ANALYTICAL METHODS FOR FOOD SAFETY
BY MASS SPECTROMETRY

农药兽药多组分残留质谱分析技术

U0248764

ANALYTICAL METHODS FOR FOOD SAFETY BY MASS SPECTROMETRY
VOLUME I PESTICIDES

农药兽药多组分残留
质谱分析技术
第1卷 农药

GUO-FANG PANG
庞国芳 著

Academician, Chinese Academy of Engineering, Beijing, China
中国工程院院士

Chairman of the Academic Committee and a Chief Scientist of Beijing
Advanced Innovation Center for Food Nutrition and Human Health,
Beijing, China
北京食品营养与人类健康高精尖创新中心学术委员会主任

化学工业出版社
·北京·

ELSEVIER

ACADEMIC PRESS
An imprint of Elsevier

本书是《农药兽药多组分残留质谱分析技术》的第1卷农药篇，介绍了农药多残留分析技术和千余种农药不同色谱-质谱条件下的质谱参数，实现了一次样品制备，GC-MS和LC-MS/MS同时测定，扩大了检测目标的范围。书中包括20项适用于10余类农产品（包括水果、蔬菜、粮谷、茶叶、中草药、食用菌、果蔬汁、蜂蜜、动物组织、水产品、牛奶、奶粉及饮用水）中800多种农药化学污染物残留的色谱-质谱分析方法。

本书可供从事食品安全、农业环境保护及农药开发利用等技术研究与应用的技术人员参考，也可作为大学教学参考用书。

图书在版编目（CIP）数据

农药兽药多组分残留质谱分析技术. 第1卷，农药＝
Analytical Methods for Food Safety by Mass Spectrometry：
Volume Ⅰ Pesticides：英文/庞国芳著. —北京：化学工
业出版社，2019.11
　　ISBN 978-7-122-34918-7

Ⅰ.①农… Ⅱ.①庞… Ⅲ.①农药残留量分析-质谱法-
英文 Ⅳ.①X592.02

中国版本图书馆CIP数据核字（2019）第151854号

责任编辑：成荣霞　　　　　　　　　　　　　　装帧设计：关　飞
责任校对：杜杏然

出版发行：化学工业出版社（北京市东城区青年湖南街13号　邮政编码100011）
印　　装：中煤（北京）印务有限公司
710mm×1000mm　1/16　印张55　字数775千字　2019年11月北京第1版第1次印刷

购书咨询：010-64518888　　　　　　　　　　　　售后服务：010-64518899
网　　址：http：//www.cip.com.cn
凡购买本书，如有缺损质量问题，本社销售中心负责调换。

Contents

Main Researchers/Editors
主要研究者 / 著者

Guo-Fang Pang
庞国芳

Qiao-Ying Chang
常巧英

Chun-Lin Fan
范春林

Hui-Qin Wu
吴惠勤

Xiao-Lan Huang
黄晓兰

Yan-Zhong Cao
曹彦忠

Yong-Ming Liu
刘永明

Preface

Today more and more requirements and restrictions are being placed on the residues of pesticides and veterinary drugs in edible animal and plant-derived agricultural products all over the world, and more and more kinds of pesticides must be controlled. Meanwhile, the index of maximum residue limit (MRL) for pesticides is getting lower, so the barriers to international trade of edible agricultural products are becoming higher.

The research team of Guo-Fang Pang took the lead in research focusing on the strategic development of food safety. They met the challenge, and a series of low-cost and highly efficient detection methods have been established.

In the field of pesticide residues determination, the authors conducted a systematic study of the detection of multiclasses and multitypes of pesticide residues in different edible agricultural products, and 20 high-throughput sample preparation and analytical methods were developed for simultaneous extraction, separation, enrichment and determination of 400–500 pesticide residues with a once-for-all sample preparation. These methods are applicable to the simultaneous determination of multipesticide residues in animal and plant-derived agricultural products, such as fruit juices, vegetable juices, fruit wines, fruits, vegetables, cereals, tea, Chinese medicinal herbs (such as ramulus mori, honeysuckle, Chinese wolfberry, and lotus leaf), edible fungi (such as mushrooms), honey, milk, milk powder, aquatic products (such as globefish, eel, and prawn), animal muscle tissues, and drinking water. A high-selectivity, high-sensitivity, high-resolution, and high-throughput analysis methodology for the detection of more than 1000 pesticide residues has been established. These methods are considered to be the international leaders in the type and quantity of pesticides that can be detected simultaneously.

In veterinary drug residue detection, the authors have established 65 high-selectivity, high-sensitivity veterinary drug residue detection methods for the determination of 20 types of nearly 200 commonly used veterinary drugs that may be residual in edible animal-derived agricultural products, and the methods are applicable to samples with complicated bases, such as edible animal tissue (such as muscle, liver, kidney, and fat), milk, milk powder, bee products (such as honey, royal jelly, and its lyophilized powder), aquatic products (such as globefish, eel, and prawn), and animal urine.

This book is a systemic summary of the research of Guo-Fang Pang's team on the theory and applied practice of detection technology of pesticide residues over the past 20 years. The innovative research achievements at an internationally advanced level in this field are fully demonstrated in this work. In addition, the progress of food safety detection technology has been promoted. This is mainly reflected in the following aspects:

1. High-throughput sample pretreatment technology: A series of technical problems such as the extraction of pesticide and veterinary drug residues at the level of micrograms per kilogram in various complex matrix samples and the effective purification of the CO-extracted disrupting chemicals have been overcome, and high-throughput sample preparation and purification technology at an international leading level has been developed. Implementation of the detection of 400–500 pesticide residues simultaneously using one sample preparation has been achieved.
2. The most comprehensive database available of pesticide mass spectrometry has been constructed: gas chromatography-(tandem) mass spectrometry and liquid chromatography-(tandem) mass spectrometry characteristics of 1000 commonly used pesticides worldwide were systematically studied and a database with tens of thousands of mass spectra has been constructed. The database can provide a qualitative and quantitative basis and has laid the foundation for research and development of high-throughput detection technology.
3. A new technique of group detection by time intervals using chromatography and mass spectrometry is proposed: pesticides with similar chemical properties and retention times are divided into several groups in turn; thus, the selectivity of the method can be enhanced. Each group of pesticides is detected according to the peak order and the time intervals, so the selectivity of the sensitivity of the methods can be improved and the monitoring range expanded.

Furthermore, academician Pang's team also cooperated fully with units from the quality inspection system of China, institutions of higher learning, and scientific research institutes for collaborative verification of the reliability and applicability of the methods, thus leading to the formation of Chinese national standards (GB/T series standards). These methods have been widely used in the detection of pesticide and veterinary drug residues in edible animal and plant-derived agricultural products and they have made a great contribution to the guarantee of food safety in China. We believe these methods will provide a useful reference value for edible animal and plant-derived agricultural product safety detection all over the world.

The book is suitable for scientific researchers in the residues testing areas and technicians working in agricultural products testing and inspection, as well as teachers and scholars of higher learning institutions as their reading material or reference literature.

Wei Fusheng

October 10, 2017

Brief Introduction

Food safety is a major security issue necessary to ensure the sound development of human society. The widespread use of pesticides, veterinary drugs, and other agricultural chemicals have contributed to widespread contamination in agricultural products. The long-term consumption of food with high residues of pesticides and veterinary drugs will cause both acute and chronic toxicity in humans and induce resistant strains, thus resulting in allergies, cancer, mutation, and teratogenicity. In order to ensure human health and food safety, countries all over the world have issued strict food safety and hygiene standards. The People's Republic of China (PRC) has implemented strict monitoring of pesticide and veterinary drug residues, such as maximum residue limits for veterinary drugs in food of animal origin (Announcement No. 235 of the Ministry of Agriculture of the People's Republic of China) and national food safety standards: Maximum residue limits for pesticides in food (GB 2763-2016). The commonly used pesticides and veterinary drugs approved in China and the main agricultural products of urban residents' daily consumption are almost all covered.

The book is a systematic summary of the high-throughput chromatography-mass spectrometry technique for the analysis of multipesticides, veterinary drugs, and other agricultural chemical residues in agriculture products of plant and animal origin. Its technical characteristics are mainly embodied in the simultaneous determination of hundreds of compounds, as well as low cost and high efficiency.

This book is divided into two volumes. In Volume 1, pesticides and related chemicals are selected as the main research object, while gas chromatography-mass spectrometry (GC-MS), gas chromatography-tandem mass spectrometry (GC-MS-MS), and liquid chromatography-tandem mass spectrometry (LC-MS-MS) are concerned. This volume includes more than 20 analytical methods for the high-throughput analysis of multipesticide residues, and more than 1000 pesticides and related chemicals in more than 10 kinds of agricultural products and drinking water can be detected, and 400–500 kinds of pesticide residues can be detected simultaneously with the pretreatment undertaken just once. The methods for the analysis of 793 pesticides and related chemical residues in foods of plant origin applicable to fruit juices, vegetable juices, fruit wines, fruits, vegetables, cereals, tea, Chinese medicinal herbs (ramulus mori, honey-suckle, Chinese wolfberry, and lotus leaf), and edible fungi (mushroom) are introduced in Part 1; the methods for the analysis of 790 pesticides and related chemical residues in foods of animal origin applicable to honey, milk, milk powder, aquatic products (globefish, eel, and prawn), and animal muscle tissues are presented in Part 2; while the methods for the analysis of 450 pesticides and related chemical residues in drinking water are introduced in Part 3. Furthermore, the three major parameter databases adopted in the analysis of more than 1000 pesticides and related chemicals are also included in this volume: ①chromatography-mass spectrometry characteristic parameters include retention time, quantitative and qualitative ions, fragment voltage and collision energy, etc.; ②performance parameters of the methods include linear equation, linear range and correlation coefficient, etc.; and ③gel permeation chromatography (GPC) purification analysis parameters.

In Volume 2, the methods for the analysis of 20 species (nearly 200 kinds) of veterinary drug residues in edible animal tissues (muscle, liver, kidney, and fat), dairies (milk and milk powder), bee products (honey, royal jelly, and its lyophilized powder), aquatic products (globefish, eel, and roasted eel) and animal urine are selected as main subjects of study. The 65 analytical methods for the analysis of multiveterinary drug residues are described in different chapters by category (sulfonamides, β-adrenergic agonists, aminoglycosides, chloramphenicols, β-lactams, macrolides, nitrofurans, anabolic steroids, nonsteroidal anabolic steroids, glucocorticoids, fluoroquinolones, tetracyclines, sedatives, pyrazolones, quinoxalines, nitromidazoles, benzimidazoles, levamisole, thiourea pyrimidines, and polyethers), and 90% of the methods are LC-MS-MS. Meanwhile the physical and chemical properties, efficacy, side effects, and maximum allowable residual limit of all the compounds are also provided.

In short, this book is the summary of the work of the author team who, for more than 20 years, engaged in the research and practice of detection technology of pesticides and veterinary drug residues. These methods for the analysis of pesticides and veterinary drugs are innovative research results based on the international frontier of pesticides and veterinary drug residue analysis, and the analytical techniques adopted are new technologies widely concerned in the field of residue analysis in the world today. The performance indexes of the methods can meet the requirements of the Codex Alimentarius Commission and the world's major developed countries; meanwhile the methods are in conformity with the developing trend of international residue analysis and they are advanced in the world. In addition, the methods established in this book have become the current effective national standard methods in China and they can be used as the detection basis of relevant testing institutions and the law basis of the relevant government departments. Nevertheless, due to the limitations of the level, there may be unavoidable errors. We would kindly ask the users of this publication to provide feedback to the authors so that subsequent editions may be improved upon.

Guo-Fang Pang

Chapter 1

Introduction

Pesticides are chemical and biological products used to control, destroy, repel, or attack organisms such as pests, mites, eelworms, pathogens, weeds, and mice. According to their functions, pesticides can be categorized as insecticides, fungicides, nematocides, molluscicides, bacteriocides, attractants, fumigants, rodenticides, defoliants, desiccants, insect growth and plant growth regulators, etc. Alternatively, based on the functional groups of their molecular structures, pesticides can be categorized as inorganic pesticides, organochlorine insecticides, organophosphate pesticides, carbamate pesticides, triazine herbicides, etc.

Pesticides are important materials in modern agriculture production because they play a key role in preventing and controlling infectious diseases and enhancing production yields at harvest. However, the long-term use of pesticides has also resulted in significant contamination of the soil, the air we breathe, and the environment.

1.1 INSECTICIDES

Insecticides are used in agriculture, medicine, industry, and our homes against insects at all developmental stages. Many insecticides are also used against eggs and larvae of insects, such as omethoate, monocrotophos and fenpropathrin, etc. Insecticides fall into two types: inorganic and organic. Inorganic insecticides are compounds containing arsenic, copper, lead, or mercury. They are highly persistent in terrestrial environments, and are slowly dispersed through leaching and erosion by wind and water. Inorganic insecticides are used much less now than previously, having been widely replaced by synthetic organic insecticides. Organic insecticides can be categorized as natural and synthetic. Natural insecticides include botanical and biological insecticides. There are many types of synthetic organic insecticides, including organophosphates, carbamates, organochlorines, and pyrethroids.

1.1.1 Organophosphate Insecticides

The function of organophosphate insecticides is to inhibit the activity of cholinesterase of nerve tissue in insects and eventually to kill them. Usage of organophosphate insecticides has a long history. Today, many organophosphate insecticides, including parathion, dichlorvos, fenitrothion, fenthion, cyanofenphos, trichlorfon, tetrachlorvinphos, dimethoate, phorate, phosphamidon, monocrotophos, menazon, pirimiphos-ethyl, diazinon, trichloronate, chlorpyrifos, phoxim, malathion, methamidophos, temephos, fenchlorphos, profenofos, etc., are still widely used.

1.1.2 Carbamate Insecticides

The hydrogen atom that connects with carbon in a carbamate ester derivative is substituted by an aryl group. Carbamate insecticides are not used as widely as organophosphate insecticides, but their effect is more rapid and more selective. These insecticides include carbaryl, isoprocarb, carbafuran, aldicarb, etc.

1.1.3 Organochlorine Insecticides

This group of insecticides was used widely in agricultural production earlier than other insecticides. In particular, the invention of DDT is considered the beginning of modern pesticide use. These insecticides have road-spectrum activity and are very persistent. There are two kinds of organochlorine insecticides. One type is made with benzene, such as DDT and hexachlorobenzene; the other is made without benzene, such as cyclopentadiene and camphechlor, etc.

1.1.4 Pyrethroid Insecticides

Pyrethroid insecticides were invented in the 1970s. The appearance of pyrethroid insecticides changed the features of insecticides in not only production technology and farming practice, but also in practices for increasing production. Pyrethroid insecticides currently used are called "super-efficiency" insecticides. They require the application of less than $1.5\,g$, or at most $30\,g$ per $1000\,m^2$, of land. Thus, with only a small amount, pyrethyroid insecticides can kill pests and stimulate growth, and so have a marked effect on increasing production. They mainly include resmethrin, tetramethrin, phenothrin, cyphenothrin, fenpropathrin, permethrin, cypermethrin, deltamethrin, and fenvalerate.

1.1.5 Neonicotinoid Insecticides

Neonicotinoid insecticides are of high killing activity and attract the attention of different industries owing to their influences on many aspects such as the reproductive system of pollinating insects and avian as well as the respiratory system of rats. Neonictonioid insecticides are classified as three generations such as chloronicotine and thionicotine and furan nicotine including imidacloprid, acetamiprid, thiamethoxam, furosemide, fluidoximide, chlorothiazoline, enipramine, thiacloprid, thiamethoxam. In 2013, EU announced a moratorium on the use of three staple neonicotinoid insecticides for certain crops, while the draft of the new regulation proclaimed the ban on the use of neonicotinoid insecticides on all the field crops except the greenhouse crops. In 2016, officials from Department of Health in Canada conveyed the message that their Federal Pesticides Supervisory Institutions planned to eliminate a disputative insecticide-imidacloprid. In 2018, France announced a ban on the neonicotinoid insecticide, which would be prohibited from use on all the crops in due time including on the treatment of seeds. At present, there are already literatures on the analytical methods for determination of neonicotinoid insecticides in red wine, white wine, environmental water, honey orange leaves, fruits and vegetables, sunflower seeds, tea soups, pollens, atmospheric particulate samples, soil samples and hematuria, and the cleanup techniques adopted are SPE, DLLME, solid-liquid extraction, QuEchERS, etc., while LC-MS-MS are adopted mostly.

1.2 HERBICIDES

Chemical herbicides can increase output of crops markedly by selectively killing weeds and stimulating growth. With chemical herbicides, an increase of 10% in output can be obtained compared to manual work. The employment of chemical herbicides could save the use of 75–300 laborers in the paddy field and 30–45 in the glebe. The application of chemical herbicides promotes good agricultural practices, including planting techniques and methods, boosting production of soya, corn, wheat, and cotton, but their residues have been detected on soya, corn, wheat, rice, and cotton plants in China. There are several kinds of synthetic herbicides, including phenolic, phenoxy acid, benzoic acid, diphenyl ether, dinitroaniline, amide, carbamate, urea, sulfonyl urea, s-triazine, quarternary ammonium, organophosphate herbicides, and others.

1.2.1 Phenolic Herbicides

The earliest applied in the field, phenolic herbicides are synthetic organic herbicides. There are two types based on structural differences. One group has a substituted phenol and the other group has a substituted chlorophenol group. All phenolic herbicides have low water solubility and high fat solubility and most have contact poisoning and bad selectivity. The function of phenolic herbicides is to kill plants by inhibiting respiration. Sodium pentachlorophenol is the most widely used.

1.2.2 Phenoxy Acid Herbicides

The main molecular structure of phenoxy acid herbicides contains phenoxyacetic acid, phenoxy propionic acid, and aryloxy-phenoxy propionic acid. Most effective on caulines and leaves, the common characteristics of these herbicides include high selectivity, broad spectrum, low dosage, low cost, and safety. Phenoxy acid herbicide is safe for humans, livestock, and the environment when normal dosages are used. However, the synthesis of 2,4,5-trichlorophenoxyacetic acid (2,4,5-T) generates dioxin, which is a strong teratogen. 2,4,5-T is, therefore, restricted or banned in many countries. Phenoxy acid herbicides mainly include 2,4-D, MCPA, diclofop-methyl, and fluazifop-butyl.

1.2.3 Benzoic Acid Herbicides

Except for chloramben, benzoic acid herbicides have activity both on leaves and soil. They not only have weeding activity, but also have the action of auxin. Benzoic acid herbicides can be absorbed and transmitted through the root, stem, and leaves. Benzoic acid herbicides mainly include chloramben and dicamba.

1.2.4 Diphenyl Ether Herbicides

These herbicides have low water solubility and tend to be adsorbed by soil colloid. They are hard to alleviate, and have a moderate persistency. Most diphenyl ether herbicides are used on soil before seeding to prevent annual weeding and have a function of contact poisoning. When substitution is at the ortho position, it has weeding activity only in illumination, such as chlornitrofen, poly tetramethylene oxide, bifenox, oxyfluorfen, etc. Most are used in the paddy fields, due to low poisonous effect on fish and shellfish. Diphenyl ether herbicides include nitrofen, acifluorfen sodium, fomesafen, oxyfluorfen, fluoroglycofen, attackweed, farmaid, lactofen, and chloromethoxynil.

1.2.5 Dinitroaniline Herbicides

Dinitroaniline herbicides were screened starting from 1953. Trifluralin, with a high activity and selectivity, was successfully selected from 80 compounds in 1959. Since then, dinitroaniline herbicides have become a very important herbicide. Their common characteristics are broad spectrum, high efficiency, tendency to volatilize and photolyze, steady effect, and safety. They include trifluralin, butralin, pendimethalin, etc.

1.2.6 Amide Herbicides

In 1956, P. C. Hamm reported that pethoxamide could prevent the growth of annual gramineous species and some broadleaf weeds. In the same year, pethoxamide was popularized by Monsanto Corporation. This was the beginning of the development of amide herbicides. Since then, amide herbicides have developed rapidly. Among them, propanil, a selective herbicide that was invented in 1960, is the most frequently used amide herbicide in the paddy fields. The main compounds of amide herbicides include propanil, alachlor, metolachlor, pretilachlor, napropamide, mefluidide and diflufenican.

1.2.7   Carbamate Herbicides

Carbamate herbicide use started in 1945 when Templeman and Sexton discovered the weed characteristics of propham. After the report of the weeding effect of chlorpropham in 1951, barban, swep, phenmedipham, asulam, desmedipham, and latifolinine were synthesized and utilized subsequently.

1.2.8   Thiocarbamate Herbicides

Thiocarbamate herbicides were developed from 1945 when the Stauffe Corporation discovered the weeding characteristic of EPTC. In 1960 and 1961, Monsanto Company invented diallate and triallate to prevent wild oats successively. In 1965, a Japanese combinatorial chemistry company produced thiobencarb, a highly efficient herbicide for killing barnyard grass. This was a great advancement for this kind of herbicide. The main compounds of thiocarbamate herbicides include thiobencarb, molinate, and vernolate.

1.2.9   Urea Herbicides

It was reported in the 1940s that urea compounds could restrain the growth of plants, but their weeding function was not found. Until 1951, when H. C. Bucha and C. W. Todd discovered the weeding function of monuron, urea herbicides had been developed gradually and became popular in use with different compounds. Urea herbicides mainly include diuron, linuron, and chlortoluron.

1.2.10   Sulfonylurea Herbicides

Sulfonylurea herbicides are a new kind of herbicide, which has been developed rapidly, with the highest weeding function. G. Levitt in 1978 and D. W. Finnerty in 1979 gave the earliest reports that chlorsulfuron possessed weeding activity. From then on, the study of sulfonylurea herbicides entered into a new era. They represent a great advancement in the history of herbicides, bringing herbicides into the era of superefficiency. Sulfonylurea herbicides mainly include chlorsulfuron, metsulfuron-methyl, and tribenuron-methyl.

1.2.11   S-Triazine Herbicides

S-Triazine herbicides were developed from 1932 when A. Gast et al. found the weeding activity of chlorazine for the first time. In particular, after the high weeding effect of simazine and trietazine was found and brought to industrialized production, the development of s-Triazine herbicides proceeded more quickly and the herbicide became one of the most important modern herbicides. Today s-Triazine herbicides stay in the lead in the aspects of applicability, selectivity, output, and sale. This type of herbicide mainly includes simazine, atrazine, and prometryne.

1.2.12 Quaternary Ammoniums Herbicides

Synthesized in the 1950s, these herbicides are also called bipyridine herbicides. The ICI Company is the pioneer in researching the biologic activity of bipyridine compounds and brought the first quaternary ammoniums herbicide, diquat, into industrialized production in 1957. The second variety was carried out by the British Plant Protecting Company in 1958. Afterward, some similar compounds were found and produced, but most of them had no large output. Quaternary ammoniums herbicides mainly include paraquat and diquat.

1.2.13 Organophosphate Herbicides

The production of organophosphate herbicides began in 1958 when Uniroyal Company found the first organophosphate herbicide, 2,4-DEP. Subsequently, some varieties were employed successively for use on dry land, paddy fields, and infield. Among them, glyphosate drew the most attention and was used most widely because of its high efficiency, broad spectrum, low level of toxicity, low levels of residual contamination, and safety to the environment. There are many varieties of organophosphate herbicides, but the most important ones are bensulide, amiprophos, 2,4-DEP, and glyphosate.

1.2.14 Other Herbicides

There are other herbicides besides the 13 described here, including nitrile herbicide containing dichlorobenil, ioxynil, ioxynil-octanoate and bromoxynil, aliphatic herbicide containing MCA, TCA, and dalapon, imidazolinones herbicide containing imazaquin, irnazamethabenz, and imazathapyr, cyclohexenone herbicide containing clethodim, cycloxydim, tralkoxydim, sethoxydim, and alloxydim, pyridine herbicide containing picloram, clopyralid, trichlopyr, haloxydine, fluridone, and fluroxypyr, pyrazole herbicide containing pyrazolate, pyrazoxyfen, benzofenap, metazachlor, and difenzoquat, pyrimidine herbicide containing isocil, bromacil, lenacil, and bentazon, quinolinecarboxylic acids herbicide containing quinmerac and quinclorac, and other heterocyclic herbicides containing oxadiazon, dimethazone, and cinmethylin.

1.3 BACTERICIDES

In order to increase food production, it is important to prevent diseases in addition to killing insects and clearing weeds. Bactericides are the oldest medications in human history and they include sulfur, lime sulfur, bordeaux mixture, mercurial, anilinopyrimidines, and strobins. Compared to insecticides and herbicides, bactericides have fewer varieties and a more stable market.

However, great advancements have been made since the 1980s in the development of bactericides such as triazoles, amide, anilinopyrimidines, and strobins.

1.3.1　Triazole Bactericides

Triazole bactericide drew much attention when Bayer Company produced triadimefon, which is the first commercial bactericide with the chiral carbon atom. Currently, there are more than 40 commercial varieties of triazoles. They can affect the formation of the cell wall by blocking the synthesis of ergosterin in epiphyte. Also, they can prevent and cure most harmful originated epiphytes for crops. Most of the triazoles bactericides have the characteristics of high efficiency, broad spectrum, persistency, strong systemic effect, and stereoselectivity. They mainly include tebuconazole, propiconazol, cyprodinil, epoxiconazole, metconazole, and tetraconazole.

1.3.2　Amide Bactericides

Amide bactericides have a history of several decades and have more than 30 varieties. As the special medicament for preventing and curing peronosporales fungal, this kind of bactericide has remarkable capabilities of protecting, curing, and eradicating. It is widely used in preventing late blight of tomatoes and potatoes. The main compounds of this bactericide include metalaxyl, oxadixyl anchor, benalaxyl, mefloamide, fenhexamid, and ethaboxam.

1.3.3　Pyridinamine Bactericides

Pyridinamine is an important bactericide developed in the early 1990s. It has low power when unorganized, but exhibits a strong prevention effect when used on host plants. It can prevent pathogen inbreaking to the host by restraining the synthesis of methionine and exudation of cell wall-degrading enzyme. Pyridinamine bactericides include pyrimethanil, mepanipyrim, cymoxanil, and diflumetorim.

1.3.4　Strobin Bactericides

Strobins are derived from the crude antibiotic strobilurins A, which possess sterilization activity. Strobins have a unique ability to block the synthesis of cellular ATP by locking the electron transfer between cytochrome b and c1, rendering their bactericidal capacity by restraining the breath of the mitochondrion. They mainly include azoxystrobin, kresoxim methyl, trifloxystrobin, kresoximmethyl, picoxystrobin, pyraclostrobin, fluoxastrobin and 1-octyl alcohol.

1.3.5   Oxazo Bactericides

Currently, scientists are paying attention to the research on isoxazolyl bactericides. There are three varieties reported: famoxadone, cyazofamid, and fenamidone.

1.3.6   Pyrrole Bactericides

Pyrrole bactericides are derived from crude pyrrolnitrin. They have specific effects on botrytis and they include fenpiclonil and fludioxonil.

1.3.7   Amino Acid Bactericides

Amino acid bactericides are also drawing a lot of current attention because they are safe to humans and the environment. They include mainly benthiavalicarb and iprovalicarb.

1.3.8   Derivants of Cinnamic Acid

The sterilization activity of 3,4-dimethoxycinamicacid was reported in the 1970s. Since then, a series of derivatives of cinnamic acid with sterilization activity have been developed. They include mainly dimethomorph and flumorph.

1.3.9   Others

Other kinds of bactericides include SYP-Z048, activated ester, spiroxamine, and benzoquinate.

1.4   PESTICIDES BANNED IN THE WORLD

1.4.1   Pesticides Banned in China

It is prescribed in bulletin 199 of the Ministry of Agriculture of the People's Republic of China that the following virulent pesticides are banned or prohibited for use on vegetables, fruit trees, tea, and Chinese traditional medicines.

Banned pesticides: HCH, DDT, camphechlor, dibromochloropane, chlordimeform, EDB, nitrofen, aldrin, dieldrin, mercury compounds, arsena, lead, fluoroacetamide, gliftor, tetramine, sodium fluoroacetamide, and silatrane.

Pesticides prohibited and restricted for use on vegetables, fruit trees, tea, and Chinese traditional medicines: methamidophos, parathion-methyl, parathion, monocrotophos, phosphamidon, phorate, isofenphos-methyl, terbufos, phosfolan-methyl, sulfotep, demeton, carbofuran, aldicarb, ethoprophos, phosfolan, coumaphos, fonofos, isazofos, and fenamiphos. Pesticides prohibited for use on tea trees: dicofol, fenvalerate. In addition, GB 2763-2016 China National

Food Safety Standard-Maximum Residue Limits for Pesticides in Food has stipulated 4140 MRLs for 433 Pesticides.

1.4.2 Pesticides Banned in Japan

The Japanese positive listing system has established a maximum residue limit (MRL) for all of the agricultural chemicals. Under this system, 15 pesticides and chemicals should not be found in food: cyhexatin, azocyclotin, propham, captafol, anitrole, daminozide, diethylstilbestrol, coumaphos, ronidazole, metronidazole, dimetridazole, chloramphenicol, chloramphenicol, nitrofurans, carbadox, including QCA. Another 8 pesticides and chemicals should not be detected in some foods, involving 84 kinds of foods and 166 standard limitations. These eight compounds are triazophos, trenbolone acetate, dexamethasone, parathion, ethylene dibromide, dieldrin, aldrin, and clenbuterol. In addition, the Japanese government's Positive List System implemented in 2006 put forward 53862 limit standards for over 300 kinds of agricultural products, 799 pesticides, veterinary drugs and feeds additives, of which there were 428 inspection items for pork, 581 for rice and 276 for teas.

1.4.3 Pesticides Banned in European Union

The European Union (EU) prohibits the sale of 320 pesticides, which include 63 pesticides produced, used, and sold in China. These pesticides, used widely on fruits, vegetables, tea, and in paddys, include 30 acaricides, 20 herbicides, 9 antiseptics, 3 plant growth regulators, and 1 molluscicide. The acaricides referred to are cartap, fenpropathrin, ethion, tetramethrin, δ-endotoxin from *Bacillus thuringiensis*, isazofos, omethoate, profenofos, triazophos, phorate, quinalphos, temephos, pirimiphos-ethyl, terbufos, chlorfluazuron, sulfotep, monocrotophos, phosphamidon, bromopropylate, phenthoate, propoxur, flucythrinate, diafenthiuron, tralomethrin, benzoximate, allethrin, minoctadine, tetradifon, thiocyclam, and fonofos. The herbicides are prometryn, hexazinone, fluazifop-butyl, ametryn, sethoxydi, carbine, butachlor, vernam, mefenacet, dimepiperate, metolachlor, difenzoquat, fenoxaprop ethyl, cyanazine, acifluorfen, benazolin-ethyl, fomesafen, fluoroglycofen, quizalofop-ethyl, and haloxyfop. The antiseptics are thiophanate, isoprothiolane, methanearsonate, mepronil, validamycin, iminoctadine, anilazine, fenaminsulf, and oxadixyl. The plant growth regulators are lumetralin, triapenthenol, and 2,4,5-T. The molluscicide is trifenmorph. EU has currently stipulated 162248 MRLs for 460 pesticides in 2016.

Chapter 2

Analytical Methods for 793 Pesticides and Related Chemical Residues in Products of Plant Origin

2.1 DETERMINATION OF 512 PESTICIDES AND RELATED CHEMICAL RESIDUES IN FRUIT JUICE AND WINE: LC-MS-MS METHOD (GB/T 23206-2008)

2.1.1 Scope

This method is applicable to the quantitative determination of 490 pesticides and related chemical residues and qualitative determination of 512 pesticides and related chemical residues in orange juice, apple juice, grape juice, cabbage juice, carrot juice, dry wine, semidry wine, semisweet wine, and sweet wine.

The limit of quantitative determination for this method of 490 pesticides and related chemicals is 0.01 µg/kg to 1.60 mg/kg (see Table 2.1).

2.1.2 Principle

Samples are extracted with acetonitrile (containing 1% acetate acid), and the extracts are cleaned up with Sep-Pak Vac cartridges. The pesticides and related chemicals are eluted with acetonitrile-toluene (3 + 1), and the solutions are analyzed by LC-MS-MS.

2.1.3 Reagents and Materials

"Water" is the first grade of GB/T6682 specified.

Acetonitrile: HPLC Grade
Acetone: HPLC Grade
Isooctane: HPLC Grade
Methanol: HPLC Grade
Toluene: HPLC Grade

Analytical Methods for Food Safety by Mass Spectrometry. https://doi.org/10.1016/B978-0-12-814167-0.00002-8
Copyright © 2018 Chemical Industry Press.
Published by Elsevier Inc. under an exclusive license with Chemical Industry Press.

TABLE 2.1 The LOD, LOQ of 512 Pesticides and Related Chemical Residues Determined by LC-MS-MS

No.	Pesticides	LOD (μg/kg)	LOQ (μg/kg)	No.	Pesticides	LOD (μg/kg)	LOQ (μg/kg)
	Group A			31	Terrbucarb	0.35	0.70
1	Propham	18.33	36.67	32	Penconazole	0.33	0.67
2	Isoprocarb	0.38	0.77	33	Myclobutanil	0.17	0.33
3	3,4,5-Trimethacarb	0.06	0.11	34	Paclobutrazol	0.10	0.19
4	Cycluron	0.03	0.07	35	Fenthion sulfoxide	0.05	0.10
5	Carbaryle	1.72	3.44	36	Triadimenol	1.76	3.52
6	Propachlor	0.05	0.09	37	Butralin	0.32	0.63
7	Rabenzazole	0.22	0.44	38	Spiroxamine	0.01	0.02
8	Simetryn	0.02	0.05	39	Tolclofos methyl	11.09	22.19
9	Monolinuron	0.59	1.19	40	Desmedipham	0.67	1.34
10	Mevinphos	0.26	0.52	41	Methidathion	1.78	3.55
11	Aziprotryne	0.23	0.46	42	Allethrin	10.07	20.13
12	Secbumeton	0.01	0.02	43	Diazinon	0.12	0.24
13	Cyprodinil	0.12	0.25	44	Edifenphos	0.13	0.25
14	Buturon	1.49	2.99	45	Pretilachlor	0.06	0.11
15	Carbetamide	0.61	1.21	46	Flusilazole	0.10	0.19
16	Pirimicarb	0.03	0.05	47	Iprovalicarb	0.39	0.77
17	Clomazone	0.07	0.14	48	Benodanil	0.58	1.16
18	Cyanazine	0.03	0.05	49	Flutolanil	0.19	0.38
19	Prometryne	0.03	0.05	50	Famphur	0.60	1.20
20	Paraoxon methyl	0.13	0.25	51	Benalyxyl	0.21	0.41
21	4,4-dichloro-benzophenone	2.27	4.53	52	Diclobutrazole	0.08	0.16
				53	Etaconazole	0.30	0.59
22	Thiacloprid	0.06	0.12	54	Fenarimol	0.10	0.20
23	Imidacloprid	3.67	7.33	55	Phthalic acid, dicyclobexyl ester	0.33	0.67
24	Ethidimuron	0.25	0.50				
25	Isomethiozin	0.18	0.36	56	Tetramethirn	0.30	0.61
26	Cis and trans diallate	14.87	29.73	57	Dichlofluanid	0.43	0.87
				58	Cloquintocet mexyl	0.31	0.63
27	Acetochlor	7.90	15.80				
28	Nitenpyram	2.85	5.71	59	Bitertanol	5.57	11.13
				60	Chlorprifos methyl	2.67	5.33
29	Methoprotryne	0.04	0.08				
30	Dimethenamid	0.72	1.43	61	Tepraloxydim	2.03	4.07

TABLE 2.1 The LOD, LOQ of 512 Pesticides and Related Chemical Residues Determined by LC-MS-MS—cont'd

No.	Pesticides	LOD (μg/kg)	LOQ (μg/kg)	No.	Pesticides	LOD (μg/kg)	LOQ (μg/kg)
62	Thiophanate methyl	3.33	6.67	91	Propoxur	4.07	8.13
				92	Isouron	0.07	0.14
63	Azinphos ethyl	18.15	36.31	93	Chlorotoluron	0.10	0.21
64	Clodinafop propargyl	0.41	0.81	94	Thiofanox	26.17	52.33
65	Triflumuron	0.65	1.31	95	Chlorbufam	30.50	61.00
66	Isoxaflutole	0.65	1.30	96	Bendiocarb	0.53	1.06
67	Thiophanat ethyl	3.36	6.72	97	Propazine	0.05	0.11
68	Quizalofop-ethyl	0.11	0.23	98	Terbuthylazine	0.08	0.16
69	Haloxyfop-methyl	0.44	0.88	99	Diuron	0.26	0.52
70	Fluazifop butyl	0.04	0.09	100	Chlormephos	74.67	149.33
71	Bromophos-ethyl	94.62	189.23	101	Carboxin	0.09	0.19
72	Bensulide	5.70	11.40	102	Difenzoquat-methyl sulfate	0.14	0.27
73	Triasulfuron	0.27	0.54	103	Clothianidin	10.50	21.00
74	Bromfenvinfos	0.50	1.01	104	Pronamide	2.56	5.13
75	Azoxystrobin	0.08	0.15	105	Dimethachloro	0.32	0.63
76	Pyrazophos	0.27	0.54	106	Methobromuron	2.81	5.61
77	Flufenoxuron	0.53	1.06	107	Phorate	52.33	104.67
78	Indoxacarb	1.26	2.51	108	Aclonifen	4.03	8.07
79	Emamectin benzoate	0.05	0.11	109	Mephosfolan	0.39	0.77
	Group B			110	Imibenzonazole-des-benzyl	1.04	2.07
80	Ethylene Thiourea	8.70	17.40	111	Neburon	1.18	2.37
81	Daminozide	0.43	0.87	112	Mefenoxam	0.26	0.51
82	Dazomet	21.17	42.33	113	Prothoate	0.41	0.82
83	Nicotine	0.37	0.73	114	Ethofume sate	62.00	124.00
84	Fenuron	0.17	0.34	115	Iprobenfos	1.38	2.76
85	Cyromazine	1.21	2.41	116	TEPP	1.73	3.47
86	Crimidine	0.26	0.52	117	Cyproconazole	0.12	0.24
87	Acephate	2.22	4.45	118	Thiamethoxam	5.50	11.00
88	Molinate	0.35	0.70	119	Crufomate	0.09	0.17
89	Carbendazim	0.08	0.16	120	Etrimfos	3.13	6.25
90	6-chloro-4-hydroxy-3-phenyl-pyridazin	0.28	0.55	121	Coumatetralyl	0.23	0.45
				122	Cythioate	13.33	26.67

Continued

TABLE 2.1 The LOD, LOQ of 512 Pesticides and Related Chemical Residues Determined by LC-MS-MS—cont'd

No.	Pesticides	LOD (µg/kg)	LOQ (µg/kg)	No.	Pesticides	LOD (µg/kg)	LOQ (µg/kg)
123	Phosphamidon	0.65	1.29	153	Propargite	11.43	22.87
124	Phenmedipham	0.75	1.49	154	Bromuconazole	0.52	1.05
125	Bifenazate	3.80	7.60	155	Picolinafen	0.12	0.24
126	Fenhexamid	0.16	0.32	156	Fluthiacet methyl	0.88	1.77
127	Flutriafol	1.43	2.86				
128	Furalaxyl	0.13	0.26	157	Trifloxystrobin	0.33	0.67
129	Bioallethrin	33.00	66.00	158	Chlorimuron ethyl	5.07	10.13
130	Cyanofenphos	3.47	6.93	159	Hexaflumuron	4.20	8.40
131	Pirimiphos methyl	0.03	0.07	160	Novaluron	1.34	2.68
				161	Hydramethylnon	0.29	0.57
132	Buprofezin	0.15	0.29		Group C		
133	Disulfoton sulfone	0.41	0.82	162	Maleic hydrazide	13.33	26.67
134	Fenazaquin	0.05	0.11	163	Methamidophos	0.82	1.64
135	Triazophos	0.11	0.23	164	EPTC	6.22	12.45
136	DEF	0.27	0.54	165	Diethyltolu-amide	0.09	0.18
137	Pyriftalid	0.10	0.21				
138	Metconazole	0.22	0.44	166	Monuron	5.79	11.58
139	Pyriproxyfen	0.07	0.14	167	Pyrimethanil	0.11	0.23
140	Cycloxydim	0.42	0.85	168	Fenfuram	0.13	0.26
141	Isoxaben	0.03	0.06	169	Quinoclamine	1.32	2.64
142	Flurtamone	0.07	0.15	170	Fenobucarb	0.98	1.97
143	Trifluralin	55.80	111.60	171	Ethirimol	0.09	0.19
144	Flamprop methyl	3.37	6.73	172	Propanil	3.60	7.20
145	Bioresmethrin	1.24	2.47	173	Carbofuran	2.18	4.35
146	Propiconazole	0.29	0.59	174	Acetamiprid	0.24	0.48
147	Chlorpyrifos	8.97	17.93	175	Mepanipyrim	0.05	0.11
148	Fluchloralin	81.33	162.67	176	Prometon	0.02	0.04
149	Chlorsulfuron	0.46	0.91	177	Methiocarb	6.87	13.73
150	Clethodim	0.35	0.69	178	Metoxuron	0.11	0.21
151	Flamprop isopropyl	0.07	0.14	179	Dimethoate	1.27	2.53
				180	Methfuroxam	0.05	0.09
152	Tetrachlorvinphos	0.37	0.74	181	Fluometuron	0.15	0.31

TABLE 2.1 The LOD, LOQ of 512 Pesticides and Related Chemical Residues Determined by LC-MS-MS—cont'd

No.	Pesticides	LOD (µg/kg)	LOQ (µg/kg)	No.	Pesticides	LOD (µg/kg)	LOQ (µg/kg)
182	Dicrotophos	0.19	0.38	213	Pyrimitate	0.03	0.06
183	Monalide	0.20	0.40	214	Fensulfothin	0.33	0.67
184	Diphenamid	0.02	0.05	215	Fluorochloridone	2.30	4.59
185	Ethoprophos	0.46	0.92	216	Butachlor	3.34	6.69
186	Fonofos	1.24	2.49	217	Kresoxim-methyl	16.76	33.53
187	Etridiazol	16.74	33.47	218	Triticonazole	0.50	1.01
188	Furmecyclox	0.14	0.28	219	Fenamiphos sulfoxide	0.12	0.25
189	Hexazinone	0.02	0.04	220	Thenylchlor	4.02	8.05
190	Dimethametryn	0.02	0.04	221	Fenoxanil	6.57	13.13
191	Trichlorphon	0.19	0.37	222	Fluridone	0.03	0.06
192	Demeton(O+S)	1.13	2.26	223	Epoxiconazole	0.68	1.35
193	Benoxacor	1.15	2.30	224	Chlorphoxim	12.93	25.86
194	Bromacil	3.93	7.87	225	Fenamiphos sulfone	0.07	0.15
195	Phorate sulfoxide	61.38	122.76	226	Fenbuconazole	0.27	0.55
196	Brompyrazon	0.60	1.20	227	Isofenphos	36.45	72.89
197	Oxycarboxin	0.15	0.30	228	Phenothrin	56.53	113.07
198	Mepronil	0.06	0.13	229	Fentin-chloride	2.88	5.75
199	Disulfoton	78.28	156.57	230	Piperophos	1.54	3.08
200	Fenthion	8.67	17.33	231	Piperonyl butoxide	0.19	0.38
201	Metalaxyl	0.08	0.17	232	Oxyflurofen	9.76	19.52
202	Ofurace	0.17	0.33	233	Coumaphos	0.35	0.70
203	Dodemorph	0.07	0.13	234	Flufenacet	0.88	1.77
204	Fosthiazate	0.09	0.19	235	Phosalone	8.01	16.01
205	Imazamethabenz-methyl	0.03	0.05	236	Methoxyfenozide	0.62	1.23
206	Disulfoton-sulfoxide	0.47	0.95	237	Prochloraz	0.34	0.69
207	Isoprothiolane	0.31	0.62	238	Aspon	0.29	0.58
208	Imazalil	0.33	0.67	239	Ethion	0.49	0.99
209	Phoxim	13.80	27.60	240	Thifensulfuron-methyl	3.57	7.13
210	Quinalphos	0.33	0.67	241	Dithiopyr	1.73	3.47
211	Ditalimfos	11.20	22.40	242	Spirodiclofen	1.65	3.30
212	Fenoxycarb	3.05	6.09	243	Fenpyroximate	0.23	0.45

Continued

TABLE 2.1 The LOD, LOQ of 512 Pesticides and Related Chemical Residues Determined by LC-MS-MS—cont'd

No.	Pesticides	LOD (µg/kg)	LOQ (µg/kg)	No.	Pesticides	LOD (µg/kg)	LOQ (µg/kg)
244	Flumiclorac-pentyl	1.77	3.54	274	Thiofanox-sulfoxide	1.38	2.76
245	Temephos	0.20	0.41	275	Cartap hydrochloride	346.67	693.33
246	Butafenacil	1.58	3.17	276	Methacrifos	403.95	807.90
247	Spinosad	0.09	0.19	277	Terbutryn	0.00	0.01
	Group D			278	Thionazin	3.78	7.56
248	Mepiquat chloride	0.15	0.30	279	Linuron(a)	1.94	3.88
249	Allidochlor	6.84	13.68	280	Heptanophos	0.97	1.95
250	Propamocarb	0.01	0.03	281	Prosulfocarb	0.06	0.12
251	Tricyclazole	0.21	0.42	282	Dipropetryn	0.05	0.09
252	Thiabendazole	0.08	0.16	283	Thiobencarb	0.55	1.10
253	Metamitron	1.06	2.12	284	Tri-iso-butyl phosphate	0.60	1.19
254	Isoproturon	0.02	0.05	285	Tri-n-butyl phosphate	0.06	0.12
255	Atratone	0.03	0.06				
256	Oesmetryn	0.03	0.06	286	Diethofencarb	0.33	0.67
257	Metribuzin	0.09	0.18	287	Alachlor	1.23	2.47
258	DMST	6.67	13.33	289	Cadusafos	0.19	0.38
259	Cycloate	0.74	1.48	299	Metazachlor	0.16	0.33
260	Atrazine	0.06	0.12	290	Propetamphos	9.00	18.00
261	Butylate	50.33	100.67	291	Terbufos	373.33	746.67
262	Pymetrozin	5.71	11.43	292	Simeconazole	0.49	0.98
263	Chloridazon	0.39	0.78	293	Triadimefon	1.31	2.63
264	Sulfallate	34.53	69.07	294	Phorate sulfone	7.00	14.00
265	Ethiofencarb	0.82	1.64	295	Tridemorph	0.43	0.87
266	Terbumeton	0.02	0.03	296	Mefenacet	0.37	0.74
267	Cyprazine	0.01	0.01	297	Fenamiphos	0.03	0.07
268	Ametryn	0.16	0.32	298	Fenpropimorph	0.03	0.06
269	Tebuthiuron	0.04	0.07	299	Tebuconazole	0.37	0.74
270	Trietazine	0.10	0.20	300	Isopropalin	5.00	10.00
271	Sebutylazine	0.05	0.10	301	Nuarimol	0.17	0.33
272	Dibutyl succinate	37.07	74.13	302	Bupirimate	0.12	0.23
273	Tebutam	0.02	0.05	303	Azinphos-methyl	184.06	368.11

TABLE 2.1 The LOD, LOQ of 512 Pesticides and Related Chemical Residues Determined by LC-MS-MS—cont'd

No.	Pesticides	LOD (µg/kg)	LOQ (µg/kg)	No.	Pesticides	LOD (µg/kg)	LOQ (µg/kg)
304	Tebupirimfos	0.02	0.04		Group E		
305	Phenthoate	15.39	30.78	335	4-Aminopyridine	0.14	0.29
306	Sulfotep	0.43	0.87	336	Chlormequat	0.02	0.04
307	Sulprofos	0.97	1.95	337	Methomyl	1.59	3.19
308	EPN	5.50	11.00	338	Pyroquilon	0.58	1.16
309	Azamethiphos	0.13	0.27	339	Fuberidazole	0.32	0.63
310	Diniconazole	0.22	0.45	340	Isocarbamid	0.28	0.57
311	Flumetsulam	0.05	0.10	341	Butocarboxim	0.26	0.52
312	Sethoxydim	14.93	29.87	342	Chlordimeform	0.22	0.44
313	Pencycuron	0.05	0.09	343	Cymoxanil	9.27	18.53
314	Mecarbam	3.27	6.53	344	Vernolate	0.04	0.09
315	Tralkoxydim	0.05	0.11	345	Chlorthiamid	1.47	2.94
316	Malathion	0.94	1.88	346	Aminocarb	2.74	5.47
317	Pyributicarb	0.06	0.11	347	Dimethirimol	0.02	0.04
318	Pyridaphenthion	0.15	0.29	348	Omethoate	1.61	3.22
319	Pirimiphos-ethyl	0.01	0.02	349	Ethoxyquin	0.59	1.17
				350	Dichlorvos	0.09	0.18
320	Thiodicarb	6.56	13.12	351	Aldicarb Sulfone	3.56	7.12
321	Pyraclofos	0.17	0.33	352	Dioxacarb	0.56	1.12
322	Picoxystrobin	1.41	2.81	353	Benzyladenine	11.80	23.60
323	Tetraconazole	0.29	0.57	354	Demeton-S-Methyl	0.88	1.77
324	Mefenpyr-diethyl	2.09	4.19	355	Ethiofencarb-Sulfoxide	37.33	74.67
325	Profenefos	0.34	0.67	356	Cyanohos	0.00	0.00
326	Pyraclostrobin	0.08	0.17	357	Thiometon	96.33	192.67
327	Dimethomorph	0.06	0.12	358	Folpet	23.10	46.20
328	Kadethrin	0.55	1.11	359	Demeton-S-Methyl Sulfone	3.29	6.59
329	Thiazopyr	0.33	0.65				
330	Benfuracarb-methyl	2.73	5.46	360	Dimepiperate	630.00	1 260.00
331	Cinosulfuron	0.19	0.37	361	Fenpropidin	0.03	0.06
332	Pyrazosulfuron-ethyl	1.14	2.28	362	Imazapic	0.98	1.97
				363	Paraoxon-Ethyl	0.08	0.16
333	Metosulam	0.73	1.47	364	Aldimorph	0.53	1.05
334	Chlorfluazuron	1.45	2.89	365	Vinclozolin	0.42	0.85

Continued

TABLE 2.1 The LOD, LOQ of 512 Pesticides and Related Chemical Residues Determined by LC-MS-MS—cont'd

No.	Pesticides	LOD (µg/kg)	LOQ (µg/kg)	No.	Pesticides	LOD (µg/kg)	LOQ (µg/kg)
366	Uniconazole	0.40	0.80	396	Pyrazoxyfen	0.05	0.11
367	Pyrifenox	0.04	0.09	397	Flubenzimine	1.30	2.59
368	Chlorthion	22.27	44.53	398	Zeta Cypermethrin	0.11	0.23
369	Dicapthon	0.04	0.08	399	Haloxyfop-2-Ethoxyethyl	0.42	0.83
370	Clofentezine	0.13	0.25				
371	Norflurazon	0.04	0.09	400	Esfenvalerate	69.33	138.66
372	Triallate	7.70	15.40	401	Fluoroglycofen-Ethyl	0.83	1.67
373	Quinoxyphen	25.57	51.13	402	Tau-Fluvalinate	38.33	76.67
374	Fenthion Sulfone	2.91	5.82		Group F		
375	Flurochloridone	0.22	0.43	403	Acrylamide	2.97	5.93
376	Phthalic Acid, Benzyl Butyl Ester	105.33	210.67	404	Tert-Butylamine	6.49	12.98
				405	Hymexazol	37.36	74.71
377	Isazofos	0.03	0.06	406	Chlormequat Chloride	0.12	0.23
378	Dichlofenthion	4.99	9.98	407	Phthalimide	7.17	14.33
379	Vamidothion Sulfone	79.33	158.67	408	Dimefox	11.37	22.73
				409	Metolcarb	4.23	8.47
380	Terbufos Sulfone	14.77	29.53	410	Diphenylamin	0.07	0.14
381	Dinitramine	0.30	0.60	411	1-Naphthy Acetamide	0.14	0.27
382	Cyazofamid	0.75	1.50	412	Atrazine-Desethyl	0.10	0.21
383	Trichloronat	11.13	22.27	413	2,6-Dichloro-benzamide	0.75	1.50
384	Resmethrin-2	0.05	0.10				
385	Boscalid	0.79	1.59	414	Aldicarb	43.50	87.00
386	Nitralin	5.73	11.47	415	Dimethyl Phthalate	2.20	4.40
387	Fenpropathrin	40.83	81.67	416	Chlordimeform Hydrochloride	0.44	0.88
388	Hexythiazox	3.93	7.87				
389	Florasulam	2.90	5.80	417	Simeton	0.18	0.37
390	Benzoximate	3.28	6.55	418	Dinotefuran	1.70	3.39
391	Benzoylprop-Ethyl	51.33	102.67	419	Pebulate	0.57	1.13
392	Pyrimidifen	2.33	4.67	420	Acibenzolar-S-Methyl	0.51	1.03
393	Furathiocarb	0.32	0.64	421	Dioxabenzofos	2.31	4.61
394	Trans-Permethin	0.80	1.60	422	Oxamyl	91.34	182.69
395	Etofenprox	380.05	760.09	423	Thidiazuron	0.05	0.10

TABLE 2.1 The LOD, LOQ of 512 Pesticides and Related Chemical Residues Determined by LC-MS-MS—cont'd

No.	Pesticides	LOD (µg/kg)	LOQ (µg/kg)	No.	Pesticides	LOD (µg/kg)	LOQ (µg/kg)
424	Methabenzthiazuron	0.01	0.02	453	Cinidon-Ethyl	2.43	4.86
425	Butoxycarboxim	4.43	8.87	454	Imibenconazole	1.71	3.42
426	Mexacarbate	0.16	0.31	455	Propaquiafop	0.21	0.41
427	Demeton-S-Methyl Sulfoxide	0.65	1.31	456	Lactofen	10.33	20.67
				457	Benzofenap	0.01	0.03
428	Thiofanox Sulfone	4.01	8.03	458	Dinoseb Acetate	6.88	13.76
429	Phosfolan	0.08	0.16	459	Propisochlor	0.13	0.27
430	Demeton-S	13.33	26.67	460	Silafluofen,	101.33	202.67
431	Fenthionoxon	0.20	0.40	461	Etobenzanid	0.13	0.27
432	Napropamide	0.21	0.42	462	Fentrazamide	2.07	4.13
433	Fenitrothion	4.47	8.93	463	Pentachloro-aniline	0.62	1.25
434	Phthalic Acid, Dibutyl Ester	6.60	13.20	464	Cyphenothrin	2.80	5.60
435	Metolachlor	0.07	0.13	465	Dieldrin	26.93	53.87
436	Procymidone	14.43	28.87	466	Malaoxon	0.78	1.56
437	Vamidothion	0.76	1.52	467	Dodine	1.33	2.67
438	Triamiphos	0.00	0.00	468	Propylene Thiourea	5.01	10.03
439	Prallethrin	0.00	0.00		Group G		
440	Cumyluron	0.22	0.44	469	Dalapon	38.46	76.91
441	Imazamox	0.30	0.60	470	Flupropanate	3.83	7.66
442	Warfarin	0.45	0.89	471	2-Phenylphenol	28.31	56.63
443	Phosmet	2.95	5.91	472	3-Phenylphenol	0.67	1.33
444	Ronnel	2.19	4.38	473	Clopyralld	46.67	93.33
445	Pyrethrin	5.97	11.93	474	DNOC	0.43	0.87
446	Phthalic Acid, Biscyclohexyl Ester	0.11	0.23	475	Cloprop	1.90	3.80
447	Carpropamid	0.87	1.73	476	Dicloran	8.09	16.19
448	Tebufenpyrad	0.04	0.08	477	Aminopyralid	61.00	122.00
449	Tebufenozide	4.63	9.27	478	Chlorpropham	2.63	5.26
450	Dialifos	26.17	52.33	479	Mecoprop	0.82	1.63
451	Chlorthiophos	5.30	10.60	480	Terbacil	0.15	0.29
452	Rotenone	0.39	0.77	481	Dicamba	210.99	421.97

Continued

TABLE 2.1 The LOD, LOQ of 512 Pesticides and Related Chemical Residues Determined by LC-MS-MS—cont'd

No.	Pesticides	LOD (µg/kg)	LOQ (µg/kg)	No.	Pesticides	LOD (µg/kg)	LOQ (µg/kg)
482	MCPB	2.36	4.73	497	Gibberellic Acid	11.06	22.11
483	Dichlorprop	0.25	0.49	498	Acifluorfen	19.67	39.33
484	Bentazone	0.17	0.34	499	Heptachlor	0.05	0.10
485	Dinoseb	0.07	0.13	500	Famoxadone	7.55	15.10
486	Dinoterb	0.04	0.08	501	Sulfentrazone	14.93	29.87
487	Fludioxonil	10.36	20.72	502	Diflufenican	4.71	9.42
488	Chlorfenethol	27.38	54.77	503	Ethiprole	6.64	13.28
489	Isocarbophos	0.01	0.01	504	Flusulfamide	0.07	0.14
490	Naptalam	0.32	0.65	505	Cyclosulfamuron	57.28	114.56
491	Chlorobenzuron	3.40	6.80	506	Triforine	70.15	140.30
492	Chloramphenicolum	0.65	1.29	507	Halosulfuron-Methyl	1.63	3.27
493	Alloxydim-Sodium	0.03	0.07	508	Fluazinam	11.77	23.53
				509	Fluazuron	0.00	0.01
494	Pyrithlobac Sodium	230.33	460.67	510	Lufenuron	0.00	0.01
				511	Kelevan	1 607.14	3 214.27
495	Sulfanitran	0.51	1.01	512	Acrinathrin	1.35	2.69
496	Oryzalin	0.82	1.64				

Acetic Acid: Guaranteed Reagent

Formic Acid: Guaranteed Reagent

Ammonium Acetate: Guaranteed Reagent

Anhydrous Sodium Acetate: Analytically Pure

Membrane Filters: 13 mm, 0.2 µm

Waters Sep-Pak Vac Cartridge: 6 mL, 1 g, or Equivalent

0.1% Formic Acid-Water (V/V)

5 mmol/L Ammonium Acetate: Weigh 0.375 g Ammonium Acetate, Dilute With Water to 1000 mL

Acetonitrile + Toluene (3+1, V/V)

1% Acetic Acid in Acetonitrile (V/V)

Acetonitrile + Water(3+2, V/V)

Anhydrous Sodium Sulfate, Anhydrous Magnesium Sulfate: Analytically Pure, Heated at 650°C for 4 h and Kept in a Desiccator

Pesticide and Related Chemicals Standard: Purity ≥95%

Stock Standard Solution: Accurately weigh 5–10 mg of individual pesticide and related chemical standards (accurate to 0.1 mg) into a 10-mL volumetric flask. Dissolve and dilute to volume with methanol, toluene, acetone, acetonitrile, or isooctane, etc., depending on each individual compound's solubility (for diluting solvent refer to Appendix B). Standard solutions are stored in the dark below 4°C and are used for one month.

Mixed Standard Solution: Depending on properties and retention times of compounds, all compounds are divided into seven groups: A, B, C, D, E, F, and G. The concentration of each compound is determined by sensitivity of the instrument for analysis. For 512 pesticides and related chemicals, grouping and concentration are listed in Appendix B.

Depending on group number, mixed standard solution concentration, and stock standard solution concentration, appropriate amounts of individual stock standard solution are pipetted into a 100-mL volumetric flask, being diluted to volume with methanol. Mixed standard solutions are stored in the dark below 4°C and are used for 1 month.

Working Standard Mixed Solution in Matrix: The working standard mixture solution in matrix is prepared by diluting an appropriate amount of mixed standard solution with blank extract, which has gone through the method with the rest of the samples. Mix thoroughly. They are used for plotting the standard curve. The working standard mixture solution in matrix must be freshly prepared.

2.1.4   Apparatus

LC-MS-MS: Equipped With ESI
Analytical Balance: Capable Weighing From 0.1 mg to 0.01 g
Pear-Shaped Flask: 200 mL
Pipette: 1 mL
Screw Vial: 2.0 mL, With Screw Caps and PTFE septa
Glass Mason Jar With Cap: 50 mL
Vortex
Nitrogen Evaporator
Low Speed Centrifuge: 4200 rpm
Rotary Evaporator

2.1.5   Sample Pretreatment

2.1.5.1   Preparation of Test Sample

For fruit juice and wine samples, mix thoroughly by stirring in clean enamel bins. Pour 0.5 kg of the prepared test sample into a clean and dry sample bottle. Seal and label.

The test samples should be stored in a frozen state.

2.1.5.2 Extraction

Weigh 15-g test sample (accurate to 0.01 g) (the wine sample is 15 mL) into a 50-mL glass mason jar, add 15 mL 1% of acetate acid in acetonitrile, and vortex mixture for 15 min. Then add 1.5 g anhydrous sodium acetate to the jar and then vortex for 1 min. Then add 6 g anhydrous magnesium sulfate to the jar and vortex for 2 min. After centrifugation at 4200 rpm for 5 min, 7.5 mL of the supernatant is transferred to another jar to await clean-up.

2.1.5.3 Clean-up

Add sodium sulfate into the Sep-Pak Vac cartridge to ca. 2 cm. Fix the cartridge into a support to which a pear-shaped flask is connected. Condition the cartridge with 5 mL acetonitrile-toluene (3 + 1) before adding the sample. Once the solution gets to the top of the sodium sulfate, pipette the supernatant into the cartridge immediately. Insert a reservoir into the cartridges. Elute the pesticides with 25 mL acetonitrile-toluene (3 + 1). Evaporate the eluate to ca. 0.5 mL using rotary evaporator at 40°C, evaporate the solutions to dryness by nitrogen evaporate at 35°C. Reconstitute the sample to 1 mL with acetonitrile-water (3 + 2). Finally the extract is filtered through a 0.2 μm filter into a glass vial for LC-MS-MS determination.

2.1.6 Determination

2.1.6.1 LC-MS-MS Operating Condition

Conditions for the Pesticides and Related Chemicals of A, B, C, D, E, and F Groups

(a) Chromatography column: ZORBAX SB-C_{18}, 3.5 μm, 100 × 2.1 mm (diameter), or equivalent
(b) Mobile phase program and flow rate: as Table 2.2
(c) Column temperature: 40°C
(d) Injection volume: 10 μL
(e) Ion source: electrospray ionization source
(f) Ion mode: positive
(g) Nebulizer gas: nitrogen
(h) Nebulizer gas pressure: 0.28 MPa
(i) Ion spray voltage: 4000 V
(j) Nebulizer gas temperature: 350°C
(k) Nebulizer gas flow rate: 10 L/min
(l) Monitoring ion pairs, collision energy and fragmentor (see Table 5.10)

Conditions for the Pesticides and Related Chemicals of G Group

(a) Chromatography column: ZORBAX SB-C_{18}, 3.5 μm, 100 × 2.1 mm (diameter), or equivalent

TABLE 2.2 Mobile Phase Program and the Flow Rate

Step	Time (min)	Flow Rate (μL/min)	Mobile Phase A 0.1% Formic Acid-Water (%)	Mobile Phase B Acetonitrile (%)
0	0.00	400	99.0	1.0
1	3.00	400	70.0	30.0
2	6.00	400	60.0	40.0
3	9.00	400	60.0	40.0
4	15.00	400	40.0	60.0
5	19.00	400	1.0	99.0
6	23.00	400	1.0	99.0
7	23.01	400	99.0	1.0

TABLE 2.3 Mobile Phase Program and Flow Rate

Step	Time (min)	Flow Rate (μL/min)	Mobile Phase A 5 mmol/L Ammonium Acetate-Water (%)	Mobile Phase B Acetonitrile (%)
0	0.00	400	99.0	1.0
1	3.00	400	70.0	30.0
2	6.00	400	60.0	40.0
3	9.00	400	60.0	40.0
4	15.00	400	40.0	60.0
5	19.00	400	1.0	99.0
6	23.00	400	1.0	99.0
7	23.01	400	99.0	1.0

(b) Mobile phase program and flow rate: as Table 2.3
(c) Column temperature: 40°C
(d) Injection volume: 10 μL
(e) Ion source: electrospray ionization source
(f) Ion mode: negative mode
(g) Nebulizer gas: nitrogen

(h) Nebulizer gas pressure: 0.28 MPa
 (i) Ion spray voltage: 4000 V
 (j) Nebulizer gas temperature; 350°C
(k) Nebulizer gas flow rate: 10 L/min
 (l) Monitoring ion pairs, collision energy and fragmentor (see Table 5.10)

2.1.6.2 Qualitative Determination

If the retention times of peaks of the sample solution are the same as those of the peaks from the working standard mixed solution, the selected ions appearing in the background-subtracted mass spectrum, and the abundance ratios of the selected ions are within the expected limits (abundance ratios >50%, permitted tolerances are ±10%; abundance ratios >20% to 50%, permitted tolerances are ±15%; abundance ratios >10% to 20%, permitted tolerances are ±20%; abundance ratios ≤10%, permitted tolerances are ±50%), the sample is confirmed to contain this pesticide compound.

2.1.6.3 Quantitative Determination

The external standard method is used for quantitation with standard curves for LC-MS-MS. To compensate for the matrix effect, quantitation is based on a series of working standard solutions prepared in blank matrix extract. The standard curves are established by injection of different concentrations of working standard mixed solutions in matrix separately. The responses of pesticides in the sample solution should be in the linear range of the instrumental detection.

2.1.6.4 Parallel Test

A parallel test is carried out for the same testing sample.

2.1.6.5 Blank Test

The operation of the blank test is the same as that described in the method of determination, but without the addition of sample.

2.1.7 Precision

The precision data of the method for this standard is according to the stipulations of GB/T 6379.1 and GB/T 6379.2. The values of repeatability and reproducibility are obtained and calculated at a 95% confidence level.

RESEARCHERS

Guo-Fang Pang, Yan Li, Chun-Lin Fan, Yu-Jing Lian, Xin-xin Ji, Cui-cui-Yao, Shu-jun Zhao, Jun-Hong Zheng, Yong-ming Liu, Yan-Zhong Cao, Jin-Jie Zhang, Xue-Min Li.

Qinhuangdao Entry-Exit Inspection and Quarantine Bureau, 39 Haibin Rd, Qinhuangdao, Hebei, PC 066002, People's Republic of China.

2.2   DETERMINATION OF 500 PESTICIDES AND RELATED CHEMICAL RESIDUES IN FRUITS AND VEGETABLES: GC-MS METHOD (GB/T 19648-2006)

2.2.1   Scope

This method is applicable to the determination of the 500 pesticides and related chemicals residues in apple, orange, grape, cabbage, celery, tomato.

The limits of detection of this method are 0.0063 mg/kg to 0.8000 mg/kg (see Table 2.4).

2.2.2   Principle

The samples are homogenized with acetonitrile and sodium chloride. The solutions are centrifuged, and the supernatants of the acetonitrile phase are cleaned up with SPE cartridges. The pesticides and related chemicals are eluted with acetonitrile + toluene (3 + 1), and the solutions are analyzed by GC-MS.

2.2.3   Reagents and Materials

Acetonitrile: HPLC Grade
Sodium Chloride: G.R.
Sodium Sulfate: Anhydrous, Analytically Pure. Ignited at 650°C for 4 h and Kept in a Desiccator
Toluene: G.R.
Acetone: Analytically Pure, All Glass Distilled
Dichloromethane: Pesticide Grade
Hexane: Analytically Pure, All Glass Distilled
Envi-18 SPE Cartridge: 12 mL, 2.0 g, or Equivalent
Envi-Carb SPE Cartridge: 6 mL, 0.5 g, or Equivalent
Sep-Pak NH_2 Cartridge: 12 mL, 2.0 g, or Equivalent
Pesticide and Related Chemicals Standard: Purity ≥95%
   Stock Standard Solution: Accurately weigh 5–10 mg of individual pesticide and related chemical standards (accurate to 0.1 mg) into a 10-mL volumetric flask. Dissolve and dilute to volume with toluene, toluene + acetone combination, dichloromethane, etc., depending on each individual compound's solubility. (For diluting solvent refer to Appendix B)
   Mixed Standard Solution (Mixed Standard Solution A, B, C, D and E): Depending on properties and retention times of compounds, all compounds are divided into five groups of A, B, C, D, and E. The mixed standard solution concentrations are determined by their sensitivity on the

TABLE 2.4 The LOD, LOQ of 500 Pesticides and Related Chemical Residues Determined by GC-MS

No.	Pesticides	LOD (mg/kg)	LOQ (mg/kg)	No.	Pesticides	LOD (mg/kg)	LOQ (mg/kg)
	Group A			29	Vinclozolin	0.006 3	0.012 6
1	Allidochlor	0.012 5	0.025 0	30	beta-HCH	0.006 3	0.012 6
2	Dichlormid	0.012 5	0.025 0	31	Metalaxyl	0.018 8	0.037 6
3	Etridiazol	0.018 8	0.037 6	32	Chlorpyrifos (-ethyl)	0.006 3	0.012 6
4	Chlormephos	0.012 5	0.025 0				
5	Propham	0.006 3	0.012 6	33	Methyl-Parathion	0.025 0	0.050 0
6	Cycloate	0.006 3	0.012 6	34	Anthraquinone	0.006 3	0.012 6
7	Diphenylamine	0.006 3	0.012 6	35	Delta-HCH	0.012 5	0.025 0
8	Chlordimeform	0.006 3	0.012 6	36	Fenthion	0.006 3	0.012 6
9	Ethalfluralin	0.025 0	0.050 0	37	Malathion	0.025 0	0.050 0
10	Phorate	0.006 3	0.012 6	38	Fenitrothion	0.012 5	0.025 0
11	Thiometon	0.006 3	0.012 6	39	Paraoxon-ethyl	0.200 0	0.400 0
12	Quintozene	0.012 5	0.025 0	40	Triadimefon	0.012 5	0.025 0
13	Atrazine-desethyl	0.006 3	0.012 6	41	Parathion	0.025 0	0.050 0
				42	Pendimethalin	0.025 0	0.050 0
14	Clomazone	0.006 3	0.012 6	43	Linuron	0.025 0	0.050 0
15	Diazinon	0.006 3	0.012 6	44	Chlorbenside	0.0 125	0.025 0
16	Fonofos	0.006 3	0.012 6	45	Bromophos-ethyl	0.006 3	0.012 6
17	Etrimfos	0.006 3	0.012 6	46	Quinalphos	0.006 3	0.012 6
18	Simazine	0.006 3	0.012 6	47	trans-Chlordane	0.006 3	0.012 6
19	Propetamphos	C.006 3	0.012 6	48	Phenthoate	0.012 5	0.025 0
20	Secbumeton	0.006 3	0.012 6	49	Metazachlor	0.018 8	0.037 6
21	Dichlofenthion	0.006 3	0.012 6	50	Fenothiocarb	0.012 5	0.025 0
22	Pronamide	0.006 3	0.012 6	51	Prothiophos	0.006 3	0.012 6
23	Mexacarbate	0.018 8	0.037 6	52	Chlorfurenol	0.018 8	0.037 6
24	Aldrin	0.012 5	0.025 0	53	Dieldrin	0.012 5	0.025 0
25	Dinitramine	0.025 0	0.050 0	54	Procymidone	0.006 3	0.012 6
26	Ronnel	0.012 5	0.025 0	55	Methidathion	0.012 5	0.025 0
27	Prometryne	0.006 3	0.012 6	56	Cyanazine	0.018 8	0.037 6
28	Cyprazine	0.006 3	0.012 6	57	Napropamide	0.018 8	0.037 6

TABLE 2.4 The LOD, LOQ of 500 Pesticides and Related Chemical Residues Determined by GC-MS—cont'd

No.	Pesticides	LOD (mg/kg)	LOQ (mg/kg)	No.	Pesticides	LOD (mg/kg)	LOQ (mg/kg)
58	Oxadiazone	0.006 3	0.012 6	87	cis-Permethrin	0.006 3	0.012 6
59	Fenamiphos	0.018 8	0.037 6	88	Trans-Permethrin	0.006 3	0.012 6
60	Tetrasul	0.006 3	0.012 6	89	Pyrazophos	0.012 5	0.025 0
61	Aramite	0.006 3	0.012 6	90	Cypermethrin	0.018 8	0.037 6
62	Bupirimate	0.006 3	0.012 6	91	Fenvalerate	0.025 0	0.050 0
63	Carboxin	0.150 0	0.300 0	92	Deltamethrin	0.037 5	0.075 0
64	Flutolanil	0.006 3	0.012 6		Group B		
65	4,4'-DDD	0.006 3	0.012 6	93	EPTC	0.018 8	0.037 6
66	Ethion	0.012 5	0.025 0	94	Butylate	0.018 8	0.037 6
67	Sulprofos	0.012 5	0.025 0	95	Dichlobenil	0.001 3	0.002 6
68	Etaconazole-1	0.018 8	0.037 6	96	Pebulate	0.018 8	0.037 6
69	Etaconazole-2	0.018 8	0.037 6	97	Nitrapyrin	0.018 8	0.037 6
70	Myclobutanil	0.006 3	0.012 6	98	Mevinphos	0.012 5	0.025 0
71	Diclofop-methyl	0.006 3	0.012 6	99	Chloroneb	0.006 3	0.012 6
				100	Tecnazene	0.012 5	0.025 0
72	Propiconazole	0.018 8	0.037 6	101	Heptenophos	0.018 8	0.037 6
73	Fensulfothion	0.012 5	0.025 0	102	Hexachlorobenzene	0.006 3	0.012 6
74	Bifenthrin	0.006 3	0.012 6	103	Ethoprophos	0.018 8	0.037 6
75	Mirex	0.006 3	0.012 6	104	cis -Diallate	0.012 5	0.025 0
76	Benodanil	0.018 8	0.037 6	105	Propachlor	0.018 8	0.037 6
77	Nuarimol	0.012 5	0.025 0	106	trans-Diallate	0.012 5	0.025 0
78	Methoxychlor	0.050 0	0.100 0	107	Trifluralin	0.012 5	0.025 0
79	Oxadixyl	0.006 3	0.012 6	108	Chlorpropham	0.012 5	0.025 0
80	Tetramethirn	0.012 5	0.025 0	109	Sulfotep	0.006 3	0.012 6
81	Tebuconazole	0.018 8	0.037 6	110	Sulfallate	0.012 5	0.025 0
82	Norflurazon	0.006 3	0.012 6	111	Alpha-HCH	0.006 3	0.012 6
83	Pyridaphenthion	0.006 3	0.012 6	112	Terbufos	0.012 5	0.025 0
84	Phosmet	0.012 5	0.025 0	113	Terbumeton	0.018 8	0.037 6
85	Tetradifon	0.006 3	0.012 6	114	Profluralin	0.025 0	0.050 0
86	Oxycarboxin	0.037 5	0.075 0				

Continued

TABLE 2.4 The LOD, LOQ of 500 Pesticides and Related Chemical Residues Determined by GC-MS—cont'd

No.	Pesticides	LOD (mg/kg)	LOQ (mg/kg)	No.	Pesticides	LOD (mg/kg)	LOQ (mg/kg)
115	Dioxathion	0.025 0	0.050 0	143	Crufomate	0.037 5	0.075 0
116	Propazine	0.006 3	0.012 6	144	Chlorfenvinphos	0.018 8	0.037 6
117	Chlorbufam	0.012 5	0.025 0	145	Cis-Chlordane	0.012 5	0.025 0
118	Dicloran	0.012 5	0.025 0	146	Tolylfluanide	0.150 0	0.300 0
119	Terbuthylazine	0.006 3	0.012 6	147	4,4′-DDE	0.006 3	0.012 6
120	Monolinuron	0.025 0	0.050 0	148	Butachlor	0.012 5	0.025 0
121	Flufenoxuron	0.018 8	0.037 6	149	Chlozolinate	0.012 5	0.025 0
122	Cyanophos	0.012 5	0.025 0	150	Crotoxyphos	0.037 5	0.075 0
123	Chlorpyrifos-methyl	0.006 3	0.012 6	151	Iodofenphos	0.012 5	0.025 0
				152	Tetrachlorvinphos	0.018 8	0.037 6
124	Desmetryn	0.006 3	0.012 6	153	Chlorbromuron	0.150 0	0.300 0
125	Dimethachlor	0.018 8	0.037 6	154	Profenofos	0.037 5	0.075 0
126	Alachlor	0.018 8	0.037 6	155	Fluorochloridone	0.012 5	0.025 0
127	Pirimiphos-methyl	0.006 3	0.012 6	156	Buprofezin	0.012 5	0.025 0
128	Terbutryn	0.012 5	0.025 0	157	2,4′-DDD	0.006 3	0.012 6
129	Thiobencarb	0.012 5	0.025 0	158	Endrin	0.075 0	0.150 0
130	Aspon	0.012 5	0.025 0	159	Hexaconazole	0.037 5	0.075 0
131	Dicofol	0.012 5	0.025 0	160	Chlorfenson	0.012 5	0.025 0
132	Metolachlor	0.006 3	0.012 6	161	2,4′-DDT	0.012 5	0.025 0
133	Oxy-chlordane	0.006 3	0.012 6	162	Paclobutrazol	0.018 8	0.037 6
134	Pirimiphos-ethyl	0.012 5	0.025 0	163	Methoprotryne	0.018 8	0.037 6
				164	Erbon	0.012 5	0.025 0
135	Methoprene	0.025 0	0.050 0	165	Chloro-propylate	0.006 3	0.012 6
136	Bromofos	0.012 5	0.025 0	166	Flamprop-methyl	0.006 3	0.012 6
137	Dichlofluanid	0.300 0	0.600 0	167	Nitrofen	0.037 5	0.075 0
138	Ethofumesate	0.012 5	0.025 0	168	Oxyfluorfen	0.025 0	0.050 0
139	Isopropalin	0.012 5	0.025 0	169	Chlorthiophos	0.018 8	0.037 6
140	Endosulfan-1	0.037 5	0.075 0	170	Endosulfan-II	0.037 5	0.075 0
141	Propanil	0.012 5	0.025 0	171	Flamprop-Isopropyl	0.006 3	0.012 6
142	Isofenphos	0.012 5	0.025 0				

TABLE 2.4 The LOD, LOQ of 500 Pesticides and Related Chemical Residues Determined by GC-MS—cont'd

No.	Pesticides	LOD (mg/kg)	LOQ (mg/kg)	No.	Pesticides	LOD (mg/kg)	LOQ (mg/kg)
172	4,4'-DDT	0.012 5	0.025 0	201	Cis-1,2,3,6-Tetrahydrophthalimide	0.018 8	0.037 6
173	Carbofenothion	0.012 5	0.025 0	202	Fenobucarb	0.012 5	0.025 0
174	Benalaxyl	0.006 3	0.012 6	203	Benfluralin	0.006 3	0.012 6
175	Edifenphos	0.012 5	0.025 0	204	Hexaflumuron	0.037 5	0.075 0
176	Triazophos	0.018 8	0.037 6	205	Prometon	0.018 8	0.037 6
177	Cyanofenphos	0.006 3	0.012 6	206	Triallate	0.012 5	0.025 0
178	Chlorbenside sulfone	0.012 5	0.025 0	207	Pyrimethanil	0.006 3	0.012 6
179	Endosulfan-Sulfate	0.018 8	0.037 6	208	Gamma-HCH	0.012 5	0.025 0
180	Bromopropylate	0.012 5	0.025 0	209	Disulfoton	0.006 3	0.012 6
181	Benzoylprop-ethyl	0.018 8	0.037 6	210	Atrizine	0.006 3	0.012 6
182	Fenpropathrin	0.012 5	0.025 0	211	Heptachlor	0.018 8	0.037 6
183	Leptophos	0.012 5	0.025 0	212	Iprobenfos	0.018 8	0.037 6
184	EPN	0.025 0	0.050 0	213	Isazofos	0.012 5	0.025 0
185	Hexazinone	0.018 8	0.037 6	214	Plifenate	0.012 5	0.025 0
186	Phosalone	0.012 5	0.025 0	215	Fenpropimorph	0.006 3	0.012 6
187	Azinphos-methyl	0.037 5	0.075 0	216	Transfluthrin	0.006 3	0.012 6
188	Fenarimol	0.012 5	0.025 0	217	Fluchloralin	0.025 0	0.050 0
189	Azinphos-ethyl	0.012 5	0.025 0	218	Tolclofos-methyl	0.006 3	0.012 6
190	Prochloraz	0.037 5	0.075 0	219	Propisochlor	0.006 3	0.012 6
191	Coumaphos	0.037 5	0.075 0	220	Ametryn	0.018 8	0.037 6
192	Cyfluthrin	0.075 0	0.150 0	221	Simetryn	0.012 5	0.025 0
193	Fluvalinate	0.075 0	0.150 0	222	Metobromuron	0.037 5	0.075 0
	Group C			223	Metribuzin	0.018 8	0.037 6
194	Dichlorvos	0.037 5	0.075 0	224	Dimethipin	0.018 8	0.037 6
195	Biphenyl	0.006 3	0.012 6	225	HCH, epsilon-	0.012 5	0.025 0
196	Vernolate	0.006 3	0.012 6	226	Dipropetryn	0.006 3	0.012 6
197	3,5-Dichloroaniline	0.050 0	0.100 0	227	Formothion	0.012 5	0.025 0
198	Molinate	0.006 3	0.012 6	228	Diethofencarb	0.037 5	0.075 0
199	Methacrifos	0.006 3	0.012 6	229	Dimepiperate	0.012 5	0.025 0
200	2-Phenylphenol	0.006 3	0.012 6	230	Bioallethrin-1	0.025 0	0.050 0

Continued

TABLE 2.4 The LOD, LOQ of 500 Pesticides and Related Chemical Residues Determined by GC-MS—cont'd

No.	Pesticides	LOD (mg/kg)	LOQ (mg/kg)	No.	Pesticides	LOD (mg/kg)	LOQ (mg/kg)
231	Bioallethrin-2	0.025 0	0.050 0	261	Sethoxydim	0.450 0	0.900 0
232	2,4'-DDE	0.006 3	0.012 6	262	Anilofos	0.012 5	0.025 0
233	Fenson	0.006 3	0.012 6	263	Acrinathrin	0.012 5	0.025 0
234	Diphenamid	0.006 3	0.012 6	264	Lambda-Cyhalothrin	0.006 3	0.012 6
235	Chlorthion	0.012 5	0.025 0				
236	Prallethrin	0.018 8	0.037 6	265	Mefenacet	0.018 8	0.037 6
237	Penconazole	0.018 8	0.037 6	266	Permethrin	0.012 5	0.025 0
238	Mecarbam	0.025 0	0.050 0	267	Pyridaben	0.006 3	0.012 6
239	Tetraconazole	0.018 8	0.037 6	268	Fluoroglycofen-ethyl	0.075 0	0.150 0
240	Propaphos	0.012 5	0.025 0	269	Bitertanol	0.018 8	0.037 6
241	Flumetralin	0.012 5	0.025 0	270	Etofenprox	0.006 3	0.012 6
242	Triadimenol	0.018 8	0.037 6	271	Cycloxydim	0.600 0	1.200 0
243	Pretilachlor	0.012 5	0.025 0	272	Alpha-Cypermethrin	0.012 5	0.025 0
244	Kresoxim-methyl	0.006 3	0.012 6	273	Flucythrinate	0.012 5	0.025 0
245	Fluazifop-butyl	0.006 3	0.012 6	274	Esfenvalerate	0.025 0	0.050 0
246	Chlorfluazuron	0.018 8	0.037 6	275	Difenonazole	0.037 5	0.075 0
247	Chlorobenzilate	0.006 3	0.012 6	276	Flumioxazin	0.012 5	0.025 0
248	Uniconazole	0.012 5	0.025 0	277	Flumiclorac-pentyl	0.012 5	0.025 0
249	Flusilazole	0.018 8	0.037 6		**Group D**		
250	Fluorodifen	0.006 3	0.012 6	278	Dimefox	0.018 8	0.037 6
251	Diniconazole	0.018 8	0.037 6	279	Disulfoton-sulfoxide	0.012 5	0.025 0
252	Piperonyl butoxide	0.006 3	0.012 6	280	Pentachlorobenzene	0.006 3	0.012 6
253	Propargite	0.012 5	0.025 0	281	Tri-iso-butyl phosphate	0.006 3	0.012 6
254	Mepronil	0.006 3	0.012 6	282	Crimidine	0.006 3	0.012 6
255	Dimefuron	0.025 0	0.050 0	283	BDMC-1	0.012 5	0.025 0
256	Diflufenican	0.006 3	0.012 6	284	Chlorfenprop-methyl	0.006 3	0.012 6
257	Fenazaquin	0.006 3	0.012 6	285	Thionazin	0.006 3	0.012 6
258	Phenothrin	0.006 3	0.012 6	286	2,3,5,6-tetrachloroaniline	0.006 3	0.012 6
259	Fludioxonil	0.006 3	0.012 6				
260	Fenoxycarb	0.037 5	0.075 0	287	Tri-n-butyl phosphate	0.012 5	0.025 0

TABLE 2.4 The LOD, LOQ of 500 Pesticides and Related Chemical Residues Determined by GC-MS—cont'd

No.	Pesticides	LOD (mg/kg)	LOQ (mg/kg)	No.	Pesticides	LOD (mg/kg)	LOQ (mg/kg)
288	2,3,4,5-tetrachloroanisole	0.006 3	0.012 6	315	Prosulfocarb	0.006 3	0.012 6
289	Pentachloroanisole	0.006 3	0.012 6	316	Dimethenamid	0.006 3	0.012 6
290	Tebutam	0.012 5	0.025 0	317	Fenchlorphos oxon	0.012 5	0.025 0
291	Dioxabenzofos	0.062 5	0.125 0	318	BDMC-2	0.025 0	0.050 0
292	Methabenzthiazuron	0.062 5	0.125 0	319	Paraoxon-methyl	0.012 5	0.025 0
293	Simetone	0.012 5	0.025 0	320	Monalide	0.012 5	0.025 0
294	Atratone	0.006 3	0.012 6	321	Musk tibeten	0.006 3	0.012 6
295	Desisopropyl-atrazine	0.050 0	0.100 0	322	Isobenzan	0.006 3	0.012 6
296	Terbufos sulfone	0.006 3	0.012 6	323	Octachlorostyrene	0.006 3	0.012 6
297	Tefluthrin	0.006 3	0.012 6	324	Pyrimitate	0.006 3	0.012 6
298	Bromocylen	0.006 3	0.012 6	325	Isodrin	0.006 3	0.012 6
299	Trietazine	0.006 3	0.012 6	326	Isomethiozin	0.012 5	0.025 0
300	Etrimfos oxon	0.006 3	0.012 6	327	Trichloronat	0.006 3	0.012 6
301	Cycluron	0.018 8	0.037 6	328	Dacthal	0.006 3	0.012 6
302	2,6-dichlorobenzamide	0.012 5	0.025 0	329	4,4-dichlorobenzophenone	0.006 3	0.012 6
303	DE-PCB 28	0.006 3	0.012 6	330	Nitrothal-isopropyl	0.012 5	0.025 0
304	DE-PCB 31	0.006 3	0.012 6	331	Musk ketone	0.006 3	0.012 6
305	Desethyl-sebuthylazine	0.012 5	0.025 0	332	Rabenzazole	0.006 3	0.012 6
306	2,3,4,5-tetrachloroaniline	0.012 5	0.025 0	333	Cyprodinil	0.006 3	0.012 6
307	Musk ambrette	0.006 3	0.012 6	334	Fuberidazole	0.031 3	0.062 6
308	Musk xylene	0.006 3	0.012 6	335	Isofenphos oxon	0.012 5	0.025 0
309	Pentachloroaniline	0.006 3	0.012 6	336	Dicapthon	0.031 3	0.062 6
310	Aziprotryne	0.050 0	0.100 0	337	DE-PCB 101	0.006 3	0.012 6
311	Sebutylazine	0.006 3	0.012 6	338	MCPA-butoxyethyl ester	0.006 3	0.012 6
312	Isocarbamid	0.031 3	0.062 6	339	Isocarbophos	0.012 5	0.025 0
313	DE-PCB 52	0.006 3	0.012 6	340	Phorate sulfone	0.006 3	0.012 6
314	Musk moskene	0.006 3	0.012 6	341	Chlorfenethol	0.006 3	0.012 6
				342	Trans-nonachlor	0.006 3	0.012 6
				343	Dinobuton	0.062 5	0.125 0

Continued

TABLE 2.4 The LOD, LOQ of 500 Pesticides and Related Chemical Residues Determined by GC-MS—cont'd

No.	Pesticides	LOD (mg/kg)	LOQ (mg/kg)	No.	Pesticides	LOD (mg/kg)	LOQ (mg/kg)
344	DEF	0.012 5	0.025 0	372	Tebufenpyrad	0.006 3	0.012 6
345	Flurochloridone	0.012 5	0.025 0	373	Cloquintocet-mexyl	0.006 3	0.012 6
346	Bromfenvinfos	0.006 3	0.012 6	374	Lenacil	0.062 5	0.125 0
347	Perthane	0.006 3	0.012 6	375	Bromuconazole-1	0.012 5	0.025 0
348	Ditalimfos	0.006 3	0.012 6	376	Desbrom- leptophos	0.006 3	0.012 6
349	DE-PCB 118	0.006 3	0.012 6	377	Bromuconazole-2	0.012 5	0.025 0
350	4,4-dibromobenzophenone	0.006 3	0.012 6	378	Nitralin	0.062 5	0.125 0
				379	Fenamiphos sulfoxide	0.200 0	0.400 0
351	Flutriafol	0.012 5	0.025 0	380	Fenamiphos sulfone	0.025 0	0.050 0
352	Mephosfolan	0.012 5	0.025 0	381	Fenpiclonil	0.025 0	0.050 0
353	Athidathion	0.012 5	0.025 0	382	Fluquinconazole	0.006 3	0.012 6
354	DE-PCB 153	0.006 3	0.012 6	383	Fenbuconazole	0.012 5	0.025 0
355	Diclobutrazole	0.025 0	0.050 0		**Group E**		
356	Disulfoton sulfone	0.012 5	0.025 0	384	Propoxur-1	0.012 5	0.025 0
357	Hexythiazox	0.050 0	0.100 0	385	Isoprocarb -1	0.012 5	0.025 0
358	DE-PCB 138	0.006 3	0.012 6	386	Methamidophos	0.200 0	0.400 0
359	Triamiphos	0.012 5	0.025 0	387	Acenaphthene	0.006 3	0.012 6
360	Resmethrin-1	0.100 0	0.200 0	388	Dibutyl succinate	0.012 5	0.025 0
361	Cyproconazole	0.006 3	0.012 6	389	Phthalimide	0.012 5	0.025 0
362	Resmethrin-2	0.100 0	0.200 0	390	Chlorethoxyfos	0.012 5	0.025 0
363	Phthalic acid, benzyl butyl ester	0.006 3	0.012 6	391	Isoprocarb -2	0.012 5	0.025 0
364	Clodinafop-propargyl	0.012 5	0.025 0	392	Pencycuron	0.012 5	0.025 0
365	Fenthion sulfoxide	0.025 0	0.050 0	393	Tebuthiuron	0.025 0	0.050 0
366	Fluotrimazole	0.006 3	0.012 6	394	demeton-S-methyl	0.025 0	0.050 0
367	Fluroxypr-1-methylheptyl ester	0.006 3	0.012 6	395	Cadusafos	0.025 0	0.050 0
				396	Propoxur-2	0.012 5	0.025 0
368	Fenthion sulfone	0.025 0	0.050 0	397	Phenanthrene	0.006 3	0.012 6
369	Triphenyl phosphate	0.006 3	0.012 6	398	Spiroxamine -1	0.012 5	0.025 0
370	Metamitron	0.062 5	0.125 0	399	Fenpyroximate	0.050 0	0.100 0
371	DE-PCB 180	0.006 3	0.012 6	400	Tebupirimfos	0.012 5	0.025 0

TABLE 2.4 The LOD, LOQ of 500 Pesticides and Related Chemical Residues Determined by GC-MS—cont'd

No.	Pesticides	LOD (mg/kg)	LOQ (mg/kg)	No.	Pesticides	LOD (mg/kg)	LOQ (mg/kg)
401	prohydrojasmon	0.025 0	0.050 0	431	Quinoclamine	0.025 0	0.050 0
402	Fenpropidin	0.012 5	0.025 0	432	Methothrin-1	0.012 5	0.025 0
403	Dichloran	0.012 5	0.025 0	433	Flufenacet	0.050 0	0.100 0
404	Pyroquilon	0.006 3	0.012 6	434	Methothrin-2	0.012 5	0.025 0
405	Spiroxamine -2	0.012 5	0.025 0	435	Pyrifenox -2	0.050 0	0.100 0
406	Propyzamide	0.012 5	0.025 0	436	Fenoxanil	0.012 5	0.025 0
407	Pirimicarb	0.012 5	0.025 0	437	Phthalide	0.025 0	0.050 0
408	Phosphamidon -1	0.050 0	0.100 0	438	Furalaxyl	0.012 5	0.025 0
409	Benoxacor	0.012 5	0.025 0	439	Thiamethoxam	0.025 0	0.050 0
410	Bromobutide	0.006 3	0.012 6	440	Mepanipyrim	0.006 3	0.012 6
411	Acetochlor	0.012 5	0.025 0	441	Captan	0.400 0	0.800 0
412	Tridiphane	0.025 0	0.050 0	442	Bromacil	0.050 0	0.100 0
413	Terbucarb	0.012 5	0.025 0	443	Picoxystrobin	0.012 5	0.025 0
414	Esprocarb	0.012 5	0.025 0	444	Butamifos	0.006 3	0.012 6
415	Fenfuram	0.012 5	0.025 0	445	Imazamethabenz-methyl	0.018 8	0.037 6
416	Acibenzolar-S-Methyl	0.012 5	0.025 0	446	Metominostrobin-1	0.025 0	0.050 0
417	Benfuresate	0.012 5	0.025 0	447	TCMTB	0.100 0	0.200 0
418	Dithiopyr	0.006 3	0.012 6	448	Methiocarb Sulfone	0.800 0	1.600 0
419	Mefenoxam	0.012 5	0.025 0	449	Imazalil	0.025 0	0.050 0
420	Malaoxon	0.100 0	0.200 0	450	Isoprothiolane	0.012 5	0.025 0
421	Phosphamidon -2	0.050 0	0.100 0	451	Cyflufenamid	0.100 0	0.200 0
422	Simeconazole	0.012 5	0.025 0	452	Pyriminobac-Methyl	0.025 0	0.050 0
423	Chlorthal-dimethyl	0.012 5	0.025 0	453	Isoxathion	0.050 0	0.100 0
424	Thiazopyr	0.012 5	0.025 0	454	Metominostrobin-2	0.025 0	0.050 0
425	Dimethylvinphos	0.012 5	0.025 0	455	Diofenolan -1	0.012 5	0.025 0
426	Butralin	0.025 0	0.050 0	456	Thifluzamide	0.050 0	0.100 0
427	Zoxamide	0.012 5	0.025 0	457	Diofenolan -2	0.012 5	0.025 0
428	Pyrifenox -1	0.050 0	0.100 0	458	Quinoxyphen	0.006 3	0.012 6
429	Allethrin	0.025 0	0.050 0	459	Chlorfenapyr	0.050 0	0.100 0
430	Dimethametryn	0.006 3	0.012 6				

Continued

TABLE 2.4 The LOD, LOQ of 500 Pesticides and Related Chemical Residues Determined by GC-MS—cont'd

No.	Pesticides	LOD (mg/kg)	LOQ (mg/kg)	No.	Pesticides	LOD (mg/kg)	LOQ (mg/kg)
460	Trifloxystrobin	0.025 0	0.050 0	481	Bifenazate	0.050 0	0.100 0
461	Imibenconazole-des-benzyl	0.025 0	0.050 0	482	Endrin ketone	0.025 0	0.050 0
				483	Clomeprop	0.006 3	0.012 6
462	Isoxadifen-Ethyl	0.012 5	0.025 0	484	Fenamidone	0.006 3	0.012 6
463	Fipronil	0.050 0	0.100 0	485	Naproanilide	0.006 3	0.012 6
464	Imiprothrin-1	0.012 5	0.025 0	486	Pyraclostrobin	0.150 0	0.300 0
465	Carfentrazone-Ethyl	0.012 5	0.025 0	487	Lactofen	0.050 0	0.100 0
466	Imiprothrin-2	0.012 5	0.025 0	488	Tralkoxydim	0.050 0	0.100 0
467	Epoxiconazole -1	0.050 0	0.100 0	489	Pyraclofos	0.050 0	0.100 0
468	Pyraflufen Ethyl	0.012 5	0.025 0	490	Dialifos	0.050 0	0.100 0
469	Pyributicarb	0.012 5	0.025 0	491	Spirodiclofen	0.050 0	0.100 0
470	Thenylchlor	0.012 5	0.025 0	492	Halfenprox	0.025 0	0.050 0
471	Clethodim	0.025 0	0.050 0	493	Flurtamone	0.025 0	0.050 0
472	Mefenpyr-diethyl	0.018 8	0.037 6	494	Pyriftalid	0.006 3	0.012 6
473	Famphur	0.025 0	0.050 0	495	Silafluofen	0.006 3	0.012 6
474	Etoxazole	0.037 5	0.075 0	496	Pyrimidifen	0.025 0	0.050 0
475	Pyriproxyfen	0.006 3	0.012 6	497	Acetamiprid	0.200 0	0.400 0
476	Epoxiconazole-2	0.050 0	0.100 0	498	Butafenacil	0.006 3	0.012 6
477	Picolinafen	0.006 3	0.012 6	499	Cafenstrole	0.075 0	0.150 0
478	Iprodione	0.025 0	0.050 0	500	Fluridone	0.050 0	0.100 0
479	Piperophos	0.018 8	0.037 6				
480	Ofurace	0.018 8	0.037 6				

instrument used for analysis. For 500 pesticides and related chemical grouping and mixed standard solution concentration of this standard, reference is made to Appendix B.

Depending on group number, mixed standard solution concentration, and stock standard solution concentration, appropriate amounts of individual stock standard solution are pipetted into a 100-mL volumetric flask, being diluted to volume with toluene. Mixed standard solutions are stored in the dark below 4°C and are used for one month.

Internal Standard Solution: Accurately weigh 3.5 mg heptachlor epoxide into a 100-mL volumetric flask. Dissolve and dilute to volume with toluene.

Working Standard Mixed Solution in Matrix: Working standard mixed solutions in matrix of A, B, C, D, and E group pesticides are prepared by diluting 40 µL internal standard solution and an appropriate amount of mixed standard solution to 1.0 mL with blank extract that has been taken through the method with the rest of the samples. Mix thoroughly. Working standard mixed solutions in matrix must be freshly prepared.

2.2.4   Apparatus

GC-MS: Equipped With EI
Analytical Balances: Capable of Weighing From 0.1 mg to 0.01 g
Homogenizer: Not Less Than 20,000 rpm
Pear-Shaped Flask: 200 mL
Pipette: 1 mL
Nitrogen Evaporator

2.2.5   Sample Pretreatment

2.2.5.1   Preparation of Test Sample

Sampling is done according to GB/T8855 and samples are cut up, mixed thoroughly, sealed, and labeled.

The test samples are stored in the dark at 4°C.

2.2.5.2   Extraction

Weigh 20 g of test sample (accurate to 0.01 g) into 80-mL centrifuge tube. Add 40 mL of acetonitrile, and homogenize at 15,000 rpm for 1 min. Add 5 g sodium chloride to the centrifuge tube and homogenize at 15,000 rpm for 1 min again, then centrifuge at 3000 rpm for 5 min. Pipette 20 mL of the top acetonitrile layer of extracts (corresponds to 10 g test sample) for clean-up.

2.2.5.3   Clean-up

Fix the Envi-18 cartridge in a support. Condition the cartridge with 10 mL acetonitrile before adding sample. Connect a pear-shaped flask to the cartridge. Decant the preceding extracts into the cartridge. Elute with 15 mL acetonitrile. Evaporate the eluate to 1 mL using rotary evaporator at 40°C. Add sodium sulfate into Envi-Carb cartridge to a height of ca. 2 cm. Connect the cartridge to the top of the Sep-Pak NH_2 cartridge in series. Fix the cartridges into a support to which a pear-shaped flask is connected. Condition the cartridges with 4 mL acetonitrile + toluene (3 + 1) before adding the sample. Once the solution gets to the top of the sodium sulfate, pipet the eluate into the cartridges immediately. Rinse the pear-shaped flask with 3×2 mL acetonitrile + toluene (3 + 1) and decant it into the cartridges. Insert a reservoir into the cartridges. Elute the pesticides with 25 mL acetonitrile +

toluene (3 + 1). Evaporate the eluate to ca. 0.5 mL using a rotary evaporator at 40°C. Exchange with 2×5 mL hexane twice and make up to ca. 1 mL. Add 40 μL internal standard solution and mix thoroughly. The solution is ready for GC-MS determination.

2.2.6 GC-MS Determination

2.2.6.1 GC-MS Operating Condition

(a) Column: DB-1701 ($30 \text{ m} \times 0.25 \text{ mm} \times 0.25 \text{ μm}$) capillary column, or equivalent
(b) Column temperature: 40°C hold 1 min, at 30°C/min to 130°C, at 5°C/min to 250°C, at 10°C/min to 300°C, hold 5 min
(c) Carrier gas: Helium, purity ≥99.999%, flow rate: 1.2 mL/min
(d) Injection port temperature: 290°C
(e) Injection volume: 1 μL
(f) Injection mode: Splitless, purge on after 1.5 min
(g) Ionization voltage: 70 eV
(h) Ion source temperature: 230°C
(i) GC-MS interface temperature: 280°C
(j) Selected ion monitoring mode: Each compound selects 1 quantifying ion and 2–3 qualifying ions. All of the detected ions of each group are detected according to programmed time and sequence of peaking. The retention times, quantifying ions, qualifying ions, and the abundance ratios of quantifying ion and qualifying ions for each compound are listed in Table 5.3. The programmed time and dwell time for the ions detected for each compound in each group are listed in Table 5.4

2.2.6.2 Qualitative Determination

In the samples determined, four injections are required to analyze for all pesticides according to GC-MS operating conditions; if the retention time in the peaks of the sample solution are the same as for the peaks of the working standard mixed solution, and the selected ions of the background-subtracted mass spectrum appear, and the abundance ratios of selected ions are within the expected limits (abundance ratios >50%, permitted tolerances are ±10%; abundance ratios >20% to 50%, permitted tolerances are ±15%; abundance ratios >10% to 20%, permitted tolerances are ±20%; abundance ratios ≤10%, permitted tolerances are ±50%), the sample is confirmed to contain this pesticide compound. In the case where results are still not definitive, the sample should be reinjected with acquisition in scan mode (sufficient sensitivity) or with additional confirmatory ions or using other instruments that have higher sensitivity.

2.2.6.3   Quantitative Determination

The results are quantitated using heptachlor epoxide for an internal standard and the quantitation ion response for each analyte. To compensate for the matrix effect, quantitation is based on a mixed standard prepared in blank matrix extract. The concentration of standard solution and the detected sample solution must be similar.

2.2.6.4   Parallel Test

A parallel test is carried out for the same testing sample.

2.2.6.5   Blank Test

The operation of the blank test is the same as that described in the method of determination, but without addition of sample.

2.2.7   Precision

The precision data of the method for this standard have been determined according to the stipulations of GB/T 6379.1 and GB/T 6379.2. The values of repeatability and reproducibility are obtained and calculated at a 95% confidence level.

RESEARCHERS

Guo-Fang Pang, Yong-Ming Liu, Chun-Lin Fan, Yan-Zhong Cao, Jin-Jie Zhang, Xue-Min Li, Yan-Ping Wu.

Qinhuangdao Entry-Exit Inspection and Quarantine Bureau, 39 Haibin Rd, Qinhuangdao, Hebei, PC 066002, People's Republic of China.

2.3   DETERMINATION OF 450 PESTICIDES AND RELATED CHEMICAL RESIDUES IN FRUITS AND VEGETABLES: LC-MS-MS METHOD (GB/T 20769-2008)

2.3.1   Scope

The method is applicable to the quantitative determination of the residues of 381 pesticides and related chemicals and the qualitative determination of the residues of 69 pesticides and related chemicals in fruits and vegetables, such as apples, oranges, cabbages, celery, and tomatoes. The limits of quantitative determination of this method of 381 pesticides and related chemicals are 0.01 µg/kg to 0.606 mg/kg (see Table 2.5).

TABLE 2.5 The LOD, LOQ of 450 Pesticides and Related Chemical Residues Determined by LC-MS-MS

No.	Pesticides	LOD (µg/kg)	LOQ (µg/kg)	No.	Pesticides	LOD (µg/kg)	LOQ (µg/kg)
	Group A			31	penconazole	0.50	1.00
1	Isoprocarb	0.58	1.16	32	myclobutanil	0.25	0.50
2	3,4,5-trimethacarb	0.09	0.18	33	paclobutrazol	0.14	0.28
3	Cycluron	0.05	0.10	34	fenthion sulfoxide	0.08	0.16
4	Carbaryl	2.58	5.16	35	triadimenol	2.64	5.28
5	Propachlor	0.07	0.14	36	butralin	0.48	0.96
6	Rabenzazole	0.33	0.66	37	spiroxamine	0.01	0.02
7	Simetryn	0.03	0.06	38	tolclofos methyl	16.64	33.28
8	monolinuron	0.89	1.78	39	desmedipham	1.01	2.02
9	Mevinphos	0.39	0.78	40	methidathion	2.67	5.34
10	Aziprotryne	0.35	0.70	41	allethrin	15.10	30.20
11	Secbumeton	0.02	0.04	42	triallate	5.05	10.10
12	Cyprodinil	0.18	0.36	43	diazinon	0.18	0.36
13	Buturon	2.24	4.48	44	edifenphos	0.19	0.38
14	carbetamide	0.91	1.82	45	pretilachlor	0.08	0.16
15	Pirimicarb	0.04	0.08	46	flusilazole	0.15	0.30
16	clomazone dimethazone	0.11	0.22	47	iprovalicarb	0.58	1.16
17	cyanazine	0.04	0.08	48	benodanil	0.87	1.74
18	prometryne	0.04	0.08	49	flutolanil	0.29	0.58
19	paraoxon methyl	0.19	0.38	50	famphur	0.90	1.80
20	4,4-dichlorobenzo-phenone	3.40	6.80	51	benalyxyl	0.31	0.62
				52	diclobutrazole	0.12	0.24
21	thiacloprid	0.09	0.18	53	etaconazole	0.45	0.90
22	imidacloprid	5.50	11.00	54	fenarimol	0.15	0.30
23	ethidimuron	0.38	0.76	55	phthalic acid, dicyclobexyl ester	0.50	1.00
24	isomethiozin	0.27	0.54				
25	diallate	22.30	44.60	56	tetramethirn	0.46	0.92
26	acetochlor	11.85	23.70	57	dichlofluanid	0.65	1.30
27	nitenpyram	4.28	8.56	58	cloquintocet mexyl	0.47	0.94
28	methoprotryne	0.06	0.12	59	bitertanol	8.35	16.70
29	dimethenamid	1.08	2.16	60	azinphos ethyl	27.23	54.46
30	terrbucarb	0.53	1.06	61	clodinafop propargyl	0.61	1.22

TABLE 2.5 The LOD, LOQ of 450 Pesticides and Related Chemical Residues Determined by LC-MS-MS—cont'd

No.	Pesticides	LOD (μg/kg)	LOQ (μg/kg)	No.	Pesticides	LOD (μg/kg)	LOQ (μg/kg)
62	triflumuron	0.98	1.96	94	chlormephos	112.00	224.00
63	isoxaflutole	0.98	1.96	95	carboxin	0.14	0.28
64	anilofos	0.18	0.36	96	difenzoquat-methyl sulfate	0.20	0.40
65	quizalofop-ethyl	0.17	0.34	97	clothianidin	15.75	31.50
66	haloxyfop-methyl	0.66	1.32	98	pronamide	3.85	7.70
67	fluazifop butyl	0.07	0.14	99	dimethachloro	0.48	0.96
68	bromophos-ethyl	141.92	283.84	100	methobromuron	4.21	8.42
69	dialifos	15.00	30.00	101	phorate	78.50	157.00
70	bensulide	8.55	17.10	102	aclonifen	6.05	12.10
71	triasulfuron	0.40	0.80	103	mephosfolan	0.58	1.16
72	bromfenvinfos	0.76	1.52	104	imibenzonazole-des-benzyl	1.56	3.12
73	pyrazophos	0.41	0.82	105	neburon	1.78	3.56
74	flufenoxuron	0.79	1.58	106	mefenoxam	0.38	0.76
75	indoxacarb	1.89	3.78	107	prothoate	0.62	1.24
76	emamectin benzoate	0.08	0.16	108	ethofume sate	93.00	186.00
	Group B			109	iprobenfos	2.07	4.14
77	ethylene thiourea	13.05	26.10	110	TEPP	2.60	5.20
78	dazomet	31.75	63.50	111	cyproconazole	0.18	0.36
79	nicotine	0.55	1.10	112	thiamethoxam	8.25	16.50
80	fenuron	0.26	0.52	113	crufomate	0.13	0.26
81	cyromazine	1.81	3.62	114	etrimfos	4.69	9.38
82	crimidine	0.39	0.78	115	cythioate	20.00	40.00
83	molinate	0.53	1.06	116	phosphamidon	0.97	1.94
84	carbendazim	0.12	0.24	117	phenmedipham	1.12	2.24
85	propoxur	6.10	12.20	118	bifenazate	5.70	11.40
86	isouron	0.10	0.20	119	flutriafol	2.15	4.30
87	chlorotoluron	0.16	0.32	120	furalaxyl	0.19	0.38
88	thiofanox	39.25	78.50	121	bioallethrin	49.50	99.00
89	chlorbufam	45.75	91.50	122	cyanofenphos	5.20	10.40
90	bendiocarb	0.80	1.60	123	pirimiphos methyl	0.05	0.10
91	propazine	0.08	0.16	124	buprofezin	0.22	0.44
92	terbuthylazine	0.12	0.24	125	disulfoton sulfone	0.62	1.24
93	diuron	0.39	0.78				

Continued

TABLE 2.5 The LOD, LOQ of 450 Pesticides and Related Chemical Residues Determined by LC-MS-MS—cont'd

No.	Pesticides	LOD (µg/kg)	LOQ (µg/kg)	No.	Pesticides	LOD (µg/kg)	LOQ (µg/kg)
126	fenazaquin	0.08	0.16	158	ethirimol	0.14	0.28
127	triazophos	0.17	0.34	159	propanil	5.40	10.80
128	DEF	0.40	0.80	160	carbofuran	3.27	6.54
129	pyriftalid	0.16	0.32	161	acetamiprid	0.36	0.72
130	metconazole	0.33	0.66	162	mepanipyrim	0.08	0.16
131	pyriproxyfen	0.11	0.22	163	prometon	0.03	0.06
132	isoxaben	0.05	0.10	164	metoxuron	0.16	0.32
133	flurtamone	0.11	0.22	165	dimethoate	1.90	3.80
134	trifluralin	130.50	261.00	166	fluometuron	0.23	0.46
135	flamprop methyl	5.05	10.10	167	dicrotophos	0.29	0.58
136	bioresmethrin	1.86	3.72	168	monalide	0.30	0.60
137	propiconazole	0.44	0.88	169	diphenamid	0.04	0.08
138	chlorpyrifos	13.45	26.90	170	ethoprophos	0.69	1.38
139	fluchloralin	122.00	244.00	171	fonofos	1.86	3.72
140	chlorsulfuron	0.69	1.38	172	furmecyclox	0.21	0.42
141	tetrachlorvinphos	0.56	1.12	173	hexazinone	0.03	0.06
142	propargite	17.15	34.30	174	dimethametryn	0.03	0.06
143	bromuconazole	0.79	1.58	175	trichlorphon	0.28	0.56
144	picolinafen	0.18	0.36	176	demeton(o+s)	1.69	3.38
145	fluthiacet methyl	1.33	2.66	177	benoxacor	1.73	3.46
146	trifloxystrobin	0.50	1.00	178	bromacil	5.90	11.80
147	hexaflumuron	6.30	12.60	179	phorate sulfoxide	92.07	184.14
148	novaluron	2.01	4.02	180	brompyrazon	0.90	1.80
149	flurazuron	6.70	13.40	181	oxycarboxin	0.22	0.44
	Group C			182	mepronil	0.09	0.18
150	methamidophos	1.23	2.46	183	disulfoton	117.42	234.84
151	EPTC	9.33	18.66	184	fenthion	13.00	26.00
152	diethyltoluamide	0.14	0.28	185	metalaxyl	0.13	0.26
153	monuron	8.68	17.36	186	ofurace	0.25	0.50
154	pyrimethanil	0.17	0.34	187	dodemorph	0.10	0.20
155	fenfuram	0.20	0.40	188	fosthiazate	0.14	0.28
156	quinoclamine	1.98	3.96	189	imazamethabenz-methyl	0.04	0.08
157	fenobucarb	1.48	2.96				

TABLE 2.5 The LOD, LOQ of 450 Pesticides and Related Chemical Residues Determined by LC-MS-MS—cont'd

No.	Pesticides	LOD (µg/kg)	LOQ (µg/kg)	No.	Pesticides	LOD (µg/kg)	LOQ (µg/kg)
190	disulfoton-sulfoxide	0.71	1.42	223	aspon	0.43	0.86
191	isoprothiolane	0.46	0.92	224	ethion	0.74	1.48
192	imazalil	0.50	1.00	225	dithiopyr	2.60	5.20
193	phoxim	20.70	41.40	226	spirodiclofen	2.48	4.96
194	quinalphos	0.50	1.00	227	fenpyroximate	0.34	0.68
195	ditalimfos	16.80	33.60	228	flumiclorac-pentyl	2.65	5.30
196	fenoxycarb	4.57	9.14	229	temephos	0.30	0.60
197	pyrimitate	0.04	0.08	230	butafenacil	2.38	4.76
198	fensulfothin	0.50	1.00	231	spinosad	0.14	0.28
199	fluorochloridone	3.45	6.90		Group D		
200	butachlor	5.02	10.04	232	mepiquat chloride	0.23	0.46
201	kresoxim-methyl	25.15	50.30	233	allidochlor	10.26	20.52
202	triticonazole	0.76	1.52	234	propamocarb	0.02	0.04
203	fenamiphos sulfoxide	0.18	0.36	235	tricyclazole	0.31	0.62
204	thenylchlor	6.04	12.08	236	thiabendazole	0.12	0.24
205	pyrethrin	88.16	176.32	237	metamitron	1.59	3.18
206	fenoxanil	9.85	19.70	238	isoproturon	0.03	0.06
207	fluridone	0.05	0.10	239	atratone	0.05	0.10
208	epoxiconazole	1.01	2.02	240	oesmetryn	0.04	0.08
209	chlorphoxim	19.39	38.78	241	metribuzin	0.14	0.28
210	fenamiphos sulfone	0.11	0.22	242	DMST	10.00	20.00
211	fenbuconazole	0.41	0.82	243	cycloate	1.11	2.22
212	isofenphos	54.67	109.34	244	atrazine	0.09	0.18
213	phenothrin	84.80	169.60	245	butylate	75.50	151.00
214	fentin-chloride	4.31	8.62	246	chloridazon	0.58	1.16
215	piperophos	2.31	4.62	247	sulfallate	51.80	103.60
216	piperonyl butoxide	0.28	0.56	248	ethiofencarb	1.23	2.46
217	oxyflurofen	14.64	29.28	249	terbumeton	0.02	0.04
218	coumaphos	0.53	1.06	250	cyprazine	0.01	0.02
219	flufenacet	1.33	2.66	251	ametryn	0.24	0.48
220	phosalone	12.01	24.02	252	tebuthiuron	0.05	0.10
221	methoxyfenozide	0.93	1.86	253	trietazine	0.15	0.30
222	prochloraz	0.52	1.04	254	sebutylazine	0.08	0.16

Continued

TABLE 2.5 The LOD, LOQ of 450 Pesticides and Related Chemical Residues Determined by LC-MS-MS—cont'd

No.	Pesticides	LOD (μg/kg)	LOQ (μg/kg)	No.	Pesticides	LOD (μg/kg)	LOQ (μg/kg)
255	dibutyl succinate	55.60	111.20	288	bupirimate	0.18	0.36
256	tebutam	0.03	0.06	289	azinphos-methyl	276.08	552.16
257	thiofanox-sulfoxide	2.07	4.14	290	tebupirimfos	0.03	0.06
258	cartap hydrochloride	520.00	1 040.00	291	phenthoate	23.09	46.18
259	methacrifos	605.92	1 211.84	292	sulfotep	0.65	1.30
260	terbutryn	5.72	11.44	293	sulprofos	1.46	2.92
261	triazoxide	2.00	4.00	294	EPN	8.25	16.50
262	thionazin	5.67	11.34	295	azamethiphos	0.20	0.40
263	linuron	2.91	5.82	296	diniconazole	0.34	0.68
264	heptanophos	1.46	2.92	297	pencycuron	0.07	0.14
265	prosulfocarb	0.09	0.18	298	mecarbam	4.90	9.80
266	propyzamide	1.74	3.48	299	tralkoxydim	0.08	0.16
267	dipropetryn	0.07	0.14	300	malathion	1.41	2.82
268	thiobencarb	0.83	1.66	301	pyributicarb	0.08	0.16
269	tri-iso-butyl phosphate	0.89	1.78	302	pyridaphenthion	0.22	0.44
270	tri-n-butyl phosphate	0.09	0.18	303	pirimiphos-ethyl	0.01	0.02
271	diethofencarb	0.50	1.00	304	pyraclofos	0.25	0.50
272	alachlor	1.85	3.70	305	picoxystrobin	2.11	4.22
273	cadusafos	0.29	0.58	306	tetraconazole	0.43	0.86
274	metazachlor	0.25	0.50	307	mefenpyr-diethyl	3.14	6.28
275	propetamphos	13.50	27.00	308	profenefos	0.50	1.00
276	terbufos	560.00	1 120.00	309	pyraclostrobin	0.13	0.26
277	simeconazole	0.74	1.48	310	dimethomorph	0.09	0.18
278	triadimefon	1.97	3.94	311	kadethrin	0.83	1.66
279	phorate sulfone	10.50	21.00	312	thiazopyr	0.49	0.98
280	tridemorph	0.65	1.30	313	chlorfluazuron	2.17	4.34
281	mefenacet	0.55	1.10		Group E		
282	azaconazole	0.20	0.40	314	chlormequat	0.03	0.06
283	fenamiphos	0.05	0.10	315	methomyl	2.39	4.78
284	fenpropimorph	0.05	0.10	316	pyroquilon	0.87	1.74
285	tebuconazole	0.56	1.12	317	fuberidazole	0.47	0.94
286	isopropalin	7.50	15.00	318	chlordimeform	0.33	0.66
287	nuarimol	0.25	0.50	319	cymoxanil	13.90	27.80

TABLE 2.5 The LOD, LOQ of 450 Pesticides and Related Chemical Residues Determined by LC-MS-MS—cont'd

No.	Pesticides	LOD (µg/kg)	LOQ (µg/kg)	No.	Pesticides	LOD (µg/kg)	LOQ (µg/kg)
320	vernolate	0.06	0.12	351	isazofos	0.04	0.08
321	promecarb	2.14	4.28	352	dichlofenthion	7.55	15.10
322	aminocarb	4.11	8.22	353	vamidothion sulfone	119.00	238.00
323	dimethirimol	0.03	0.06	354	terbufos sulfone	22.15	44.30
324	chlortoluron	0.09	0.18	355	dinitramine	0.45	0.90
325	omethoate	2.41	4.82	356	trichloronat	16.70	33.40
326	dichlorvos	0.14	0.28	357	resmethrin-2	0.08	0.16
327	aldicarb sulfone	5.35	10.70	358	boscalid	1.19	2.38
328	dioxacarb	0.84	1.68	359	nitralin	8.60	17.20
329	benzyladenine	17.7	35.40	360	hexythiazox	5.90	11.80
330	oxabetrinil	10.00	20.00	361	florasulam	4.35	8.70
331	ethiofencarb-sulfoxide	56.00	112.00	362	benzoximate	4.92	9.84
332	cyanohos	2.53	5.06	363	pyridaben	3.04	6.08
333	etridiazole	0.26	0.52	364	benzoylprop-ethyl	77.00	154.00
334	thiometon	144.5	289.00	365	pyrimidifen	3.50	7.00
335	folpet	34.65	69.30	366	pyridate	19.95	39.90
336	demeton-s-methyl sulfone	4.94	9.88	367	*trans*-permethin	1.20	2.40
337	fenpropidin	0.05	0.10	368	pyrazoxyfen	0.08	0.16
338	paraoxon-ethyl	0.12	0.24	369	flubenzimine	1.95	3.90
339	aldimorph	0.79	1.58	370	*zeta* cypermethrin	0.17	0.34
340	uniconazole	0.60	1.20	371	haloxyfop-2-ethoxyethyl	0.63	1.26
341	pyrifenox	0.07	0.14		Group F		
342	chlorthion	33.40	66.80	372	acrylamide	4.45	8.90
343	dicapthon	0.06	0.12	373	*tert*-butylamine	9.74	19.48
344	clofentezine	0.19	0.38	374	chlormequat chloride	0.18	0.36
345	norflurazon	0.06	0.12	375	phthalimide	10.75	21.50
346	quinoxyphen	38.35	76.70	376	dimefox	17.05	34.10
347	fenthion sulfone	4.37	8.74	377	metolcarb	6.35	12.70
348	methoprene	1.31	2.62	378	diphenylamin	0.10	0.20
349	flurochloridone	0.32	0.64	379	*1*-naphthy acetamide	0.20	0.40
350	phthalic acid, benzyl butyl ester	158.00	316.00	380	atrazine-desethyl	0.16	0.32

Continued

TABLE 2.5 The LOD, LOQ of 450 Pesticides and Related Chemical Residues Determined by LC-MS-MS—cont'd

No.	Pesticides	LOD (µg/kg)	LOQ (µg/kg)	No.	Pesticides	LOD (µg/kg)	LOQ (µg/kg)
381	2,6-dichlorobenzamide	1.13	2.26	411	phthalic acid, biscyclohexyl ester	0.17	0.34
382	dimethyl phthalate	3.30	6.60	412	carpropamid	1.30	2.60
383	chlordimeform hydrochloride	0.66	1.32	413	tebufenpyrad	0.06	0.12
384	simeton	0.28	0.56	414	zoxamide	1.12	2.24
385	dinotefuran	2.55	5.10	415	tebufenozide	6.95	13.90
386	pebulate	0.85	1.70	416	chlorthiophos	7.95	15.90
387	acibenzolar-s-methyl	0.77	1.54	417	naled	37.05	74.10
388	dioxabenzofos	3.46	6.92	418	tolfenpyrad	0.02	0.04
389	methabenz-thiazuron	0.02	0.04	419	dicofol	0.45	0.90
390	butoxycarboxim	6.65	13.30	420	cinidon-ethyl	3.65	7.30
391	mexacarbate	0.24	0.48	421	rotenone	0.58	1.16
392	demeton-s-methyl sulfoxide	0.98	1.96	422	propaquiafop	0.31	0.62
393	thiofanox sulfone	6.02	12.04	423	lactofen	15.50	31.00
394	phosfolan	0.12	0.24		Group G		
395	triclopyr	0.05	0.10	424	dalapon	57.69	115.38
396	demeton-s	20.00	40.00	425	2-phenylphenol	42.47	84.94
397	imazapyr	2.57	5.14	426	3-phenylphenol	1.00	2.00
398	napropamide	0.32	0.64	427	DNOC	0.65	1.30
399	fenitrothion	6.70	13.40	428	dicloran	12.14	24.28
400	phthalic acid, dibutyl ester	9.90	19.80	429	chlorpropham	3.94	7.88
401	metolachlor	0.10	0.20	430	terbacil	0.22	0.44
402	procymidone	21.65	43.30	431	dicamba	316.48	632.96
403	vamidothion	1.14	2.28	432	bentazone	0.26	0.52
404	chloroxuron	0.11	0.22	433	dinoseb	0.10	0.20
405	triamiphos	1.15	2.30	434	dinoterb	0.06	0.12
406	dithianon	2.14	4.28	435	2,4-DB	534.94	1 069.88
407	prallethrin	14.70	29.40	436	fludioxonil	15.54	31.08
408	cumyluron	0.33	0.66	437	trinexapac-ethyl	17.67	35.34
409	phosmet	4.43	8.86	438	chlorfenethol	41.08	82.16
410	ronnel	3.28	6.56	439	chlorobenzuron	5.10	10.20
				440	chloramphenicolum	0.97	1.94

TABLE 2.5 The LOD, LOQ of 450 Pesticides and Related Chemical Residues Determined by LC-MS-MS—cont'd

No.	Pesticides	LOD (µg/kg)	LOQ (µg/kg)	No.	Pesticides	LOD (µg/kg)	LOQ (µg/kg)
441	sulfanitran	0.76	1.52	446	diflufenican	7.07	14.14
442	mesotrion	575.14	1 150.28	447	ethiprole	9.96	19.92
443	oryzalin	1.23	2.46	448	flusulfamide	0.10	0.20
444	ioxynil	0.15	0.30	449	fomesafen	0.51	1.02
445	famoxadone	11.32	22.64	450	lufenuron	0.01	0.02

2.3.2   Principle

The samples are homogenized with acetonitrile and sodium chloride. The solutions are centrifuged, and the supernatants of the acetonitrile phase are cleaned up with Sep-Pak Vac cartridges. The pesticides and related chemicals are eluted with acetonitrile + toluene (3+1), and the solutions are analyzed by LC-MS-MS, using an external standard method for quantification.

2.3.3   Reagents and Materials

"Water" is the first grade of GB/T6682 specified.

Acetonitrile: HPLC Grade
Hexane: HPLC Grade
Isooctane: HPLC Grade
Toluene: G.R.
Acetone: HPLC Grade
Dichloromethane: HPLC Grade
Methanol: HPLC Grade
Filter Membrane (Nylon): 13 mm × 0.2 µm
Aminopropyl Sep-Pak Vac Cartridges: 1 g, 6 mL, or Equivalent
Acetonitrile + Toluene (3+1, V/V)
Acetonitrile + Water (3+2, V/V)
0.05% Formic Acid-Water (V/V)
5 mmol/L Ammonium Acetate: Weigh 0.375 g Ammonium Acetate, Dilute With Water to 1000 mL
Sodium Sulfate: Anhydrous, Analytically Pure. Ignited at 650°C for 4 h and Keep in a Desiccator
Sodium Chloride: G.R.
Pesticide and Related Chemicals Standard: Purity ≥95%
    Stock Standard Solution: Accurately weigh 5–10 mg of individual pesticide and related chemical standards (accurate to 0.1 mg) into a 10-mL

volumetric flask. Dissolve and dilute to volume with methanol, toluene, acetone, acetonitrile, or isooctane, etc., depending on each individual compound's solubility. (For diluting solvent refer to Appendix B) Standard solutions are stored in the dark below 0°C to 4°C and are used for 1 year. Mixed Standard Solution (Mixed Standard Solution A, B, C, D, E, F, and G): Depending on properties and retention time of each compound, all compounds are divided into seven groups A, B, C, D, E, F, and G. Depending on instrument sensitivity of each compound, the concentrations of mixed standard solutions are decided. For 450 pesticides and related chemicals grouping and concentration of mixed standard solutions for this standard, reference is made to Appendix B.

Depending on group number, mixed standard solution concentration, and stock standard solution concentration, appropriate amounts of individual stock standard solutions are pipetted into a 100-mL volumetric flask, with seven group pesticides and related chemicals being diluted to volume with methanol. Mixed standard solutions are stored in the dark below 0°C to 4°C and are used for 1 month.

Working Standard Mixed Solution in Matrix: Working standard mixed solutions in matrix of A, B, C, D, E, F and G groups of pesticides and related chemicals are prepared by mixing different concentrations of working standard mixed solutions with sample blank extract that has been taken through the method with the rest of the samples. These solutions are used to construct calibration plots. Working standard mixed solutions in matrix must be freshly prepared.

2.3.4 Apparatus

LC-MS-MS: Equipped With ESI Source
Analytical Balances: Capable of Weighing From 0.1 mg to 0.01 g
Homogenizer: Max. rpm 20,000 rpm
Centrifuge Tube: 80 mL
Centrifuge: Max. rpm 4200 rpm
Evaporate
Pear-Shaped Flask
Pipette: 1 mL
Screw Vial: 2.0 mL, With Screw Caps and PTFE Septa
Nitrogen Evaporator

2.3.5 Sample Pretreatment

2.3.5.1 Preparation of Test Sample

Sampling is done according to GB/T8855 and the samples are cut up, mixed thoroughly, sealed, and labeled.

The test samples are stored in the dark at 0°C to 4°C.

2.3.5.2 Extraction

Weigh 20-g test sample (accurate to 0.01 g) into 80-mL centrifuge tube. Add 40 mL acetonitrile, and homogenize at 15,000 rpm for 1 min. Add 5 g sodium chloride to the centrifuge tube and homogenize at 15,000 rpm for 1 min again, then centrifuge at 3800 rpm for 5 min. Pipette 20 mL of the top acetonitrile layer of extracts (corresponding to a 10-g test sample), and concentrate the extracts to ca. 1 mL using the rotary evaporator at 40°C for clean-up.

2.3.5.3 Clean-up

Add sodium sulfate into the Sep-Pak Vac cartridge to a height of 2 cm. Fix the cartridge into a support to which a pear-shaped flask is connected. Condition the cartridge with 4 mL acetonitrile-toluene (3 + 1) before adding the sample. Once the solution gets to the top of the sodium sulfate, pipet the eluate into the cartridge. Rinse the pear-shaped flask with 3 × 2 mL acetonitrile-toluene (3 + 1) and decant it to the cartridge. Insert a reservoir into the cartridges. Elute the pesticides with 25 mL acetonitrile-toluene (3 + 1). Evaporate the eluate to 0.5 mL using a rotary evaporator at 40°C. Then evaporate the eluate to dryness using a nitrogen evaporator. Make up to 1 mL with acetonitrile-water (3 + 2) and mix thoroughly. Finally, filter the extract through a 0.2-μm filter into a glass vial for LC-MS-MS determination.

2.3.6 Determination

2.3.6.1 LC-MS-MS Operating Conditions

Operating Conditions for the Pesticides and Related Chemicals of A, B, C, D, E, and F Group

   (a) Chromatography column: Atlantis T3, 3 μm, 150 × 2.1 mm, or equivalent
   (b) Mobile phase program and the flow rate: as Table 2.6
   (c) Column temperature: 40°C
   (d) Injection volume: 20 μL
   (e) Ion source: ESI
   (f) Scan mode: positive ion
   (g) Monitor mode: multiple reaction monitor
   (h) Ion spray voltage: 5000 V
   (i) Nebulizer gas: 0.483 MPa
   (j) Curtain gas: 0.138 MPa
   (k) Auxiliary gas: 0.379 MPa
   (l) Ion source temperature: 725°C
   (m) Monitoring ion pairs, collision energy, and declustering potentials: see Table 5.10

TABLE 2.6 Mobile Phase and Flow Rate

Time (min)	Flow Rate (μL/min)	Mobile Phase A 0.05% Formic Acid Solution (%)	Mobile Phase B Acetonitrile (%)
0.00	200	90.0	10.0
4.00	200	50.0	50.0
15.00	200	40.0	60.0
23.00	200	20.0	80.0
30.00	200	5.0	95.0
35.00	200	5.0	95.0
35.01	200	90.0	10.0
50.00	200	90.0	10.0

TABLE 2.7 Mobile Phase and Flow Rate

Time (min)	Flow Rate (μL/min)	Mobile Phase A 5 mmol/L Ammonium Acetate-Water (%)	Mobile Phase B Acetonitrile (%)
0.00	200	90.0	10.0
4.00	200	50.0	50.0
15.00	200	40.0	60.0
20.00	200	20.0	80.0
25.00	200	5.0	95.0
32.00	200	5.0	95.0
32.01	200	90.0	10.0
40.00	200	90.0	10.0

Operating Conditions for the Pesticides and Related Chemicals of G Group

(a) Chromatography column: Inertsil C8, 5 μm, 150×2.1 mm, or equivalent
(b) Mobile phase program and the flow rate: as Table 2.7
(c) Column temperature: 40°C
(d) Injection volume: 20 μL

(e) Ion source: ESI
(f) Scan mode: negative ion
(g) Monitor mode: multiple reaction monitor
(h) Ion spray voltage: $-4200\,V$
(i) Nebulizer gas: 0.42 MPa
(j) Curtain gas: 0.32 MPa
(k) Auxiliary gas: 0.35 MPa
(l) Ion source temperature: 700°C
(m) Monitoring ion pairs, collision energy, and declustering potentials: see Table 5.10

2.3.6.2   Qualitative Determination

If the retention times of peaks of the sample solution are the same as those of the peaks from the working standard mixed solution, and the selected ions appeared in the background-subtracted mass spectrum, and the abundance ratios of the selected ions are within the expected limits (abundance ratios $>50\%$, permitted tolerances are $\pm 10\%$; abundance ratios $>20\%$ to 50%, permitted tolerances are $\pm 15\%$; abundance ratios $>10\%$ to 20%, permitted tolerances are $\pm 20\%$; abundance ratios $\leq 10\%$, permitted tolerances are $\pm 50\%$), the sample is confirmed to contain this pesticide compound.

2.3.6.3   Quantitative Determination

External standard method is used for quantitation with standard curves for LC-MS-MS. To compensate for the matrix effect, quantitation is based on a series of working standard solutions prepared in blank matrix extract. The standard curves are established by injection of different concentrations of working standard mixed solutions in matrix separately. The responses of pesticides in the sample solution should be in the linear range of the instrumental detection.

2.3.6.4   Parallel Test

A parallel test is carried out for the same testing sample.

2.3.6.5   Blank Test

The operation of the blank test is the same as that described in the method of determination, but without the addition of sample.

2.3.7   Precision

The precision data of the method for this standard have been determined according to the stipulations of GB/T 6379.1 and GB/T 6379.2. The values of repeatability and reproducibility are obtained and calculated at a 95% confidence level.

RESEARCHERS

Guo-Fang Pang, Yan Li, Chun-Lin Fan, Yu-Jing Lian, Jing Cao,Wen-Wen Wang,Jun-Hong Zheng, Yong-Ming Liu, Yan-Zhong Cao, Jin-Jie Zhang, Xue-Min Li.

Qinhuangdao Entry-Exit Inspection and Quarantine Bureau, 39 Haibin Rd, Qinhuangdao, Hebei, PC 066002, People's Republic of China.

2.4 DETERMINATION OF 475 PESTICIDES AND RELATED CHEMICAL RESIDUES IN GRAINS: GC-MS METHOD (GB/T 19649-2006)

2.4.1 Scope

This method is applicable to the determination of residues of 475 pesticides and related chemicals in barley, wheat, oat, rice, and maize.

The limits of detection of method for this standard are 0.0025 to 0.8000 mg/kg (refer to Table 2.8).

2.4.2 Principle

The samples are extracted with acetonitrile by accelerated solvent extractor, the extracts are cleaned up with a solid phase extraction cartridge, pesticides and related chemicals are eluted with acetonitrile + toluene (3 + 1), and the solutions are analyzed by GC-MS.

2.4.3 Reagents and Materials

Acetonitrile: HPLC Grade
Celit 545: G.R.
Sodium Sulfate: Anhydrous, Analytically Pure. Ignited at 650°C for 4 h and Kept in a Desiccator
Toluene: G.R.
Acetone: Analytically Pure. Redistilled
Dichloromethane: HPLC Grade
Envi-18 Cartridge: 12 mL, 2000 mg, or Equivalent
Envi-Carb Cartridge: 6 mL, 500 mg, or Equivalent
Sep-Pak NH$_2$ Cartridge: 3 mL, 500 mg, or Equivalent
Pesticide and Related Chemicals Standard: Purity ≥95%
Stock Standard Solution: Accurately weigh 5–10 mg of individual pesticide and related chemicals standard (accurate to 0.1 mg) into a 10-mL volumetric flask. Dissolve and dilute to volume with toluene, toluene + acetone combination, dichloromethane, etc., depending on each individual compound's solubility. (For diluting solvent, refer to Appendix B)

TABLE 2.8 The LOD, LOQ of 475 Pesticides and Related Chemical Residues Determined by GC-MS

No.	Pesticides	LOD (mg/kg)	LOQ (mg/kg)	No.	Pesticides	LOD (mg/kg)	LOQ (mg/kg)
	Group A			32	Methyl-Parathion	0.050 0	0.100 0
1	Allidochlor	0.025 0	0.050 0	33	Anthraquinone	0.012 5	0.025 0
2	Dichlormid	0.025 0	0.050 0	34	Delta-HCH	0.025 0	0.050 0
3	Etridiazol	0.037 5	0.075 0	35	Fenthion	0.012 5	0.025 0
4	Chlormephos	0.025 0	0.050 0	36	Malathion	0.050 0	0.100 0
5	Propham	0.012 5	0.025 0	37	Fenitrothion	0.025 0	0.050 0
6	Cycloate	0.012 5	0.025 0	38	Paraoxon-ethyl	0.050 0	0.100 0
7	Diphenylamine	0.012 5	0.025 0	39	Triadimefon	0.025 0	0.050 0
8	Ethalfluralin	0.050 0	0.100 0	40	Parathion	0.050 0	0.100 0
9	Phorate	0.012 5	0.025 0	41	Pendimethalin	0.050 0	0.100 0
10	Thiometon	0.012 5	0.025 0	42	Linuron	0.050 0	0.100 0
11	Quintozene	0.025 0	0.050 0	43	Chlorbenside	0.025 0	0.050 0
12	Atrazine-desethyl	0.012 5	0.025 0	44	Bromophos-ethyl	0.012 5	0.025 0
13	Clomazone	0.012 5	0.025 0	45	Quinalphos	0.012 5	0.025 0
14	Diazinon	0.012 5	0.025 0	46	trans-Chlordane	0.012 5	0.025 0
15	Fonofos	0.012 5	0.025 0	47	Phenthoate	0.025 0	0.050 0
16	Etrimfos	0.012 5	0.025 0	48	Metazachlor	0.037 5	0.075 0
17	Simazine	0.012 5	0.025 0	49	Fenothiocarb	0.025 0	0.050 0
18	Propetamphos	0.012 5	0.025 0	50	Prothiophos	0.012 5	0.025 0
19	Secbumeton	0.012 5	0.025 0	51	Chlorflurenol	0.037 5	0.075 0
20	Dichlofenthion	0.012 5	0.025 0	52	Dieldrin	0.025 0	0.050 0
21	Pronamide	0.012 5	0.025 0	53	Procymidone	0.012 5	0.025 0
22	Mexacarbate	0.037 5	0.075 0	54	Methidathion	0.025 0	0.050 0
23	Aldrin	0.025 0	0.050 0	55	Napropamide	0.037 5	0.075 0
24	Dinitramine	0.050 0	0.100 0	56	Oxadiazone	0.012 5	0.025 0
25	Ronnel	0.025 0	0.050 0	57	Fenamiphos	0.037 5	0.075 0
26	Prometryne	0.012 5	0.025 0	58	Tetrasul	0.012 5	0.025 0
27	Cyprazine	0.100 0	0.200 0	59	Aramite	0.012 5	0.025 0
28	Vinclozolin	0.012 5	0.025 0	60	Bupirimate	0.012 5	0.025 0
29	Beta-HCH	0.012 5	0.025 0	61	Carboxin	0.037 5	0.075 0
30	Metalaxyl	0.037 5	0.075 0	62	Flutolanil	0.012 5	0.025 0
31	Chlorpyrifos (-ethyl)	0.012 5	0.025 0	63	4,4'-DDD	0.012 5	0.025 0
				64	Ethion	0.025 0	0.050 0

Continued

TABLE 2.8 The LOD, LOQ of 475 Pesticides and Related Chemical Residues Determined by GC-MS—cont'd

No.	Pesticides	LOD (mg/kg)	LOQ (mg/kg)	No.	Pesticides	LOD (mg/kg)	LOQ (mg/kg)
65	Sulprofos	0.025 0	0.050 0	97	Chloroneb	0.012 5	0.025 0
66	Etaconazole-1	0.037 5	0.075 0	98	Tecnazene	0.025 0	0.050 0
67	Etaconazole-2	0.037 5	0.075 0	99	Heptanophos	0.037 5	0.075 0
68	Myclobutanil	0.012 5	0.025 0	100	Hexachlorobenzene	0.012 5	0.025 0
69	Diclofop-methyl	0.012 5	0.025 0	101	Ethoprophos	0.037 5	0.075 0
70	Propiconazole	0.037 5	0.075 0	102	cis-Diallate	0.025 0	0.050 0
71	Fensulfothion	0.025 0	0.050 0	103	Propachlor	0.037 5	0.075 0
72	Bifenthrin	0.012 5	0.025 0	104	trans-Diallate	0.025 0	0.050 0
73	Mirex	0.012 5	0.025 0	105	Trifluralin	0.025 0	0.050 0
74	Benodanil	0.037 5	0.075 0	106	Chlorpropham	0.025 0	0.050 0
75	Nuarimol	0.025 0	0.050 0	107	Sulfotep	0.012 5	0.025 0
76	Methoxychlor	0.012 5	0.025 0	108	Sulfallate	0.025 0	0.050 0
77	Oxadixyl	0.012 5	0.025 0	109	Alpha-HCH	0.012 5	0.025 0
78	Tetramethirn	0.025 0	0.050 0	110	Terbufos	0.025 0	0.050 0
79	Tebuconazole	0.037 5	0.075 0	111	Terbumeton	0.037 5	0.075 0
80	Norflurazon	0.012 5	0.025 0	112	Profluralin	0.050 0	0.100 0
81	Pyridaphenthion	0.012 5	0.025 0	113	Dioxathion	0.050 0	0.100 0
82	Phosmet	0.025 0	0.050 0	114	Propazine	0.012 5	0.025 0
83	Tetradifon	0.012 5	0.025 0	115	Chlorbufam	0.025 0	0.050 0
84	Oxycarboxin	0.075 0	0.150 0	116	Dicloran	0.0250	0.050 0
85	cis-Permethrin	0.012 5	0.025 0	117	Terbuthylazine	0.012 5	0.025 0
86	trans-Permethrin	0.012 5	0.025 0	118	Monolinuron	0.050 0	0.100 0
87	Pyrazophos	0.025 0	0.050 0	119	Flufenoxuron	0.037 5	0.075 0
88	Cypermethrin	0.037 5	0.075 0	120	Cyanophos	0.025 0	0.050 0
89	Fenvalerate	0.050 0	0.100 0	121	Chlorpyrifos-methyl	0.012 5	0.025 0
90	Deltamethrin	0.075 0	0.150 0	122	Desmetryn	0.012 5	0.025 0
	Group B			123	Dimethachlor	0.037 5	0.075 0
91	EPTC	0.037 5	0.075 0	124	Alachlor	0.037 5	0.075 0
92	Butylate	0.037 5	0.075 0	125	Pirimiphos-methyl	0.012 5	0.025 0
93	Dichlobenil	0.002 5	0.005 0	126	Terbutryn	0.025 0	0.050 0
94	Pebulate	0.037 5	0.075 0	127	Thiobencarb	0.025 0	0.050 0
95	Nitrapyrin	0.037 5	0.075 0	128	Aspon	0.025 0	0.050 0
96	Mevinphos	0.025 0	0.050 0	129	Dicofol	0.200 0	0.400 0

TABLE 2.8 The LOD, LOQ of 475 Pesticides and Related Chemical Residues Determined by GC-MS—cont'd

No.	Pesticides	LOD (mg/kg)	LOQ (mg/kg)	No.	Pesticides	LOD (mg/kg)	LOQ (mg/kg)
130	Metolachlor	0.012 5	0.0250	162	Oxyfluorfen	0.050 0	0.100 0
131	Oxy-chlordane	0.012 5	0.025 0	163	Chlorthiophos	0.037 5	0.075 0
132	Methoprene	0.050 0	0.100 0	164	Endosulfan II	0.075 0	0.150 0
133	Bromofos	0.025 0	0.050 0	165	Flamprop-Isopropyl	0.012 5	0.025 0
134	Ethofumesate	0.025 0	0.050 0	166	4,4′-DDT	0.025 0	0.050 0
135	Isopropalin	0.025 0	0.050 0	167	Carbofenothion	0.025 0	0.050 0
136	Endosulfan I	0.075 0	0.150 0	168	Benalaxyl	0.012 5	0.025 0
137	Propanil	0.025 0	0.050 0	169	Edifenphos	0.025 0	0.050 0
138	Isofenphos	0.025 0	0.050 0	170	Triazophos	0.037 5	0.075 0
139	Crufomate	0.075 0	0.150 0	171	Cyanofenphos	0.012 5	0.025 0
140	Chlorfenvinphos	0.037 5	0.075 0	172	Chlorbenside sulfone	0.025 0	0.050 0
141	cis-Chlordane	0.025 0	0.050 0	173	Endosulfan-Sulfate	0.037 5	0.075 0
142	Tolylfluanide	0.037 5	0.075 0	174	Bromopropylate	0.025 0	0.050 0
143	4,4′-DDE	0.012 5	0.025 0	175	Benzoylprop-ethyl	0.037 5	0.075 0
144	Butachlor	0.025 0	0.050 0	176	Fenpropathrin	0.025 0	0.050 0
145	Chlozolinate	0.025 0	0.050 0	177	Leptophos	0.025 0	0.050 0
146	Crotoxyphos	0.075 0	0.150 0	178	EPN	0.050 0	0.100 0
147	Iodofenphos	0.025 0	0.050 0	179	Hexazinone	0.037 5	0.075 0
148	Tetrachlorvinphos	0.037 5	0.075 0	180	Phosalone	0.025 0	0.050 0
149	Chlorbromuron	0.300 0	0.600 0	181	Azinphos-methyl	0.075 0	0.150 0
150	Profenofos	0.075 0	0.150 0	182	Fenarimol	0.025 0	0.050 0
151	Fluorochloridone	0.025 0	0.050 0	183	Azinphos-ethyl	0.025 0	0.050 0
152	2,4′-DDD	0.012 5	0.025 0	184	Coumaphos	0.075 0	0.1500
153	Endrin	0.150 0	0.300 0	185	Cyfluthrin	0.150 0	0.300 0
154	Hexaconazole	0.075 0	0.150 0	186	Fluvalinate	0.150 0	0.300 0
155	Chlorfenson	0.025 0	0.050 0		Group C		
156	2,4′-DDT	0.025 0	0.050 0	187	Dichlorvos	0.600 0	1.200 0
157	Paclobutrazol	0.037 5	0.075 0	188	Biphenyl	0.012 5	0.025 0
158	Methoprotryne	0.037 5	0.075 0	189	Vernolate	0.012 5	0.025 0
159	Chloropropylate	0.012 5	0.025 0	190	3,5-Dichloroaniline	0.012 5	0.025 0
160	Flamprop-methyl	0.012 5	0.025 0	191	Molinate	0.012 5	0.025 0
161	Nitrofen	0.075 0	0.150 0	192	Methacrifos	0.012 5	0.025 0

Continued

TABLE 2.8 The LOD, LOQ of 475 Pesticides and Related Chemical Residues Determined by GC-MS—cont'd

No.	Pesticides	LOD (mg/kg)	LOQ (mg/kg)	No.	Pesticides	LOD (mg/kg)	LOQ (mg/kg)
193	2-Phenylphenol	0.012 5	0.025 0	225	Mecarbam	0.050 0	0.100 0
194	Tetrahydro phthalimide	0.037 5	0.075 0	226	Tetraconazole	0.037 5	0.075 0
				227	Propaphos	0.025 0	0.050 0
195	Fenobucarb	0.025 0	0.050 0	228	Flumetralin	0.025 0	0.050 0
196	Benfluralin	0.012 5	0.025 0	229	Triadimenol	0.037 5	0.075 0
197	Hexaflumuron	0.075 0	0.150 0	230	Pretilachlor	0.025 0	0.050 0
198	Triallate	0.025 0	0.050 0	231	Kresoxim-methyl	0.012 5	0.025 0
199	Pyrimethanil	0.012 5	0.025 0	232	Fluazifop-butyl	0.012 5	0.025 0
200	Gamma-HCH	0.025 0	0.050 0	233	Chlorfluazuron	0.037 5	0.075 0
201	Disulfoton	0.012 5	0.025 0	234	Chlorobenzilate	0.012 5	0.025 0
202	Atrizine	0.012 5	0.025 0	235	Uniconazole	0.025 0	0.050 0
203	Heptachlor	0.037 5	0.075 0	236	Flusilazole	0.037 5	0.075 0
204	Iprobenfos	0.037 5	0.075 0	237	Fluorodifen	0.012 5	0.025 0
205	Isazofos	0.025 0	0.050 0	238	Diniconazole	0.037 5	0.075 0
206	Plifenate	0.025 0	0.050 0	239	Piperonyl butoxide	0.012 5	0.025 0
207	Transfluthrin	0.012 5	0.025 0	240	Propargite	0.025 0	0.050 0
208	Fluchloraline	0.050 0	0.100 0	241	Mepronil	0.012 5	0.025 0
209	Tolclofos-methyl	0.012 5	0.025 0	242	Dimefuron	0.050 0	0.100 0
210	Propisochlor	0.012 5	0.025 0	243	Diflufenican	0.012 5	0.025 0
211	Metobromuron	0.075 0	0.150 0	244	Phenothrin	0.012 5	0.025 0
212	Metribuzin	0.037 5	0.075 0	245	Fludioxonil	0.012 5	0.025 0
213	HCH, epsilon-	0.025 0	0.050 0	246	Fenoxycarb	0.075 0	0.150 0
214	Formothion	0.025 0	0.050 0	247	Sethoxydim	0.112 5	0.225 0
215	Diethofencarb	0.075 0	0.150 0	248	Anilofos	0.025 0	0.050 0
216	Dimepiperate	0.025 0	0.050 0	249	Acrinathrin	0.025 0	0.050 0
217	Bioallethrin-1	0.050 0	0.100 0	250	Lambda-Cyhalothrin	0.012 5	0.025 0
218	Bioallethrin-2	0.050 0	0.100 0	251	Mefenacet	0.037 5	0.075 0
219	2,4'-DDE	0.012 5	0.025 0	252	Permethrin	0.025 0	0.050 0
220	Fenson	0.012 5	0.025 0	253	Pyridaben	0.012 5	0.025 0
221	Diphenamid	0.012 5	0.025 0	254	Fluoroglycofen-ethyl	0.150 0	0.300 0
222	Chlorthion	0.025 0	0.050 0	255	Bitertanol	0.037 5	0.075 0
223	Prallethrin	0.037 5	0.075 0	256	Etofenprox	0.012 5	0.025 0
224	Penconazole	0.037 5	0.075 0	257	Cycloxydim	1.200 0	2.400 0

TABLE 2.8 The LOD, LOQ of 475 Pesticides and Related Chemical Residues Determined by GC-MS—cont'd

No.	Pesticides	LOD (mg/kg)	LOQ (mg/kg)	No.	Pesticides	LOD (mg/kg)	LOQ (mg/kg)
318	Rabenzazole	0.012 5	0.025 0	349	Triphenyl phosphate	0.012 5	0.025 0
319	Cyprodinil	0.012 5	0.025 0	350	Metamitron	0.125 0	0.250 0
320	Isofenphos oxon	0.025 0	0.050 0	351	DE-PCB 180	0.012 5	0.025 0
321	Dicapthon	0.062 5	0.125 0	352	Tebufenpyrad	0.012 5	0.025 0
322	DE-PCB 101	0.012 5	0.025 0	353	Lenacil	0.125 0	0.250 0
323	MCPA-Butoxyethyl ester	0.012 5	0.025 0	354	Bromuconazole-1	0.025 0	0.050 0
				355	Desbrom- leptophos	0.012 5	0.025 0
324	Isocarbophos	0.025 0	0.050 0	356	Bromuconazole-2	0.025 0	0.050 0
325	Phorate sulfone	0.012 5	0.025 0	357	Nitralin	0.125 0	0.250 0
326	Chlorfenethol	0.012 5	0.025 0	358	Fenamiphos sulfoxide	0.400 0	0.800 0
327	Trans-Nonachlor	0.012 5	0.025 0	359	Fenamiphos sulfone	0.050 0	0.100 0
328	Dinobuton	0.125 0	0.250 0	360	Fenpiclonil	0.050 0	0.100 0
329	DEF	0.025 0	0.050 0	361	Fluquinconazole	0.012 5	0.025 0
330	Flurochloridone	0.025 0	0.050 0	362	Fenbuconazole	0.025 0	0.050 0
331	Bromfenvinfos	0.012 5	0.025 0		Group E		
332	Perthane	0.012 5	0.025 0	363	Propoxur-1	0.025 0	0.050 0
333	DE-PCB 118	0.012 5	0.025 0	364	Isoprocarb -1	0.025 0	0.050 0
334	4,4-Dibromoben zophenone	0.012 5	0.025 0	365	Methamidophos	0.400 0	0.800 0
335	Flutriafol	0.025 0	0.050 0	366	Acenaphthene	0.012 5	0.025 0
336	Mephosfolan	0.025 0	0.050 0	367	Dibutyl succinate	0.025 0	0.050 0
337	Athidathion	0.025 0	0.050 0	368	Phthalimide	0.025 0	0.050 0
338	DE-PCB 153	0.012 5	0.025 0	369	Chlorethoxyfos	0.025 0	0.050 0
339	Diclobutrazole	0.050 0	0.100 0	370	Isoprocarb -2	0.025 0	0.050 0
340	Disulfoton sulfone	0.025 0	0.050 0	371	Pencycuron	0.100 0	0.200 0
				372	Tebuthiuron	0.050 0	0.100 0
341	Hexythiazox	0.100 0	0.200 0	373	demeton-S-methyl	0.050 0	0.100 0
342	DE-PCB 138	0.012 5	0.025 0	374	Cadusafos	0.050 0	0.100 0
343	Cyproconazole	0.012 5	0.025 0	375	Propoxur-2	0.025 0	0.050 0
344	Clodinafop-propargyl	0.025 0	0.050 0	376	Phenanthrene	0.012 5	0.025 0
345	Fenthion sulfoxide	0.050 0	0.100 0	377	Fenpyroximate	0.100 0	0.200 0
346	Fluotrimazole	0.012 5	0.025 0	378	Tebupirimfos	0.025 0	0.050 0
347	Fluroxypr-1-methylheptyl ester	0.012 5	0.025 0	379	Prohydrojasmon	0.050 0	0.100 0
348	Fenthion sulfone	0.050 0	0.100 0	380	Dichloran	0.025 0	0.050 0

TABLE 2.8 The LOD, LOQ of 475 Pesticides and Related Chemical Residues Determined by GC-MS—cont'd

No.	Pesticides	LOD (mg/kg)	LOQ (mg/kg)	No.	Pesticides	LOD (mg/kg)	LOQ (mg/kg)
318	Rabenzazole	0.012 5	0.025 0	349	Triphenyl phosphate	0.012 5	0.025 0
319	Cyprodinil	0.012 5	0.025 0	350	Metamitron	0.125 0	0.250 0
320	Isofenphos oxon	0.025 0	0.050 0	351	DE-PCB 180	0.012 5	0.025 0
321	Dicapthon	0.062 5	0.125 0	352	Tebufenpyrad	0.012 5	0.025 0
322	DE-PCB 101	0.012 5	0.025 0	353	Lenacil	0.125 0	0.250 0
323	MCPA-Butoxyethyl ester	0.012 5	0.025 0	354	Bromuconazole-1	0.025 0	0.050 0
				355	Desbrom- leptophos	0.012 5	0.025 0
324	Isocarbophos	0.025 0	0.050 0	356	Bromuconazole-2	0.025 0	0.050 0
325	Phorate sulfone	0.012 5	0.025 0	357	Nitralin	0.125 0	0.250 0
326	Chlorfenethol	0.012 5	0.025 0	358	Fenamiphos sulfoxide	0.400 0	0.800 0
327	Trans-Nonachlor	0.012 5	0.025 0	359	Fenamiphos sulfone	0.050 0	0.100 0
328	Dinobuton	0.125 0	0.250 0	360	Fenpiclonil	0.050 0	0.100 0
329	DEF	0.025 0	0.050 0	361	Fluquinconazole	0.012 5	0.025 0
330	Flurochloridone	0.025 0	0.050 0	362	Fenbuconazole	0.025 0	0.050 0
331	Bromfenvinfos	0.012 5	0.025 0		Group E		
332	Perthane	0.012 5	0.025 0	363	Propoxur-1	0.025 0	0.050 0
333	DE-PCB 118	0.012 5	0.025 0	364	Isoprocarb -1	0.025 0	0.050 0
334	4,4-Dibromoben zophenone	0.012 5	0.025 0	365	Methamidophos	0.400 0	0.800 0
335	Flutriafol	0.025 0	0.050 0	366	Acenaphthene	0.012 5	0.025 0
336	Mephosfolan	0.025 0	0.050 0	367	Dibutyl succinate	0.025 0	0.050 0
337	Athidathion	0.025 0	0.050 0	368	Phthalimide	0.025 0	0.050 0
338	DE-PCB 153	0.012 5	0.025 0	369	Chlorethoxyfos	0.025 0	0.050 0
339	Diclobutrazole	0.050 0	0.100 0	370	Isoprocarb -2	0.025 0	0.050 0
340	Disulfoton sulfone	0.025 0	0.050 0	371	Pencycuron	0.100 0	0.200 0
				372	Tebuthiuron	0.050 0	0.100 0
341	Hexythiazox	0.100 0	0.200 0	373	demeton-S-methyl	0.050 0	0.100 0
342	DE-PCB 138	0.012 5	0.025 0	374	Cadusafos	0.050 0	0.100 0
343	Cyproconazole	0.012 5	0.025 0	375	Propoxur-2	0.025 0	0.050 0
344	Clodinafop-propargyl	0.025 0	0.050 0	376	Phenanthrene	0.012 5	0.025 0
345	Fenthion sulfoxide	0.050 0	0.100 0	377	Fenpyroximate	0.100 0	0.200 0
346	Fluotrimazole	0.012 5	0.025 0	378	Tebupirimfos	0.025 0	0.050 0
347	Fluroxypr-1-methylheptyl ester	0.012 5	0.025 0	379	Prohydrojasmon	0.050 0	0.100 0
348	Fenthion sulfone	0.050 0	0.100 0	380	Dichloran	0.025 0	0.050 0

TABLE 2.8 The LOD, LOQ of 475 Pesticides and Related Chemical Residues Determined by GC-MS—cont'd

No.	Pesticides	LOD (mg/kg)	LOQ (mg/kg)	No.	Pesticides	LOD (mg/kg)	LOQ (mg/kg)
381	Pyroquilon	0.012 5	0.025 0	414	Furalaxyl	0.025 0	0.050 0
382	Propyzamide	0.025 0	0.050 0	415	Thiamethoxam	0.050 0	0.100 0
383	Pirimicarb	0.050 0	0.100 0	416	Mepanipyrim	0.012 5	0.025 0
384	Phosphamidon -1	0.100 0	0.200 0	417	Bromacil	0.100 0	0.200 0
385	Benoxacor	0.025 0	0.050 0	418	Picoxystrobin	0.025 0	0.050 0
386	Bromobutide	0.012 5	0.025 0	419	Butamifos	0.012 5	0.025 0
387	Acetochlor	0.025 0	0.050 0	420	Imazamethabenz-methyl	0.037 5	0.075 0
388	Tridiphane	0.050 0	0.100 0	421	Metominostrobin-1	0.050 0	0.100 0
389	Terbucarb	0.025 0	0.050 0	422	TCMTB	0.200 0	0.400 0
390	Esprocarb	0.025 0	0.050 0	423	Methiocarb Sulfone	0.800 0	1.600 0
391	Fenfuram	0.025 0	0.050 0	424	Imazalil	0.100 0	0.200 0
392	Acibenzolar-S-Methyl	0.025 0	0.050 0	425	Isoprothiolane	0.025 0	0.050 0
393	Benfuresate	0.025 0	0.050 0	426	Cyflufenamid	0.200 0	0.400 0
394	Dithiopyr	0.012 5	0.025 0	427	Pyriminobac-Methyl	0.050 0	0.100 0
395	Mefenoxam	0.025 0	0.050 0	428	Isoxathion	0.100 0	0.200 0
396	Malaoxon	0.200 0	0.400 0	429	Metominostrobin-2	0.050 0	0.100 0
397	Phosphamidon -2	0.100 0	0.200 0	430	Diofenolan -1	0.025 0	0.050 0
398	Simeconazole	0.025 0	0.050 0	431	Thifluzamide	0.100 0	0.200 0
399	Chlorthal-dimethyl	0.025 0	0.050 0	432	Diofenolan -2	0.025 0	0.050 0
400	Thiazopyr	0.025 0	0.050 0	433	Quinoxyphen	0.012 5	0.025 0
401	Dimethylvinphos	0.025 0	0.050 0	434	Chlorfenapyr	0.100 0	0.200 0
402	Butralin	0.050 0	0.100 0	435	Trifloxystrobin	0.050 0	0.100 0
403	Zoxamide	0.025 0	0.050 0	436	Imibenconazole-des-benzyl	0.050 0	0.100 0
404	Pyrifenox -1	0.100 0	0.200 0	437	Isoxadifen-Ethyl	0.025 0	0.050 0
405	Allethrin	0.050 0	0.100 0	438	Fipronil	0.100 0	0.200 0
406	Dimethametryn	0.012 5	0.025 0	439	Imiprothrin-1	0.025 0	0.050 0
407	Quinoclamine	0.050 0	0.100 0	440	Carfentrazone-Ethyl	0.025 0	0.050 0
408	Methothrin-1	0.025 0	0.050 0	441	Imiprothrin-2	0.025 0	0.050 0
409	Flufenacet	0.100 0	0.200 0	442	Epoxiconazole -1	0.100 0	0.200 0
410	Methothrin-2	0.025 0	0.050 0	443	Pyraflufen Ethyl	0.025 0	0.050 0
411	Pyrifenox -2	0.100 0	0.200 0	444	Pyributicarb	0.025 0	0.050 0
412	Fenoxanil	0.025 0	0.050 0				
413	Phthalide	0.050 0	0.100 0				

Continued

TABLE 2.8 The LOD, LOQ of 475 Pesticides and Related Chemical Residues Determined by GC-MS—cont'd

No.	Pesticides	LOD (mg/kg)	LOQ (mg/kg)	No.	Pesticides	LOD (mg/kg)	LOQ (mg/kg)
445	Thenylchlor	0.025 0	0.050 0	461	Pyraclostrobin	0.300 0	0.600 0
446	Clethodim	0.100 0	0.200 0	462	Lactofen	0.100 0	0.200 0
447	Mefenpyr-diethyl	0.037 5	0.075 0	463	Tralkoxydim	0.400 0	0.800 0
448	Famphur	0.050 0	0.100 0	464	Pyraclofos	0.100 0	0.200 0
449	Etoxazole	0.075 0	0.150 0	465	Dialifos	0.100 0	0.200 0
450	Pyriproxyfen	0.012 5	0.025 0	466	Spirodiclofen	0.100 0	0.200 0
451	Epoxiconazole-2	0.100 0	0.200 0	467	Halfenprox	0.050 0	0.100 0
452	Picolinafen	0.012 5	0.025 0	468	Flurtamone	0.050 0	0.100 0
453	Iprodione	0.050 0	0.100 0	469	Pyriftalid	0.012 5	0.025 0
454	Piperophos	0.037 5	0.075 0	470	Silafluofen	0.012 5	0.025 0
455	Ofurace	0.037 5	0.075 0	471	Pyrimidifen	0.100 0	0.200 0
456	Bifenazate	0.100 0	0.200 0	472	Acetamiprid	0.150 0	0.300 0
457	Endrin ketone	0.050 0	0.100 0	473	Butafenacil	0.012 5	0.025 0
458	Clomeprop	0.012 5	0.025 0	474	Cafenstrole	0.150 0	0.300 0
459	Fenamidone	0.012 5	0.025 0	475	Fluridone	0.025 0	0.050 0
460	Naproanilide	0.012 5	0.025 0				

Mixed Standard Solution (Mixed Standard Solution A, B, C, D, and E): Depending on properties and retention time of each compound, all compounds are divided into five groups: A, B, C, D, and E. Depending upon instrument sensitivity of each compound, the concentrations of mixed standard solutions are decided. For 475 pesticides and related chemicals grouping and concentration of mixed standard solution of this standard, reference is made to Appendix B.

Depending on group number, mixed standard solution concentration, and stock standard solution concentration, appropriate amounts of individual stock standard solutions are pipetted into a 100-mL volumetric flask, with five group pesticides and related chemicals being diluted to volume with toluene. Mixed standard solutions are stored in the dark below 4°C and are used for 1 month.

Internal Standard Solution: Accurately weigh 3.5 mg heptachlor epoxide into a 100-mL volumetric flask. Dissolve and dilute to volume with toluene.

Working Standard Mixed Solution in Matrix: Working standard mixed solutions in matrix of A, B, C, D, and E group pesticides and related chemicals are prepared by diluting 40 μL internal standard solution and appropriate amounts of mixed standard solution to 1.0 mL with blank extract that has been taken through the method with the rest of the samples, mixed thoroughly.

Working standard mixed solutions in matrix must be freshly prepared.

2.4.4 Apparatus

GC-MS: Equipped With EI
Analytical Balances: Capable of Weighing From 0.1 mg to 0.01 g
Accelerated Solvent Extractor: Equipped With 34-mL Cell
Nitrogen Evaporator
Pear-Shaped Flask: 200 mL
Pipette: 1 mL

2.4.5 Sample Pretreatment

2.4.5.1 Preparation of Test Sample

According to GB/T 5491, take the representative grains samples. Grind the samples thoroughly with a grinder and pass through a 20-mesh sieve. Mix thoroughly, seal, and label.

The test samples are stored at room temperature.

2.4.5.2 Extraction

A 10-g test sample (accurate to 0.01 g) and 10 g Celit 545 are weighed, mixed thoroughly, placed in a 34-mL cell of accelerated solvent extractor, and extracted with acetonitrile by heating for 5 min and in the static state for 3 min and then by cycling twice at 10.34 MPa and 80°C. The sample is then rinsed with acetonitrile of 60% of cell volume, and purged with nitrogen for 100 s. The extracts are then mixed thoroughly. For the least oily samples, half of the extracts (corresponding to a 5-g test sample) are used for clean-up. For the oiliest samples, a quarter of the extracts (corresponding to a 2.5-g test sample) are used for clean-up.

2.4.5.3 Clean-up

The Envi-18 cartridge is fixed to a support and conditioned with 10 mL acetonitrile before addition of the sample. A pear-shaped flask is connected to the cartridge, and the extracts obtained as described previously are decanted into the cartridge. The cartridge is eluted with 15 mL acetonitrile and the eluate is concentrated to ca. 1 mL by rotary evaporator at 40°C. The concentrate is used for the next stage of clean-up.

Sodium sulfate (ca. 2 cm) is placed on top of an Envi-Carb cartridge and this is connected in series to the top of an Sep-Pak NH_2 cartridge. The cartridges are fixed to a support and conditioned with 4 mL acetonitrile + toluene (3 + 1).

When the conditioning solution reaches the top of the sodium sulfate, the cartridges are connected to a pear-shaped flask and the concentrated sample obtained as described previously is added to the cartridges. The pear-shaped flask is rinsed with $3 \times 2\,mL$ acetonitrile + toluene $(3+1)$, and the washings are also applied to the cartridges. A reservoir is attached to the cartridges and the pesticides and related chemicals are eluted with $25\,mL$ acetonitrile + toluene $(3+1)$. The eluate is concentrated to ca. $0.5\,mL$ by rotary evaporator at 40°C, exchanged with $2 \times 5\,mL$ hexane twice and diluted to ca. $1\,mL$. Internal standard solution $(40\,\mu L)$ is added and the solution, after thorough mixing, is then ready for GC-MS analysis.

2.4.6   GC-MS Method Determination

2.4.6.1   GC-MS Operating Condition

(a) Column: DB-1701 $(30\,m \times 0.25\,mm \times 0.25\,\mu m)$ capillary column, or equivalent
(b) Column temperature: 40°C hold 1 min, at 30°C/min to 130°C, at 5°C/min to 250°C, at 10°C/min to 300°C, hold 5 min
(c) Carrier gas: Helium, purity $\geq 99.999\%$, flow rate: $1.2\,mL/min$
(d) Injection port temperature: 290°C
(e) Injection volume: $1\,\mu L$
(f) Injection mode: Splitless, purge on after 1.5 min
(g) Ionization voltage: 70 eV
(h) Ion source temperature: 230°C
(i) GC-MS interface temperature: 280°C
(j) Selected ion monitoring mode: Each compound selects 1 quantifying ion and 2–3 qualifying ions. All of the detected ions of each group are detected according to programmed time and sequence of peaking. The retention times, quantifying ions, qualifying ions, and the abundance ratios of quantifying ion and qualifying ions for each compound are listed in Table 5.3. The programmed time and dwell time for the ions detected for each compound in each group are listed in Table 5.4

2.4.6.2   Qualitative Determination

In the samples determined, four injections are required to analyze all pesticides and related chemicals by GC-MS under the operating condition. If the retention times of peaks of the sample solution are the same as those of the peaks from the working standard mixed solution, and the selected ions appeared in the background-subtracted mass spectrum, and the abundance ratios of the selected ions are within the expected limits (abundance ratios $>50\%$, permitted tolerances are $\pm 10\%$; abundance ratios $>20\%$ to 50%, permitted tolerances are $\pm 15\%$; abundance ratios $>10\%$ to 20%, permitted tolerances are $\pm 20\%$; abundance ratios $\leq 10\%$, permitted tolerances are $\pm 50\%$), the sample is confirmed to contain this pesticide compound. If the results are not definitive, the sample is reinjected with acquisition in scan mode (sufficient sensitivity) or with additional confirmatory ions or using other instruments of higher sensitivity.

2.4.6.3   Quantitative Determination

Quantification is performed using an internal standard and the quantifying ion for GC-MS. The internal standard is heptachlor epoxide. To compensate for the matrix effects, quantitation is based on a mixed standard prepared in blank matrix extract. The concentrations in the standard solution and in the sample solution analyzed must be close.

2.4.6.4   Parallel Test

A parallel test is carried out for the same testing sample.

2.4.6.5   Blank Test

The operation of the blank test is the same as that described in the method of determination, but without the addition of sample.

2.4.7   Precision

The precision data of the method for this standard have been determined according to the stipulations of GB/T 6379.1 and GB/T 6379.2. The values of repeatability and reproducibility are obtained and calculated at the 95% confidence level.

RESEARCHERS

Guo-Fang Pang, Yong-Ming Liu, Chun-Lin Fan, Yan-Ping Wu, Yan-Zhong Cao, Jin-Jie Zhang, Xue-Min Li, Jin Li.

Qinhuangdao Entry-Exit Inspection and Quarantine Bureau, 39 Haibin Rd, Qinhuangdao, Hebei, PC 066002, People's Republic of China.

2.5   DETERMINATION OF 486 PESTICIDES AND RELATED CHEMICAL RESIDUES IN GRAINS: LC-MS-MS METHOD (GB/T 20770-2008)

2.5.1   Scope

This method is applicable to the determination of residues of 486 pesticides and related chemicals in barley, wheat, oat, rice, and maize. The standard is applicable to the quantitative determination of residues of 376 pesticides and related chemicals.

The limits of quantitative determination of this method of 376 pesticides and related chemicals are 0.02 µg/kg to 0.96 mg/kg (refer to Table 2.9).

TABLE 2.9 The LOD, LOQ of 486 Pesticides and Related Chemical Residues Determined by LC-MS-MS

No.	Pesticides	LOD (µg/kg)	LOQ (µg/kg)	No.	Pesticides	LOD (µg/kg)	LOQ (µg/kg)
	Group A			32.	penconazole	1.00	2.00
1.	propham	55.00	110.00	33.	myclobutanil	0.50	1.00
2.	isoprocarb	1.15	2.30	34.	imazethapyr	0.56	1.12
3.	3,4,5-trimethacarb	0.17	0.34	35.	paclobutrazol	0.29	0.58
4.	cycluron	0.10	0.20	36.	fenthion sulfoxide	0.16	0.32
5.	carbaryl	5.16	10.32	37.	triadimenol	5.28	10.56
6.	propachlor	0.14	0.28	38.	butralin	0.95	1.90
7.	rabenzazole	0.67	1.34	39.	spiroxamine	0.03	0.06
8.	simetryn	0.07	0.14	40.	tolclofos methyl	33.28	66.56
9.	monolinuron	1.78	3.56	41.	desmedipham	2.01	4.02
10.	mevinphos	0.78	1.56	42.	methidathion	5.33	10.66
11.	aziprotryne	0.69	1.38	43.	allethrin	30.20	60.40
12.	secbumeton	0.04	0.08	44.	diazinon	0.38	0.76
13.	cyprodinil	0.37	0.74	45.	edifenphos	0.17	0.34
14.	buturon	4.48	8.96	46.	pretilachlor	0.36	0.72
15.	carbetamide	1.82	3.64	47.	flusilazole	0.29	0.58
16.	pirimicarb	0.08	0.16	48.	iprovalicarb	1.16	2.32
17.	clomazone	0.21	0.42	49.	benodanil	1.74	3.48
18.	cyanazine	0.08	0.16	50.	flutolanil	0.57	1.14
19.	prometryne	0.08	0.16	51.	famphur	1.80	3.60
20.	paraoxon methyl	0.38	0.76	52.	benalyxyl	0.62	1.24
21.	4,4-dichloro-benzophenone	6.80	13.60	53.	diclobutrazole	0.23	0.46
22.	thiacloprid	0.19	0.38	54.	etaconazole	0.89	1.78
23.	imidacloprid	11.00	22.00	55.	fenarimol	0.30	0.60
24.	ethidimuron	0.75	1.50	56.	tetramethirn	0.91	1.82
25.	isomethiozin	0.53	1.06	57.	cloquintocet mexyl	0.94	1.88
26.	diallate	44.60	89.20	58.	bitertanol	16.70	33.40
27.	acetochlor	23.70	47.40	59.	tepraloxydim	6.10	12.20
28.	nitenpyram	8.56	17.12	60.	thiophanate methyl	10.00	20.00
29.	methoprotryne	0.12	0.24	61.	azinphos ethyl	54.46	108.92
30.	dimethenamid	2.15	4.30	62.	triflumuron	1.96	3.92
31.	terrbucarb	1.05	2.10	63.	isoxaflutole	1.95	3.90
				64.	anilofos	0.36	0.72

TABLE 2.9 The LOD, LOQ of 486 Pesticides and Related Chemical Residues Determined by LC-MS-MS—cont'd

No.	Pesticides	LOD (µg/kg)	LOQ (µg/kg)	No.	Pesticides	LOD (µg/kg)	LOQ (µg/kg)
65.	thiophanat ethyl	10.08	20.16	96.	methobromuron	8.42	16.84
66.	quizalofop-ethyl	0.34	0.68	97.	phorate	157.00	314.00
67.	haloxyfop-methyl	1.32	2.64	98.	aclonifen	12.10	24.20
68.	fluazifop butyl	0.13	0.26	99.	mephosfolan	1.16	2.32
69.	bromophos-ethyl	283.85	567.70	100.	imibenzon-azole-des-benzyl	3.11	6.22
70.	bensulide	17.10	34.20				
71.	triasulfuron	0.80	1.60	101.	neburon	3.55	7.10
72.	bromfenvinfos	1.51	3.02	102.	mefenoxam	0.77	1.54
73.	azoxystrobin	0.23	0.46	103.	prothoate	1.23	2.46
74.	indoxacarb	3.77	7.54	104.	ethofume sate	186.00	372.00
				105.	iprobenfos	4.14	8.28
	Group B			106.	TEPP	5.20	10.40
75.	dazomet	63.50	127.00	107.	cyproconazole	0.37	0.74
76.	nicotine	1.10	2.20	108.	thiamethoxam	16.50	33.00
77.	fenuron	0.52	1.04	109.	crufomate	0.26	0.52
78.	crimidine	0.78	1.56	110.	etrimfos	9.38	18.76
79.	acephate	6.67	13.34	111.	coumatetralyl	0.68	1.36
80.	molinate	1.05	2.10	112.	cythioate	40.00	80.00
81.	carbendazim	0.23	0.46	113.	phosphamidon	1.94	3.88
82.	6-chloro-4-hydroxy-3-phenyl-pyridazin	0.83	1.66	114.	phenmedipham	2.24	4.48
83.	propoxur	12.20	24.40	115.	fenhexamid	0.47	0.94
84.	isouron	0.20	0.40	116.	flutriafol	4.29	8.58
85.	chlorotoluron	0.31	0.62	117.	furalaxyl	0.39	0.78
86.	thiofanox	78.50	157.00	118.	bioallethrin	99.00	198.00
87.	chlorbufam	91.50	183.00	119.	cyanofenphos	10.40	20.80
88.	bendiocarb	1.59	3.18	120.	pirimiphos methyl	0.10	0.20
89.	propazine	0.16	0.32	121.	buprofezin	0.44	0.88
90.	terbuthylazine	0.23	0.46	122.	disulfoton sulfone	1.23	2.46
91.	diuron	0.78	1.56	123.	fenazaquin	0.16	0.32
92.	chlormephos	224.00	448.00	124.	triazophos	0.34	0.68
93.	clothianidin	31.50	63.00	125.	DEF	0.81	1.62
94.	pronamide	7.69	15.38	126.	pyriftalid	0.31	0.62
95.	dimethachloro	0.95	1.90	127.	metconazole	0.66	1.32

Continued

TABLE 2.9 The LOD, LOQ of 486 Pesticides and Related Chemical Residues Determined by LC-MS-MS—cont'd

No.	Pesticides	LOD (µg/kg)	LOQ (µg/kg)	No.	Pesticides	LOD (µg/kg)	LOQ (µg/kg)
128.	pyriproxyfen	0.22	0.44	160.	methiocarb	20.60	41.20
129.	cycloxydim	1.27	2.54	161.	metoxuron	0.32	0.64
130.	isoxaben	0.09	0.18	162.	dimethoate	3.80	7.60
131.	flurtamone	0.22	0.44	163.	methfuroxam	0.14	0.28
132.	trifluralin	167.40	334.80	164.	fluometuron	0.46	0.92
133.	flamprop methyl	10.10	20.20	165.	dicrotophos	0.57	1.14
134.	bioresmethrin	3.71	7.42	166.	monalide	0.60	1.20
135.	propiconazole	0.88	1.76	167.	diphenamid	0.07	0.14
136.	chlorpyrifos	26.90	53.80	168.	ethoprophos	1.38	2.76
137.	fluchloralin	244.00	488.00	169.	fonofos	3.73	7.46
138.	clethodim	1.04	2.08	170.	etridiazol	50.21	100.42
139.	flamprop isopropyl	0.22	0.44	171.	hexazinone	0.06	0.12
140.	tetra-chlorvinphos	1.11	2.22	172.	dimethametryn	0.06	0.12
141.	propargite	34.30	68.60	173.	trichlorphon	0.56	1.12
142.	bromuconazole	1.57	3.14	174.	demeton(o+s)	3.39	6.78
143.	picolinafen	0.36	0.72	175.	benoxacor	3.45	6.90
144.	fluthiacet methyl	2.65	5.30	176.	bromacil	11.80	23.60
145.	trifloxystrobin	1.00	2.00	177.	phorate sulfoxide	184.14	368.28
146.	chlorimuron ethyl	15.20	30.40	178.	brompyrazon	1.80	3.60
	Group C			179.	oxycarboxin	0.45	0.90
147.	methamidophos	2.47	4.94	180.	mepronil	0.19	0.38
148.	EPTC	18.67	37.34	181.	disulfoton	234.85	469.70
149.	diethyltoluamide	0.28	0.56	182.	fenthion	26.00	52.00
150.	monuron	17.37	34.74	183.	metalaxyl	0.25	0.50
151.	pyrimethanil	0.34	0.68	184.	ofurace	0.50	1.00
152.	fenfuram	0.39	0.78	185.	dodemorph	0.20	0.40
153.	quinoclamine	3.96	7.92	186.	imaza-methabenz-methyl	0.08	0.16
154.	fenobucarb	2.95	5.90	187.	disulfoton-sulfoxide	1.42	2.84
155.	propanil	10.80	21.60	188.	isoprothiolane	0.92	1.84
156.	carbofuran	6.53	13.06	189.	imazalil	1.00	2.00
157.	acetamiprid	0.72	1.44	190.	phoxim	41.40	82.80
158.	mepanipyrim	0.16	0.32	191.	quinalphos	1.00	2.00
159.	prometon	0.07	0.14	192.	ditalimfos	33.61	67.22

TABLE 2.9 The LOD, LOQ of 486 Pesticides and Related Chemical Residues Determined by LC-MS-MS—cont'd

No.	Pesticides	LOD (µg/kg)	LOQ (µg/kg)	No.	Pesticides	LOD (µg/kg)	LOQ (µg/kg)
193.	fenoxycarb	9.14	18.28	226.	flumiclorac-pentyl	5.30	10.60
194.	pyrimitate	0.09	0.18	227.	temephos	0.61	1.22
195.	fensulfothin	1.00	2.00	228.	butafenacil	4.75	9.50
196.	fluorochloridone	6.89	13.78		Group D		
197.	butachlor	10.03	20.06	229.	mepiquat chloride	0.45	0.90
198.	kresoxim-methyl	50.29	100.58	230.	allidochlor	20.52	41.04
199.	triticonazole	1.51	3.02	231.	propamocarb	0.04	0.08
200.	fenamiphos sulfoxide	0.37	0.74	232.	tricyclazole	0.62	1.24
201.	thenylchlor	12.07	24.14	233.	thiabendazole	0.24	0.48
202.	fenoxanil	19.70	39.40	234.	metamitron	3.18	6.36
203.	fluridone	0.09	0.18	235.	isoproturon	0.07	0.14
204.	epoxiconazole	2.03	4.06	236.	atratone	0.09	0.18
205.	chlorphoxim	38.79	77.58	237.	oesmetryn	0.09	0.18
206.	fenamiphos sulfone	0.22	0.44	238.	metribuzin	0.27	0.54
207.	fenbuconazole	0.82	1.64	239.	DMST	20.00	40.00
208.	isofenphos	109.34	218.68	240.	cycloate	2.22	4.44
209.	phenothrin	169.60	339.20	241.	atrazine	0.18	0.36
210.	piperophos	4.62	9.24	242.	butylate	151.00	302.00
211.	piperonyl butoxide	0.57	1.14	243.	pymetrozin	17.14	34.28
212.	oxyflurofen	29.27	58.54	244.	chloridazon	1.16	2.32
213.	coumaphos	1.05	2.10	245.	sulfallate	103.60	207.20
214.	flufenacet	2.65	5.30	246.	ethiofencarb	2.46	4.92
215.	phosalone	24.02	48.04	247.	terbumeton	0.05	0.10
216.	methoxy fenozide	1.85	3.70	248.	cyprazine	0.05	0.10
217.	prochloraz	1.03	2.06	249.	ametryn	0.48	0.96
218.	aspon	0.87	1.74	250.	tebuthiuron	0.11	0.22
219.	ethion	1.48	2.96	251.	trietazine	0.30	0.60
220.	diafenthiuron	0.14	0.28	252.	sebutylazine	0.16	0.32
221.	thifensulfuron-methyl	10.70	21.40	253.	dibutyl succinate	111.20	222.40
222.	ethoxysulfuron	2.29	4.58	254.	tebutam	0.07	0.14
223.	dithiopyr	5.20	10.40	255.	thiofanox-sulfoxide	4.15	8.30
224.	spirodiclofen	4.95	9.90	256.	cartap hydrochloride	1040.00	2080.00
225.	fenpyroximate	0.68	1.36	257.	methacrifos	242.37	484.74

Continued

TABLE 2.9 The LOD, LOQ of 486 Pesticides and Related Chemical Residues Determined by LC-MS-MS—cont'd

No.	Pesticides	LOD (µg/kg)	LOQ (µg/kg)	No.	Pesticides	LOD (µg/kg)	LOQ (µg/kg)
258.	terbutryn	11.35	22.70	290.	sulprofos	2.92	5.84
259.	triazoxide	4.00	8.00	291.	EPN	16.50	33.00
260.	thionazin	11.34	22.68	292.	azamethiphos	0.40	0.80
261.	linuron	5.82	11.64	293.	diniconazole	0.67	1.34
262.	heptanophos	2.92	5.84	294.	flumetsulam	0.15	0.30
263.	prosulfocarb	0.18	0.36	295.	sethoxydim	44.80	89.60
264.	dipropetryn	0.14	0.28	296.	pencycuron	0.14	0.28
265.	thiobencarb	1.65	3.30	297.	mecarbam	9.80	19.60
266.	tri-iso-butyl phosphate	1.79	3.58	298.	tralkoxydim	0.16	0.32
				299.	malathion	2.82	5.64
267.	tri-n-butyl phosphate	0.19	0.38	300.	pyributicarb	0.17	0.34
268.	diethofencarb	1.00	2.00	301.	pyridaphenthion	0.44	0.88
269.	alachlor	3.70	7.40	302.	pirimiphos-ethyl	0.05	0.10
270.	cadusafos	0.58	1.16	303.	thiodicarb	19.68	39.36
271.	metazachlor	0.49	0.98	304.	pyraclofos	0.50	1.00
272.	propetamphos	27.00	54.00	305.	picoxystrobin	4.22	8.44
273.	terbufos	1120.00	2240.00	306.	tetraconazole	0.86	1.72
274.	simeconazole	1.47	2.94	307.	mefenpyr-diethyl	6.28	12.56
275.	triadimefon	3.94	7.88	308.	profenefos	1.01	2.02
276.	phorate sulfone	21.00	42.00	309.	pyraclostrobin	0.25	0.50
277.	tridemorph	1.30	2.60	310.	dimethomorph	0.18	0.36
278.	mefenacet	1.10	2.20	311.	kadethrin	1.66	3.32
279.	azaconazole	0.40	0.80	312.	thiazopyr	0.98	1.96
280.	fenamiphos	0.10	0.20	313.	benfuracarb-methyl	8.19	16.38
281.	fenpropimorph	0.09	0.18	314.	cinosulfuron	0.56	1.12
282.	tebuconazole	1.12	2.24	315.	pyrazosulfuron-ethyl	3.42	6.84
283.	isopropalin	15.00	30.00	316.	metosulam	2.20	4.40
284.	nuarimol	0.50	1.00		Group E		
285.	bupirimate	0.35	0.70	317.	4-amino-pyridine	0.43	0.86
286.	azinphos-methyl	552.17	1104.34	318.	methomyl	4.78	9.56
287.	tebupirimfos	0.06	0.12	319.	pyroquilon	1.74	3.48
288.	phenthoate	46.18	92.36	320.	fuberidazole	0.95	1.90
289.	sulfotep	1.30	2.60	321.	isocarbamid	0.85	1.70

TABLE 2.9 The LOD, LOQ of 486 Pesticides and Related Chemical Residues Determined by LC-MS-MS—cont'd

No.	Pesticides	LOD (μg/kg)	LOQ (μg/kg)	No.	Pesticides	LOD (μg/kg)	LOQ (μg/kg)
322.	butocarboxim	0.79	1.58	354.	quinoxyphen	76.70	153.40
323.	chlordimeform	0.67	1.34	355.	fenthion sulfone	8.73	17.46
324.	cymoxanil	27.80	55.60	356.	fluro-chloridone	0.65	1.30
325.	vernolate	0.13	0.26	357.	phthalic acid,benzyl butyl ester	316.00	632.00
326.	chlorthiamid	4.41	8.82	358.	isazofos	0.09	0.18
327.	aminocarb	8.21	16.42	359.	dichlofenthion	15.10	30.20
328.	dimethirimol	0.06	0.12	360.	vamidothion sulfone	238.00	476.00
329.	omethoate	4.83	9.66	361.	terbufos sulfone	44.30	88.60
330.	ethoxyquin	1.76	3.52	362.	dinitramine	0.90	1.80
331.	dichlorvos	0.27	0.54	363.	cyazofamid	2.25	4.50
332.	aldicarb sulfone	10.70	21.40	364.	trichloronat	33.40	66.80
333.	dioxacarb	1.68	3.36	365.	resmethrin-2	0.15	0.30
334.	benzyladenine	35.40	70.80	366.	boscalid	2.38	4.76
335.	demeton-s-methyl	2.65	5.30	367.	nitralin	17.20	34.40
336.	ethiofencarb-sulfoxide	112.00	224.00	368.	fenpropathrin	122.50	245.00
337.	cyanohos	5.06	10.12	369.	hexythiazox	11.80	23.60
338.	thiometon	289.00	578.00	370.	florasulam	8.70	17.40
339.	folpet	69.30	138.60	371.	benzoximate	9.83	19.66
340.	dimepiperate	1 890.00	3 780.00	372.	benzoylprop-ethyl	154.00	308.00
341.	fenpropidin	0.09	0.18	373.	pyrimidifen	7.00	14.00
342.	amidithion	329.00	658.00	374.	furathiocarb	0.96	1.92
343.	imazapic	2.95	5.90	375.	trans-permethin	2.40	4.80
344.	paraoxon-ethyl	0.24	0.48	376.	etofenprox	114.00	228.00
345.	aldimorph	1.58	3.16	377.	pyrazoxyfen	0.16	0.32
346.	vinclozolin	1.27	2.54	378.	flubenzimine	3.89	7.78
347.	uniconazole	1.20	2.40	379.	*zeta* cypermethrin	0.34	0.68
348.	pyrifenox	0.13	0.26	380.	haloxyfop-2-ethoxyethyl	1.25	2.50
349.	chlorthion	66.80	133.60	381.	esfenvalerate	208.00	416.00
350.	dicapthon	0.12	0.24	382.	fluoroglycofen-ethyl	2.50	5.00
351.	clofentezine	0.38	0.76		Group F		
352.	norflurazon	0.13	0.26	383.	acrylamide	8.90	17.80
353.	triallate	23.10	46.20	384.	*tert*-butylamine	19.48	38.96

Continued

TABLE 2.9 The LOD, LOQ of 486 Pesticides and Related Chemical Residues Determined by LC-MS-MS—cont'd

No.	Pesticides	LOD (µg/kg)	LOQ (µg/kg)	No.	Pesticides	LOD (µg/kg)	LOQ (µg/kg)
385.	hymexazol	112.07	224.14	416.	chloroxuron	0.22	0.44
386.	phthalimide	21.50	43.00	417.	triamiphos	2.30	4.60
387.	dimefox	34.10	68.20	418.	prallethrin	0.65	1.30
388.	metolcarb	12.70	25.40	419.	cumyluron	0.66	1.32
389.	diphenylamin	0.21	0.42	420.	imazamox	0.90	1.80
390.	1-naphthy acetamide	0.41	0.82	421.	warfarin	1.34	2.68
391.	atrazine-desethyl	0.31	0.62	422.	phosmet	8.86	17.72
392.	2,6-dichloro-benzamide	2.25	4.50	423.	ronnel	6.57	13.14
393.	aldicarb	130.50	261.00	424.	phthalic acid, biscyclohexyl ester	0.34	0.68
394.	dimethyl phthalate	6.60	13.20	425.	carpropamid	2.60	5.20
395.	chlordimeform hydrochloride	1.32	2.64	426.	tebufenpyrad	0.13	0.26
396.	simeton	0.55	1.10	427.	tebufenozide	13.90	27.80
397.	dinotefuran	5.09	10.18	428.	chlorthiophos	15.90	31.80
398.	pebulate	1.70	3.40	429.	dialifos	78.50	157.00
399.	acibenzolar-s-methyl	1.54	3.08	430.	cinidon-ethyl	7.29	14.58
400.	dioxabenzofos	6.92	13.84	431.	rotenone	1.16	2.32
401.	oxamyl	274.03	548.06	432.	imibenconazole	5.13	10.26
402.	thidiazuron pestanal	0.15	0.30	433.	propaquiafop	0.62	1.24
403.	metha-benzthiazuron	0.04	0.08	434.	2,3,4,5-tetrachloro-aniline	26.80	53.60
404.	butoxycarboxim	13.30	26.60	435.	benzofenap	0.04	0.08
405.	demeton-s-methyl sulfoxide	1.96	3.92	436.	dinoseb acetate	20.64	41.28
406.	thiofanox sulfone	12.04	24.08	437.	propisochlor	0.40	0.80
407.	phosfolan	0.24	0.48	438.	silafluofen,	304.00	608.00
408.	demeton-s	40.00	80.00	439.	etobenzanid	0.40	0.80
409.	fenthion oxon	0.59	1.18	440.	fentrazamide	6.20	12.40
410.	napropamide	0.64	1.28	441.	pentachloro-aniline	1.87	3.74
411.	fenitrothion	13.40	26.80	442.	cyphenothrin	8.40	16.80
412.	phthalic acid, dibutyl ester	19.80	39.60	443.	dieldrin	80.80	161.60
413.	metolachlor	0.20	0.40	444.	etoxazole	0.44	0.88
414.	procymidone	43.30	86.60	445.	malaoxon	2.34	4.68
415.	vamidothion	2.28	4.56	446.	dodine	4.00	8.00
				447.	dalapon	115.37	230.74

TABLE 2.9 The LOD, LOQ of 486 Pesticides and Related Chemical Residues Determined by LC-MS-MS—cont'd

No.	Pesticides	LOD (μg/kg)	LOQ (μg/kg)	No.	Pesticides	LOD (μg/kg)	LOQ (μg/kg)
448.	flupropanate	11.49	22.98	468.	chlorfenethol	82.15	164.30
449.	2,6-difluoro-benzoic acid	852.04	1704.08	469.	bromoxynil	0.90	1.80
				470.	chlorobenzuron	10.20	20.40
450.	trichloroacetic acid sodium salt	140.79	281.58	471.	chlorampheni-colum	1.94	3.88
451.	2-phenylphenol	84.94	169.88	472.	alloxydim-sodium	0.10	0.20
452.	3-phenylphenol	2.00	4.00	473.	sulfanitran	1.52	3.04
453.	clopyralld	140.00	280.00	474.	mesotrion	1 150.28	2 300.56
454.	DNOC	1.30	2.60	475.	oryzalin	2.46	4.92
455.	cloprop	5.70	11.40	476.	acifluorfen	59.00	118.00
456.	dicloran	24.28	48.56	477.	ioxynil	0.31	0.62
457.	chlorpropham	7.88	15.76	478.	famoxadone	22.64	45.28
458.	mecoprop	2.45	4.90	479.	diflufenican	14.14	28.28
459.	terbacil	0.44	0.88	480.	ethiprole	19.93	39.86
460.	dicamba	632.96	1 265.92	481.	flusulfamide	0.21	0.42
461.	MCPB	7.09	14.18	482.	cyclosulfamuron	171.84	343.68
462.	bentazone	0.52	1.04	483.	fomesafen	1.01	2.02
463.	dinoseb	0.20	0.40	484.	iodosulfuron-methyl sodium	10.60	21.20
464.	dinoterb	0.12	0.24	485.	kelevan	964.00	1 928.00
465.	forchlorfenuron	5.70	11.40	486.	iodosulfuron-methyl	33.30	66.60
466.	fludioxonil	31.08	62.16				
467.	fluroxypyr	96.03	192.06				

2.5.2 Principle

The samples are extracted with acetonitrile by homogenizing, the extracts are cleaned up by GPC, the solutions are analyzed by LC-MS-MS, using an external standard method for quantification.

2.5.3 Reagents and Materials

"Water" is first grade of GB/T6682 specified.

Acetonitrile: HPLC Grade
Methanol: HPLC Grade

Cyclohexane: P.R. Grade
Ethyl Acetate: P.R. Grade
Hexane: P.R. Grade
Toluene: G.R.
Acetone: HPLC Grade
Isooctane: HPLC Grade
0.1% Formic Acid-Water (V/V)
5 mmol/L Ammonium Acetate: Weigh 0.375 g Ammonium Acetate, Dilute
With Water to 1000 mL
Acetonitrile + Toluene (3 + 1, V/V)
Acetonitrile + Water (3 + 2, V/V)
Ethyl acetate + Cyclohexane (1 + 1, V/V)
Sodium Sulfate: Anhydrous, Analytically Pure. Ignited at 650°C for 4 h and
Kept in a Desiccator
Filter Membrane (Nylon): 13 mm × 0.2 μm and 13 mm × 0.45 μm
Pesticide and Related Chemicals Standard: Purity ≥95%

  Stock Standard Solution: Accurately weigh 5–10 mg of individual pesticide and related chemicals standards (accurate to 0.1 mg) into a 10-mL
  volumetric flask. Dissolve and dilute to volume with methanol, acetonitrile, toluene, toluene + acetone combination, isooctane, etc., depending
  on each individual compound's solubility (for diluting solvent refer to
  Appendix B). Standard solutions are stored in the dark below 0°C to
  4°C and are used for 1 year.

  Mixed Standard Solution (Mixed Standard Solution A, B, C, D, E, F,
  and G): Depending on properties and retention time of each compound, all compounds are divided into seven groups: A, B, C, D, E,
  F, and G. The concentrations of mixed standard solutions are
  decided depending on instrument sensitivity for each compound.
  For 486 pesticides and related chemicals groupings and concentration
  of mixed standard solutions of this standard, reference is made to
  Appendix B.

  Depending on group number, mixed standard solution concentration, and
  stock standard solution concentration, appropriate amounts of individual
  stock standard solutions are pipetted into a 100-mL volumetric flask,
  with seven group pesticides and related chemicals being diluted to volume with methanol. Mixed standard solutions are stored in the dark
  below 0°C to 4°C and are used for 1 month.

  Working Standard Mixed Solution in Matrix: Working standard mixed
  solutions in matrix of A, B, C, D, E, F, and G groups pesticides and
  related chemicals are prepared by mixing different concentrations of
  working standard mixed solutions with sample blank extract that has
  been taken through the method with the rest of the samples. These solutions are used to construct calibration plots. Working standard mixed
  solutions in matrix must be freshly prepared.

2.5.4   Apparatus

LC-MS-MS: Equipped With ESI Source
GPC: Equipped 400×25 mm, BIO-Beads S-X3 Column or Equivalent
Analytical Balances: Capable of Weighing From 0.1 mg to 0.01 g
Homogenizer: Max. rpm: 24,000 rpm
Pear-Shaped Flask
Nitrogen Evaporator
Centrifuge Tube: 80 mL
Pipette: 1 mL
Screw Vial: 2.0 mL, With Screw Caps and PTFE Septa
Centrifuge: Max. rpm 4200 rpm
Rotary Evaporate

2.5.5   Sample Pretreatment

2.5.5.1   Preparation of Test Sample

According to GB/T 5491, take the representative grain samples. Grind the samples thoroughly with a grinder and pass all through a 20-mesh sieve. Mix thoroughly, seal, and label.

The test samples are stored at room temperature.

2.5.5.2   Extraction

Weigh 10 g (accurate to 0.01 g) of the test sample into a centrifuge tube filled with 15 g anhydrous sodium sulfate. Add 35 mL acetonitrile, and homogenize with homogenizer for 1 min, and centrifuge at 3800 rpm for 5 min. The supernatants are made to pass through a glass funnel containing anhydrous sodium sulfate and collected in a pear-shaped flask. Rehomogenize sample plug in centrifuge tube with 30 mL acetonitrile, recentrifuge and then transfer to the previously mentioned glass funnel before the extracts are combined, which are then placed in a waterbath of 40°C and evaporated to 0.5 mL on a rotary evaporator for clean-up. Add 5 mL ethyl acetate plus cyclohexane (1+1) twice for solvent exchange, and finally evaporate the sample to a 5-mL solution for clean-up.

2.5.5.3   GPC Clean-up

2.5.5.3.1   Conditions

(a) GPC Column: 400×25 mm (i.d.), filled with BIO-Beads S-X3 or equivalent
(b) Detection wavelength: 254 nm
(c) Mobile phase: cychexane+ethyl acetate (1+1)
(d) Flow rate: 5 mL/min
(e) Injection volume: 5 mL
(f) Start collect time: 22 min
(g) Stop collect time: 40 min

2.5.5.3.2 Clean-up

Transfer the concentrated extracts to a 10-mL volumetric flask with cychexane +ethyl acetate (1 + 1). Rinse pear-shaped flask twice with 5 mL cychexane + ethyl acetate (1 + 1) and transfer them to the previously mentioned 10-mL volumetric flask. Dilute to volume with cychexane +ethyl acetate (1 + 1) and mix thoroughly. Filter the solution to 10-mL test tube with 0.45-μm filter membrane for GPC clean-up. The fractions of 22 min to 40 min are collected in a 200-mL pear-shaped flask and concentrated to 0.5 mL on a rotary evaporator at 40°C. Evaporate it to dryness using a nitrogen evaporator. Dissolve residues with 1.0 mL acetonitrile + water (3 + 2) and pass through the 0.20-μm filter membrane. The solution is ready for LC-MS-MS determination.

2.5.6 Determination

2.5.6.1 LC-MS-MS Operation Condition

Conditions for the Pesticides and Related Chemicals of A, B, C, D, E, and F Group (ESI Positive Ion Source)

- **(a)** Column: ZORBOX SB-C$_{18}$, 3.5 μm, 100 × 2.1 mm (i.d.) or equivalent
- **(b)** Mobile phase and flow rate, see Table 2.2
- **(c)** Column temperature: 40°C
- **(d)** Injection volume: 10 μL
- **(e)** Ion source: electrospray ionization source Scan mode: Positive ion scan
- **(f)** Nebulizer gas: nitrogen
- **(g)** Nebulizer gas: 0.28 MPa
- **(h)** Ionspray voltage: 4000 V
- **(i)** Nebulizer gas temperature: 350°C
- **(j)** Dryness gas Flow rate: 10 L/min
- **(k)** Monitoring ion pairs, collision energy, and fragmentor, see Table 5.10

Conditions for the Pesticides and Related Chemicals of G Group (ESI Positive Ion Source)

- **(a)** Column: ZORBOX SB-C$_{18}$, 3.5 μm, 100 × 2.1 mm (i.d.) or equivalent
- **(b)** Mobile phase and flow rate, see Table 2.3
- **(c)** Column temperature: 40°C
- **(d)** Injection volume: 10 μL
- **(e)** Ion source: electrospray ionization source
- **(f)** Ion mode: negative mode
- **(g)** Nebulizer gas: nitrogen
- **(h)** Nebulizer gas: 0.28 MPa
- **(i)** Ionspray voltage: 4000 V
- **(j)** Nebulizer gas temperature: 350°C
- **(k)** Dryness gas flow rate: 10 L/min
- **(l)** Monitoring ion pairs, collision energy, and fragmentor, see Table 5.10

2.5.6.2   Qualitative Determination

If the retention times of peaks of the sample solution are the same as those of the peaks from the working standard mixed solution, and the selected ions appeared in the background-subtracted mass spectrum, and the abundance ratios of the selected ions are within the expected limits (abundance ratios >50%, permitted tolerances are ±10%; abundance ratios >20% to 50%, permitted tolerances are ±15%; abundance ratios >10% to 20%, permitted tolerances are ±20%; abundance ratios ≤10%, permitted tolerances are ±50%), the sample is confirmed to contain this pesticide compound.

2.5.6.3   Quantitative Determination

An external standard method is used for quantitation with standard curves for LC-MS-MS. In order to compensate for the matrix effect, quantitation is based on a series of working standard solutions prepared in blank matrix extract. The standard curves are established by injection of different concentrations of working standard mixed solutions in matrix separately. The responses of pesticides in the sample solution should be in the linear range of the instrumental detection.

2.5.6.4   Parallel Test

A parallel test is carried out for the same testing sample.

2.5.6.5   Blank Test

The operation of the blank test is the same as that described in the method of determination, but without the addition of sample.

2.5.7   Precision

The precision data of the method for this standard have been determined according to the stipulations of GB/T 6379.1 and GB/T 6379.2. The values of repeatability and reproducibility are obtained and calculated at the 95% confidence level.

RESEARCHERS

Guo-Fang Pang, Yong-Ming Liu, Chun-Lin Fan, Yan-Zhong Cao, Jing Cao, Feng Zheng, Yu-Jing Lian, Guang-Qun Jia, Jin-Jie Zhang, Xue-Min Li, Yan-Ping Wu, Jin Li.

Qinhuangdao Entry-Exit Inspection and Quarantine Bureau, 39 Haibin Rd, Qinhuangdao, Hebei, PC 066002, People's Republic of China.

2.6 DETERMINATION OF 519 PESTICIDES AND RELATED CHEMICAL RESIDUES IN TEA LEAVES: GC-MS METHOD (GB/T 23204-2008)

2.6.1 Scope

This method is applicable to the qualitative determination of residues of 490 pesticides and related chemicals in tea leaves, including green tea, black tea, puer tea, and oolong; the quantitative determination of residues of 453 pesticides and related chemicals; and the determination of 29 acidic herbicide (clopyralid, cloprop, 4-CPA, dicamba, MCPA, dichlorprop, bromoxynil, 2,4-D, triclopyr, NAA, pentachlorphenol, fenoprop, chloramben, MCPB,2,4,5-T, fluroxypyr, 2,4-DB, bentazone, OH-ioxynil, picloram, quinclorac, fluazifop, haloxyfop, flamprop acid, acifluorfen, pyritiobacsodium, fenhxamid, quizalofop, and bispyribacsodium) residues in tea leaves, including green tea, black tea, puer tea, and oolong.

This limits of detection of the method of the quantitative determination of 453 pesticides and related chemicals are: 0.001 mg/kg to 0.500 mg/kg (see Table 2.10); the limit of detection of the method of 29 acidic herbicides is: 0.01 mg/kg.

2.6.2 Determination of 490 Pesticides and Related Chemical Residues in Tea: GC-MS Method

2.6.2.1 Principle

The samples are extracted with acetonitrile, and the extracts are cleaned up with the Cleanert TPT cartridges. The pesticides and related chemicals are eluted with acetonitrile-toluene (3+1), and the solutions are analyzed by GC-MS, using the internal standard method of quantitative analysis.

2.6.2.2 Reagents and Materials

Acetonitrile: HPLC Grade
Toluene: G.R.
Acetone: Analytically Pure, Redistillation
Dichloromethane: HPLC Grade
Hexane: Analytically Pure, Redistillation
Methanol: HPLC Grade
Sodium Sulfate: Anhydrous, Analytically Pure. Ignited at 650°C for 4 h and Kept in a Desiccator
Acetonitrile-Toluene (3+1, V/V)
Milli Pore Filter Membrane (Nylon): 13 mm × 0.2 μm
Internal Standard Solution: Accurately Weigh 3.5 mg Heptachlor Epoxide into a 100-mL Volumetric Flask. Dissolve and Dilute to Volume With Toluene

TABLE 2.10 The LOD, LOQ of 490 Pesticides and Related Chemical Residues Determined by GC-MS

No.	Pesticides	LOD (mg/kg)	LOQ (mg/kg)	No.	Pesticides	LOD (mg/kg)	LOQ (mg/kg)
	Group A			35	anthraquinone	0.0125	0.0250
1	allidochlor	0.0100	0.0200	36	fenthion	0.0050	0.0100
2	dichlormid	0.0100	0.0200	37	malathion	0.0200	0.0400
3	etridiazol	0.0150	0.0300	38	paraoxon-ethyl	0.1600	0.3200
4	chlormephos	0.0100	0.0200	39	fenitrothion	0.0100	0.0200
5	propham	0.0050	0.0100	40	triadimefon	0.0100	0.0200
6	cycloate	0.0050	0.0100	41	linuron	0.0200	0.0400
7	diphenylamine	0.0050	0.0100	42	pendimethalin	0.0200	0.0400
8	chlordimeform	0.0050	0.0100	43	chlorbenside	0.0100	0.0200
9	ethalfluralin	0.0200	0.0400	44	bromophos-ethyl	0.0050	0.0100
10	phorate	0.0050	0.0100	45	quinalphos	0.0050	0.0100
11	thiometon	0.0050	0.0100	46	*trans*-chlordane	0.0050	0.0100
12	quintozene	0.0100	0.0200	47	phenthoate	0.0100	0.0200
13	atrazine-desethyl	0.0050	0.0100	48	metazachlor	0.0150	0.0300
14	clomazone	0.0050	0.0100	49	prothiophos	0.0050	0.0100
15	diazinon	0.0050	0.0100	50	chlorfurenol	0.0150	0.0300
16	fonofos	0.0050	0.0100	51	procymidone	0.0050	0.0100
17	etrimfos	0.0050	0.0100	52	dieldrin	0.0100	0.0200
18	propetamphos	0.0050	0.0100	53	methidathion	0.0250	0.0500
19	secbumeton	0.0050	0.0100	54	napropamide	0.0150	0.0300
20	pronamide	0.0050	0.0100	55	cyanazine	0.0150	0.0300
21	dichlofenthion	0.0050	0.0100	56	oxadiazone	0.0050	0.0100
22	mexacarbate	0.0150	0.0300	57	fenamiphos	0.0150	0.0300
23	dimethoate	0.0200	0.0400	58	tetrasul	0.0050	0.0100
24	dinitramine	0.0200	0.0400	59	bupirimate	0.0050	0.0100
25	aldrin	0.0100	0.0200	60	flutolanil	0.0050	0.0100
26	ronnel	0.0100	0.0200	61	carboxin	0.0150	0.0300
27	prometryne	0.0050	0.0100	62	*p,p'*-DDD	0.0050	0.0100
28	cyprazine	0.0050	0.0100	63	ethion	0.0100	0.0200
29	vinclozolin	0.0050	0.0100	64	etaconazole-1	0.0150	0.0300
30	*beta*-HCH	0.0050	0.0100	65	sulprofos	0.0100	0.0200
31	metalaxyl	0.0150	0.0300	66	etaconazole-2	0.0150	0.0300
32	methyl-parathion	0.0200	0.0400	67	myclobutanil	0.0050	0.0100
33	chlorpyrifos (-ethyl)	0.0050	0.0100	68	fensulfothion	0.0250	0.0500
34	*delta*-HCH	0.0100	0.0200	69	diclofop-methyl	0.0050	0.0100

Continued

TABLE 2.10 The LOD, LOQ of 490 Pesticides and Related Chemical Residues Determined by GC–MS—cont'd

No.	Pesticides	LOD (mg/kg)	LOQ (mg/kg)	No.	Pesticides	LOD (mg/kg)	LOQ (mg/kg)
70	propiconazole-1	0.0150	0.0300	105	*trans*-diallate	0.0100	0.0200
71	propiconazole-2	0.0150	0.0300	106	chlorpropham	0.0100	0.0200
72	bifenthrin	0.0050	0.0100	107	sulfotep	0.0050	0.0100
73	mirex	0.0050	0.0100	108	sulfallate	0.0100	0.0200
74	carbosulfan	0.0150	0.0300	109	*alpha*-HCH	0.0050	0.0100
75	nuarimol	0.0100	0.0200	110	terbufos	0.0100	0.0200
76	benodanil	0.0150	0.0300	111	profluralin	0.0200	0.0400
77	methoxychlor	0.0050	0.0100	112	dioxathion	0.0500	0.1000
78	oxadixyl	0.0050	0.0100	113	propazine	0.0050	0.0100
79	tebuconazole	0.0375	0.0750	114	chlorbufam	0.0250	0.0500
80	tetramethirn	0.0125	0.0250	115	dicloran	0.0100	0.0200
81	norflurazon	0.0050	0.0100	116	terbuthylazine	0.0125	0.0250
82	pyridaphenthion	0.0050	0.0100	117	monolinuron	0.0200	0.0400
83	tetradifon	0.0050	0.0100	118	cyanophos	0.0100	0.0200
84	*cis*-permethrin	0.0050	0.0100	119	flufenoxuron	0.0150	0.0300
85	pyrazophos	0.0100	0.0200	120	chlorpyrifos-methyl	0.0050	0.0100
86	*trans*-permethrin	0.0125	0.0250	121	desmetryn	0.0050	0.0100
87	cypermethrin	0.0150	0.0300	122	dimethachlor	0.0150	0.0300
88	fenvalerate-1	0.0200	0.0400	123	alachlor	0.0150	0.0300
89	fenvalerate-2	0.0200	0.0400	124	pirimiphos-methyl	0.0050	0.0100
90	deltamethrin	0.0750	0.1500	125	terbutryn	0.0100	0.0200
	Group B			126	aspon	0.0100	0.0200
91	EPTC	0.0150	0.0300	127	thiobencarb	0.0100	0.0200
92	butylate	0.0150	0.0300	128	dicofol	0.0100	0.0200
93	dichlobenil	0.0010	0.0020	129	metolachlor	0.0050	0.0100
94	pebulate	0.0150	0.0300	130	pirimiphos-ethyl	0.0100	0.0200
95	nitrapyrin	0.0150	0.0300	131	oxy-chlordane	0.0125	0.0250
96	mevinphos	0.0100	0.0200	132	dichlofluanid	0.0300	0.0600
97	chloroneb	0.0050	0.0100	133	methoprene	0.0200	0.0400
98	tecnazene	0.0100	0.0200	134	bromofos	0.0100	0.0200
99	heptanophos	0.0150	0.0300	135	ethofumesate	0.0100	0.0200
100	ethoprophos	0.0150	0.0300	136	isopropalin	0.0100	0.0200
101	hexachlorobenzene	0.0050	0.0100	137	propanil	0.0100	0.0200
102	propachlor	0.0150	0.0300	138	crufomate	0.0300	0.0600
103	*cis*-diallate	0.0100	0.0200	139	isofenphos	0.0100	0.0200
104	trifluralin	0.0100	0.0200	140	endosulfan -1	0.0300	0.0600

TABLE 2.10 The LOD, LOQ of 490 Pesticides and Related Chemical Residues Determined by GC-MS—cont'd

No.	Pesticides	LOD (mg/kg)	LOQ (mg/kg)	No.	Pesticides	LOD (mg/kg)	LOQ (mg/kg)
141	chlorfenvinphos	0.0150	0.0300	176	EPN	0.0200	0.0400
142	tolylfluanide	0.0150	0.0300	177	hexazinone	0.0150	0.0300
143	cis- chlordane	0.0100	0.0200	178	leptophos	0.0100	0.0200
144	butachlor	0.0100	0.0200	179	bifenox	0.0100	0.0200
145	chlozolinate	0.0100	0.0200	180	phosalone	0.0100	0.0200
146	p,p'-DDE	0.0050	0.0100	181	azinphos-methyl	0.0750	0.1500
147	iodofenphos	0.0100	0.0200	182	fenarimol	0.0100	0.0200
148	tetrachlorvinphos	0.0150	0.0300	183	azinphos-ethyl	0.0250	0.0500
149	profenofos	0.0300	0.0600	184	cyfluthrin	0.1200	0.2400
150	buprofezin	0.0100	0.0200	185	prochloraz	0.0600	0.1200
151	hexaconazole	0.0300	0.0600	186	coumaphos	0.0300	0.0600
152	o,p'-DDD	0.0050	0.0100	187	fluvalinate	0.0600	0.1200
153	chlorfenson	0.0100	0.0200		Group C		
154	fluorochloridone	0.0100	0.0200	188	dichlorvos	0.0300	0.0600
155	endrin	0.0600	0.1200	189	biphenyl	0.0050	0.0100
156	paclobutrazol	0.0150	0.0300	190	propamocarb	0.0150	0.0300
157	o,p'-DDT	0.0100	0.0200	191	vernolate	0.0050	0.0100
158	methoprotryne	0.0150	0.0300	192	3,5-dichloroaniline	0.0125	0.0250
159	chloropropylate	0.0050	0.0100	193	methacrifos	0.0050	0.0100
160	flamprop-methyl	0.0050	0.0100	194	molinate	0.0125	0.0250
161	nitrofen	0.0300	0.0600	195	2-phenylphenol	0.0050	0.0100
162	oxyfluorfen	0.0200	0.0400	196	cis-1,2,3,6-tetrahydrophthalimide	0.0150	0.0300
163	chlorthiophos	0.0150	0.0300	197	fenobucarb	0.0100	0.0200
164	flamprop-isopropyl	0.0050	0.0100	198	benfluralin	0.0050	0.0100
165	carbofenothion	0.0100	0.0200	199	hexaflumuron	0.0300	0.0600
166	p,p'-DDT	0.0100	0.0200	200	prometon	0.0150	0.0300
167	benalaxyl	0.0050	0.0100	201	triallate	0.0100	0.0200
168	edifenphos	0.0100	0.0200	202	pyrimethanil	0.0050	0.0100
169	triazophos	0.0150	0.0300	203	gamma-HCH	0.0100	0.0200
170	cyanofenphos	0.0050	0.0100	204	disulfoton	0.0050	0.0100
171	chlorbenside sulfone	0.0100	0.0200	205	atrizine	0.0050	0.0100
172	endosulfan-sulfate	0.0150	0.0300	206	iprobenfos	0.0150	0.0300
173	bromopropylate	0.0100	0.0200	207	heptachlor	0.0150	0.0300
174	benzoylprop-ethyl	0.0150	0.0300	208	isazofos	0.0100	0.0200
175	fenpropathrin	0.0100	0.0200				

Continued

TABLE 2.10 The LOD, LOQ of 490 Pesticides and Related Chemical Residues Determined by GC-MS—cont'd

No.	Pesticides	LOD (mg/kg)	LOQ (mg/kg)	No.	Pesticides	LOD (mg/kg)	LOQ (mg/kg)
209	plifenate	0.0050	0.0100	245	mepronil	0.0050	0.0100
210	fluchloralin	0.0200	0.0400	246	diflufenican	0.0050	0.0100
211	transfluthrin	0.0050	0.0100	247	fludioxonil	0.0125	0.0250
212	fenpropimorph	0.0100	0.0200	248	fenazaquin	0.0050	0.0100
213	tolclofos-methyl	0.0050	0.0100	249	phenothrin	0.0125	0.0250
214	propisochlor	0.0050	0.0100	250	amitraz	0.0150	0.0300
215	metobromuron	0.0300	0.0600	251	anilofos	0.0100	0.0200
216	ametryn	0.0150	0.0300	252	lambda-cyhalothrin	0.0050	0.0100
217	metribuzin	0.0150	0.0300	253	mefenacet	0.0150	0.0300
218	dipropetryn	0.0050	0.0100	254	permethrin	0.0100	0.0200
219	formothion	0.0250	0.0500	255	pyridaben	0.0050	0.0100
220	diethofencarb	0.0300	0.0600	256	fluoroglycofen-ethyl	0.0600	0.1200
221	dimepiperate	0.0100	0.0200	257	bitertanol	0.0150	0.0300
222	bioallethrin-1	0.0500	0.1000	258	etofenprox	0.0125	0.0250
223	bioallethrin-2	0.0500	0.1000	259	*alpha*-cypermethrin	0.0250	0.0500
224	fenson	0.0050	0.0100	260	flucythrinate-1	0.0100	0.0200
225	*o,p'*-DDE	0.0125	0.0250	261	flucythrinate-2	0.0100	0.0200
226	diphenamid	0.0050	0.0100	262	esfenvalerate	0.0200	0.0400
227	penconazole	0.0150	0.0300	263	difenconazole-2	0.0300	0.0600
228	tetraconazole	0.0150	0.0300	264	difenonazole-1	0.0300	0.0600
229	mecarbam	0.0200	0.0400	265	flumioxazin	0.0100	0.0200
230	propaphos	0.0100	0.0200	266	flumiclorac-pentyl	0.0100	0.0200
231	flumetralin	0.0100	0.0200		Group D		
232	triadimenol-1	0.0150	0.0300	267	dimefox	0.0150	0.0300
233	triadimenol-2	0.0375	0.0750	268	disulfoton-sulfoxide	0.0100	0.0200
234	pretilachlor	0.0100	0.0200	269	pentachlorobenzene	0.0050	0.0100
235	kresoxim-methyl	0.0050	0.0100	270	crimidine	0.0050	0.0100
236	fluazifop-butyl	0.0050	0.0100	271	BDMC-1	0.0100	0.0200
237	chlorfluazuron	0.0150	0.0300	272	chlorfenprop-methyl	0.0050	0.0100
238	chlorobenzilate	0.0050	0.0100	273	thionazin	0.0050	0.0100
239	flusilazole	0.0150	0.0300	274	2,3,5,6-tetrachloroaniline	0.0050	0.0100
240	fluorodifen	0.0050	0.0100	275	*tri-N*-butyl phosphate	0.0100	0.0200
241	diniconazole	0.0150	0.0300	276	2,3,4,5-tetrachloroanisole	0.5000	1.0000
242	piperonyl butoxide	0.0050	0.0100	277	pentachloroanisole	0.0050	0.0100
243	dimefuron	0.0200	0.0400	278	tebutam	0.0100	0.0200
244	propargite	0.0125	0.0250	279	methabenzthiazuron	0.0500	0.1000

TABLE 2.10 The LOD, LOQ of 490 Pesticides and Related Chemical Residues Determined by GC-MS—cont'd

No.	Pesticides	LOD (mg/kg)	LOQ (mg/kg)	No.	Pesticides	LOD (mg/kg)	LOQ (mg/kg)
280	desisopropyl-atrazine	0.0400	0.0800	314	isofenphos oxon	0.0100	0.0200
281	simetone	0.0100	0.0200	315	fuberidazole	0.0250	0.0500
282	atratone	0.0125	0.0250	316	dicapthon	0.0250	0.0500
283	tefluthrin	0.0050	0.0100	317	mcpa-butoxyethyl ester	0.0050	0.0100
284	bromocylen	0.0050	0.0100	318	de-PCB 101	0.0050	0.0100
285	trietazine	0.0050	0.0100	319	isocarbophos	0.0100	0.0200
286	2,6-dichlorobenzamide	0.0100	0.0200	320	phorate sulfone	0.0050	0.0100
287	cycluron	0.0150	0.0300	321	chlorfenethol	0.0050	0.0100
288	de-PCB 28	0.0050	0.0100	322	trans-nonachlor	0.0050	0.0100
289	de-PCB 31	0.0050	0.0100	323	DEF	0.0100	0.0200
290	desethyl-sebuthylazine	0.0100	0.0200	324	flurochloridone	0.0100	0.0200
				325	bromfenvinfos	0.0050	0.0100
291	2,3,4,5-tetrachloroaniline	0.0100	0.0200	326	perthane	0.0125	0.0250
292	musk ambrette	0.0050	0.0100	327	de-PCB 118	0.0050	0.0100
293	musk xylene	0.0050	0.0100	328	mephosfolan	0.0100	0.0200
294	pentachloroaniline	0.0050	0.0100	329	4,4-dibromobenzophenone	0.0050	0.0100
295	aziprotryne	0.0400	0.0800				
296	isocarbamid	0.0250	0.0500	330	flutriafol	0.0100	0.0200
297	sebutylazine	0.0050	0.0100	331	de-PCB 153	0.0050	0.0100
298	musk moskene	0.0050	0.0100	332	diclobutrazole	0.0200	0.0400
299	de-PCB 52	0.0050	0.0100	333	disulfoton sulfone	0.0250	0.0500
300	prosulfocarb	0.0050	0.0100	334	hexythiazox	0.0400	0.0800
301	dimethenamid	0.0050	0.0100	335	de-PCB 138	0.0125	0.0250
302	BDMC-2	0.0250	0.0500	336	cyproconazole	0.0125	0.0250
303	monalide	0.0100	0.0200	337	resmethrin-1	0.0250	0.0500
304	musk tibeten	0.0050	0.0100	338	resmethrin-2	0.0250	0.0500
305	isobenzan	0.0050	0.0100	339	phthalic acid,benzyl butyl ester	0.0050	0.0100
306	octachlorostyrene	0.0050	0.0100				
307	isodrin	0.0050	0.0100	340	clodinafop-propargyl	0.0100	0.0200
308	isomethiozin	0.0100	0.0200	341	fenthion sulfoxide	0.0500	0.1000
309	dacthal	0.0050	0.0100	342	fluotrimazole	0.0050	0.0100
310	4,4-dichlorobenzophenone	0.0050	0.0100	343	fluroxypr-1-methylheptyl ester	0.0050	0.0100
311	nitrothal-isopropyl	0.0100	0.0200	344	fenthion sulfone	0.0200	0.0400
312	rabenzazole	0.0050	0.0100	345	metamitron	0.0500	0.1000
313	cyprodinil	0.0050	0.0100	346	triphenyl phosphate	0.0050	0.0100
				347	de-PCB 180	0.0050	0.0100

Continued

TABLE 2.10 The LOD, LOQ of 490 Pesticides and Related Chemical Residues Determined by GC-MS—cont'd

No.	Pesticides	LOD (mg/kg)	LOQ (mg/kg)	No.	Pesticides	LOD (mg/kg)	LOQ (mg/kg)
348	tebufenpyrad	0.0050	0.0100	382	esprocarb	0.0100	0.0200
349	cloquintocet-mexyl	0.0500	0.1000	383	terbucarb-2	0.0100	0.0200
350	lenacil	0.0050	0.0100	384	acibenzolar-s-methyl	0.0250	0.0500
351	bromuconazole-1	0.0100	0.0200	385	mefenoxam	0.0100	0.0200
352	bromuconazole-2	0.0100	0.0200	386	malaoxon	0.0250	0.0500
353	nitralin	0.0500	0.1000	387	chlorthal-dimethyl	0.0100	0.0200
354	fenamiphos sulfoxide	0.0500	0.1000	388	simeconazole	0.0100	0.0200
355	fenamiphos sulfone	0.0200	0.0400	389	terbacil	0.0100	0.0200
356	fenpiclonil	0.0200	0.0400	390	thiazopyr	0.0100	0.0200
357	fluquinconazole	0.0050	0.0100	391	dimethylvinphos	0.0250	0.0500
358	fenbuconazole	0.0250	0.0500	392	zoxamide	0.0100	0.0200
	Group E			393	allethrin	0.0200	0.0400
359	propoxur-1	0.1000	0.2000	394	quinoclamine	0.0200	0.0400
360	isoprocarb -1	0.0100	0.0200	395	flufenacet	0.1000	0.2000
361	terbucarb-1	0.0100	0.0200	396	fenoxanil	0.0100	0.0200
362	dibutyl succinate	0.0100	0.0200	397	furalaxyl	0.0100	0.0200
363	chlorethoxyfos	0.0100	0.0200	398	bromacil	0.0250	0.0500
364	isoprocarb -2	0.0100	0.0200	399	picoxystrobin	0.0100	0.0200
365	tebuthiuron	0.0200	0.0400	400	butamifos	0.0050	0.0100
366	pencycuron	0.0200	0.0400	401	imazamethabenz-methyl	0.0150	0.0300
367	demeton-s-methyl	0.0500	0.1000	402	methiocarb sulfone	0.1600	0.3200
368	propoxur-2	0.0400	0.0800	403	metominostrobin	0.0200	0.0400
369	phenanthrene	0.0050	0.0100	404	imazalil	0.0200	0.0400
370	fenpyroximate	0.0400	0.0800	405	isoprothiolane	0.0100	0.0200
371	tebupirimfos	0.0100	0.0200	406	cyflufenamid	0.0800	0.1600
372	prohydrojasmon	0.0250	0.0500	407	isoxathion	0.1000	0.2000
373	fenpropidin	0.0100	0.0200	408	quinoxyphen	0.0050	0.0100
374	dichloran	0.0125	0.0250	409	trifloxystrobin	0.0200	0.0400
375	pyroquilon	0.0050	0.0100	410	imibenconazole-des-benzyl	0.0200	0.0400
376	propyzamide	0.0100	0.0200	411	imiprothrin-1	0.0100	0.0200
377	pirimicarb	0.0250	0.0500	412	fipronil	0.1000	0.2000
378	benoxacor	0.0100	0.0200	413	imiprothrin-2	0.0100	0.0200
379	phosphamidon -1	0.0125	0.0250	414	epoxiconazole -1	0.1000	0.2000
380	acetochlor	0.0100	0.0200	415	pyributicarb	0.0250	0.0500
381	tridiphane	0.0200	0.0400				

TABLE 2.10 The LOD, LOQ of 490 Pesticides and Related Chemical Residues Determined by GC-MS—cont'd

No.	Pesticides	LOD (mg/kg)	LOQ (mg/kg)	No.	Pesticides	LOD (mg/kg)	LOQ (mg/kg)
416	pyraflufen ethyl	0.0100	0.0200	450	dicrotophos	0.0400	0.0800
417	thenylchlor	0.0100	0.0200	451	3,4,5-trimethacarb	0.0400	0.0800
418	mefenpyr-diethyl	0.0150	0.0300	452	2,4,5-T	0.1000	0.2000
419	etoxazole	0.0300	0.0600	453	3-phenylphenol	0.0300	0.0600
420	epoxiconazole-2	0.1000	0.2000	454	furmecyclox	0.0150	0.0300
421	pyriproxyfen	0.0050	0.0100	455	spiroxamine -2	0.0100	0.0200
422	iprodione	0.0200	0.0400	456	DMSA	0.0400	0.0800
423	ofurace	0.0150	0.0300	457	sobutylazine	0.0100	0.0200
424	piperophos	0.0150	0.0300	458	cinmethylin	0.0250	0.0500
425	clomeprop	0.0050	0.0100	459	monocrotophos	0.1000	0.2000
426	fenamidone	0.0125	0.0250	460	s421(octachlorodipropyl ether)-1	0.1000	0.2000
427	pyraclostrobin	0.1500	0.3000				
428	lactofen	0.0400	0.0800	461	s421(octachlorodipropyl ether)-2	0.1000	0.2000
429	pyraclofos	0.0400	0.0800	462	dodemorph	0.0150	0.0300
430	dialifos	0.4000	0.8000	463	fenchlorphos	0.0200	0.0400
431	spirodiclofen	0.1000	0.2000	464	difenoxuron	0.0400	0.0800
432	flurtamone	0.0250	0.0500	465	butralin	0.0200	0.0400
433	pyriftalid	0.0125	0.0250	466	pyrifenox-1	0.0400	0.0800
434	silafluofen	0.0125	0.0250	467	thiabendazole	0.1000	0.2000
435	pyrimidifen	0.0250	0.0500	468	iprovalicarb-1	0.0200	0.0400
436	butafenacil	0.0050	0.0100	469	azaconazole	0.0200	0.0400
437	cafenstrole	0.0200	0.0400	470	iprovalicarb-2	0.0200	0.0400
	Group F			471	diofenolan -1	0.0100	0.0200
438	tribenuron-methyl	0.0050	0.0100	472	diofenolan -2	0.0100	0.0200
439	ethiofencarb	0.0500	0.1000	473	aclonifen	0.2500	0.5000
440	dioxacarb	0.0400	0.0800	474	chlorfenapyr	0.1000	0.2000
441	dimethyl phthalate	0.0200	0.0400	475	bioresmethrin	0.0100	0.0200
442	4-chlorophenoxy acetic acid	0.0063	0.0126	476	isoxadifen-ethyl	0.0100	0.0200
				477	carfentrazone-ethyl	0.0100	0.0200
443	phthalimide	0.0250	0.0500	478	fenhexamid	0.2500	0.5000
444	diethyltoluamide	0.0040	0.0080	479	spiromesifen	0.0500	0.1000
445	2,4-D	0.1000	0.2000	480	fluazinam	0.1000	0.2000
446	carbaryl	0.0150	0.0300	481	bifenazate	0.0400	0.0800
447	cadusafos	0.0200	0.0400	482	endrin ketone	0.0800	0.1600
448	demetom-s	0.0200	0.0400	483	norflurazon-desmethyl	0.0500	0.1000
449	spiroxamine -1	0.0100	0.0200				

Continued

TABLE 2.10 The LOD, LOQ of 490 Pesticides and Related Chemical Residues Determined by GC-MS—cont'd

No.	Pesticides	LOD (mg/kg)	LOQ (mg/kg)	No.	Pesticides	LOD (mg/kg)	LOQ (mg/kg)
484	gamma-cyhaloterin-1	0.0040	0.0080	488	halfenprox	0.0250	0.0500
485	metoconazole	0.0200	0.0400	489	boscalid	0.0200	0.0400
486	cyhalofop-butyl	0.0100	0.0200	490	dimethomorph	0.0100	0.0200
487	gamma-cyhalothrin-2	0.0040	0.0080				

Pesticide and Related Chemicals Standard: Purity ≥95%

Stock Standard Solution: Accurately weigh 5–10 mg of individual pesticide and related chemical standards (accurate to 0.1 mg) into a 10-mL volumetric flask. Dissolve and dilute to volume with toluene, toluene + acetone combination, dichloromethane, etc., depending on each individual compound's solubility (for diluting solvent refer to Appendix B); all standard mixtures are stored in the dark at 0°C to 4°C and can be used for 1 year.

Mixed Standard Solution (Mixed Standard Solution A, B, C, D, E, and F): Depending on properties and retention times of compounds, 490 compounds are divided into six groups: A, B, C, D, E, and F. The mixed standard solution concentration is determined by its sensitivity on the instrument used for analysis. For 490 pesticides and related chemical grouping and mixed standard solution concentration of this standard, reference is made to Appendix B.

Depending on group number, mixed standard solution concentration, and stock standard solution concentration, appropriate amounts of individual stock standard solution are pipetted into a 100-mL volumetric flask, being diluted to volume with toluene. Mixed standard solutions are stored in the dark at 0°C to 4°C and are used for 1 month.

Working Standard Mixed Solution in Matrix: Working standard mixed solutions in matrix of A, B, C, D, E, and F group pesticides and related chemicals are prepared by diluting 40 μL internal standard solution and an appropriate amount of mixed standard solution to 1.0 mL with blank extract that has been taken through the method with the rest of the samples. Mix thoroughly. Working standard mixed solutions in matrix must be freshly prepared.

Solid Phase Extraction Cartridge: Cleanert TPT, 10 mL, 2.0 g or Equivalent

2.6.2.3 Apparatus

GC-MS: Equipped With EI

Analytical Balances: Capable of Weighing From 0.1 mg to 0.01 g

Homogenizer: Rotational Speed Not Lower Than 120,000 rpm
Rotary Evaporator
Pear-Shaped Flask: 200 mL
Pipette: 1 mL
Centrifuge: Rotational Speed no Lower Than 4200 rpm

2.6.2.4 Sample Pretreatment

2.6.2.4.1 Preparation of Test Sample

Tea samples are comminuted by a crushing mill, passed through a 20-mesh sieve, mixed, sealed, and labeled.

The test samples should be stored at ambient temperature.

2.6.2.4.2 Extraction

Weigh 5-g test sample (accurate to 0.01 g) into a 80-mL centrifuge tube, add 15 mL acetonitrile, and homogenize at 15,000 rpm for 1 min; then centrifuge at 4200 rpm for 5 min. Pipet the preceding acetonitrile layer of the extracts into a pear-shaped flask. The residue is once again extracted with 15 mL acetonitrile and the previously described procedure repeated. The two portions collected are combined and the extract concentrated to ca. 1 mL with a rotary evaporator at 40°C for clean-up.

2.6.2.4.3 Clean-up

Add sodium sulfate into the Cleanert TPT cartridge to ca. 2 cm. Activate the cartridge with 10 mL acetonitrile-toluent (3+1) before adding the sample. Fix the cartridge into a support to which a pear-shaped flask is connected. Once the solution gets to the top of the sodium sulfate, pipette the eluate into the cartridge immediately. Rinse the pear-shaped flask with 3×2 mL acetonitrile-toluent (3+1) and decant it into the cartridge. Insert a reservoir into the cartridge. Elute the pesticides with 25 mL acetonitrile-toluene (3+1). Evaporate the eluate to ca. 0.5 mL using a rotary evaporator at 40°C. Exchange with 2×5 mL hexane twice and make up to ca. 1 mL. Add 40 μL internal standard solution. Mix thoroughly. The solution is ready for GC-MS determination.

2.6.2.5 Determination

2.6.2.5.1 GC-MS Operating Conditions

(a) Column: DB-1701 capillary column (30 m × 0.25 mm × 0.25 μm), or equivalent
(b) Column temperature: 40°C hold 1 min, at 30°C/min to 130°C, at 5°C/min to 250°C, at 10°C/min to 300°C, hold 5 min
(c) Carrier gas: Helium, purity ≥99.999%, flow rate: 1.2 mL/min
(d) Injection port temperature: 290°C

(e) Injection volume: 1 μL
(f) Injection mode: Splitless, purge on after 1.5 min
(g) Ionization voltage: 70 eV
(h) Ion source temperature: 230°C
(i) GC-MS interface temperature: 280°C
(j) Solvent delay: group A 8.30 min, group B 7.80 min, group C 7.30 min, group D 5.50 min, group E 6.10 min, group F 5.50 min
(k) Selected ion monitoring mode: Each compound selects 1 quantifying ion and 2–3 qualifying ions. All of the detected ions of each group are detected according to programmed time and sequence of peaking. The retention times, quantifying ions, qualifying ions, and the abundance ratios of quantifying ion and qualifying ions for each compound are listed in Table 5.3. The programmed times and dwell times for the ions detected for each compound in each group are listed in Table 5.4

2.6.2.5.2   Qualitative Determination

In the samples determined, five injections are required to analyze for all pesticides according to GC-MS operating conditions; if the retention times in the peaks of the sample solution are the same as for the peaks of the working standard mixed solution, and the selected ions of the background-subtracted mass spectrum appear, and the abundance ratios of selected ions are within the expected limits (abundance ratios >50%, permitted tolerances are ±10%; abundance ratios >20% to 50%, permitted tolerances are ±15%; abundance ratios >10% to 20%, permitted tolerances are ±20%; abundance ratios ≤10%, permitted tolerances are ±50%), the sample is confirmed to contain this pesticide compound. In the case where the results are still not definitive, the sample should be reinjected with acquisition in scan mode (sufficient sensitivity) or with additional confirmatory ions or by other instruments that have higher sensitivity.

2.6.2.5.3   Quantitative Determination

The results are quantitated using heptachlor epoxide for an internal standard and the quantitation ion response for each analyte. In order to compensate for the matrix effect, quantitation is based on a mixed standard prepared in blank matrix extract. The concentration of the standard solution and the detected sample solution must be similar.

2.6.2.5.4   Parallel Test

A parallel test is carried out for the same testing sample.

2.6.2.5.5   Blank Test

The operation of the blank test is the same as that described in the method of determination, but without the addition of sample.

2.6.2.6   *Precision*

The precision data of the method for this standard have been determined according to the stipulations of GB/T 6379.1 and GB/T 6379.2. The values of repeatability and reproducibility are obtained and calculated at the 95% confidence level.

2.6.3   Determination of 29 Acidic Herbicides in Tea: GC-MS Method

2.6.3.1   *Principle*

The samples are extracted with acetonitrile by ultrasonator, the extracts are cleaned up with graphitized carbon black cartridges, the eluants are derived with trimethylsilylation diazomethane and are cleaned up with florisil cartridges, and the solutions are analyzed by GC-MS, using an external standard method of quantitative analysis.

2.6.3.2   *Reagents and Materials*

Unless otherwise specified, all reagents are analytically pure.

Acetonitrile: HPLC Grade
Acetone: HPLC Grade
Hexane: G.R.
Methanol: HPLC Grade
Benzene
Acetic Acid
Toluene: HPLC Grade
Sodium sulfate: anhydrous, ignited at 650°C for 4 h and kept in a desiccator
Trimethylsilylation diazomethane of n-Hexane Solution (2.0 mol/L)
Acetonitrile-Toluene-Acetic Acid Solution (75+25+1, V/V/V): Accurately take 75 mL Acetonitrile, 25 mL Toluene, and 1 mL Acetic Acid and Mix Thoroughly
Acetone-Hexane Solution (2+8, V/V): accurately take 20 mL acetone and 80 mL hexane and mix thoroughly
Methanol-Benzene Solution (2+8, V/V): accurately take 20 mL methanol and 80 mL benzene and mix thoroughly
Pesticide Standard: clopyralid, cloprop, 4-CPA, dicamba, MCPA, dichlorprop, bromoxynil, 2,4-D, triclopyr, NAA, pentachlorphenol, fenoprop, chloramben, MCPB,2,4,5-T, fluroxypyr, 2,4-DB, bentazone, OH-ioxynil, picloram, quinclorac, fluazifop, haloxyfop, flamprop acid,

acifluorfen, pyritiobacsodium, fenhexamid, quizalofop, bispyribacsodium, purity ≥95%

Stock Standard Solution: Accurately weigh individual pesticide and related chemical standards (accurate to 0.1 mg) into a 50-mL brown volumetric flask. Dissolve and dilute to volume with acetone; the concentration of each stock standard solution is 500 μg/mL. The stock standard solutions are stored in the dark at 0°C to 4°C and can be used for 3 months.

Working Standard Mixed Median Solution: Accurately take 2.0 mL of individual pesticide and related chemical standards (accurate to 0.1 mg) into a 100-mL brown volumetric flask. Dissolve and dilute to volume with acetone; the concentrations of each mixed standard median solution are 10 μg/mL. The standard median solutions are stored in the dark at 0°C to 4°C and can be used for 1 month.

Working standard mixed solution: According to detection levels required, take the needed volume of mixed standard median solution to dilute to mixed standard solution of appropriate concentration. Working standard mixed solutions must be freshly prepared.

Solid phase extraction Cartridge: Graphitized Carbon Black, 1.0 g, 12 mL, or Equivalent

Solid phase extraction Cartridge: Florisil, 250 mg, 3 mL, or Equivalent

2.6.3.3 Apparatus

GC-MS: Equipped With EI
Solid Phase Extraction Device
Centrifuge: 5000 rpm
Vortex Mix
Rotary Evaporator
Nitrogen Evaporator
Analytical Balances: Capable of Weighing From 0.1 mg to 0.01 g
Pear-Shaped Flask: 100 mL

2.6.3.4 Sample Pretreatment

2.6.3.4.1 Preparation of Test Sample

According to SN/T 0918-2000, 500 g representative tea samples are comminuted by a crushing mill, passed through a 40-mesh sieve, mixed, and are then divided into two as test samples to subpackage into a clean container, seal, and label.

The test samples should be stored at 0°C to 4°C.

During the storage process, samples should be prevented from being contaminated or experiencing any change of residue content.

2.6.3.4.2 Extraction

Weigh 2.5-g test sample (accurate to 0.01 g) into a 50-mL centrifuge tube with plug, add 20 mL acetonitrile, and perform ultrasound extraction for 30 min; add 2 g anhydrous sodium sulfate, then place on vortex mix to extract for 5 min, centrifuge at 5000 rpm for 5 min, and pipet the acetonitrile layer of the extracts into a 100-mL pear-shaped flask. Extract the residue once again with 20 mL acetonitrile and repeat previously described procedure. Combine the two portions collected and concentrate the extract to ca. 1 mL with a rotary evaporator at 40°C for clean-up.

2.6.3.4.3 Clean-up With Graphitized Carbon Black Solid Phase Extraction

Fix the graphitized carbon black cartridge into a support to which a pear-shaped flask is connected.

Add sodium sulfate into the graphitized carbon black cartridge to ca. 1 cm. Activate the cartridge with 10 mL acetonitrile-toluent-acetic acid (75+25+1, V/V/V) before adding the sample and disposable eluent. Once the solution gets to the top of the sodium sulfate, pipette the eluate into the graphitized carbon black cartridge immediately. Rinse the pear-shaped flask with 3×2 mL acetonitrile-toluent-acetic acid (75+25+1, V/V/V) and decant it into the cartridge. Insert a reservoir into the cartridges. Elute the pesticides with 25 mL acetonitrile-toluent-acetic acid (75+25+1, V/V/V) and collect the eluant in another 100-mL pear-shaped flask.

2.6.3.4.4 Derivatization

Concentrate the eluent to ca. 1 mL with a rotary evaporator at 40°C. Then evaporate the eluate to dryness using a nitrogen evaporator, and dissolve to 2 mL with benzene-methanol. Add to 0.2 mL trimethylsilylation diazomethane of n-hexane solution, cap plug, and mix thoroughly. Place in a water bath to maintain 30 min at 30°C. Then evaporate the eluate to dryness using a nitrogen evaporator, and dissolve the residual with 5 mL n-hexane.

2.6.3.4.5 Clean-up With Forisil Solid Phase Extraction

Activate the forisil cartridge with 3 mL acetone and 6 mL n-hexane before adding the sample and disposable eluent. Pipette the dissolved residua with 5 mL n-hexane into the forisil cartridge and dispose of the loading solution; then elute the pesticides with 6 mL acetone n-hexane and collect the eluant in a 10-mL test tube with scale. Evaporate the eluate to dryness using a nitrogen evaporator at 45°C. Make up to 0.5 mL with acetone for GC-MS determination.

2.6.3.5 Determination

2.6.3.5.1 GC-MS Operating Condition

(a) Column: DB-1701 capillary column (30 m × 0.25 mm × 0.25 μm), or equivalent
(b) Column temperature: 40°C hold 1 min, at 40°C/min to 130°C, at 5°C/min to 250°C, at 10°C/min to 300°C, hold 5 min
(c) Injection port temperature: 290°C
(d) GC-MS interface temperature: 280°C
(e) Carrier gas: Helium, purity ≥99.999%, flow rate: 1.2 mL/min
(f) Injection volume: 1 μL
(g) Injection mode: Splitless, purge on after 1.0 min
(h) Ionization mode: EI
(i) Ionization voltage: 70 eV
(j) Selected ion monitoring mode (SIM): Each compound selects 1 quantifying ion and 2–3 qualifying ions. All of the detected ions of each group are detected according to programmed time and sequence of peaking. The retention times, quantifying ions, qualifying ions, and the abundance ratios of quantifying ion and qualifying ions for each compound are listed in Table 2.11
(k) Solvent delay time: 9 min

2.6.3.5.2 Qualitative Determination

In the samples determined, if the retention times in the peaks of the sample solution are the same as those for the peaks of the working standard mixed solution, and the selected ions of the background-subtracted mass spectrum appear, and the abundance ratios of selected ions are within the expected limits (abundance ratios >50%, permitted tolerances are ±10%; abundance ratios >20% to 50%, permitted tolerances are ±15%; abundance ratios >10% to 20%, permitted tolerances are ±20%; abundance ratios ≤10%, permitted tolerances are ±50%), the sample is confirmed to contain this pesticide compound. In cases where the results are still not definitive, the sample should be reinjected with acquisition in scan mode (sufficient sensitivity) or with additional confirmatory ions or by using other instruments that have higher sensitivity. Monitoring ions for SIM acquisition for 29 acidic herbicides, see Table 2.12.

2.6.3.5.3 Quantitative Determination

Based on the content of acidic herbicides in the sample solution, a quantitative determination method similar to that for the standard working solution is adopted. The response of the standard solution and the detected sample solution must be in a linear range of the detection apparatus. The same volumes of the working standard mixed solution and the sample solution are detected by turns.

TABLE 2.11 The Retention Times, Quantifying Ions, Qualifying Ions and The Abundance Ratios for 29 Acidic Herbicides

No.	Pesticides	CAS	Retention Time/min	Quantifying ion	Qualifying ion 1	Qualifying ion 2	LOD (mg/kg)
1	cloprop	101-10-0	11.59	155(100)	157(32)	214(37)	0.01
2	clopyralid	1702-17-6	10.93	147(100)	149(57)	146(65)	0.01
3	4-CPA	122-88-3	11.89	200(100)	141(96)	111(60)	0.01
4	dicamba	1918-00-9	12.01	203(100)	205(64)	234(25)	0.01
5	MCPA	94-74-6	13.16	214(100)	141(94)	155(65)	0.01
6	dichlorprop	120-36-5	13.95	162(100)	189(54)	248(45)	0.01
7	bromoxynil	1689-84-5	14.87	291(100)	276(51)	289(53)	0.01
8	2,4-D	120-36-5	14.87	199(100)	234(62)	175(37)	0.01
9	pentachlorphenol	87-86-5	15.93	265(100)	280(95)	237(90)	0.01
10	NAA	86-87-3	15.41	141(100)	200(40)	210(18)	0.01
11	triclopyr	64700-56-7	15.04	210(100)	269(31)	212(45)	0.01
12	fenoprop	93-72-1	16.65	196(100)	198(97)	223(38)	0.01
13	MCPB	94-74-6	17.77	101(100)	101(100)	59(68)	0.01
14	2,4,5-T	93-76-5	18.78	233(100)	235(66)	268(50)	0.01
15	2,4-DB	94-82-6	20.22	101(100)	59(60)	162(23)	0.01
16	chloramben	133-90-4	17.27	188(100)	219(76)	160(40)	0.01
17	fluroxypyr	69377-81-7	18.80	209(100)	211(65)	268(48)	0.01

Continued

TABLE 2.11 The Retention Times, Quantifying Ions, Qualifying Ions and The Abundance Ratios for 29 Acidic Herbicides—cont'd

No.	Pesticides	CAS	Retention Time/min	Quantifying ion	Qualifying ion 1	Qualifying ion 2	LOD (mg/kg)
18	OH-ioxynil	1689-83-4	20.72	385(100)	370(35)	243(38)	0.01
19	bentazone	25057-89-0	20.42	212(100)	105(65)	254(27)	0.01
20	quinclorac	84087-01-4	21.99	224(100)	226(65)	197(51)	0.01
21	fluazifop	79241-46-6	22.88	341(100)	282(97)	254(90)	0.01
22	picloram	1918-2-1	21.95	196(100)	197(80)	198(95)	0.01
23	haloxyfop	69806-34-4	23.36	316(100)	288(94)	375(81)	0.01
24	flamprop acid	58667-63-3	26.18	105(100)	106(90)	77(36)	0.01
25	pyritiobacsodium	123343-16-8	27.21	281(100)	283(38)	282(15)	0.01
26	acifluorfen	50594-66-6	26.63	375(100)	344(55)	223(44)	0.01
27	fenhxamid	126833-17-8	29.17	97(100)	55(38)	191(27)	0.01
28	quizalofop	76578-14-8	33.09	299(100)	243(87)	163(35)	0.01
29	bispyribacsodium	125401-92-5	34.40	385(100)	384(30)	386(22)	0.01

TABLE 2.12 Monitoring Ions for SIM Acquisition for 29 Acidic Herbicides Determined by GC-MS

No.	Time/min	Ion (amu)	Dwell time/ms
1	11.31	110,111,113,141,147,149,174,188,200,203, 205,234,128,155,157,214	50
2	13.64	162,164,189,248,125,141,214	100
3	14.53	175,199,210,212,234,237,265,267,269,271, 276,289,291,293	50
4	17.04	196,198,223,282,115,141,200,59,101,107	100
5	18.42	59,101,160,162,188,115,141,200,59,101,107	50
6	19.69	105,133,181,209,211,212243,254,268,385	50
7	23.18	288,290,316,375,196,198,225,254,161,197, 224,226,227,254,282,341	50
8	28.40	55,97,176,191,223,344,375,377,77,105, 230,276,281,283,309	50
9	33.91	385,386,387,413,163,243,299,372	100

2.6.3.5.4   Parallel Test

A parallel test is carried out for the same testing sample.

2.6.3.5.5   Blank Test

The operation of the blank test is the same as that described in the method of determination, but without the addition of sample.

2.6.3.6   *Precision*

The precision data of the method for this standard have been determined according to the stipulations of GB/T 6379.1 and GB/T 6379.2. The values of repeatability and reproducibility are obtained and calculated at the 95% confidence level.

RESEARCHERS

Researcher of determination of 490 pesticides and related chemical residues in tea GC-MS method: Zong-Mao Chen, Guo-Fang Pang, Yan-Zhong Cao, Xue-Yan Hu, Chun-Lin Fan, Ping Liang, Qun-Jie Wang, Zheng-Yun Lou, Peng-Jian Luo, Fu-Bin Tang, Guang-Ming Liu, Jun-Yan Zhang.

Researcher of determination of 29 acidic herbicides in tea GC-MS method: Hong-Fei Yan, Ying Zhang, Zhi-Qiang Huang, Ping Huang, Yong-Jun Li, Mei-Ling Wang, Guo-Fang Pang.

Qinhuangdao Entry-Exit Inspection and Quarantine Bureau, 39 Haibin Rd, Qinhuangdao, Hebei, PC 066002, People's Republic of China.

2.7 DETERMINATION OF 448 PESTICIDES AND RELATED CHEMICAL RESIDUES IN TEA LEAVES: LC-MS-MS METHOD (GB/T23205-2008)

2.7.1 Scope

The method is applicable to the quantitative determination of 418 pesticides and related chemical residues and qualitative determination of 30 pesticides and related chemical residues in green tea, black tea, puer tea, and oolong.

The limit of quantitative determination of this method of 418 pesticides and related chemicals is 0.03 µg/kg to 1.21 mg/kg (see Table 2.13).

2.7.2 Principle

The samples are extracted with acetonitrile, and the extracts are cleaned up with Cleanert TPT cartridges. The pesticides and related chemicals are eluted with acetonitrile-toluene (3+1), and the solutions are analyzed by LC-MS-MS, using an external standard method for quantitative analysis.

2.7.3 Reagents and Materials

"Water" is the first grade of GB/T6682 specified.

Acetonitrile: HPLC Grade
Toluene: G.R.
Acetone: HPLC Grade
Isooctane: HPLC Grade
Methanol: HPLC Grade
Acetic Acid: G.R.
Sodium Chloride: Analytically Pure
Membrane Filters: 13 mm × 0.2 µm
Solid Phase Extraction Column: Cleanert TPT, 10 mL, 2.0 g or Equivalent
0.1% Formic Acid Solution (Volume Fraction)
5 mmol/Lmmol/L Ammonium Acetate Solution: Accurately Weigh 0.375 g of Ammonium Acetate Into a 1000-mL Volumetric Flask. Dissolve and Dilute to Volume With Water
Acetonitrile-Toluene (3+1, V/V)
Acetonitrile-Water (3+2, V/V)

TABLE 2.13 The LOD, LOQ of 448 Pesticides and Related Chemical Residues Determined by LC-MS-MS

No.	Pesticides	LOD (µg/kg)	LOQ (µg/kg)	No.	Pesticides	LOD (µg/kg)	LOQ (µg/kg)
	Group A			33	myclobutanil	0.50	1.00
1	propham	55.00	110.00	34	imazethapyr	0.56	1.12
2	isoprocarb	1.15	2.30	35	paclobutrazol	0.29	0.58
3	3,4,5-trimethacarb	0.17	0.34	36	fenthion sulfoxide	0.16	0.32
4	cycluron	0.10	0.20	37	triadimenol	5.28	10.56
5	carbaryl	5.16	10.32	38	butralin	0.95	1.90
6	propachlor	0.14	0.28	39	spiroxamine	0.03	0.06
7	rabenzazole	0.67	1.34	40	tolclofos methyl	33.28	66.56
8	simetryn	0.07	0.14	41	methidathion	5.33	10.66
9	monolinuron	1.78	3.56	42	allethrin	30.20	60.40
10	mevinphos	0.78	1.56	43	diazinon	0.36	0.72
11	aziprotryne	0.69	1.38	44	edifenphos	0.38	0.76
12	secbumeton	0.04	0.08	45	Pretilachlor	0.17	0.34
13	cyprodinil	0.37	0.74	46	flusilazole	0.29	0.58
14	buturon	4.48	8.96	47	iprovalicarb	1.16	2.32
15	carbetamide	1.82	3.64	48	benodanil	1.74	3.48
16	pirimicarb	0.08	0.16	49	flutolanil	0.57	1.14
17	clomazone	0.21	0.42	50	famphur	1.80	3.60
18	cyanazine	0.08	0.16	51	benalyxyl	0.62	1.24
19	prometryne	0.08	0.16	52	diclobutrazole	0.23	0.46
20	paraoxon methyl	0.38	0.76	53	etaconazole	0.89	1.78
21	4,4-dichloro-benzophenone	6.80	13.60	54	fenarimol	0.30	0.60
22	thiacloprid	0.19	0.38	55	tetramethirn	0.91	1.82
23	imidacloprid	11.00	22.00	56	dichlofluanid	1.30	2.60
24	ethidimuron	0.75	1.50	57	cloquintocet mexyl	0.94	1.88
25	isomethiozin	0.53	1.06	58	bitertanol	16.70	33.40
26	diallate	44.60	89.20	59	chlorprifos methyl	8.00	16.00
27	acetochlor	23.70	47.40	60	azinphos ethyl	54.46	108.92
28	nitenpyram	8.56	17.12	61	clodinafop propargyl	1.22	2.44
29	methoprotryne	0.12	0.24	62	triflumuron	1.96	3.92
30	dimethenamid	2.15	4.30	63	isoxaflutole	1.95	3.90
31	terrbucarb	1.05	2.10	64	anilofos	0.36	0.72
32	penconazole	1.00	2.00	65	quizalofop-ethyl	0.34	0.68

Continued

TABLE 2.13 The LOD, LOQ of 448 Pesticides and Related Chemical Residues Determined by LC-MS-MS—cont'd

No.	Pesticides	LOD (µg/kg)	LOQ (µg/kg)	No.	Pesticides	LOD (µg/kg)	LOQ (µg/kg)
66	haloxyfop-methyl	1.32	2.64	98	methobromuron	8.42	16.84
67	fluazifop butyl	0.13	0.26	99	phorate	157.00	314.00
68	bromophos-ethyl	283.85	567.70	100	aclonifen	12.10	24.20
69	bensulide	17.10	34.20	101	mephosfolan	1.16	2.32
70	bromfenvinfos	1.51	3.02	102	imibenzonazole-des-benzyl	3.11	6.22
71	azoxystrobin	0.23	0.46				
72	pyrazophos	0.81	1.62	103	neburon	3.55	7.10
73	flufenoxuron	1.58	3.16	104	mefenoxam	0.77	1.54
74	indoxacarb	3.77	7.54	105	prothoate	1.23	2.46
	Group B			106	ethofume sate	186.00	372.00
75	ethylene thiourea	26.10	52.20	107	iprobenfos	4.14	8.28
76	daminozide	1.30	2.60	108	TEPP	5.20	10.40
77	dazomet	63.50	127.00	109	cyproconazole	0.37	0.74
78	nicotine	1.10	2.20	110	thiamethoxam	16.50	33.00
79	fenuron	0.52	1.04	111	crufomate	0.26	0.52
80	crimidine	0.78	1.56	112	etrimfos	18.76	37.52
81	molinate	1.05	2.10	113	coumatetralyl	0.68	1.36
82	carbendazim	0.23	0.46	114	cythioate	40.00	80.00
83	6-chloro-4-hydroxy-3-phenyl-pyridazin	0.83	1.66	115	phosphamidon	1.94	3.88
				116	phenmedipham	2.24	4.48
84	propoxur	12.20	24.40	117	bifenazate	11.40	22.80
85	isouron	0.20	0.40	118	fenhexamid	0.47	0.94
86	chlorotoluron	0.31	0.62	119	flutriafol	4.29	8.58
87	thiofanox	78.50	157.00	120	furalaxyl	0.39	0.78
88	chlorbufam	91.50	183.00	121	bioallethrin	99.00	198.00
89	bendiocarb	1.59	3.18	122	cyanofenphos	10.40	20.80
90	propazine	0.16	0.32	123	pirimiphos methyl	0.10	0.20
91	terbuthylazine	0.23	0.46	124	buprofezin	0.44	0.88
92	diuron	0.78	1.56	125	disulfoton sulfone	1.23	2.46
93	chlormephos	224.00	448.00	126	fenazaquin	0.16	0.32
94	carboxin	0.28	0.56	127	triazophos	0.34	0.68
95	clothianidin	31.50	63.00	128	DEF	0.81	1.62
96	pronamide	7.69	15.38	129	pyriftalid	0.31	0.62
97	dimethachloro	0.95	1.90	130	metconazole	0.66	1.32

TABLE 2.13 The LOD, LOQ of 448 Pesticides and Related Chemical Residues Determined by LC-MS-MS—cont'd

No.	Pesticides	LOD (µg/kg)	LOQ (µg/kg)	No.	Pesticides	LOD (µg/kg)	LOQ (µg/kg)
131	pyriproxyfen	0.22	0.44	163	mepanipyrim	0.16	0.32
132	isoxaben	0.09	0.18	164	prometon	0.07	0.14
133	flurtamone	0.22	0.44	165	methiocarb	20.60	41.20
134	trifluralin	167.40	334.80	166	metoxuron	0.32	0.64
135	flamprop methyl	10.10	20.20	167	dimethoate	3.80	7.60
136	bioresmethrin	3.71	7.42	168	fluometuron	0.46	0.92
137	propiconazole	0.88	1.76	169	dicrotophos	0.57	1.14
138	chlorpyrifos	26.90	53.80	170	monalide	0.60	1.20
139	fluchloralin	244.00	488.00	171	diphenamid	0.14	0.28
140	chlorsulfuron	1.37	2.74	172	ethoprophos	1.38	2.76
141	flamprop isopropyl	0.22	0.44	173	fonofos	3.73	7.46
142	tetrachlorvinphos	1.11	2.22	174	etridiazol	50.21	100.42
143	propargite	34.30	68.60	175	hexazinone	0.06	0.12
144	bromuconazole	1.57	3.14	176	dimethametryn	0.06	0.12
145	picolinafen	0.36	0.72	177	trichlorphon	0.56	1.12
146	fluthiacet methyl	2.65	5.30	178	demeton(o+s)	3.39	6.78
147	trifloxystrobin	1.00	2.00	179	benoxacor	3.45	6.90
148	hexaflumuron	12.60	25.20	180	bromacil	11.80	23.60
149	novaluron	4.02	8.04	181	phorate sulfoxide	184.14	368.28
150	Flurazuron	13.40	26.80	182	brompyrazon	1.80	3.60
	Group C			183	oxycarboxin	0.45	0.90
151	maleic hydrazide	40.00	80.00	184	mepronil	0.19	0.38
152	methamidophos	2.47	4.94	185	disulfoton	234.85	469.70
153	EPTC	18.67	37.34	186	fenthion	26.00	52.00
154	diethyltoluamide	0.28	0.56	187	metalaxyl	0.25	0.50
155	monuron	17.37	34.74	188	ofurace	0.50	1.00
156	pyrimethanil	0.34	0.68	189	fosthiazate	0.28	0.56
157	fenfuram	0.39	0.78	190	imazamethabenz-methyl	0.08	0.16
158	quinoclamine	3.96	7.92	191	disulfoton-sulfoxide	1.42	2.84
159	fenobucarb	2.95	5.90	192	isoprothiolane	0.92	1.84
160	propanil	10.80	21.60	193	imazalil	1.00	2.00
161	carbofuran	6.53	13.06	194	phoxim	41.40	82.80
162	acetamiprid	0.72	1.44				

Continued

TABLE 2.13 The LOD, LOQ of 448 Pesticides and Related Chemical Residues Determined by LC-MS-MS—cont'd

No.	Pesticides	LOD (μg/kg)	LOQ (μg/kg)	No.	Pesticides	LOD (μg/kg)	LOQ (μg/kg)
195	quinalphos	1.00	2.00	227	butafenacil	4.75	9.50
196	fenoxycarb	9.14	18.28	228	spinosad	0.28	0.56
197	pyrimitate	0.09	0.18		Group D		
198	fensulfothin	1.00	2.00	229	mepiquat chloride	0.45	0.90
199	fluorochloridone	6.89	13.78	230	allidochlor	20.52	41.04
200	butachlor	10.03	20.06	231	tricyclazole	0.62	1.24
201	kresoxim-methyl	50.29	100.58	232	metamitron	3.18	6.36
202	triticonazole	1.51	3.02	233	isoproturon	0.07	0.14
203	fenamiphos sulfoxide	0.37	0.74	234	atratone	0.09	0.18
				235	oesmetryn	0.09	0.18
204	thenylchlor	12.07	24.14	236	metribuzin	0.27	0.54
205	fenoxanil	19.70	39.40	237	DMST	20.00	40.00
206	fluridone	0.09	0.18	238	cycloate	2.22	4.44
207	epoxiconazole	2.03	4.06	239	atrazine	0.18	0.36
208	chlorphoxim	38.79	77.58	240	butylate	302.00	604.00
209	fenamiphos sulfone	0.22	0.44	241	pymetrozin	17.14	34.28
210	fenbuconazole	0.82	1.64	242	chloridazon	1.16	2.32
211	isofenphos	109.34	218.68	243	sulfallate	103.60	207.20
212	phenothrin	169.60	339.20	244	ethiofencarb	2.46	4.92
213	piperophos	4.62	9.24	245	terbumeton	0.05	0.10
214	piperonyl butoxide	0.57	1.14	246	cyprazine	0.03	0.06
215	oxyflurofen	29.27	58.54	247	ametryn	0.48	0.96
216	flufenacet	2.65	5.30	248	tebuthiuron	0.11	0.22
217	phosalone	24.02	48.04	249	trietazine	0.30	0.60
218	methoxyfenozide	1.85	3.70	250	sebutylazine	0.16	0.32
219	aspon	0.87	1.74	251	dibutyl succinate	111.20	222.40
220	ethion	1.48	2.96	252	tebutam	0.07	0.14
221	diafenthiuron	0.14	0.28	253	thiofanox-sulfoxide	4.15	8.30
222	dithiopyr	5.20	10.40	254	cartap hydrochloride	1040.00	2080.00
223	spirodiclofen	4.95	9.90	255	methacrifos	1211.85	2423.70
224	fenpyroximate	0.68	1.36	256	thionazin	11.34	22.68
225	flumiclorac-pentyl	5.30	10.60	257	linuron	5.82	11.64
226	temephos	0.61	1.22				

TABLE 2.13 The LOD, LOQ of 448 Pesticides and Related Chemical Residues Determined by LC-MS-MS—cont'd

No.	Pesticides	LOD (µg/kg)	LOQ (µg/kg)	No.	Pesticides	LOD (µg/kg)	LOQ (µg/kg)
258	heptanophos	2.92	5.84	290	tralkoxydim	0.16	0.32
259	prosulfocarb	0.18	0.36	291	malathion	2.82	5.64
260	dipropetryn	0.14	0.28	292	pyributicarb	0.17	0.34
261	thiobencarb	1.65	3.30	293	pyridaphenthion	0.44	0.88
262	tri-iso-butyl phosphate	1.79	3.58	294	pirimiphos-ethyl	0.03	0.06
263	tri-n-butyl phosphate	0.19	0.38	295	thiodicarb	19.68	39.36
				296	pyraclofos	0.50	1.00
264	diethofencarb	1.00	2.00	297	picoxystrobin	4.22	8.44
265	cadusafos	0.58	1.16	298	tetraconazole	0.86	1.72
266	metazachlor	0.49	0.98	299	mefenpyr-diethyl	6.28	12.56
267	propetamphos	27.00	54.00	300	profenofos	1.01	2.02
268	terbufos	1120.00	2240.00	301	pyraclostrobin	0.25	0.50
269	simeconazole	1.47	2.94	302	dimethomorph	0.18	0.36
270	triadimefon	3.94	7.88	303	kadethrin	3.33	6.66
271	phorate sulfone	21.00	42.00	304	thiazopyr	0.98	1.96
272	tridemorph	1.30	2.60	305	chlorfluazuron	4.34	8.68
273	mefenacet	1.10	2.20		Group E		
274	fenamiphos	0.10	0.20	306	4-aminopyridine	0.43	0.86
275	fenpropimorph	0.09	0.18	307	methomyl	4.78	9.56
276	tebuconazole	1.12	2.24	308	pyroquilon	1.74	3.48
277	isopropalin	15.00	30.00	309	fuberidazole	0.95	1.90
278	nuarimol	0.50	1.00	310	isocarbamid	0.85	1.70
279	bupirimate	0.35	0.70	311	butocarboxim	0.79	1.58
280	azinphos-methyl	552.17	1104.34	312	chlordimeform	0.67	1.34
281	tebupirimfos	0.06	0.12	313	cymoxanil	27.80	55.60
282	phenthoate	46.18	92.36	314	chlorthiamid	4.41	8.82
283	sulfotep	1.30	2.60	315	aminocarb	8.21	16.42
284	sulprofos	2.92	5.84	316	omethoate	4.83	9.66
285	EPN	16.50	33.00	317	ethoxyquin	1.76	3.52
286	diniconazole	0.67	1.34	318	aldicarb sulfone	10.68	21.36
287	sethoxydim	44.80	89.60	319	dioxacarb	1.68	3.36
288	pencycuron	0.14	0.28	320	demeton-s-methyl	2.65	5.30
289	mecarbam	9.80	19.60	321	cyanophos	5.05	10.10

Continued

TABLE 2.13 The LOD, LOQ of 448 Pesticides and Related Chemical Residues Determined by LC-MS-MS—cont'd

No.	Pesticides	LOD (µg/kg)	LOQ (µg/kg)	No.	Pesticides	LOD (µg/kg)	LOQ (µg/kg)
322	thiometon	289.00	578.00	352	fenpropathrin	122.50	245.00
323	folpet	69.30	138.60	353	hexythiazox	11.80	23.60
324	demeton-s-methyl sulfone	9.88	19.76	354	benzoximate	9.83	19.66
325	fenpropidin	0.09	0.18	355	benzoylprop-ethyl	154.00	308.00
326	amidithion	95.20	190.40	356	pyrimidifen	7.00	14.00
327	imazapic	2.95	5.90	357	furathiocarb	0.96	1.92
328	paraoxon-ethyl	0.24	0.48	358	trans-permethin	2.40	4.80
329	aldimorph	1.58	3.16	359	etofenprox	114.01	228.02
330	vinclozolin	1.27	2.54	360	pyrazoxyfen	0.16	0.32
331	uniconazole	1.20	2.40	361	flubenzimine	3.89	7.78
332	pyrifenox	0.13	0.26	362	zeta- cypermethrin	0.34	0.68
333	chlorthion	66.80	133.60	363	haloxyfop-2-ethoxyethyl	1.25	2.50
334	dicapthon	0.12	0.24	364	esfenvalerate	208.00	416.00
335	clofentezine	0.38	0.76	365	fluoroglycofen-ethyl	2.50	5.00
336	norflurazon	0.13	0.26	366	tau-fluvalinate	115.00	230.00
337	triallate	23.10	46.20		Group F		
338	quinoxyphen	76.70	153.40	367	acrylamide	17.80	35.60
339	fenthion sulfone	8.73	17.46	368	tert-butylamine	19.48	38.96
340	flurochloridone	0.65	1.30	369	hymexazol	112.07	224.14
341	phthalic acid, benzyl butyl ester	316.00	632.00	370	phthalimide	21.50	43.00
342	isazofos	0.09	0.18	371	dimefox	34.10	68.20
343	dichlofenthion	14.98	29.96	372	metolcarb	12.70	25.40
344	vamidothion sulfone	238.00	476.00	373	diphenylamin	0.21	0.42
345	terbufos sulfone	44.30	88.60	374	1-naphthy acetamide	0.41	0.82
346	dinitramine	0.90	1.80	375	atrazine-desethyl	0.62	1.24
347	cyazofamid	2.25	4.50	376	2,6-dichloro-benzamide	2.25	4.50
348	trichloronat	33.40	66.80	377	aldicarb	130.50	261.00
349	resmethrin-2	0.15	0.30	378	dimethyl phthalate	6.60	13.20
350	boscalid	2.38	4.76	379	chlordimeform hydrochloride	2.64	5.28
351	nitralin	17.20	34.40	380	simeton	1.10	2.20

TABLE 2.13 The LOD, LOQ of 448 Pesticides and Related Chemical Residues Determined by LC-MS-MS—cont'd

No.	Pesticides	LOD (μg/kg)	LOQ (μg/kg)	No.	Pesticides	LOD (μg/kg)	LOQ (μg/kg)
381	dinotefuran	5.09	10.18	409	phthalic acid, biscyclohexyl ester	0.34	0.68
382	pebulate	1.70	3.40	410	carpropamid	2.60	5.20
383	acibenzolar-s-methyl	1.54	3.08	411	tebufenpyrad	0.13	0.26
384	dioxabenzofos	6.92	13.84	412	chlorthiophos	31.80	63.60
385	oxamyl	274.03	548.06	413	dialifos	78.50	157.00
386	methabenz-thiazuron	0.07	0.14	414	cinidon-ethyl	14.58	29.16
				415	rotenone	2.32	4.64
387	butoxycarboxim	26.60	53.20	416	imibenconazole	5.13	10.26
388	mexacarbate	0.47	0.94	417	propaquiafop	0.62	1.24
389	demeton-s-methyl sulfoxide	1.96	3.92	418	lactofen	62.00	124.00
390	thiofanox sulfone	24.08	48.16	419	benzofenap	0.04	0.08
391	phosfolan	0.24	0.48	420	dinoseb acetate	20.64	41.28
392	demeton-s	80.00	160.00	421	propisochlor	0.40	0.80
393	fenthion oxon	0.59	1.18	422	silafluofen,	304.00	608.00
394	napropamide	1.27	2.54	423	etobenzanid	0.40	0.80
395	fenitrothion	26.80	53.60	424	fentrazamide	6.20	12.40
396	phthalic acid, dibutyl ester	19.80	39.60	425	pentachloroaniline	1.87	3.74
				426	carbosulfan	0.40	0.80
397	metolachlor	0.20	0.40	427	cyphenothrin	8.40	16.80
398	procymidone	43.30	86.60	428	dimefuron	2.00	4.00
399	vamidothion	4.56	9.12	429	malaoxon	2.34	4.68
400	chloroxuron	0.22	0.44	430	chlorbenside sulfone	0.40	0.80
401	triamiphos	0.03	0.06				
402	prallethrin	0.10	0.20	431	dodine	8.00	16.00
403	cumyluron	1.32	2.64		Group G		
404	imazamox	0.90	1.80	432	dalapon	115.37	230.74
405	warfarin	1.34	2.68	433	2-phenylphenol	84.94	169.88
406	phosmet	8.86	17.72	434	3-phenylphenol	2.00	4.00
407	ronnel	6.57	13.14	435	dicloran	24.28	48.56
408	pyrethrin	17.90	35.80	436	chlorpropham	7.88	15.76

Continued

TABLE 2.13 The LOD, LOQ of 448 Pesticides and Related Chemical Residues Determined by LC-MS-MS—cont'd

No.	Pesticides	LOD (µg/kg)	LOQ (µg/kg)	No.	Pesticides	LOD (µg/kg)	LOQ (µg/kg)
437	terbacil	0.44	0.88	443	chloramphenicolum	1.94	3.88
438	2,4-D	5.93	11.86	444	famoxadone	22.64	45.28
439	fludioxonil	31.08	62.16	445	diflufenican	14.14	28.28
440	chlorfenethol	82.15	164.30	446	ethiprole	19.93	39.86
441	naptalam	0.97	1.94	447	fluazinam	35.30	70.60
442	chlorobenzuron	10.20	20.40	448	kelevan	4820.00	9640.00

Sodium Sulfate: Anhydrous, Analytically Pure. Ignited at 650°C for 4 h and Kept in a Desiccator, Cooling for Use

Pesticide and Related Chemical Standard: Purity ≥95%

Stock Standard Solution: Accurately weigh 5–10 mg of individual pesticide and related chemical standard (accurate to 0.1 mg) into a 10-mL volumetric flask. Dissolve and dilute to volume with methanol, toluene, acetone, acetonitrile, isooctane, etc., depending on each individual compound's solubility. (For diluting solvent refer to Appendix B) All standard mixtures are stored in the dark at 0°C to 4°C and can be used for 1 year.

Mixed Standard Solution (Mixed Standard Solution A, B, C, D, E, F, and G): Depending on the properties and retention times of compounds, all compounds are divided into seven groups: A, B, C, D, E, F, and G. The concentration of each compound is determined by its sensitivity on the instrument for analysis. For 448 pesticides and related chemical standards, grouping and concentration are listed in Appendix B.

Depending on group number, mixed standard solution concentration, and stock standard solution concentration, appropriate amounts of individual stock standard solution are pipetted into a 100-mL volumetric flask and diluted to volume with methanol. Mixed standard solutions are stored in the dark below 4°C and used for one month.

Working Standard Mixed Solution in Matrix: Working standard mixture solutions in matrix of A, B, C, D, E, F, and G group pesticides and related chemicals are prepared by diluting an appropriate amount of mixed standard solution with blank extract that has been taken through the method with the rest of the samples. Mix thoroughly. They are used for plotting the standard curve. Working standard mixture solution in matrix must be freshly prepared.

2.7.4 Apparatus

LC-MS-MS: Equipped With ESI
Analytical Balances: Capable of Weighing From 0.1 mg to 0.01 g
Pear-Shaped Flask: 200 mL
Pipette: 1 mL
Sample Bottle: 2 mL With PTFE Screw-Cap
Centrifuge Tube With Plug: 50 mL
Nitrogen Evaporator
Low-Speed Centrifuge: 4200 rpm
Rotary Evaporators
Homogenizer

2.7.5 Sample Pretreatment

2.7.5.1 Preparation of Test Sample

The samples are comminuted by crushing mill, packaged, and stored.
The samples should be stored in a dry location.

2.7.5.2 Extraction

Weigh 10-g test sample (accurate to 0.01 g) into a 50-mL centrifuge tube with plug, add 30 mL acetonitrile, and homogenize at 15,000 rpm for 1 min, then centrifuge at 4200 rpm for 5 min. Pipette the acetonitrile layer of the extracts into a pear-shaped flask. The residues are extracted with 30 mL acetonitrile and the previously described procedure is repeated. The residues are once again extracted with 20 mL acetonitrile. The three portions collected are combined. Then concentrate the extract to nearly dry with a rotary evaporator at 45°C with a nitrogen evaporator blowing dry; then add 5 mL acetonitrile to dissolve the residue. Take 1 mL for clean-up.

2.7.5.3 Clean-up

Add sodium sulfate into the Cleanert TPT cartridge to ca. 2 cm. Fix the cartridge into a support to which a pear-shaped flask is connected. Activate the cartridge with 5 mL acetonitrile-toluent (3 + 1) before adding the sample. Once the solution gets to the top of the sodium sulfate, pipette the eluate into the cartridge immediately. Change the pear-shaped flask to accept the loading solution. Insert a reservoir into the cartridges. Elute the pesticides with 25 mL acetonitrile-toluene (3 + 1). Evaporate the eluate to ca. 0.5 mL using rotary evaporator at 45°C, and then evaporate the eluate to dryness using a nitrogen evaporator at 35°C and make up to 1 mL with acetonitrile-water (3 + 2); mix thoroughly. Finally, filter the extract through a 0.2-μm filter into a glass vial for LC-MS-MS determination.

2.7.6   Determination

2.7.6.1   LC-MS-MS Operating Condition

Determination of Conditions of Group A, B, C, D, E, F Pesticides and Related Chemicals

(a) Chromatography column: ZORBAX SB-C$_{18}$, 3.5 μm, 100 × 2.1 mm or equivalent
(b) Mobile phase program and the flow rate: refer to Table 2.2
(c) Column temperature: 40°C
(d) Injection volume: 10 μL
(e) Scan mode: ESI
(f) Scan polarity: positive ion
(g) Nebulizer gas: nitrogen gas
(h) Nebulizer gas pressure: 0.28 MPa
(i) Ion spray voltage: 4000 V
(j) Dry gas temperature: 350°C
(k) Dry gas flow rate: 10 L/min
(l) Monitoring ion pairs, collision energy, and declustering potentials: see Table 5.10

Determination Conditions of Group G Pesticides and Related Chemicals

(a) Chromatography column: ZORBAX SB-C$_{18}$, 3.5 μm, 100 × 2.1 mm or equivalent
(b) Mobile phase program and the flow rate: refer to Table 2.3
(c) Column temperature: 40°C
(d) Injection volume: 10 μL
(e) Scan mode: ESI
(f) Scan polarity: negative ion
(g) Nebulizer gas: nitrogen gas
(h) Nebulizer gas pressure: 0.28 MPa
(i) Ion spray voltage: 4000 V
(j) Dry gas temperature: 350°C
(k) Dry gas flow rate: 10 L/min
(l) Monitoring ion pairs, collision energy and declustering potentials: refer to Table 5.10

2.7.6.2   Qualitative Determination

In the samples determined, if the retention times of the peaks of the sample solution are the same as those of the peaks of the working standard mixed solution, and the selected ions of the background-subtracted mass spectrum appear, and also the abundance ratios of selected ions are within the expected limits (abundance ratios >50%, permitted tolerances are ±20%; abundance ratios >20% to

50%, permitted tolerances are ±25%; abundance ratios >10% to 20%, permitted tolerances are ±30%; abundance ratios ≤10%, permitted tolerances are ±50%), the sample is confirmed to contain this pesticide compound.

2.7.6.3   Quantitative Determination

The external standard method is used for quantitation with standard curves for LC-MS-MS. To compensate for the matrix effect, quantitation is based on a series of working standard solutions prepared in blank matrix extract. The standard curves are established by injection of different concentrations of working standard mixed solutions in matrix separately. The responses of pesticides in the sample solution should be in the linear range of the instrumental detection.

2.7.6.4   Parallel Test

A parallel test is carried out for the same testing sample.

2.7.6.5   Blank Test

The operation of the blank test is the same as that described in the method of determination, but without the addition of sample.

2.7.7   Precision

The precision data of the method for this standard is according to the stipulations of GB/T 6379.1 and GB/T 6379.2. The values of repeatability and reproducibility are obtained and calculated at 95% confidence level.

RESEARCHERS

Guo-Fang Pang, Zong-Mao Chen, Yan-Zhong Cao, Xue-Yan Hu, Chun-Lin Fan, Yan-Zhong Cao, Xue-Yan Hu, Chun-Lin Fan, Ping Liang, Peng-Jian Luo, Zheng-Yun Lou, Yan Li, Guang-Ming Liu, Fu-Bin Tang, Jun-Yan Zhang.

Qinhuangdao Entry-Exit Inspection and Quarantine Bureau, 39 Haibin Rd, Qinhuangdao, Hebei, PC 066002, People's Republic of China.

2.8   DETERMINATION OF 488 PESTICIDES AND RELATED CHEMICAL RESIDUES IN MULBERRY TWIG, HONEYSUCKLE, BARBARY WOLFBERRY FRUIT, AND LOTUS LEAF: GC-MS METHOD (GB/T 23200-2008)

2.8.1   Scope

This method is applicable to the qualitative determination of 488 and quantitative determination of 431 pesticides and related chemical residues in mulberry twig, honeysuckle, barbary wolfberry fruit, and lotus leaf.

The limits of detection of the 431 pesticides and related chemical residues of this method are: 0.002 mg/kg to 0.960 mg/kg (see Table 2.14).

2.8.2   Principle

The samples are homogenized with acetonitrile and sodium chloride. The solutions are centrifuged and the supernatants of acetonitrile phase are cleaned up with SPE cartridges. The pesticides and related chemicals are eluted with hexane-acetone. The solutions are analyzed by GC-MS and quantitated with an internal standard.

2.8.3   Reagents and Materials

Acetonitrile: HPLC Grade
Toluene: HPLC Grade
Hexane: HPLC Grade
Acetone: HPLC Grade
Dichloromethane: HPLC Grade
Hexane + Acetone (2+3, V/V)
Sodium Chloride: G.R.
Sodium Sulfate: Anhydrous, Analytically Pure. Ignited at 650°C for 4 h and Kept in a Desiccator
Pesticide and Related Chemicals Standard: Purity ≥95%
  Stock Standard Solution: Accurately weigh 5–10 mg of individual pesticide and related chemical standards (accurate to 0.1 mg) into a 10-mL volumetric flask. Dissolve and dilute to volume with toluene, toluene + acetone combination, dichloromethane, etc., depending on each individual compound's solubility. (For diluting solvent, refer to Appendix B) Stock standard solutions are stored in the dark below 4°C and are used for 1 year.
  Internal Standard Solution: Accurately weigh 3.5 mg heptachlor epoxide into a 100-mL volumetric flask. Dissolve and dilute to volume with toluene.
  Mixed Standard Solution (Mixed Standard Solution A, B, C, D, E, and F): Depending on properties and retention times of the compounds, all compounds are divided into five groups: A, B, C, D, E, and F. The mixed standard solution concentrations are determined by the sensitivity of the instrument used for analysis. For 488 pesticides and related chemical groupings and mixed standard solution concentrations of this standard, reference is made to Appendix B.
  Depending on group number, mixed standard solution concentration, and stock standard solution concentration, appropriate amounts of individual stock standard solution are pipetted into a 100-mL volumetric flask, being diluted to volume with toluene. Mixed standard solutions are stored in the dark below 4°C and used for 1 month.

TABLE 2.14 The LOD, LOQ of 488 Pesticides and Related Chemical Residues Determined by GC-MS

No.	Pesticides	LOD (mg/kg)	LOQ (mg/kg)	No.	Pesticides	LOD (mg/kg)	LOQ (mg/kg)
	Group A			33	chlorpyrifos (-ethyl)	0.012 5	0.025 0
1	allidochlor	0.025 0	0.050 0	34	delta-HCH	0.025 0	0.050 0
2	dichlormid	0.025 0	0.050 0	35	fenthion	0.012 5	0.025 0
3	etridiazol	0.037 5	0.075 0	36	malathion	0.050 0	0.100 0
4	chlormephos	0.025 0	0.050 0	37	paraoxon-ethyl	0.400 0	0.800 0
5	propham	0.012 5	0.025 0	38	fenitrothion	0.025 0	0.050 0
6	cycloate	0.012 5	0.025 0	39	triadimefon	0.025 0	0.050 0
7	diphenylamine	0.012 5	0.025 0	40	linuron	0.050 0	0.100 0
8	chlordimeform	0.012 5	0.025 0	41	pendimethalin	0.050 0	0.100 0
9	ethalfluralin	0.050 0	0.100 0	42	chlorbenside	0.025 0	0.050 0
10	phorate	0.012 5	0.025 0	43	bromophos-ethyl	0.012 5	0.025 0
11	thiometon	0.012 5	0.025 0	44	quinalphos	0.012 5	0.025 0
12	quintozene	0.025 0	0.050 0	45	trans-chlordane	0.012 5	0.025 0
13	atrazine-desethyl	0.012 5	0.025 0	46	phenthoate	0.025 0	0.050 0
14	clomazone	0.012 5	0.025 0	47	metazachlor	0.037 5	0.075 0
15	diazinon	0.012 5	0.025 0	48	prothiofos	0.012 5	0.025 0
16	fonofos	0.012 5	0.025 0	49	chlorfurenol	0.037 5	0.075 0
17	etrimfos	0.012 5	0.025 0	50	procymidone	0.012 5	0.025 0
18	propetamphos	0.012 5	0.025 0	51	dieldrin	0.025 0	0.050 0
19	secbumeton	0.012 5	0.025 0	52	methidathion	0.025 0	0.050 0
20	pronamide	0.012 5	0.025 0	53	napropamide	0.037 5	0.075 0
21	dichlofenthion	0.012 5	0.025 0	54	cyanazine	0.037 5	0.075 0
22	mexacarbate	0.037 5	0.075 0	55	oxadiazone	0.012 5	0.025 0
23	dimethoate	0.050 0	0.100 0	56	fenamiphos	0.037 5	0.075 0
24	dinitramine	0.050 0	0.100 0	57	tetrasul	0.012 5	0.025 0
25	aldrin	0.025 0	0.050 0	58	bupirimate	0.012 5	0.025 0
26	ronnel	0.025 0	0.050 0	59	flutolanil	0.012 5	0.025 0
27	prometryne	0.012 5	0.025 0	60	carboxin	0.300 0	0.600 0
28	cyprazine	0.012 5	0.025 0	61	p,p'-DDD	0.012 5	0.025 0
29	vinclozolin	0.012 5	0.025 0	62	ethion	0.025 0	0.050 0
30	beta-HCH	0.012 5	0.025 0	63	etaconazole-1	0.037 5	0.075 0
31	metalaxyl	0.037 5	0.075 0	64	sulprofos	0.025 0	0.050 0
32	methyl-parathion	0.050 0	0.100 0	65	etaconazole-2	0.037 5	0.075 0

Continued

TABLE 2.14 The LOD, LOQ of 488 Pesticides and Related Chemical Residues Determined by GC-MS—cont'd

No.	Pesticides	LOD (mg/kg)	LOQ (mg/kg)	No.	Pesticides	LOD (mg/kg)	LOQ (mg/kg)
66	myclobutanil	0.012 5	0.025 0	98	heptanophos	0.037 5	0.075 0
67	fensulfothion	0.025 0	0.050 0	99	ethoprophos	0.037 5	0.075 0
68	diclofop-methyl	0.012 5	0.025 0	100	hexachlorobenzene	0.012 5	0.025 0
69	propiconazole-1	0.037 5	0.075 0	101	propachlor	0.037 5	0.075 0
70	propiconazole-2	0.037 5	0.075 0	102	cis-diallate	0.025 0	0.050 0
71	bifenthrin	0.012 5	0.025 0	103	trifluralin	0.025 0	0.050 0
72	mirex	0.012 5	0.025 0	104	trans-diallate	0.025 0	0.050 0
73	carbosulfan	0.037 5	0.075 0	105	chlorpropham	0.025 0	0.050 0
74	nuarimol	0.025 0	0.050 0	106	sulfotep	0.012 5	0.025 0
75	benodanil	0.037 5	0.075 0	107	sulfallate	0.025 0	0.050 0
76	methoxychlor	0.100 0	0.200 0	108	alpha-HCH	0.012 5	0.025 0
77	oxadixyl	0.012 5	0.025 0	109	terbufos	0.025 0	0.050 0
78	tebuconazole	0.037 5	0.075 0	110	profluralin	0.050 0	0.100 0
79	tetramethirn	0.025 0	0.050 0	111	dioxathion	0.050 0	0.100 0
80	norflurazon	0.012 5	0.025 0	112	propazine	0.012 5	0.025 0
81	pyridaphenthion	0.012 5	0.025 0	113	chlorbufam	0.025 0	0.050 0
82	tetradifon	0.012 5	0.025 0	114	dicloran	0.025 0	0.050 0
83	cis-permethrin	0.012 5	0.025 0	115	terbuthylazine	0.012 5	0.025 0
84	pyrazophos	0.025 0	0.050 0	116	monolinuron	0.050 0	0.100 0
85	trans-permethrin	0.012 5	0.025 0	117	flufenoxuron	0.037 5	0.075 0
86	cypermethrin	0.037 5	0.075 0	118	chlorpyrifos-methyl	0.012 5	0.025 0
87	fenvalerate-1	0.050 0	0.100 0	119	desmetryn	0.012 5	0.025 0
88	fenvalerate-2	0.050 0	0.100 0	120	dimethachlor	0.037 5	0.075 0
89	deltamethrin	0.075 0	0.150 0	121	alachlor	0.037 5	0.075 0
	Group B			122	pirimiphos-methyl	0.012 5	0.025 0
90	EPTC	0.037 5	0.075 0	123	terbutryn	0.025 0	0.050 0
91	butylate	0.037 5	0.075 0	124	aspon	0.025 0	0.050 0
92	dichlobenil	0.002 5	0.005 0	125	thiobencarb	0.025 0	0.050 0
93	pebulate	0.037 5	0.075 0	126	dicofol	0.025 0	0.050 0
94	nitrapyrin	0.037 5	0.075 0	127	metolachlor	0.012 5	0.025 0
95	mevinphos	0.025 0	0.050 0	128	pirimiphos-ethyl	0.025 0	0.050 0
96	chloroneb	0.012 5	0.025 0	129	dichlofluanid	0.600 0	1.200 0
97	tecnazene	0.025 0	0.050 0	130	methoprene	0.050 0	0.100 0

TABLE 2.14 The LOD, LOQ of 488 Pesticides and Related Chemical Residues Determined by GC-MS—cont'd

No.	Pesticides	LOD (mg/kg)	LOQ (mg/kg)	No.	Pesticides	LOD (mg/kg)	LOQ (mg/kg)
131	bromofos	0.025 0	0.050 0	164	carbofenothion	0.025 0	0.050 0
132	ethofumesate	0.025 0	0.050 0	165	p,p'-DDT	0.025 0	0.050 0
133	isopropalin	0.025 0	0.050 0	166	benalaxyl	0.012 5	0.025 0
134	propanil	0.025 0	0.050 0	167	edifenphos	0.025 0	0.050 0
135	crufomate	0.075 0	0.150 0	168	triazophos	0.037 5	0.075 0
136	isofenphos	0.025 0	0.050 0	169	cyanofenphos	0.012 5	0.025 0
137	endosulfan-1	0.075 0	0.150 0	170	chlorbenside sulfone	0.025 0	0.050 0
138	chlorfenvinphos	0.037 5	0.075 0	171	endosulfan-sulfate	0.037 5	0.075 0
139	tolylfluanide	0.300 0	0.600 0	172	bromopropylate	0.025 0	0.050 0
140	cis-chlordane	0.025 0	0.050 0	173	benzoylprop-ethyl	0.037 5	0.075 0
141	butachlor	0.025 0	0.050 0	174	fenpropathrin	0.025 0	0.050 0
142	chlozolinate	0.025 0	0.050 0	175	EPN	0.050 0	0.100 0
143	p,p'-DDE	0.012 5	0.025 0	176	hexazinone	0.037 5	0.075 0
144	iodofenphos	0.025 0	0.050 0	177	leptophos	0.025 0	0.050 0
145	tetrachlorvinphos	0.037 5	0.075 0	178	bifenox	0.025 0	0.050 0
146	chlorbromuron	0.300 0	0.600 0	179	phosalone	0.025 0	0.050 0
147	profenofos	0.075 0	0.150 0	180	azinphos-methyl	0.075 0	0.150 0
148	buprofezin	0.025 0	0.050 0	181	fenarimol	0.025 0	0.050 0
149	hexaconazole	0.075 0	0.150 0	182	azinphos-ethyl	0.025 0	0.050 0
150	o,p'-DDD	0.012 5	0.025 0	183	cyfluthrin	0.150 0	0.300 0
151	chlorfenson	0.025 0	0.050 0	184	prochloraz	0.075 0	0.150 0
152	fluorochloridone	0.025 0	0.050 0	185	coumaphos	0.075 0	0.150 0
153	endrin	0.150 0	0.300 0	186	fluvalinate	0.150 0	0.300 0
154	paclobutrazol	0.037 5	0.075 0		Group C		
155	o,p'-DDT	0.025 0	0.050 0	187	dichlorvos	0.075 0	0.150 0
156	methoprotryne	0.037 5	0.075 0	188	biphenyl	0.012 5	0.025 0
157	chloropropylate	0.012 5	0.025 0	189	propamocarb	0.037 5	0.075 0
158	flamprop-methyl	0.012 5	0.025 0	190	vernolate	0.012 5	0.025 0
159	nitrofen	0.075 0	0.150 0	191	3,5-dichloroaniline	0.012 5	0.025 0
160	oxyfluorfen	0.050 0	0.100 0	192	methacrifos	0.012 5	0.025 0
161	chlorthiophos	0.037 5	0.075 0	193	molinate	0.012 5	0.025 0
162	flamprop-isopropyl	0.012 5	0.025 0	194	2-phenylphenol	0.012 5	0.025 0
163	endosulfan -2	0.075 0	0.150 0				

Continued

TABLE 2.14 The LOD, LOQ of 488 Pesticides and Related Chemical Residues Determined by GC-MS—cont'd

No.	Pesticides	LOD (mg/kg)	LOQ (mg/kg)	No.	Pesticides	LOD (mg/kg)	LOQ (mg/kg)
195	cis-1,2,3,6-tetrahydrophthalimide	0.037 5	0.075 0	227	diphenamid	0.012 5	0.025 0
				228	penconazole	0.037 5	0.075 0
196	fenobucarb	0.025 0	0.050 0	229	tetraconazole	0.037 5	0.075 0
197	benfluralin	0.012 5	0.025 0	230	mecarbam	0.050 0	0.100 0
198	hexaflumuron	0.075 0	0.150 0	231	propaphos	0.025 0	0.050 0
199	prometon	0.037 5	0.075 0	232	flumetralin	0.025 0	0.050 0
200	triallate	0.025 0	0.050 0	233	triadimenol-1	0.037 5	0.075 0
201	pyrimethanil	0.012 5	0.025 0	234	triadimenol-2	0.037 5	0.075 0
202	gamma-HCH	0.025 0	0.050 0	235	pretilachlor	0.025 0	0.050 0
203	disulfoton	0.012 5	0.025 0	236	kresoxim-methyl	0.012 5	0.025 0
204	atrizine	0.012 5	0.025 0	237	fluazifop-butyl	0.012 5	0.025 0
205	iprobenfos	0.037 5	0.075 0	238	chlorfluazuron	0.037 5	0.075 0
206	heptachlor	0.037 5	0.075 0	239	chlorobenzilate	0.012 5	0.025 0
207	isazofos	0.025 0	0.050 0	240	flusilazole	0.037 5	0.075 0
208	plifenate	0.025 0	0.050 0	241	fluorodifen	0.012 5	0.025 0
209	fluchloralin	0.050 0	0.100 0	242	diniconazole	0.037 5	0.075 0
210	transfluthrin	0.012 5	0.025 0	243	piperonyl butoxide	0.012 5	0.025 0
211	fenpropimorph	0.012 5	0.025 0	244	dimefuron	0.050 0	0.100 0
212	tolclofos-methyl	0.012 5	0.025 0	245	propargite	0.025 0	0.050 0
213	propisochlor	0.012 5	0.025 0	246	mepronil	0.012 5	0.025 0
214	metobromuron	0.075 0	0.150 0	247	diflufenican	0.012 5	0.025 0
215	ametryn	0.037 5	0.075 0	248	fludioxonil	0.012 5	0.025 0
216	simetryn	0.025 0	0.050 0	249	fenazaquin	0.012 5	0.025 0
217	metribuzin	0.037 5	0.075 0	250	phenothrin	0.012 5	0.025 0
218	dimethipin	0.037 5	0.075 0	251	sethoxydim	0.900 0	1.800 0
219	dipropetryn	0.012 5	0.025 0	252	anilofos	0.025 0	0.050 0
220	formothion	0.025 0	0.050 0	253	acrinathrin	0.025 0	0.050 0
221	diethofencarb	0.075 0	0.150 0	254	lambda-cyhalothrin	0.012 5	0.025 0
222	dimepiperate	0.025 0	0.050 0	255	mefenacet	0.037 5	0.075 0
223	bioallethrin-1	0.050 0	0.100 0	256	permethrin	0.025 0	0.050 0
224	bioallethrin-2	0.050 0	0.100 0	257	pyridaben	0.012 5	0.025 0
225	fenson	0.012 5	0.025 0	258	fluoroglycofen-ethyl	0.150 0	0.300 0
226	o,p'-DDE	0.012 5	0.025 0	259	bitertanol	0.037 5	0.075 0

TABLE 2.14 The LOD, LOQ of 488 Pesticides and Related Chemical Residues Determined by GC-MS—cont'd

No.	Pesticides	LOD (mg/kg)	LOQ (mg/kg)	No.	Pesticides	LOD (mg/kg)	LOQ (mg/kg)
260	etofenprox	0.012 5	0.025 0	291	2,3,4,5-tetrachloroaniline	0.025 0	0.050 0
261	cycloxydim	1.200 0	2.400 0				
262	*alpha*-cypermethrin	0.025 0	0.050 0	292	musk ambrette	0.012 5	0.025 0
263	flucythrinate-1	0.025 0	0.050 0	293	musk xylene	0.012 5	0.025 0
264	flucythrinate-2	0.025 0	0.050 0	294	pentachloroaniline	0.012 5	0.025 0
265	esfenvalerate	0.050 0	0.100 0	295	aziprotryne	0.100 0	0.200 0
266	difenconazole-2	0.075 0	0.150 0	296	isocarbamid	0.062 5	0.125 0
267	difenonazole-1	0.075 0	0.150 0	297	sebutylazine	0.012 5	0.025 0
268	flumioxazin	0.025 0	0.050 0	298	musk moskene	0.012 5	0.025 0
269	flumiclorac-pentyl	0.025 0	0.050 0	299	*de*-PCB 52	0.012 5	0.025 0
	Group D			300	prosulfocarb	0.012 5	0.025 0
270	dimefox	0.037 5	0.075 0	301	dimethenamid	0.012 5	0.025 0
271	disulfoton-sulfoxide	0.025 0	0.050 0	302	BDMC-2	0.025 0	0.050 0
272	pentachlorobenzene	0.012 5	0.025 0	303	monalide	0.025 0	0.050 0
273	crimidine	0.012 5	0.025 0	304	isobenzan	0.012 5	0.025 0
274	BDMC-1	0.025 0	0.050 0	305	octachlorostyrene	0.012 5	0.025 0
275	chlorfenprop-methyl	0.012 5	0.025 0	306	isodrin	0.012 5	0.025 0
276	thionazin	0.012 5	0.025 0	307	isomethiozin	0.025 0	0.050 0
277	2,3,5,6-tetrachloroaniline	0.012 5	0.025 0	308	trichloronat	0.012 5	0.025 0
				309	dacthal	0.012 5	0.025 0
278	*tri-n*-butyl phosphate	0.025 0	0.050 0	310	4,4-dichlorobenzophenone	0.012 5	0.025 0
279	2,3,4,5-tetrachloroanisole	0.012 5	0.025 0				
280	pentachloroanisole	0.012 5	0.025 0	311	nitrothal-isopropyl	0.025 0	0.050 0
281	tebutam	0.025 0	0.050 0	312	musk ketone	0.012 5	0.025 0
282	methabenzthiazuron	0.125 0	0.250 0	313	rabenzazole	0.012 5	0.025 0
283	simetone	0.025 0	0.050 0	314	cyprodinil	0.012 5	0.025 0
284	atratone	0.012 5	0.025 0	315	fuberidazole	0.062 5	0.125 0
285	tefluthrin	0.012 5	0.025 0	316	dicapthon	0.062 5	0.125 0
286	bromocylen	0.012 5	0.025 0	317	*mcpa*-butoxyethyl ester	0.012 5	0.025 0
287	trietazine	0.012 5	0.025 0	318	*de*-PCB 101	0.012 5	0.025 0
288	cycluron	0.037 5	0.075 0	319	isocarbophos	0.025 0	0.050 0
289	*de*-PCB 28	0.012 5	0.025 0	320	phorate sulfone	0.012 5	0.025 0
290	*de*-PCB 31	0.012 5	0.025 0	321	chlorfenethol	0.012 5	0.025 0
				322	*trans*-nonachlor	0.012 5	0.025 0

Continued

TABLE 2.14 The LOD, LOQ of 488 Pesticides and Related Chemical Residues Determined by GC-MS—cont'd

No.	Pesticides	LOD (mg/kg)	LOQ (mg/kg)	No.	Pesticides	LOD (mg/kg)	LOQ (mg/kg)
323	DEF	0.025 0	0.050 0	354	fenamiphos sulfoxide	0.400 0	0.800 0
324	flurochloridone	0.025 0	0.050 0	355	fenamiphos sulfone	0.050 0	0.100 0
325	bromfenvinfos	0.012 5	0.025 0				
326	perthane	0.012 5	0.025 0	356	fenpiclonil	0.050 0	0.100 0
327	de-PCB 118	0.012 5	0.025 0	357	fluquinconazole	0.012 5	0.025 0
328	mephosfolan	0.025 0	0.050 0	358	fenbuconazole	0.025 0	0.050 0
329	4,4-dibromobenzophenone	0.012 5	0.025 0	359	propoxur-1	0.025 0	0.050 0
330	flutriafol	0.025 0	0.050 0		Group E		
331	de-PCB 153	0.012 5	0.025 0	360	XMC	0.025 0	0.050 0
332	diclobutrazole	0.050 0	0.100 0	361	isoprocarb-1	0.025 0	0.050 0
333	disulfoton sulfone	0.025 0	0.050 0	362	acenaphthene	0.012 5	0.025 0
334	hexythiazox	0.100 0	0.200 0	363	terbucarb-1	0.025 0	0.050 0
335	de-PCB 138	0.012 5	0.025 0	364	chlorethoxyfos	0.025 0	0.050 0
336	cyproconazole	0.012 5	0.025 0	365	isoprocarb-2	0.025 0	0.050 0
337	resmethrin-1	0.200 0	0.400 0	366	tebuthiuron	0.050 0	0.100 0
338	resmethrin-2	0.200 0	0.400 0	367	pencycuron	0.050 0	0.100 0
339	phthalic acid,benzyl butyl ester	0.012 5	0.025 0	368	demeton-s-methyl	0.050 0	0.100 0
340	clodinafop-propargyl	0.025 0	0.050 0	369	naled	0.200 0	0.400 0
341	fenthion sulfoxide	0.050 0	0.100 0	370	phenanthrene	0.012 5	0.025 0
342	fluotrimazole	0.012 5	0.025 0	371	fenpyroximate	0.100 0	0.200 0
343	fluroxypr-1-methylheptyl ester	0.012 5	0.025 0	372	tebupirimfos	0.025 0	0.050 0
				373	prohydrojasmon	0.050 0	0.100 0
344	fenthion sulfone	0.050 0	0.100 0	374	fenpropidin	0.025 0	0.050 0
345	metamitron	0.125 0	0.250 0	375	dichloran	0.025 0	0.050 0
346	triphenyl phosphate	0.012 5	0.025 0	376	pyroquilon	0.012 5	0.025 0
347	de-PCB 180	0.012 5	0.025 0	377	propyzamide	0.025 0	0.050 0
348	tebufenpyrad	0.012 5	0.025 0	378	pirimicarb	0.025 0	0.050 0
349	cloquintocet-mexyl	0.012 5	0.025 0	379	bromobutide	0.012 5	0.025 0
350	lenacil	0.125 0	0.250 0	380	tridiphane	0.050 0	0.100 0
351	bromuconazole-1	0.025 0	0.050 0	381	esprocarb	0.025 0	0.050 0
352	bromuconazole-2	0.025 0	0.050 0	382	terbucarb-2	0.025 0	0.050 0
353	nitralin	0.125 0	0.250 0	383	fenfuram	0.025 0	0.050 0

TABLE 2.14 The LOD, LOQ of 488 Pesticides and Related Chemical Residues Determined by GC-MS—cont'd

No.	Pesticides	LOD (mg/kg)	LOQ (mg/kg)	No.	Pesticides	LOD (mg/kg)	LOQ (mg/kg)
384	acibenzolar-s-methyl	0.025 0	0.050 0	416	pyraflufen ethyl	0.025 0	0.050 0
385	benfuresate	0.025 0	0.050 0	417	thenylchlor	0.025 0	0.050 0
386	mefenoxam	0.025 0	0.050 0	418	clethodim	0.050 0	0.100 0
387	malaoxon	0.200 0	0.400 0	419	mefenpyr-diethyl	0.037 5	0.075 0
388	phosphamidon -2	0.100 0	0.200 0	420	etoxazole	0.075 0	0.150 0
389	chlorthal-dimethyl	0.025 0	0.050 0	421	epoxiconazole-2	0.100 0	0.200 0
390	simeconazole	0.025 0	0.050 0	422	famphur	0.050 0	0.100 0
391	terbacil	0.025 0	0.050 0	423	pyriproxyfen	0.025 0	0.050 0
392	thiazopyr	0.025 0	0.050 0	424	iprodione	0.050 0	0.100 0
393	dimethylvinphos	0.025 0	0.050 0	425	ofurace	0.037 5	0.075 0
394	zoxamide	0.025 0	0.050 0	426	piperophos	0.037 5	0.075 0
395	allethrin	0.050 0	0.100 0	427	clomeprop	0.012 5	0.025 0
396	quinoclamine	0.050 0	0.100 0	428	fenamidone	0.012 5	0.025 0
397	fenoxanil	0.025 0	0.050 0	429	tralkoxydim	0.100 0	0.200 0
398	furalaxyl	0.025 0	0.050 0	430	pyraclofos	0.100 0	0.200 0
399	bromacil	0.025 0	0.050 0	431	spirodiclofen	0.100 0	0.200 0
400	picoxystrobin	0.025 0	0.050 0	432	flurtamone	0.025 0	0.050 0
401	butamifos	0.012 5	0.025 0	433	pyriftalid	0.012 5	0.025 0
402	imazamethabenz-methyl	0.037 5	0.075 0	434	silafluofen	0.012 5	0.025 0
				435	pyrimidifen	0.025 0	0.050 0
403	methiocarb sulfone	0.400 0	0.800 0	436	butafenacil	0.012 5	0.025 0
404	TCMTB	0.200 0	0.400 0	437	fluridone	0.025 0	0.050 0
405	metominostrobin	0.050 0	0.100 0	438	tribenuron-methyl	0.012 5	0.025 0
406	imazalil	0.050 0	0.100 0				
407	isoprothiolane	0.025 0	0.050 0		Group F		
408	cyflufenamid	0.200 0	0.400 0	439	ethiofencarb	0.125 0	0.250 0
409	isoxathion	0.100 0	0.200 0	440	dioxacarb	0.100 0	0.200 0
410	quinoxyphen	0.012 5	0.025 0	441	dimethyl phthalate	0.050 0	0.100 0
411	trifloxystrobin	0.050 0	0.100 0	442	4-chlorophenoxy acetic acid	0.006 3	0.012 6
412	imibenconazole-des-benzyl	0.050 0	0.100 0				
				443	phthalimide	0.025 0	0.050 0
413	fipronil	0.100 0	0.200 0	444	diethyltoluamide	0.010 0	0.020 0
414	epoxiconazole-1	0.100 0	0.200 0	445	2,4-D	0.250 0	0.500 0
415	pyributicarb	0.025 0	0.050 0	446	carbaryl	0.037 5	0.075 0

Continued

TABLE 2.14 The LOD, LOQ of 488 Pesticides and Related Chemical Residues Determined by GC-MS—cont'd

No.	Pesticides	LOD (mg/kg)	LOQ (mg/kg)	No.	Pesticides	LOD (mg/kg)	LOQ (mg/kg)
447	cadusafos	0.050 0	0.100 0	467	iprovalicarb-2	0.050 0	0.100 0
448	spiroxamine-1	0.025 0	0.050 0	468	diofenolan-1	0.025 0	0.050 0
449	dicrotophos	0.100 0	0.200 0	469	diofenolan-2	0.025 0	0.050 0
450	2,4,5-T	0.250 0	0.500 0	470	aclonifen	0.250 0	0.500 0
451	3-phenylphenol	0.075 0	0.150 0	471	chlorfenapyr	0.100 0	0.200 0
452	furmecyclox	0.037 5	0.075 0	472	bioresmethrin	0.025 0	0.050 0
453	spiroxamine-2	0.025 0	0.050 0	473	isoxadifen-ethyl	0.025 0	0.050 0
454	dmsa	0.100 0	0.200 0	474	carfentrazone-ethyl	0.025 0	0.050 0
455	sobutylazine	0.025 0	0.050 0	475	halosulfuran-methyl	0.250 0	0.500 0
456	s421 (octachlorodipropyl ether)-1	0.250 0	0.500 0	476	tricyclazole	0.075 0	0.150 0
				477	fenhexamid	0.250 0	0.500 0
457	s421 (octachlorodipropyl ether)-2	0.250 0	0.500 0	478	spiromesifen	0.125 0	0.250 0
				479	bifenazate	0.100 0	0.200 0
458	dodemorph	0.037 5	0.075 0	480	endrin ketone	0.200 0	0.400 0
459	desmedipham	0.250 0	0.500 0	481	gamma-cyhaloterin-1	0.010 0	0.020 0
460	fenchlorphos	0.050 0	0.100 0	482	metoconazole	0.050 0	0.100 0
461	difenoxuron	0.100 0	0.200 0	483	cyhalofop-butyl	0.025 0	0.050 0
462	butralin	0.050 0	0.100 0	484	gamma-cyhalothrin-2	0.010 0	0.020 0
463	dimethametryn	0.012 5	0.025 0	485	halfenprox	0.025 0	0.050 0
464	pyrifenox-1	0.100 0	0.200 0	486	acetamiprid	0.050 0	0.100 0
465	iprovalicarb-1	0.050 0	0.100 0	487	boscalid	0.050 0	0.100 0
466	azaconazole	0.050 0	0.100 0	488	dimethomorph	0.025 0	0.050 0

Working Standard Mixed Solution in Matrix: Working standard mixed solutions in matrix of A, B, C, D, E, and F group pesticides are prepared by diluting 40 μL internal standard solution and an appropriate amount of mixed standard solution to 1.0 mL with blank extract that has been taken through the method with the rest of the samples. Mix thoroughly.

Working standard mixed solutions in matrix must be freshly prepared. Solid Phase Extraction Cartridge: Cleanert TPH 10 mL, 2.0 g, or Equivalent Microporous Filtration Membrane: 13 mm × 0.2 μm

2.8.4   Apparatus

GC-MS: Equipped With EI
Analytical Balances: Capable of Weighing From 0.1 mg to 0.01 g
Homogenizer: Not Less Than 12,000 rpm
Centrifuge: Not Less Than 4200 rpm
Pear-Shaped Flask: 150 mL
Pipette: 1 mL
Rotary Evaporator

2.8.5   Sample Pretreatment

2.8.5.1   Preparation of Test Sample

Samples are cut up and pulverized, sealed, and labeled.

The mulberry twig, honeysuckle, and lotus leaf are stored at ambient temperature. The barbary wolfberry fruit is stored in the dark at 4°C.

2.8.5.2   Extraction

Weigh 5 g of honeysuckle, barbary wolfberry fruit, or 2.5 g of mulberry twig, lotus leaf (accurate to 0.01 g) into a 50-mL centrifuge tube. Add 15 mL acetonitrile (add 5 mL into centrifuge tube with barbary wolfberry fruit), and homogenize at 15,000 rpm for 1 min. Add 2 g sodium chloride to the centrifuge tube and homogenize at 15,000 rpm for 1 min again, then centrifuge at 4200 rpm for 5 min. Pipette 20 mL of the top acetonitrile layer of extracts (corresponding to 10-g test sample) for clean-up. Collect extracts in 150-mL pear-shaped flask. Add 15 mL acetonitrile into the centrifuge tube and homogenize for 1 min again, then centrifuge at 4200 rpm for 5 min. Combine the extracts in the pear-shaped flask. Evaporate the eluate to 1.0 to 2.0 mL using a rotary evaporator at 40°C for clean-up.

2.8.5.3   Clean-up

Add sodium sulfate into Cleanert TPH cartridge to a height of 2 cm. Fix the cartridges into a support to which a pear-shaped flask is connected. Condition the cartridges with 10 mL hexane-acetone before adding the sample. Once the solution gets to the top of the sodium sulfate, pipette the eluate into the cartridges immediately. Rinse the pear-shaped flask with 3×2 mL acetonitrile + toluene (3 + 1, V/V), and decant it into the cartridges. Insert a reservoir into the cartridges. Elute the pesticides with 25 mL acetonitrile + toluene (3 + 1, V/V). Evaporate the eluate to ca. 0.5 mL using a rotary evaporator at 40°C. Exchange with 2×5 mL hexane twice and make up to ca. 1 mL. Add 40 μL internal standard solution and mix thoroughly. The solution is ready for GC-MS determination.

2.8.6   Determination

2.8.6.1   GC-MS Operating Condition

(a) Column: DB-1701 (14% cyanopropyl-phenyl) (30 m × 0.25 mm × 0.25 μm) capillary column, or equivalent
(b) Column temperature: 40°C hold 1 min, at 30°C/min to 130°C, at 5°C/min to 250°C, at 10°C/min to 300°C, hold 5 min
(c) Carrier gas: Helium, purity ≥99.999%, flow rate: 1.2 mL/min
(d) Injection port temperature: 290°C
(e) Injection volume: 1 μL
(f) Injection mode: Splitless, purge on after 1.5 min
(g) Ionization voltage: 70 eV
(h) Ion source temperature: 230°C
(i) GC-MS interface temperature: 80°C
(j) Solvent delay time: group A: 8.30 min, group B: 7.80 min, group C: 7.30 min, group D: 5.50 min, group E: 6.10 min, group F: 5.50 min
(k) Selected ion monitoring mode: For each compound select 1 quantifying ion and 2–3 qualifying ions. All of the detected ions of each group are detected according to programmed time and sequence of peaking. The retention times, quantifying ions, qualifying ions, and the abundance ratios of quantifying ion and qualifying ions for each compound are listed in Table 5.3. The programmed time and dwell time for the ions detected for each compound in each group are listed in Table 5.4

2.8.6.2   Qualitative Determination

In the samples determined, injections are required to analyze for all pesticides according to GC-MS operating conditions. If the retention times in the peaks of the sample solution are the same as those for the peaks of the working standard mixed solution, and the selected ions of the background-subtracted mass spectrum appear, and the abundance ratios of the selected ions are within the expected limits (abundance ratios >50%, permitted tolerances are ±10%; abundance ratios >20% to 50%, permitted tolerances are ±15%; abundance ratios >10% to 20%, permitted tolerances are ±20%; abundance ratios ≤10%, permitted tolerances are ±50%), the sample is confirmed to contain this pesticide compound. In cases where the results are still not definitive, the sample should be reinjected with acquisition in scan mode (sufficient sensitivity) or with additional confirmatory ions or using other instruments that have higher sensitivity.

2.8.6.3   Quantitative Determination

The results are quantitated using heptachlor epoxide for an internal standard and the quantitation ion response for each analyte. To compensate for the matrix effect, quantitation is based on a mixed standard prepared in blank matrix extract. The concentration of standard solution and the detected sample solution must be similar.

2.8.6.4 Parallel Test

A parallel test is carried out for the same testing sample.

2.8.6.5 Blank Test

The operation of the blank test is the same as that described in the method of determination, but without the addition of sample.

2.8.7 Precision

The precision data of the method for this standard have been determined according to the stipulations of GB/T 6379.1 and GB/T 6379.2. The values of repeatability and reproducibility are obtained and calculated at the 95% confidence level.

RESEARCHERS

Guo-Fang Pang, Chun-Lin Fan, Qun-Jie Wang, Wei Huang, Ping Liang, Cui-Cui Yao, Wan Wang, Xun-dong Zhu.

Qinhuangdao Entry-Exit Inspection and Quarantine Bureau, 39 Haibin Rd, Qinhuangdao, Hebei, PC 066002, People's Republic of China.

2.9 DETERMINATION OF 413 PESTICIDES AND RELATED CHEMICAL RESIDUES IN MULBERRY TWIG, HONEYSUCKLE, BARBARY WOLFBERRY FRUIT, AND LOTUS LEAF: LC-MS-MS METHOD (GB/T 23201-2008)

2.9.1 Scope

This method is applicable to the determination of 413 pesticides and related chemical residues in mulberry twig, honeysuckle, barbary wolfberry fruit, and lotus leaf.

The limits of quantitative determination of this method of 369 pesticides and related chemicals are: 0.01 µg/kg to 1.26 mg/kg (refer to Table 2.15).

2.9.2 Principle

The samples are extracted with acetonitrile and sodium chloride by homogenizing. The solutions are centrifuged, and the supernatants of the acetonitrile phase are cleaned up with SPE cartridges. The pesticides and related chemicals are eluted with acetonitrile + toluene (3+1), and the solutions are analyzed by LC-MS-MS, using the external standard method for quantification.

2.9.3   Reagents and Materials

"Water" is first grade of GB/T6682 specified.

Acetonitrile: HPLC Grade
Toluene: HPLC Grade
Methanol: HPLC Grade
Cyclohexane: HPLC Grade
Isooctane: HPLC Grade
Sodium Chloride: Analytically Pure
Cleanert TPH Cartridge: 10 mL, 2.0 g, or Equivalent
Micropore Membrane Filters (nylon): 13 mm × 0.2 μm
0.1% Formic Acid Solution (V/V)
5 mmol/L Ammonium Acetate Solution
Acetonitrile + Toluene (3+1, V/V)
Acetonitrile + Water (3+2, V/V)
Sodium Sulfate: Anhydrous, Analytically Pure. Ignited at 650°C for 4 h and
Kept in a Desiccator
Pesticide and Related Chemicals Standard: Purity ≥95%

Stock Standard Solution: Accurately weigh 5–10 mg of individual pesticide and related chemicals standards (accurate to 0.1 mg) into a 10-mL
volumetric flask. Dissolve and dilute to volume with methanol, toluene,
cyclohexane, isooctane, etc., depending on each individual compound's
solubility. (For diluting solvent refer to Appendix B) Standard solution is
stored in the dark at 0°C to 4°C and is used for 1 year.

Mixed Standard Solution (Mixed Standard Solution A, B, C, D, E, F,
and G): Depending on the properties and retention times of each compound, all compounds are divided into seven groups: A, B, C, D, E, F,
and G. The concentrations of mixed standard solutions are decided
depending on the instrument sensitivity of each compound. For 413 pesticides and related chemical groupings and concentrations of mixed standard solutions of this standard, reference is made to Appendix B.

Depending on group number, mixed standard solution concentration, and
stock standard solution concentration, appropriate amounts of individual
stock standard solutions are pipetted into a 100-mL volumetric flask,
with five groups of pesticides and related chemicals being diluted to volume with methanol. Mixed standard solutions are stored in the dark at 0°
C to 4°C and are used for 1 month.

Working Standard Mixed Solution in Matrix: Working standard mixed
solutions in matrix of A, B, C, D, E, F, and G group pesticides and related
chemicals are prepared by mixing different concentrations of working
standard mixed solutions with sample blank extract that has been taken
through the method with the rest of the samples. These solutions are used
to construct calibration plots. Working standard mixed solutions in
matrix must be freshly prepared.

TABLE 2.15 The LOD, LOQ of 413 Pesticides and Related Chemical Residues Determined by LC-MS-MS

No.	Pesticides	LOD (μg/kg)	LOQ (μg/kg)	No.	Pesticides	LOD (μg/kg)	LOQ (μg/kg)
	Group A			32	myclobutanil	1.00	2.00
1	propham	110.00	220.00	33	paclobutrazol	0.57	1.14
2	isoprocarb	2.30	4.60	34	fenthion sulfoxide	0.31	0.62
3	3,4,5-trimethacarb	0.34	0.68	35	triadimenol	10.55	21.10
				36	butralin	1.90	3.80
4	cycluron	0.21	0.42	37	spiroxamine	0.05	0.10
5	carbaryl	10.32	20.64	38	tolclofos methyl	66.56	133.12
6	propachlor	0.27	0.54	39	desmedipham	4.03	8.06
7	rabenzazole	1.33	2.66	40	methidathion	10.66	21.32
8	simetryn	0.14	0.28	41	allethrin	60.40	120.80
9	monolinuron	3.56	7.12	42	diazinon	0.71	1.42
10	mevinphos	1.57	3.14	43	edifenphos	0.75	1.50
11	aziprotryne	1.38	2.76	44	flusilazole	0.58	1.16
12	secbumeton	0.07	0.14	45	iprovalicarb	2.32	4.64
13	cyprodinil	0.74	1.48	46	benodanil	3.48	6.96
14	buturon	8.96	17.92	47	flutolanil	1.15	2.30
15	carbetamide	3.64	7.28	48	famphur	3.60	7.20
16	pirimicarb	0.15	0.30	49	benalyxyl	1.24	2.48
17	clomazone dimethazone	0.42	0.84	50	diclobutrazole	0.47	0.94
18	cyanazine	0.16	0.32	51	etaconazole	1.78	3.56
19	prometryne	0.16	0.32	52	fenarimol	0.61	1.22
20	paraoxon methyl	0.76	1.52	53	tetramethirn	1.82	3.64
21	thiacloprid	0.37	0.74	54	cloquintocet mexyl	1.88	3.76
22	imidacloprid	22.00	44.00	55	bitertanol	33.40	66.80
23	ethidimuron	1.50	3.00	56	chlorprifos methyl	16.00	32.00
24	isomethiozin	1.07	2.14	57	azinphos ethyl	108.93	217.86
25	diallate	89.20	178.40	58	clodinafop propargyl	2.44	4.88
26	acetochlor	47.40	94.80	59	triflumuron	3.92	7.84
27	nitenpyram	17.12	34.24	60	isoxaflutole	3.90	7.80
28	methoprotryne	0.24	0.48	61	quizalofop-ethyl	0.68	1.36
29	dimethenamid	4.30	8.60	62	haloxyfop-methyl	2.64	5.28
30	terbucarb	2.10	4.20	63	fluazifop butyl	0.26	0.52
31	penconazole	2.00	4.00	64	bromophos-ethyl	567.69	1135.38

Continued

TABLE 2.15 The LOD, LOQ of 413 Pesticides and Related Chemical Residues Determined by LC-MS-MS—cont'd

No.	Pesticides	LOD (μg/kg)	LOQ (μg/kg)	No.	Pesticides	LOD (μg/kg)	LOQ (μg/kg)
65	bensulide	34.20	68.40	96	imibenzonazole-des-benzyl	6.22	12.44
66	bromfenvinfos	3.02	6.04	97	neburon	7.10	14.20
67	azoxystrobin	0.45	0.90	98	mefenoxam	1.54	3.08
68	pyrazophos	1.62	3.24	99	ethofume sate	372.00	744.00
69	flufenoxuron	3.17	6.34	100	iprobenfos	8.28	16.56
70	indoxacarb	7.54	15.08	101	cyproconazole	0.73	1.46
	Group B			102	thiamethoxam	33.00	66.00
71	ethylene thiourea	52.20	104.40	103	crufomate	0.52	1.04
72	dazomet	127	254.00	104	etrimfos	18.76	37.52
73	nicotine	2.20	4.40	105	cythioate	80.00	160.00
74	fenuron	1.03	2.06	106	phosphamidon	3.88	7.76
75	crimidine	1.56	3.12	107	phenmedipham	8.96	17.92
76	molinate	2.10	4.20	108	fenhexamid	0.95	1.90
77	6-chloro-4-hydroxy-3-phenyl-pyridazin	1.65	3.30	109	flutriafol	8.58	17.16
78	propoxur	24.40	48.80	110	furalaxyl	0.77	1.54
79	isouron	0.41	0.82	111	bioallethrin	198.00	396.00
80	chlorotoluron	0.62	1.24	112	cyanofenphos	20.80	41.60
81	thiofanox	157.00	314.00	113	pirimiphos methyl	0.20	0.40
82	chlorbufam	183.00	366.00	114	buprofezin	0.88	1.76
83	bendiocarb	3.18	6.36	115	disulfoton sulfone	2.46	4.92
84	propazine	0.32	0.64	116	fenazaquin	0.32	0.64
85	terbuthylazine	0.47	0.94	117	triazophos	0.68	1.36
86	diuron	1.56	3.12	118	DEF	1.61	3.22
87	chlormephos	448.00	896.00	119	pyriftalid	0.62	1.24
88	carboxin	0.56	1.12	120	metconazole	1.32	2.64
89	clothianidin	63.00	126.00	121	pyriproxyfen	0.43	0.86
90	pronamide	15.38	30.76	122	isoxaben	0.19	0.38
91	dimethachloro	1.90	3.80	123	flurtamone	0.44	0.88
92	methobromuron	16.84	33.68	124	trifluralin	334.80	669.60
93	phorate	314.00	628.00	125	flamprop methyl	20.20	40.40
94	aclonifen	24.20	48.40	126	propiconazole	1.76	3.52
95	mephosfolan	2.32	4.64	127	chlorpyrifos	53.80	107.60

TABLE 2.15 The LOD, LOQ of 413 Pesticides and Related Chemical Residues Determined by LC-MS-MS—cont'd

No.	Pesticides	LOD (μg/kg)	LOQ (μg/kg)	No.	Pesticides	LOD (μg/kg)	LOQ (μg/kg)
128	fluchloralin	488.00	976.00	160	ethoprophos	2.76	5.52
129	flamprop isopropyl	0.43	0.86	161	fonofos	7.46	14.92
130	tetrachlorvinphos	2.22	4.44	162	etridiazol	100.42	200.84
131	propargite	68.60	137.20	163	hexazinone	0.12	0.24
132	bromuconazole	3.14	6.28	164	dimethametryn	0.11	0.22
133	picolinafen	0.73	1.46	165	demeton(o+s)	6.77	13.54
134	fluthiacet methyl	5.30	10.60	166	benoxacor	6.90	13.80
135	trifloxystrobin	2	4.00	167	bromacil	23.60	47.20
136	hexaflumuron	25.20	50.40	168	phorate sulfoxide	368.28	736.56
137	novaluron	8.04	16.08	169	brompyrazon	3.60	7.20
138	flurazuron	26.80	53.60	170	mepronil	0.38	0.76
	Group C			171	disulfoton	469.70	939.40
139	maleic hydrazide	80.00	160.00	172	fenthion	52.00	104.00
140	methamidophos	4.93	9.86	173	metalaxyl	0.50	1.00
141	EPTC	37.34	74.68	174	ofurace	1.00	2.00
142	diethyltoluamide	0.55	1.10	175	dodemorph	0.40	0.80
143	monuron	34.74	69.48	176	imazamethabenz-methyl	0.16	0.32
144	pyrimethanil	0.68	1.36	177	isoprothiolane	1.85	3.70
145	fenfuram	0.78	1.56	178	imazalil	2.00	4.00
146	quinoclamine	7.92	15.84	179	phoxim	82.80	165.60
147	fenobucarb	5.90	11.80	180	quinalphos	2.00	4.00
148	propanil	21.59	43.18	181	fenoxycarb	18.27	36.54
149	carbofuran	13.06	26.12	182	pyrimitate	0.17	0.34
150	acetamiprid	1.44	2.88	183	fensulfothin	2.00	4.00
151	mepanipyrim	0.32	0.64	184	fluorochloridone	13.78	27.56
152	prometon	0.13	0.26	185	butachlor	20.07	40.14
153	methiocarb	41.20	82.40	186	kresoxim-methyl	100.58	201.16
154	metoxuron	0.64	1.28	187	triticonazole	3.02	6.04
155	dimethoate	7.60	15.20	188	fenamiphos sulfoxide	0.74	1.48
156	fluometuron	0.92	1.84	189	thenylchlor	24.14	48.28
157	dicrotophos	1.14	2.28	190	fenoxanil	39.40	78.80
158	monalide	1.20	2.40	191	fluridone	0.18	0.36
159	diphenamid	0.14	0.28	192	epoxiconazole	4.06	8.12

Continued

TABLE 2.15 The LOD, LOQ of 413 Pesticides and Related Chemical Residues Determined by LC-MS-MS—cont'd

No.	Pesticides	LOD (µg/kg)	LOQ (µg/kg)	No.	Pesticides	LOD (µg/kg)	LOQ (µg/kg)
193	chlorphoxim	77.57	155.14	225	chloridazon	2.33	4.66
194	fenamiphos sulfone	0.45	0.90	226	sulfallate	207.20	414.40
195	fenbuconazole	1.65	3.30	227	ethiofencarb	4.92	9.84
196	isofenphos	218.67	437.34	228	terbumeton	0.10	0.20
197	phenothrin	339.20	678.40	229	cyprazine	0.04	0.08
198	piperophos	9.24	18.48	230	ametryn	0.96	1.92
199	piperonyl butoxide	1.13	2.26	231	tebuthiuron	0.22	0.44
200	oxyflurofen	58.55	117.10	232	trietazine	0.60	1.20
201	coumaphos	2.10	4.20	233	sebutylazine	0.31	0.62
202	flufenacet	5.30	10.60	234	dibutyl succinate	222.40	444.80
203	phosalone	48.04	96.08	235	tebutam	0.14	0.28
204	methoxyfenozide	3.70	7.40	236	thiofanox-sulfoxide	8.29	16.58
205	prochloraz	2.07	4.14	237	methacrifos	242.37	484.74
206	aspon	1.73	3.46	238	terbutryn	0.02	0.04
207	ethion	2.96	5.92	239	thionazin	22.68	45.36
208	diafenthiuron	0.28	0.56	240	linuron	11.63	23.26
209	dithiopyr	10.40	20.80	241	heptanophos	5.84	11.68
210	fenpyroximate	1.36	2.72	242	prosulfocarb	0.37	0.74
211	flumiclorac-pentyl	10.61	21.22	243	dipropetryn	6.96	13.92
212	temephos	1.22	2.44	244	thiobencarb	0.27	0.54
213	butafenacil	9.50	19.00	245	tri-n-butyl phosphate	0.37	0.74
	Group D			246	diethofencarb	2.00	4.00
214	thiabendazole	0.49	0.98	247	alachlor	7.40	14.80
215	metamitron	6.36	12.72	248	cadusafos	1.15	2.30
216	isoproturon	0.14	0.28	249	metazachlor	0.98	1.96
217	atratone	0.18	0.36	250	propetamphos	54.00	108.00
218	oesmetryn	0.17	0.34	251	simeconazole	2.94	5.88
219	metribuzin	0.54	1.08	252	triadimefon	7.88	15.76
220	DMST	40.00	80.00	253	phorate sulfone	42.00	84.00
221	cycloate	4.44	8.88	254	tridemorph	2.60	5.20
222	atrazine	0.36	0.72	255	mefenacet	2.21	4.42
223	butylate	302.00	604.00	256	fenpropimorph	0.18	0.36
224	pymetrozin	34.28	68.56	257	tebuconazole	2.23	4.46

TABLE 2.15 The LOD, LOQ of 413 Pesticides and Related Chemical Residues Determined by LC-MS-MS—cont'd

No.	Pesticides	LOD (μg/kg)	LOQ (μg/kg)	No.	Pesticides	LOD (μg/kg)	LOQ (μg/kg)
258	isopropalin	30.00	60.00	290	vernolate	0.26	0.52
259	nuarimol	1.00	2.00	291	chlorthiamid	8.82	17.64
260	bupirimate	0.70	1.40	292	aminocarb	16.42	32.84
261	azinphos-methyl	110.433	220.87	293	dimethirimol	0.12	0.24
262	tebupirimfos	0.13	0.26	294	omethoate	9.65	19.30
263	phenthoate	92.35	184.70	295	ethoxyquin	3.52	7.04
264	sulfotep	2.60	5.20	296	dichlorvos	0.55	1.10
265	sulprofos	5.84	11.68	297	aldicarb sulfone	21.40	42.80
266	EPN	33.00	66.00	298	dioxacarb	3.36	6.72
267	diniconazole	1.34	2.68	299	ethiofencarb-sulfoxide	224.00	448.00
268	pencycuron	0.27	0.54	300	cyanophos	10.12	20.24
269	mecarbam	19.60	39.20	301	thiometon	578.00	1156.00
270	tralkoxydim	0.32	0.64	302	folpet	138.60	277.20
271	malathion	5.64	11.28	303	demeton-s-methyl sulfone	19.76	39.52
272	pyributicarb	0.34	0.68	304	dimepiperate	1260.00	2520.00
273	pyridaphenthion	0.87	1.74	305	fenpropidin	0.18	0.36
274	pirimiphos-ethyl	0.05	0.10	306	amidithion	658.00	1316.00
275	pyraclofos	1.00	2.00	307	paraoxon-ethyl	0.47	0.94
276	picoxystrobin	8.44	16.88	308	aldimorph	3.16	6.32
277	tetraconazole	1.72	3.44	309	vinclozolin	2.54	5.08
278	mefenpyr-diethyl	12.56	25.12	310	uniconazole	2.40	4.80
279	profenefos	2.02	4.04	311	pyrifenox	0.27	0.54
280	pyraclostrobin	0.51	1.02	312	dicapthon	0.24	0.48
281	thiazopyr	1.96	3.92	313	clofentezine	0.76	1.52
	Group E			314	norflurazon	0.26	0.52
282	4-aminopyridine	0.87	1.74	315	triallate	46.20	92.40
283	methomyl	9.56	19.12	316	quinoxyphen	153.40	306.80
284	pyroquilon	3.48	6.96	317	fenthion sulfone	17.46	34.92
285	fuberidazole	1.89	3.78	318	flurochloridone	1.29	2.58
286	isocarbamid	1.70	3.40	319	phthalic acid,benzyl butyl ester	632.00	1264.00
287	butocarboxim	1.57	3.14	320	isazofos	0.18	0.36
288	chlordimeform	1.33	2.66	321	dichlofenthion	30.20	60.40
289	cymoxanil	55.60	111.20				

Continued

TABLE 2.15 The LOD, LOQ of 413 Pesticides and Related Chemical Residues Determined by LC-MS-MS—cont'd

No.	Pesticides	LOD (µg/kg)	LOQ (µg/kg)	No.	Pesticides	LOD (µg/kg)	LOQ (µg/kg)
322	vamidothion sulfone	476.00	952.00	353	simeton	1.10	2.20
323	terbufos sulfone	88.60	177.20	354	dinotefuram	10.18	20.36
324	dinitramine	1.79	3.58	355	pebulate	3.40	6.80
325	trichloronate	66.80	133.60	356	acibenzolar-s-methyl	3.08	6.16
326	resmethrin-2	0.30	0.60	357	dioxabenzofos	13.84	27.68
327	boscalid	4.76	9.52	358	oxamyl-oxime	548.06	1096.12
328	nitralin	34.40	68.80	359	metha-benzthiazuron	0.07	0.14
329	fenpropathrin	245.00	490.00	360	butoxycarboxim	26.60	53.20
330	hexythiazox	23.60	47.20	361	demeton-s-methyl sulfoxide	3.92	7.84
331	benzoximate	19.66	39.32	362	thiofanox sulfone	24.08	48.16
332	benzoylprop-ethyl	308.00	616.00	363	phosfolan	0.49	0.98
333	pyrimidifen	14.00	28.00	364	demeton-s	80.00	160.00
334	furathiocarb	1.92	3.84	365	fenthion oxon	1.19	2.38
335	trans-permethin	4.80	9.60	366	napropamide	1.27	2.54
336	etofenprox	228.00	456.00	367	fenitrothion	26.80	53.60
337	pyrazoxyfen	0.33	0.66	368	phthalic acid, dibutyl ester	39.60	79.20
338	zeta cypermethrin	0.68	1.36	369	metolachlor	0.39	0.78
339	haloxyfop-2-ethoxyethyl	2.50	5.00	370	procymidone	86.60	173.20
340	tau-fluvalinate	230.00	460.00	371	vamidothion	4.56	9.12
	Group F			372	triamiphos	0.01	0.02
341	acrylamide	17.80	35.60	373	cumyluron	1.32	2.64
342	tert-butylamine	38.95	77.90	374	phosmet	17.72	35.44
343	phthalimide	43.00	86.00	375	ronnel (fenchlorphos)	13.13	26.26
344	dimefox	68.20	136.40	376	pyrethrin	35.80	71.60
345	metolcarb	25.40	50.80	377	phthalic acid, biscyclohexyl ester	0.68	1.36
346	diphenylamin	0.41	0.82	378	carpropamid	5.20	10.40
347	1-naphthy acetamide	0.81	1.62	379	tebufenpyrad	0.25	0.50
348	atrazine-desethyl	0.62	1.24	380	tebufenozide	27.80	55.60
349	2,6-dichloro benzamide	4.50	9.00	381	chlorthiophos	31.80	63.60
350	aldicarb	261.00	522.00	382	dialifos	157.00	314.00
351	dimethyl phthalate	13.20	26.40	383	cinidon-ethyl	14.58	29.16
352	chlordimeform hydrochloride	2.64	5.28				

TABLE 2.15 The LOD, LOQ of 413 Pesticides and Related Chemical Residues Determined by LC-MS-MS—cont'd

No.	Pesticides	LOD (µg/kg)	LOQ (µg/kg)	No.	Pesticides	LOD (µg/kg)	LOQ (µg/kg)
384	rotenone	2.32	4.64	399	2-phenylphenol	169.88	339.76
385	imibenconazole	10.26	20.52	400	3-phenylphenol	4.00	8.00
386	propaquiafop	1.24	2.48	401	dicloran	48.56	97.12
387	lactofen	62.00	124.00	402	chlorpropham	15.77	31.54
388	benzofenap	0.08	0.16	403	terbacil	0.88	1.76
389	dinoseb acetate	41.28	82.56	404	chlorfenethol	164.30	328.60
390	imazapic	1.68	3.36	405	chlorobenzuron	20.40	40.80
391	propisochlor	0.80	1.60	406	chloramph-enicolum	3.88	7.76
392	silafluofen	608.00	1216.00	407	oryzalin	4.91	9.82
393	fentrazamide	12.40	24.80	408	famoxadone	45.29	90.58
394	pentachloro-aniline	3.74	7.48	409	diflufenican	28.27	56.54
395	cyphenothrin	16.80	33.60	410	ethiprole	39.85	79.70
396	dimefuron	4.00	8.00	411	fluazinam	70.60	141.20
397	malaoxon	4.69	9.38	412	kelevan	964.282	1928.56
	Group G			413	acrinathrin	8.08	16.16
398	dalapon	230.74	461.48				

2.9.4   Apparatus

LC-MS-MS: equipped with ESI source
Analytical balances: Capable of weighing From 0.1 mg to 0.01 g
Homogenizer: Max. rpm 24,000 rpm
Centrifuge: Max. rpm 42,000 rpm
Rotary Evaporator
Pear-Shaped Flask
Pipette: 1 mL
Centrifuge Tube: 50 mL With Stopper
Vial: 2 mL With PTFE Screw Cap
Nitrogen Evaporator

2.9.5   Sample Pretreatment

2.9.5.1   Preparation of Test Sample

Grind the mulberry twig, honeysuckle, and lotus leaf thoroughly with a grinder until it is pulverized to a fine powder. Barbary wolfberry fruit can be tested directly.

The test samples are stored at room temperature. Be sure to seal carefully and protect from dampness.

2.9.5.2 Extraction

Weigh 2 g (accurate to 0.01 g) of the mulberry twig, honeysuckle, barbary wolfberry fruit, and lotus leaf into 50 mL centrifuge tubes. Add 15 mL acetonitrile (mix an extra 5 mL water in the barbary wolfberry fruit), and homogenize at 15,000 rpm for 1 min. Add 5 g sodium chloride to the centrifuge tube and homogenize 1 min again, then centrifuge at 4200 rpm for 5 min. The supernatants are then all moved into a 150-mL pear-shaped flask. Add 15 mL acetonitrile into the centrifuge tube again, homogenize, centrifuge, and then transfer to the previously mentioned glass funnel before the extracts are combined, which are then placed in a waterbath of 40°C and evaporated to 1 mL to 2 mL using a rotary evaporator for clean-up.

2.9.5.3 Clean-up

Add sodium sulfate into the Cleanert TPH cartridge to a height of 2 cm. Fix the cartridge into a support to which a pear-shaped flask is connected. Condition the cartridge with 10 mL acetonitrile-toluene (3+1) before adding the sample. Once the solution gets to the top of the sodium sulfate, transfer the concentrated solution into the cartridge quickly. Rinse the pear-shaped flask with 2 mL acetonitrile-toluene (3+1) twice and transfer it into the cartridge too. Insert a 50-mL reservoir into the cartridges. Elute the pesticides with 25 mL acetonitrile-toluene (3+1). Evaporate the eluate to 1 mL using a rotary evaporator at 40°C. Then evaporate the eluate to dryness using a nitrogen evaporator. Make up to 1 mL with acetonitrile-water (3+2) and mix thoroughly. Finally, filter the extract through a 0.2-μm filter into a glass vial for LC-MS-MS determination.

2.9.6 Determination

2.9.6.1 LC-MS-MS Operating Conditions

Operating Conditions for Groups A, B, C, D, E, and F (ESI⁺)

(a) Column: ZORBOX SB-C_{18}, 3.5 μm, 100 × 2.1 mm, or equivalent
(b) Mobile phase and the condition of gradient elution, see Table 2.2
(c) Column temperature: 40°C
(d) Injection volume: 10 μL
(e) Ion source: ESI
(f) Scan mode: Positive ion scan
(g) Monitor mode: Multiple reaction monitor

(h) Ionspray voltage: 4000 V
(i) Nebulizer gas: 0.28 MPa
(j) Dry gas temperature: 350°C
(k) Dry gas flow rate: 10 L/min
(l) MRM transitions for precursor/product ion, quantifying for precursor/ product ion, declustering potential, collision energy, and voltage-source fragmentation: see Table 5.10

Operating Conditions of G Group (ESI⁻)

(a) Column: ZORBOX SB-C$_{18}$, 3.5 µm, 100 × 2.1 mm, or equivalent
(b) Mobile phase and the condition of gradient elution, see Table 2.3
(c) Column temperature: 40°C
(d) Injection volume: 10 µL
(e) Ion source: ESI
(f) Scan mode: Negative ion scan
(g) Monitor mode: Multiple reaction monitor
(h) Ionspray voltage: 4000 V
(i) Nebulizer gas: 0. 28 MPa
(j) Dry gas temperature: 350°C
(k) Dry gas flow rate: 10 L/min
(l) MRM transitions for precursor/product ion, quantifying for precursor/ product ion, declustering potential, collision energy, and voltage-source fragmentation: see Table 5.10

2.9.6.2 Qualitative Determination

Sample solutions are analyzed using the LC-MS-MS operating conditions. For the samples determined, if the retention times of peaks found in the sample solution chromatogram are the same as those for the peaks in the standard in the blank matrix extract chromatogram, and the abundance ratios of MRM transitions for the precursor/product ion are within the expected limits (relative abundance >50%, permitted tolerances ±20%; relative abundance >20% to 50%, permitted tolerances ±25%; relative abundance >10% to 20%, permitted tolerances ±30%; relative abundance ≤10%, permitted tolerances ±50%), the sample is confirmed to contain this pesticide compound.

2.9.6.3 Quantitative Determination

The external standard method is used for quantitation with standard curves for LC-MS-MS. In order to compensate for the matrix effect, quantitation is based on a series of working standard solutions prepared in blank matrix extract. The standard curves are established by injection of different concentrations of working standard mixed solutions in matrix separately. The responses of pesticides in the sample solution should be in the linear range of the instrumental detection.

2.9.6.4 Parallel Test

A parallel test is carried out for the same testing sample.

2.9.6.5 Blank Test

The operation of the blank test is the same as that described in the method of determination, but without the addition of sample.

2.9.7 Precision

The precision data of the method for this standard have been determined according to the stipulations of GB/T 6379.1 and GB/T 6379.2. The values of repeatability and reproducibility are obtained and calculated at the 95% confidence level.

RESEARCHERS

Guo-Fang Pang, Chun-Lin Fan, Ping Liang, Wei Huang, Qun-Jie Wang, Xin-Xin Ji, Xu-Dong Zhu, Wan Wang.

Qinhuangdao Entry-Exit Inspection and Quarantine Bureau, 39 Haibin Rd, Qinhuangdao, Hebei, PC 066002, People's Republic of China.

2.10 DETERMINATION OF 503 PESTICIDES AND RELATED CHEMICAL RESIDUES IN MUSHROOMS: GC-MS METHOD (GB/T 23216-2008)

2.10.1 Scope

This method is applicable to the qualitative determination of the 503 pesticides and related chemicals and quantitative determination of 478 pesticides and related chemical residues in nameko, *Flammulina*, and moodear mushrooms.

The limits of detection of the 478 pesticides and related chemical residues of this method: 0.0013 mg/kg to 1.6000 mg/kg (see Table 2.16).

2.10.2 Principle

The samples are homogenized with acetonitrile and sodium chloride. The solutions are centrifuged and the supernatants of the acetonitrile phase are cleaned up with SPE cartridges. The pesticides and related chemicals are eluted with acetonitrile + toluene (3 + 1). The solutions are analyzed by GC-MS and quantitated using the internal standard method.

2.10.3 Reagents and Materials

Acetonitrile: HPLC Grade

Toluene: HPLC Grade

Hexane: HPLC Grade

Acetone: HPLC Grade

Cyclohexane: HPLC Grade

Isooctane: HPLC Grade

Ethyl Acetate: HPLC Grade

Acetonitrile + Toluene (3+1, V/V)

Sodium Chloride: G.R.

Sodium Sulfate: Anhydrous, Analytically Pure. Ignited at 650°C for 4 h and Kept in a Desiccator

Pesticide and Related Chemicals Standard: Purity ≥95%

Stock Standard Solution: Accurately weigh 5–10 mg of individual pesticide and related chemical standards (accurate to 0.1 mg) into a 10-mL volumetric flask. Dissolve and dilute to volume with toluene, toluene + acetone combination, dichloromethane, etc., depending on each individual compound's solubility. (For diluting solvent, refer to Appendix B) Stock standard solutions are stored in the dark below 4°C and are used for 1 year.

Internal Standard Solution: Accurately weigh 3.5 mg heptachlor epoxide into a 100-mL volumetric flask. Dissolve and dilute to volume with toluene.

Mixed Standard Solution (Mixed Standard Solution A, B, C, D, E, and F): Depending on properties and retention times of compounds, all compounds are divided into five groups: A, B, C, D, E, and F. The mixed standard solution concentrations are determined by the sensitivity of the instrument used for analysis. For 503 pesticides and related chemicals grouping and mixed standard solution concentrations of this standard, reference is made to Appendix B.

Depending on group number, mixed standard solution concentration, and stock standard solution concentration, appropriate amounts of individual stock standard solution are pipetted into a 100-mL volumetric flask, being diluted to volume with toluene. Mixed standard solutions are stored in the dark below 4°C and are used for 1 month.

Working Standard Mixed Solution in Matrix: Working standard mixed solutions in matrix of A, B, C, D, E and F group pesticides are prepared by diluting 40 μL internal standard solution and an appropriate amount of mixed standard solution to 1.0 mL with blank extract that has been taken through the method with the rest of the samples. Mix thoroughly. Working standard mixed solutions in matrix must be freshly prepared.

Solid Phase Extraction Cartridge: Sep-Pak Carbon NH$_2$, 6 mL, 1 g, or Equivalent.

TABLE 2.16 The LOD, LOQ of 503 Pesticides and Related Chemical Residues Determined by GC-MS

No.	Pesticides	LOD (mg/kg)	LOQ (mg/kg)	No.	Pesticides	LOD (mg/kg)	LOQ (mg/kg)
	Group A			33	delta-HCH	0.012 5	0.025 0
1	allidochlor	0.012 5	0.025 0	34	anthraquinone	0.006 3	0.012 6
2	etridiazol	0.018 8	0.037 6	35	fenthion	0.006 3	0.012 6
3	chlormephos	0.012 5	0.025 0	36	malathion	0.025 0	0.050 0
4	propham	0.006 3	0.012 6	37	paraoxon-ethyl	0.200 0	0.400 0
5	cycloate	0.006 3	0.012 6	38	fenitrothion	0.012 5	0.025 0
6	diphenylamine	0.006 3	0.012 6	39	triadimefon	0.012 5	0.025 0
7	chlordimeform	0.006 3	0.012 6	40	linuron	0.025 0	0.050 0
8	ethalfluralin	0.025 0	0.050 0	41	pendimethalin	0.025 0	0.050 0
9	phorate	0.006 3	0.012 6	42	chlorbenside	0.012 5	0.025 0
10	thiometon	0.006 3	0.012 6	43	bromophos-ethyl	0.006 3	0.012 6
11	quintozene	0.012 5	0.025 0	44	quinalphos	0.006 3	0.012 6
12	atrazine-desethyl	0.006 3	0.012 6	45	trans-chlordane	0.006 3	0.012 6
13	clomazone	0.006 3	0.012 6	46	phenthoate	0.012 5	0.025 0
14	diazinon	0.006 3	0.012 6	47	metazachlor	0.018 8	0.037 6
15	fonofos	0.006 3	0.012 6	48	prothiophos	0.006 3	0.012 6
16	etrimfos	0.006 3	0.012 6	49	chlorfurenol	0.018 8	0.037 6
17	propetamphos	0.006 3	0.012 6	50	procymidone	0.006 3	0.012 6
18	secbumeton	0.006 3	0.012 6	51	dieldrin	0.012 5	0.025 0
19	pronamide	0.006 3	0.012 6	52	methidathion	0.012 5	0.025 0
20	dichlofenthion	0.006 3	0.012 6	53	napropamide	0.018 8	0.037 6
21	mexacarbate	0.018 8	0.037 6	54	cyanazine	0.018 8	0.037 6
22	dimethoate	0.025 0	0.050 0	55	oxadiazone	0.006 3	0.012 6
23	dinitramine	0.025 0	0.050 0	56	fenamiphos	0.018 8	0.037 6
24	aldrin	0.012 5	0.025 0	57	tetrasul	0.006 3	0.012 6
25	ronnel	0.012 5	0.025 0	58	bupirimate	0.006 3	0.012 6
26	prometryne	0.006 3	0.012 6	59	flutolanil	0.006 3	0.012 6
27	cyprazine	0.006 3	0.012 6	60	carboxin	0.150 0	0.300 0
28	vinclozolin	0.006 3	0.012 6	61	p,p'-DDD	0.006 3	0.012 6
29	beta-HCH	0.006 3	0.012 6	62	ethion	0.012 5	0.025 0
30	metalaxyl	0.018 8	0.037 6	63	etaconazole-1	0.018 8	0.037 6
31	methyl-parathion	0.025 0	0.050 0	64	sulprofos	0.012 5	0.025 0
32	chlorpyrifos (-ethyl)	0.006 3	0.012 6	65	etaconazole-2	0.018 8	0.037 6

TABLE 2.16 The LOD, LOQ of 503 Pesticides and Related Chemical Residues Determined by GC-MS—cont'd

No.	Pesticides	LOD (mg/kg)	LOQ (mg/kg)	No.	Pesticides	LOD (mg/kg)	LOQ (mg/kg)
66	myclobutanil	0.006 3	0.012 6	97	tecnazene	0.012 5	0.025 0
67	fensulfothion	0.012 5	0.025 0	98	heptanophos	0.018 8	0.037 6
68	diclofop-methyl	0.006 3	0.012 6	99	ethoprophos	0.018 8	0.037 6
69	propiconazole-1	0.018 8	0.037 6	100	hexachlorobenzene	0.006 3	0.012 6
70	propiconazole-2	0.018 8	0.037 6	101	propachlor	0.018 8	0.037 6
71	bifenthrin	0.006 3	0.012 6	102	cis-diallate	0.012 5	0.025 0
72	mirex	0.006 3	0.012 6	103	trifluralin	0.012 5	0.025 0
73	nuarimol	0.012 5	0.025 0	104	trans-diallate	0.012 5	0.025 0
74	benodanil	0.018 8	0.037 6	105	chlorpropham	0.012 5	0.025 0
75	methoxychlor	0.050 0	0.100 0	106	sulfotep	0.006 3	0.012 6
76	oxadixyl	0.006 3	0.012 6	107	sulfallate	0.012 5	0.025 0
77	tebuconazole	0.018 8	0.037 6	108	alpha-HCH	0.006 3	0.012 6
78	tetramethirn	0.012 5	0.025 0	109	terbufos	0.012 5	0.025 0
79	norflurazon	0.006 3	0.012 6	110	terbumeton	0.018 8	0.037 6
80	pyridaphenthion	0.006 3	0.012 6	111	profluralin	0.025 0	0.050 0
81	phosmet	0.012 5	0.025 0	112	dioxathion	0.025 0	0.050 0
82	tetradifon	0.006 3	0.012 6	113	propazine	0.006 3	0.012 6
83	cis-permethrin	0.006 3	0.012 6	114	chlorbufam	0.012 5	0.025 0
84	pyrazophos	0.012 5	0.025 0	115	dicloran	0.012 5	0.025 0
85	trans-permethrin	0.006 3	0.012 6	116	terbuthylazine	0.006 3	0.012 6
86	cypermethrin	0.018 8	0.037 6	117	monolinuron	0.025 0	0.050 0
87	fenvalerate-1	0.025 0	0.050 0	118	flufenoxuron	0.018 8	0.037 6
88	fenvalerate-2	0.025 0	0.050 0	119	chlorpyrifos-methyl	0.006 3	0.012 6
89	deltamethrin	0.037 5	0.075 0	120	desmetryn	0.006 3	0.012 6
	Group B			121	dimethachlor	0.018 8	0.037 6
90	EPTC	0.018 8	0.037 6	122	alachlor	0.018 8	0.037 6
91	butylate	0.018 8	0.037 6	123	pirimiphos-methyl	0.006 3	0.012 6
92	dichlobenil	0.001 3	0.002 6	124	terbutryn	0.012 5	0.025 0
93	pebulate	0.018 8	0.037 6	125	aspon	0.012 5	0.025 0
94	nitrapyrin	0.018 8	0.037 6	126	thiobencarb	0.012 5	0.025 0
95	mevinphos	0.012 5	0.025 0	127	dicofol	0.012 5	0.025 0
96	chloroneb	0.006 3	0.012 6	128	metolachlor	0.006 3	0.012 6

Continued

TABLE 2.16 The LOD, LOQ of 503 Pesticides and Related Chemical Residues Determined by GC-MS—cont'd

No.	Pesticides	LOD (mg/kg)	LOQ (mg/kg)	No.	Pesticides	LOD (mg/kg)	LOQ (mg/kg)
129	pirimiphos-ethyl	0.012 5	0.025 0	162	flamprop-isopropyl	0.006 3	0.012 6
130	dichlofluanid	0.300 0	0.600 0	163	endosulfan -2	0.037 5	0.075 0
131	methoprene	0.025 0	0.050 0	164	carbofenothion	0.012 5	0.025 0
132	bromofos	0.012 5	0.025 0	165	p,p'-DDT	0.012 5	0.025 0
133	ethofumesate	0.012 5	0.025 0	166	benalaxyl	0.006 3	0.012 6
134	isopropalin	0.012 5	0.025 0	167	edifenphos	0.012 5	0.025 0
135	propanil	0.012 5	0.025 0	168	triazophos	0.018 8	0.037 6
136	crufomate	0.037 5	0.075 0	169	cyanofenphos	0.006 3	0.012 6
137	isofenphos	0.012 5	0.025 0	170	chlorbenside sulfone	0.012 5	0.025 0
138	endosulfan -1	0.037 5	0.075 0	171	endosulfan-sulfate	0.018 8	0.037 6
139	chlorfenvinphos	0.018 8	0.037 6	172	bromopropylate	0.012 5	0.025 0
140	tolylfluanide	0.150 0	0.300 0	173	benzoylprop-ethyl	0.018 8	0.037 6
141	cis-chlordane	0.012 5	0.025 0	174	fenpropathrin	0.012 5	0.025 0
142	butachlor	0.012 5	0.025 0	175	EPN	0.025 0	0.050 0
143	chlozolinate	0.012 5	0.025 0	176	captafol	0.112 5	0.225 0
144	p,p'-DDE	0.006 3	0.012 6	177	hexazinone	0.018 8	0.037 6
145	iodofenphos	0.012 5	0.025 0	178	leptophos	0.012 5	0.025 0
146	tetrachlorvinphos	0.018 8	0.037 6	179	bifenox	0.012 5	0.025 0
147	chlorbromuron	0.150 0	0.300 0	180	phosalone	0.012 5	0.025 0
148	profenofos	0.037 5	0.075 0	181	azinphos-methyl	0.037 5	0.075 0
149	buprofezin	0.012 5	0.025 0	182	fenarimol	0.012 5	0.025 0
150	o,p'-DDD	0.006 3	0.012 6	183	azinphos-ethyl	0.012 5	0.025 0
151	chlorfenson	0.012 5	0.025 0	184	cyfluthrin	0.075 0	0.150 0
152	fluorochloridone	0.012 5	0.025 0	185	prochloraz	0.037 5	0.075 0
153	endrin	0.075 0	0.150 0	186	coumaphos	0.037 5	0.075 0
154	paclobutrazol	0.018 8	0.037 6	187	tau-fluvalinate	0.075 0	0.150 0
155	o,p'-DDT	0.012 5	0.025 0		Group C		
156	methoprotryne	0.018 8	0.037 6	188	dichlorvos	0.037 5	0.075 0
157	chloropropylate	0.006 3	0.012 6	189	biphenyl	0.006 3	0.012 6
158	flamprop-methyl	0.006 3	0.012 6	190	propamocarb	0.018 8	0.037 6
159	nitrofen	0.037 5	0.075 0	191	vernolate	0.006 3	0.012 6
160	oxyfluorfen	0.025 0	0.050 0	192	3,5-dichloroaniline	0.006 3	0.012 6
161	chlorthiophos	0.018 8	0.037 6	193	methacrifos	0.006 3	0.012 6

TABLE 2.16 The LOD, LOQ of 503 Pesticides and Related Chemical Residues Determined by GC-MS—cont'd

No.	Pesticides	LOD (mg/kg)	LOQ (mg/kg)	No.	Pesticides	LOD (mg/kg)	LOQ (mg/kg)
194	molinate	0.006 3	0.012 6	226	diphenamid	0.006 3	0.012 6
195	2-phenylphenol	0.006 3	0.012 6	227	penconazole	0.018 8	0.037 6
196	cis-1,2,3,6-tetrahydrophthalimide	0.018 8	0.037 6	228	tetraconazole	0.018 8	0.037 6
				229	mecarbam	0.025 0	0.050 0
197	fenobucarb	0.012 5	0.025 0	230	propaphos	0.012 5	0.025 0
198	benfluralin	0.006 3	0.012 6	231	flumetralin	0.012 5	0.025 0
199	hexaflumuron	0.037 5	0.075 0	232	triadimenol-1	0.018 8	0.037 6
200	prometon	0.018 8	0.037 6	233	triadimenol-2	0.018 8	0.037 6
201	triallate	0.012 5	0.025 0	234	pretilachlor	0.012 5	0.025 0
202	pyrimethanil	0.006 3	0.012 6	235	kresoxim-methyl	0.006 3	0.012 6
203	gamma-HCH	0.012 5	0.025 0	236	fluazifop-butyl	0.006 3	0.012 6
204	disulfoton	0.006 3	0.012 6	237	chlorfluazuron	0.018 8	0.037 6
205	atrizine	0.006 3	0.012 6	238	chlorobenzilate	0.006 3	0.012 6
206	iprobenfos	0.018 8	0.037 6	239	flusilazole	0.018 8	0.037 6
207	isazofos	0.012 5	0.025 0	240	fluorodifen	0.006 3	0.012 6
208	plifenate	0.012 5	0.025 0	241	diniconazole	0.018 8	0.037 6
209	fluchloralin	0.025 0	0.050 0	242	piperonyl butoxide	0.006 3	0.012 6
210	transfluthrin	0.006 3	0.012 6	243	dimefuron	0.025 0	0.050 0
211	fenpropimorph	0.006 3	0.012 6	244	propargite	0.012 5	0.025 0
212	tolclofos-methyl	0.006 3	0.012 6	245	mepronil	0.006 3	0.012 6
213	propisochlor	0.006 3	0.012 6	246	diflufenican	0.006 3	0.012 6
214	metobromuron	0.037 5	0.075 0	247	fludioxonil	0.006 3	0.012 6
215	ametryn	0.018 8	0.037 6	248	fenazaquin	0.006 3	0.012 6
216	simetryn	0.012 5	0.025 0	249	phenothrin	0.006 3	0.012 6
217	metribuzin	0.018 8	0.037 6	250	fenoxycarb	0.037 5	0.075 0
218	dimethipin	0.018 8	0.037 6	251	sethoxydim	0.450 0	0.900 0
219	dipropetryn	0.006 3	0.012 6	252	amitraz	0.018 8	0.037 6
220	diethofencarb	0.037 5	0.075 0	253	anilofos	0.012 5	0.025 0
221	dimepiperate	0.012 5	0.025 0	254	acrinathrin	0.012 5	0.025 0
222	bioallethrin-1	0.025 0	0.050 0				
223	bioallethrin-2	0.025 0	0.050 0	255	lambda-cyhalothrin	0.006 3	0.012 6
224	fenson	0.006 3	0.012 6	256	mefenacet	0.018 8	0.037 6
225	o,p'-DDE	0.006 3	0.012 6	257	permethrin	0.012 5	0.025 0

Continued

TABLE 2.16 The LOD, LOQ of 503 Pesticides and Related Chemical Residues Determined by GC-MS—cont'd

No.	Pesticides	LOD (mg/kg)	LOQ (mg/kg)	No.	Pesticides	LOD (mg/kg)	LOQ (mg/kg)
258	pyridaben	0.006 3	0.012 6	288	tefluthrin	0.006 3	0.012 6
259	fluoroglycofen-ethyl	0.075 0	0.150 0	289	bromocylen	0.006 3	0.012 6
260	bitertanol	0.018 8	0.037 6	290	trietazine	0.006 3	0.012 6
261	etofenprox	0.006 3	0.012 6	291	2,6-dichlorobenzamide	0.012 5	0.025 0
262	cycloxydim	0.600 0	1.200 0	292	cycluron	0.018 8	0.037 6
263	alpha-cypermethrin	0.012 5	0.025 0	293	DE-PCB 28	0.006 3	0.012 6
				294	DE-PCB 31	0.006 3	0.012 6
264	flucythrinate-1	0.012 5	0.025 0	295	desethyl-sebuthylazine	0.012 5	0.025 0
265	flucythrinate-2	0.012 5	0.025 0	296	2,3,4,5-tetrachloroaniline	0.012 5	0.025 0
266	esfenvalerate	0.025 0	0.050 0				
267	difenconazole-2	0.037 5	0.075 0	297	musk ambrette	0.006 3	0.012 6
268	difenonazole-1	0.037 5	0.075 0	298	pentachloroaniline	0.006 3	0.012 6
269	flumioxazin	0.012 5	0.025 0	299	aziprotryne	0.050 0	0.100 0
270	flumiclorac-pentyl	0.012 5	0.025 0	300	isocarbamid	0.031 3	0.062 6
	Group D			301	sebutylazine	0.006 3	0.012 6
271	dimefox	0.018 8	0.037 6	302	musk moskene	0.006 3	0.012 6
272	disulfoton-sulfoxide	0.100 0	0.200 0	303	DE-PCB 52	0.006 3	0.012 6
273	pentachlorobenzene	0.006 3	0.012 6	304	prosulfocarb	0.006 3	0.012 6
274	tri-iso-butyl phosphate	0.006 3	0.012 6	305	dimethenamid	0.006 3	0.012 6
275	crimidine	0.006 3	0.012 6	306	BDMC-2	0.012 5	0.025 0
276	BDMC-1	0.012 5	0.025 0	307	monalide	0.012 5	0.025 0
277	chlorfenprop-methyl	0.006 3	0.012 6	308	musk tibeten	0.006 3	0.012 6
278	thionazin	0.006 3	0.012 6	309	octachlorostyrene	0.006 3	0.012 6
279	2,3,5,6-tetrachloroaniline	0.006 3	0.012 6	310	isodrin	0.006 3	0.012 6
				311	isomethiozin	0.012 5	0.025 0
280	tri-n-butyl phosphate	0.012 5	0.025 0	312	trichloronat	0.006 3	0.012 6
281	2,3,4,5-tetrachloroanisole	0.006 3	0.012 6	313	dacthal	0.006 3	0.012 6
282	pentachloroanisole	0.006 3	0.012 6	314	4,4-dichlorobenzophenone	0.006 3	0.012 6
283	tebutam	0.012 5	0.025 0	315	nitrothal-isopropyl	0.012 5	0.025 0
284	dioxabenzofos	0.062 5	0.125 0	316	rabenzazole	0.006 3	0.012 6
285	methabenzthiazuron	0.062 5	0.125 0	317	cyprodinil	0.006 3	0.012 6
286	simetone	0.012 5	0.025 0	318	fuberidazole	0.031 3	0.062 6
287	atratone	0.006 3	0.012 6	319	dicapthon	0.031 3	0.062 6

TABLE 2.16 The LOD, LOQ of 503 Pesticides and Related Chemical Residues Determined by GC-MS—cont'd

No.	Pesticides	LOD (mg/kg)	LOQ (mg/kg)	No.	Pesticides	LOD (mg/kg)	LOQ (mg/kg)
320	MCPA-butoxyethyl ester	0.006 3	0.012 6	350	metamitron	0.062 5	0.125 0
321	DE-PCB 101	0.006 3	0.012 6	351	triphenyl phosphate	0.006 3	0.012 6
322	methfuroxam	0.006 3	0.012 6	352	DE-PCB 180	0.006 3	0.012 6
323	isocarbophos	0.012 5	0.025 0	353	tebufenpyrad	0.006 3	0.012 6
324	phorate sulfone	0.006 3	0.012 6	354	cloquintocet-mexyl	0.006 3	0.012 6
325	chlorfenethol	0.006 3	0.012 6	355	lenacil	0.062 5	0.125 0
326	trans-nonachlor	0.006 3	0.012 6	356	bromuconazole-1	0.012 5	0.025 0
327	DEF	0.012 5	0.025 0	357	bromuconazole-2	0.012 5	0.025 0
328	flurochloridone	0.012 5	0.025 0	358	nitralin	0.062 5	0.125 0
329	bromfenvinfos	0.006 3	0.012 6	359	fenamiphos sulfoxide	0.200 0	0.400 0
330	perthane	0.006 3	0.012 6	360	fenamiphos sulfone	0.025 0	0.050 0
331	ditalimfos	0.006 3	0.012 6	361	fenpiclonil	0.025 0	0.050 0
332	DE-PCB 118	0.006 3	0.012 6	362	fluquinconazole	0.006 3	0.012 6
333	mephosfolan	0.012 5	0.025 0	363	fenbuconazole	0.012 5	0.025 0
334	4,4-dibromobenzophenone	0.006 3	0.012 6		Group E		
				364	propoxur-1	0.012 5	0.025 0
335	flutriafol	0.012 5	0.025 0	365	XMC	0.012 5	0.025 0
336	athidathion	0.100 0	0.200 0	366	isoprocarb -1	0.012 5	0.025 0
337	DE-PCB 153	0.006 3	0.012 6	367	terbucarb-1	0.012 5	0.025 0
338	diclobutrazole	0.025 0	0.050 0	368	dibutyl succinate	0.012 5	0.025 0
339	disulfoton sulfone	0.012 5	0.025 0	369	chlorethoxyfos	0.100 0	0.200 0
340	DE-PCB 138	0.006 3	0.012 6	370	isoprocarb -2	0.012 5	0.025 0
341	cyproconazole	0.006 3	0.012 6	371	tebuthiuron	0.025 0	0.050 0
342	resmethrin-1	0.100 0	0.200 0	372	pencycuron	0.025 0	0.050 0
343	resmethrin-2	0.100 0	0.200 0	373	demeton-s-methyl	0.025 0	0.050 0
344	phthalic acid, benzyl butyl ester	0.006 3	0.012 6	374	propoxur-2	0.012 5	0.025 0
				375	naled	0.100 0	0.200 0
345	clodinafop-propargyl	0.012 5	0.025 0	376	phenanthrene	0.006 3	0.012 6
346	fenthion sulfoxide	0.025 0	0.050 0	377	fenpyroximate	0.050 0	0.100 0
347	fluotrimazole	0.006 3	0.012 6	378	tebupirimfos	0.012 5	0.025 0
348	fluroxypr-1-methylheptyl ester	0.006 3	0.012 6	379	fenpropidin	0.012 5	0.025 0
				380	dichloran	0.012 5	0.025 0
349	fenthion sulfone	0.025 0	0.050 0				

Continued

TABLE 2.16 The LOD, LOQ of 503 Pesticides and Related Chemical Residues Determined by GC-MS—cont'd

No.	Pesticides	LOD (mg/kg)	LOQ (mg/kg)	No.	Pesticides	LOD (mg/kg)	LOQ (mg/kg)
381	pyroquilon	0.050 4	0.100 8	414	TCMTB	0.800 0	1.600 0
382	propyzamide	0.100 0	0.200 0	415	metominostrobin	0.025 0	0.050 0
383	pirimicarb	0.012 5	0.025 0	416	imazalil	0.025 0	0.050 0
384	benoxacor	0.012 5	0.025 0	417	isoprothiolane	0.012 5	0.025 0
385	phosphamidon -1	0.400 0	0.800 0	418	cyflufenamid	0.100 0	0.200 0
386	acetochlor	0.012 5	0.025 0	419	quinoxyphen	0.006 3	0.012 6
387	tridiphane	0.100 0	0.200 0	420	trifloxystrobin	0.025 0	0.050 0
388	esprocarb	0.012 5	0.025 0	421	imibenconazole-des-benzyl	0.025 0	0.050 0
389	terbucarb-2	0.012 5	0.025 0	422	imiprothrin-1	0.012 5	0.025 0
390	fenfuram	0.012 5	0.025 0	423	fipronil	0.050 0	0.100 0
391	acibenzolar-s-methyl	0.012 5	0.025 0	424	imiprothrin-2	0.012 5	0.025 0
392	benfuresate	0.100 0	0.200 0	425	epoxiconazole -1	0.050 0	0.100 0
393	mefenoxam	0.012 5	0.025 0	426	pyributicarb	0.012 5	0.025 0
394	malaoxon	0.100 0	0.200 0	427	pyraflufen ethyl	0.012 5	0.025 0
395	phosphamidon -2	0.400 0	0.800 0	428	thenylchlor	0.012 5	0.025 0
396	chlorthal-dimethyl	0.012 5	0.025 0	429	clethodim	0.025 0	0.050 0
397	simeconazole	0.012 5	0.025 0	430	mefenpyr-diethyl	0.018 8	0.037 6
398	terbacil	0.012 5	0.025 0	431	etoxazole	0.037 5	0.075 0
399	thiazopyr	0.012 5	0.025 0	432	epoxiconazole-2	0.050 0	0.100 0
400	dimethylvinphos	0.012 5	0.025 0	433	famphur	0.025 0	0.050 0
401	zoxamide	0.012 5	0.025 0	434	pyriproxyfen	0.012 5	0.025 0
402	allethrin	0.025 0	0.050 0	435	iprodione	0.025 0	0.050 0
403	quinoclamine	0.025 0	0.050 0	436	ofurace	0.018 8	0.037 6
404	flufenacet	0.050 0	0.100 0	437	piperophos	0.018 8	0.037 6
405	fenoxanil	0.012 5	0.025 0	438	fenamidone	0.006 3	0.012 6
406	furalaxyl	0.012 5	0.025 0	439	pyraclostrobin	1.200 0	2.400 0
407	mepanipyrim	0.050 0	0.100 0	440	tralkoxydim	0.050 0	0.100 0
408	thiamethoxam	0.200 0	0.400 0	441	pyraclofos	0.400 0	0.800 0
409	captan	0.800 0	1.600 0	442	dialifos	1.600 0	3.200 0
410	bromacil	0.012 5	0.025 0	443	spirodiclofen	0.050 0	0.100 0
411	picoxystrobin	0.012 5	0.025 0	444	flurtamone	0.012 5	0.025 0
412	butamifos	0.006 3	0.012 6	445	pyriftalid	0.006 3	0.012 6
413	imazamethabenz-methyl	0.018 8	0.037 6	446	silafluofen	0.006 3	0.012 6

TABLE 2.16 The LOD, LOQ of 503 Pesticides and Related Chemical Residues Determined by GC-MS—cont'd

No.	Pesticides	LOD (mg/kg)	LOQ (mg/kg)	No.	Pesticides	LOD (mg/kg)	LOQ (mg/kg)
447	pyrimidifen	0.012 5	0.025 0	474	desmedipham	0.125 0	0.250 0
448	butafenacil	0.006 3	0.012 6	475	fenchlorphos	0.025 0	0.050 0
449	fluridone	0.012 5	0.025 0	476	butralin	0.025 0	0.050 0
	Group F			477	pyrifenox -2	0.050 0	0.100 0
450	tribenuron-methyl	0.006 3	0.012 6	478	dimethametryn	0.006 3	0.012 6
451	ethiofencarb	0.062 5	0.125 0	479	pyrifenox-1	0.050 0	0.100 0
452	dioxacarb	0.050 0	0.100 0	480	thiabendazole	0.125 0	0.250 0
453	dimethyl phthalate	0.025 0	0.050 0	481	iprovalicarb-1	0.025 0	0.050 0
454	4-chlorophenoxy acetic acid	0.003 1	0.006 2	482	azaconazole	0.025 0	0.050 0
				483	iprovalicarb-2	0.025 0	0.050 0
455	triflumizole	0.200 0	0.400 0	484	diofenolan -1	0.012 5	0.025 0
456	phthalimide	0.012 5	0.025 0	485	diofenolan -2	0.012 5	0.025 0
457	acephate	0.125 0	0.250 0	486	aclonifen	0.125 0	0.250 0
458	diethyltoluamide	0.005 0	0.010 0	487	chlorfenapyr	0.050 0	0.100 0
459	2,4-D	0.125 0	0.250 0	488	bioresmethrin	0.012 5	0.025 0
460	carbaryl	0.018 8	0.037 6	489	carfentrazone-ethyl	0.012 5	0.025 0
461	cadusafos	0.025 0	0.050 0	490	tricyclazole	0.037 5	0.075 0
462	demetom-s	0.200 0	0.400 0	491	spiromesifen	0.062 5	0.125 0
463	spiroxamine -1	0.012 5	0.025 0	492	phenkapton	0.300 0	0.600 0
464	dicrotophos	0.050 0	0.100 0	493	bifenazate	0.050 0	0.100 0
465	2,4,5-T	0.125 0	0.250 0	494	endrin ketone	0.100 0	0.200 0
466	3-phenylphenol	0.037 5	0.075 0	495	norflurazon-desmethyl	0.025 0	0.050 0
467	furmecyclox	0.018 8	0.037 6				
468	spiroxamine -2	0.012 5	0.025 0	496	metoconazole	0.025 0	0.050 0
469	DMSA	0.050 0	0.100 0	497	cyhalofop-butyl	0.012 5	0.025 0
470	sobutylazine	0.100 0	0.200 0	498	halfenprox	0.012 5	0.025 0
				499	acetamiprid	0.025 0	0.050 0
471	s421 (octachlorodipropyl ether)-1	0.125 0	0.250 0	500	boscalid	0.025 0	0.050 0
				501	tralomethrin-1	0.050 0	0.100 0
472	s421 (octachlorodipropyl ether)-2	0.125 0	0.250 0	502	tralomethrin-2	0.050 0	0.100 0
				503	dimethomorph	0.012 5	0.025 0
473	dodemorph	0.018 8	0.037 6				

2.10.4 Apparatus

GC-MS: Equipped With EI
Analytical Balances: Capable of Weighing From 0.1 mg to 0.01 g
Centrifuge Tube: 80 mL
Homogenizer: Not Less Than 12,000 rpm
Centrifuge: Not Less Than 4200 rpm
Pear-Shaped Flask: 200 mL
Pipette: 1 mL and 10 mL
Rotary Evaporator
Vials: 2 mL, With PTFE Screw Cap

2.10.5 Sample Pretreatment

2.10.5.1 Preparation of Test Sample

The samples are cut up, mixed thoroughly, sealed, and labeled.
The test samples are stored in the dark at 4°C.

2.10.5.2 Extraction

Weigh 20 g of test sample (accurate to 0.01 g) into 80 mL centrifuge tube. Add 40 mL acetonitrile, and homogenize at 15,000 rpm for 1 min. Add 5 g sodium chloride to the centrifuge tube and homogenize at 15,000 rpm for 1 min again, then centrifuge at 4200 rpm for 5 min. Pipette 20 mL of the top acetonitrile layer of extracts (corresponding to 10 g test sample). Evaporate the eluate to 1 mL using rotary evaporator at 40°C for clean-up.

2.10.5.3 Clean-up

Add sodium sulfate into the cartridge to a height of 2 cm. Connect the cartridge to the top of the Sep-Pak Carbon NH_2 cartridge in series. Fix the cartridges into a support to which a pear-shaped flask is connected. Condition the cartridges with 4 mL acetonitrile + toluene (3+1, V/V) before adding the sample. Once the solution gets to the top of the sodium sulfate, pipette the eluate into the cartridges immediately. Rinse the pear-shaped flask with 3×2 mL acetonitrile + toluene (3+1, V/V), and decant it into the cartridges. Insert a reservoir into the cartridges. Elute the pesticides with 25 mL acetonitrile + toluene (3+1, V/V). Evaporate the eluate to ca. 0.5 mL using a rotary evaporator at 40°C. Exchange with 2×5 mL hexane twice and make up to ca. 1 mL. Add 40 μL internal standard solution and mix thoroughly. The solution is ready for GC-MS determination.

2.10.6   Determination

2.10.6.1   GC-MS Operating Condition

(a) Column: DB-1701 (14% cyanopropyl-phenyl) (30 m × 0.25 mm × 0.25 µm) capillary column, or equivalent

(b) Column temperature: 40°C hold 1 min, at 30°C/min to 130°C, at 5°C/min to 250°C, at 10°C/min to 300°C, hold 5 min

(c) Carrier gas: Helium, purity ≥99.999%, flow rate: 1.2 mL/min

(d) Injection port temperature: 290°C

(e) Injection volume: 1 µL

(f) Injection mode: Splitless, purge on after 1.5 min

(g) Ionization voltage: 70 eV

(h) Ion source temperature: 230°C

(i) GC-MS interface temperature: 80°C

(j) Solvent delay time: group A: 8.30 min, group B: 7.80 min, group C: 7.30 min, group D: 5.50 min, group E: 6.10 min, group F: 5.50 min

(k) Selected ion monitoring mode: Each compound selects 1 quantifying ion and 2–3 qualifying ions. All of the detected ions of each group are detected according to programmed time and sequence of peaking. The retention times, quantifying ions, qualifying ions, and the abundance ratios of quantifying ion and qualifying ions for each compound are listed in Table 5.3. The programmed time and dwell time for the ions detected for each compound in each group are listed in Table 5.4

2.10.6.2   Qualitative Determination

In the samples determined, injections are required to analyze for all pesticides according to GC-MS operating conditions. If the retention times in the peaks of the sample solution are the same as those for the peaks of the working standard mixed solution, and the selected ions of the background-subtracted mass spectrum appear, and the abundance ratios of selected ions are within the expected limits (abundance ratios >50%, permitted tolerances are ±10%; abundance ratios >20% to 50%, permitted tolerances are ±15%; abundance ratios >10% to 20%, permitted tolerances are ±20%; abundance ratios ≤10%, permitted tolerances are ±50%), the sample is confirmed to contain this pesticide compound. In the cases where results are still not definitive, the sample should be reinjected with acquisition in scan mode (sufficient sensitivity) or with additional confirmatory ions or using other instruments that have higher sensitivity.

2.10.6.3   Quantitative Determination

The results are quantitated using heptachlor epoxide for an internal standard and the quantitation ion response for each analyte. To compensate for the matrix effect, quantitation is based on a mixed standard prepared in blank matrix extract. The concentration of standard solution and the detected sample solution must be similar.

2.10.6.4 Parallel Test

A parallel test is carried out for the same testing sample.

2.10.6.5 Blank Test

The operation of the blank test is the same as that described in the method of determination, but without the addition of sample.

2.10.7 Precision

The precision data of the method for this standard have been determined according to the stipulations of GB/T 6379.1 and GB/T 6379.2. The values of repeatability and reproducibility are obtained and calculated at the 95% confidence level.

RESEARCHERS

Guo-Fang Pang, Jun-Hong Zheng, Feng Zheng, Ming-Lin Wang, Cui-Cui Yao.
Qinhuangdao Entry-Exit Inspection and Quarantine Bureau, 39 Haibin Rd, Qinhuangdao, Hebei, PC 066002, People's Republic of China.

2.11 DETERMINATION OF 440 PESTICIDES AND RELATED CHEMICAL RESIDUES IN EDIBLE FUNGUS: LC-MS-MS METHOD (GB/T 23202-2008)

2.11.1 Scope

The method is applicable to the qualitative determination of 440 pesticides and related chemical residues and quantitative determination of 364 pesticides and related chemical residues in nameko, *Flammulina*, wood ear, and shiitake mushroom.

The limits of quantitative determination of this method of 364 pesticides and related chemicals is 0.06 µg/kg to 0.61 mg/kg (see Table 2.17).

2.11.2 Principle

The samples are homogenized with acetonitrile. The solutions are salted out and centrifuged, and the supernatants of the acetonitrile phase are cleaned up with SPE cartridges. The pesticides and related chemicals are eluted with acetonitrile + toluene (3 + 1), and the solutions are analyzed by LC-MS-MS, using an external standard method for quantification.

TABLE 2.17 The LOD, LOQ of 440 Pesticides and Related Chemical Residues Determined by LC-MS-MS

No.	Pesticides	LOD (µg/kg)	LOQ (µg/kg)	No.	Pesticides	LOD (µg/kg)	LOQ (µg/kg)
	Group A			33	myclobutanil	0.25	0.50
1	propham	27.50	55.00	34	paclobutrazol	0.15	0.30
2	isoprocarb	0.58	1.16	35	fenthion sulfoxide	0.08	0.16
3	3,4,5-trimethacarb	0.08	0.16	36	triadimenol	2.65	5.30
4	cycluron	0.05	0.10	37	butralin	0.48	0.96
5	carbaryl	2.58	5.16	38	spiroxamine	0.03	0.06
6	propachlor	0.08	0.16	39	tolclofos methyl	16.65	33.30
7	rabenzazole	0.33	0.66	40	desmedipham	1.00	2.00
8	simetryn	0.03	0.06	41	methidathion	2.68	5.36
9	monolinuron	0.90	1.80	42	allethrin	15.10	30.20
10	mevinphos	0.40	0.80	43	diazinon	0.18	0.36
11	aziprotryne	0.35	0.70	44	edifenphos	0.20	0.40
12	secbumeton	0.03	0.06	45	pretilachlor	0.08	0.16
13	cyprodinil	0.18	0.36	46	flusilazole	0.15	0.30
14	buturon	2.25	4.50	47	iprovalicarb	0.58	1.16
15	carbetamide	0.90	1.80	48	benodanil	0.88	1.76
16	pirmicarb	0.04	0.08	49	flutolanil	0.28	0.56
17	clomazone	0.10	0.20	50	famphur	0.90	1.80
18	cyanazine	0.05	0.10	51	benalyxyl	0.30	0.60
19	prometryne	0.05	0.10	52	diclobutrazole	0.13	0.26
20	paraoxon methyl	0.20	0.40	53	etaconazole	0.45	0.90
21	4,4-dichlorobenzophenone	3.40	6.80	54	fenarimol	0.15	0.30
22	thiacloprid	0.10	0.20	55	phthalic acid, dicyclobexyl ester	0.50	1.00
23	imidacloprid	5.50	11.00				
24	ethidimuron	0.38	0.76	56	tetramethirn	0.45	0.90
25	isomethiozin	0.28	0.56	57	dichlofluanid	0.65	1.30
26	cis and trans diallate	22.30	44.60	58	cloquintocet mexyl	0.48	0.96
27	acetochlor	11.85	23.70	59	bitertanol	8.35	16.70
28	nitenpyram	4.28	8.56	60	azinphos ethyl	27.23	54.46
29	methoprotryne	0.05	0.10	61	clodinafop propargyl	0.60	1.20
30	dimethenamid	1.08	2.16	62	triflumuron	0.98	1.96
31	terrbucarb	0.53	1.06	63	isoxaflutole	0.98	1.96
32	penconazole	0.50	1.00	64	anilofos	0.18	0.36

Continued

TABLE 2.17 The LOD, LOQ of 440 Pesticides and Related Chemical Residues Determined by LC-MS-MS—cont'd

No.	Pesticides	LOD (µg/kg)	LOQ (µg/kg)	No.	Pesticides	LOD (µg/kg)	LOQ (µg/kg)
65	thiophanat ethyl	5.05	10.10	97	difenzoquat-methyl sulfate	0.20	0.40
66	quizalofop-ethyl	0.18	0.36	98	clothianidin	15.75	31.50
67	haloxyfop-methyl	0.65	1.30	99	pronamide	3.85	7.70
68	fluazifop butyl	0.08	0.16	100	dimethachloro	0.48	0.96
69	bromophos-ethyl	141.93	283.86	101	methobromuron	4.20	8.40
70	bensulide	8.55	17.10	102	phorate	78.50	157.00
71	bromfenvinfos	0.75	1.50	103	aclonifen	6.05	12.10
72	azoxystrobin	0.13	0.26	104	mephosfolan	0.58	1.16
73	pyrazophos	0.40	0.80	105	imibenzonazole-des-benzyl	1.55	3.10
74	flufenoxuron	0.80	1.60	106	neburon	1.78	3.56
75	indoxacarb	1.88	3.76	107	mefenoxam	0.38	0.76
76	emamectin benzoate	0.08	0.16	108	ethofume sate	93.00	186.00
	Group B			109	iprobenfos	2.08	4.16
77	dazomet	31.75	63.50	110	cyproconazole	0.18	0.36
78	nicotine	0.55	1.10	111	thiamethoxam	8.25	16.50
79	fenuron	0.25	0.50	112	etrimfos	4.70	9.40
80	cyromazine	1.80	3.60	113	cythioate	20.00	40.00
81	crimidine	0.40	0.80	114	phosphamidon	0.98	1.96
82	acephate	3.33	6.66	115	phenmedipham	1.13	2.26
83	molinate	0.53	1.06	116	bifenazate	5.70	11.40
84	carbendazim	0.13	0.26	117	fenhexamid	0.23	0.46
85	6-chloro-4-hydroxy-3-phenyl-pyridazin	0.43	0.86	118	flutriafol	2.15	4.30
86	propoxur	6.10	12.20	119	furalaxyl	0.20	0.40
87	isouron	0.10	0.20	120	bioallethrin	49.50	99.00
88	chlorotoluron	0.15	0.30	121	cyanofenphos	5.20	10.40
89	thiofanox	39.25	78.50	122	pirimiphos methyl	0.05	0.10
90	chlorbufam	45.75	91.50	123	buprofezin	0.23	0.46
91	bendiocarb	0.80	1.60	124	disulfoton sulfone	0.63	1.26
92	propazine	0.08	0.16	125	fenazaquin	0.08	0.16
93	terbuthylazine	0.13	0.26	126	triazophos	0.18	0.36
94	diuron	0.40	0.80	127	DEF	0.40	0.80
95	chlormephos	112.00	224.00	128	pyriftalid	0.15	0.30
96	carboxin	0.15	0.30	129	metconazole	0.33	0.66

TABLE 2.17 The LOD, LOQ of 440 Pesticides and Related Chemical Residues Determined by LC-MS-MS—cont'd

No.	Pesticides	LOD (µg/kg)	LOQ (µg/kg)	No.	Pesticides	LOD (µg/kg)	LOQ (µg/kg)
130	pyriproxyfen	0.10	0.20	162	propanil	5.40	10.80
131	cycloxydim	0.63	1.26	163	carbofuran	3.28	6.56
132	isoxaben	0.05	0.10	164	acetamiprid	0.35	0.70
133	flurtamone	0.10	0.20	165	mepanipyrim	0.08	0.16
134	trifluralin	83.70	167.40	166	prometon	0.03	0.06
135	flamprop methyl	5.05	10.10	167	methiocarb	41.20	82.40
136	bioresmethrin	1.85	3.70	168	metoxuron	0.15	0.30
137	propiconazole	0.45	0.90	169	dimethoate	1.90	3.80
138	chlorpyrifos	13.45	26.90	170	fluometuron	0.23	0.46
139	fluchloralin	122.00	244.00	171	dicrotophos	0.28	0.56
140	clethodim	0.53	1.06	172	monalide	0.30	0.60
141	flamprop isopropyl	0.10	0.20	173	diphenamid	0.03	0.06
142	tetrachlorvinphos	0.55	1.10	174	ethoprophos	0.70	1.40
143	propargite	17.15	34.30	175	fonofos	1.88	3.76
144	bromuconazole	0.78	1.56	176	etridiazol	25.10	50.20
145	picolinafen	0.18	0.36	177	hexazinone	0.03	0.06
146	fluthiacet methyl	1.33	2.66	178	dimethametryn	0.03	0.06
147	trifloxystrobin	0.50	1.00	179	trichlorphon	0.28	0.56
148	chlorimuron ethyl	7.60	15.20	180	demeton(o+s)	1.70	3.40
149	hexaflumuron	6.30	12.60	181	benoxacor	1.73	3.46
150	novaluron	2.00	4.00	182	bromacil	5.90	11.80
151	flurazuron	6.70	13.40	183	phorate sulfoxide	86.63	173.26
	Group C			184	brompyrazon	0.90	1.80
152	maleic hydrazide	20.00	40.00	185	oxycarboxin	0.23	0.46
153	methamidophos	1.23	2.46	186	mepronil	0.10	0.20
154	EPTC	9.33	18.66	187	disulfoton	117.43	234.86
155	diethyltoluamide	0.15	0.30	188	fenthion	13.00	26.00
156	monuron	8.68	17.36	189	metalaxyl	0.13	0.26
157	pyrimethanil	0.18	0.36	190	ofurace	0.25	0.50
158	fenfuram	0.20	0.40	191	dodemorph	0.10	0.20
159	quinoclamine	1.98	3.96	192	imazamethabenz-methyl	0.05	0.10
160	fenobucarb	1.48	2.96	193	isoprothiolane	0.45	0.90
161	ethirimol	0.15	0.30	194	imazalil	0.50	1.00

Continued

TABLE 2.17 The LOD, LOQ of 440 Pesticides and Related Chemical Residues Determined by LC-MS-MS—cont'd

No.	Pesticides	LOD (µg/kg)	LOQ (µg/kg)	No.	Pesticides	LOD (µg/kg)	LOQ (µg/kg)
195	phoxim	20.70	41.40	228	temephos	0.30	0.60
196	quinalphos	0.50	1.00	229	butafenacil	2.38	4.76
197	fenoxycarb	4.58	9.16	230	spinosad	0.15	0.30
198	pyrimitate	0.05	0.10		Group D		
199	fensulfothin	0.50	1.00	231	allidochlor	10.25	20.50
200	fluorochloridone	3.45	6.90	232	propamocarb	0.03	0.06
201	butachlor	5.03	10.06	233	thiabendazole	0.13	0.26
202	kresoxim-methyl	25.15	50.30	234	metamitron	1.60	3.20
203	triticonazole	0.75	1.50	235	isoproturon	0.03	0.06
204	fenamiphos sulfoxide	0.18	0.36	236	atratone	0.05	0.10
205	thenylchlor	6.03	12.06	237	metribuzin	0.13	0.26
206	fenoxanil	9.85	19.70	238	DMST	10.00	20.00
207	fluridone	0.05	0.10	239	cycloate	1.10	2.20
208	epoxiconazole	1.03	2.06	240	atrazine	0.10	0.20
209	chlorphoxim	19.40	38.80	241	butylate	75.50	151.00
210	fenamiphos sulfone	0.10	0.20	242	pymetrozin	8.58	17.16
211	fenbuconazole	0.40	0.80	243	chloridazon	0.58	1.16
212	isofenphos	54.68	109.36	244	sulfallate	51.80	103.60
213	phenothrin	84.80	169.60	245	ethiofencarb	1.23	2.46
214	piperophos	2.30	4.60	246	terbumeton	0.03	0.06
215	piperonyl butoxide	0.28	0.56	247	cyprazine	0.06	0.12
216	oxyflurofen	14.63	29.26	248	ametryn	0.25	0.50
217	coumaphos	0.53	1.06	249	tebuthiuron	0.05	0.10
218	flufenacet	1.33	2.66	250	trietazine	0.15	0.30
219	phosalone	12.00	24.00	251	sebutylazine	0.08	0.16
220	methoxyfenozide	0.93	1.86	252	dibutyl succinate	55.60	111.20
221	prochloraz	0.53	1.06	253	tebutam	0.03	0.06
222	aspon	0.43	0.86	254	thiofanox-sulfoxide	2.08	4.16
223	ethion	0.75	1.50	255	methacrifos	605.93	1211.86
224	dithiopyr	2.60	5.20	256	terbutryn	0.05	0.10
225	spirodiclofen	2.48	4.96	257	triazoxide	2.00	4.00
226	fenpyroximate	0.35	0.70				
227	flumiclorac-pentyl	2.65	5.30	258	thionazin	5.68	11.36

TABLE 2.17 The LOD, LOQ of 440 Pesticides and Related Chemical Residues Determined by LC-MS-MS—cont'd

No.	Pesticides	LOD (µg/kg)	LOQ (µg/kg)	No.	Pesticides	LOD (µg/kg)	LOQ (µg/kg)
259	linuron	2.90	5.80	292	tralkoxydim	0.08	0.16
260	heptanophos	1.45	2.90	293	malathion	1.40	2.80
261	prosulfocarb	0.10	0.20	294	pyributicarb	0.08	0.16
262	dipropetryn	0.08	0.16	295	pyridaphenthion	0.23	0.46
263	thiobencarb	0.83	1.66	296	pirimiphos-ethyl	0.05	0.10
264	tri-iso-butyl phosphate	1.00	2.00	297	thiodicarb	9.85	19.70
265	tri-n-butyl phosphate	0.10	0.20	298	pyraclofos	0.25	0.50
266	diethofencarb	0.50	1.00	299	picoxystrobin	2.10	4.20
267	alachlor	1.85	3.70	300	tetraconazole	0.43	0.86
268	cadusafos	0.30	0.60	301	mefenpyr-diethyl	3.15	6.30
269	metazachlor	0.25	0.50	302	profenefos	0.50	1.00
270	propetamphos	13.50	27.00	303	pyraclostrobin	0.13	0.26
271	simeconazole	0.73	1.46	304	dimethomorph	0.10	0.20
272	triadimefon	1.98	3.96	305	kadethrin	0.83	1.66
273	phorate sulfone	10.50	21.00	306	thiazopyr	0.50	1.00
274	tridemorph	0.65	1.30	307	chlorfluazuron	2.18	4.36
275	mefenacet	0.55	1.10		Group E		0.00
276	fenpropimorph	0.05	0.10	308	4-aminopyridine	0.23	0.46
277	tebuconazole	0.55	1.10	309	chlormequat	0.03	0.06
278	isopropalin	7.50	15.00	310	methomyl	2.40	4.80
279	nuarimol	0.25	0.50	311	pyroquilon	0.88	1.76
280	bupirimate	0.18	0.36	312	fuberidazole	0.48	0.96
281	azinphos-methyl	276.08	552.16	313	isocarbamid	0.43	0.86
282	tebupirimfos	0.03	0.06	314	butocarboxim	0.40	0.80
283	phenthoate	23.10	46.20	315	chlordimeform	0.33	0.66
284	sulfotep	0.65	1.30	316	vernolate	0.07	0.14
285	sulprofos	1.45	2.90	317	aminocarb	4.10	8.20
286	EPN	8.25	16.50	318	dimethirimol	0.03	0.06
287	azamethiphos	0.20	0.40	319	omethoate	2.43	4.86
288	diniconazole	0.33	0.66	320	dichlorvos	0.13	0.26
289	sethoxydim	24.90	49.80	321	aldicarb sulfone	5.35	10.70
290	pencycuron	0.08	0.16	322	dioxacarb	0.85	1.70
291	mecarbam	4.90	9.80	323	demeton-s-methyl	1.33	2.66

Continued

TABLE 2.17 The LOD, LOQ of 440 Pesticides and Related Chemical Residues Determined by LC-MS-MS—cont'd

No.	Pesticides	LOD (µg/kg)	LOQ (µg/kg)	No.	Pesticides	LOD (µg/kg)	LOQ (µg/kg)
324	ethiofencarb-sulfoxide	56.00	112.00	356	hexythiazox	5.90	11.80
325	thiometon	144.50	289.00	357	benzoximate	2.15	4.30
326	folpet	34.65	69.30	358	benzoylprop-ethyl	77.00	154.00
327	demeton-s-methyl sulfone	4.95	9.90	359	pyrimidifen	3.50	7.00
328	fenpropidin	0.05	0.10	360	furathiocarb	0.48	0.96
329	amidithion	164.50	329.00	361	*trans*-permethin	1.20	2.40
330	imazapic	1.48	2.96	362	etofenprox	570.00	1140.00
331	paraoxon-ethyl	0.13	0.26	363	pyrazoxyfen	0.08	0.16
332	aldimorph	0.80	1.60	364	flubenzimine	1.95	3.90
333	vinclozolin	0.63	1.26	365	haloxyfop-2-ethoxyethyl	0.63	1.26
334	uniconazole	0.60	1.20	366	*tau*-fluvalinate	57.50	115.00
335	pyrifenox	0.08	0.16		Group F		
336	chlorthion	33.40	66.80	367	acrylamide	4.45	8.90
337	dicapthon	0.05	0.10	368	*tert*-butylamine	9.75	19.50
338	clofentezine	0.20	0.40	369	phthalimide	10.75	21.50
339	norflurazon	0.08	0.16	370	dimefox	17.05	34.10
340	triallate	11.55	23.10	371	diphenylamin	0.10	0.20
341	quinoxyphen	38.35	76.70	372	*1*-naphthy acetamide	0.20	0.40
342	fenthion sulfone	4.38	8.76	373	atrazine-desethyl	0.15	0.30
343	flurochloridone	0.33	0.66	374	*2,6*-dichlorobenzamide	1.13	2.26
344	phthalic acid,benzyl butyl ester	158.00	316.00	375	aldicarb	65.25	130.50
				376	simeton	0.28	0.56
345	isazofos	0.05	0.10	377	dinotefuran	2.55	5.10
346	dichlofenthion	7.55	15.10	378	pebulate	0.85	1.70
347	vamidothion sulfone	119.00	238.00	379	acibenzolar-s-methyl	0.78	1.56
348	terbufos sulfone	22.15	44.30	380	dioxabenzofos	3.45	6.90
349	dinitramine	0.45	0.90	381	oxamyl	137.03	274.06
350	cyazofamid	1.13	2.26	382	methabenzthiazuron	0.03	0.06
351	trichloronat	16.70	33.40	383	butoxycarboxim	6.65	13.30
352	resmethrin-2	0.08	0.16	384	demeton-s-methyl sulfoxide	0.98	1.96
353	boscalid	1.20	2.40	385	thiofanox sulfone	6.03	12.06
354	nitralin	8.60	17.20	386	phosfolan	0.13	0.26
355	fenpropathrin	61.25	122.50	387	demeton-s	20.00	40.00

TABLE 2.17 The LOD, LOQ of 440 Pesticides and Related Chemical Residues Determined by LC-MS-MS—cont'd

No.	Pesticides	LOD (µg/kg)	LOQ (µg/kg)	No.	Pesticides	LOD (µg/kg)	LOQ (µg/kg)
388	fenthion oxon	0.30	0.60	416	propylene thiourea	7.53	15.06
389	napropamide	0.33	0.66		Group G		
390	fenitrothion	6.70	13.40	417	dalapon	57.68	115.36
391	phthalic acid, dibutyl ester	9.90	19.80	418	2-phenylphenol	42.48	84.96
392	metolachlor	0.10	0.20	419	3-phenylphenol	1.00	2.00
393	procymidone	21.65	43.30	420	DNOC	0.65	1.30
394	vamidothion	1.15	2.30	421	dicloran	12.15	24.30
395	cumyluron	0.33	0.66	422	chlorpropham	3.95	7.90
396	phosmet	4.43	8.86	423	terbacil	0.23	0.46
397	ronnel	3.28	6.56	424	bentazone	0.25	0.50
398	pyrethrin	8.95	17.90	425	dinoseb	0.10	0.20
399	phthalic acid, biscyclohexyl ester	0.18	0.36	426	dinoterb	0.05	0.10
				427	fludioxonil	15.55	31.10
400	carpropamid	1.30	2.60	428	trinexapac-ethyl	17.67	35.34
401	tebufenpyrad	0.08	0.16				
402	tebufenozide	6.95	13.90	429	chlorfenethol	41.08	82.16
403	chlorthiophos	7.95	15.90	430	chlorobenzuron	5.10	10.20
404	dialifos	39.25	78.50	431	chlorampheni-colum	0.98	1.96
405	rotenone	0.58	1.16				
406	imibenconazole	2.58	5.16	432	oryzalin	1.23	2.46
407	propaquiafop	0.30	0.60	433	famoxadone	11.33	22.66
408	lactofen	15.50	31.00	434	diflufenican	7.08	14.16
409	benzofenap	0.03	0.06	435	ethiprole	9.98	19.96
410	dinoseb acetate	10.33	20.66	436	flusulfamide	0.10	0.20
411	propisochlor	0.20	0.40	437	fomesafen	0.50	1.00
412	silafluofen	152.00	304.00	438	fluazinam	17.65	35.30
413	etobenzanid	0.20	0.40	439	kelevan	2410.71	4821.42
414	fentrazamide	3.10	6.20	440	acrinathrin	2.02	4.04
415	cyphenothrin	4.20	8.40				

2.11.3   Reagents and Materials

"Water" is first grade of GB/T6682 specified.

Acetonitrile: HPLC Grade
Acetone: HPLC Grade
Toluene: HPLC Grade
Isooctane: HPLC Grade
Dichloromethane: HPLC Grade
Hexane: HPLC Grade
Methanol: HPLC Grade
Acetonitrile-Toluene (3+1, V/V)
0.1% Formic Acid Solution (Volume Fraction)
5 mmol/L Ammonium Acetate Solution
Acetonitrile-Water (3+2, V/V)
Sodium Chloride: G.R.
Sodium Sulfate: Anhydrous, Analytically Pure. Ignited at 650°C for 4 h and Kept in a Desiccator
Pesticide and Related Chemicals Standard: Purity ≥95%

Stock Standard Solution: Accurately weigh 5–10 mg of individual pesticide and related chemical standards (accurate to 0.1 mg) into a 10-mL volumetric flask. Dissolve and dilute to volume with methanol, hexane, acetone, acetonitrile, isooctane, etc., depending on each individual compound's solubility. (For diluting solvent, refer to Appendix B) Solutions are stored in the dark from 0°C to 4°C and are used for 1 year.

Mixed Standard Solution (Mixed Standard Solution A, B, C, D, E, F, and G): Depending on properties and retention times of compounds, all compounds are divided into seven groups: A, B, C, D, E, F, and G. The concentration of each compound is determined by the sensitivity of the instrument for analysis. For 440 pesticides and related chemicals, groupings and concentrations are listed in Appendix B.

Depending on group number, mixed standard solution concentration, and stock standard solution concentration, appropriate amounts of individual stock standard solution are pipetted into a 100-mL volumetric flask, being diluted to volume with methanol. Mixed standard solutions are stored in the dark below 4°C and used for 1 month.

Working Standard Mixed Solution in Matrix: Working standard mixture solution in matrix is prepared by diluting an appropriate amount of mixed standard solution with blank extract that has been taken through the method with the rest of the samples. Mix thoroughly. These are used for plotting the standard curve. Working standard mixture solution in matrix must be prepared freshly.

SPE-Cartridge: Sep-Pak Carbon NH$_2$, 6 mL, 1 g or Equivalent
Filter Membrane (Nylon): 13 mm × 0.2 μm

2.11.4 Apparatus

LC-MS-MS: Equipped With ESI
Analytical Balances: Capable of Weighing From 0.1 mg to 0.01 g
Centrifuge Tube: 80 mL
Pipette: 1 mL, 10 mL
Pear-Shaped Flasks: 100 mL
Homogenizer: Not Less Than 12,000 rpm
Centrifuge: Not Less Than 4500 rpm
Rotary Evaporator
Nitrogen Evaporator
Vials: 2 mL, With PTFE Screw-Cap

2.11.5 Sample Pretreatment

2.11.5.1 Preparation of Test Sample

Part of edible sample is cut up, mixed, labeled, and sealed.
The test samples are stored in the dark at 0°C to 4°C.

2.11.5.2 Extraction

Weigh 20 g test sample (accurate to 0.01 g) into 80-mL centrifuge tube. Add 40 mL acetonitrile, and homogenize at 15, 000 rpm for 1 min. Add 5 g sodium chloride to the centrifuge tube and homogenize at 15,000 rpm for 1 min again, then centrifuge at 4200 rpm for 5 min. Pipette 20 mL of the top acetonitrile layer of extracts (corresponding to a 10-g test sample), and concentrate the extracts to ca. 1 mL using the rotary evaporator at 40°C for clean-up.

2.11.5.3 Clean-up

Add sodium sulfate into the Envi-Carb cartridge to a height of ca. 2 cm. Connect the cartridge to the top of the aminopropyl Sep-Pak cartridge in series. Fix the cartridge into a support to which a pear-shaped flask is connected. Condition the cartridge with 4 mL acetonitrile-toluene (3+1) before adding the sample. Once the solution gets to the top of the sodium sulfate, pipette the eluate into the cartridge. Rinse the pear-shaped flask with 3 × 2 mL acetonitrile-toluene (3+1) and decant it into the cartridge. Insert a 50-mL reservoir into the cartridges. Elute the pesticides with 25 mL acetonitrile-toluene (3+1). Evaporate the eluate to ca. 0.5 mL using a rotary evaporator at 40°C. Then evaporate the eluate to dryness using a nitrogen evaporator at 45°C. Make up to 1 mL with acetonitrile-water (3 +2) and mix thoroughly. Finally, filter the extract through a 0.2-μm filter into a glass vial for LC-MS-MS determination.

Extract and cleanup blank edible fungus samples as that described above.

2.11.6 Determination

2.11.6.1 LC-MS-MS Operating Condition

Operating Conditions for Group A, B, C, D, E and F

(a) Column: ZORBAX SB-C_{18}, 3.5 μm, 100 × 2.1 mm (i.d) or equivalent
(b) Mobile phase and the condition of gradient elution, see Table 2.2
(c) Column temperature: 40°C
(d) Injection volume: 10 μL
(e) Ion source: ESI
(f) Scan mode: Positive ion scan
(g) Atomization gas: Nitrogen gas
(h) Nebulizer gas: 0.28 MPa
(i) Ionspray voltage: 4000 V
(j) Drying temperature: 350°C
(k) Drying air flow rate: 10 L/min
(l) MRM transitions for precursor/production, quantifying for precursor/production, declustering potential, collision energy, and collision cell exit potential: see Table 5.10

Operating Conditions for Group G

(a) Column: ZORBAX SB-C_{18}, 3.5 μm, 100 × 2.1 mm (i.d) or equivalent
(b) Mobile phase and the condition of gradient elution, see Table 2.3
(c) Column temperature: 40°C
(d) Injection volume: 10 μL
(e) Ion source: ESI
(f) Scan mode: Negative ion scan
(g) Atomization gas: Nitrogen gas
(h) Nebulizer gas: 0.28 MPa
(i) Ionspray voltage: 4000 V
(j) Drying temperature: 350°C
(k) Drying air flow rate: 10 L/min
(l) MRM transitions for precursor/product ion, quantifying for precursor/product ion, declustering potential, collision energy and collision cell exit potential: see Table 5.10

2.11.6.2 Qualitative Determination

Sample solutions are analyzed using LC-MS-MS operating conditions. For the samples determined, if the retention times of peaks found in the sample solution chromatogram are the same as those for the peaks in the standard in blank matrix extract chromatogram, and the abundance ratios of MRM transitions for precursor/product ions are within the expected limits (relative abundance >50%, permitted tolerances ±20%; relative abundance >20% to 50%, permitted tolerances ±25%; relative abundance >10% to 20%, permitted tolerances ±30%; relative abundance ≤10%, permitted tolerances ±50%), the sample is confirmed to contain this pesticide compound.

2.11.6.3   Quantitative Determination

xternal standard method is used for quantitation with standard curves for LC-MS-MS. To compensate for the matrix effect, quantitation is based on a series of working standard solutions prepared in blank matrix extract. The standard curves are established by injection of different concentrations of working standard mixed solutions in matrix separately. The responses of pesticides in the sample solution should be in the linear range of the instrumental detection.

2.11.6.4   Parallel Test

A parallel test is carried out for the same testing sample.

2.11.6.5   Blank Test

The operation of the blank test is the same as that described in the method of determination, but without the addition of sample.

2.11.7   Precision

The precision data of the method for this standard is according to the stipulations of GB/T 6379.1 and GB/T 6379.2. The values of repeatability and reproducibility are obtained and calculated at the 95% confidence level.

RESEARCHERS

Guo-Fang Pang, Chun-Lin Fan, Jun-Hong Zheng, Ming-Lin Wang, Wen-Wen Wang.

Qinhuangdao Entry-Exit Inspection and Quarantine Bureau, 39 Haibin Rd, Qinhuangdao, Hebei, PC 066002, People's Republic of China.

Chapter 3

Analytical Methods for 790 Pesticides and Related Chemical Residues in Products of Animal Origin

3.1 DETERMINATION OF 511 PESTICIDES AND RELATED CHEMICAL RESIDUES IN MILK AND MILK POWDER: GC-MS METHOD (GB/T 23210-2008)

3.1.1 Scope

This method is applicable to the quantitative determination of 504 pesticides and related chemical residues and qualitative determination of 487 pesticides and related chemical residues in milk; it is also applicable to the quantitative determination of 498 pesticides and related chemical residues and qualitative determination of 489 pesticides and related chemical residues in milk powder.

The limit of quantitative determination of the method as applied to 487 pesticides and related chemicals is 0.0008 mg/L to 0.4 mg/L. The limit of quantitative determination of the method as applied to 489 pesticides and related chemicals is 0.0042 mg/kg to 2.0 mg/kg, See Table 3.1.

3.1.2 Principle

The samples are extracted with acetonitrile, cleaned up with a C_{18} SPE cartridge. The pesticides and related chemicals are eluted with acetonitrile and the solutions are analyzed by GC-MS.

Quantitation is performed by an external method.

3.1.3 Reagents and Materials

Acetonitrile: HPLC Grade
Hexane: HPLC Grade
Toluene: Guaranteed Reagent
Acetone: HPLC Grade
Isooctane: HPLC Grade

Analytical Methods for Food Safety by Mass Spectrometry. https://doi.org/10.1016/B978-0-12-814167-0.00003-X
Copyright © 2018 Chemical Industry Press.
Published by Elsevier Inc. under an exclusive license with Chemical Industry Press.

TABLE 3.1 The LOD, LOQ of 511 Pesticides and Related Chemical Residues Determined by GC-MS

No.	Pesticides	LOD for Milk / (mg/ mL)	LOD for Milk Powder / (mg/kg)	No.	Pesticides	LOD for Milk / (mg/ mL)	LOD for Milk Powder / (mg/kg)
	Group A			28	cyprazine	0.0042	0.0208
1	allidochlor	0.0083	0.0417	29	vinclozolin	0.0042	0.0208
2	dichlormid	0.0083	0.0417	30	beta-HCH	0.0042	0.0208
3	etridiazol	0.0500	0.0625	31	metalaxyl	0.0125	0.0625
4	chlormephos	0.0332	0.0417	32	methyl-parathion	0.0167	0.0833
5	propham	0.0042	0.0208	33	chlorpyrifos (ethyl)	0.0042	0.0208
6	cycloate	0.0042	0.0208	34	delta- HCH	0.0083	0.0417
7	diphenylamine	0.0042	0.0208	35	fenthion	0.0042	0.0208
8	chlordimeform	0.0168	0.0832	36	malathion	0.0167	0.0833
9	ethalfluralin	0.0167	0.0833	37	paraoxon-ethyl	0.1333	0.6667
10	phorate	0.0042	0.0208	38	fenitrothion	0.0083	0.0417
11	thiometon	0.0042	0.0208	39	triadimefon	0.0083	0.0417
12	quintozene	0.0083	0.0417	40	parathion	0.0668	0.0833
13	atrazine-desethyl	0.0042	0.0208	41	linuron	0.0167	—
14	clomazone	0.0042	0.0208	42	pendimethalin	0.0167	0.0833
15	diazinon	0.0042	0.0208	43	chlorbenside	0.0083	0.0417
16	fonofos	0.0042	0.0208	44	bromophos-ethyl	0.0168	0.0208
17	etrimfos	0.0042	0.0208	45	quinalphos	0.0042	0.0208
18	propetamphos	0.0042	0.0208	46	trans-chlordane	0.0042	0.0208
19	secbumeton	0.0042	0.0208	47	metazachlor	0.0125	0.0625
20	pronamide	0.0042	0.0208	48	prothiophos	0.0042	0.0208
21	dichlofenthion	0.0042	0.0208	49	folpet	0.2000	0.2500
22	mexacarbate	0.0125	0.2500	50	chlorfurenol	0.0125	0.0625
23	dinitramine	0.0167	0.0833	51	procymidone	0.0042	0.0208
24	dimethoate	0.0167	0.0833	52	dieldrin	0.0083	0.0417
25	aldrin	0.0083	0.0417	53	methidathion	0.0083	—
26	ronnel	0.0083	0.0417	54	cyanazine	0.0125	0.0625
27	prometryne	0.0042	0.0208	55	napropamide	0.0125	0.0625

TABLE 3.1 The LOD, LOQ of 511 Pesticides and Related Chemical Residues Determined by GC-MS—cont'd

No.	Pesticides	LOD for Milk / (mg/ mL)	LOD for Milk Powder / (mg/kg)	No.	Pesticides	LOD for Milk / (mg/ mL)	LOD for Milk Powder / (mg/kg)
56	oxadiazone	0.0042	0.0208	84	*trans*-permethrin	0.0042	0.0832
57	tetrasul	0.0168	0.0208	85	cypermethrin	0.0125	0.2500
58	fenamiphos	0.0125	0.0625	86	fenvalerate-1	0.0167	0.0833
59	bupirimate	0.0042	0.0208	87	fenvalerate-2	0.0167	0.0833
60	flutolanil	0.0042	0.0208	88	deltamethrin	0.0250	0.1250
61	carboxin	0.1000	0.5000		Group B		
62	*p,p'*-DDD	0.0042	0.0208	89	EPTC	0.0125	0.0625
63	ethion	0.0083	0.0417	90	butylate	0.0125	0.0625
64	etaconazole-1	0.0500	0.0625	91	dichlobenil	0.0008	0.0042
65	sulprofos	0.0083	0.0417	92	pebulate	0.0125	0.0625
66	etaconazole-2	0.0125	0.0625	93	nitrapyrin	0.0125	0.0625
67	myclobutanil	0.0042	0.0208	94	mevinphos	0.0083	0.0417
68	fensulfothion	0.0083	0.0417	95	chloroneb	0.0042	0.0208
69	propiconazole-1	0.0125	0.0625	96	tecnazene	0.0083	0.0417
70	propiconazole-2	0.0125	0.0625	97	heptanophos	0.0125	0.0625
71	bifenthrin	0.0042	0.0208	98	ethoprophos	0.0125	0.0625
72	mirex	0.0168	0.0208	99	hexachlorobenzene	0.0042	0.0208
73	nuarimol	0.0083	0.0417	100	propachlor	0.0125	0.0625
74	benodanil	0.0125	0.0625	101	*cis-* diallate	0.0083	0.0417
75	methoxychlor	0.0333	0.1667	102	trifluralin	0.0083	0.0417
76	oxadixyl	0.0168	0.0208	103	*trans*-diallate	0.0083	0.0417
77	tebuconazole	0.0125	0.0625	104	chlorpropham	0.0083	0.0417
78	tetramethirn	0.0083	0.0417	105	sulfotep	0.0042	0.0208
79	norflurazon	0.0042	0.0208	106	sulfallate	0.0083	0.0417
80	pyridaphenthion	0.0042	0.0208	107	*alpha*-HCH	0.0042	0.0208
81	phosmet	0.0083	0.0417	108	terbufos	0.0083	0.0417
82	tetradifon	0.0042	0.0208	109	terbumeton	0.0500	0.0625
83	pyrazophos	0.0083	0.0417	110	profluralin	0.0167	0.0833

Continued

TABLE 3.1 The LOD, LOQ of 511 Pesticides and Related Chemical Residues Determined by GC-MS—cont'd

No.	Pesticides	LOD for Milk / (mg/ mL)	LOD for Milk Powder / (mg/kg)	No.	Pesticides	LOD for Milk / (mg/ mL)	LOD for Milk Powder / (mg/kg)
111	dioxathion	0.0167	0.0833	139	endosulfan-i	0.0250	0.1250
112	propazine	0.0042	0.0208	140	chlorfenvinphos	0.0125	0.0625
113	chlorbufam	0.0083	0.0417	141	tolylfluanide	0.1000	0.5000
114	dicloran	0.0083	0.0417	142	cis- chlordane	0.0083	0.0417
115	terbuthylazine	0.0042	0.0208	143	butachlor	0.0083	0.0417
116	monolinuron	0.0167	0.0833	144	chlozolinate	0.0083	0.0417
117	cyanophos	0.0083	0.0417	145	p,p'-DDE	0.0168	0.0208
118	flufenoxuron	0.0500	0.0625	146	iodofenphos	0.0083	0.0417
119	chlorpyrifos-methyl	0.0042	0.0208	147	tetrachlorvinphos	0.0125	0.0625
120	desmetryn	0.0042	0.0208	148	chlorbromuron	0.1000	0.5000
121	dimethachlor	0.0125	0.0625	149	profenofos	0.0250	0.1250
122	alachlor	0.0125	0.0625	150	buprofezin	0.0083	0.0417
123	pirimiphos-methyl	0.0042	0.0208	151	hexaconazole	0.0250	0.1250
124	terbutryn	0.0083	0.0417	152	4,4,-DDD	0.0042	0.0208
125	aspon	0.0083	0.0417	153	chlorfenson	0.0083	0.0417
126	thiobencarb	0.0083	0.0417	154	fluorochloridone	0.0083	0.0417
127	dicofol	0.0083	0.0417	155	endrin	0.0500	0.2500
128	metolachlor	0.0168	0.0208	156	paclobutrazol	0.0125	0.0625
129	pirimiphos-ethyl	0.0083	0.0417	157	2,4'-DDT	0.0332	0.0417
130	oxy-chlordane	0.0168	0.0208	158	methoprotryne	0.0125	0.0625
131	dichlofluanid	0.2000	1.0000	159	erbon	0.0332	0.0417
132	methoprene	0.0167	0.0833	160	chloropropylate	0.0042	0.0208
133	bromofos	0.0083	0.0417	161	flamprop-methyl	0.0042	0.0208
134	ethofumesate	0.0083	0.0417	162	nitrofen	0.0250	0.1250
135	isopropalin	0.0083	0.0417	163	oxyfluorfen	0.0167	0.0833
136	propanil	0.0083	0.0417	164	chlorthiophos	0.0125	0.0625
137	crufomate	0.0250	0.1250	165	flamprop-isopropyl	0.0042	0.0208
138	isofenphos	0.0083	0.0417	166	endosulfan-ii	0.0250	0.1250

TABLE 3.1 The LOD, LOQ of 511 Pesticides and Related Chemical Residues Determined by GC-MS—cont'd

No.	Pesticides	LOD for Milk / (mg/mL)	LOD for Milk Powder / (mg/kg)	No.	Pesticides	LOD for Milk / (mg/mL)	LOD for Milk Powder / (mg/kg)
167	carbofenothion	0.0083	0.0417	194	methacrifos	0.0042	0.0208
168	p,p'-DDT	0.0083	0.0417	195	molinate	0.0042	0.0208
169	benalaxyl	0.0042	0.0208	196	2-phenylphenol	0.0042	0.0208
170	edifenphos	0.0083	0.0417	197	cis-1,2,3,6-tetrahydrophthalimide	0.0125	0.0625
171	triazophos	0.0500	0.0625				
172	cyanofenphos	0.0042	0.0208	198	fenobucarb	0.0083	0.0417
173	chlorbenside sulfone	0.0083	0.0417	199	benfluralin	0.0042	0.0208
174	endosulfan-sulfate	0.0125	0.0625	200	hexaflumuron	0.0250	0.1250
175	bromopropylate	0.0083	0.0417	201	prometon	0.0125	0.0625
176	benzoylprop-ethyl	0.0125	0.0625	202	triallate	0.0083	0.0417
177	fenpropathrin	0.0083	0.0417	203	pyrimethanil	0.0042	0.0208
178	leptophos	0.0083	0.0417	204	gamma-HCH	0.0083	0.0417
179	EPN	0.0167	0.0833	205	disulfoton	0.0042	0.0208
180	captafol	0.0750	1.5000	206	atrizine	0.0042	0.0208
181	hexazinone	0.0125	0.0625	207	iprobenfos	0.0125	0.0625
182	bifenox	0.0083	0.0417	208	heptachlor	0.0125	0.0625
183	phosalone	0.0083	0.0417	209	isazofos	0.0083	0.0417
184	fenarimol	0.0083	0.0417	210	plifenate	0.0083	0.0417
185	azinphos-methyl	0.0250	0.1250	211	fluchloralin	0.0167	0.0833
186	azinphos-ethyl	0.0083	0.0417	212	transfluthrin	0.0042	0.0208
187	coumaphos	0.0250	0.5000	213	fenpropimorph	0.0042	0.0208
188	cyfluthrin	0.0500	0.2500	214	tolclofos-methyl	0.0042	0.0208
189	(tau-)fluvalinate	0.5000	0.2500	215	propisochlor	0.0042	0.0208
	Group C			216	ametryn	0.0125	0.0625
190	dichlorvos	0.0250	0.1250	217	metribuzin	0.0125	0.0625
191	biphenyl	0.0042	0.0208	218	dimethipin	0.0500	—
192	vernolate	0.0042	0.0208	219	dipropetryn	0.0042	0.0208
193	3,5-dichloroaniline	0.0042	0.0208	220	formothion	0.0083	0.0417

Continued

TABLE 3.1 The LOD, LOQ of 511 Pesticides and Related Chemical Residues Determined by GC-MS—cont'd

No.	Pesticides	LOD for Milk / (mg/mL)	LOD for Milk Powder / (mg/kg)	No.	Pesticides	LOD for Milk / (mg/mL)	LOD for Milk Powder / (mg/kg)
221	diethofencarb	0.0250	0.1250	249	fenazaquin	0.0042	0.0208
222	dimepiperate	0.0083	0.0417	250	phenothrin	0.0042	0.0208
223	bioallethrin-1	0.0167	0.0833	251	fenoxycarb	0.0250	0.1250
224	bioallethrin-2	0.0167	0.0833	252	propamocarb	0.0125	0.2500
225	fenson	0.0042	0.0208	253	anilofos	0.0083	0.0417
226	2.4'- DDE	0.0042	0.0208	254	acrinathrin	0.0083	0.0417
227	diphenamid	0.0042	0.0208	255	lambda-cyhalothrin	0.0042	0.0208
228	penconazole	0.0125	0.0625	256	mefenacet	0.0500	0.0625
229	tetraconazole	0.0125	0.0625	257	permethrin	0.0083	0.0417
230	mecarbam	0.0167	0.0833	258	pyridaben	0.0042	0.0208
231	propaphos	0.0083	0.0417	259	fluoroglycofen-ethyl	0.0500	0.2500
232	flumetralin	0.0083	0.0417	260	bitertanol	0.0125	0.0625
233	triadimenol-1	0.0125	0.0625	261	etofenprox	0.0042	0.0208
234	triadimenol-2	0.0125	0.0625	262	cycloxydim	0.4000	2.0000
235	kresoxim-methyl	0.0042	0.0208	263	alpha-cypermethrin	0.0083	0.0417
236	fluazifop-butyl	0.0042	0.0208	264	flucythrinate-1	0.0083	0.0417
237	chlorfluazuron	0.0125	0.2500	265	flucythrinate-2	0.0083	0.0417
238	chlorobenzilate	0.0042	0.0208	266	esfenvalerate	0.0167	0.0833
239	uniconazole	0.0083	0.0417	267	difenonazole-1	0.0250	0.1250
240	flusilazole	0.0125	0.0625	268	difenconazole-2	0.0250	0.1250
241	fluorodifen	0.0042	0.0208	269	flumioxazin	0.0083	0.0417
242	diniconazole	0.0125	0.0625	270	flumiclorac-pentyl	0.0083	0.0417
243	piperonyl butoxide	0.0042	0.0208		Group D		
244	dimefuron	0.0668	0.3332	271	dimefox	0.0125	0.0625
245	propargite	0.0332	0.1668	272	disulfoton-sulfoxide	0.0083	0.0417
246	mepronil	0.0042	0.0208	273	pentachlorobenzene	0.0042	0.0832
247	diflufenican	0.0042	0.0208	274	crimidine	0.0042	0.0208
248	fludioxonil	0.0042	0.0208	275	BDMC-1	0.0083	0.0417

TABLE 3.1 The LOD, LOQ of 511 Pesticides and Related Chemical Residues Determined by GC-MS—cont'd

No.	Pesticides	LOD for Milk / (mg/mL)	LOD for Milk Powder / (mg/kg)	No.	Pesticides	LOD for Milk / (mg/mL)	LOD for Milk Powder / (mg/kg)
276	chlorfenprop-methyl	0.0042	0.0208	302	DE-PCB 52	0.0042	0.0208
277	thionazin	0.0042	0.0208	303	prosulfocarb	0.0042	0.0208
278	2,3,5,6-tetrachloroaniline	0.0042	0.0208	304	dimethenamid	0.0042	0.0208
				305	BDMC-2	0.0083	0.0417
279	tri-n-butyl phosphate	0.0083	0.0417	306	monalide	0.0083	0.0417
280	2,3,4,5-tetrachloroanisole	0.0042	0.0208	307	octachlorostyrene	0.0042	0.0208
281	pentachloroanisole	0.0042	0.0208	308	isodrin	0.0042	0.0208
282	tebutam	0.0083	0.0417	309	isomethiozin	0.0083	0.0417
283	methabenzthiazuron	0.0417	0.2083	310	trichloronat	0.0042	0.0208
284	simetone	0.0083	0.0417	311	dacthal	0.0042	0.0208
285	atratone	0.0042	0.0208	312	4,4-dichlorobenzophenone	0.0042	0.0208
286	bromocylen	0.0042	0.0208	313	nitrothal-isopropyl	0.0083	0.0417
287	tefluthrin	0.0042	0.0208	314	musk ketone	0.0042	0.0208
288	trietazine	0.0042	0.0208	315	rabenzazole	0.0042	0.0208
289	2,6-dichlorobenzamide	0.0083	0.0417	316	cyprodinil	0.0042	0.0208
290	cycluron	0.0125	0.0625	317	fuberidazole	—	0.4168
291	DE-PCB 28	0.0042	0.0208	318	dicapthon	0.0208	0.1042
292	DE-PCB 31	0.0042	0.0208	319	methfuroxam	0.0168	—
293	desethyl-sebuthylazine	0.0083	0.0417	320	mcpa-butoxyethyl ester	0.0042	0.0208
294	2,3,4,5-tetrachloroaniline	0.0083	0.0417	321	DE-PCB 101	0.0042	0.0208
				322	isocarbophos	0.0083	0.0417
295	musk ambrette	0.0042	0.0208	323	phorate sulfone	0.0042	0.0208
296	musk xylenez	0.0042	0.0208	324	chlorfenethol	0.0042	0.0208
297	pentachloroaniline	0.0042	0.0208	325	trans-nonachlor	0.0042	0.0208
298	aziprotryne	0.0333	0.1667	326	dinobuton	0.1668	—
299	isocarbamid	0.0208	0.1042	327	DEF	0.0083	0.0417
300	sebutylazine	0.0042	0.0208	328	bromfenvinfos	0.0042	0.0208
301	musk moskene	0.0042	0.0208	329	perthane	0.0042	0.0208

Continued

TABLE 3.1 The LOD, LOQ of 511 Pesticides and Related Chemical Residues Determined by GC-MS—cont'd

No.	Pesticides	LOD for Milk / (mg/mL)	LOD for Milk Powder / (mg/kg)	No.	Pesticides	LOD for Milk / (mg/mL)	LOD for Milk Powder / (mg/kg)
330	ditalimfos	0.0042	0.0208	356	nitralin	0.0417	0.2083
331	DE-PCB 118	0.0042	0.0832	357	fenamiphos sulfoxide	0.1333	0.6667
332	mephosfolan	0.0083	0.0417	358	fenamiphos sulfone	0.0167	0.0833
333	4,4-dibromobenzophenone	0.0042	0.0208	359	fenpiclonil	0.0167	0.0833
				360	fluquinconazole	0.0042	0.0208
334	flutriafol	0.0083	0.0417	361	fenbuconazole	0.0083	0.0417
335	DE-PCB 153	0.0042	0.0208		Group E		
336	diclobutrazole	0.0167	0.0833	362	propoxur-1	0.0083	0.0417
337	disulfoton sulfone	0.0083	0.0417	363	XMC	0.0332	0.0417
338	hexythiazox	0.0333	0.1667	364	isoprocarb -1	0.0083	0.0417
339	DE-PCB 138	0.0042	0.0208	365	acenaphthene	0.0042	0.0208
340	cyproconazole	0.0042	0.0208	366	dibutyl succinate	0.0083	0.0417
341	resmethrin-1	0.2668	1.3332	367	chlorethoxyfos	0.0083	0.1668
342	resmethrin-2	0.0667	0.3333	368	isoprocarb -2	0.0083	0.0417
343	phthalic acid,benzyl butyl ester	0.0042	0.0208	369	tebuthiuron	0.0167	0.0833
344	clodinafop-propargyl	0.0083	0.0417	370	pencycuron	0.0167	0.0833
345	fenthion sulfoxide	0.0167	0.0833	371	demeton-s-methyl	0.0167	0.0833
346	fluotrimazole	0.0042	0.0208	372	propoxur-2	0.0083	0.0417
347	fluroxypr-1-methylheptyl ester	0.0042	0.0208	373	phenanthrene	0.0042	0.0208
				374	spiroxamine -1	0.0083	0.0417
348	fenthion sulfone	0.0167	0.0833	375	fenpyroximate	0.0333	0.1667
349	triphenyl phosphate	0.0042	0.0208	376	tebupirimfos	0.0083	0.1668
350	DE-PCB 180	0.0042	0.0832	377	prohydrojasmon	0.0167	0.0833
351	tebufenpyrad	0.0042	0.0208	378	fenpropidin	0.0083	—
352	cloquintocet-mexyl	0.0042	0.0208	379	pyroquilon	0.0042	0.0208
353	lenacil	0.0417	0.2083	380	dinoterb	—	0.3332
354	bromuconazole-1	0.0083	0.0417	381	propyzamide	0.0083	0.0417
355	bromuconazole-2	0.0083	0.0417	382	benoxacor	0.0083	0.0417

TABLE 3.1 The LOD, LOQ of 511 Pesticides and Related Chemical Residues Determined by GC-MS—cont'd

No.	Pesticides	LOD for Milk / (mg/mL)	LOD for Milk Powder / (mg/kg)	No.	Pesticides	LOD for Milk / (mg/mL)	LOD for Milk Powder / (mg/kg)
383	phosphamidon -1	0.1332	0.1667	411	imazamethabenz-methyl	0.0500	—
384	acetochlor	0.0083	0.0417	412	methiocarb sulfone	0.1333	0.6667
385	tridiphane	0.0167	0.0833	413	tcmtb	0.0667	0.3333
386	acibenzolar-s-methyl	0.0083	0.0417	414	metominostrobin	0.0167	0.0833
387	terbucarb-1	0.0083	0.0417	415	imazalil	0.0668	0.0833
388	terbucarb-2	0.0083	0.0417	416	isoprothiolane	0.0083	0.0417
389	fenfuram	0.0083	0.0417	417	cyflufenamid	0.0667	0.3333
390	benfuresate	0.0083	0.0417	418	pyriminobac-methyl	0.0167	—
391	mefenoxam	0.0083	0.0417	419	quinoxyphen	0.0042	0.0832
392	malaoxon	0.0667	0.3333	420	trifloxystrobin	0.0167	0.0833
393	phosphamidon -2	0.0333	0.6668	421	imibenconazole-des-benzyl	0.0668	0.3332
394	chlorthal-dimethyl	0.0083	0.0417	422	pyraflufen ethyl	0.0083	0.0417
395	simeconazole	0.0083	0.0417	423	imiprothrin-1	0.0083	0.0417
396	terbacil	0.0083	0.0417	424	epoxiconazole -1	0.0333	0.1667
397	thiazopyr	0.0083	0.0417	425	imiprothrin-2	0.0083	0.0417
398	dimethylvinphos	0.0083	0.0417	426	pyributicarb	0.0083	0.0417
399	zoxamide	0.0083	0.0417	427	thenylchlor	0.0083	0.0417
400	allethrin	0.0167	0.0833	428	clethodim	0.0167	0.0833
401	quinoclamine	0.0167	0.0833	429	chrysene	0.0042	0.0208
402	methothrin-1	0.0083	0.0417	430	mefenpyr-diethyl	0.0125	0.0625
403	methothrin-2	0.0083	0.0417	431	etoxazole	0.0250	0.1250
404	phenthoate	0.0083	0.0417	432	epoxiconazole-2	0.0333	0.1667
405	fenoxanil	0.0083	0.0417	433	pyriproxyfen	0.0083	0.0417
406	furalaxyl	0.0083	0.0417	434	chromafenozide	0.0333	0.1667
407	thiamethoxam	0.0668	0.0833	435	piperophos	0.0125	0.0625
408	bromacil	0.0083	0.0417	436	fenamidone	0.0042	0.0208
409	picoxystrobin	0.0083	0.0417	437	cis-permethrin	0.0168	0.0208
410	butamifos	0.0042	0.0208				

Continued

TABLE 3.1 The LOD, LOQ of 511 Pesticides and Related Chemical Residues Determined by GC-MS—cont'd

No.	Pesticides	LOD for Milk / (mg/ mL)	LOD for Milk Powder / (mg/kg)	No.	Pesticides	LOD for Milk / (mg/ mL)	LOD for Milk Powder / (mg/kg)
438	pyraclostrobin	0.4000	2.0000	465	DMSA	—	0.1667
439	tralkoxydim	0.0333	0.1667	466	sobutylazine	0.0083	0.0417
440	pyraclofos	0.0333	0.1667	467	cinmethylin	0.0083	—
441	dialifos	0.1333	0.6667	468	pirimicarb	0.0083	—
442	spirodiclofen	0.0333	0.1667	469	s421 (octachlorodipropyl ether)-1	0.0833	0.4167
443	flurtamone	0.0332	0.0417				
444	pyrimidifen	0.0083	0.0417	470	s421 (octachlorodipropyl ether)-2	0.0833	0.4167
445	silafluofen	0.0168	0.0832				
446	butafenacil	0.0042	0.0208	471	dodemorph	0.0125	—
	Group F			472	desmedipham	0.0833	0.4167
447	tribenuron-methyl	0.0042	0.0208	473	fenchlorphos	0.0167	0.0833
448	ethiofencarb	0.0417	0.2083	474	esprocarb	0.0083	0.0417
449	dioxacarb	0.0333	0.1667	475	difenoxuron	0.0333	0.1667
450	dimethyl phthalate	0.0167	0.0833	476	butralin	0.0167	0.0833
451	4-chlorophenoxy acetic acid	0.0084	0.0104	477	dimethametryn	0.0042	0.0208
452	phthalimide	0.0083	0.0417	478	flufenacet	0.0333	0.6668
453	acephate	0.0833	0.4167	479	pyrifenox-1	0.0333	—
454	diethyltoluamide	0.0033	0.0167	480	bensulide	0.0333	0.1667
455	2,4-D	0.0833	0.4167	481	thiabendazole	0.0833	1.6668
456	carbaryl	0.0125	0.0625	482	iprovalicarb-1	0.0167	0.0833
457	cadusafos	0.0167	0.0833	483	azaconazole	0.0167	0.0833
458	endothal	—	0.4167	484	iprovalicarb-2	0.0167	0.0833
459	demetom-s	0.0167	0.0833	485	diofenolan -1	0.0083	0.0417
460	dicrotophos	0.0333	—	486	diofenolan -2	0.0083	0.0417
461	3.4.5-trimethacarb	0.0333	0.1667	487	aclonifen	0.0833	0.4167
462	2,4,5-T	0.0833	1.6668	488	bioresmethrin	0.0332	0.0417
463	3-phenylphenol	0.0250	0.1250	489	carfentrazone-ethyl	0.0083	0.0417
464	furmecyclox	0.0500	0.0625	490	endrin aldehyde	0.3332	0.4167

TABLE 3.1 The LOD, LOQ of 511 Pesticides and Related Chemical Residues Determined by GC-MS—cont'd

No.	Pesticides	LOD for Milk / (mg/ mL)	LOD for Milk Powder / (mg/kg)	No.	Pesticides	LOD for Milk / (mg/ mL)	LOD for Milk Powder / (mg/kg)
491	halosulfuran-methyl	0.3332	0.4167	502	*gamma*-cyhalothrin-2	0.0033	0.0167
492	tricyclazole	0.0250	0.1250	503	cythioate	0.0833	0.4167
493	fenhexamid	0.0833	0.4167	504	halfenprox	0.0083	0.0417
494	spiromesifen	0.0417	0.2083	505	pyriftalid	0.0042	0.0208
495	phenkapton	0.0250	0.1250	506	acetamiprid	0.0167	0.0833
496	famphur	0.0167	0.3332	507	boscalid	0.0167	0.0833
497	fluazinam	0.1332	0.6668	508	tralomethrin-1	0.0042	0.0208
498	bifenazate	0.1332	0.1667	509	tralomethrin-2	0.0042	0.0832
499	endrin ketone	0.0667	0.3333	510	dimethomorph	0.0083	0.0417
500	*gamma*-cyhaloterin-1	0.0033	0.0668	511	azoxystrobin	—	0.2083
501	cyhalofop-butyl	0.0083	0.0417				

Ethyl Acetate: HPLC Grade

NaCl: Analytically Pure

Magnesium Sulfate (MgSO$_4$·7H$_2$O): Analytical Reagent

Pesticides and Related Chemicals Standard: Purity ≥95%

Pesticides and Related Chemicals Standard Solution

Stock Standard Solution: Accurately weigh 5–10 mg of individual pesticide and related chemical standards (accurate to 0.1 mg) into a 10-mL volumetric flask. Dissolve and dilute to volume with toluene, toluene-acetone combination, dichloromethane, etc., depending on each individual compound's solubility. (For diluting solvent, refer to Appendix B.) Mixed Standard Solution (Mixed Standard Solutions A, B, C, D, E, and F): Depending on properties and retention time of each compound, all compounds are divided into sixd groups: A, B, C, D, E, and F. The concentrations of the mixed standard solutions are decided based on instrument sensitivity of each compound. For 511 pesticides and related chemical groupings and concentrations of mixed standard solutions of this standard, reference is made to Appendix B.

Depending on group number, mixed standard solution concentration, and stock standard solution concentration, appropriate amounts of individual stock standard solutions are pipetted into a 100-mL volumetric flask, with five group pesticides and related chemicals being diluted to volume with toluene. Mixed standard solutions are stored in the dark below 4°C and are used for 1 month.

Internal Standard Solution: Accurately weigh 3.5 mg heptachlor epoxide into a 100-mL volumetric flask. Dissolve and dilute to volume with toluene. Working Standard Mixed Solution in Matrix: Working standard mixed solutions in matrix of A, B, C, D, E, and F group pesticides and related chemicals are prepared by diluting 40 µL internal standard solution and appropriate amounts of mixed standard solution to 1.0 mL with blank extract that has been taken through the method with the rest of the samples, and mixing thoroughly.

Working standard mixed solutions in matrix must be freshly prepared. ENVI™-18 SPE Cartridge: 12 mL, 2000 mg, or Equivalent

3.1.4 Apparatus

GC-MS: Equipped With EI
Analytical Balances: Capable of Weighing From 0.1 mg to 0.01 g
Centrifuge Tube: 50 mL
Oscillator
Homogenizer: Not Less Than 15,000 rpm
Pear-Shaped Flask: 100 mL
Pipette: 1 mL and 10 mL
Centrifuge: Not Less Than 12,000 rpm
Rotary Evaporator
Sample Vial: 2 mL, With PTFE Screw-Cap

3.1.5 Sample Pretreatment

3.1.5.1 Storage of Test Sample

The milk powder samples are stored at normal temperature. The milk samples are stored at 0°C to 4°C.

3.1.5.2 Extraction

Milk powder: Weigh 3.0 g test sample (accurate to 0.01 g) into 50-mL centrifuge tube. Add 20 mL acetonitrile and 4 g $MgSO_4$ and homogenize at 15,000 rpm for 1 min; then centrifuge for 5 min at 4200 rpm. Collect the extracts into 100-mL pear-shaped flask and rehomogenize sample in centrifuge tube with 20 mL acetonitrile; recentrifuge and then transfer the extracts to the pear-shaped flask. Concentrate the extracts to ca. 1 mL using the rotary evaporator at 40°C for clean-up.

Milk: measure 10 mL test sample into 50-mL centrifuge tube. Add 20 mL acetonitrile, 4 g $MgSO_4$, and 1 g NaCl, shake for 1 min, then centrifuge for 8 min at 4200 rpm. Collect the top acetonitrile layer of extracts to 100 mL in pear-shaped flask, reshake sample in centrifuge tube with 20 mL acetonitrile, recentrifuge, and then transfer the extracts to the pear-shaped flask. Concentrate the extracts to ca. 1 mL using the rotary evaporator at 40°C for clean-up.

3.1.5.3 Clean-up

Condition the cartridge with 10 mL acetonitrile before adding the sample. Fix the cartridge into a support to which a pear-shaped flask is connected. Pipette the concentrated extracts into the cartridge. Rinse the pear-shaped flask with 2×5 mL acetonitrile and decant it to the cartridge. Collect the eluate to a 100-mL pear-shaped flask and elute the pesticides with 10 mL acetonitrile. Evaporate the eluate to ca. 0.5 mL using a rotary evaporator at 40°C. Exchange with 2×5 mL hexane twice and diluted to ca. 1 mL. Add internal standard solution (40 μL) and the solution, after thorough mixing, is then ready for GC-MS analysis.

Extract and cleanup blank milk and milk powder samples as that described above.

3.1.6 Determination

3.1.6.1 GC-MS Operating Conditions

(a) Column: DB-170 (14% n-propyl cyanide-phenyl)-methyl polysiloxane ($30 m \times 0.25 mm \times 0.25 μm$) capillary column, or equivalent

(b) Column temperature: 40°C hold 1 min, at 30°C/min to 130°C, at 5°C/min to 250°C, at 10°C/min to 300°C, hold 5 min

(c) Carrier gas: Helium, purity ≥99.999%, flow rate: 1.2 mL/min

(d) Injection port temperature: 290°C

(e) Injection volume: 1 μL

(f) Injection mode: Splitless, purge on after 1.5 min

(g) Ionization voltage: 70 eV

(h) Ion source temperature: 230°C

(i) GC-MS interface temperature: 280°C

(j) Solvent delay: A Group 8.30 min, B Group 7.80 min, C Group 7.30 min, D Group 5.50 min, E Group 6.10 min, F Group 5.50 min

(k) Selected ion monitoring mode: For each compound select 1 quantifying ion and 2–3 qualifying ions. All of the detected ions of each group are detected according to programmed time and sequence of peaking. The retention times, quantifying ions, qualifying ions, and the abundance ratios of quantifying ion and qualifying ions for each compound are listed in Table 5.3. The programmed time and dwell time for the ions detected for each compound in each group are listed in Table 5.4.

3.1.6.2 Qualitative Determination

In the samples determined, four injections are required to analyze all pesticides and related chemicals by GC-MS under the operating conditions. If the retention times of peaks of the sample solution are the same as those of the peaks from the working standard mixed solution, and the selected ions appear in the background-subtracted mass spectrum, and the abundance ratios of the selected ions are within the expected limits (abundance ratios >50%, permitted tolerances are ±10%; abundance ratios >20% to 50%, permitted tolerances are ±15%; abundance ratios >10% to 20%, permitted tolerances are ±20%; abundance ratios ≤10%, permitted tolerances are ±50%), the sample is confirmed to contain this pesticide compound. If the results are not definitive, the sample is reinjected with acquisition in scan mode (sufficient sensitivity) or with additional confirmatory ions or using other instruments of higher sensitivity.

3.1.6.3 Quantitative Determination

Quantitation is performed using an internal standard and the quantifying ion for GC-MS. The internal standard is heptachlor epoxide. To compensate for the matrix effects, quantitation is based on a mixed standard prepared in blank matrix extract. The concentrations in the standard solution and in the sample solution analyzed must be close.

3.1.6.4 Parallel Test

A parallel test is carried out for the same testing sample.

3.1.6.5 Blank Test

The operation of the blank test is the same as that described in the determination method, but without the addition of sample.

3.1.7 Precision

The precision data of the method for this standard have been determined according to the stipulations of GB/T 6379.1 and GB/T 6379.2. The values of repeatability and reproducibility are obtained and calculated at the 95% confidence level.

RESEARCHERS

Guo-Fang Pang, Jun-Hong Zheng, Chun-Lin Fan, Feng Zheng.
　Qinhuangdao Entry-Exit Inspection and Quarantine Bureau, 39 Haibin Rd, Qinhuangdao, Hebei, PC 066002, People's Republic of China.

3.2 DETERMINATION OF 493 PESTICIDES AND RELATED CHEMICAL RESIDUES IN MILK AND MILK POWDER: LC-MS-MS METHOD (GB/T 23211-2008)

3.2.1 Scope

This method is applicable to the quantitative determination of 482 pesticides and related chemical residues and qualitative determination of 441 pesticides and related chemical residues in milk. It is applicable to the quantitative determination of 481 pesticides and related chemical residues and qualitative determination of 427 pesticides and related chemical residues in milk powder.

The limit of quantitative determination of this method of 441 pesticides and related chemicals is 0.01 µg/L to 2.41 mg/L. The limit of quantitative determination of this method of 427 pesticides and related chemicals is 0.04 µg/kg to 8.04 mg/kg. See Table 3.2.

3.2.2 Principle

The samples were extracted with acetonitrile and cleaned up with a C_{18} SPE cartridge. The pesticides and related chemicals are eluted with acetonitrile, and the solutions are analyzed by LC-MS-MS.

Quantitation with internal method.

3.2.3 Reagents and Materials

"Water" is first grade of GB/T6682 as specified.

Acetonitrile: HPLC Grade
Acetone: HPLC Grade
Isooctane: HPLC Grade
Toluene: Guaranteed Reagent
Hexane: HPLC Grade
Methanol: HPLC Grade
Formic Acid: Guaranteed Reagent
Ammonium Acetate: Guaranteed Reagent
0.1 % Formic Acid Water
5 mmol/L Ammonium Acetate
Acetonitrile + water (3+2, V/V)
NaCl: Analytical Reagent
Magnesium Sulfate ($MgSO_4 \cdot 7H_2O$): Analytical Reagent
Pesticide and Related Chemicals Standard: Purity $\geq 95\%$
   Stock Standard Solution: Accurately weigh 5–10 mg of individual pesticide and related chemical standards (accurate to 0.1 mg) into a 10-mL volumetric flask. Dissolve and dilute to volume with methanol, toluene,

TABLE 3.2 The LOD, LOQ of 493 Pesticides and Related Chemical Residues Determined by LC-MS-MS

No.	Pesticides	LOD for Milk (μg/L)	LOD for Milk Powder (μg/kg)	No.	Pesticides	LOD for Milk (μg/L)	LOD for Milk Powder (μg/kg)
	Group A			27	nitenpyram	4.28	14.25
1	propham	27.50	91.67	28	methoprotryne	0.05	0.17
2	isoprocarb	0.58	1.92	29	dimethenamid	4.32	3.58
3	3,4,5-trimethacarb	0.08	0.25	30	terrbucarb	0.53	1.75
4	cycluron	0.05	0.17	31	penconazole	0.50	1.67
5	carbaryl	2.58	8.58	32	myclobutanil	0.25	0.83
6	propachlor	0.08	0.25	33	imazethapyr	0.28	0.92
7	rabenzazole	0.33	1.08	34	paclobutrazol	0.15	0.50
8	simetryn	0.03	0.08	35	fenthion sulfoxide	0.08	0.25
9	monolinuron	0.90	3.00	36	triadimenol	2.65	8.83
10	mevinphos	0.40	1.33	37	butralin	0.48	1.58
11	aziprotryne	0.35	1.17	38	tolclofos methyl	16.65	55.50
12	secbumeton	0.03	0.08	39	desmedipham	1.00	3.33
13	cyprodinil	0.18	0.58	40	allethrin	15.10	50.33
14	buturon	2.25	7.50	41	pretilachlor	0.08	0.25
15	carbetamide	0.90	3.00	42	flusilazole	0.15	0.50
16	pirimicarb	0.05	0.17	43	benodanil	0.88	2.92
17	clomazone dimethazone	0.10	0.33	44	flutolanil	0.28	0.92
				45	famphur	0.90	3.00
18	prometryne	0.05	0.17	46	benalyxyl	0.30	1.00
19	paraoxon methyl	0.20	0.67	47	Diclobutrazole	0.13	0.42
20	4,4-dichlorobenzophenone	3.40	11.33	48	etaconazole	0.45	1.50
21	thiacloprid	0.10	0.33	49	phthalic acid, dicyclohexyl ester	0.50	1.67
22	imidacloprid	5.50	18.33	50	tetramethirn	0.45	1.50
23	ethidimuron	0.38	1.25	51	dichlofluanid	0.65	2.17
24	isomethiozin	0.28	0.92	52	chlorprifos methyl	4.00	13.33
25	diallate	22.30	74.33	53	bitertanol	8.35	27.83
26	acetochlor	11.85	39.50	54	tepraloxydim	3.05	10.17

TABLE 3.2 The LOD, LOQ of 493 Pesticides and Related Chemical Residues Determined by LC-MS-MS—cont'd

No.	Pesticides	LOD for Milk (µg/L)	LOD for Milk Powder (µg/kg)	No.	Pesticides	LOD for Milk (µg/L)	LOD for Milk Powder (µg/kg)
55	thiophanate methyl	5.00	16.67	82	isouron	0.10	0.33
56	azinphos ethyl	27.23	—	83	chlorotoluron	0.15	0.50
57	triflumuron	0.98	3.25	84	thiofanox	39.25	130.83
58	anilofos	0.18	0.58	85	chlorbufam	45.75	152.50
59	thiophanat ethyl	5.05	16.83	86	bendiocarb	0.80	2.67
60	quizalofop-ethyl	0.18	0.58	87	propazine	0.08	0.25
61	haloxyfop-methyl	0.65	2.17	88	terbuthylazine	0.13	0.42
62	fluazifop butyl	0.08	0.25	89	diuron	0.40	1.33
63	bromophos-ethyl	141.93	473.08	90	chlormephos	112.00	373.33
64	bensulide	8.55	28.50	91	carboxin	0.15	0.50
65	bromfenvinfos	0.75	2.50	92	difenzoquat-methyl sulfate	0.20	0.67
66	azoxystrobin	0.13	0.42				
67	pyrazophos	0.40	1.33	93	clothianidin	15.75	52.50
68	flufenoxuron	0.80	2.67	94	pronamide	3.85	12.83
69	indoxacarb	1.88	6.25	95	dimethachloro	0.48	1.58
70	emamectin benzoate	0.08	0.25	96	methobromuron	4.20	14.00
	Group B			97	phorate	78.50	261.67
71	ethylene thiourea	13.05	—	98	aclonifen	6.05	20.17
72	dazomet	—	105.83	99	mephosfolan	0.58	1.92
73	nicotine	0.55	1.83	100	imibenzonazole-des-benzyl	1.55	5.17
74	fenuron	0.25	0.83				
75	cyromazine	1.80	6.00	101	neburon	1.78	5.92
76	crimidine	0.40	1.33	102	mefenoxam	0.38	1.25
77	acephate	3.33	11.08	103	ethofume sate	93.00	310.00
78	molinate	0.53	1.75	104	iprobenfos	2.08	6.92
79	carbendazim	0.13	0.42	105	TEPP	2.60	8.67
80	6-chloro-4-hydroxy-3-phenyl-pyridazin	0.43	1.42	106	cyproconazole	0.18	0.58
81	propoxur	6.10	20.33	107	thiamethoxam	8.25	27.50
				108	crufomate	0.13	0.42

TABLE 3.2 The LOD, LOQ of 493 Pesticides and Related Chemical Residues Determined by LC-MS-MS—cont'd

No.	Pesticides	LOD for Milk (μg/L)	LOD for Milk Powder (μg/kg)	No.	Pesticides	LOD for Milk (μg/L)	LOD for Milk Powder (μg/kg)
109	etrimfos	4.70	15.67	138	chlorsulfuron	0.68	2.25
110	coumatetralyl	0.35	1.17	139	clethodim	0.53	1.75
111	cythioate	20.00	66.67	140	flamprop isopropyl	0.10	0.33
112	phosphamidon	0.98	3.25	141	tetrachlorvinphos	0.55	1.83
113	phenmedipham	1.13	3.75	142	propargite	17.15	57.17
114	bifenazate	5.70	19.00	143	bromuconazole	0.78	2.58
115	fenhexamid	0.23	—	144	picolinafen	0.18	0.58
116	flutriafol	2.15	7.17	145	fluthiacet methyl	1.33	4.42
117	furalaxyl	0.20	0.67	146	trifloxystrobin	0.50	1.67
118	bioallethrin	49.50	165.00	147	chlorimuron ethyl	7.60	25.33
119	cyanofenphos	5.20	17.33	148	hexaflumuron	6.30	21.00
120	pirimiphos methyl	0.05	0.17	149	novaluron	2.00	6.67
121	buprofezin	0.23	0.75	150	hydramethylnon	0.43	1.42
122	disulfoton sulfone	0.63	2.08	151	flurazuron	6.70	22.33
123	fenazaquin	0.08	0.25	152	maleic hydrazide	20.00	66.67
124	triazophos	0.18	0.58		Group C		
125	DEF	0.40	1.33	153	methamidophos	1.23	4.08
126	pyriftalid	0.15	0.50	154	EPTC	9.33	31.08
127	metconazole	0.33	1.08	155	diethyltoluamide	0.15	0.50
128	pyriproxyfen	0.10	0.33	156	monuron	8.68	28.92
129	cycloxydim	0.63	2.08	157	pyrimethanil	0.18	0.58
130	isoxaben	0.05	0.17	158	fenfuram	0.20	0.67
131	flurtamone	0.10	0.33	159	quinoclamine	1.98	6.58
132	trifluralin	310.00	1033.33	160	fenobucarb	1.48	4.92
133	flamprop methyl	5.05	16.83	161	ethirimol	0.15	0.50
134	bioresmethrin	1.85	6.17	162	propanil	5.40	18.00
135	propiconazole	0.45	1.50	163	carbofuran	3.28	10.92
136	chlorpyrifos	13.45	44.83	164	acetamiprid	0.35	1.17
137	fluchloralin	122.00	406.67	165	mepanipyrim	0.08	0.25

TABLE 3.2 The LOD, LOQ of 493 Pesticides and Related Chemical Residues Determined by LC-MS-MS—cont'd

No.	Pesticides	LOD for Milk (µg/L)	LOD for Milk Powder (µg/kg)	No.	Pesticides	LOD for Milk (µg/L)	LOD for Milk Powder (µg/kg)
166	prometon	0.03	0.08	194	isoprothiolane	0.45	1.50
167	metoxuron	0.15	0.50	195	imazalil	0.50	1.67
168	dimethoate	1.90	6.33	196	phoxim	20.70	69.00
169	methfuroxam	0.08	—	197	quinalphos	0.50	1.67
170	fluometuron	0.23	0.75	198	ditalimfos	16.80	56.00
171	dicrotophos	0.28	0.92	199	fenoxycarb	4.58	15.25
172	monalide	0.30	1.00	200	pyrimitate	0.05	0.17
173	diphenamid	0.03	0.08	201	fensulfothin	0.50	1.67
174	ethoprophos	0.70	2.33	202	fluorochloridone	3.45	11.50
175	fonofos	1.88	6.25	203	butachlor	5.03	16.75
176	etridiazole	25.10	83.67	204	imazaquin	0.73	2.42
177	furmecyclox	0.20	0.67	205	kresoxim-methyl	25.15	83.83
178	hexazinone	0.03	0.08	206	triticonazole	0.75	2.50
179	dimethametryn	0.03	0.08	207	fenamiphos sulfoxide	0.18	0.58
180	trichlorphon	0.28	0.92	208	thenylchlor	6.03	20.08
181	demeton(o+s)	1.70	5.67	209	fenoxanil	9.85	32.83
182	benoxacor	1.73	5.75	210	fluridone	0.05	0.17
183	bromacil	5.90	19.67	211	epoxiconazole	1.03	3.42
184	phorate sulfoxide	86.63	288.75	212	chlorphoxim	19.40	64.67
185	brompyrazon	0.90	3.00	213	fenamiphos sulfone	0.10	0.33
186	oxycarboxin	0.23	0.75	214	fenbuconazole	0.40	1.33
187	mepronil	0.10	0.33	215	isofenphos	54.68	182.25
188	disulfoton	117.43	391.42	216	oryzalin	16.00	53.33
189	fenthion	13.00	43.33	217	fentin-chloride	4.33	14.42
190	metalaxyl	0.13	0.42	218	piperophos	2.30	7.67
191	ofurace	0.25	0.83	219	piperonyl butoxide	0.28	0.92
192	dodemorph	0.10	0.33	220	oxyflurofen	14.63	48.75
193	imazamethabenz-methyl	0.05	0.17	221	coumaphos	0.53	1.75
				222	flufenacet	1.33	4.42

Continued

TABLE 3.2 The LOD, LOQ of 493 Pesticides and Related Chemical Residues Determined by LC-MS-MS—cont'd

No.	Pesticides	LOD for Milk (µg/L)	LOD for Milk Powder (µg/kg)	No.	Pesticides	LOD for Milk (µg/L)	LOD for Milk Powder (µg/kg)
223	phosalone	12.00	40.00	251	pymetrozin	—	28.58
224	methoxyfenozide	0.93	3.08	252	chloridazon	0.58	1.92
225	prochloraz	0.53	1.75	253	sulfallate	51.80	172.67
226	aspon	0.43	1.42	254	ethiofencarb	1.23	4.08
227	ethion	0.75	2.50	255	terbumeton	0.03	0.08
228	diafenthiuron	0.08	1.00	256	cyanazine	0.05	0.17
229	thifensulfuron-methyl	5.35	17.83	257	ametryn	0.25	0.83
230	ethoxysulfuron	1.15	3.83	258	tebuthiuron	0.05	0.17
231	dithiopyr	2.60	8.67	259	trietazine	0.15	0.50
232	spirodiclofen	2.48	8.25	260	sebutylazine	0.08	0.25
233	fenpyroximate	0.35	1.17	261	dibutyl succinate	55.60	185.33
234	flumiclorac-pentyl	2.65	8.83	262	tebutam	0.03	0.08
235	temephos	0.30	1.00	263	thiofanox-sulfoxide	2.08	6.92
236	butafenacil	2.38	7.92	264	terbutryn	0.05	0.17
237	spinosad	0.15	0.50	265	triazoxide	2.00	6.67
	Group D			266	thionazin	5.68	18.92
238	mepiquat chloride	0.23	0.75	267	linuron	2.90	9.67
239	allidochlor	10.25	34.17	268	heptanophos	1.45	4.83
240	propamocarb	0.03	0.08	269	prosulfocarb	0.10	0.33
241	thiabendazole	0.13	0.42	270	dipropetryn	0.08	0.25
242	metamitron	1.60	5.33	271	thiobencarb	0.83	2.75
243	isoproturon	0.03	0.08	272	tri-iso-butyl phosphate	1.00	—
244	atratone	0.05	0.17	273	tri-n-butyl phosphate	0.10	0.33
245	desmetryn (oesmetryn)	0.05	0.17	274	diethofencarb	0.50	1.67
246	metribuzin	0.13	0.42	275	alachlor	1.85	6.17
247	DMST	10.00	33.33	276	cadusafos	0.30	1.00
248	cycloate	1.10	3.67	277	metazachlor	0.25	0.83
249	atrazine	0.10	0.33	278	propetamphos	13.50	45.00
250	butylate	75.50	251.67	279	terbufos	560.00	1866.67

TABLE 3.2 The LOD, LOQ of 493 Pesticides and Related Chemical Residues
Determined by LC-MS-MS—cont'd

No.	Pesticides	LOD for Milk (µg/L)	LOD for Milk Powder (µg/kg)	No.	Pesticides	LOD for Milk (µg/L)	LOD for Milk Powder (µg/kg)
280	simeconazole	0.73	2.42	309	tetraconazole	0.43	1.42
281	triadimefon	1.98	6.58	310	mefenpyr-diethyl	3.15	10.50
282	phorate sulfone	10.50	35.00	311	profenefos	0.50	1.67
283	tridemorph	0.65	2.17	312	pyraclostrobin	0.13	0.42
284	mefenacet	0.55	1.83	313	dimethomorph	0.10	0.33
285	fenpropimorph	0.05	0.17	314	kadethrin	0.83	2.75
286	tebuconazole	0.55	1.83	315	thiazopyr	0.50	1.67
287	isopropalin	7.50	25.00	316	benfuracarb-methyl	4.10	13.67
288	nuarimol	0.25	0.83	317	cinosulfuron	0.28	0.92
289	bupirimate	0.18	0.58	318	pyrazosulfuron-ethyl	1.70	5.67
290	azinphos-methyl	276.08	920.25	319	metosulam	1.10	3.67
291	tebupirimfos	0.03	0.08	320	chlorfluazuron	2.18	7.25
292	phenthoate	23.10	77.00		Group E		
293	sulfotep	0.65	2.17	321	chlormequat	0.03	0.08
294	sulprofos	1.45	4.83	322	methomyl	2.40	8.00
295	EPN	8.25	27.50	323	pyroquilon	0.88	2.92
296	azamethiphos	0.20	0.67	324	fuberidazole	0.48	1.58
297	diniconazole	0.33	1.08	325	isocarbamid	0.43	1.42
298	flumetsulam	0.08	0.25	326	butocarboxim	0.40	1.33
299	pencycuron	0.08	0.25	327	chlordimeform	0.33	1.08
300	mecarbam	4.90	16.33	328	cymoxanil	13.90	46.33
301	tralkoxydim	0.08	0.25	329	vernolate	0.08	0.25
302	malathion	1.40	4.67	330	chlorthiamid	2.20	7.33
303	pyributicarb	0.08	0.25	331	aminocarb	4.10	13.67
304	pirimiphos-ethyl	0.01	0.04	332	dimethirimol	—	0.08
305	pyridaphenthion	0.23	0.75	333	chlortoluron	0.08	0.25
306	thiodicarb	9.85	32.83	334	omethoate	2.43	8.08
307	pyraclofos	0.25	0.83	335	ethoxyquin	0.88	2.92
308	picoxystrobin	2.10	7.00	336	dichlorvos	0.13	0.42

Continued

TABLE 3.2 The LOD, LOQ of 493 Pesticides and Related Chemical Residues Determined by LC-MS-MS—cont'd

No.	Pesticides	LOD for Milk (µg/L)	LOD for Milk Powder (µg/kg)	No.	Pesticides	LOD for Milk (µg/L)	LOD for Milk Powder (µg/kg)
337	aldicarb sulfone	5.35	17.83	364	isazofos	0.05	0.17
338	dioxacarb	0.85	2.83	365	dichlofenthion	7.55	25.17
339	benzyladenine	17.70	59.00	366	vamidothion sulfone	119.00	396.67
340	demeton-s-methyl	1.33	4.42	367	terbufos sulfone	22.15	73.83
341	ethiofencarb-sulfoxide	56.00	186.67	368	dinitramine	0.45	1.50
342	thiometon	144.50	481.67	369	cyazofamid	1.13	3.75
343	folpet	34.65	—	370	trichloronat	16.70	55.67
344	demeton-s-methyl sulfone	4.95	16.50	371	resmethrin-2	0.08	0.25
				372	boscalid	1.20	4.00
345	dimepiperate	945.00	3150.00	373	nitralin	8.60	28.67
346	fenpropidin	—	0.17	374	fenpropathrin	61.25	204.17
347	amidithion	164.50	548.33	375	hexythiazox	5.90	19.67
348	imazapic	1.48	4.92	376	florasulam	4.35	14.50
349	paraoxon-ethyl	0.13	0.42	377	benzoximate	4.93	16.42
350	aldimorph	0.80	2.67	378	benzoylprop-ethyl	77.00	256.67
351	vinclozolin	0.63	2.08	379	pyrimidifen	—	11.67
352	uniconazole	0.20	0.67	380	furathiocarb	0.48	1.58
353	pyrifenox	0.08	0.25	381	trans-permethin	1.20	4.00
354	chlorthion	33.40	111.33	382	etofenprox	570.00	1900.00
355	dicapthon	0.05	0.17	383	pyrazoxyfen	0.08	0.25
356	clofentezine	0.20	0.67	384	flubenzimine	1.95	6.50
357	norflurazon	0.08	0.25	385	zeta cypermethrin	0.18	0.58
358	triallate	5.05	16.83	386	haloxyfop-2-ethoxyethyl	0.63	2.08
359	ziram	19.60	—				
360	quinoxyphen	38.35	127.83	387	fluoroglycofen-ethyl	1.25	4.17
361	fenthion sulfone	4.38	14.58	388	tau-fluvalinate	57.50	191.67
362	flurochloridone	0.33	1.08		Group F		
363	phthalic acid,benzyl butyl ester	158.00	526.67	389	acrylamide	4.45	14.83
				390	tert-butylamine	9.75	32.50

TABLE 3.2 The LOD, LOQ of 493 Pesticides and Related Chemical Residues Determined by LC-MS-MS—cont'd

No.	Pesticides	LOD for Milk (μg/L)	LOD for Milk Powder (μg/kg)	No.	Pesticides	LOD for Milk (μg/L)	LOD for Milk Powder (μg/kg)
391	hymexazol	56.03	186.75	417	procymidone	21.65	72.17
392	phthalimide	10.75	35.83	418	vamidothion	1.15	30.67
393	dimefox	17.05	56.83	419	cumyluron	0.33	1.08
394	metolcarb	6.35	21.17	420	ronnel	3.28	10.92
395	diphenylamin	0.10	0.33	421	pyrethrin	88.15	293.83
396	1-naphthy acetamide	0.20	0.67	422	phthalic acid, biscyclohexyl ester	0.18	0.58
397	atrazine-desethyl	0.15	0.50	423	carpropamid	1.30	4.33
398	2,6-dichlorobenzamide	1.13	3.75	424	tebufenozide	6.95	23.17
399	aldicarb	65.25	217.50	425	chlorthiophos	7.95	26.50
400	dimethyl phthalate	3.30	11.00	426	dialifos	39.25	130.83
401	chlordimeform hydrochloride	0.65	2.17	427	cinidon-ethyl	3.65	12.17
				428	rotenone	0.58	1.92
402	simeton	0.28	0.92	429	imibenconazole	2.58	8.58
403	dinotefuran	2.55	8.50	430	propaquiafop	0.30	1.00
404	pebulate	0.85	2.83	431	lactofen	15.50	51.67
405	acibenzolar-s-methyl	0.78	2.58	432	benzofenap	0.03	0.08
406	dioxabenzofos	3.45	11.50	433	dinoseb acetate	10.33	34.42
407	oxamyl	137.03	456.75	434	propisochlor	0.20	0.67
408	methabenzthiazuron	0.03	0.08	435	etobenzanid	0.20	0.67
409	butoxycarboxim	6.65	22.17	436	fentrazamide	3.10	10.33
410	demeton-s-methyl sulfoxide	—	3.25	437	cyphenothrin	4.20	14.00
411	phosfolan	0.13	0.42	438	dieldrin	40.40	538.68
412	demeton-s	—	66.67	439	malaoxon	1.18	3.92
413	napropamide	0.33	1.08	440	dodine	2.00	—
414	fenitrothion	6.70	22.33	441	propylene thiourea	7.53	25.08
415	phthalic acid, dibutyl ester	9.90	33.00		Group G		
				442	dalapon	57.68	192.25
416	metolachlor	0.10	0.33	443	flupropanate	5.75	19.17

Continued

TABLE 3.2 The LOD, LOQ of 493 Pesticides and Related Chemical Residues Determined by LC-MS-MS—cont'd

No.	Pesticides	LOD for Milk (µg/L)	LOD for Milk Powder (µg/kg)	No.	Pesticides	LOD for Milk (µg/L)	LOD for Milk Powder (µg/kg)
444	2-phenylphenol	42.48	141.58	470	naptalam	—	1.58
445	3-phenylphenol	1.00	3.33	471	chlorobenzuron	5.10	17.00
446	clopyralld	70.00	233.33	472	chloramphenicolum	0.98	3.25
447	DNOC	0.65	2.17	473	alloxydim-sodium	0.05	0.17
448	cloprop	2.85	9.50	474	pyrithlobac sodium	345.50	1151.67
449	dicloran	12.15	40.50	475	dimehypo	100.05	333.50
450	aminopyralid	—	305.00	476	sulfanitran	0.75	2.50
451	chlorpropham	3.95	13.17	477	oryzalin	1.23	4.08
452	mecoprop	1.23	4.08	478	gibberellic acid	16.58	55.25
453	terbacil	0.23	0.75	479	acifluorfen	29.50	98.33
454	2,4-D	2.98	9.92	480	ioxynil	0.15	0.50
455	dicamba	316.48	1054.92	481	famoxadone	11.33	37.75
456	MCPB	3.55	11.83	482	sulfentrazone	22.40	74.67
457	fenaminosulf	56.35	187.83	483	diflufenican	7.08	23.58
458	picloram	133.53	445.09	484	ethiprole	9.98	33.25
459	bentazone	0.25	0.83	485	flusulfamide	0.10	0.33
460	dinoseb	0.10	0.33	486	cyclosulfamuron	85.93	286.42
461	dinoterb	0.05	0.17	487	fomesafen	0.50	1.67
462	forchlorfenuron	2.85	9.50	488	fluazinam	17.65	58.83
463	fludioxonil	15.55	51.83	489	fluazuron	0.05	0.17
464	2,4,5-T	—	14.58	490	iodosulfuron-methyl sodium	5.30	17.67
465	fluroxypyr	—	160.08				
466	chlorfenethol	41.08	136.92	491	kelevan	2410.70	8035.67
467	fenoxaprop	1.63	5.42	492	acrinathrin	2.03	6.75
468	cyclanilide	0.85	2.83	493	iodosulfuron-methyl	16.65	55.50
469	bromoxynil	0.45	1.50				

acetone, acetonitrile, isooctane, etc., depending on each individual compound's solubility (for diluting solvent refer to Appendix B).

Mixed Standard Solution (Mixed Standard Solution A, B, C, D, E, F, and G Groups): Depending on properties and retention times of compounds, all compounds are divided into five groups: A, B, C, D, E, F, and G. The concentration of each compound is determined by its sensitivity on the instrument for analysis. For 493 pesticides and related chemicals, groupings and concentrations are listed in Appendix B.

Depending on group number, mixed standard solution concentration, and stock standard solution concentration, appropriate amounts of individual stock standard solution are pipetted into a 100-mL volumetric flask, being diluted to volume with methanol. Mixed standard solutions are stored in the dark below 4°C and used for one month.

Working Standard Mixed Solution in Matrix: Working standard mixture solution in matrix is prepared by diluting an appropriate amount of mixed standard solution of A, B, C, D, E, F, and G groups with blank extract that has been taken through the method with the rest of the samples. Mix thoroughly. These are used for plotting the standard curve.

Working standard mixture solution in matrix must be freshly prepared.

Membrane Filters (Nylon): 13 mm, 0.2 μm

ENVI^{TM}-18 SPE Cartridge: 12 mL, 2000 mg, or Equivalent

3.2.4   Apparatus

LC-MS-MS: Equipped With ESI

Analytical Balances: Capable of Weighing From 0.1 mg to 0.01 g

Centrifuge Tube: 50 mL

Homogenizer: Not Less Than 15,000 rpm

Centrifuge: Not Less Than 12,000 rpm

Pear-Shaped Flask: 100 mL

Pipette: 1 mL and 10 mL

Rotary Evaporator

Nitrogen Evaporator

Sample Vial: 2 mL, With PTFE Screw-Cap

3.2.5   Sample Pretreatment

3.2.5.1   Storage of Test Sample

The milk powder samples are stored at normal temperature; the milk samples are stored at 0°C to 4°C.

3.2.5.2   Extraction

Milk powder: Weigh 3.0 g test sample (accurate to 0.01 g) into 50-mL centrifuge tube. Add 20 mL acetonitrile and 4 g $MgSO_4$, homogenize at 15,000 rpm for

1 min, then centrifuge for 5 min at 4200 rpm. Collect extracts into 100 mL pear-shaped flask, rehomogenize sample plug in centrifuge tube with 20 mL acetonitrile, recentrifuge, and then transfer the extracts to the pear-shaped flask. Concentrate the extracts to ca. 1 mL using the rotary evaporator at 40°C for clean-up.

Milk: measure 10 mL of test sample into 50-mL centrifuge tube. Add 20 mL acetonitrile, 4 g MgSO$_4$, and 1 g NaCl, shake for 1 min, then centrifuge for 8 min at 4200 rpm. Collect the top acetonitrile layer of extracts into 100 mL pear-shaped flask, reshake sample in centrifuge tube with 20 mL acetonitrile, recentrifuge, and then transfer the top layer to the pear-shaped flask. Concentrate the extracts to ca. 1 mL using the rotary evaporator at 40°C for clean-up.

3.2.5.3 Clean-up

Condition the cartridge with 10 mL acetonitrile before adding the sample. Fix the cartridge into a support to which a pear-shaped flask is connected. Pipette the concentrated extracts into the cartridge. Rinse the pear-shaped flask with 5 × 2 mL acetonitrile and decant it to the cartridge. Collect the eluate into a 100-mL pear-shaped flask. Elute the pesticides with 10 mL acetonitrile. Evaporate the eluate to ca. 0.5 mL using a rotary evaporator at 40°C. Then evaporate the eluate to dryness using a nitrogen evaporator. Make up to 1 mL with acetonitrile-water (3+2) and mix thoroughly. Finally, filter the extract through a 0.2-μm filter into a sample vial for LC-MS-MS determination.

Extract and cleanup blank milk and milk powder samples as that described above.

3.2.6 Determination

3.2.6.1 LC-MS-MS Operating Condition

Conditions for A, B, C, D, E, and F Groups

(a) Chromatography column: ZORBAX SB-C$_{18}$, 3.5 μm, 100 mm × 2.1 mm (i.d.), or equivalent
(b) Mobile phase program and the flow rate: as in Table 3.3
(c) Column temperature: 40°C
(d) Injection volume: 10 μL
(e) Ionization mode: electrospray ionization
(f) Scan mode: positive ion
(g) Atomization gas: nitrogen gas
(h) Atomization gas pressure: 0.28 MPa
(i) Ion spray voltage: 4000 V
(j) Drying air Temperature: 350°C
(k) Drying air flow speed: 10 L/min
(l) Monitoring ion pairs, collision energy and fragmentor: see Table 5.10

TABLE 3.3 Mobile Phase Program and the Flow Rate

Program	Time (min)	Flow Rate (µL/min)	Mobile phase A 0.1% Formic Acid-Water (%)	Mobile Phase B Acetonitrile (%)
0	0.00	400	99.0	1.0
1	3.00	400	70.0	30.0
2	6.00	400	60.0	40.0
3	9.00	400	60.0	40.0
4	15.00	400	40.0	60.0
5	19.00	400	1.0	99.0
6	23.00	400	1.0	99.0
7	23.01	400	99.0	1.0

Conditions for G Group

(a) Chromatography column: ZORBAX SB-C$_{18}$, 3.5 µm, 100 mm × 2.1 mm (i.d.), or equivalent
(b) Mobile phase program and the flow rate: as in Table 3.4
(c) Column temperature: 40°C
(d) Injection volume: 10 µL

TABLE 3.4 Mobile Phase Program and the Flow Rate

Program	Time (min)	Flow Rate (µL/min)	Mobile Phase A 0.1% Ammonium Acetate-Water (%)	Mobile Phase B Acetonitrile (%)
0	0.00	400	99.0	1.0
1	3.00	400	70.0	30.0
2	6.00	400	60.0	40.0
3	9.00	400	60.0	40.0
4	15.00	400	40.0	60.0
5	19.00	400	1.0	99.0
6	23.00	400	1.0	99.0
7	23.01	400	99.0	1.0

(e) Ionization mode: electrospray ionization
(f) Scan mode: negative ion
(g) Atomization gas: nitrogen gas
(h) Atomization gas pressure: 0.28 MPa
(i) Ion spray voltage: 4000 V
(j) Drying air Temperature: 350°C
(k) Drying air flow speed: 10 L/min
(l) Monitoring ions pairs, collision energy and fragmentor: see Table 5.10

3.2.6.2 Qualitative Determination

For the samples determined, if the retention times of peaks found in the sample solution chromatogram are the same as the peaks in the standard in blank matrix extract chromatogram, and the abundance ratios of MRM transitions for precursor/product ion are within the expected limits (relative abundance >50%, ±20% deviation permitted; relative abundance of >20%–50%, allowing ±25% deviation; relative abundance of >10%–20%, allowing ±30% deviation; relative abundance ≤10%, ±50% deviation permitted), the sample is confirmed to contain this compound.

3.2.6.3 Quantitative Determination

An external standard method is used for quantitation with standard curves for LC-MS-MS. In order to compensate for the matrix effect, quantitation is based on a series of working standard solutions prepared in blank matrix extract. The standard curves are established by injection of different concentrations of working standard mixed solutions in matrix separately. The responses of pesticides in the sample solution should be in the linear range of the instrumental detection.

3.2.6.4 Parallel Test

A parallel test is carried out for the same testing sample.

3.2.6.5 Blank Test

The operation of the blank test is the same as that described in the method of determination, but without the addition of sample.

3.2.7 Precision

The precision data of the method for this standard is according to the stipulations of GB/T 6379.1 and GB/T 6379.2. The values of repeatability and reproducibility are obtained and calculated at the 95% confidence level.

RESEARCHERS

Guo-Fang Pang, Jun-Hong Zheng, Chun-Lin Fan, Yan Li.
    Qinhuangdao Entry-Exit Inspection and Quarantine Bureau, 39 Haibin Rd, Qinhuangdao, Hebei, PC 066002, People's Republic of China.

3.3 DETERMINATION OF 485 PESTICIDES AND RELATED CHEMICAL RESIDUES IN FUGU, EEL, AND PRAWN: GC-MS METHOD (GB/T 23207-2008)

3.3.1 Scope

This method is applicable to the determination of 485 pesticides and related chemicals residue in fugu, eel, and prawn.

The limits of quantification of 427 pesticides and related chemical residues for this standard are 0.0025 mg/kg to 0.6000 mg/kg (refer to Table 3.5).

TABLE 3.5 The LOD, LOQ of 485 Pesticides and Related Chemical Residues Determined by GC-MS

No.	Pesticides	LOD (mg/kg)	LOQ (mg/kg)	No.	Pesticides	LOD (mg/kg)	LOQ (mg/kg)
	Group A			20	pronamide	0.0125	0.0250
1	allidochlor	0.0250	0.0500	21	dichlofenthion	0.0125	0.0250
2	dichlormid	0.0250	0.0500	22	mexacarbate	0.0375	0.0750
3	etridiazol	0.0375	0.0750	23	dimethoate	0.0500	0.1000
4	chlormephos	0.0250	0.0500	24	dinitramine	0.0500	0.1000
5	propham	0.0125	0.0250	25	aldrin	0.0250	0.0500
6	cycloate	0.0125	0.0250	26	ronnel	0.0250	0.0500
7	diphenylamine	0.0125	0.0250	27	prometryne	0.0125	0.0250
8	chlordimeform	0.0125	0.0250	28	cyprazine	0.0125	0.0250
9	ethalfluralin	0.0500	0.1000	29	vinclozolin	0.0125	0.0250
10	phorate	0.0125	0.0250	30	*beta*-HCH	0.0125	0.0250
11	thiometon	0.0125	0.0250	31	metalaxyl	0.0375	0.0750
12	quintozene	0.0250	0.0500	32	methyl-parathion	0.0500	0.1000
13	atrazine-desethyl	0.0125	0.0250	33	chlorpyrifos (-ethyl)	0.0125	0.0250
14	clomazone	0.0125	0.0250	34	*delta*-HCH	0.0250	0.0500
15	diazinon	0.0125	0.0250	35	fenthion	0.0125	0.0250
16	fonofos	0.0125	0.0250	36	malathion	0.0500	0.1000
17	etrimfos	0.0125	0.0250	37	paraoxon-ethyl	0.4000	0.8000
18	propetamphos	0.0125	0.0250	38	fenitrothion	0.0250	0.0500
19	secbumeton	0.0125	0.0250	39	triadimefon	0.0250	0.0500

Continued

TABLE 3.5 The LOD, LOQ of 485 Pesticides and Related Chemical Residues Determined by GC-MS—cont'd

No.	Pesticides	LOD (mg/kg)	LOQ (mg/kg)	No.	Pesticides	LOD (mg/kg)	LOQ (mg/kg)
40	linuron	0.2000	0.4000	70	propiconazole-2	0.0375	0.0750
41	pendimethalin	0.0500	0.1000	71	bifenthrin*	0.0125	0.0250
42	chlorbenside	0.0250	0.0500	72	mirex	0.0125	0.0250
43	bromophos-ethyl	0.0125	0.0250	73	carbosulfan	0.0375	0.0750
44	quinalphos	0.0125	0.0250	74	nuarimol	0.0250	0.0500
45	trans-chlordane	0.0125	0.0250	75	benodanil	0.0375	0.0750
46	phenthoate	0.0250	0.0500	76	methoxychlor	0.1000	0.2000
47	metazachlor	0.0375	0.0750	77	oxadixyl	0.0125	0.0250
48	prothiophos	0.0125	0.0250	78	tebuconazole	0.0375	0.0750
49	chlorfurenol	0.0375	0.0750	79	tetramethirn	0.0250	0.0500
50	folpet	0.6000	1.2000	80	norflurazon	0.0125	0.0250
51	procymidone	0.0125	0.0250	81	pyridaphenthion	0.0125	0.0250
52	dieldrin	0.0250	0.0500	82	phosmet	0.0250	0.0500
53	methidathion	0.0250	0.0500	83	tetradifon	0.0125	0.0250
54	napropamide	0.0375	0.0750	84	oxycarboxin	0.0750	0.1500
55	cyanazine	0.1500	0.3000	85	cis-permethrin	0.0125	0.0250
56	oxadiazone	0.0125	0.0250	86	pyrazophos	0.0250	0.0500
57	fenamiphos	0.0375	0.0750	87	trans-permethrin	0.0125	0.0250
58	tetrasul	0.0125	0.0250	88	cypermethrin	0.0375	0.0750
59	bupirimate	0.0125	0.0250	89	fenvalerate-1	0.0500	0.1000
60	flutolanil	0.0125	0.0250	90	fenvalerate-2	0.0500	0.1000
61	carboxin*	0.3000	0.6000	91	deltamethrin*	0.0750	0.1500
62	p,p'-DDD	0.0125	0.0250		Group B		0.0000
63	ethion	0.0250	0.0500	92	EPTC	0.0375	0.0750
64	etaconazole-1	0.0375	0.0750	93	butylate	0.0375	0.0750
65	sulprofos	0.0250	0.0500	94	dichlobenil	0.0025	0.0050
66	etaconazole-2	0.0375	0.0750	95	pebulate	0.0375	0.0750
67	myclobutanil	0.0125	0.0250	96	nitrapyrin	0.0375	0.0750
68	diclofop-methyl	0.0125	0.0250	97	mevinphos	0.0250	0.0500
69	propiconazole-1	0.0375	0.0750	98	chloroneb	0.0125	0.0250

TABLE 3.5 The LOD, LOQ of 485 Pesticides and Related Chemical Residues Determined by GC-MS—cont'd

No.	Pesticides	LOD (mg/kg)	LOQ (mg/kg)	No.	Pesticides	LOD (mg/kg)	LOQ (mg/kg)
99	tecnazene	0.0250	0.0500	129	thiobencarb	0.0250	0.0500
100	heptanophos	0.0375	0.0750	130	dicofol	0.0250	0.0500
101	ethoprophos	0.0375	0.0750	131	metolachlor	0.0125	0.0250
102	hexachlorobenzene	0.0125	0.0250	132	pirimiphos-ethyl	0.0250	0.0500
103	propachlor	0.0375	0.0750	133	oxy-chlordane	0.0125	0.0250
104	cis- diallate	0.0250	0.0500	134	dichlofluanid*	0.6000	1.2000
105	trifluralin*	0.0250	0.0500	135	methoprene*	0.0500	0.1000
106	trans-diallate*	0.0250	0.0500	136	bromofos	0.0250	0.0500
107	chlorpropham	0.0250	0.0500	137	ethofumesate	0.0250	0.0500
108	sulfotep	0.0125	0.0250	138	isopropalin*	0.0250	0.0500
109	sulfallate	0.0250	0.0500	139	propanil	0.0250	0.0500
110	alpha-HCH	0.0125	0.0250	140	crufomate	0.0750	0.1500
111	terbufos	0.0250	0.0500	141	isofenphos	0.1000	0.2000
112	terbumeton	0.1500	0.3000	142	endosulfan -1	0.0750	0.1500
113	profluralin*	0.0500	0.1000	143	chlorfenvinphos	0.0375	0.0750
114	dioxathion	0.0500	0.1000	144	tolylfluanide	0.3000	0.6000
115	propazine*	0.0125	0.0250	145	cis-chlordane*	0.0250	0.0500
116	chlorbufam	0.1000	0.2000	146	butachlor	0.0250	0.0500
117	dicloran	0.0250	0.0500	147	chlozolinate	0.0250	0.0500
118	terbuthylazine	0.0500	0.1000	148	p,p′-DDE	0.0125	0.0250
119	monolinuron	0.0500	0.1000	149	iodofenphos	0.0250	0.0500
120	cyanophos	0.0250	0.0500	150	tetrachlorvinphos	0.0375	0.0750
121	flufenoxuron*	0.0375	0.0750	151	profenofos	0.0750	0.1500
122	chlorpyrifos-methyl	0.0125	0.0250	152	buprofezin	0.0250	0.0500
123	desmetryn	0.0125	0.0250	153	hexaconazole	0.0750	0.1500
124	dimethachlor	0.0375	0.0750	154	o,p′-DDD	0.0125	0.0250
125	alachlor	0.0375	0.0750	155	chlorfenson	0.0250	0.0500
126	pirimiphos-methyl	0.0125	0.0250	156	fluorochloridone	0.0250	0.0500
127	terbutryn	0.0250	0.0500	157	endrin	0.1500	0.3000
128	aspon	0.0250	0.0500	158	paclobutrazol	0.1500	0.3000

Continued

TABLE 3.5 The LOD, LOQ of 485 Pesticides and Related Chemical Residues Determined by GC-MS—cont'd

No.	Pesticides	LOD (mg/kg)	LOQ (mg/kg)	No.	Pesticides	LOD (mg/kg)	LOQ (mg/kg)
159	o,p'-DDT	0.0250	0.0500	189	coumaphos*	0.0750	0.1500
160	methoprotryne	0.0375	0.0750		Group C		0.0000
161	chloropropylate	0.0125	0.0250	190	dichlorvos	0.0750	0.1500
162	flamprop-methyl	0.0125	0.0250	191	biphenyl	0.0125	0.0250
163	nitrofen	0.0750	0.1500	192	vernolate	0.0125	0.0250
164	oxyfluorfen	0.2000	0.4000	193	3,5-dichloroaniline	0.0125	0.0250
165	chlorthiophos	0.0375	0.0750	194	methacrifos	0.0125	0.0250
166	flamprop-isopropyl	0.0125	0.0250	195	molinate	0.0125	0.0250
167	endosulfan -2	0.0750	0.1500	196	2-phenylphenol	0.0125	0.0250
168	carbofenothion	0.0250	0.0500	197	cis-1,2,3,6-tetrahydrophthalimide	0.0375	0.0750
169	p,p'-DDT*	0.0250	0.0500				
170	benalaxyl	0.0125	0.0250	198	fenobucarb	0.0250	0.0500
171	edifenphos	0.0250	0.0500	199	benfluralin*	0.0125	0.0250
172	triazophos	0.0375	0.0750	200	prometon	0.0375	0.0750
173	cyanofenphos	0.0125	0.0250	201	triallate	0.0250	0.0500
174	chlorbenside sulfone	0.0250	0.0500	202	pyrimethanil	0.0125	0.0250
175	endosulfan-sulfate	0.0375	0.0750	203	gamma-HCH	0.0250	0.0500
176	bromopropylate	0.0250	0.0500	204	disulfoton	0.0125	0.0250
177	benzoylprop-ethyl	0.0375	0.0750	205	atrizine*	0.0125	0.0250
178	fenpropathrin	0.1000	0.2000	206	iprobenfos*	0.0375	0.0750
179	EPN	0.0500	0.1000	207	heptachlor	0.0375	0.0750
180	hexazinone	0.0375	0.0750	208	isazofos	0.0250	0.0500
181	leptophos	0.0250	0.0500	209	plifenate	0.0250	0.0500
182	bifenox	0.0250	0.0500	210	fluchloralin	0.0500	0.1000
183	phosalone	0.0250	0.0500	211	transfluthrin*	0.0125	0.0250
184	azinphos-methyl	0.0750	0.1500	212	fenpropimorph*	0.0125	0.0250
185	fenarimol	0.0250	0.0500	213	tolclofos-methyl*	0.0125	0.0250
186	azinphos-ethyl	0.0250	0.0500	214	propisochlor	0.0125	0.0250
187	cyfluthrin	0.6000	1.2000	215	ametryn	0.0375	0.0750
188	prochloraz	0.0750	0.1500	216	simetryn	0.0250	0.0500

TABLE 3.5 The LOD, LOQ of 485 Pesticides and Related Chemical Residues Determined by GC-MS—cont'd

No.	Pesticides	LOD (mg/kg)	LOQ (mg/kg)	No.	Pesticides	LOD (mg/kg)	LOQ (mg/kg)
217	metribuzin	0.0375	0.0750	247	phenothrin	0.0125	0.0250
218	dipropetryn*	0.0125	0.0250	248	anilofos	0.0250	0.0500
219	formothion	0.0250	0.0500	249	*lambda*-cyhalothrin*	0.0125	0.0250
220	diethofencarb	0.0750	0.1500	250	mefenacet	0.0375	0.0750
221	dimepiperate	0.0250	0.0500	251	permethrin	0.0250	0.0500
222	bioallethrin-1	0.0500	0.1000	252	pyridaben	0.0125	0.0250
223	bioallethrin-2	0.0500	0.1000	253	fluoroglycofen-ethyl*	0.1500	0.3000
224	fenson	0.0125	0.0250	254	bitertanol*	0.0375	0.0750
225	*o,p'*-DDE	0.0125	0.0250	255	etofenprox	0.0125	0.0250
226	diphenamid	0.0125	0.0250	256	cycloxydim	0.1500	0.3000
227	penconazole	0.0375	0.0750	257	alpha-cypermethrin*	0.0250	0.0500
228	tetraconazole*	0.0375	0.0750	258	flucythrinate-1*	0.0250	0.0500
229	mecarbam	0.0500	0.1000	259	flucythrinate-2*	0.0250	0.0500
230	propaphos	0.0250	0.0500	260	esfenvalerate*	0.0500	0.1000
231	flumetralin*	0.0250	0.0500	261	difenconazole-2	0.0750	0.1500
232	triadimenol-1*	0.0375	0.0750	262	difenonazole-1	0.0750	0.1500
233	triadimenol-2*	0.0375	0.0750	263	flumioxazin	0.1000	0.2000
234	pretilachlor	0.0250	0.0500	264	flumiclorac-pentyl	0.0250	0.0500
235	kresoxim-methyl	0.0125	0.0250		Group D		0.0000
236	fluazifop-butyl*	0.0125	0.0250	265	dimefox	0.0375	0.0750
237	chlorobenzilate	0.0125	0.0250	266	disulfoton-sulfoxide	0.0250	0.0500
238	flusilazole	0.0375	0.0750	267	pentachlorobenzene	0.0125	0.0250
239	fluorodifen*	0.0125	0.0250	268	crimidine	0.0125	0.0250
240	diniconazole	0.0375	0.0750	269	BDMC-1	0.0250	0.0500
241	piperonyl butoxide	0.0125	0.0250	270	chlorfenprop-methyl	0.0125	0.0250
242	propargite	0.1000	0.2000	271	thionazin	0.0125	0.0250
243	mepronil	0.0125	0.0250	272	2,3,5,6-tetrachloroaniline	0.0125	0.0250
244	diflufenican*	0.0125	0.0250	273	tri-n-butyl phosphate*	0.0250	0.0500
245	fludioxonil*	0.0125	0.0250	274	2,3,4,5-tetrachloroanisole	0.0125	0.0250
246	fenazaquin	0.0125	0.0250	275	pentachloroanisole	0.0125	0.0250

Continued

TABLE 3.5 The LOD, LOQ of 485 Pesticides and Related Chemical Residues Determined by GC-MS—cont'd

No.	Pesticides	LOD (mg/kg)	LOQ (mg/kg)	No.	Pesticides	LOD (mg/kg)	LOQ (mg/kg)
276	tebutam	0.0250	0.0500	306	isomethiozin	0.0250	0.0500
277	methabenzthiazuron	0.1250	0.2500	307	trichloronat	0.0125	0.0250
278	desisopropyl-atrazine	0.1000	0.2000	308	dacthal	0.0125	0.0250
279	simetone	0.0250	0.0500	309	4,4-dichlorobenzophenone	0.0125	0.0250
280	atratone	0.0125	0.0250				
281	tefluthrin*	0.0125	0.0250	310	nitrothal-isopropyl*	0.0250	0.0500
282	bromocylen	0.0125	0.0250	311	musk ketone	0.0125	0.0250
283	trietazine	0.0125	0.0250	312	rabenzazole	0.0125	0.0250
284	2,6-dichlorobenzamide	0.0250	0.0500	313	cyprodinil	0.0125	0.0250
285	cycluron	0.0375	0.0750	314	isofenphos oxon	0.0250	0.0500
286	DE-PCB 28	0.0125	0.0250	315	fuberidazole	0.0625	0.1250
287	DE-PCB 31	0.0125	0.0250	316	dicapthon	0.0625	0.1250
288	desethyl-sebuthylazine*	0.0250	0.0500	317	MCPA-butoxyethyl ester	0.0125	0.0250
289	2,3,4,5-tetrachloroaniline	0.0250	0.0500	318	DE-PCB 101	0.0125	0.0250
290	musk ambrette	0.0125	0.0250	319	isocarbophos	0.0250	0.0500
291	musk xylene	0.0125	0.0250	320	phorate sulfone	0.0125	0.0250
292	pentachloroaniline	0.0125	0.0250	321	chlorfenethol	0.0125	0.0250
293	aziprotryne	0.1000	0.2000	322	trans-nonachlor*	0.0125	0.0250
294	isocarbamid	0.0625	0.1250	323	DEF	0.0250	0.0500
295	sebutylazine*	0.0125	0.0250	324	flurochloridone*	0.0250	0.0500
296	musk moskene	0.0125	0.0250	325	bromfenvinfos	0.0125	0.0250
297	DE-PCB 52	0.0125	0.0250	326	perthane	0.0125	0.0250
298	prosulfocarb	0.0125	0.0250	327	ditalimfos	0.0125	0.0250
299	dimethenamid	0.0125	0.0250	328	DE-PCB 118	0.0125	0.0250
300	BDMC-2	0.0250	0.0500	329	mephosfolan	0.0250	0.0500
301	monalide*	0.0250	0.0500	330	4,4-dibromobenzophenone	0.0125	0.0250
302	musk tibeten	0.0125	0.0250				
303	isobenzan	0.0125	0.0250	331	flutriafol	0.0250	0.0500
304	octachlorostyrene	0.0125	0.0250	332	DE-PCB 153	0.0125	0.0250
305	isodrin	0.0125	0.0250	333	diclobutrazole*	0.0500	0.1000
				334	disulfoton sulfone	0.1000	0.2000

TABLE 3.5 The LOD, LOQ of 485 Pesticides and Related Chemical Residues Determined by GC-MS—cont'd

No.	Pesticides	LOD (mg/kg)	LOQ (mg/kg)	No.	Pesticides	LOD (mg/kg)	LOQ (mg/kg)
335	hexythiazox	0.1000	0.2000	363	dibutyl succinate*	0.0250	0.0500
336	DE-PCB 138	0.0125	0.0250	364	chlorethoxyfos	0.0250	0.0500
337	cyproconazole	0.0125	0.0250	365	isoprocarb -2	0.0250	0.0500
338	resmethrin-1	0.2000	0.4000	366	tebuthiuron	0.0500	0.1000
339	resmethrin-2	0.2000	0.4000	367	pencycuron	0.0500	0.1000
340	phthalic acid,benzyl butyl ester	0.0125	0.0250	368	demeton-s-methyl	0.0500	0.1000
				369	propoxur-2	0.0250	0.0500
341	clodinafop-propargyl	0.0250	0.0500	370	phenanthrene	0.0125	0.0250
342	fenthion sulfoxide	0.0500	0.1000	371	fenpyroximate	0.1000	0.2000
343	fluotrimazole	0.0125	0.0250	372	tebupirimfos*	0.0250	0.0500
344	fluroxypr-1-methylheptyl ester*	0.0125	0.0250	373	prohydrojasmon	0.0500	0.1000
345	fenthion sulfone	0.0500	0.1000	374	fenpropidin*	0.0250	0.0500
346	metamitron	0.1250	0.2500	375	dichloran	0.0250	0.0500
347	triphenyl phosphate	0.0125	0.0250	376	propyzamide*	0.0250	0.0500
348	DE-PCB 180	0.0125	0.0250	377	pirimicarb	0.0250	0.0500
349	tebufenpyrad	0.0125	0.0250	378	benoxacor	0.0250	0.0500
350	cloquintocet-mexyl	0.0125	0.0250	379	phosphamidon -1	0.1000	0.2000
351	lenacil	0.1250	0.2500	380	acetochlor	0.0250	0.0500
352	bromuconazole-1	0.0250	0.0500	381	tridiphane	0.0500	0.1000
353	bromuconazole-2	0.0250	0.0500	382	esprocarb	0.0250	0.0500
354	nitralin*	0.1250	0.2500	383	fenfuram	0.0250	0.0500
355	fenamiphos sulfone	0.0500	0.1000	384	acibenzolar-s-methyl*	0.0250	0.0500
356	fenpiclonil	0.0500	0.1000	385	benfuresate	0.0250	0.0500
357	fluquinconazole	0.0125	0.0250	386	mefenoxam	0.0250	0.0500
358	fenbuconazole	0.0250	0.0500	387	malaoxon	0.2000	0.4000
	Group E		0.0000	388	chlorthal-dimethyl	0.0250	0.0500
359	propoxur-1*	0.0250	0.0500	389	simeconazole*	0.0250	0.0500
360	XMC	0.0250	0.0500	390	terbacil	0.0250	0.0500
361	isoprocarb-1*	0.0250	0.0500	391	thiazopyr*	0.0250	0.0500
362	acenaphthene	0.0500	0.1000	392	zoxamide*	0.0250	0.0500

Continued

TABLE 3.5 The LOD, LOQ of 485 Pesticides and Related Chemical Residues Determined by GC-MS—cont'd

No.	Pesticides	LOD (mg/kg)	LOQ (mg/kg)	No.	Pesticides	LOD (mg/kg)	LOQ (mg/kg)
393	allethrin*	0.0500	0.1000	422	iprodione	0.0500	0.1000
394	quinoclamine	0.0500	0.1000	423	ofurace	0.0375	0.0750
395	furalaxyl	0.0250	0.0500	424	piperophos*	0.0375	0.0750
396	thiamethoxam *	0.0500	0.1000	425	clomeprop	0.0125	0.0250
397	bromacil	0.0250	0.0500	426	fenamidone	0.0125	0.0250
398	picoxystrobin	0.0250	0.0500	427	tralkoxydim	0.1000	0.2000
399	butamifos	0.0125	0.0250	428	pyraclofos	0.1000	0.2000
400	imazamethabenz-methyl	0.0375	0.0750	429	dialifos	0.4000	0.8000
401	TCMTB	0.2000	0.4000	430	spirodiclofen	0.1000	0.2000
402	metominostrobin	0.0500	0.1000	431	flurtamone*	0.0250	0.0500
403	imazalil	0.0500	0.1000	432	pyriftalid	0.0125	0.0250
404	isoprothiolane	0.0250	0.0500	433	silafluofen *	0.0125	0.0250
405	cyflufenamid*	0.2000	0.4000	434	pyrimidifen	0.0250	0.0500
406	isoxathion	0.1000	0.2000	435	butafenacil *	0.0125	0.0250
407	quinoxyphen	0.0125	0.0250	436	cafenstrole*	0.0500	0.1000
408	trifloxystrobin	0.0500	0.1000		Group F		
409	imibenconazole-des-benzyl*	0.0500	0.1000	437	tribenuron-methyl	0.0125	0.0250
				438	ethiofencarb	0.1250	0.2500
410	imiprothrin-1	0.0250	0.0500	439	dioxacarb	0.1000	0.2000
411	imiprothrin-2*	0.0250	0.0500	440	dimethyl phthalate	0.0500	0.1000
412	epoxiconazole -1	0.1000	0.2000	441	phthalimide	0.0250	0.0500
413	pyributicarb	0.0250	0.0500	442	diethyltoluamide	0.0100	0.0200
414	pyraflufen ethyl	0.0250	0.0500	443	2,4-D	0.2500	0.5000
415	thenylchlor	0.0250	0.0500	444	carbaryl	0.0375	0.0750
416	clethodim	0.0500	0.1000	445	cadusafos	0.0500	0.1000
417	mefenpyr-diethyl	0.0375	0.0750	446	demetom-s	0.0500	0.1000
418	etoxazole*	0.0750	0.1500	447	spiroxamine-1	0.0250	0.0500
419	epoxiconazole-2	0.1000	0.2000	448	dicrotophos	0.1000	0.2000
420	famphur	0.0500	0.1000	449	3.4.5-trimethacarb	0.1000	0.2000
421	pyriproxyfen	0.0250	0.0500	450	3-phenylphenol	0.0750	0.1500

TABLE 3.5 The LOD, LOQ of 485 Pesticides and Related Chemical Residues Determined by GC-MS—cont'd

No.	Pesticides	LOD (mg/kg)	LOQ (mg/kg)	No.	Pesticides	LOD (mg/kg)	LOQ (mg/kg)
451	furmecyclox	0.0375	0.0750	469	diofenolan -1	0.0250	0.0500
452	spiroxamine -2	0.0250	0.0500	470	diofenolan -2	0.0250	0.0500
453	sobutylazine*	0.0250	0.0500	471	aclonifen	0.2500	0.5000
454	monocrotophos	0.1000	0.2000	472	bioresmethrin*	0.0250	0.0500
455	S421(octachlorodipropyl ether)-1	0.2500	0.5000	473	isoxadifen-ethyl	0.0250	0.0500
				474	carfentrazone-ethyl	0.0250	0.0500
456	S421(octachlorodipropyl ether)-2	0.2500	0.5000	475	endrin aldehyde*	0.2500	0.5000
457	dodemorph	0.0375	0.0750	476	halosulfuran-methyl	0.2500	0.5000
458	desmedipham*	0.2500	0.5000	477	tricyclazole	0.3000	0.6000
459	fenchlorphos	0.0500	0.1000	478	fenhexamid*	0.2500	0.5000
460	difenoxuron*	0.1000	0.2000	479	bifenazate	0.1000	0.2000
461	butralin	0.0500	0.1000	480	endrin ketone	0.2000	0.4000
462	pyrifenox -2	0.1000	0.2000	481	metoconazole	0.0500	0.1000
463	dimethametryn	0.0125	0.0250	482	cyhalofop-butyl*	0.0250	0.0500
464	pyrifenox-1	0.1000	0.2000	483	halfenprox*	0.0250	0.0500
465	thiabendazole	0.2500	0.5000	484	boscalid	0.0500	0.1000
466	iprovalicarb-1*	0.0500	0.1000	485	dimethomorph	0.0250	0.0500
467	azaconazole	0.0500	0.1000				
468	iprovalicarb-2*	0.0500	0.1000				

*The compounds of qualitative determination only.

3.3.2 Principle

The test samples are extracted with cyclohexane-ethyl acetate (1+1) mixed solution. The extracts are cleaned up by GPC with Bio-beads S-X3 column, and the solutions are analyzed by GC-MS.

3.3.3 Reagents and Materials

Toluene: HPLC Grade
Hexane: HPLC Grade
Cyclohexane: HPLC Grade
Ethyl acetate: HPLC Grade

Acetone: HPLC Grade

Dichloromethane: HPLC Grade

Filter Membrane (Nylon): 13 mm × 0.45 μm

Anhydrous Sodium Sulfate: Ignite at 650°C for 4 h; Store in a Desiccator

Cyclohexane-Ethyl Acetate (1+1)

Pesticide Standards: Purity ≥95%

Stock Standard Solution: Accurately weigh 5–10 mg of individual pesticide and related chemical standards (accurate to 0.1 mg) into a 10-mL volumetric flask. Dissolve and dilute to volume with toluene, toluene-acetone combination, dichloromethane, etc., depending on each individual compound's solubility (for diluting solvent refer to Appendix B).

Mixed Standard Solution (Mixed Standard Solution A, B, C, D, E, and F): Depending on properties and retention times of each compound, all compounds are divided into five groups: A, B, C, D, E. Depending upon instrument sensitivity of each compound, the concentrations of mixed standard solutions are decided. For 485 pesticides and related chemicals grouping and concentration of mixed standard solution of this standard, reference is made to Appendix B.

Depending on group number, mixed standard solution concentration, and stock standard solution concentration, appropriate amounts of individual stock standard solutions are pipetted into a 100-mL volumetric flask, with five group pesticides and related chemicals being diluted to volume with toluene. Mixed standard solutions are stored in the dark below 4°C and are used for one month.

Internal Standard Solution: Accurately weigh 3.5 mg heptachlor epoxide into a 100-mL volumetric flask. Dissolve and dilute to volume with toluene.

Working Standard Mixed Solution in Matrix: Working standard mixed solutions in matrix of A, B, C, D, E, and F group pesticides and related chemicals are prepared by diluting 40 μL internal standard solution and appropriate amounts of mixed standard solution to 1.0 mL with blank extract that has been taken through the method with the rest of the samples and mixing thoroughly. Working standard mixed solutions in matrix must be freshly prepared.

3.3.4 Apparatus

GC-MS: Equipped With EI

GPC: Equipped 360 mm × 25 mm BIO-Beads S-X3 Column or Other Equivalence In-Line Concentrator

Analytical Balances: Capable of Weighing From 0.1 mg to 0.01 g

Rotary Evaporator

Homogenizer: Max. 24,000 rpm

Centrifuge: Max. 4200 rpm

Pear-Shaped Flasks: 150 mL
Pipette: 1 mL

3.3.5 Sample Pretreatment

3.3.5.1 Preparation of Test Sample

Part of the representative sample is taken from each bag of the primary sample and homogenized by grinding in a meat grinder. The homogenized sample is thoroughly mixed and reduced to at least 500 g by quartering and then placed in a clean container as the test sample. Label and seal.

The test samples should be stored below $-18°C$.

3.3.5.2 Extraction

Weigh 10 g (accurate to 0.01 g) of the test sample into a 50-mL centrifuge tube filled with 20 g anhydrous sodium sulfate. Add 35 mL volume of cyclohexane-ethyl acetate (1+1) mixed solution. The mixture is homogenized for 1.5 min at 15,000 rpm and then centrifuged for 3 min at 3000 rpm. Pass the supernatant through a glass funnel containing a glass wool plug and ca. 15 g Na_2SO_4. Collect extracts in 100-mL pear-shaped flask. Proceed twice with 35 mL cyclohexane-ethyl acetate mixed solution. Combine the extracts in the pear-shaped flask. Evaporate to ca. 5 mL on rotary evaporator at 40°C and then proceed to clean-up. If the residue content in the fat is to be calculated, collect the extracts in a weighted pear-shaped flask. Evaporate the extracts to ca. 5 mL on a rotary evaporator at 40°C and evaporate the solvent to dryness with a nitrogen evaporator. Weigh the pear-shaped flask and proceed to clean-up.

3.3.5.3 GPC Clean-up

3.3.5.3.1 Conditions

(a) GPC Column: 360 mm × 25 mm, filled with Bio-Beads S-X3
(b) Detection wavelength: 254 nm
(c) Mobile phase: cychexane-ethyl acetate (1+1) mixed solution
(d) Flow rate: 4.7 mL/min
(e) Injection volume: 5 mL
(f) Start collect time: 26 min
(g) Stop collect time: 44 min
(h) In-line concentration temperature and vacuum degree: 1 zone: 45°C 33.3 kPa; 2 zone: 49°C 29.3 kPa; 3 zone: 52°C 26.6 kPa
(i) Concentration endpoint mode: Level sensor mode
(j) Concentration endpoint temperature and vacuum degree: 1 zone: 51°C 26.60 kPa; 2 zone: 50°C 23.94 kPa

3.3.5.3.2 Clean-up

Transfer the concentrated extracts into a 10-mL volumetric flask with cychexane+ethyl acetate (1+1) mixed solution. Rinse pear-shaped flask twice with 5 mL cychexane-ethyl acetate (1+1) mixed solution and transfer into the 10-mL volumetric flask. Dilute to volume with cychexane-ethyl acetate (1+1) mixed solution. Mix thoroughly. Filter the solution to a 10-mL tube. Inject 5 mL of the filtered extract solution into GPC. Collect the fraction of 22 min to 40 min in a 100-mL pear-shaped flask. Evaporate it to ca. 0.5 mL on rotary evaporator at 40°C. Exchange with 2 × 5 mL hexane twice and make up to 1 mL. Add 40 μL internal standard solution for GC-MS determination.

Extract and cleanup blank fugu, eel, and prawn samples as that described above.

3.3.6 Determination

3.3.6.1 GC-MS Operating Conditions

(a) Column: DB-1701 (30 m × 0.25 mm × 0.25 μm) capillary column, or equivalent
(b) Column temperature: 40°C hold 1 min, at 30°C/min to 130°C, at 5°C/min to 250°C, at 10°C/min to 300°C, hold 5 min
(c) Carrier gas: Helium, purity ≥99.999%, flow rate: 1.2 mL/min
(d) Injection port temperature: 290°C
(e) Injection volume: 1 μL
(f) Injection mode: Splitless, purge on after 1.5 min
(g) Ionization voltage: 70 eV
(h) Ion source temperature: 230°C
(i) GC-MS interface temperature: 280°C
(j) Solvent delay time: group A is 8.30 min; group B is 7.80 min; group C is 7.30 min; group D is 5.50 min; group E is 5.50 min; group F is 5.50 min
(k) Selected ion monitoring mode: Select 1 quantifying ion and 2–3 qualifying ions from each compound. All of the detected ions of each group are detected according to the period of time and order of compounds. For the retention times, quantifying ions, qualifying ions, and the abundance ratios of quantifying ion and qualifying ions of each compound, refer to Table 5.3. For the period of time and dwell time of detected ions of each group, refer to Table 5.4

3.3.6.2 Qualitative Determination

In the samples determined, four injections are required to analyze all pesticides and related chemicals by GC-MS under the operating conditions. If the retention times of peaks of the sample solution are the same as those of the peaks from the working standard mixed solution, and the selected ions appeared in the background-subtracted mass spectrum, and the abundance ratios of the selected ions are within the expected limits (abundance ratios >50%, permitted tolerances are ±10%; abundance ratios >20% to 50%, permitted tolerances are ±15%; abundance ratios >10% to 20%, permitted tolerances are ±20%; abundance ratios ≤10%, permitted tolerances are ±50%), the sample is confirmed to contain this pesticide compound. If the results are not definitive, the sample is reinjected with acquisition in scan mode (sufficient sensitivity) or with additional confirmatory ions, or other instruments of higher sensitivity are used.

3.3.6.3 Quantitative Determination

Quantitation is performed using an internal standard and the quantifying ion for GC-MS. The internal standard is heptachlor epoxide. To compensate for the matrix effects, quantitation is based on a mixed standard prepared in blank matrix extract. The concentrations in the standard solution and in the sample solution analyzed must be close.

3.3.6.4 Parallel Test

A parallel test is carried out for the same testing sample.

3.3.6.5 Blank Test

The operation of the blank test is the same as that described in the method of determination, but without the addition of sample.

3.3.7 Precision

The precision data of the method for this standard have been determined according to the stipulations of GB/T 6379.1 and GB/T 6379.2. The values of repeatability and reproducibility are obtained and calculated at the 95% confidence level.

RESEARCHERS

Guo-Fang Pang, Feng Zheng, Chun-Lin Fan, Ming-Lin Wang, Yong-Ming Liu, Jing Cao.

Qinhuangdao Entry-Exit Inspection and Quarantine Bureau, 39 Haibin Rd, Qinhuangdao, Hebei, PC 066002, People's Republic of China.

3.4 DETERMINATION OF 450 PESTICIDES AND RELATED CHEMICAL RESIDUES IN FUGU, EEL, AND PRAWN: LC-MS-MS METHOD (GB/T 23208-2008)

3.4.1 Scope

This method is applicable to the determination of 450 pesticides and related chemical residues in fugu, eel, and prawn; it is also applicable to the quantification of 380 pesticides and related chemical residues in fugu, eel, and prawn.

The limits of quantification of 380 pesticides and related chemical residues for this standard are 0.02 μg/kg to 0.195 mg/kg (refer to Table 3.6).

TABLE 3.6 The LOD, LOQ of 450 Pesticides and Related Chemical Residues Determined by LC-MS-MS

No.	Pesticides	LOD (μg/kg)	LOQ (μg/kg)	No.	Pesticides	LOD (μg/kg)	LOQ (μg/kg)
	Group A			20	paraoxon methyl	0.31	0.62
1	propham	44.00	88.00	21	4,4-dichlorobenzophenone	5.44	10.88
2	isoprocarb	0.92	1.84				
3	3,4,5-trimethacarb	0.14	0.28	22	thiacloprid	0.15	0.30
4	cycluron	0.08	0.16	23	imidacloprid	8.80	17.60
5	carbaryl	4.13	8.26	24	ethidimuron	0.60	1.20
6	propachlor	0.11	0.22	25	isomethiozin	0.43	0.86
7	rabenzazole	0.53	1.06	26	diallate	35.68	71.36
8	simetryn	0.05	0.10	27	acetochlor	18.96	37.92
9	monolinuron	1.42	2.84	28	nitenpyram	6.85	13.70
10	mevinphos	0.63	1.26	29	methoprotryne	0.10	0.20
11	aziprotryne	0.55	1.10	30	dimethenamid	1.72	3.44
12	secbumeton	0.03	0.06	31	terrbucarb	0.84	1.68
13	cyprodinil	0.30	0.60	32	penconazole	0.80	1.60
14	buturon	3.58	7.16	33	myclobutanil	0.40	0.80
15	carbetamide	1.46	2.92	34	imazethapyr	0.45	0.90
16	pirimicarb	0.06	0.12	35	paclobutrazol	0.23	0.46
17	clomazone	0.17	0.34	36	fenthion sulfoxide	0.13	0.26
18	cyanazine	0.07	0.14	37	triadimenol	4.22	8.44
19	prometryne	0.07	0.14	38	butralin	0.76	1.52

TABLE 3.6 The LOD, LOQ of 450 Pesticides and Related Chemical Residues Determined by LC-MS-MS—cont'd

No.	Pesticides	LOD (µg/kg)	LOQ (µg/kg)	No.	Pesticides	LOD (µg/kg)	LOQ (µg/kg)
39	spiroxamine	0.02	0.04	68	quizalofop-ethyl	0.27	0.54
40	tolclofos methyl	26.62	53.24	69	haloxyfop-methyl	1.06	2.12
41	desmedipham	1.61	3.22	70	fluazifop butyl	0.11	0.22
42	methidathion	4.26	8.52	71	bromophos-ethyl	908.31	1816.62
43	allethrin	24.16	48.32	72	bensulide	13.68	27.36
44	diazinon	0.29	0.58	73	triasulfuron	0.64	1.28
45	edifenphos	0.30	0.60	74	bromfenvinfos	1.21	2.42
46	pretilachlor	0.13	0.26	75	azoxystrobin	0.18	0.36
47	flusilazole	0.23	0.46	76	pyrazophos	0.65	1.30
48	iprovalicarb	0.93	1.86	77	flufenoxuron	1.27	2.54
49	benodanil	1.39	2.78	78	indoxacarb	3.02	6.04
50	flutolanil	0.46	0.92		Group B		
51	famphur	1.44	2.88	79	dazomet	50.80	101.60
52	benalyxyl	0.50	1.00	80	nicotine	0.88	1.76
53	diclobutrazole	0.19	0.38	81	fenuron	0.41	0.82
54	etaconazole	0.71	1.42	82	cyromazine	2.90	5.80
55	fenarimol	0.24	0.48	83	crimidine	0.62	1.24
56	phthalic acid, dicyclobexyl ester	0.80	1.60	84	acephate	5.34	10.68
				85	molinate	0.84	1.68
57	tetramethirn	0.73	1.46	86	carbendazim	0.19	0.38
58	dichlofluanid	1.04	2.08	87	6-chloro-4-hydroxy-3-phenyl-pyridazin	0.66	1.32
59	cloquintocet mexyl	0.75	1.50				
60	bitertanol	13.36	26.72	88	propoxur	9.76	19.52
61	chlorprifos methyl	6.40	12.80	89	isouron	0.16	0.32
62	tepraloxydim	4.88	9.76	90	chlorotoluron	0.25	0.50
63	azinphos ethyl	43.57	87.14	91	thiofanox	62.80	125.60
64	clodinafop propargyl	0.98	1.96	92	chlorbufam	73.20	146.40
65	triflumuron	1.57	3.14	93	bendiocarb	1.27	2.54
66	isoxaflutole	1.56	3.12	94	propazine	0.13	0.26
67	anilofos	0.29	0.58	95	terbuthylazine	0.19	0.38

Continued

TABLE 3.6 The LOD, LOQ of 450 Pesticides and Related Chemical Residues Determined by LC-MS-MS—cont'd

No.	Pesticides	LOD (µg/kg)	LOQ (µg/kg)	No.	Pesticides	LOD (µg/kg)	LOQ (µg/kg)
96	diuron	0.62	1.24	124	flutriafol	3.43	6.86
97	chlormephos	179.2	358.40	125	furalaxyl	0.31	0.62
98	arboxin	0.22	0.44	126	bioallethrin	79.20	158.40
99	difenzoquat-methyl sulfate	1.30	2.60	127	cyanofenphos	8.32	16.64
100	clothianidin	25.20	50.40	128	pirimiphos methyl	0.08	0.16
101	pronamide	6.15	12.30	129	buprofezin	0.35	0.70
102	dimethachloro	0.76	1.52	130	disulfoton sulfone	0.98	1.96
103	methobromuron	6.74	13.48	131	fenazaquin	0.13	0.26
104	phorate	125.60	251.20	132	triazophos	0.27	0.54
105	aclonifen	9.68	19.36	133	DEF	0.65	1.30
106	mephosfolan	0.93	1.86	134	pyriftalid	0.25	0.50
107	imibenconazole-des-benzyl	2.49	4.98	135	metconazole	0.53	1.06
108	neburon	2.84	5.68	136	pyriproxyfen	0.17	0.34
109	mefenoxam	0.62	1.24	137	isoxaben	0.07	0.14
110	prothoate	0.98	1.96	138	flurtamone	0.18	0.36
111	ethofume sate	148.80	297.60	139	trifluralin	133.92	267.84
112	iprobenfos	3.31	6.62	140	flamprop methyl	8.08	16.16
113	TEPP	4.16	8.32	141	bioresmethrin	2.97	5.94
114	cyproconazole	0.29	0.58	142	propiconazole	0.70	1.40
115	thiamethoxam	13.20	26.40	143	chlorpyrifos	21.52	43.04
116	crufomate	0.21	0.42	144	fluchloralin	195.20	390.40
117	etrimfos	30.02	60.04	145	chlorsulfuron	4.38	8.76
118	coumatetralyl	0.54	1.08	146	clethodim	0.83	1.66
119	cythioate	32.00	64.00	147	flamprop isopropyl	0.69	1.38
120	phosphamidon	1.55	3.10	148	tetrachlorvinphos	0.89	1.78
121	phenmedipham	1.79	3.58	149	propargite	27.44	54.88
122	bifenazate	9.12	18.24	150	bromuconazole	1.26	2.52
123	fenhexamid	0.38	0.76	151	picolinafen	0.29	0.58
				152	fluthiacet methyl	2.12	4.24

TABLE 3.6 The LOD, LOQ of 450 Pesticides and Related Chemical Residues Determined by LC-MS-MS—cont'd

No.	Pesticides	LOD (µg/kg)	LOQ (µg/kg)	No.	Pesticides	LOD (µg/kg)	LOQ (µg/kg)
153	trifloxystrobin	0.80	1.60	181	fonofos	2.98	5.96
154	chlorimuron ethyl	12.16	24.32	182	etridiazol	40.17	80.34
155	hexaflumuron	10.08	20.16	183	furmecyclox	1.33	2.66
156	hydramethylnon	0.69	1.38	184	hexazinone	0.05	0.10
157	flurazuron	10.72	21.44	185	dimethametryn	0.04	0.08
	Group C		0.00	186	trichlorphon	0.45	0.90
158	methamidophos	1.97	3.94	187	benoxacor	2.76	5.52
159	EPTC	14.94	29.88	188	bromacil	9.44	18.88
160	diethyltoluamide	0.22	0.44	189	phorate sulfoxide	147.31	294.62
161	monuron	13.89	27.78	190	brompyrazon	1.44	2.88
162	pyrimethanil	0.27	0.54	191	oxycarboxin	0.36	0.72
163	fenfuram	0.31	0.62	192	mepronil	0.15	0.30
164	quinoclamine	3.17	6.34	193	disulfoton	187.88	375.76
165	fenobucarb	2.36	4.72	194	metalaxyl	0.20	0.40
166	ethirimol	0.22	0.44	195	ofurace	0.40	0.80
167	propanil	8.64	17.28	196	dodemorph	0.16	0.32
168	carbofuran	5.22	10.44	197	imazamethabenz-methyl	0.07	0.14
169	acetamiprid	0.58	1.16	198	disulfoton-sulfoxide	1.14	2.28
170	mepanipyrim	0.13	0.26	199	isoprothiolane	0.74	1.48
171	prometon	0.05	0.10	200	imazalil	0.80	1.60
172	methiocarb	16.48	32.96	201	phoxim	33.12	66.24
173	metoxuron	0.26	0.52	202	quinalphos	0.80	1.60
174	dimethoate	3.04	6.08	203	ditalimfos	26.88	53.76
175	methfuroxam	0.11	0.22	204	fenoxycarb	7.31	14.62
176	fluometuron	0.37	0.74	205	pyrimitate	0.07	0.14
177	dicrotophos	0.46	0.92	206	fensulfothin	0.80	1.60
178	monalide	0.48	0.96	207	fluorochloridone	22.05	44.10
179	diphenamid	0.06	0.12	208	butachlor	8.03	16.06
180	ethoprophos	1.11	2.22	209	imazaquin	1.16	2.32

Continued

TABLE 3.6 The LOD, LOQ of 450 Pesticides and Related Chemical Residues Determined by LC-MS-MS—cont'd

No.	Pesticides	LOD (μg/ kg)	LOQ (μg/kg)	No.	Pesticides	LOD (μg/ kg)	LOQ (μg/kg)
210	kresoxim-methyl	40.23	80.46		Group D		
211	triticonazole	1.21	2.42	239	mepiquat chloride	0.36	0.72
212	fenamiphos sulfoxide	0.30	0.60	240	allidochlor	16.42	32.84
213	thenylchlor	9.66	19.32	241	thiabendazole	0.20	0.40
214	fenoxanil	15.76	31.52	242	metamitron	2.54	5.08
215	fluridone	0.07	0.14	243	isoproturon	0.05	0.10
216	epoxiconazole	1.62	3.24	244	atratone	0.07	0.14
217	chlorphoxim	31.03	62.06	245	desmetryn	0.07	0.14
218	fenamiphos sulfone	0.18	0.36	246	metribuzin	0.22	0.44
219	fenbuconazole	0.66	1.32	247	DMST	16.00	32.00
220	isofenphos	87.47	174.94	248	cycloate	1.78	3.56
221	phenothrin	135.68	271.36	249	atrazine	0.14	0.28
222	fentin-chloride	6.90	13.80	250	butylate	120.80	241.60
223	piperophos	3.70	7.40	251	pymetrozin	13.71	27.42
224	piperonyl butoxide	0.45	0.90	252	chloridazon	0.93	1.86
225	oxyflurofen	23.42	46.84	253	sulfallate	82.88	165.76
226	coumaphos	0.84	1.68	254	ethiofencarb	7.87	15.74
227	flufenacet	2.12	4.24	255	terbumeton	0.04	0.08
228	phosalone	19.22	38.44	256	ametryn	0.38	0.76
229	methoxyfenozide	1.48	2.96	257	tebuthiuron	0.09	0.18
230	prochloraz	0.83	1.66	258	trietazine	0.24	0.48
231	aspon	0.69	1.38	259	sebutylazine	0.13	0.26
232	ethion	1.18	2.36	260	dibutyl succinate	88.96	177.92
233	thifensulfuron-methyl	8.56	17.12	261	tebutam	0.05	0.10
234	ethoxysulfuron	1.83	3.66	262	thiofanox-sulfoxide	3.32	6.64
235	spirodiclofen	3.96	7.92	263	methacrifos	969.48	1938.96
236	fenpyroximate	0.54	1.08	264	triazoxide	3.20	6.40
237	flumiclorac-pentyl	4.24	8.48	265	thionazin	9.07	18.14
238	temephos	0.49	0.98	266	linuron	4.65	9.30

TABLE 3.6 The LOD, LOQ of 450 Pesticides and Related Chemical Residues Determined by LC-MS-MS—cont'd

No.	Pesticides	LOD (µg/kg)	LOQ (µg/kg)	No.	Pesticides	LOD (µg/kg)	LOQ (µg/kg)
267	heptanophos	2.34	4.68	296	diniconazole	0.54	1.08
268	prosulfocarb	0.15	0.30	297	flumetsulam	0.48	0.96
269	dipropetryn	0.11	0.22	298	pencycuron	0.11	0.22
270	thiobencarb	1.32	2.64	299	mecarbam	7.84	15.68
271	tri-n-butyl phosphate	0.15	0.30	300	tralkoxydim	0.13	0.26
272	diethofencarb	0.80	1.60	301	malathion	2.26	4.52
273	alachlor	2.96	5.92	302	pyributicarb	0.14	0.28
274	cadusafos	0.46	0.92	303	pyridaphenthion	0.35	0.70
275	metazachlor	0.39	0.78	304	thiodicarb	15.75	31.50
276	propetamphos	21.60	43.20	305	pyraclofos	0.40	0.80
277	terbufos	896.00	1792.00	306	picoxystrobin	3.38	6.76
278	simeconazole	1.18	2.36	307	tetraconazole	0.69	1.38
279	triadimefon	3.15	6.30	308	mefenpyr-diethyl	5.02	10.04
280	phorate sulfone	16.80	33.60	309	profenefos	3.23	6.46
281	tridemorph	1.04	2.08	310	pyraclostrobin	0.20	0.40
282	mefenacet	0.88	1.76	311	dimethomorph	0.14	0.28
283	fenamiphos	0.08	0.16	312	kadethrin	1.33	2.66
284	fenpropimorph	0.07	0.14	313	thiazopyr	0.78	1.56
285	tebuconazole	0.89	1.78	314	benfuracarb-methyl	6.55	13.10
286	isopropalin	12.00	24.00	315	cinosulfuron	0.45	0.90
287	nuarimol	0.40	0.80	316	pyrazosulfuron-ethyl	10.94	21.88
288	bupirimate	0.28	0.56	317	metosulam	1.76	3.52
289	azinphos-methyl	441.73	883.46		Group E		
290	tebupirimfos	0.05	0.10	318	4-aminopyridine	0.35	0.70
291	phenthoate	36.94	73.88	319	methomyl	3.82	7.64
292	sulfotep	1.04	2.08	320	pyroquilon	1.39	2.78
293	sulprofos	2.34	4.68	321	fuberidazole	0.76	1.52
294	EPN	13.20	26.40	322	isocarbamid	0.68	1.36
295	azamethiphos	0.32	0.64	323	butocarboxim	0.63	1.26

Continued

TABLE 3.6 The LOD, LOQ of 450 Pesticides and Related Chemical Residues Determined by LC-MS-MS—cont'd

No.	Pesticides	LOD (µg/ kg)	LOQ (µg/kg)	No.	Pesticides	LOD (µg/ kg)	LOQ (µg/kg)
324	chlordimeform	0.53	1.06	353	flurochloridone	0.52	1.04
325	cymoxanil	22.24	44.48	354	phthalic acid,benzyl butyl ester	252.80	505.60
326	vernolate	0.10	0.20	355	isazofos	0.07	0.14
327	aminocarb	6.57	13.14	356	dichlofenthion	12.08	24.16
328	dimethirimol	0.05	0.10	357	vamidothion sulfone	190.40	380.80
329	omethoate	3.86	7.72	358	terbufos sulfone	35.44	70.88
330	ethoxyquin	1.41	2.82	359	cyazofamid	1.80	3.60
331	dichlorvos	0.22	0.44	360	trichloronat	26.72	53.44
332	aldicarb sulfone	8.56	17.12	361	resmethrin	0.12	0.24
333	dioxacarb	1.34	2.68	362	boscalid	1.90	3.80
334	benzyladenine	28.32	56.64	363	nitralin	13.76	27.52
335	demeton-s-methyl	2.12	4.24	364	fenpropathrin	98.00	196.00
336	ethiofencarb-sulfoxide	89.60	179.20	365	hexythiazox	9.44	18.88
337	thiometon	231.20	462.40	366	benzoylprop-ethyl	123.20	246.40
338	folpet	55.44	110.88	367	pyrimidifen	5.60	11.20
339	demeton-s-methyl sulfone	7.90	15.80	368	furathiocarb	0.77	1.54
340	fenpropidin	0.07	0.14	369	trans-permethin	1.92	3.84
341	paraoxon-ethyl	0.19	0.38	370	etofenprox	912.00	1824.00
342	aldimorph	1.26	2.52	371	haloxyfop-2-ethoxyethyl	1.00	2.00
343	vinclozolin	1.02	2.04	372	esfenvalerate	166.4	332.80
344	uniconazole	0.96	1.92	373	fluoroglycofen-ethyl	2.00	4.00
345	pyrifenox	0.11	0.22		Group F		
346	chlorthion	53.44	106.88	374	acrylamide	7.12	14.24
347	dicapthon	0.10	0.20	375	tert-butylamine	15.58	31.16
348	clofentezine	0.31	0.62	376	phthalimide	17.20	34.40
349	norflurazon	0.10	0.20	377	metolcarb	10.16	20.32
350	triallate	18.48	36.96	378	1-naphthy acetamide	0.32	0.64
351	quinoxyphen	61.36	122.72	379	2,6-dichlorobenzamide	1.80	3.60
352	fenthion sulfone	6.98	13.96	380	aldicarb	104.40	208.80

TABLE 3.6 The LOD, LOQ of 450 Pesticides and Related Chemical Residues
Determined by LC-MS-MS—cont'd

No.	Pesticides	LOD (µg/kg)	LOQ (µg/kg)	No.	Pesticides	LOD (µg/kg)	LOQ (µg/kg)
381	dimethyl phthalate	5.28	10.56	409	rotenone	0.93	1.86
382	simeton	0.44	0.88	410	imibenconazole	4.10	8.20
383	dinotefuran	4.07	8.14	411	propaquizafop	0.49	0.98
384	pebulate	1.36	2.72	412	benzofenap	0.03	0.06
385	acibenzolar-s-methyl	1.23	2.46	413	benzoximate	13.70	27.40
386	oxamyl	219.22	438.44	414	dinoseb acetate	16.51	33.02
387	thidiazuron	0.12	0.24	415	imazapic	0.67	1.34
388	methabenzthiazuron	0.03	0.06	416	propisochlor	0.32	0.64
389	butoxycarboxim	10.64	21.28	417	silafluofen	243.20	486.40
390	demeton-s-methyl sulfoxide	1.57	3.14	418	fentrazamide	4.96	9.92
				419	pentachloroaniline	1.50	3.00
391	thiofanox sulfone	9.63	19.26	420	cyphenothrin	6.72	13.44
392	phosfolan	0.19	0.38	421	dimefuron	1.60	3.20
393	demeton-s	32.00	64.00	422	etoxazole	0.35	0.70
394	fenthion oxon	1.90	3.80	423	dodine	3.20	6.40
395	napropamide	0.51	1.02	424	propylene thiourea	12.03	24.06
396	fenitrothion	10.72	21.44		Group G		
397	metolachlor	0.16	0.32	425	dalapon	92.30	184.60
398	procymidone	34.64	69.28	426	2-phenylphenol	67.95	135.90
399	cumyluron	0.53	1.06	427	3-phenylphenol	1.60	3.20
400	warfarin	4.29	8.58	428	DNOC	1.04	2.08
401	phosmet	7.09	14.18	429	cloprop	4.56	9.12
402	ronnel	5.25	10.50	430	dicloran	19.42	38.84
403	pyrethrin	14.32	28.64	431	chlorpropham	6.31	12.62
404	phthalic acid, biscyclohexyl ester	0.27	0.54	432	mecoprop	1.96	3.92
405	carpropamid	8.32	16.64	433	terbacil	0.35	0.70
406	tebufenpyrad	0.10	0.20	434	dicamba	506.37	1012.74
407	chlorthiophos	12.72	25.44	435	MCPB	5.67	11.34
408	dialifos	62.80	125.60	436	dinoseb	0.16	0.32

Continued

TABLE 3.6 The LOD, LOQ of 450 Pesticides and Related Chemical Residues Determined by LC-MS-MS—cont'd

No.	Pesticides	LOD (µg/kg)	LOQ (µg/kg)	No.	Pesticides	LOD (µg/kg)	LOQ (µg/kg)
437	dinoterb	0.10	0.20	445	oryzalin	7.86	15.72
438	forchlorfenuron	4.56	9.12	446	famoxadone	18.12	36.24
439	fludioxonil	24.86	49.72	447	diflufenican	11.31	22.62
440	chlorfenethol	65.72	131.44	448	ethiprole	15.94	31.88
441	chlorobenzuron	8.16	16.32	449	cyclosulfamuron	137.47	274.94
442	chloramphenicolum	1.55	3.10	450	iodosulfuron-methyl sodium	33.92	67.84
443	alloxydim-sodium	0.08	0.16				
444	sulfanitran	1.22	2.44				

3.4.2 Principle

The test samples are extracted with cyclohexane-ethyl acetate (1+1) mixed solution. The extracts are cleaned up by GPC with a Bio-beads S-X3 column, diluting with acetonitrile-water (3+2), and the solutions are analyzed by LC-MS-MS, using an external standard method for quantification.

3.4.3 Reagents and Materials

"Water" is first grade of GB/T6682 specified.

Acetonitrile: HPLC Grade
Cyclohexane: HPLC Grade
Ethyl Acetate: HPLC Grade
Formic Acid: P.R. Grade
Ammonium Acetate: Analytically Pure
Methanol: HPLC Grade
Toluene: HPLC Grade
Acetone: HPLC Grade
Sodium Acetate: Anhydrous, Analytically Pure
Membrane Filters (Nylon): 13 mm × 0.2 µm, 13 mm × 0.45 µm
Sodium Sulfate, Magnesium Sulfate: Anhydrous, Analytically Pure. Ignited at 650°C for 4 h and Kept in a Desiccator
0.1% Formic Acid (V/V)
5 mmol/L Ammonium Acetate Solution
Pesticide and Related Chemical Standards: Purity ≥95%
Cyclohexane+Ethyl Acetate: 1:1, (V/V)

Stock Standard Solution: Accurately weigh 5 mg of individual pesticide and related chemical standards (accurate to 0.1 mg) into a 10-mL volumetric flask. Dissolve and dilute to volume with methanol, toluene, acetone, isooctane, etc., depending on each individual compound's solubility. (For diluting solvent, refer to Appendix B.)

Mixed Standard Solution (Mixed Standard Solution A, B, C, D, E, F, and G): Depending on properties and retention times of each compound, all compounds are divided into seven groups: A, B, C, D, E, F, and G. Depending on the instrument sensitivity of each compound, the concentrations of mixed standard solutions are decided. For 450 pesticides and related chemical groupings and concentrations of mixed standard solution of this standard, reference is made to Appendix B.

Depending on group number, mixed standard solution concentration, and stock standard solution concentration, appropriate amounts of individual stock standard solutions are pipetted into a 100-mL volumetric flask, with five group pesticides and related chemicals being diluted to volume with methanol. Mixed standard solutions are stored in the dark below 4°C and are used for 1 month.

Working Standard Mixed Solution in Matrix: Working standard mixed solutions in matrix of A, B, C, D, E, F, and G group pesticides and related chemicals are prepared by mixing different concentrations of working standard mixed solutions with sample blank extract that has been taken through the method with the rest of the samples. These solutions are used to construct calibration plots. Working standard mixed solutions in matrix must be freshly prepared.

3.4.4 Apparatus

LC-MS-MS: Equipped With ESI Source
GPC: Equipped 400 mm × 25 mm, BIO-Beads S-X3 Column
Analytical Balances: Capable of Weighing From 0.1 mg to 0.01 g
Rotary Evaporator
Homogenizer: Max. 25,000 rpm
Centrifuge: Max. 4200 rpm
Nitrogen Evaporator
Pear-Shaped Flasks: 100 mL
Pipette: 1 mL, 5 mL

3.4.5 Sample Pretreatment

3.4.5.1 *Preparation of Test Sample*

Part of the representative sample is taken from each bag of the primary sample and homogenized by grinding in a meat grinder. The homogenized sample is thoroughly mixed and reduced to at least 500 g by quartering and is then placed in a clean container as the test sample. Label and seal.

The test samples should be stored below −18°C.

3.4.5.2   Extraction

Weigh 10 g (accurate to 0.01 g) of the test sample into a 50-mL centrifuge tube filled with 20 g anhydrous sodium sulfate. Add 35 mL volume of cyclohexane-ethyl acetate (1+1) mixed solution. The mixture is homogenized for 1.5 min at 15,000 rpm, and then centrifuged for 3 min at 3000 rpm. Pass the supernatant through a glass funnel containing a glass wool plug and ca. 15 g Na_2SO_4. Collect extracts in a 100-mL pear-shaped flask. Proceed twice with 35 mL cyclohexane-ethyl acetate mixed solution. Combine the extracts in the pear-shaped flask. Evaporate them to ca. 5 mL on rotary evaporator at 40°C and proceed to clean-up. If the residue content in the fat is calculated, collect the extracts in a weighted pear-shaped flask. Evaporate the extracts to ca. 5 mL on a rotary evaporator at 40°C and evaporate the solvent to dryness with a nitrogen evaporator. Weigh the pear-shaped flask and proceed to clean-up.

3.4.5.3   GPC Clean-up

3.4.5.3.1   Conditions

(a) GPC Column: 360 mm × 25 mm, filled with Bio-Beads S-X3
(b) Detection wavelength: 254 nm
(c) Mobile phase: cychexane-ethyl acetate (1+1)
(d) Flow rate: 4.7 mL/min
(e) Injection volume: 5 mL
(f) Start collect time: 26 min
(g) Stop collect time: 44 min

3.4.5.3.2   Clean-up

Transfer the concentrated extracts into a 10-mL volumetric flask with cychexane-ethyl acetate (1+1) mixed solution. Rinse pear-shaped flask twice with 5 mL cychexane-ethyl acetate (1+1) mixed solution and transfer into the 10-mL volumetric flask. Dilute to volume with cychexane-ethyl acetate (1+1) mixed solution. Mix thoroughly. Filter the solution into 10-mL test tube with 0.20-μm filter membrane for GPC clean-up. The fractions from 22 min to 40 min are collected in a 100-mL pear-shaped flask and concentrated to ca. 0.5 mL on a rotary evaporator at 40°C. Evaporate to dryness using a nitrogen evaporator. Dissolve residues with 1.0 mL acetonitrile-water (3+2) and pass through the 0.20-μm filter membrane. The solution is ready for LC-MS-MS determination.

Extract blank fugu, eel, and prawn samples as described that described above. Prepare working standard mixed solutions in matrix.

3.4.6 Determination

3.4.6.1 LC-MS-MS Operating Conditions

Conditions for the Pesticides and Related Chemicals of Group A-F

(a) Column: ZORBAX SB-C$_{18}$, 3.5 μm, 100 mm × 2.1 mm, or equivalent
(b) Mobile phase and flow rate: see Table 3.3
(c) Column temperature: 40°C
(d) Injection volume: 10 μL
(e) Ion source: electrospray ionization source
(f) Ion mode: positive
(g) Nebulizer gas: nitrogen
(h) Nebulizer gas pressure: 0.28 MPa
(i) Ion spray voltage: 4000 V
(j) Dryness gas temperature: 350°C
(k) Dryness gas flow rate: 10 L/min
(l) Monitoring ion pairs, collision energy and fragmentor: see Table 5.10

3.4.6.2 Conditions for the Pesticides and Related Chemicals of Group G

(a) Chromatography column: ZORBAX SB-C$_{18}$, 3.5 μm, 100 mm × 2.1 mm (diameter), or equivalent
(b) Mobile phase program and flow rate: as in Table 3.4
(c) Column temperature: 40°C
(d) Injection volume: 10 μL
(e) Ion source: electrospray ionization source
(f) Ion mode: negative mode
(g) Nebulizer gas: nitrogen
(h) Nebulizer gas pressure: 0.28 MPa
(i) Ion spray voltage: 4000 V
(j) Dryness gas temperature: 350°C
(k) Dryness gas flow rate: 10 L/min
(l) Monitoring ion pairs, collision energy, and fragmentor: see Table 5.10

3.4.6.3 Qualitative Determination

Sample solutions are analyzed using to LC-MS-MS operating conditions. For the samples determined, if the retention times of peaks found in the sample solution chromatogram are the same as the peaks in the standard in the blank matrix extract chromatogram, and the abundance ratios of MRM transitions for the precursor/product ion are within the expected limits (relative abundance >50%, permitted ±20% deviation; relative abundance >20% to 50%, permitted ±25% deviation; relative abundance >10% to 20%, permitted ±30% deviation; relative abundance ≤10%, permitted ±50% deviation), the sample is confirmed to contain this pesticide compound.

3.4.6.4 Quantitative Determination

An external standard method is used for quantitation with standard curves for LC-MS-MS. In order to compensate for the matrix effect, quantitation is based on a series of working standard solutions prepared in blank matrix extract. The standard curves are established by injection of different concentrations of working standard mixed solutions in matrix separately. The responses of pesticides in the sample solution should be in the linear range of the instrumental detection.

3.4.6.5 Parallel Test

A parallel test is carried out for the same testing sample.

3.4.6.6 Blank Test

The operation of the blank test is the same as that described in the method of determination, but without the addition of sample.

3.4.7 Precision

The precision data of the method for this standard have been determined according to the stipulations of GB/T 6379.1 and GB/T 6379.2. The values of repeatability and reproducibility are obtained and calculated at the 95% confidence level.

RESEARCHERS

Guo-Fang Pang, Feng Zheng, Chun-Lin Fan, Yan Li, Min-Lin Wang.

Qinhuangdao Entry-Exit Inspection and Quarantine Bureau, 39 Haibin Rd, Qinhuangdao, Hebei, PC 066002, People's Republic of China.

3.5 DETERMINATION OF 497 PESTICIDES AND RELATED CHEMICAL RESIDUES IN HONEY, FRUIT JUICE, AND WINE: GC-MS METHOD (GB/T 19426-2006)

3.5.1 Scope

This method is applicable to the determination of 497 pesticides and related chemical residues in honey, fruit juice, and wine.

The limits of detection of this method: 0.001 mg/kg to 0.300 mg/kg (see Table 3.7)

3.5.2 Principle

The samples are extracted with dichloromethane, the extracts are cleaned up with Envi-Carb and Sep-Pak cartridges. The pesticides and related chemicals are eluted with acetonitrile-toluene (3+1), and the solutions are analyzed by GC-MS.

TABLE 3.7 The LOD, LOQ of 497 Pesticides and Related Chemical Residues Determined by GC-MS

No.	Pesticides	LOD (mg/kg)	LOQ (mg/kg)	No.	Pesticides	LOD (mg/kg)	LOQ (mg/kg)
	Group A			29	Cyprazine	0.017	0.0340
1	Allidochlor	0.033	0.0660	30	Chlorothalonil	0.033	0.0660
2	Dichlormid	0.017	0.0340	31	Vinclozolin	0.017	0.0340
3	Etridiazol	0.050	0.1000	32	Beta-HCH	0.017	0.0340
4	Chlormephos	0.033	0.0660	33	Metalaxyl	0.050	0.1000
5	Propham	0.017	0.0340	34	Chlorpyrifos (-ethyl)	0.017	0.0340
6	Cycloate	0.017	0.0340	35	Methyl-Parathion	0.066	0.1320
7	Diphenylamine	0.017	0.0340	36	Anthraquinone	0.017	0.0340
8	Chlordimeform	0.017	0.0340	37	Delta-HCH	0.033	0.0660
9	Ethalfluralin	0.066	0.1320	38	Fenthion	0.017	0.0340
10	Phorate	0.017	0.0340	39	Malathion	0.066	0.1320
11	Thiometon	0.017	0.0340	40	Fenitrothion	0.033	0.0660
12	Quintozene	0.033	0.0660	41	Paraoxon-ethyl	0.033	0.0660
13	Atrazine-desethyl	0.017	0.0340	42	Triadimefon	0.017	0.0340
14	Clomazone	0.017	0.0340	43	Parathion	0.033	0.0660
15	Diazinon	0.017	0.0340	44	Pendimethalin	0.011	0.0220
16	Fonofos	0.017	0.0340	45	Linuron	0.033	0.0660
17	Etrimfos	0.017	0.0340	46	Chlorbenside	0.017	0.0340
18	Simazine	0.080	0.1600	47	Bromophos-ethyl	0.008	0.0160
19	Propetamphos	0.017	0.0340	48	Quinalphos	0.008	0.0160
20	Secbumeton	0.017	0.0340	49	trans-Chlordane	0.006	0.0120
21	Dichlofenthion	0.017	0.0340	50	Phenthoate	0.017	0.0340
22	Pronamide	0.017	0.0340	51	Metazachlor	0.010	0.0200
23	Mexacarbate	0.050	0.1000	52	fenothiocarb	0.006	0.0120
24	Dimethoate	0.066	0.1320	53	Prothiophos	0.008	0.0160
25	Aldrin	0.033	0.0660	54	Folpet	0.100	0.2000
26	Dinitramine	0.066	0.1320	55	Chlorflurenol	0.005	0.0100
27	Ronnel	0.033	0.0660	56	Dieldrin	0.017	0.0340
28	Prometryne	0.017	0.0340	57	Procymidone	0.008	0.0160

Continued

TABLE 3.7 The LOD, LOQ of 497 Pesticides and Related Chemical Residues Determined by GC-MS—cont'd

No.	Pesticides	LOD (mg/kg)	LOQ (mg/kg)	No.	Pesticides	LOD (mg/kg)	LOQ (mg/kg)
58	Methidathion	0.011	0.0220	87	Phosmet	0.008	0.0160
59	Cyanazine	0.013	0.0260	88	Tetradifon	0.006	0.0120
60	Napropamide	0.010	0.0200	89	Oxycarboxin	0.012	0.0240
61	Oxadiazone	0.008	0.0160	90	cis-Permethrin	0.008	0.0160
62	Fenamiphos	0.017	0.0340	91	trans-Permethrin	0.008	0.0160
63	Tetrasul	0.004	0.0080	92	Pyrazophos	0.007	0.0140
64	Aramite	0.004	0.0080	93	Cypermethrin	0.025	0.0500
65	Bupirimate	0.006	0.0120	94	Fenvalerate	0.017	0.0340
66	Carboxin	0.005	0.0100	95	Deltamethrin	0.050	0.1000
67	Flutolanil	0.004	0.0080		Group B		
68	4,4'-DDD	0.004	0.0080	96	EPTC	0.012	0.0240
69	Ethion	0.008	0.0160	97	Butylate	0.012	0.0240
70	Sulprofos	0.007	0.0140	98	Dichlobenil	0.001	0.0020
71	Etaconazole	0.012	0.0240	99	Pebulate	0.012	0.0240
72	Myclobutanil	0.008	0.0160	100	Nitrapyrin	0.025	0.0500
73	Diclofop-methyl	0.004	0.0080	101	Mevinphos	0.017	0.0340
74	Propiconazole	0.012	0.0240	102	Chloroneb	0.008	0.0160
75	Fensulfothion	0.011	0.0220	103	Tecnazene	0.017	0.0340
76	Bifenthrin	0.006	0.0120	104	Heptanophos	0.025	0.0500
77	Carbosulfan	0.010	0.0200	105	Hexachlorobenzene	0.008	0.0160
78	Mirex	0.004	0.0080	106	Ethoprophos	0.025	0.0500
79	Benodanil	0.008	0.0160	107	Propachlor	0.012	0.0240
80	Nuarimol	0.007	0.0140	108	cis and trans-Diallate	0.017	0.0340
81	Methoxychlor	0.008	0.0160	109	Trifluralin	0.017	0.0340
82	Oxadixyl	0.008	0.0160	110	Chlorpropham	0.017	0.0340
83	Tetramethirn	0.007	0.0140	111	Sulfotep	0.008	0.0160
84	Tebuconazole	0.012	0.0240	112	Sulfallate	0.017	0.0340
85	Norflurazon	0.008	0.0160	113	Alpha-HCH	0.017	0.0340
86	Pyridaphenthion	0.008	0.0160	114	Terbufos	0.017	0.0340

TABLE 3.7 The LOD, LOQ of 497 Pesticides and Related Chemical Residues Determined by GC-MS—cont'd

No.	Pesticides	LOD (mg/kg)	LOQ (mg/kg)	No.	Pesticides	LOD (mg/kg)	LOQ (mg/kg)
115	Terbumeton	0.012	0.0240	144	Isofenphos	0.017	0.0340
116	Profluralin	0.033	0.0660	145	Crufomate	0.050	0.1000
117	Dioxathion	0.068	0.1360	146	cis and trane-Chlorfenvinphos	0.025	0.0500
118	Propazine	0.008	0.0160				
119	Chlorbufam	0.033	0.0660	147	cis-Chlordane	0.017	0.0340
120	Dicloran	0.033	0.0660	148	Tolylfluanide	0.025	0.0500
121	Terbuthylazine	0.008	0.0160	149	4,4'-DDE	0.008	0.0160
122	Monolinuron	0.033	0.0660	150	Butachlor	0.017	0.0340
123	Flufenoxuron	0.050	0.1000	151	Chlozolinate	0.017	0.0340
124	Cyanophos	0.033	0.0660	152	Crotoxyphos	0.050	0.1000
125	Chlorpyrifos-methyl	0.008	0.0160	153	Iodofenphos	0.017	0.0340
126	Desmetryn	0.008	0.0160	154	Tetrachlorvinphos	0.025	0.0500
127	Dimethachlor	0.010	0.0200	155	Chlorbromuron	0.204	0.4080
128	Alachlor	0.025	0.0500	156	Profenofos	0.050	0.1000
129	Pirimiphos-methyl	0.008	0.0160	157	Fluorochloridone	0.017	0.0340
130	Terbutryn	0.017	0.0340	158	Buprofezin	0.017	0.0340
131	Thiobencarb	0.017	0.0340	159	2,4'-DDD	0.008	0.0160
132	Aspon	0.017	0.0340	160	Endrin	0.100	0.2000
133	Dicofol	0.017	0.0340	161	Hexaconazole	0.050	0.1000
134	Metolachlor	0.008	0.0160	162	Chlorfenson	0.017	0.0340
135	Oxy-chlordane	0.017	0.0340	163	2,4'-DDT	0.017	0.0340
136	Pirimiphos-ethyl	0.017	0.0340	164	Paclobutrazol	0.025	0.0500
137	Methoprene	0.033	0.0660	165	Methoprotryne	0.025	0.0500
138	Bromofos	0.017	0.0340	166	Erbon	0.017	0.0340
139	Dichlofluanid	0.050	0.1000	167	Chloropropylate	0.008	0.0160
140	Ethofumesate	0.017	0.0340	168	Flamprop-methyl	0.008	0.0160
141	Isopropalin	0.017	0.0340	169	Nitrofen	0.050	0.1000
142	Endosulfan I	0.050	0.1000	170	Oxyfluorfen	0.033	0.0660
143	Propanil	0.017	0.0340	171	Chlorthiophos	0.025	0.0500

Continued

TABLE 3.7 The LOD, LOQ of 497 Pesticides and Related Chemical Residues Determined by GC-MS—cont'd

No.	Pesticides	LOD (mg/ kg)	LOQ (mg/ kg)	No.	Pesticides	LOD (mg/ kg)	LOQ (mg/ kg)
172	Flamprop-Isopropyl	0.008	0.0160	200	Vernolate	0.008	0.0160
173	4,4'-DDT	0.017	0.0340	201	3,5-Dichloroaniline	0.008	0.0160
174	Carbofenothion	0.017	0.0340	202	Molinate	0.008	0.0160
175	Benalaxyl	0.008	0.0160	203	Methacrifos	0.008	0.0160
176	Edifenphos	0.017	0.0340	204	2-Phenylphenol	0.004	0.0080
177	Triazophos	0.025	0.0500	205	Tetrahydrophthalimide	0.025	0.0500
178	Cyanofenphos	0.008	0.0160	206	Fenobucarb	0.008	0.0160
179	Chlorbenside sulfone	0.017	0.0340	207	Benfluralin	0.008	0.0160
180	Endosulfan-Sulfate	0.025	0.0500	208	Hexaflumuron	0.050	0.1000
181	Bromopropylate	0.017	0.0340	209	Prometon	0.008	0.0160
182	Benzoylprop-ethyl	0.025	0.0500	210	Triallate	0.008	0.0160
183	Fenpropathrin	0.017	0.0340	211	Pyrimethanil	0.004	0.0080
184	Captafol	0.300	0.6000	212	Gamma-HCH	0.008	0.0160
185	Leptophos	0.017	0.0340	213	Disulfoton	0.008	0.0160
186	EPN	0.033	0.0660	214	Atrizine	0.008	0.0160
187	Hexazinone	0.012	0.0240	215	Heptachlor	0.025	0.0500
188	Bifenox	0.017	0.0340	216	Iprobenfos	0.017	0.0340
189	Phosalone	0.017	0.0340	217	Isazofos	0.017	0.0340
190	Azinphos-methyl	0.050	0.1000	218	Plifenate	0.017	0.0340
191	Fenarimol	0.017	0.0340	219	Fenpropimorph	0.006	0.0120
192	Azinphos-ethyl	0.017	0.0340	220	Transfluthrin	0.008	0.0160
193	Prochloraz	0.050	0.1000	221	Fluchloralin	0.033	0.0660
194	Coumaphos	0.050	0.1000	222	Tolclofos-methyl	0.008	0.0160
195	Cyfluthrin	0.100	0.2000	223	Propisochlor	0.006	0.0120
196	Fluvalinate	0.050	0.1000	224	Ametryn	0.017	0.0340
	Group C			225	Simetryn	0.008	0.0160
197	Dichlorvos	0.017	0.0340	226	Metobromuron	0.050	0.1000
198	Biphenyl	0.004	0.0080	227	Metribuzin	0.017	0.0340
199	Propamocarb	0.050	0.1000	228	Dimethipin	0.050	0.1000

TABLE 3.7 The LOD, LOQ of 497 Pesticides and Related Chemical Residues Determined by GC–MS—cont'd

No.	Pesticides	LOD (mg/kg)	LOQ (mg/kg)	No.	Pesticides	LOD (mg/kg)	LOQ (mg/kg)
229	Epsilon-HCH	0.017	0.0340	258	Mepronil	0.006	0.0120
230	Dipropetryn	0.008	0.0160	259	Dimefuron	0.033	0.0660
231	Formothion	0.017	0.0340	260	Diflufenican	0.006	0.0120
232	Terbacil	0.017	0.0340	261	Fenazaquin	0.006	0.0120
233	Diethofencarb	0.025	0.0500	262	Phenothrin	0.006	0.0120
234	Dimepiperate	0.017	0.0340	263	Fludioxonil	0.008	0.0160
235	Bioallethrin	0.033	0.0660	264	Fenoxycarb	0.020	0.0400
236	2,4'-DDE	0.006	0.0120	265	Sethoxydim	0.076	0.1520
237	Fenson	0.006	0.0120	266	Amitraz	0.010	0.0200
238	Diphenamid	0.006	0.0120	267	Anilofos	0.017	0.0340
239	Chlorthion	0.017	0.0340	268	Acrinathrin	0.017	0.0340
240	Prallethrin	0.017	0.0340	269	Lambda-Cyhalothrin	0.008	0.0160
241	Penconazole	0.017	0.0340	270	Mefenacet	0.025	0.0500
242	Mecarbam	0.017	0.0340	271	Permethrin	0.008	0.0160
243	Tetraconazole	0.025	0.0500	272	Pyridaben	0.006	0.0120
244	Propaphos	0.017	0.0340	273	Fluoroglycofen-ethyl	0.050	0.1000
245	Flumetralin	0.017	0.0340	274	Bitertanol	0.010	0.0200
246	Triadimenol	0.025	0.0500	275	Etofenprox	0.004	0.0080
247	Pretilachlor	0.017	0.0340	276	Cycloxydim	0.040	0.0800
248	Kresoxim-methyl	0.006	0.0120	277	Alpha-Cypermethrin	0.008	0.0160
249	Fluazifop-butyl	0.006	0.0120	278	Flucythrinate	0.017	0.0340
250	Chlorfluazuron	0.025	0.0500	279	Esfenvalerate	0.033	0.0660
251	Chlorobenzilate	0.006	0.0120	280	Difenoconazole	0.033	0.0660
252	Uniconazole	0.017	0.0340	281	Flumioxazin	0.017	0.0340
253	Flusilazole	0.025	0.0500	282	Flumiclorac-pentyl	0.017	0.0340
254	Fluorodifen	0.033	0.0660		Group D		
255	Diniconazole	0.025	0.0500	283	Dimefox	0.013	0.0260
256	Piperonyl butoxide	0.006	0.0120	284	Disulfoton-Sulfoxide	0.008	0.0160
257	Propargite	0.017	0.0340	285	Pentachlorobenzene	0.004	0.0080

Continued

TABLE 3.7 The LOD, LOQ of 497 Pesticides and Related Chemical Residues Determined by GC-MS—cont'd

No.	Pesticides	LOD (mg/kg)	LOQ (mg/kg)	No.	Pesticides	LOD (mg/kg)	LOQ (mg/kg)
286	Tri-Iso-Butyl Phosphate	0.004	0.0080	315	Aziprotryne	0.033	0.0660
287	Crimidine	0.004	0.0080	316	Sebutylazine	0.004	0.0080
288	BDMC-1	0.008	0.0160	317	Isocarbamid	0.021	0.0420
289	Chlorfenprop-Methyl	0.004	0.0080	318	DE-PCB 52	0.004	0.0080
290	Thionazin	0.004	0.0080	319	Musk Moskene	0.004	0.0080
291	2,3,5,6-Tetrachloroaniline	0.004	0.0080	320	Prosulfocarb	0.004	0.0080
292	Tri-N-Butyl Phosphate	0.008	0.0160	321	Dimethenamid	0.004	0.0080
293	2,3,4,5-Tetrachloroanisole	0.004	0.0080	322	Fenchlorphos Oxon	0.008	0.0160
294	Pentachloroanisole	0.004	0.0080	323	BDMC-2	0.008	0.0160
295	Tebutam	0.008	0.0160	324	Paraoxon-Methyl	0.008	0.0160
296	Dioxabenzofos	0.042	0.0840	325	Monalide	0.008	0.0160
297	Methabenzthiazuron	0.042	0.0840	326	Musk Tibeten	0.004	0.0080
298	Simeton	0.008	0.0160	327	Isobenzan	0.004	0.0080
299	Atratone	0.004	0.0080	328	Octachlorostyrene	0.004	0.0080
300	Desisopropyl-Atrazine	0.033	0.0660	329	Pyrimitate	0.004	0.0080
301	Terbufos Sulfone	0.004	0.0080	330	Isodrin	0.004	0.0080
302	Tefluthrin	0.004	0.0080	331	Isomethiozin	0.008	0.0160
303	Bromocylen	0.004	0.0080	332	Trichloronat	0.004	0.0080
304	Trietazine	0.004	0.0080	333	Dacthal	0.004	0.0080
305	Etrimfos oxon	0.004	0.0080	334	4,4-Dichlorobenzophenone	0.004	0.0080
306	Cycluron	0.013	0.0260				
307	2,6-dichlorobenzamide	0.008	0.0160	335	Nitrothal-Isopropyl	0.008	0.0160
308	DE-PCB 28	0.004	0.0080	336	Musk Ketone	0.004	0.0080
309	DE-PCB 31	0.004	0.0080	337	Rabenzazole	0.004	0.0080
310	Desethyl-Sebuthylazine	0.008	0.0160	338	Cyprodinil	0.004	0.0080
311	2,3,4,5-Tetrachloroaniline	0.008	0.0160	339	Fuberidazole	0.021	0.0420
312	Musk Ambrette	0.004	0.0080	340	Dicapthon	0.021	0.0420
313	Musk Xylene	0.004	0.0080	341	DE-PCB 101	0.004	0.0080
314	Pentachloroaniline	0.004	0.0080	342	MCPA-butoxyethyl ester	0.004	0.0080

TABLE 3.7 The LOD, LOQ of 497 Pesticides and Related Chemical Residues Determined by GC-MS—cont'd

No.	Pesticides	LOD (mg/kg)	LOQ (mg/kg)	No.	Pesticides	LOD (mg/kg)	LOQ (mg/kg)
343	Isocarbophos	0.008	0.0160	370	Fenthion sulfone	0.017	0.0340
344	Phorate sulfone	0.004	0.0080	371	Triphenyl phosphate	0.004	0.0080
345	Chlorfenethol	0.004	0.0080	372	Metamitron	0.042	0.0840
346	Trans-nonachlor	0.004	0.0080	373	DE-PCB 180	0.004	0.0080
347	DEF	0.008	0.0160	374	Tebufenpyrad	0.004	0.0080
348	Flurochloridone	0.008	0.0160	375	Cloquintocet-mexyl	0.004	0.0080
349	Bromfenvinfos	0.004	0.0080	376	Lenacil	0.042	0.0840
350	Perthane	0.004	0.0080	377	Bromuconazole-1	0.008	0.0160
351	DE-PCB 118	0.004	0.0080	378	Desbrom- leptophos	0.004	0.0080
352	4,4-Dibromobenzophenone	0.004	0.0080	379	Bromuconazole-2	0.008	0.0160
				380	Nitralin	0.042	0.0840
353	Flutriafol	0.008	0.0160	381	Fenamiphos sulfoxide	0.017	0.0340
354	Mephosfolan	0.008	0.0160	382	Fenamiphos sulfone	0.017	0.0340
355	Athidathion	0.008	0.0160	383	Fenpiclonil	0.017	0.0340
356	DE-PCB 153	0.004	0.0080	384	Fluquinconazole	0.004	0.0080
357	Diclobutrazole	0.017	0.0340	385	Fenbuconazole	0.008	0.0160
358	Disulfoton sulfone	0.008	0.0160		Group E		
359	Hexythiazox	0.033	0.0660	386	Propoxur -1	0.008	0.0160
360	DE-PCB 138	0.004	0.0080	387	Isoprocarb -1	0.008	0.0160
361	Triamiphos	0.008	0.0160	388	Acenaphthene	0.004	0.0080
362	Resmethrin-1	0.008	0.0160	389	Dibutyl Succinate	0.008	0.0160
363	Cyproconazole	0.004	0.0080	390	Phthalimide	0.008	0.0160
364	Resmethrin-2	0.008	0.0160	391	Chlorethoxyfos	0.008	0.0160
365	Phthalic acid,benzyl butyl ester	0.004	0.0080	392	Isoprocarb -2	0.008	0.0160
366	Clodinafop-propargyl	0.008	0.0160	393	Pencycuron	0.008	0.0160
367	Fenthion sulfoxide	0.017	0.0340	394	Tebuthiuron	0.017	0.0340
368	Fluotrimazole	0.004	0.0080	395	Demeton-S-Methyl	0.017	0.0340
369	Fluroxypr-1-methylheptyl ester	0.004	0.0080	396	Cadusafos	0.017	0.0340
				397	Propoxur -2	0.008	0.0160

Continued

TABLE 3.7 The LOD, LOQ of 497 Pesticides and Related Chemical Residues Determined by GC-MS—cont'd

No.	Pesticides	LOD (mg/kg)	LOQ (mg/kg)	No.	Pesticides	LOD (mg/kg)	LOQ (mg/kg)
398	Phenanthrene	0.004	0.0080	427	Butralin	0.017	0.0340
399	Spiroxamine -1	0.008	0.0160	428	Zoxamide	0.008	0.0160
400	Fenpyroximate	0.033	0.0660	429	Pyrifenox -1	0.033	0.0660
401	Tebupirimfos	0.008	0.0160	430	Allethrin	0.017	0.0340
402	Prohydrojasmon	0.017	0.0340	431	Dimethametryn	0.004	0.0080
403	Fenpropidin	0.008	0.0160	432	Quinoclamine	0.017	0.0340
404	Dichloran	0.008	0.0160	433	Methothrin-1	0.008	0.0160
405	Pyroquilon	0.004	0.0080	434	Flufenacet	0.033	0.0660
406	Spiroxamine -2	0.008	0.0160	435	Methothrin-2	0.008	0.0160
407	Propyzamide	0.008	0.0160	436	Pyrifenox -2	0.033	0.0660
408	Pirimicarb	0.008	0.0160	437	Fenoxanil	0.008	0.0160
409	Phosphamidon -1	0.033	0.0660	438	Phthalide	0.017	0.0340
410	Benoxacor	0.008	0.0160	439	Furalaxyl	0.008	0.0160
411	Bromobutide	0.004	0.0080	440	Mepanipyrim	0.004	0.0080
412	Acetochlor	0.008	0.0160	441	Bromacil	0.033	0.0660
413	Tridiphane	0.017	0.0340	442	Picoxystrobin	0.008	0.0160
414	Terbucarb	0.008	0.0160	443	Butamifos	0.004	0.0080
415	Esprocarb	0.008	0.0160	444	Imazamethabenz-methyl	0.013	0.0260
416	Fenfuram	0.008	0.0160	445	Metominostrobin-1	0.017	0.0340
417	Acibenzolar-S-Methyl	0.008	0.0160	446	TCMTB	0.067	0.1340
418	Benfuresate	0.008	0.0160	447	Methiocarb Sulfone	0.033	0.0660
419	Dithiopyr	0.004	0.0080	448	Imazalil	0.017	0.0340
420	Mefenoxam	0.008	0.0160	449	Isoprothiolane	0.008	0.0160
421	Malaoxon	0.067	0.1340	450	Cyflufenamid	0.067	0.1340
422	Phosphamidon -2	0.033	0.0660	451	Pyriminobac-Methyl	0.017	0.0340
423	Simeconazole	0.008	0.0160	452	Isoxathion	0.033	0.0660
424	Chlorthal-dimethyl	0.008	0.0160	453	Metominostrobin-2	0.017	0.0340
425	Thiazopyr	0.008	0.0160	454	Diofenolan -1	0.008	0.0160
426	Dimethylvinphos	0.008	0.0160	455	Diofenolan -2	0.008	0.0160

TABLE 3.7 The LOD, LOQ of 497 Pesticides and Related Chemical Residues Determined by GC-MS—cont'd

No.	Pesticides	LOD (mg/ kg)	LOQ (mg/ kg)	No.	Pesticides	LOD (mg/ kg)	LOQ (mg/ kg)
456	Quinoxyphen	0.004	0.0080	477	Piperophos	0.013	0.0260
457	Chlorfenapyr	0.033	0.0660	478	Ofurace	0.013	0.0260
458	Trifloxystrobin	0.017	0.0340	479	Bifenazate	0.033	0.0660
459	Imibenconazole-des-benzyl	0.017	0.0340	480	Endrin Ketone	0.017	0.0340
				481	Clomeprop	0.004	0.0080
460	Isoxadifen-ethyl	0.008	0.0160	482	Fenamidone	0.004	0.0080
461	Fipronil	0.033	0.0660	483	Naproanilide	0.004	0.0080
462	Imiprothrin-1	0.008	0.0160	484	Pyraclostrobin	0.100	0.2000
463	Carfentrazone-ethyl	0.008	0.0160	485	Lactofen	0.033	0.0660
464	Imiprothrin-2	0.008	0.0160	486	Tralkoxydim	0.033	0.0660
465	Epoxiconazole -1	0.033	0.0660	487	Pyraclofos	0.033	0.0660
466	Pyraflufen ethyl	0.008	0.0160	488	Dialifos	0.033	0.0660
467	Pyributicarb	0.008	0.0160	489	Spirodiclofen	0.033	0.0660
468	Thenylchlor	0.008	0.0160	490	Halfenprox	0.017	0.0340
469	Clethodim	0.017	0.0340	491	Flurtamone	0.017	0.0340
470	Mefenpyr-diethyl	0.013	0.0260	492	Pyriftalid	0.004	0.0080
471	Famphur	0.017	0.0340	493	Silafluofen	0.004	0.0080
472	Etoxazole	0.025	0.0500	494	Pyrimidifen	0.017	0.0340
473	Pyriproxyfen	0.004	0.0080	495	Butafenacil	0.004	0.0080
474	Epoxiconazole -2	0.033	0.0660	496	Cafenstrole	0.050	0.1000
475	Picolinafen	0.004	0.0080	497	Fluridone	0.008	0.0160
476	Iprodione	0.017	0.0340				

3.5.3 Reagents and Materials

Acetonitrile: HPLC Grade
Acetone: HPLC Grade
Dichloromethane: Pesticide Grade
Sodium Sulfate: Anhydrous, Analytically Pure. Ignited at 650°C for 4 h and Kept in a Desiccator
Toluene: G.R.

Hexane: Pesticide Grade
Envi-Carb SPE Cartridge: 6 mL, 0.5 g, or Equivalent
Sep-Pak NH$_2$ Cartridge: 3 mL, 0.5 g, or Equivalent
Pesticide and Related Chemicals Standard: Purity ≥95%

Stock Standard Solution: Accurately weigh 5–10 mg of individual pesticide and related chemical standards (accurate to 0.1 mg) into a 10-mL volumetric flask. Dissolve and dilute to volume with toluene, toluene + acetone combination, dichloromethane, etc., depending on each individual compound's solubility (for diluting solvent refer to Appendix B).

Mixed Standard Solution (Mixed Standard Solution A, B, C, D, and E): Depending on properties and retention times of the compounds, all compounds are divided into five groups: A, B, C, D, and E. The mixed standard solution concentration is determined by its sensitivity on the instrument used for analysis. For 497 pesticides and related chemical groupings and mixed standard solution concentrations of this standard, reference is made to Appendix B.

Depending on group number, mixed standard solution concentration, and stock standard solution concentration, appropriate amounts of individual stock standard solution are pipetted into a 100-mL volumetric flask, being diluted to volume with toluene. Mixed standard solutions are stored in the dark below 4°C and are used for one month.

Internal Standard Solution: Accurately weigh 3.5 mg heptachlor epoxide into a 100-mL volumetric flask. Dissolve and dilute to volume with toluene.

Working Standard Mixed Solution in Matrix: Working standard mixed solutions in matrix of A, B, C, D, and E group pesticides and related chemicals are prepared by diluting 40 μL internal standard solution and an appropriate amount of mixed standard solution to 1.0 mL with blank extract that has been taken through the method with the rest of the samples. Mix thoroughly. Working standard mixed solutions in matrix must be prepared fresh.

3.5.4 Apparatus

GC-MS: Equipped With EI
Analytical Balances: Capable of Weighing From 0.1 mg to 0.01 g
Pear-Shaped Flask: 200 mL
Pipette: 1 mL
Glass Mason Jar With Cap: 250 mL
Separatory Funnel: 250 mL

3.5.5 Sample Pretreatment

3.5.5.1 Preparation of Test Sample

For honey sample, if the sample is not crystallized, mix it thoroughly by stirring. If crystallized, place the closed container in a water bath, warm at ≤60°C with occasional shaking until liquefied, mix thoroughly, and promptly cool to room temperature. Pour 0.5 kg of the prepared test sample into a clean and dry sample bottle. Seal and label.

For fruit juice and wine samples, mix thoroughly by stirring. Pour 0.5 kg of the prepared test sample into a clean and dry sample bottle. Seal and label.

The test samples should be stored at ambient temperature.

3.5.5.2 Extraction

Weigh 15 g test sample (accurate to 0.01 g) into a 250-mL glass mason jar, add 30 mL water and place it in a shaking water bath at 40°C to shake for 15 min, ensuring the honey sample has completely dissolved in the water. Add 10 mL acetone to the jar and transfer the jar contents to a 250-mL separatory funnel. Rinse the jar with 40 mL dichloromethane several times and transfer this rinse to the separatory funnel for partitioning. Partition by using eight strokes of the separatory funnel. Pass the bottom layer through a funnel containing anhydrous sodium sulfate into a 200-mL pear-shaped flask. Add 5 mL acetone and 40 mL dichloromethane into the separatory funnel; repeat steps twice. Concentrate the extract to ca. 1 mL with a rotary evaporator at 40°C for clean-up.

3.5.5.3 Clean-up

Add sodium sulfate into the Envi-Carb cartridge to ca. 2 cm. Connect the cartridge to the top of the aminopropyl Sep-Pak cartridge in series. Fix the cartridge into a support to which a pear-shaped flask is connected. Condition the cartridge with 4 mL acetonitrile-toluene (3+1) before adding the sample. Once the solution gets to the top of the sodium sulfate, pipette the eluate into the cartridge immediately. Rinse the pear-shaped flask with 3×2 mL acetonitrile-toluent (3+1) and decant it into the cartridge. Insert a reservoir into the cartridge. Elute the pesticide with 25 mL acetonitrile-toluene (3+1). Evaporate the eluate to ca. 0.5 mL using a rotary evaporator at 40°C. Exchange with 2×5 mL hexane twice and make up to ca. 1 mL. Add 40 μL internal standard solution. Mix thoroughly. The solution is ready for GC-MS determination.

3.5.6 Determination

3.5.6.1 GC-MS Operating Condition

(a) Column: DB-1701 capillary column (30 m × 0.25 mm × 0.25 μm), or equivalent
(b) Column temperature: 40°C hold 1 min, at 30°C/min to 130°C, at 5°C/min to 250°C, at 10°C/min to 300°C, hold 5 min
(c) Carrier gas: Helium, purity ≥99.999%, flow rate: 1.2 mL/min
(d) Injection port temperature: 290°C
(e) Injection volume: 1 μL
(f) Injection mode: Splitless, purge on after 1.5 min
(g) Ionization voltage: 70 eV

(h) Ion source temperature: 230°C

(i) GC-MS interface temperature: 280°C

(j) Selected ion monitoring mode: Each compound selects 1 quantifying ion and 2–3 qualifying ions. All of the detected ions of each group are detected according to programmed time and sequence of peaking. The retention times, quantifying ions, qualifying ions and the abundance ratios of quantifying ion and qualifying ions for each compound are listed in Table 5.3. The programmed time and dwell time for the ions detected for each compound in each group are listed in Table 5.4

3.5.6.2   Qualitative Determination

In the samples determined, five injections are required to analyze for all pesticides according to GC-MS operating conditions, if the retention time in the peaks of the sample solution are the same as for the peaks of the working standard mixed solution, and the selected ions of the background-subtracted mass spectrum appear, and the abundance ratios of selected ions are within the expected limits (abundance ratios >50%, permitted tolerances are ±10%; abundance ratios >20% to 50%, permitted tolerances are ±15%; abundance ratios >10% to 20%, permitted tolerances are ±20%; abundance ratios ≤10%, permitted tolerances are ±50%), the sample is confirmed to contain this pesticide compound. In the case where results are still not definitive, the sample should be reinjected with acquisition in scan mode (sufficient sensitivity) or with additional confirmatory ions or by use of other instruments that have higher sensitivity.

3.5.6.3   Quantitative Determination

The results are quantitated using heptachlor epoxide for an internal standard and the quantitation ion response for each analyte. To compensate for the matrix effect, quantitation is based on a mixed standard prepared in blank matrix extract. The concentrations of the standard solution and the detected sample solution must be similar.

3.5.6.4   Parallel Test

A parallel test is carried out for the same testing sample.

3.5.6.5   Blank Test

The operation of the blank test is the same as that described in the method of determination, but without the addition of sample.

3.5.7   Precision

The precision data of the method for this standard have been determined according to the stipulations of GB/T 6379.1 and GB/T 6379.2. The values of repeatability and reproducibility are obtained and calculated at the 95% confidence level.

RESEARCHERS

Guo-Fang Pang, Chun-Lin Fan, Yong-Ming Liu, Yan-Zhong Cao, Jin-Jie Zhang, Bao-Lian Fu, Guang-Qun Jia, Xue-Min Li, Yan-Ping Wu.

Qinhuangdao Entry-Exit Inspection and Quarantine Bureau, 39 Haibin Rd, Qinhuangdao, Hebei, PC 066002, People's Republic of China.

3.6 DETERMINATION OF 486 PESTICIDES AND RELATED CHEMICAL RESIDUES IN HONEY: LC-MS-MS METHOD (GB/T 20771-2008)

3.6.1 Scope

This method is applicable to the quantitative determination of 461 pesticides and related chemical residues and qualitative determination of 486 pesticides and related chemical residues in yanghuai honey, youcai honey, duanshu honey, qiaomai honey, and zaohua honey. The limit of quantitative determination of this method of 461 pesticides and related chemicals is 0.01 µg/kg to 3.34 mg/kg (see Table 3.8).

3.6.2 Principle

Samples are extracted with dichloromethane, and the extracts are cleaned up with Sep-Pak Vac cartridges. The pesticides and related chemicals are eluted with acetonitrile-toluene (3 + 1), and the solutions are analyzed by LC-MS-MS.

3.6.3 Reagents and Materials

"Water" is first grade of GB/T6682 specified.

Acetonitrile: HPLC Grade
Acetone: HPLC Grade
Toluene: HPLC Grade
Isooctane: HPLC Grade
Hexane: HPLC Grade
Methanol: HPLC Grade
Dichloromethane: HPLC Grade
Membrane Filters: 13 mm, 0.2 µm
Waters Sep-Pak Vac Cartridge: 6 mL, 1 g, or Equivalent
0.05% Formic Acid-Water (V/V)
5 mmol/L Ammonium Acetate: Weigh 0.375 g Ammonium Acetate, Dilute With Water to 1000 mL
Acetonitrile + Toluene (3 + 1, V/V)
Acetonitrile + Water (3 + 2, V/V)

TABLE 3.8 The LOD, LOQ of 486 Pesticides and Related Chemical Residues Determined by LC-MS-MS

No.	Pesticides	LOD (μg/kg)	LOQ (μg/kg)	No.	Pesticides	LOD (μg/kg)	LOQ (μg/kg)
	Group A			29	dimethenamid	0.24	0.48
1	propham	13.70	27.40	30	terrbucarb	0.13	0.26
2	isoprocarb	0.14	0.28	31	penconazole	0.16	0.32
3	3,4,5-trimethacarb	0.03	0.06	32	myclobutanil	0.09	0.18
4	cycluron	0.04	0.08	33	paclobutrazol	0.05	0.10
5	carbaryl	0.34	0.68	34	fenthion sulfoxide	0.10	0.20
6	propachlor	0.04	0.08	35	triadimenol	0.75	1.50
7	rabenzazole	0.22	0.44	36	butralin	0.17	0.34
8	simetryn	0.02	0.04	37	spiroxamine	0.01	0.02
9	monolinuron	0.20	0.40	38	tolclofos methyl	8.29	16.58
10	mevinphos	0.15	0.30	39	desmedipham	0.08	0.16
11	aziprotryne	0.19	0.38	40	methidathion	0.33	0.66
12	secbumeton	0.01	0.02	41	allethrin	3.20	6.40
13	cyprodinil	0.08	0.16	42	triallate	1.32	2.64
14	buturon	0.67	1.34	43	diazinon	0.06	0.12
15	carbetamide	0.57	1.14	44	edifenphos	0.08	0.16
16	pirimicarb	0.03	0.06	45	pretilachlor	0.03	0.06
17	clomazone dimethazone	0.04	0.08	46	flusilazole	0.06	0.12
18	cyanazine	0.02	0.04	47	iprovalicarb	0.15	0.30
19	prometryne	0.02	0.04	48	benodanil	0.30	0.60
20	4,4-dichlorobenzophenone	1.19	2.38	49	flutolanil	0.14	0.28
				50	famphur	0.37	0.74
21	thiacloprid	0.09	0.18	51	benalyxyl	0.11	0.22
22	imidacloprid	2.49	4.98	52	diclobutrazole	0.07	0.14
23	ethidimuron	0.50	1.00	53	etaconazole	0.17	0.34
24	isomethiozin	0.07	0.14	54	fenarimol	0.08	0.16
25	diallate	4.17	8.34	55	phthalic acid, dicyclobexyl ester	0.10	0.20
26	acetochlor	3.40	6.80				
27	nitenpyram	31.70	63.40	56	tetramethirn	0.24	0.48
28	methoprotryne	0.02	0.04	57	dichlofluanid	9.89	19.78

TABLE 3.8 The LOD, LOQ of 486 Pesticides and Related Chemical Residues Determined by LC-MS-MS—cont'd

No.	Pesticides	LOD (µg/kg)	LOQ (µg/kg)	No.	Pesticides	LOD (µg/kg)	LOQ (µg/kg)
58	cloquintocet mexyl	0.02	0.04	87	propoxur	0.73	1.46
59	bitertanol	0.87	1.74	88	isouron	0.04	0.08
60	chlorprifos methyl	9.15	18.30	89	chlorotoluron	0.07	0.14
61	thiophanate methyl	3.05	6.10	90	thiofanox	8.90	17.80
62	azinphos ethyl	15.80	31.60	91	chlorbufam	9.74	19.48
63	clodinafop propargyl	0.36	0.72	92	bendiocarb	0.07	0.14
64	triflumuron	0.34	0.68	93	propazine	0.03	0.06
65	isoxaflutole	0.39	0.78	94	terbuthylazine	0.03	0.06
66	anilofos	0.06	0.12	95	diuron	0.12	0.24
67	thiophanat ethyl	4.42	8.84	96	chlormephos	490.00	980.00
68	quizalofop-ethyl	0.09	0.18	97	carboxin	0.08	0.16
69	haloxyfop-methyl	0.22	0.44	98	difenzoquat-methyl sulfate	0.08	0.16
70	fluazifop butyl	0.03	0.06	99	clothianidin	14.50	29.00
71	bromophos-ethyl	13.10	26.20	100	pronamide	0.47	0.94
72	dialifos	9.32	18.64	101	dimethachloro	0.10	0.20
73	bensulide	1.52	3.04	102	methobromuron	1.25	2.50
74	bromfenvinfos	0.32	0.64	103	phorate	3.49	6.98
75	azoxystrobin	0.04	0.08	104	aclonifen	1.61	3.22
76	pyrazophos	0.08	0.16	105	mephosfolan	0.07	0.14
77	flufenoxuron	0.27	0.54	106	imibenzonazole-des-benzyl	0.50	1.00
78	indoxacarb	0.72	1.44	107	neburon	0.31	0.62
79	emamectin benzoate	0.03	0.06	108	mefenoxam	0.08	0.16
	Group B			109	prothoate	0.11	0.22
80	dazomet	33.60	67.20	110	ethofume sate	41.70	83.40
81	nicotine	0.18	0.36	111	iprobenfos	0.65	1.30
82	fenuron	0.17	0.34	112	TEPP	0.79	1.58
83	crimidine	0.20	0.40	113	cyproconazole	0.07	0.14
84	acephate	1.38	2.76	114	thiamethoxam	5.42	10.84
85	molinate	0.19	0.38	115	crufomate	0.04	0.08
86	carbendazim	0.06	0.12				

Continued

TABLE 3.8 The LOD, LOQ of 486 Pesticides and Related Chemical Residues Determined by LC-MS-MS—cont'd

No.	Pesticides	LOD (µg/kg)	LOQ (µg/kg)	No.	Pesticides	LOD (µg/kg)	LOQ (µg/kg)
116	etrimfos	19.00	38.00	146	propargite	3.99	7.98
117	cythioate	16.00	32.00	147	bromuconazole	0.23	0.46
118	phosphamidon	0.18	0.36	148	picolinafen	0.10	0.20
119	phenmedipham	0.06	0.12	149	fluthiacet methyl	0.43	0.86
120	bifenazate	4.51	9.02	150	trifloxystrobin	0.52	1.04
121	flutriafol	0.31	0.62	151	hexaflumuron	1.38	2.76
122	furalaxyl	0.03	0.06	152	novaluron	0.52	1.04
123	bioallethrin	5.61	11.22	153	flurazuron	0.50	1.00
124	cyanofenphos	0.75	1.50		Group C		
125	pirimiphos methyl	0.02	0.04	154	methamidophos	0.73	1.46
126	buprofezin	0.05	0.10	155	diethyltoluamide	0.01	0.02
127	disulfoton sulfone	0.09	0.18	156	monuron	2.25	4.50
128	fenazaquin	0.03	0.06	157	pyrimethanil	0.09	0.18
129	triazophos	0.04	0.08	158	fenfuram	0.05	0.10
130	DEF	0.13	0.26	159	quinoclamine	0.63	1.26
131	pyriftalid	0.05	0.10	160	fenobucarb	1.42	2.84
132	metconazole	0.06	0.12	161	propanil	0.89	1.78
133	pyriproxyfen	0.02	0.04	162	carbofuran	0.43	0.86
134	cycloxydim	0.21	0.42	163	acetamiprid	0.24	0.48
135	isoxaben	0.02	0.04	164	mepanipyrim	0.04	0.08
136	flurtamone	0.04	0.08	165	prometon	0.01	0.02
137	trifluralin	150.00	300.00	166	methiocarb	20.70	41.40
138	flamprop methyl	0.69	1.38	167	metoxuron	0.14	0.28
139	bioresmethrin	0.37	0.74	168	dimethoate	0.68	1.36
140	propiconazole	0.11	0.22	169	methfuroxam	0.03	0.06
141	chlorpyrifos	2.35	4.70	170	fluometuron	0.07	0.14
142	fluchloralin	33.70	67.40	171	dicrotophos	0.18	0.36
143	clethodim	0.29	0.58	172	monalide	0.14	0.28
144	flamprop isopropyl	0.03	0.06	173	diphenamid	0.01	0.02
145	tetrachlorvinphos	0.17	0.34	174	ethoprophos	0.17	0.34

TABLE 3.8 The LOD, LOQ of 486 Pesticides and Related Chemical Residues Determined by LC-MS-MS—cont'd

No.	Pesticides	LOD (µg/kg)	LOQ (µg/kg)	No.	Pesticides	LOD (µg/kg)	LOQ (µg/kg)
175	fonofos	0.39	0.78	205	butachlor	1.49	2.98
176	etridiazol	8.55	17.10	206	kresoxim-methyl	26.90	53.80
177	furmecyclox	0.08	0.16	207	triticonazole	0.25	0.50
178	hexazinone	0.02	0.04	208	fenamiphos sulfoxide	0.09	0.18
179	dimethametryn	0.01	0.02	209	thenylchlor	0.69	1.38
180	trichlorphon	0.52	1.04	210	pyrethrin	19.20	38.40
181	demeton(o+s)	4.85	9.70	211	fenoxanil	0.76	1.52
182	benoxacor	0.57	1.14	212	fluridone	0.03	0.06
183	bromacil	6.16	12.32	213	epoxiconazole	0.21	0.42
184	phorate sulfoxide	51.70	103.40	214	chlorphoxim	4.74	9.48
185	brompyrazon	0.80	1.60	215	fenamiphos sulfone	0.05	0.10
186	oxycarboxin	0.21	0.42	216	fenbuconazole	0.13	0.26
187	mepronil	0.03	0.06	217	isofenphos	29.30	58.60
188	disulfoton	23.00	46.00	218	oryzalin	7.91	15.82
189	fenthion	7.72	15.44	219	phenothrin	21.40	42.80
190	metalaxyl	0.04	0.08	220	piperophos	0.49	0.98
191	ofurace	0.05	0.10	221	piperonyl butoxide	0.05	0.10
192	dodemorph	0.04	0.08	222	oxyflurofen	6.75	13.50
193	fosthiazate	0.07	0.14	223	coumaphos	2.52	5.04
194	imazamethabenz-methyl	0.03	0.06	224	flufenacet	0.40	0.80
195	disulfoton-sulfoxide	0.10	0.20	225	phosalone	0.53	1.06
196	isoprothiolane	0.10	0.20	226	methoxyfenozide	0.19	0.38
197	imazalil	0.16	0.32	227	prochloraz	0.10	0.20
198	phoxim	26.30	52.60	228	aspon	0.18	0.36
199	quinalphos	0.14	0.28	229	ethion	0.11	0.22
200	ditalimfos	3.90	7.80	230	diafenthiuron pestanal	0.84	1.68
201	fenoxycarb	0.59	1.18	231	dithiopyr	0.93	1.86
202	pyrimitate	0.02	0.04	232	spirodiclofen	0.36	0.72
203	fensulfothin	0.12	0.24	233	fenpyroximate	0.06	0.12
204	fluorochloridone	1.53	3.06	234	flumiclorac-pentyl	0.71	1.42

Continued

TABLE 3.8 The LOD, LOQ of 486 Pesticides and Related Chemical Residues Determined by LC-MS-MS—cont'd

No.	Pesticides	LOD (μg/kg)	LOQ (μg/kg)	No.	Pesticides	LOD (μg/kg)	LOQ (μg/kg)
235	temephos	0.12	0.24	264	terbutryn	0.10	0.20
236	butafenacil	0.45	0.90	265	triazoxide	1.13	2.26
237	spinosad	0.09	0.18	266	thionazin	1.27	2.54
	Group D			267	linuron	0.97	1.94
238	allidochlor	1.13	2.26	268	heptanophos	0.22	0.44
239	tricyclazole	0.19	0.38	269	prosulfocarb	0.05	0.10
240	thiabendazole	0.11	0.22	270	propyzamide	0.62	1.24
241	metamitron	0.86	1.72	271	dipropetryn	0.03	0.06
242	isoproturon	0.01	0.02	272	thiobencarb	0.19	0.38
243	atratone	0.03	0.06	273	tri-iso-butyl phosphate	0.35	0.70
244	oesmetryn	0.04	0.08	274	tri-n-butyl phosphate	0.03	0.06
245	metribuzin	0.04	0.08	275	diethofencarb	1.04	2.08
246	DMST	1.79	3.58	276	alachlor	0.75	1.50
247	cycloate	0.18	0.36	277	cadusafos	0.08	0.16
248	atrazine	0.05	0.10	278	metazachlor	0.10	0.20
249	butylate	3.42	6.84	279	propetamphos	6.20	12.40
250	pymetrozin	3.16	6.32	280	terbufos	18.00	36.00
251	chloridazon	0.72	1.44	281	uniconazole	0.10	0.20
252	sulfallate	14.80	29.60	282	simeconazole	0.15	0.30
253	terbumeton	0.01	0.02	283	triadimefon	0.63	1.26
254	cyprazine	0.01	0.02	284	phorate sulfone	0.85	1.70
255	ametryn	0.08	0.16	285	tridemorph	0.56	1.12
256	tebuthiuron	0.04	0.08	286	mefenacet	0.13	0.26
257	trietazine	0.08	0.16	287	azaconazole	0.10	0.20
258	sebutylazine	0.03	0.06	288	fenamiphos	0.04	0.08
259	dibutyl succinate	4.98	9.96	289	fenpropimorph	0.03	0.06
260	tebutam	0.02	0.04	290	tebuconazole	0.17	0.34
261	thiofanox-sulfoxide	0.63	1.26	291	isopropalin	2.88	5.76
262	cartap hydrochloride	806.00	1612.00	292	nuarimol	0.10	0.20
263	methacrifos	15.00	30.00	293	bupirimate	0.06	0.12

TABLE 3.8 The LOD, LOQ of 486 Pesticides and Related Chemical Residues Determined by LC-MS-MS—cont'd

No.	Pesticides	LOD (μg/kg)	LOQ (μg/kg)	No.	Pesticides	LOD (μg/kg)	LOQ (μg/kg)
294	azinphos-methyl	26.20	52.40		Group E		
295	tebupirimfos	0.02	0.04	323	4-aminopyridine	0.45	0.90
296	phenthoate	4.34	8.68	324	methomyl	3.86	7.72
297	sulfotep	0.26	0.52	325	oxamyl-oxime	3.60	7.20
298	sulprofos	0.78	1.56	326	pyroquilon	0.70	1.40
299	EPN	4.40	8.80	327	fuberidazole	2.67	5.34
300	azamethiphos	0.13	0.26	328	isocarbamid	0.16	0.32
301	diniconazole	0.18	0.36	329	butocarboxim	0.37	0.74
302	sethoxydim	5.29	10.58	330	chlordimeform	1.03	2.06
303	pencycuron	0.02	0.04	331	cymoxanil	2.39	4.78
304	mecarbam	1.90	3.80	332	vernolate	0.05	0.10
305	tralkoxydim	0.05	0.10	333	promecarb	0.79	1.58
306	malathion	0.27	0.54	334	aminocarb	11.80	23.60
307	pyributicarb	0.07	0.14	335	dimethirimol	0.12	0.24
308	pyridaphenthion	0.08	0.16	336	omethoate	0.03	0.06
309	pirimiphos-ethyl	0.01	0.02	337	ethoxyquin	7.04	14.08
310	thiodicarb	40.30	80.60	338	dichlorvos	0.18	0.36
311	pyraclofos	0.17	0.34	339	aldicarb sulfone	7.01	14.02
312	picoxystrobin	0.66	1.32	340	dioxacarb	12.50	25.00
313	tetraconazole	0.20	0.40	341	demeton-s-methyl	0.64	1.28
314	mefenpyr-diethyl	0.75	1.50	342	oxabetrinil	0.01	0.02
315	profenefos	0.38	0.76	343	ethiofencarb-sulfoxide	11.40	22.80
316	pyraclostrobin	0.04	0.08	344	cyanohos	5.77	11.54
317	dimethomorph	0.06	0.12	345	etridiazole	0.31	0.62
318	kadethrin	0.26	0.52	346	thiometon	91.00	182.00
319	thiazopyr	0.45	0.90	347	folpet	94.10	188.20
320	cinosulfuron	0.12	0.24	348	demeton-s-methyl sulfone	0.87	1.74
321	iodosulfuron-methyl	5.04	10.08	349	dimepiperate	1060.00	2120.00
322	chlorfluazuron	0.68	1.36	350	fenpropidin	0.10	0.20

Continued

TABLE 3.8 The LOD, LOQ of 486 Pesticides and Related Chemical Residues Determined by LC-MS-MS—cont'd

No.	Pesticides	LOD (µg/kg)	LOQ (µg/kg)	No.	Pesticides	LOD (µg/kg)	LOQ (µg/kg)
351	amidithion	2.80	5.60	380	pyridate	29.10	58.20
352	paraoxon-ethyl	0.09	0.18	381	furathiocarb	1.00	2.00
353	aldimorph	0.16	0.32	382	trans-permethin	0.22	0.44
354	pyrifenox	0.09	0.18	383	pyrazoxyfen	0.03	0.06
355	chlorthion	100.00	200.00	384	flubenzimine	10.50	21.00
356	dicapthon	0.12	0.24	385	zeta cypermethrin	0.18	0.36
357	clofentezine	0.32	0.64	386	haloxyfop-2-ethoxyethyl	1.20	2.40
358	norflurazon	0.07	0.14		Group F		0.00
359	quinoxyphen	31.30	62.60	387	acrylamide	12.50	25.00
360	fenthion sulfone	6.38	12.76	388	tert-butylamine	29.90	59.80
361	methoprene	1.72	3.44	389	phthalimide	25.40	50.80
362	flurochloridone	0.46	0.92	390	dimefox	2.41	4.82
363	phthalic acid,benzyl butyl ester	136.00	272.00	391	metolcarb	5.40	10.80
				392	diphenylamin	0.10	0.20
364	isazofos	0.09	0.18	393	1-naphthyl acetamide	0.20	0.40
365	dichlofenthion	2.51	5.02	394	atrazine-desethyl	0.14	0.28
366	vamidothion sulfone	33.00	66.00	395	2,6-dichlorobenzamide	1.54	3.08
367	terbufos sulfone	9.84	19.68	396	dimethyl phthalate	4.69	9.38
368	dinitramine	1.60	3.20	397	chlordimeform hydrochloride	1.21	2.42
369	cyazofamid	0.43	0.86				
370	trichloronat	3.78	7.56	398	simeton	0.08	0.16
371	resmethrin-2	0.11	0.22	399	dinotefuran	4.39	8.78
372	boscalid	1.37	2.74	400	pebulate	0.94	1.88
373	nitralin	4.51	9.02	401	acibenzolar-s-methyl	0.96	1.92
374	cafenstrol	11.40	22.80	402	dioxabenzofos	5.53	11.06
375	hexythiazox	19.30	38.60	403	oxamyl	19.40	38.80
376	benzoximate	3.63	7.26	404	methabenzthiazuron	0.02	0.04
377	pyridaben	2.40	4.80	405	butoxycarboxim	4.30	8.60
378	benzoylprop-ethyl	130.00	260.00	406	mexacarbate	0.56	1.12
379	pyrimidifen	1.99	3.98	407	demeton-s-methyl sulfoxide	1.52	3.04

TABLE 3.8 The LOD, LOQ of 486 Pesticides and Related Chemical Residues Determined by LC-MS-MS—cont'd

No.	Pesticides	LOD (µg/kg)	LOQ (µg/kg)	No.	Pesticides	LOD (µg/kg)	LOQ (µg/kg)
408	thiofanox-sulfone	0.82	1.64	437	naled	72.20	144.40
409	phosfolan	0.11	0.22	438	tolfenpyrad	0.05	0.10
410	triclopyr	0.03	0.06	439	cinidon-ethyl	9.13	18.26
411	demeton-s	3.32	6.64	440	rotenone	0.69	1.38
412	fenthion oxon	0.29	0.58	441	fluotrimazole	3340.00	6680.00
413	daimuron	1.55	3.10	442	carfentrazone—ethyl	1.15	2.30
414	napropamide	0.21	0.42	443	propaquizafop	0.30	0.60
415	fenitrothion	36.60	73.20	444	lactofen	67.10	134.20
416	phthalic acid, dibutyl ester	0.05	0.10		Group G		
417	metolachlor	0.12	0.24	445	cis-1,2,3,6-tetrahydrophthalimide	0.23	0.46
418	procymidone	62.80	125.60	446	2,6-difluorobenzoic acid	25.40	50.80
419	vamidothion	0.78	1.56	447	trichloroacetic acid sodium salt	2.41	4.82
420	chloroxuron	0.13	0.26				
421	triamiphos	0.10	0.20	448	2-phenylphenol	0.10	0.20
422	dithianon	1.33	2.66	449	3-phenylphenol	0.20	0.40
423	merphos	8.02	16.04	450	dicloran	4.69	9.38
424	prallethrin	16.50	33.00	451	aminopyralid	1.21	2.42
425	cumyluron	0.22	0.44	452	chlorpropham	0.08	0.16
426	phosmet	3.81	7.62	453	terbacil	0.94	1.88
427	ronnel	2.48	4.96	454	dicamba	5.53	11.06
428	pyrethrins	21.80	43.60	455	2,3,4,5-tetrachloroaniline	0.06	0.12
429	athidathion	3.47	6.94	456	fenaminosulf	0.02	0.04
430	phthalic acid, biscyclohexyl ester	0.07	0.14	457	picloram	0.56	1.12
431	carpropamid	1.58	3.16	458	bentazone	1.52	3.04
432	tebufenpyrad	0.09	0.18	459	dinoseb	0.82	1.64
433	zoxamide	1.31	2.62	460	dinoterb	0.11	0.22
434	tebufenozide	0.94	1.88	461	fludioxonil	1.48	2.96
435	iminoctadine triacetate	4.56	9.12	462	chlorfenethol	18.30	36.60
436	chlorthiophos	11.40	22.80	463	bromoxynil	31.40	62.80

Continued

TABLE 3.8 The LOD, LOQ of 486 Pesticides and Related Chemical Residues Determined by LC-MS-MS—cont'd

No.	Pesticides	LOD (µg/kg)	LOQ (µg/kg)	No.	Pesticides	LOD (µg/kg)	LOQ (µg/kg)
464	chlorobenzuron	0.13	0.26	477	fomesafen	35.70	71.40
465	chloramphenicolum	0.10	0.20	478	lufenuron	0.30	0.60
466	fluorodifen	0.22	0.44	479	thifluzamide	67.10	134.20
467	diflufenzopyr-sodium	3.81	7.62		Group H		
468	acifluorfen	1.58	3.16	480	bediocarb	0.22	0.44
469	famoxadone	1.31	2.62	481	bromocylen	4.50	9.00
470	sulfentrazone	9.12	18.24		Group I		
471	diflufenican	11.40	22.80	482	dimethipin	3.29	6.58
472	ethiprole	72.20	144.40	483	benfuresate	50.10	100.20
473	flusulfamide	0.12	0.24	484	nitrofen	8.55	17.10
474	endosulfan-sulfate	25.50	51.00	485	bifenox	31.80	63.60
475	triforine	0.69	1.38	486	iodofenphos	16.10	32.20
476	halosulfuran-methyl	3340.00	6680.00				

Anhydrous Sodium Sulfate: Analytically Pure, Heated at 650°C for 4 h and Kept in a Desiccator

Pesticide and Related Chemicals Standard: Purity ≥95%

Stock Standard Solution: Accurately weigh 5–10 mg of individual pesticide and related chemical standards (accurate to 0.1 mg) into a 10-mL volumetric flask. Dissolve and dilute to volume with methanol, toluene, acetone, acetonitrile, or isooctane, etc. depending on each individual compound's solubility. (For diluting solvent refer to Appendix B.) Standard solutions are stored in the dark below 4°C and are used for 1 month.

Mixed Standard Solution (Mixed Standard Solution A, B, C, D, E, F, G, H, and I): Depending on properties and retention times of the compounds, all compounds are divided into nine groups: A, B, C, D, E, F, G, H, and I. The concentration of each compound is determined by its sensitivity on the instrument for analysis. For 486 pesticides and related chemical groupings and concentrations, refer to Appendix B.

Depending on group number, mixed standard solution concentrations, and stock standard solution concentrations, appropriate amounts of individual stock standard solution are pipetted into a 100-mL volumetric flask, being diluted to volume with methanol. Mixed standard solutions are stored in the dark below 4°C and are used for 1 month.

Working Standard Mixed Solution in Matrix: Working standard mixture solution in matrix is prepared by diluting an appropriate amount of mixed standard solution with blank extract that has been taken through the method with the rest of the samples. Mix thoroughly. These are used for plotting the standard curve.

Working standard mixture solution in matrix must be freshly prepared.

3.6.4 Apparatus

LC-MS-MS: Equipped With ESI
Analytical Balance: Capable of Weighing From 0.1 mg to 0.01 g
Pear-Shaped Flask: 200 mL
Pipette: 1 mL
Screw Vial: 2.0 mL, With Screw Caps and PTFE Septa
Glass Mason Jar With Cap: 250 mL
Separatory Funnel: 250 mL
Barrel Funnel
Rotary Evaporator
Nitrogen Evaporator

3.6.5 Sample Pretreatment

3.6.5.1 Preparation of Test Sample

For honey sample, if the sample is not crystallized, mix it thoroughly by stirring. If crystallized, place the closed container in a water bath, warm at ≤60°C with occasional shaking until liquefied, mix thoroughly, and promptly cool to room temperature. Pour 0.5 kg of the prepared test sample into a clean and dry sample bottle. Seal and label.

The test samples should be stored at ambient temperature.

3.6.5.2 Extraction

Weigh 15 g test sample (accurate to 0.01 g) into a 250-mL glass mason jar, add 20 mL water, and place it in a shaking water bath at 40°C to shake for 15 min, ensuring the honey sample has completely dissolved in the water. Add 10 mL acetone to the jar and transfer the jar contents to a 250-mL separatory funnel. Rinse the jar with 40 mL dichloromethane several times, and transfer this rinse to the separatory funnel for partitioning. Partition by using eight strokes of the separatory funnel. Pass the bottom layer through a funnel containing anhydrous sodium sulfate into a 200-mL pear-shaped flask. Add 5 mL acetone and 40 mL dichloromethane into the separatory funnel; repeat steps twice. Concentrate the extract to ca. 1 mL with a rotary evaporator at 40°C for clean-up.

3.6.5.3 Clean-up

Add sodium sulfate into the Sep-Pak Vac cartridge to ca. 2 cm. Fix the cartridge into a support to which a pear-shaped flask is connected. Condition the cartridge with 4 mL acetonitrile-toluene (3+1) before adding the sample. Once the solution gets to the top of the sodium sulfate, pipette the eluate into the cartridge immediately. Rinse the pear-shaped flask with 3 × 2 mL acetonitrile-toluent (3+1) and decant it into the cartridge. Insert a reservoir into the cartridge. Elute the pesticide with 25 mL acetonitrile-toluene (3+1). Evaporate the eluate to ca. 0.5 mL using a rotary evaporator at 40°C, and evaporate the solutions to dryness using a nitrogen evaporator at 35°C. Reconstitute the sample to 1 mL with acetonitrile-water (3+2). Finally the extract is filtered through a 0.2-μm filter into a glass vial for LC-MS-MS determination.

3.6.6 Determination

3.6.6.1 LC-MS-MS Operating Condition

Conditions for the Pesticides and Related Chemicals of A, B, C, D, E and F Group

(a) Chromatography column: Atlantis dc_{18}, 3 μm, 150 mm × 2.1 mm (diameter), or equivalent
(b) Mobile phase program and flow rate: as in Table 3.9
(c) Column temperature: 40°C
(d) Injection volume: 20 μL
(e) Ion source: ESI
(f) Scan mode: Positive ion scan
(g) Monitor mode: Multiple reaction monitor

TABLE 3.9 Mobile Phase Program and the Flow Rate

Time (min)	Flow Rate (μL/min)	Mobile Phase A 0.1% Formic Acid-Water (%)	Mobile Phase B Acetonitrile (%)
0.00	200	90.0	10.0
4.00	200	50.0	50.0
15.00	200	40.0	60.0
23.00	200	20.0	80.0
30.00	200	5.0	95.0
35.00	200	5.0	95.0
35.01	200	90.0	10.0
50.00	200	90.0	10.0

(h) Ion spray voltage: 5000 V
(i) Nebulizer gas: 0.483 MPa
(j) Curtain gas: 0.138 MPa
(k) Auxiliary gas: 0.379 MPa
(l) Ion source temperature: 725°C
(m) Monitoring ion pairs, collision energy, and declustering potentials: see Table 5.10

Conditions for the Pesticides and Related Chemicals of Group G

(a) Chromatography column: Inertsil C8, 5 μm, 150 mm × 2.1 mm (diameter), or equivalent
(b) Mobile phase program and flow rate: as in Table 3.10
(c) Column temperature: 40°C
(d) Injection volume: 20 μL
(e) Ion source: ESI
(f) Scan mode: Negative ion scan
(g) Monitor mode: Multiple reaction monitor
(h) Ion spray voltage: −4200 V
(i) Nebulizer gas: 0.42 MPa
(j) Curtain gas: 0.315 MPa
(k) Auxiliary gas: 0.35 MPa
(l) Ion source temperature: 700°C
(m) Monitoring ions pairs, collision energy, and declustering potentials: see Table 5.10

TABLE 3.10 Mobile Phase Program and the Flow Rate

Time (min)	Flow Rate (μL/min)	Mobile Phase A 5 mmol/L Ammonium Acetate-Water (%)	Mobile Phase B Acetonitrile (%)
0.00	200	90.0	10.0
4.00	200	50.0	50.0
15.00	200	40.0	60.0
20.00	200	20.0	80.0
25.00	200	5.0	95.0
32.00	200	5.0	95.0
32.01	200	90.0	10.0
40.00	200	90.0	10.0

Conditions for the Pesticides and Related Chemicals of Group H

(a) Chromatography column: Atlantis T3, 5 μm, 150 mm × 4.6 mm (diameter), or equivalent

(b) Mobile phase program and flow rate: as in Table 3.11

(c) Column temperature: 40°C

(d) Injection volume: 20 μL

(e) Ion source: APCI

(f) Scan mode: Positve ion scan

(g) Monitor mode: Multiple reaction monitor

(h) Nebulizer gas: 0.56 MPa

(i) Curtain gas: 0.133 MPa

(j) Auxiliary gas: 0.28 MPa

(k) Ion source temperature: 400°C

(l) Monitoring ions pairs, collision energy, and declustering potentials: see Table 5.10

Conditions for the Pesticides and Related Chemicals of Group I

(a) Chromatography column: Atlantis T3, 5 μm, 150 mm × 4.6 mm (diameter), or equivalent

(b) Mobile phase program and flow rate: as in Table 3.12

(c) Column temperature: 40°C

(d) Injection volume: 20 μL

(e) Ion source: APCI

(f) Scan mode: Negative ion scan

(g) Monitor mode: Multiple reaction monitor

(h) Nebulizer gas: 0.42 MPa

(i) Curtain gas: 0.084 MPa

(j) Auxiliary gas: 0.28 MPa

TABLE 3.11 Mobile Phase Program and the Flow Rate

Time (min)	Flow Rate (μL/min)	Mobile Phase A 5 mmol/L Ammonium Acetate-Water (%)	Mobile Phase B Acetonitrile (%)
0.00	500	80.0	20.0
2.00	500	5.0	95.0
10.00	500	5.0	95.0
10.01	500	80.0	20.0
20.00	500	80.0	20.0

TABLE 3.12 Mobile Phase Program and the Flow Rate

Time (min)	Flow Rate (μL/min)	Mobile Phase A 5 mmol/L Ammonium Acetate-Water (%)	Mobile Phase B Acetonitrile (%)
0.00	500	80.0	20.0
2.00	500	5.0	95.0
10.00	500	5.0	95.0
10.01	500	80.0	20.0
20.00	500	80.0	20.0

(k) Ion source temperature: 425°C
(l) Monitoring ions pairs, collision energy, and declustering potentials: see Table 5.10

3.6.6.2 Qualitative Determination

If the retention times of peaks of the sample solution are the same as those of the peaks from working standard mixed solution, and the selected ions appear in the background-subtracted mass spectrum, and the abundance ratios of the selected ions are within the expected limits (abundance ratios >50%, permitted tolerances are ±10%; abundance ratios >20% to 50%, permitted tolerances are ±15%; abundance ratios >10% to 20%, permitted tolerances are ±20%; abundance ratios ≤10%, permitted tolerances are ±50%), the sample is confirmed to contain this pesticide compound.

3.6.6.3 Quantitative Determination

An external standard method is used for quantitation with standard curves for LC-MS-MS. In order to compensate for the matrix effect, quantitation is based on a series of working standard solutions prepared in blank matrix extract. The standard curves are established by injection of different concentrations of working standard mixed solutions in matrix separately. The responses of pesticides in the sample solution should be in the linear range of the instrumental detection.

3.6.6.4 Parallel Test

A parallel test is carried out for the same testing sample.

3.6.6.5 Blank Test

The operation of the blank test is the same as that described in the method of determination, but without the addition of sample.

3.6.7 Precision

The precision data of the method for this standard is according to the stipulations of GB/T 6379.1 and GB/T 6379.2. The values of repeatability and reproducibility are obtained and calculated at the 95% confidence level.

RESEARCHERS

Guo-Fang Pang, Yan Li, Chun-Lin Fan, Yu-Jing Lian, Shu-jun Zhao, Jun-Hong Zheng, Yong-ming Liu, Yan-Zhong Cao, Jin-Jie Zhang, Xue-Min Li.

Qinhuangdao Entry-Exit Inspection and Quarantine Bureau, 39 Haibin Rd, Qinhuangdao, Hebei, PC 066002, People's Republic of China.

3.7 DETERMINATION OF 478 PESTICIDES AND RELATED CHEMICAL RESIDUES IN ANIMAL MUSCLES: GC-MS METHOD (GB/T 19650-2006)

3.7.1 Scope

This method is applicable to the determination of 478 pesticides and related chemical residues in bovine, porcine, mutton, chicken, and rabbit.

The limits of detection of the method for this standard are 0.0025 mg/kg to 0.3000 mg/kg (refer to Table 3.13).

3.7.2 Principle

The test samples are extracted with cyclohexane-ethyl acetate (1+1) mixed solution. The extracts are cleaned up by GPC with Bio-Beads S-X3 column, and the solutions are analyzed by GC-MS.

3.7.3 Reagents and Materials

Acetonitrile: HPLC Grade
Cyclohexane: P.R. Grade
Ethyl Acetate: P.R. Grade
Hexane: P.R. Grade
Anhydrous Sodium Sulfate: Ignite at 650°C for 4h; Store in a Desiccator
Cyclohexane-Ethyl Acetate (1+1)
Pesticide Standards: Purity ≥95%
> Stock Standard Solution: Accurately weigh 5–10 mg of individual pesticide and related chemical standards (accurate to 0.1 mg) into a 10-mL volumetric flask. Dissolve and dilute to volume with toluene, toluene-acetone combination, dichloromethane, etc., depending on each individual compound's solubility. (For diluting solvent, refer to Appendix B.)

TABLE 3.13 The LOD, LOQ of 478 Pesticides and Related Chemical Residues Determined by GC-MS

No.	Pesticides	LOD (mg/ kg)	LOQ (mg/ kg)	No.	Pesticides	LOD (mg/ kg)	LOQ (mg/ kg)
	Group A			29	Vinclozolin	0.0125	0.0250
1	Allidochlor	0.0250	0.0500	30	Beta-HCH	0.0125	0.0250
2	Dichlormid	0.0250	0.0500	31	Metalaxyl	0.0375	0.0750
3	Etridiazol	0.0375	0.0750	32	Chlorpyrifos (-ethyl)	0.0125	0.0250
4	Chlormephos	0.0250	0.0500	33	Methyl-Parathion	0.0500	0.1000
5	Propham	0.0125	0.0250	34	Anthraquinone	0.0125	0.0250
6	Cycloate	0.0125	0.0250	35	Delta-HCH	0.0250	0.0500
7	Diphenylamine	0.0125	0.0250	36	Fenthion	0.0125	0.0250
8	Chlordimeform	0.0125	0.0250	37	Malathion	0.0500	0.1000
9	Ethalfluralin	0.0500	0.1000	38	Fenitrothion	0.0250	0.0500
10	Phorate	0.0125	0.0250	39	Paraoxon-ethyl	0.0500	0.1000
11	Thiometon	0.0125	0.0250	40	Triadimefon	0.0250	0.0500
12	Quintozene	0.0250	0.0500	41	Parathion	0.0500	0.1000
13	Atrazine-desethyl	0.0125	0.0250	42	Pendimethalin	0.0500	0.1000
14	Clomazone	0.0125	0.0250	43	Linuron	0.0500	0.1000
15	Diazinon	0.0125	0.0250	44	Chlorbenside	0.0250	0.0500
16	Fonofos	0.0125	0.0250	45	Bromophos-ethyl	0.0125	0.0250
17	Etrimfos	0.0125	0.0250	46	Quinalphos	0.0125	0.0250
18	Simazine	0.0125	0.0250	47	trans-Chlordane	0.0125	0.0250
19	Propetamphos	0.0125	0.0250	48	Phenthoate	0.0250	0.0500
20	Secbumeton	0.0125	0.0250	49	Metazachlor	0.0375	0.0750
21	Dichlofenthion	0.0125	0.0250	50	Fenothiocarb	0.0250	0.0500
22	Pronamide	0.0125	0.0250	51	Prothiophos	0.0125	0.0250
23	Mexacarbate	0.0375	0.0750	52	Chlorflurenol	0.0375	0.0750
24	Aldrin	0.0250	0.0500	53	Dieldrin	0.0250	0.0500
25	Dinitramine	0.0500	0.1000	54	Procymidone	0.0125	0.0250
26	Ronnel	0.0250	0.0500	55	Methidathion	0.0250	0.0500
27	Prometryne	0.0125	0.0250	56	Napropamide	0.0375	0.0750
28	Cyprazine	0.0125	0.0250	57	Oxadiazone	0.0125	0.0250

Continued

TABLE 3.13 The LOD, LOQ of 478 Pesticides and Related Chemical Residues Determined by GC-MS—cont'd

No.	Pesticides	LOD (mg/kg)	LOQ (mg/kg)	No.	Pesticides	LOD (mg/kg)	LOQ (mg/kg)
58	Fenamiphos	0.0375	0.0750	87	cis-Permethrin	0.0125	0.0250
59	Tetrasul	0.0125	0.0250	88	Trans-Permethrin	0.0125	0.0250
60	Aramite	0.0125	0.0250	89	Pyrazophos	0.0250	0.0500
61	Bupirimate	0.0125	0.0250	90	Cypermethrin	0.0375	0.0750
62	Carboxin	0.0375	0.0750	91	Fenvalerate	0.0500	0.1000
63	Flutolanil	0.0125	0.0250	92	Deltamethrin	0.0750	0.1500
64	p,p'-DDD	0.0125	0.0250		Group B		
65	Ethion	0.0250	0.0500	93	EPTC	0.0375	0.0750
66	Sulprofos	0.0250	0.0500	94	Butylate	0.0375	0.0750
67	Etaconazole-1	0.0375	0.0750	95	Dichlobenil	0.0025	0.0050
68	Etaconazole-2	0.0375	0.0750	96	Pebulate	0.0375	0.0750
69	Myclobutanil	0.0125	0.0250	97	Nitrapyrin	0.0375	0.0750
70	Diclofop-methyl	0.0125	0.0250	98	Mevinphos	0.0250	0.0500
71	Propiconazole	0.0375	0.0750	99	Chloroneb	0.0125	0.0250
72	Fensulfothion	0.0250	0.0500	100	Tecnazene	0.0250	0.0500
73	Bifenthrin	0.0125	0.0250	101	Heptenophos	0.0375	0.0750
74	Carbosulfan	0.0375	0.0750	102	Hexachlorobenzene	0.0125	0.0250
75	Mirex	0.0125	0.0250	103	Ethoprophos	0.0375	0.0750
76	Benodanil	0.0375	0.0750	104	cis-Diallate	0.0250	0.0500
77	Nuarimol	0.0250	0.0500	105	Propachlor	0.0375	0.0750
78	Methoxychlor	0.0125	0.0250	106	trans-Diallate	0.0250	0.0500
79	Oxadixyl	0.0125	0.0250	107	Trifluralin	0.0250	0.0500
80	Tetramethirn	0.0250	0.0500	108	Chlorpropham	0.0250	0.0500
81	Tebuconazole	0.0375	0.0750	109	Sulfotep	0.0125	0.0250
82	Norflurazon	0.0125	0.0250	110	Sulfallate	0.0250	0.0500
83	Pyridaphenthion	0.0125	0.0250	111	Alpha-HCH	0.0125	0.0250
84	Phosmet	0.0250	0.0500	112	Terbufos	0.0250	0.0500
85	Tetradifon	0.0125	0.0250	113	Terbumeton	0.0375	0.0750
86	Oxycarboxin	0.0750	0.1500	114	Profluralin	0.0500	0.1000

TABLE 3.13 The LOD, LOQ of 478 Pesticides and Related Chemical Residues Determined by GC-MS—cont'd

No.	Pesticides	LOD (mg/kg)	LOQ (mg/kg)	No.	Pesticides	LOD (mg/kg)	LOQ (mg/kg)
115	Dioxathion	0.0500	0.1000	144	Cis-Chlordane	0.0250	0.0500
116	Propazine	0.0125	0.0250	145	Tolylfluanide	0.0375	0.0750
117	Chlorbufam	0.0250	0.0500	146	p,p'-DDE	0.0125	0.0250
118	Dicloran	0.0250	0.0500	147	Butachlor	0.0250	0.0500
119	Terbuthylazine	0.0125	0.0250	148	Chlozolinate	0.0250	0.0500
120	Monolinuron	0.0500	0.1000	149	Crotoxyphos	0.0750	0.1500
121	Cyanophos	0.0250	0.0500	150	Iodofenphos	0.0250	0.0500
122	Chlorpyrifos-methyl	0.0125	0.0250	151	Tetrachlorvinphos	0.0375	0.0750
123	Desmetryn	0.0125	0.0250	152	Chlorbromuron	0.3000	0.6000
124	Dimethachlor	0.0375	0.0750	153	Profenofos	0.0750	0.1500
125	Alachlor	0.0375	0.0750	154	Fluorochloridone	0.0250	0.0500
126	Pirimiphos-methyl	0.0125	0.0250	155	Buprofezin	0.0250	0.0500
127	Terbutryn	0.0250	0.0500	156	o,p'-DDD	0.0125	0.0250
128	Thiobencarb	0.0250	0.0500	157	Endrin	0.1500	0.3000
129	Aspon	0.0250	0.0500	158	Hexaconazole	0.0750	0.1500
130	Dicofol	0.0250	0.0500	159	Chlorfenson	0.0250	0.0500
131	Metolachlor	0.0125	0.0250	160	o,p'-DDT	0.0250	0.0500
132	Oxy-chlordane	0.0125	0.0250	161	Paclobutrazol	0.0375	0.0750
133	Pirimiphos-ethyl	0.0250	0.0500	162	Methoprotryne	0.0375	0.0750
134	Methoprene	0.0500	0.1000	163	Erbon	0.0250	0.0500
135	Bromofos	0.0250	0.0500	164	Chloropropylate	0.0125	0.0250
136	Dichlofluanid	0.0750	0.1500	165	Flamprop-methyl	0.0125	0.0250
137	Ethofumesate	0.0250	0.0500	166	Nitrofen	0.0750	0.1500
138	Isopropalin	0.0250	0.0500	167	Oxyfluorfen	0.0500	0.1000
139	endosulfan I	0.0750	0.1500	168	Chlorthiophos	0.0375	0.0750
140	Propanil	0.0250	0.0500	169	endosulfan I	0.0750	0.1500
141	Isofenphos	0.0250	0.0500	170	Flamprop-Isopropyl	0.0125	0.0250
142	Crufomate	0.0750	0.1500	171	p,p'-DDT	0.0250	0.0500
143	Chlorfenvinphos	0.0375	0.0750	172	Carbofenothion	0.0250	0.0500

Continued

TABLE 3.13 The LOD, LOQ of 478 Pesticides and Related Chemical Residues Determined by GC-MS—cont'd

No.	Pesticides	LOD (mg/kg)	LOQ (mg/kg)	No.	Pesticides	LOD (mg/kg)	LOQ (mg/kg)
173	Benalaxyl	0.0125	0.0250	201	Benfluralin	0.0125	0.0250
174	Edifenphos	0.0250	0.0500	202	Prometon	0.0375	0.0750
175	Triazophos	0.0375	0.0750	203	Triallate	0.0250	0.0500
176	Cyanofenphos	0.0125	0.0250	204	Pyrimethanil	0.0125	0.0250
177	Chlorbenside sulfone	0.0250	0.0500	205	Gamma-HCH	0.0250	0.0500
178	Endosulfan-Sulfate	0.0375	0.0750	206	Disulfoton	0.0125	0.0250
179	Bromopropylate	0.0250	0.0500	207	Atrizine	0.0125	0.0250
180	Benzoylprop-ethyl	0.0375	0.0750	208	Heptachlor	0.0375	0.0750
181	Fenpropathrin	0.0250	0.0500	209	Iprobenfos	0.0375	0.0750
182	Leptophos	0.0250	0.0500	210	Isazofos	0.0250	0.0500
183	EPN	0.0500	0.1000	211	Fenpropimorph	0.0125	0.0250
184	Hexazinone	0.0375	0.0750	212	Transfluthrin	0.0125	0.0250
185	Phosalone	0.0250	0.0500	213	Tolclofos-methyl	0.0125	0.0250
186	Azinphos-methyl	0.0750	0.1500	214	Propisochlor	0.0125	0.0250
187	Fenarimol	0.0250	0.0500	215	Ametryn	0.0375	0.0750
188	Azinphos-ethyl	0.0250	0.0500	216	Simetryn	0.0250	0.0500
189	Prochloraz	0.0750	0.1500	217	Metobromuron	0.0750	0.1500
190	Coumaphos	0.0750	0.1500	218	Metribuzin	0.0375	0.0750
191	Cyfluthrin	0.1500	0.3000	219	Epsilon-HCH,	0.0250	0.0500
	Group C			220	Dipropetryn	0.0125	0.0250
192	Dichlorvos	0.0750	0.1500	221	Formothion	0.0250	0.0500
193	Biphenyl	0.0125	0.0250	222	Diethofencarb	0.0750	0.1500
194	Vernolate	0.0125	0.0250	223	Dimepiperate	0.0250	0.0500
195	3,5-Dichloroaniline	0.0125	0.0250	224	Bioallethrin-1	0.0500	0.1000
196	Molinate	0.0125	0.0250	225	Bioallethrin-2	0.0500	0.1000
197	Methacrifos	0.0125	0.0250	226	o,p'-DDE	0.0125	0.0250
198	2-Phenylphenol	0.0125	0.0250	227	Fenson	0.0125	0.0250
199	Cis-1,2,3,6-Tetrahydrophthalimide	0.0375	0.0750	228	Diphenamid	0.0125	0.0250
200	Fenobucarb	0.0250	0.0500	229	Chlorthion	0.0250	0.0500

TABLE 3.13 The LOD, LOQ of 478 Pesticides and Related Chemical Residues Determined by GC-MS—cont'd

No.	Pesticides	LOD (mg/kg)	LOQ (mg/kg)	No.	Pesticides	LOD (mg/kg)	LOQ (mg/kg)
230	Prallethrin	0.0375	0.0750	259	Bitertanol	0.0375	0.0750
231	Penconazole	0.0375	0.0750	260	Etofenprox	0.0125	0.0250
232	Mecarbam	0.0500	0.1000	261	Cycloxydim	0.1500	0.3000
233	Tetraconazole	0.0375	0.0750	262	Alpha-Cypermethrin	0.0250	0.0500
234	Flumetralin	0.0250	0.0500	263	Esfenvalerate	0.0500	0.1000
235	Triadimenol	0.0375	0.0750	264	Difenoconazole	0.0750	0.1500
236	Pretilachlor	0.0250	0.0500	265	Flumioxazin	0.0250	0.0500
237	Kresoxim-methyl	0.0125	0.0250	266	Flumiclorac-pentyl	0.0250	0.0500
238	Fluazifop-butyl	0.0125	0.0250		Group D		
239	Chlorobenzilate	0.0125	0.0250	267	Dimefox	0.0375	0.0750
240	Uniconazole	0.0250	0.0500	268	Disulfoton-sulfoxide	0.0250	0.0500
241	Flusilazole	0.0375	0.0750	269	Pentachlorobenzene	0.0125	0.0250
242	Fluorodifen	0.0125	0.0250	270	Tri-iso-butyl phosphate	0.0125	0.0250
243	Diniconazole	0.0375	0.0750	271	Crimidine	0.0125	0.0250
244	Piperonyl butoxide	0.0125	0.0250	272	BDMC-1	0.0250	0.0500
245	Propargite	0.0250	0.0500	273	Chlorfenprop-methyl	0.0125	0.0250
246	Mepronil	0.0125	0.0250	274	Thionazin	0.0125	0.0250
247	Diflufenican	0.0125	0.0250	275	2,3,5,6-tetrachloroaniline	0.0125	0.0250
248	Fenazaquin	0.0125	0.0250	276	Tri-n-butyl phosphate	0.0250	0.0500
249	Phenothrin	0.0125	0.0250	277	2,3,4,5-tetrachloroanisole	0.0125	0.0250
250	Fludioxonil	0.0125	0.0250	278	Pentachloroanisole	0.0125	0.0250
251	Fenoxycarb	0.0750	0.1500	279	Tebutam	0.0250	0.0500
252	Sethoxydim	0.1125	0.2250	280	Dioxabenzofos	0.1250	0.2500
253	Amitraz	0.0375	0.0750	281	Methabenzthiazuron	0.1250	0.2500
254	Anilofos	0.0250	0.0500	282	Simetone	0.0250	0.0500
255	Mefenacet	0.0375	0.0750	283	Atratone	0.0125	0.0250
256	Permethrin	0.0250	0.0500	284	Bromocylen	0.0125	0.0250
257	Pyridaben	0.0125	0.0250	285	Trietazine	0.0125	0.0250
258	Fluoroglycofen-ethyl	0.1500	0.3000	286	Etrimfos oxon	0.0125	0.0250

Continued

TABLE 3.13 The LOD, LOQ of 478 Pesticides and Related Chemical Residues Determined by GC-MS—cont'd

No.	Pesticides	LOD (mg/kg)	LOQ (mg/kg)	No.	Pesticides	LOD (mg/kg)	LOQ (mg/kg)
287	Cycluron	0.0375	0.0750	314	Dacthal	0.0125	0.0250
288	2,6-dichlorobenzamide	0.0250	0.0500	315	4,4-dichlorobenzophenone	0.0125	0.0250
289	DE-PCB 28 2,4,4'-Trichlorobiphenyl	0.0125	0.0250	316	Nitrothal-isopropyl	0.0250	0.0500
290	DE-PCB 31 2,4',5-Trichlorobiphenyl	0.0125	0.0250	317	Rabenzazole	0.0125	0.0250
				318	Cyprodinil	0.0125	0.0250
291	Desethyl-sebuthylazine	0.0250	0.0500	319	Fuberidazole	0.0625	0.1250
292	2,3,4,5-tetrachloroaniline	0.0250	0.0500	320	Isofenphos oxon	0.0250	0.0500
293	Musk ambrette	0.0125	0.0250	321	Methfuroxam	0.0125	0.0250
294	Musk xylene	0.0125	0.0250	322	Dicapthon	0.0625	0.1250
295	Pentachloroaniline	0.0125	0.0250	323	DE-PCB 101 2,2',4,5,5'-Pentachlorobiphenyl	0.0125	0.0250
296	Aziprotryne	0.1000	0.2000				
297	Sebutylazine	0.0125	0.0250	324	MCPA-butoxyethyl ester	0.0125	0.0250
298	Isocarbamid	0.0625	0.1250	325	Isocarbophos	0.0250	0.0500
299	DE-PCB 52 2,2',5,5'-Tetrachlorobiphenyl	0.0125	0.0250	326	Phorate sulfone	0.0125	0.0250
				327	Chlorfenethol	0.0125	0.0250
300	Musk moskene	0.0125	0.0250	328	Trans-nonachlor	0.0125	0.0250
301	Prosulfocarb	0.0125	0.0250	329	Dinobuton	0.1250	0.2500
302	Dimethenamid	0.0125	0.0250	330	DEF	0.0250	0.0500
303	Fenchlorphos oxon	0.0250	0.0500	331	Flurochloridone	0.0250	0.0500
304	BDMC-2	0.0250	0.0500	332	Bromfenvinfos	0.0125	0.0250
305	Paraoxon-methyl	0.0250	0.0500	333	Perthane	0.0125	0.0250
306	Monalide	0.0250	0.0500	334	Ditalimfos	0.0125	0.0250
307	Musk tibeten	0.0125	0.0250	335	DE-PCB 118 2,3',4,4',5-Pentachlorobiphenyl	0.0125	0.0250
308	Isobenzan	0.0125	0.0250				
309	Octachlorostyrene	0.0125	0.0250	336	4,4-dibromobenzophenone	0.0125	0.0250
310	Pyrimitate	0.0125	0.0250				
311	Isodrin	0.0125	0.0250	337	Flutriafol	0.0250	0.0500
312	Isomethiozin	0.0250	0.0500	338	Mephosfolan	0.0250	0.0500
313	Trichloronat	0.0125	0.0250	339	Athidathion	0.0250	0.0500

TABLE 3.13 The LOD, LOQ of 478 Pesticides and Related Chemical Residues Determined by GC-MS—cont'd

No.	Pesticides	LOD (mg/kg)	LOQ (mg/kg)	No.	Pesticides	LOD (mg/kg)	LOQ (mg/kg)
340	DE-PCB 153 2,2′,4,4′,5,5′-Hexachlorobiphenyl	0.0125	0.0250	365	Fenamiphos sulfone	0.0500	0.1000
				366	Fenpiclonil	0.0500	0.1000
341	Diclobutrazole	0.0500	0.1000	367	Fluquinconazole	0.0125	0.0250
342	Disulfoton sulfone	0.0250	0.0500	368	Fenbuconazole	0.0250	0.0500
343	Hexythiazox	0.1000	0.2000		Group E		
344	DE-PCB 138 2,2′,3,4,4′,5′-Hexachlorobiphenyl	0.0125	0.0250	369	Propoxur-1	0.0250	0.0500
				370	Isoprocarb -1	0.0250	0.0500
345	Triamiphos	0.0250	0.0500	371	Methamidophos	0.0500	0.1000
346	Resmethrin-1	0.0250	0.0500	372	Acenaphthene	0.0125	0.0250
347	Cyproconazole	0.0125	0.0250	373	Dibutyl succinate	0.0250	0.0500
348	Resmethrin-2	0.0250	0.0500	374	Phthalimide	0.0250	0.0500
349	Phthalic acid,benzyl butyl ester	0.0125	0.0250	375	Chlorethoxyfos	0.0250	0.0500
350	Clodinafop-propargyl	0.0250	0.0500	376	Isoprocarb -2	0.0250	0.0500
351	Fenthion sulfoxide	0.0500	0.1000	377	Pencycuron	0.0250	0.0500
352	Fluotrimazole	0.0125	0.0250	378	Tebuthiuron	0.0500	0.1000
353	Fluroxypr-1-methylheptyl ester	0.0125	0.0250	379	demeton-S-methyl	0.0500	0.1000
				380	Cadusafos	0.0500	0.1000
354	Fenthion sulfone	0.0500	0.1000	381	Propoxur-2	0.0250	0.0500
355	Triphenyl phosphate	0.0125	0.0250	382	Phenanthrene	0.0125	0.0250
356	Metamitron	0.1250	0.2500	383	Fenpyroximate	0.1000	0.2000
357	DE-PCB 180 2,2′,3,4,4′,5,5′-Heptachlorobiphenyl	0.0125	0.0250	384	Tebupirimfos	0.0250	0.0500
				385	prohydrojasmon	0.0500	0.1000
358	Tebufenpyrad	0.0125	0.0250	386	Fenpropidin	0.0250	0.0500
359	Cloquintocet-mexyl	0.0125	0.0250	387	Dichloran	0.0250	0.0500
360	Lenacil	0.1250	0.2500	388	Propyzamide	0.0250	0.0500
361	Bromuconazole	0.0250	0.0500	389	Pirimicarb	0.0250	0.0500
362	Desbrom- leptophos	0.0125	0.0250	390	Phosphamidon -1	0.1000	0.2000
363	Nitralin	0.1250	0.2500	391	Benoxacor	0.0250	0.0500
364	Fenamiphos sulfoxide	0.0500	0.1000	392	Bromobutide	0.0125	0.0250

Continued

TABLE 3.13 The LOD, LOQ of 478 Pesticides and Related Chemical Residues Determined by GC-MS—cont'd

No.	Pesticides	LOD (mg/kg)	LOQ (mg/kg)	No.	Pesticides	LOD (mg/kg)	LOQ (mg/kg)
393	Acetochlor	0.0250	0.0500	422	Mepanipyrim	0.0125	0.0250
394	Tridiphane	0.0500	0.1000	423	Bromacil	0.1000	0.2000
395	Terbucarb	0.0250	0.0500	424	Picoxystrobin	0.0250	0.0500
396	Esprocarb	0.0250	0.0500	425	Butamifos	0.0125	0.0250
397	Fenfuram	0.0250	0.0500	426	Imazamethabenz-methyl	0.0375	0.0750
398	Acibenzolar-S-Methyl	0.0250	0.0500	427	Metominostrobin-1	0.0500	0.1000
399	Benfuresate	0.0250	0.0500	428	TCMTB	0.2000	0.4000
400	Dithiopyr	0.0125	0.0250	429	Methiocarb Sulfone	0.1000	0.2000
401	Mefenoxam	0.0250	0.0500	430	Imazalil	0.0500	0.1000
402	Malaoxon	0.2000	0.4000	431	Isoprothiolane	0.0250	0.0500
403	Phosphamidon -2	0.1000	0.2000	432	Pyriminobac-methyl	0.0500	0.1000
404	Simeconazole	0.0250	0.0500	433	Isoxathion	0.1000	0.2000
405	Chlorthal-dimethyl	0.0250	0.0500	434	Metominostrobin-2	0.0500	0.1000
406	Thiazopyr	0.0250	0.0500	435	Diofenolan-1	0.0250	0.0500
407	Dimethylvinphos	0.0250	0.0500	436	Diofenolan-2	0.0250	0.0500
408	Butralin	0.0500	0.1000	437	Quinoxyphen	0.0125	0.0250
409	Zoxamide	0.0250	0.0500	438	Chlorfenapyr	0.1000	0.2000
410	Pyrifenox -1	0.1000	0.2000	439	Trifloxystrobin	0.0500	0.1000
411	Allethrin	0.0500	0.1000	440	Imibenconazole-des-benzyl	0.0500	0.1000
412	Dimethametryn	0.0125	0.0250				
413	Quinoclamine	0.0500	0.1000	441	Isoxadifen-ethyl	0.0250	0.0500
414	Methothrin-1	0.0250	0.0500	442	Imiprothrin-1	0.0250	0.0500
415	Flufenacet	0.1000	0.2000	443	Carfentrazone-ethyl	0.0250	0.0500
416	Methothrin-2	0.0250	0.0500	444	Imiprothrin-2	0.0250	0.0500
417	Pyrifenox -2	0.1000	0.2000	445	Epoxiconazole -1	0.1000	0.2000
418	Fenoxanil	0.0250	0.0500	446	Pyraflufen Ethyl	0.0250	0.0500
419	Phthalide	0.0500	0.1000	447	Pyributicarb	0.0250	0.0500
420	Furalaxyl	0.0250	0.0500	448	Thenylchlor	0.0250	0.0500
421	Thiamethoxam	0.0500	0.1000	449	Clethodim	0.1000	0.2000

TABLE 3.13 The LOD, LOQ of 478 Pesticides and Related Chemical Residues Determined by GC-MS—cont'd

No.	Pesticides	LOD (mg/ kg)	LOQ (mg/ kg)	No.	Pesticides	LOD (mg/ kg)	LOQ (mg/ kg)
450	Mefenpyr-diethyl	0.0375	0.0750	465	Lactofen	0.2000	0.4000
451	Famphur	0.0500	0.1000	466	Tralkoxydim	0.1000	0.2000
452	Etoxazole	0.0750	0.1500	467	Pyraclofos	0.1000	0.2000
453	Pyriproxyfen	0.0125	0.0250	468	Dialifos	0.1000	0.2000
454	Epoxiconazole-2	0.1000	0.2000	469	Spirodiclofen	0.1000	0.2000
455	Picolinafen	0.0125	0.0250	470	Halfenprox	0.0500	0.1000
456	Iprodione	0.0500	0.1000	471	Flurtamone	0.0500	0.1000
457	Piperophos	0.0375	0.0750	472	Pyriftalid	0.0125	0.0250
458	Ofurace	0.0375	0.0750	473	Silafluofen	0.0125	0.0250
459	Bifenazate	0.1000	0.2000	474	Pyrimidifen	0.0500	0.1000
460	Endrin ketone	0.0500	0.1000	475	Acetamiprid	0.1500	0.3000
461	Clomeprop	0.0125	0.0250	476	Butafenacil	0.0500	0.1000
462	Fenamidone	0.0125	0.0250	477	Cafenstrole	0.1500	0.3000
463	Naproanilide	0.0125	0.0250	478	Fluridone	0.0250	0.0500
464	Pyraclostrobin	0.3000	0.6000				

Mixed Standard Solution (Mixed Standard Solution A, B, C, D, and E): Depending on properties and retention times of each compound, all compounds are divided into five groups: A, B, C, D, E. Depending upon instrument sensitivity of each compound, the concentration of mixed standard solutions are decided. For 478 pesticides and related chemical groupings and concentrations of mixed standard solutions of this standard, reference is made to Appendix B.

Depending on group number, mixed standard solution concentration, and stock standard solution concentration, appropriate amounts of individual stock standard solutions are pipetted into a 100-mL volumetric flask, with five group pesticides and related chemicals being diluted to volume with toluene. Mixed standard solutions are stored in the dark below 4°C and are used for 1 month.

Internal Standard Solution: Accurately weigh 3.5 mg heptachlor epoxide into a 100-mL volumetric flask. Dissolve and dilute to volume with toluene.

Working Standard Mixed Solution in Matrix: Working standard mixed solutions in matrix of A, B, C, D, and E group pesticides and related chemicals are prepared by diluting 40 μL internal standard solution and appropriate amounts of mixed standard solution to 1.0 mL with blank extract that has been taken through the method with the rest of the samples, and mixing thoroughly. Working standard mixed solutions in matrix must be freshly prepared.

3.7.4 Apparatus

GC-MS: Equipped With EI
GPC: Equipped 400 mm × 25 mm, BIO-Beads S-X3 Column
Analytical Balances: Capable of Weighing From 0.1 mg to 0.01 g
Rotary Evaporator
Homogenizer: Max. 24,000 rpm
Centrifuge: Max. 4200 rpm
Nitrogen Evaporator
Pear-Shaped Flasks
Pipette: 1 mL

3.7.5 Sample Pretreatment

3.7.5.1 Preparation of Test Sample

Part of the representative sample is taken from each bag of the primary sample and homogenized by grinding in a meat grinder. The homogenized sample is thoroughly mixed and reduced to at least 500 g by quartering and then placed in a clean container as the test sample. Label and seal.

The test samples should be stored below −18°C.

3.7.5.2 Extraction

Weigh 10 g (accurate to 0.01 g) of the test sample into a 50-mL centrifuge tube filled with 20 g anhydrous sodium sulfate. Add 35 mL volume of cyclohexane-ethyl acetate (1 + 1) mixed solution. The mixture is homogenized for 1.5 min at 15,000 rpm, and then centrifuged for 3 min at 3000 rpm. Pass the supernatant through a glass funnel containing a glass wool plug and ca. 15 g Na_2SO_4. Collect extracts in a 100-mL pear-shaped flask. Proceed twice with 35 mL cyclohexane-ethyl acetate mixed solution. Combine the extracts in the pear-shaped flask. Evaporate to ca. 5 mL on a rotary evaporator at 40°C and proceed to clean-up. If the residue content in the fat are calculated, collect the extracts in a weighted pear-shaped flask. Evaporate the extracts to ca. 5 mL on rotary evaporator at 40°C and evaporate the solvent to dryness with a nitrogen evaporator. Weigh the pear-shaped flask and proceed to-clean up.

3.7.5.3 GPC Clean-up

3.7.5.3.1 Conditions

(a) GPC Column: 400 mm × 25 mm, filled with Bio-Beads S-X3
(b) Detection wavelength: 254 nm
(c) Mobile phase: cychexane-ethyl acetate (1+1) mixed solution
(d) Flow rate: 5 mL/min
(e) Injection volume: 5 mL
(f) Start collect time: 22 min
(g) Stop collect time: 40 min

3.7.5.3.2 Clean-up

Transfer the concentrated extracts into a 10-mL volumetric flask with cychexane+ethyl acetate (1+1) mixed solution. Rinse pear-shaped flask twice with 5 mL cychexane-ethyl acetate (1+1) mixed solution and transfer into the 10-mL volumetric flask. Dilute to volume with cychexane-ethyl acetate (1+1) mixed solution, Mix thoroughly. Filter the solution into 10-mL tube. Inject 5 mL of the filtered extract solution into GPC. Collect the fraction from 22 min to 40 min in a 100-mL pear-shaped flask. Evaporate it to ca. 0.5 mL on a rotary evaporator at 40°C. Exchange with 2 × 5 mL hexane twice and make up to 1 mL. Add 40 μL internal standard solution for GC-MS determination.

Extract five blank animal muscle samples as described that described above. Prepare working standard mixed solutions in matrix.

3.7.6 Determination

3.7.6.1 GC-MS Operating Conditions

(a) Column: DB-1701 (30 m × 0.25 mm × 0.25 μm) capillary column, or equivalent
(b) Column temperature: 40°C hold 1 min, at 30°C/min to 130°C, at 5°C/min to 250°C, at 10°C/min to 300°C, hold 5 min
(c) Carrier gas: Helium, purity ≥99.999%, flow rate: 1.2 mL/min
(d) Injection port temperature: 290°C
(e) Injection volume: 1 μL
(f) Injection mode: Splitless, purge on after 1.5 min
(g) Ionization voltage: 70 eV
(h) Ion source temperature: 230°C
(i) GC-MS interface temperature: 280°C
(j) Selected ion monitoring mode: Each compound selects 1 quantifying ion and 2–3 qualifying ions. All of the detected ions of each group are detected according to the period of time and order of compounds. For the retention time, quantifying ions, qualifying ions, and the abundance ratios of the quantifying ion and qualifying ions of each compound, refer to Table 5.3. For the period of time and dwell time of the detected ions of each group, refer to Table 5.4

3.7.6.2 Qualitative Determination

In the samples determined, four injections are required to analyze all pesticides and related chemicals by GC-MS under the operating conditions. If the retention times of the peaks of the sample solution are the same as those of the peaks from the working standard mixed solution, and the selected ions appear in the background-subtracted mass spectrum, and the abundance ratios of the selected ions are within the expected limits (abundance ratios >50%, permitted tolerances are ±10%; abundance ratios >20% to 50%, permitted tolerances are ±15%; abundance ratios >10% to 20%, permitted tolerances are ±20%; abundance ratios ≤10%, permitted tolerances are ±50%), the sample is confirmed to contain this pesticide compound. If the results are not definitive, the sample is reinjected with acquisition in scan mode (sufficient sensitivity) or with additional confirmatory ions or using other instruments of higher sensitivity.

3.7.6.3 Quantitative Determination

Quantitation is performed using an internal standard and the quantifying ion for GC-MS. The internal standard is heptachlor epoxide. To compensate for the matrix effects, quantitation is based on a mixed standard prepared in blank matrix extract. The concentrations in the standard solution and in the sample solution analyzed must be close.

3.7.6.4 Parallel Test

A parallel test is carried out for the same testing sample.

3.7.6.5 Blank Test

The operation of the blank test is the same as that described in the method of determination, but without the addition of sample.

3.7.7 Precision

The precision data of the method for this standard have been determined according to the stipulations of GB/T 6379.1 and GB/T 6379.2. The values of repeatability and reproducibility are obtained and calculated at the 95% confidence level.

RESEARCHERS

Guo-Fang Pang, Yan-Zhong Cao, Yong-Ming Liu, Chun-Lin Fan, Jin-Jie Zhang, Xue-Min Li, Guang-Qun Jia, Feng Zheng, Yu-Qiu Shi, Yan-Ping Wu.
    Qinhuangdao Entry-Exit Inspection and Quarantine Bureau, 39 Haibin Rd, Qinhuangdao, Hebei, PC 066002, People's Republic of China.

3.8   DETERMINATION OF 461 PESTICIDES AND RELATED CHEMICAL RESIDUES IN ANIMAL MUSCLES: LC-MS-MS METHOD (GB/T 20772-2008)

3.8.1   Scope

The method is applicable to the qualitative determination of 461 pesticides and related chemicals residues, as well as quantitative determination of 396 pesticides and related chemicals residues in porcine, bovine, mutton, rabbit, and chicken.

The limits of quantitative determination of this method of 396 pesticides and related chemicals are 0.04 µg/kg to 4.82 mg/kg (refer to Table 3.14).

TABLE 3.14 The LOD, LOQ of 461 Pesticides and Related Chemical Residues Determined by LC-MS-MS

No.	Pesticides	LOD (µg/kg)	LOQ (µg/kg)	No.	Pesticides	LOD (µg/kg)	LOQ (µg/kg)
	Group A			19	prometryne	0.08	0.16
1	propham	55.00	110.00	20	paraoxon methyl	0.38	0.76
2	isoprocarb	1.15	2.30	21	thiacloprid	0.19	0.38
3	3,4,5-trimethacarb	0.17	0.34	22	imidacloprid	11.00	22.00
4	cycluron	0.10	0.20	23	ethidimuron	0.75	1.50
5	carbaryl	5.16	10.32	24	isomethiozin	0.53	1.06
6	propachlor	1.10	2.20	25	diallate	44.60	89.20
7	rabenzazole	0.67	1.34	26	acetochlor	23.70	47.40
8	simetryn	0.07	0.14	27	nitenpyram	8.56	17.12
9	monolinuron	1.78	3.56	28	methoprotryne	0.12	0.24
10	mevinphos	0.78	1.56	29	dimethenamid	2.15	4.30
11	aziprotryne	0.69	1.38	30	terrbucarb	1.05	2.10
12	secbumeton	0.04	0.08	31	penconazole	1.00	2.00
13	cyprodinil	0.37	0.74	32	myclobutanil	0.50	1.00
14	buturon	4.48	8.96	33	imazethapyr	0.56	1.12
15	carbetamide	1.82	3.64	34	paclobutrazol	0.29	0.58
16	pirimicarb	0.08	0.16	35	fenthion sulfoxide	0.16	0.32
17	clomazone dimethazone	0.21	0.42	36	triadimenol	5.28	10.56
18	cyanazine	0.08	0.16	37	butralin	0.95	1.90

Continued

TABLE 3.14 The LOD, LOQ of 461 Pesticides and Related Chemical Residues Determined by LC-MS-MS—cont'd

No.	Pesticides	LOD (µg/kg)	LOQ (µg/kg)	No.	Pesticides	LOD (µg/kg)	LOQ (µg/kg)
38	spiroxamine	0.12	0.24	67	fluazifop butyl	0.13	0.26
39	tolclofos methyl	33.28	66.56	68	bromophos-ethyl	283.85	567.70
40	desmedipham	2.01	4.02	69	bensulide	13.68	27.36
41	methidathion	5.33	10.66	70	triasulfuron	0.80	1.60
42	allethrin	30.20	60.40	71	bromfenvinfos	1.51	3.02
43	diazinon	0.36	0.72	72	azoxystrobin	0.23	0.46
44	edifenphos	0.38	0.76	73	pyrazophos	0.81	1.62
45	pretilachlor	0.17	0.34	74	indoxacarb	3.77	7.54
46	flusilazole	0.29	0.58		Group B		
47	iprovalicarb	1.16	2.32	75	dazomet	63.50	127.00
48	benodanil	1.74	3.48	76	nicotine	1.10	2.20
49	flutolanil	0.57	1.14	77	fenuron	0.52	1.04
50	famphur	1.80	3.60	78	cyromazine	28.96	57.92
51	benalyxyl	0.62	1.24	79	crimidine	0.78	1.56
52	diclobutrazole	0.23	0.46	80	acephate	6.67	13.34
53	etaconazole	0.89	1.78	81	molinate	1.05	2.10
54	fenarimol	0.30	0.60	82	carbendazim	0.23	0.46
55	phthalic acid, dicyclobexyl ester	1.00	2.00	83	6-chloro-4-hydroxy-3-phenyl-pyridazin	0.83	1.66
56	tetramethirn	0.91	1.82	84	propoxur	12.20	24.40
57	cloquintocet mexyl	0.94	1.88	85	isouron	0.20	0.40
58	bitertanol	16.70	33.40	86	chlorotoluron	0.31	0.62
59	chlorprifos methyl	8.00	16.00	87	thiofanox	78.50	157.00
60	tepraloxydim	6.10	12.20	88	chlorbufam	91.50	183.00
61	azinphos ethyl	54.46	108.92	89	bendiocarb	1.59	3.18
62	triflumuron	1.96	3.92	90	propazine	0.16	0.32
63	isoxaflutole	1.95	3.90	91	terbuthylazine	0.23	0.46
64	anilofos	2.86	5.72	92	diuron	0.78	1.56
65	quizalofop-ethyl	0.34	0.68	93	chlormephos	224.00	448.00
66	haloxyfop-methyl	1.32	2.64	94	carboxin	0.28	0.56

TABLE 3.14 The LOD, LOQ of 461 Pesticides and Related Chemical Residues Determined by LC-MS-MS—cont'd

No.	Pesticides	LOD (μg/kg)	LOQ (μg/kg)	No.	Pesticides	LOD (μg/kg)	LOQ (μg/kg)
95	clothianidin	31.50	63.00	124	triazophos	0.34	0.68
96	pronamide	7.69	15.38	125	DEF	0.81	1.62
97	dimethachloro	0.95	1.90	126	pyriftalid	0.31	0.62
98	methobromuron	8.42	16.84	127	metconazole	0.66	1.32
99	phorate	157.00	314.00	128	pyriproxyfen	0.22	0.44
100	aclonifen	12.10	24.20	129	cycloxydim	1.27	2.54
101	mephosfolan	1.16	2.32	130	isoxaben	0.09	0.18
102	imibenzonazole-des-benzyl	3.11	6.22	131	flurtamone	0.22	0.44
103	neburon	3.55	7.10	132	trifluralin	167.4	334.80
104	mefenoxam	0.77	1.54	133	flamprop methyl	10.10	20.20
105	ethofume sate	186.00	372.00	134	bioresmethrin	3.71	7.42
106	iprobenfos	4.14	8.28	135	propiconazole	0.88	1.76
107	TEPP	41.60	83.20	136	chlorpyrifos	26.90	53.80
108	cyproconazole	0.37	0.74	137	fluchloralin	244.00	488.00
109	thiamethoxam	16.50	33.00	138	chlorsulfuron	1.37	2.74
110	crufomate	0.26	0.52	139	clethodim	1.04	2.08
111	cythioate	40.00	80.00	140	flamprop isopropyl	0.22	0.44
112	phosphamidon	1.94	3.88	141	tetrachlorvinphos	1.11	2.22
113	phenmedipham	2.24	4.48	142	propargite	34.30	68.60
114	bifenazate	11.40	22.80	143	bromuconazole	1.57	3.14
115	fenhexamid	3.78	7.56	144	picolinafen	0.36	0.72
116	flutriafol	4.29	8.58	145	fluthiacet methyl	2.65	5.30
117	furalaxyl	0.39	0.78	146	trifloxystrobin	1.00	2.00
118	bioallethrin	99.00	198.00	147	chlorimuron ethyl	15.20	30.40
119	cyanofenphos	10.40	20.80	148	hydramethylnon	6.86	13.72
120	pirimiphos methyl	0.10	0.20		Group C		
121	buprofezin	0.44	0.88	149	methamidophos	2.47	4.94
122	disulfoton sulfone	1.23	2.46	150	EPTC	18.67	37.34
123	fenazaquin	0.16	0.32	151	diethyltoluamide	0.28	0.56
				152	monuron	17.37	34.74

Continued

TABLE 3.14 The LOD, LOQ of 461 Pesticides and Related Chemical Residues Determined by LC-MS-MS — cont'd

No.	Pesticides	LOD (μg/kg)	LOQ (μg/kg)	No.	Pesticides	LOD (μg/kg)	LOQ (μg/kg)
153	pyrimethanil	0.34	0.68	183	mepronil	0.19	0.38
154	fenfuram	0.39	0.78	184	disulfoton	234.85	469.70
155	quinoclamine	3.96	7.92	185	fenthion	26.00	52.00
156	fenobucarb	2.95	5.90	186	metalaxyl	0.25	0.50
157	ethirimol	0.28	0.56	187	ofurace	0.50	1.00
158	propanil	10.80	21.60	188	dodemorph	0.20	0.40
159	carbofuran	6.53	13.06	189	imazamethabenz-methyl	0.08	0.16
160	acetamiprid	0.72	1.44				
161	mepanipyrim	0.16	0.32	190	disulfoton-sulfoxide	11.38	22.76
162	prometon	0.07	0.14	191	isoprothiolane	0.92	1.84
163	metoxuron	0.32	0.64	192	imazalil	1.00	2.00
164	dimethoate	3.80	7.60	193	phoxim	41.40	82.80
165	methfuroxam	0.14	0.28	194	quinalphos	1.00	2.00
166	fluometuron	0.46	0.92	195	ditalimfos	33.61	67.22
167	dicrotophos	0.57	1.14	196	fenoxycarb	9.14	18.28
168	monalide	0.60	1.20	197	pyrimitate	0.09	0.18
169	diphenamid	0.07	0.14	198	fensulfothin	1.00	2.00
170	ethoprophos	1.38	2.76	199	fluorochloridone	6.89	13.78
171	fonofos	3.73	7.46	200	butachlor	10.03	20.06
172	etridiazol	50.21	100.42	201	imazaquin	1.44	2.88
173	furmecyclox	0.42	0.84	202	kresoxim-methyl	50.29	100.58
174	hexazinone	0.06	0.12	203	triticonazole	1.51	3.02
175	dimethametryn	0.06	0.12	204	fenamiphos sulfoxide	0.37	0.74
176	trichlorphon	0.56	1.12	205	thenylchlor	12.07	24.14
177	demeton(o+s)	3.39	6.78	206	fenoxanil	19.70	39.40
178	benoxacor	3.45	6.90	207	fluridone	0.09	0.18
179	bromacil	11.80	23.60	208	epoxiconazole	2.03	4.06
180	phorate sulfoxide	184.14	368.28	209	chlorphoxim	38.79	77.58
181	brompyrazon	1.80	3.60	210	fenamiphos sulfone	0.22	0.44
182	oxycarboxin	0.45	0.90	211	fenbuconazole	0.82	1.64

TABLE 3.14 The LOD, LOQ of 461 Pesticides and Related Chemical Residues Determined by LC-MS-MS—cont'd

No.	Pesticides	LOD (µg/kg)	LOQ (µg/kg)	No.	Pesticides	LOD (µg/kg)	LOQ (µg/kg)
212	isofenphos	109.34	218.68	241	DMST	20.00	40.00
213	phenothrin	169.60	339.20	242	cycloate	2.22	4.44
214	piperophos	4.62	9.24	243	atrazine	0.18	0.36
215	piperonyl butoxide	0.57	1.14	244	butylate	151.00	302.00
216	oxyflurofen	29.27	58.54	245	pymetrozin	17.14	34.28
217	coumaphos	1.05	2.10	246	chloridazon	1.16	2.32
218	flufenacet	2.65	5.30	247	sulfallate	103.60	207.20
219	phosalone	24.02	48.04	248	ethiofencarb	2.46	4.92
220	methoxyfenozide	1.85	3.70	249	terbumeton	0.05	0.10
221	prochloraz	1.03	2.06	250	cyprazine	0.08	0.16
222	aspon	6.80	13.60	251	ametryn	0.48	0.96
223	ethion	1.48	2.96	252	tebuthiuron	0.11	0.22
224	thifensulfuron-methyl	10.70	21.40	253	trietazine	0.30	0.60
225	ethoxysulfuron	2.29	4.58	254	sebutylazine	0.16	0.32
226	dithiopyr	5.20	10.40	255	dibutyl succinate	111.20	222.40
227	spirodiclofen	4.95	9.90	256	tebutam	0.07	0.14
228	fenpyroximate	0.68	1.36	257	thiofanox-sulfoxide	4.15	8.30
229	flumiclorac-pentyl	5.30	10.60	258	cartap hydrochloride	1040.00	2080.00
230	temephos	0.61	1.22	259	methacrifos	1211.85	2423.70
231	butafenacil	4.75	9.50	260	terbutryn	0.08	0.16
232	spinosad	2.27	4.54	261	triazoxide	4.00	8.00
	Group D			262	thionazin	11.34	22.68
233	allidochlor	164.16	328.32	263	linuron	5.82	11.64
234	propamocarb	2.80	5.60	264	heptanophos	2.92	5.84
235	thiabendazole	0.24	0.48	265	prosulfocarb	0.18	0.36
236	metamitron	3.18	6.36	266	dipropetryn	0.14	0.28
237	isoproturon	0.07	0.14	267	thiobencarb	1.65	3.30
238	atratone	0.09	0.18	268	tri-n-butyl phosphate	0.19	0.38
239	oesmetryn	0.09	0.18	269	diethofencarb	1.00	2.00
240	metribuzin	0.27	0.54	270	alachlor	3.70	7.40

Continued

TABLE 3.14 The LOD, LOQ of 461 Pesticides and Related Chemical Residues Determined by LC-MS-MS — cont'd

No.	Pesticides	LOD (µg/kg)	LOQ (µg/kg)	No.	Pesticides	LOD (µg/kg)	LOQ (µg/kg)
271	cadusafos	0.58	1.16	301	picoxystrobin	4.22	8.44
272	metazachlor	0.49	0.98	302	tetraconazole	0.86	1.72
273	propetamphos	27.00	54.00	303	mefenpyr-diethyl	6.28	12.56
274	terbufos	1120.00	2240.00	304	profenefos	1.01	2.02
275	simeconazole	1.47	2.94	305	pyraclostrobin	0.25	0.50
276	triadimefon	3.94	7.88	306	dimethomorph	1.60	3.20
277	phorate sulfone	21.00	42.00	307	kadethrin	1.66	3.32
278	tridemorph	10.40	20.80	308	thiazopyr	0.98	1.96
279	mefenacet	1.10	2.20	309	benfuracarb-methyl	8.19	16.38
280	fenpropimorph	0.09	0.18	310	cinosulfuron	0.56	1.12
281	tebuconazole	1.12	2.24	311	pyrazosulfuron-ethyl	3.42	6.84
282	isopropalin	15.00	30.00	312	metosulam	2.20	4.40
283	nuarimol	4.00	8.00		Group E		
284	bupirimate	0.35	0.70	313	4-aminopyridine	0.43	0.86
285	azinphos-methyl	552.17	1104.34	314	methomyl	4.78	9.56
286	tebupirimfos	0.06	0.12	315	pyroquilon	1.74	3.48
287	phenthoate	46.18	92.36	316	fuberidazole	0.95	1.90
288	sulfotep	1.30	2.60	317	isocarbamid	0.85	1.70
289	sulprofos	2.92	5.84	318	butocarboxim	0.79	1.58
290	EPN	16.50	33.00	319	chlordimeform	0.67	1.34
291	diniconazole	0.67	1.34	320	cymoxanil	27.80	55.60
292	flumetsulam	0.15	0.30	321	chlorthiamid	35.28	70.56
293	pencycuron	0.14	0.28	322	aminocarb	8.21	16.42
294	mecarbam	9.80	19.60	323	dimethirimol	0.06	0.12
295	tralkoxydim	0.16	0.32	324	omethoate	4.83	9.66
296	malathion	2.82	5.64	325	ethoxyquin	1.75	3.50
297	pyributicarb	0.17	0.34	326	dichlorvos	0.27	0.54
298	pyridaphenthion	0.44	0.88	327	aldicarb sulfone	10.70	21.40
299	pirimiphos-ethyl	0.08	0.16	328	dioxacarb	1.68	3.36
300	pyraclofos	0.50	1.00	329	benzyladenine	35.40	70.80

TABLE 3.14 The LOD, LOQ of 461 Pesticides and Related Chemical Residues Determined by LC-MS-MS—cont'd

No.	Pesticides	LOD (µg/kg)	LOQ (µg/kg)	No.	Pesticides	LOD (µg/kg)	LOQ (µg/kg)
330	demeton-s-methyl	2.65	5.30	359	florasulam	8.70	17.40
331	cyanophos	5.06	10.12	360	benzoylprop-ethyl	154.00	308.00
332	thiometon	289.00	578.00	361	pyrimidifen	7.00	14.00
333	demeton-s-methyl sulfone	9.88	19.76	362	furathiocarb	0.96	1.92
334	fenpropidin	0.09	0.18	363	*trans*-permethin	2.40	4.80
335	paraoxon-ethyl	0.24	0.48	364	etofenprox	1140.00	2280.00
336	aldimorph	1.58	3.16	365	pyrazoxyfen	1.30	2.60
337	vinclozolin	1.27	2.54	366	*zeta* cypermethrin	2.71	5.42
338	uniconazole	1.20	2.40	367	haloxyfop-2-ethoxyethyl	1.25	2.50
339	pyrifenox	0.13	0.26	368	fluoroglycofen-ethyl	2.50	5.00
340	chlorthion	66.80	133.60		Group F		
341	dicapthon	0.12	0.24	369	acrylamide	8.90	17.80
342	clofentezine	3.20	6.40	370	tert-butylamine	19.48	38.96
343	norflurazon	0.13	0.26	371	hymexazol	112.07	224.14
344	triallate	23.10	46.20	372	phthalimide	21.50	43.00
345	quinoxyphen	76.70	153.40	373	dimefox	34.10	68.20
346	fenthion sulfone	8.73	17.46	374	diphenylamin	0.21	0.42
347	flurochloridone	0.65	1.30	375	1-naphthyl acetamide	0.41	0.82
348	phthalic acid,benzyl butyl ester	316.00	632.00	376	atrazine-desethyl	0.31	0.62
				377	2,6-dichlorobenzamide	2.25	4.50
349	isazofos	0.09	0.18	378	aldicarb	130.50	261.00
350	dichlofenthion	15.10	30.20	379	chlordimeform hydrochloride	10.56	21.12
351	vamidothion sulfone	1904.00	3808.00				
352	terbufos sulfone	44.30	88.60	380	simeton	0.55	1.10
353	dinitramine	7.17	14.34	381	dinotefuran	5.09	10.18
354	trichloronat	33.40	66.80	382	pebulate	1.70	3.40
355	resmethrin-2	0.15	0.30	383	acibenzolar-s-methyl	12.32	24.64
356	boscalid	2.38	4.76	384	dioxabenzofos	6.90	13.80
357	nitralin	17.20	34.40	385	oxamyl	274.03	548.06
358	hexythiazox	11.80	23.60	386	thidiazuron	0.15	0.30

Continued

TABLE 3.14 The LOD, LOQ of 461 Pesticides and Related Chemical Residues Determined by LC-MS-MS—cont'd

No.	Pesticides	LOD (μg/kg)	LOQ (μg/kg)	No.	Pesticides	LOD (μg/kg)	LOQ (μg/kg)
387	methabenzthiazuron	0.29	0.58	415	benzoximate	4.28	8.56
388	butoxycarboxim	13.30	26.60	416	dinoseb acetate	20.64	41.28
389	demeton-s-methyl sulfoxide	1.96	3.92	417	imazapic	0.84	1.68
				418	propisochlor	0.40	0.80
390	phosfolan	0.24	0.48	419	etobenzanid	0.40	0.80
391	fenthion oxon	0.59	1.18	420	fentrazamide	6.20	12.40
392	napropamide	0.64	1.28	421	cyphenothrin	8.40	16.80
393	fenitrothion	13.40	26.80	422	malaoxon	18.75	37.50
394	phthalic acid, dibutyl ester	19.80	39.60		Group G		
395	metolachlor	0.20	0.40	423	dalapon	115.37	230.74
396	procymidone	43.30	86.60	424	2-phenylphenol	84.94	169.88
397	vamidothion	2.28	4.56	425	3-phenylphenol	2.00	4.00
398	cumyluron	0.66	1.32	426	DNOC	1.30	2.60
399	warfarin	1.35	2.70	427	cloprop	5.70	11.40
400	phosmet	8.86	17.72	428	dicloran	24.28	48.56
401	ronnel	6.57	13.14	429	chlorpropham	7.88	15.76
402	pyrethrins	17.90	35.80	430	mecoprop	2.45	4.90
403	phthalic acid, biscyclohexyl ester	0.34	0.68	431	terbacil	0.44	0.88
				432	2,4-D	5.95	11.90
404	carpropamid	2.60	5.20	433	dicamba	632.96	1265.92
405	tebufenpyrad	0.13	0.26	434	MCPB	7.09	14.18
406	tebufenozide	13.90	27.80	435	bentazone	0.52	1.04
407	chlorthiophos	15.90	31.80	436	dinoseb	0.20	0.40
408	dialifos	78.50	157.00	437	dinoterb	0.12	0.24
409	cinidon-ethyl	7.29	14.58	438	forchlorfenuron	5.70	11.40
410	rotenone	1.16	2.32	439	fludioxonil	31.08	62.16
411	imibenconazole	5.13	10.26	440	2,4,5-T	8.75	17.50
412	propaquiafop	0.62	1.24	441	fluroxypyr	96.03	192.06
413	lactofen	31.00	62.00	442	chlorfenethol	82.15	164.30
414	benzofenap	0.04	0.08	443	bromoxynil	0.90	1.80

TABLE 3.14 The LOD, LOQ of 461 Pesticides and Related Chemical Residues Determined by LC-MS-MS—cont'd

No.	Pesticides	LOD (µg/kg)	LOQ (µg/kg)	No.	Pesticides	LOD (µg/kg)	LOQ (µg/kg)
444	chlorobenzuron	10.20	20.40	454	sulfentrazone	44.80	89.60
445	chloramphenicolum	1.94	3.88	455	diflufenican	14.14	28.28
446	alloxydim-sodium	0.10	0.20	456	ethiprole	19.93	39.86
447	fempxaprop-ethyl	2.45	4.90	457	flusulfamide	0.21	0.42
448	sulfanitran	12.00	24.00	458	cyclosulfamuron	171.84	343.68
449	mesotrion	1150.28	2300.56				
450	oryzalin	2.45	4.90	459	iodosulfuron-methyl sodium	10.60	21.20
451	acifluorfen	59.00	118.00	460	kelevan	4820.00	9640.00
452	ioxynil	0.31	0.62	461	iodosulfuron-methyl	33.30	66.60
453	famoxadone	22.64	45.28				

3.8.2 Principle

The test samples are extracted with cyclohexane-ethyl acetate (1+1) mixed solution. The extracts are cleaned up by GPC, then the solutions are analyzed by LC-MS-MS and quantified using an external standard method.

3.8.3 Reagents and Materials

"Water" is first grade of GB/T6682 specified.

Acetonitrile: HPLC Grade
Toluene: HPLC Grade
Cyclohexane: HPLC Grade
Ethyl Acetate: HPLC Grade
Methanol: HPLC Grade
Isooctane: HPLC Grade
Filter Membrane (Nylon): 13 mm × 0.45 µm
Filter Membrane (Nylon): 13 mm × 0.2 µm
0.1% Formic Acid Solution (Volume Fraction)
5 mmol/L Ammonium Acetate Solution
Anhydrous Sodium Sulfate: Analytical Reagent. Ignite at 650°C for 4 h, and Store in a Desiccator
Cyclohexane-Ethyl Acetate (1+1, V/V)
Acetonitrile-Water (3+2, V/V)

Pesticide and Related Chemicals Standard: Purity ≥95%

Stock Standard Solution: Accurately weigh 5–10 mg of individual pesticide and related chemical standards (accurate to 0.1 mg) into a 10-mL volumetric flask. Dissolve and dilute to volume with methanol, toluene, acetone, isooctane, etc., depending on each individual compound's solubility (for diluting solvent refer to Appendix B). Standard solutions are stored in the dark below 4°C and are used for 1 year.

Mixed Standard Solution (Mixed Standard Solution A, B, C, D, E, F, and G): Depending on properties and retention times of each compound, 461 pesticides and related chemical residues are divided into seven groups: A, B, C, D, E, F, and G. Depending on instrument sensitivity of each compound, the concentrations of mixed standard solutions are decided. For 461 pesticides and related chemical groupings and concentrations of mixed standard solutions of this standard, reference is made to Appendix B.

Depending on group number, mixed standard solution concentration, and stock standard solution concentration, appropriate amounts of individual stock standard solutions are pipetted into a 100-mL volumetric flask, with seven group pesticides and related chemicals being diluted to volume with methanol. Mixed standard solutions are stored in the dark below 4°C and are used for 1 month.

Working Standard Mixed Solution in Matrix: Working standard mixed solutions in matrix of A, B, C, D, E, F, and G group pesticides and related chemicals are prepared by mixing different concentrations of working standard mixed solutions with sample blank extract that has been taken through the method with the rest of the samples. These solutions are used to construct calibration plots.

Working standard mixed solutions in matrix must be freshly prepared.

3.8.4   Apparatus

LC-MS-MS: Equipped With ESI Source
Gel Permeation Chromatography (GPC)
Analytical Balances: Capable of Weighing From 0.1 mg to 0.01 g
Rotary Evaporator
Homogenizer: Max. 25,000 rpm
Centrifuge: Max. 4200 rpm
Nitrogen Evaporator
Pear-Shaped Flasks: 100 mL
Pipette: 1 mL, 5 mL
Vials: 2 mL, With PTFE Screw-Cap

3.8.5   Sample Pretreatment

3.8.5.1   Preparation of Test Sample

Part of the representative sample is taken from each bag of the primary sample and homogenized by grinding in a meat grinder. The homogenized sample is thoroughly mixed and reduced to at least 500 g by quartering, and is then placed in a clean container as the test sample, sealed, and labeled.

The test samples should be stored below −18°C.

3.8.5.2 Extraction

Weigh 10 g (accurate to 0.01 g) of the test sample into a 50-mL centrifuge tube filled with 20 g anhydrous sodium sulfate. Add 35 mL volume of cyclohexane-ethyl acetate (1+1) mixed solution. The mixture is homogenized for 1.5 min at 15,000 rpm, and then centrifuged for 3 min at 3000 rpm. Pass the supernatant through a glass funnel containing a glass wool plug and anhydrous sodium sulfate. Collect extracts in 100-mL pear-shaped flask. Proceed twice with 35 mL cyclohexane-ethyl acetate mixed solution. Combine the extracts in the pear-shaped flask. Evaporate it to ca. 5 mL on rotary evaporator at 40°C and proceed to clean-up. If the residue contents in the fat are calculated, collect the extracts in a weighted pear-shaped flask. Evaporate the extracts to ca. 5 mL on a rotary evaporator at 40°C and evaporate the solvent to dryness with a nitrogen evaporator at 50°C. Weigh the pear-shaped flask and proceed to clean-up.

3.8.5.3 GPC Clean-up

3.8.5.3.1 Conditions

(a) GPC Column: 400 mm × 25 mm, filled with Bio-Beads S-X3 or equivalent
(b) Detection wavelength: 254 nm
(c) Mobile phase: cychexane-ethyl acetate (1+1, V/V)
(d) Flow rate: 5 mL/min
(e) Injection volume: 5 mL
(f) Start collect time: 22 min
(g) Stop collect time: 40 min

3.8.5.3.2 Clean-up

Transfer the concentrated extracts into a 10-mL volumetric flask with cychexane-ethyl acetate (1+1) mixed solution. Rinse pear-shaped flask twice with 5 mL cychexane-ethyl acetate (1+1) mixed solution and transfer into the 10-mL volumetric flask. Dilute to volume with cychexane-ethyl acetate (1+1) mixed solution and mix thoroughly. Filter the solution into a 10-mL test tube with 0.45-μm filter membrane for GPC clean-up. The fractions from 22 min to 40 min are collected in a 100-mL pear-shaped flask and concentrated to ca. 0.5 mL on a rotary evaporator at 40°C. Evaporate to dryness using a nitrogen evaporator. Dissolve residues with 1.0 mL acetonitrile-water (3+2) and pass through the 0.20-μm filter membrane. The solution is now ready for LC-MS-MS determination.

Extract blank animal muscle samples as described that described above. Prepare working standard mixed solutions in matrix.

3.8.6 Determination

3.8.6.1 LC-MS-MS Operating Conditions

Operating Conditions of Group A, B, C, D, E, and F

(a) Column: ZORBAX SB-C_{18}, 3.5 μm, 100 mm × 2.1 mm (i.d.) or equivalent
(b) Mobile phase and the condition of gradient elution: see Table 3.3
(c) Column temperature: 40°C
(d) Injection volume: 10 μL
(e) Ion source: ESI
(f) Scan mode: Positive ion scan
(g) Monitor mode: Multiple reaction monitor (MRM)
(h) Ionspray voltage: 4000 V
(i) Nebulizer gas: 0.28 MPa
(j) Drying temperature: 350°C
(k) Drying air flow rate: 10 L/min
(l) MRM transitions for precursor/product ion, quantifying for precursor/product ion, declustering potential, collision energy and collision cell exit potential: see Table 5.10

Operating Conditions of Group G)

(a) Column: ZORBAX SB-C_{18}, 3.5 μm, 100 mm × 2.1 mm (i.d.) or equivalent
(b) Mobile phase and the condition of gradient elution: see Table 3.4
(c) Column temperature: 40°C
(d) Injection volume: 10 μL
(e) Ion source: ESI
(f) Scan mode: Negative ion scan
(g) Monitor mode: Multiple reaction monitor (MRM)
(h) Ionspray voltage: 4000 V
(i) Nebulizer gas: 0.28 MPa
(j) Drying temperature: 350°C
(k) Drying air flow rate: 10 L/min
(l) MRM transitions for precursor/product ion, quantifying for precursor/product ion, declustering potential, collision energy and collision cell exit potential: see Table 5.10

3.8.6.2 Qualitative Determination

Sample solutions are analyzed using LC-MS-MS operating conditions. For the samples determined, if the retention times of the peaks found in the sample solution chromatogram are the same as those of the peaks in the standard in the blank matrix extract chromatogram, and the abundance ratios of MRM transitions for precursor/product ions are within the expected limits (relative abundance >50%, permitted ±20% deviation; relative abundance >20% to 50%, permitted ±25% deviation; relative abundance >10% to 20%, permitted

±30% deviation; relative abundance ≤10%, permitted ±50% deviation), the sample is confirmed to contain this pesticide compound.

3.8.6.3   Quantitative Determination

An external standard method is used for quantitation with standard curves for LC-MS-MS. To compensate for the matrix effect, quantitation is based on a series of working standard solutions prepared in blank matrix extract. The standard curves are established by injection of different concentrations of working standard mixed solutions in matrix separately. The responses of pesticides in the sample solution should be in the linear range of the instrumental detection.

3.8.6.4   Parallel Test

A parallel test is carried out for the same testing sample.

3.8.6.5   Blank Test

The operation of the blank test is the same as that described in the method of determination, but without the addition of sample.

3.8.7   Precision

The precision data of the method for this standard have been determined according to the stipulations of GB/T 6379.1 and GB/T 6379.2. The values of repeatability and reproducibility are obtained and calculated at the 95% confidence level.

RESEARCHERS

Guo-Fang Pang, Yan-Zhong Cao, Chun-Lin Fan, Guang-Qun Jia, Feng Zheng, Wen-Wen Wang, Yu-Jing Lian, Jin-Jie Zhang, Xue-Min Li, Yong-Ming Liu, Yu-Qiu Shi.

Qinhuangdao Entry-Exit Inspection and Quarantine Bureau, 39 Haibin Rd, Qinhuangdao, Hebei, PC 066002, People's Republic of China.

Chapter 4

Determination of 450 Pesticides and Related Chemical Residues in Drinking Water: LC-MS-MS Method (GB/T 23214-2008)

4.1 SCOPE

This method is applicable to the quantitative determination of 427 pesticides and related chemical residues and qualitative determination of 450 pesticides and related chemical residues in drinking water.

The limit of quantitative determination of this method concerning 427 pesticides and related chemicals is 0.010 µg/L to 0.065 mg/L (see Table 4.1).

4.2 PRINCIPLE

The samples are extracted by 1% acetic acid acetonitrile solution and cleaned up with Sep-Pak Vac cartridges. The pesticides and related chemicals are eluted with acetonitrile + toluene (3 + 1) and the solutions are analyzed by LC-MS-MS, using an external standard method for quantification.

4.3 REAGENTS AND MATERIALS

"Water" is first grade of GB/T6682 specified.

Acetonitrile: HPLC Grade
Acetone: HPLC Grade
Isooctane: HPLC Grade
Toluene: HPLC Grade
Hexane: HPLC Grade
Methanol: HPLC Grade
Acetic Acid: P.R. Grade
Formic Acid: P.R. Grade
Ammonium Acetate: Analytically Pure
Sodium Acetate: Anhydrous, Analytically Pure
Membrane Filters (Nylon): 13 mm × 0.2 µm

Analytical Methods for Food Safety by Mass Spectrometry. https://doi.org/10.1016/B978-0-12-814167-0.00004-1
Copyright © 2018 Chemical Industry Press.
Published by Elsevier Inc. under an exclusive license with Chemical Industry Press.

TABLE 4.1 The LOD, LOQ of 450 Pesticides and Related Chemical Residues Determined by LC-MS-MS

No.	Pesticides	LOD (μg/L)	No.	Pesticides	LOD (μg/L)
	Group A		29	methoprotryne	0.02
1	propham	11.00	30	dimethenamid	0.43
2	isoprocarb	0.23	31	terrbucarb	0.21
3	3,4,5-trimethacarb	0.03	32	penconazole	0.20
4	cycluron	0.02	33	myclobutanil	0.10
5	carbaryl	1.03	34	imazethapyr	0.11
6	propachlor	0.03	35	paclobutrazol	0.06
7	rabenzazole	0.13	36	fenthion sulfoxide	0.03
8	simetryn	0.01	37	triadimenol	1.06
9	monolinuron	0.36	38	butralin	0.19
10	mevinphos	0.16	39	spiroxamine	0.01
11	aziprotryne	0.14	40	tolclofos methyl	6.66
12	secbumeton	0.01	41	desmedipham	0.40
13	cyprodinil	0.07	42	methidathion	1.07
14	buturon	0.90	43	allethrin	6.04
15	carbetamide	0.36	44	diazinon	0.07
16	pirimicarb	0.02	45	edifenphos	0.08
17	clomazone	0.04	46	pretilachlor	0.03
18	cyanazine	0.02	47	flusilazole	0.06
19	prometryne	0.02	48	iprovalicarb	0.23
20	paraoxon methyl	0.08	49	benodanil	0.35
21	4,4-dichlorobenzophenone	1.36	50	flutolanil	0.12
22	thiacloprid	0.04	51	famphur	0.36
23	imidacloprid	2.20	52	benalyxyl	0.12
24	ethidimuron	0.15	53	diclobutrazole	0.05
25	isomethiozin	0.11	54	etaconazole	0.18
26	diallate	8.92	55	fenarimol	0.06
27	acetochlor	4.74	56	tetramethirn	0.18
28	nitenpyram	1.71	57	dichlofluanid	0.26

TABLE 4.1 The LOD, LOQ of 450 Pesticides and Related Chemical Residues Determined by LC-MS-MS—cont'd

No.	Pesticides	LOD (μg/L)	No.	Pesticides	LOD (μg/L)
58	cloquintocet mexyl	0.19	87	carbendazim	0.05
59	bitertanol	3.34	88	6-chloro-4-hydroxy-3-phenyl-pyridazin	0.66
60	chlorprifos methyl	1.60	89	propoxur	2.44
61	tepraloxydim	1.22	90	isouron	0.04
62	thiophanate methyl	2.00	91	chlorotoluron	0.06
63	azinphos ethyl	10.89	92	thiofanox	15.70
64	clodinafop propargyl	0.24	93	chlorbufam	18.30
65	triflumuron	0.39	94	bendiocarb	0.32
66	isoxaflutole	0.39	95	propazine	0.03
67	anilofos	0.07	96	terbuthylazine	0.05
68	thiophanat ethyl	8.06	97	diuron	0.16
69	quizalofop-ethyl	0.07	98	chlormephos	44.8
70	haloxyfop-methyl	0.26	99	carboxin	0.06
71	fluazifop butyl	0.03	100	clothianidin	6.30
72	bromophos-ethyl	56.77	101	pronamide	1.54
73	bensulide	3.42	102	dimethachloro	0.19
74	triasulfuron	0.16	103	methobromuron	1.68
75	bromfenvinfos	0.30	104	phorate	31.40
76	azoxystrobin	0.05	105	aclonifen	2.42
77	pyrazophos	0.16	106	mephosfolan	0.23
78	flufenoxuron	0.32	107	imibenconazole-des-benzyl	0.62
79	indoxacarb	0.75	108	neburon	0.71
	Group B		109	mefenoxam	0.15
80	daminozide	1.04	110	ethofume sate	37.20
81	nicotine	0.22	111	iprobenfos	0.83
82	fenuron	0.10	112	TEPP	1.04
83	cyromazine	0.72	113	cyproconazole	0.07
84	crimidine	0.16	114	thiamethoxam	3.30
85	acephate	1.33			
86	molinate	0.21			

Continued

TABLE 4.1 The LOD, LOQ of 450 Pesticides and Related Chemical Residues Determined by LC-MS-MS—cont'd

No.	Pesticides	LOD (µg/L)	No.	Pesticides	LOD (µg/L)
115	etrimfos	1.88	145	flamprop isopropyl	0.04
116	coumatetralyl	0.54	146	tetrachlorvinphos	0.22
117	cythioate	8.00	147	propargite	6.86
118	phosphamidon	0.39	148	bromuconazole	0.31
119	phenmedipham	0.45	149	picolinafen	0.07
120	bifenazate	2.28	150	fluthiacet methyl	0.53
121	flutriafol	0.86	151	trifloxystrobin	0.20
122	furalaxyl	0.08	152	hexaflumuron	2.52
123	bioallethrin	19.80	153	novaluron	0.80
124	cyanofenphos	2.08	154	flurazuron	2.68
125	pirimiphos methyl	0.02		Group C	
126	buprofezin	0.09	155	methamidophos	0.49
127	disulfoton sulfone	0.25	156	EPTC	3.73
128	fenazaquin	0.03	157	diethyltoluamide	0.06
129	triazophos	0.07	158	monuron	3.47
130	DEF	0.16	159	pyrimethanil	0.07
131	pyriftalid	0.06	160	fenfuram	0.08
132	metconazole	0.13	161	quinoclamine	0.79
133	pyriproxyfen	0.04	162	fenobucarb	0.59
134	cycloxydim	1.02	163	ethirimol	0.06
135	isoxaben	0.02	164	propanil	2.16
136	flurtamone	0.04	165	carbofuran	1.31
137	trifluralin	33.48	166	acetamiprid	0.14
138	flamprop methyl	2.02	167	mepanipyrim	0.03
139	bioresmethrin	0.74	168	prometon	0.01
140	propiconazole	0.18	169	methiocarb	4.12
141	chlorpyrifos	5.38	170	metoxuron	0.06
142	fluchloralin	48.80	171	dimethoate	0.76
143	chlorsulfuron	0.27	172	methfuroxam	0.03
144	clethodim	0.21	173	fluometuron	0.09

TABLE 4.1 The LOD, LOQ of 450 Pesticides and Related Chemical Residues Determined by LC-MS-MS—cont'd

No.	Pesticides	LOD (μg/L)	No.	Pesticides	LOD (μg/L)
174	dicrotophos	0.11	204	pyrimitate	0.02
175	monalide	0.12	205	fensulfothin	0.20
176	diphenamid	0.01	206	fluorochloridone	1.38
177	ethoprophos	0.28	207	butachlor	2.01
178	fonofos	0.75	208	imazaquin	0.29
179	etridiazol	10.04	209	kresoxim-methyl	10.06
180	furmecyclox	0.08	210	triticonazole	0.30
181	hexazinone	0.01	211	fenamiphos sulfoxide	0.07
182	dimethametryn	0.01	212	thenylchlor	2.41
183	trichlorphon	0.11	213	fenoxanil	3.94
184	demeton(o+s)	0.68	214	fluridone	0.02
185	benoxacor	0.69	215	epoxiconazole	0.41
186	bromacil	2.36	216	chlorphoxim	7.76
187	phorate sulfoxide	36.83	217	fenamiphos sulfone	0.05
188	brompyrazon	0.36	218	fenbuconazole	0.17
189	oxycarboxin	0.09	219	isofenphos	21.87
190	mepronil	0.04	220	phenothrin	33.92
191	disulfoton	46.97	221	fentin-chloride	1.73
192	fenthion	5.20	222	piperophos	0.92
193	metalaxyl	0.05	223	piperonyl butoxide	0.11
194	ofurace	0.10	224	oxyflurofen	5.86
195	dodemorph	0.04	225	coumaphos	0.21
196	imazamethabenz-methyl	0.02	226	flufenacet	0.53
197	disulfoton-sulfoxide	0.28	227	phosalone	4.80
198	isoprothiolane	0.19	228	methoxyfenozide	0.37
199	imazalil	0.20	229	prochloraz	0.21
200	phoxim	8.28	230	aspon	0.17
201	quinalphos	0.20	231	ethion	0.30
202	ditalimfos	6.72	232	diafenthiuron	0.03
203	fenoxycarb	1.83	233	thifensulfuron-methyl	2.14

Continued

TABLE 4.1 The LOD, LOQ of 450 Pesticides and Related Chemical Residues Determined by LC-MS-MS—cont'd

No.	Pesticides	LOD (μg/L)	No.	Pesticides	LOD (μg/L)
234	ethoxysulfuron	0.46	263	dibutyl succinate	22.24
235	dithiopyr	1.04	264	tebutam	0.01
236	spirodiclofen	0.99	265	thiofanox-sulfoxide	0.83
237	fenpyroximate	0.14	266	cartap hydrochloride	208.00
238	flumiclorac-pentyl	1.06	267	methacrifos	242.37
239	temephos	0.12	268	triazoxide	0.80
240	butafenacil	0.95	269	thionazin	2.27
241	spinosad	0.06	270	linuron	1.16
	Group D		271	heptanophos	0.58
242	mepiquat chloride	0.09	272	prosulfocarb	0.04
243	allidochlor	4.10	273	dipropetryn	0.03
244	propamocarb	0.01	274	thiobencarb	0.33
245	thiabendazole	0.05	275	tri-n-butyl phosphate	0.04
246	metamitron	0.64	276	alachlor	0.74
247	isoproturon	0.01	277	cadusafos	0.12
248	atratone	0.02	278	metazachlor	0.10
249	metribuzin	0.05	279	propetamphos	5.40
250	DMST	4.00	280	terbufos	224.00
251	cycloate	0.44	281	simeconazole	0.29
252	atrazine	0.04	282	triadimefon	0.79
253	butylate	120.80	283	phorate sulfone	4.20
254	pymetrozin	3.43	284	tridemorph	0.26
255	chloridazon	0.23	285	mefenacet	0.22
256	sulfallate	20.72	286	fenpropimorph	0.02
257	ethiofencarb	0.49	287	tebuconazole	0.89
258	terbumeton	0.01	288	isopropalin	3.00
259	ametryn	0.10	289	nuarimol	0.10
260	tebuthiuron	0.02	290	bupirimate	0.28
261	trietazine	0.06	291	azinphos-methyl	110.43
262	sebutylazine	0.13	292	tebupirimfos	0.01

TABLE 4.1 The LOD, LOQ of 450 Pesticides and Related Chemical Residues Determined by LC-MS-MS—cont'd

No.	Pesticides	LOD (μg/L)	No.	Pesticides	LOD (μg/L)
293	phenthoate	9.24	322	pyroquilon	0.35
294	sulfotep	0.26	323	fuberidazole	0.19
295	sulprofos	0.58	324	isocarbamid	0.17
296	EPN	3.30	325	butocarboxim	0.63
297	azamethiphos	0.08	326	chlordimeform	0.13
298	diniconazole	0.13	327	cymoxanil	5.56
299	flumetsulam	0.03	328	aminocarb	1.64
300	pencycuron	0.11	329	dimethirimol	0.01
301	mecarbam	1.96	330	omethoate	0.97
302	tralkoxydim	0.03	331	ethoxyquin	0.35
303	malathion	2.26	332	dichlorvos	0.06
304	pyributicarb	0.03	333	aldicarb sulfone	2.14
305	pyridaphenthion	0.09	334	dioxacarb	0.34
306	thiodicarb	15.75	335	benzyladenine	7.08
307	pyraclofos	0.10	336	demeton-s-methyl	0.53
308	picoxystrobin	0.84	337	ethiofencarb-sulfoxide	22.40
309	tetraconazole	0.17	338	thiometon	57.80
310	mefenpyr-diethyl	1.26	339	folpet	13.86
311	profenefos	0.20	340	demeton-s-methyl sulfone	1.98
312	pyraclostrobin	0.05	341	fenpropidin	0.02
313	dimethomorph	0.04	342	imazapic	0.59
314	kadethrin	0.33	343	paraoxon-ethyl	0.05
315	thiazopyr	0.20	344	vinclozolin	0.25
316	cinosulfuron	0.11	345	uniconazole	0.24
317	pyrazosulfuron-ethyl	0.68	346	pyrifenox	0.03
318	metosulam	1.76	347	dicapthon	0.10
	Group E		348	clofentezine	0.08
319	4-aminopyridine	0.09	349	norflurazon	0.03
320	chlormequat	0.01	350	triallate	4.62
321	methomyl	0.96	351	ziram	7.84

Continued

TABLE 4.1 The LOD, LOQ of 450 Pesticides and Related Chemical Residues Determined by LC-MS-MS—cont'd

No.	Pesticides	LOD (µg/L)	No.	Pesticides	LOD (µg/L)
352	quinoxyphen	15.34	378	diphenylamin	0.04
353	fenthion sulfone	1.75	379	1-naphthy acetamide	0.08
354	flurochloridone	0.13	380	atrazine-desethyl	0.25
355	phthalic acid,benzyl butyl ester	63.20	381	2,6-dichlorobenzamide	0.45
356	isazofos	0.02	382	aldicarb	26.10
357	dichlofenthion	3.02	383	dimethyl phthalate	1.32
358	terbufos sulfone	8.86	384	simeton	0.11
359	dinitramine	0.18	385	dinotefuran	1.02
360	trichloronat	6.68	386	pebulate	0.34
361	resmethrin	0.03	387	dioxabenzofos	5.54
362	boscalid	0.48	388	oxamyl	54.81
363	fenpropathrin	24.50	389	methabenzthiazuron	0.01
364	hexythiazox	2.36	390	butoxycarboxim	10.64
365	pyrimidifen	1.40	391	demeton-s-methyl sulfoxide	0.39
366	furathiocarb	0.19	392	phosfolan	0.05
367	trans-permethin	0.48	393	fenthion oxon	0.12
368	pyrazoxyfen	0.03	394	napropamide	0.13
369	flubenzimine	0.78	395	fenitrothion	2.68
370	zeta cypermethrin	0.07	396	phthalic acid, dibutyl ester	15.84
371	haloxyfop-2-ethoxyethyl	0.25	397	procymidone	8.66
372	tau-fluvalinate	23.00	398	cumyluron	0.13
	Group F		399	imazamox	0.72
373	tert-butylamine	3.90	400	phosmet	1.77
374	hymexazol	22.41	401	ronnel	1.31
375	chlormequat chloride	0.07	402	phthalic acid, biscyclohexyl ester	0.07
376	phthalimide	4.30	403	tebufenpyrad	0.03
377	dimefox	6.82	404	chlorthiophos	3.18

TABLE 4.1 The LOD, LOQ of 450 Pesticides and Related Chemical Residues Determined by LC-MS-MS—cont'd

No.	Pesticides	LOD (μg/L)	No.	Pesticides	LOD (μg/L)
405	dialifos	15.70	428	MCPB	1.42
406	rotenone	0.23	429	dinoseb	0.04
407	imibenconazole	1.03	430	dinoterb	0.02
408	propaquizafop	0.12	431	chlorfenethol	65.72
409	lactofen	6.20	432	bromoxynil	0.18
410	2,3,4,5-tetrachloroaniline	5.36	433	chlorobenzuron	2.04
411	benzoximate	0.86	434	chloramphenicolum	1.55
412	dinoseb acetate	4.13	435	alloxydim-sodium	0.08
413	propisochlor	0.08	436	dimehypo	40.02
414	silafluofen	243.20	437	diflufenzopyr-sodium	3.08
415	etobenzanid	0.08	438	sulfanitran	0.30
416	fentrazamide	1.24	439	oryzalin	0.49
	Group G		440	acifluorfen	11.80
417	dalapon	23.07	441	ioxynil	0.25
418	2-phenylphenol	16.99	442	famoxadone	4.53
419	3-phenylphenol	0.40	443	diflufenican	2.83
420	DNOC	0.26	444	ethiprole	3.99
421	cloprop	1.14	445	flusulfamide	0.04
422	dicloran	19.42	446	cyclosulfamuron	34.37
423	chlorpropham	1.58	447	fomesafen	0.20
424	mecoprop	0.49	448	fluazinam	7.06
425	terbacil	0.35	449	fluazuron	0.01
426	2,4-D	1.19	450	acrinathrin	0.81
427	dicamba	126.59			

Waters Sep-Pak Vac Envi-Carb Cartridge: 6 mL, 1 g, or Equivalent
Sodium Sulfate, Magnesium Sulfate: Anhydrous, Analytically Pure. Ignited at 650°C for 4 h and Kept in a Desiccator
0.1% Formic Acid (V/V)
5 mmol/L Ammonium Acetate Solution
Acetonitrile + Toluene (3 + 1, V/V)

1% Acetic Acid Acetonitrile Solution (V/V)

Acetonitrile + Water (3 + 2, V/V)

Pesticide and Related Chemicals Standard: Purity ≥95%

Stock Standard Solution: Accurately weigh 5–10 mg of individual pesticide and related chemical standards (accurate to 0.1 mg) into a 10-mL volumetric flask. Dissolve and dilute to volume with methanol, toluene, acetone, acetonitrile, isooctane, etc., depending on each individual compound's solubility. (For diluting solvent refer to Appendix B.) Standard solutions are stored in the dark below 4°C and are used for 1 year.

Mixed Standard Solution (Mixed Standard Solution A, B, C, D, E, F, and G): Depending on properties and retention times of compounds, all compounds are divided into seven groups: A, B, C, D, E, F, and G. The concentration of each compound is determined by its sensitivity on the analytical instrument. For 450 pesticides and related chemicals, groupings and concentrations are listed in Appendix B.

Depending on group number, mixed standard solution concentration, and stock standard solution concentration, appropriate amounts of individual stock standard solution are pipetted into a 100-mL volumetric flask, being diluted to volume with methanol. Mixed standard solutions are stored in the dark below 4°C and are used for 1 month.

Working Standard Mixed Solution in Matrix: Working standard mixed solutions in matrix of A, B, C, D, E, F, and G group pesticides and related chemicals are prepared by mixing different concentrations of working standard mixed solutions with sample blank extract that has been taken through the method with the rest of the samples. These solutions are used to construct calibration plots.

Working standard mixed solutions in matrix must be freshly prepared.

4.4 APPARATUS

LC-MS-MS: Equipped With ESI Source

Analytical Balances: Capable of Weighing to 0.1 mg, 0.01 g

Pear-Shaped Flask: 200 mL

Pipette: 1 mL, 10 mL

Glass Vial: 2 mL With PTFE Screw-Cap

Centrifuge Tube: 50 mL With Stopper

Nitrogen Evaporator

Vortex Mixer

Centrifuge: Max. rpm 42,000 rpm

Rotary Evaporator

4.5   SAMPLE PRETREATMENT

4.5.1   Preparation of Test Sample

Put all of the origin sample into a purity politef barre, seal, and label.
   The test samples are stored in cold preservation.

4.5.2   Extraction

Pipette 25 mL test sample (accurate to 0.1 mL) into a 100-mL centrifuge tube.
Add 40 mL 1% acetic acid acetonitrile solution, and homogenize at 15,000 rpm
for 1 min. Add 5 g sodium chloride to the centrifuge tube and vortex mix 2 min.
Then add 4 g sodium acetate into the centrifuge tube, shake 1 min again, add
15 g magnesium sulfate, shake 5 min, and centrifuge at 4200 rpm for 5 min.
Pipette 20 mL of the top acetonitrile layer of extracts into a pear-shaped flask
and concentrate the extracts to 2 mL using the rotary evaporator at 40°C for
clean-up.

4.5.3   Clean-up

Add a height of 2 cm sodium sulfate into the Envi-Carb cartridge. Fix the cartridge into a support to which a pear-shaped flask is connected. Condition the
cartridge with 5 mL acetonitrile-toluene (3 + 1) before adding the sample. Once
the solution gets to the top of the sodium sulfate, transfer the concentrated
solution into the cartridge. Rinse the pear-shaped flask with 3 mL acetonitrile-
toluene (3 + 1) twice and transfer it to the cartridge. Insert a 50-mL reservoir into
the cartridges. Elute the pesticides with 25 mL acetonitrile-toluene (3 + 1).
Evaporate the eluate to 0.5 mL using the rotary evaporator at 40°C. Then evaporate the eluate to dryness using a nitrogen evaporator. Make up to 1 mL with
acetonitrile-water (3 + 2) and mix thoroughly. Finally, filter the extract through
a 0.2-μm filter into a glass vial for LC-MS-MS determination.

4.6   DETERMINATION

4.6.1   LC-MS-MS Operating Condition

4.6.1.1   LC-MS-MS Operating Conditions of A, B, C, D, E, F Group

(a) Column: ZORBOX SB-C_{18}, 3.5 μm, 100 × 2.1 mm, or equivalent;
(b) Mobile phase and the condition of gradient elution: see Table 4.2;
(c) Column temperature: 40°C;
(d) Injection volume: 10 μL;
(e) Ion source: ESI;
(f) Scan mode: Positive ion scan;
(g) Atomization gas: Nitrogen gas;
(h) Nebulizer gas: 0. 28 MPa;

TABLE 4.2 Mobile Phase Program and the Flow Rate

Step	Time/ Min	Flow Rate (μL/Min)	Mobile Phase A (0.1% Formic Acid Solution)/%	Mobile Phase B (Acetonitrile)/%
0	0.00	400	99.0	1.0
1	3.00	400	70.0	30.0
2	6.00	400	60.0	40.0
3	9.00	400	60.0	40.0
4	15.00	400	40.0	60.0
5	19.00	400	1.0	99.0
6	23.00	400	1.0	99.0
7	23.01	400	99.0	1.0

(i) Ionspray voltage: 4000 V;
(j) Dry gas temperature: 350°C;
(k) Dry gas flow rate: 10 L/min;
(l) MRM transitions for precursor/product ion, quantifying for precursor/ product ion, declustering potential, collision energy, and voltage-source fragmentation: see Table 5.10.

4.6.1.2 LC-MS-MS Operating Condition of G group

(a) Column: ZORBOX SB-C$_{18}$, 3.5 μm, 100 × 2.1 mm, or equivalent;
(b) Mobile phase and the condition of gradient elution: see Table 4.3;
(c) Column temperature: 40°C;
(d) Injection volume: 10 μL;
(e) Ion source: ESI;
(f) Scan mode: Negative ion scan;
(g) Atomization gas: Nitrogen gas;
(h) Ionspray voltage: 4000 V;
(i) Nebulizer gas: 0. 28 MPa;
(j) Dry gas temperature: 350°C;
(k) Dry gas flow rate: 10 L/min;
(l) MRM transitions for precursor/product ion, quantifying for precursor/ product ion, declustering potential, collision energy, and voltage-source fragmentation: see Table 5.10.

TABLE 4.3 Mobile Phase and the Condition of Gradient Elution

Step	Time/ Min	Flow Rate (μL/Min)	Mobile Phase A (5 mmol/L Ammonium Acetate Solution)/%	Mobile Phase B (Acetonitrile)/%
0	0.00	400	99.0	1.0
1	3.00	400	70.0	30.0
2	6.00	400	60.0	40.0
3	9.00	400	60.0	40.0
4	15.00	400	40.0	60.0
5	19.00	400	1.0	99.0
6	23.00	400	1.0	99.0
7	23.01	400	99.0	1.0

4.6.2 Qualitative Determination

Sample solutions are analyzed using LC-MS-MS operating conditions. For the samples determined, if the retention times of the peaks found in the sample solution chromatogram are the same as those for the peaks in the standard in the blank matrix extract chromatogram, and the abundance ratios of MRM transitions for precursor/product ions are within the expected limits (relative abundance >50%, permitted ±20% deviation; relative abundance >20% to 50%, permitted ±25% deviation; relative abundance >10% to 20%, permitted ±30% deviation; relative abundance ≤10%, permitted ±50% deviation), the sample is confirmed to contain this pesticide compound.

4.6.3 Quantitative Determination

An external standard method is used for quantitation with standard curves for LC-MS-MS. To compensate for the matrix effect, quantitation is based on a series of working standard solutions prepared in blank matrix extract. The standard curves are established by injection of different concentrations of working standard mixed solutions in matrix separately. The responses of pesticides in the sample solution should be in the linear range of the instrumental detection.

4.7   PRECISION

The precision data of the method for this standard is according to the stipulations of GB/T 6379.1 and GB/T 6379.2. The values of repeatability and reproducibility are obtained and calculated at the 95% confidence level.

RESEARCHERS

Guo-Fang Pang, Feng Zheng, Chun-Lin Fan, Yan Li, Xin-Xin Ji, Ming-Lin Wang.

Qinhuangdao Entry-Exit Inspection and Quarantine Bureau, 39 Haibin Rd, Qinhuangdao, Hebei, PC 066002, People's Republic of China.

Chapter 5

Basic Research on Chromatography-Mass Spectroscopy Characteristic Parameters of Pesticide and Chemical Pollutants

In the process of writing this book, the author's team has searched 15 international influential journals (Chromatographia, J. AOAC Int., Int. J. Environ., Anal. Chem., Food Addit. Contam., J. Agric. Food Chem., J. Sep. Sci., Rapid Commun. Mass Spectrom., Food Chem., Analyst, Talanta, Anal. Bioanal. Chem., Anal. Chim. Acta, J. Chromatogr. A, Analytical Chemistry, Trac-Trends Anal. Chem) published for the period from 1991 to 2010, and has found that there are 3505 SCI papers on analytical techniques for pesticide residues in edible agricultural products from 72 countries (regions) across the five continents, among which 1942 papers are from 35 countries in Europe, 837 papers from 13 countries in America, 641 papers from 18 countries (regions) in Asia, 38 papers from 2 countries in Oceania, and 32 papers from 6 countries in Africa.

As far as the detection techniques for pesticide residues in edible agricultural products are concerned, there are 3505 papers involving 204 analytical techniques. Techniques that rank in the top 20 in terms of paper quantities are GC-MS, LC-MS-MS, LC-UV, GC-ECD, LC-MS, LC-FLD, ELISA, LC-DAD, GC-NPD, GC-MS-MS, Sensor, GC-FPD, LC-Q-TOF-MS, TLC, CE, EIA, GC-FID, CE-UV, LC-ED, GC-ITD AND IA. It is found statistically that, over the 20-year period examined, and especially in the last 10 years of that period, the technique that developed the fastest was LC-MS/MS, which rose to rank No. 1 in the last 10 years from No. 9 in the 10 years prior. GC-MS-MS has also risen from No. 19 in the first 10 years of that period to No. 8 in the last 10 years, as shown in Table 5.1. It is apparent that the period from 2001 to 2010 was an era for unprecedented development of mass spectrometric techniques.

In our country, the scientists and technologists devoted to this area are closely keeping pace with these techniques and have brought about a great leap, from No. 14 in the field during the 10-year period of 1991 to 2000, to No. 2 during the period of 2001 to 2010, ranking among the top 20 countries, as shown in Table 5.2.

Analytical Methods for Food Safety by Mass Spectrometry. https://doi.org/10.1016/B978-0-12-814167-0.00005-3
Copyright © 2018 Chemical Industry Press.
Published by Elsevier Inc. under an exclusive license with Chemical Industry Press.

TABLE 5.1 Comparison of the Top 20 Analytical Techniques Developed From 1991 to 2010

1991–2000

No.	1	2	3	4	5	6	7	8	9	10
Anal. Tech.	LC-UV	GC-MS	GC-ECD	LC-FLD	LC-MS	ELISA	GC-NPD	LC-DAD	LC-MSMS	GC-FPD
Literature	209	193	153	131	100	60	56	47	36	32
No.	11	12	13	14	15	16	17	18	19	20
Anal. Tech.	TLC	EIA	GC-FID	LC-ED	GC-ITD	CE-UV	Sensor	CE	GC-MS-MS	IA
Literature	26	23	15	14	10	8	7	7	6	6

2001–2010

No.	1	2	3	4	5	6	7	8	9	10
Anal. Tech.	LC-MSMS	GC-MS	LC-MS	GC-ECD	ELISA	LC-UV	LC-FLD	GC-MS-MS	LC-DAD	Sensor
Literature	439	291	161	126	116	104	97	84	75	10
No.	11	12	13	14	15	16	17	18	19	20
Anal. Tech.	GC-NPD	LC-QTOF	GC-FPD	CE	CE-UV	IA	GC-FID	GC-ITD	TLC	LC-ED
Literature	41	40	19	18	9	8	5	4	3	3

TABLE 5.2 Comparison of Analytical Techniques Developed Among the Top 20 Countries for 20 Years (1991–2010)

1991–2000

No.	1	2	3	4	5	6	7	8	9	10
Country	USA	Spain	GK	Canada	France	Italy	Holland	Japan	Germany	Belgium
Literature	87	46	40	28	25	18	14	14	14	13
No.	11	12	13	14	15	16	17	18	19	20
Country	Switzerland	Ireland	Australia	P.R. China	India	Portugal	Greece	Brazil	Poland	Czech
Literature	5	5	5	3	2	1	1	1	0	0

2001–2010

No.	1	2	3	4	5	6	7	8	9	10
Country	Spain	PR China	USA	Italy	Belgium	Germany	G.K.	Canada	France	Japan
Literature	273	129	117	72	48	40	40	40	38	37
No.	11	12	13	14	15	16	17	18	19	20
Country	Greece	Switzerland	Holland	Czech	Brazil	Portugal	Poland	India	Australia	Ireland
Literature	26	20	20	19	15	15	12	11	6	5

In the previous decade, there was not even a mass spectrometer available for the author's team laboratory, but the team seemed to smell the importance and urgency for investigation of LC-MS-MS starting around 2001. In 2004, the author's team made a final decision, through borrowing the instruments, to begin the study of high-throughput analytical techniques for pesticide and chemical multiresidues in edible agricultural products. All the MS analyses involving high-throughput analytical techniques for pesticide and chemical multiresidues in edible agricultural products in the 10 chapters mentioned previously were carried out with borrowed instruments. The basic research of MS characteristic parameters for pesticide and chmical residues is especially presented in this chapter, and a corresponding database has been established.

The author here is particularly grateful to Beijing AB Company and Agilent (Beijing) Technologies, as the former graciously lent us an AB-3200 LC-MS-MS (2005–2006), and the latter also, without any payment, generously provided us many instruments, such as the Agilent 6430QQQ LC-MS-MS (2009–2010), Agilent 7890A5975C GC-MS-MS (2010–2011), and Agilent 1290+6530 LC-TOF-MASS (2012–2013) LC-MS-MS. With their generous assistance, our study on high-throughput analytical techniques for pesticide and chemical residues in edible agricultural products could thus be accomplished systematically.

5.1 MASS SPECTROMETRY DATA OF 1200 PESTICIDES AND CHEMICAL POLLUTANTS DETERMINED BY GC-MS, GC-MS-MS, AND LC-MS-MS

5.1.1 Retention Times, Quantifier and Qualifier Ions of 567 Pesticides and Chemical Pollutants Determined by GC-MS

See Table 5.3.

TABLE 5.3 Retention Times, Quantifier and Qualifier Ions of 567 Pesticides and Chemical Pollutants Determined by GC-MS-MS

No.	Compound	CAS	Retention Time (min)	Qualifier Ion/1	Qualifier Ion/2	Qualifier Ion/3	Quantifier Ion
Group A							
1	Allidochlor	93-71-0	8.78	158(10)	173(15)		138(100)
2	Dichlormid	37764-25-3	9.74	166(41)	124(79)		172(100)
3	Etridiazol	2593-15-9	10.42	183(73)	140(19)		211(100)
4	Chlormephos	24934-91-6	10.53	234(70)	154(70)		121(100)
5	Propham	122-42-9	11.36	137(66)	120(51)		179(100)
6	Cycloate	1134-23-2	13.56	186(5)	215(12)		154(100)
7	Diphenylamine	122-39-4	14.55	168(58)	167(29)		169(100)
8	Chlordimeform	6164-98-3	14.93	198(30)	195(18)	183(23)	196(100)
9	Ethalfluralin	55283-68-6	15.00	316(81)	292(42)		276(100)
10	Phorate	298-02-2	15.46	121(160)	231(56)	153(3)	260(100)

Continued

TABLE 5.3 Retention Times, Quantifier and Qualifier Ions of 567 Pesticides and Chemical Pollutants Determined by GC-MS-MS—cont'd

No.	Compound	CAS	Retention Time (min)	Qualifier Ion/1	Qualifier Ion/2	Qualifier Ion/3	Quantifier Ion
11	Thiometon	640-15-3	16.20	125(55)	246(9)		88(100)
12	Quintozene	82-68-8	16.75	237(159)	249(114)		295(100)
13	Atrazine-desethyl	6190-65-4	16.76	187(32)	145(17)		172(100)
14	Clomazone	81777-89-1	17.00	138(4)	205(13)		204(100)
15	Diazinon	333-41-5	17.14	179(192)	137(172)		304(100)
16	Fonofos	944-22-9	17.31	137(141)	174(15)	202(6)	246(100)
17	Etrimfos	38260-54-7	17.92	181(40)	277(31)		292(100)
18	Propetamphos	31218-83-4	17.97	194(49)	236(30)		138(100)
19	Secbumeton	26259-45-0	18.36	210(38)	225(39)		196(100)
20	Pronamide	23950-58-5	18.72	175(62)	255(22)		173(100)
21	Dichlofenthion	97-17-6	18.80	223(78)	251(38)		279(100)
22	Mexacarbate	315-18-4	18.83	150(66)	222(27)		165(100)
23	Dimethoate	60-51-5	19.25	143(16)	229(11)		125(100)
24	Dinitramine	29091-05-2	19.35	307(38)	261(29)		305(100)
25	Aldrin	309-00-2	19.67	265(65)	293(40)	329(8)	263(100)

No.	Name	CAS	RT				
26	Ronnel	83-79-4	19.80	287(67)	125(32)		285(100)
27	Prometryne	7287-19-6	20.13	184(78)	226(60)		241(100)
28	Cyprazine	22936-86-3	20.18	227(58)	170(29)		212(100)
29	Vinclozolin	50471-44-8	20.29	212(109)	198(96)		285(100)
30	Beta-HCH	319-85-7	20.31	217(78)	181(94)	254(12)	219(100)
31	Metalaxyl	57837-19-1	20.67	249(53)	234(38)		206(100)
32	Methyl-Parathion	34388-29-9	20.82	233(66)	246(8)	200(6)	263(100)
33	Chlorpyrifos (-ethyl)	2921-88-2	20.96	258(57)	286(42)		314(100)
34	Delta-HCH	319-86-8	21.16	217(80)	181(99)	254(10)	219(100)
35	Anthraquinone	84-65-1	21.49	180(84)	152(69)		208(100)
36	Fenthion	55-38-9	21.53	169(16)	153(9)		278(100)
37	Malathion	121-75-5	21.54	158(36)	143(15)		173(100)
38	Paraoxon-ethyl	311-45-5	21.57	220(60)	247(58)	263(11)	275(100)
39	Fenitrothion	122-14-5	21.62	260(52)	247(60)		277(100)
40	Triadimefon	43121-43-3	22.22	210(50)	181(74)		208(100)
41	Parathion	56-38-2	22.32	186(23)	235(35)		291(100)
42	Linuron	330-55-2	22.44	248(30)	160(12)		61(100)
43	Pendimethalin	40318-45-4	22.59	220(22)	162(12)		252(100)
44	Chlorbenside	103-17-3	22.96	270(41)	143(11)		268(100)
45	Bromophos-ethyl	4824-78-6	23.06	303(77)	357(74)		359(100)

Continued

TABLE 5.3 Retention Times, Quantifier and Qualifier Ions of 567 Pesticides and Chemical Pollutants Determined by GC-MS-MS—cont'd

No.	Compound	CAS	Retention Time (min)	Qualifier Ion/1	Qualifier Ion/2	Qualifier Ion/3	Quantifier Ion
46	Quinalphos	84087-01-4	23.10	298(28)	157(66)		146(100)
47	trans-Chlordane	5103-74-2	23.29	375(96)	377(51)		373(100)
48	Phenthoate	2597-03-7	23.30	246(24)	320(5)		274(100)
49	Metazachlor	67129-08-2	23.32	133(120)	211(32)		209(100)
50	Fenothiocarb	62850-32-2	23.79	160(37)	253(15)		72(100)
51	Prothiophos	34643-46-4	24.04	267(88)	162(55)		309(100)
52	Chlorfurenol	2464-37-1	24.15	152(40)	274(11)		215(100)
53	Folpet	133-07-3	24.16	104(158)	297(17)		260(100)
54	Procymidone	32809-16-8	24.36	285(70)	255(15)		283(100)
55	Dieldrin	60-57-1	24.43	277(82)	380(30)	345(35)	263(100)
56	Methidathion	950-37-8	24.49	157(2)	302(4)		145(100)
57	Napropamide	15299-99-7	24.84	128(111)	171(34)		271(100)
58	Cyanazine	21725-46-2	24.94	240(56)	198(61)		225(100)
59	Oxadiazone	19666-30-9	25.06	258(62)	302(37)		175(100)
60	Fenamiphos	22224-92-6	25.29	154(56)	288(31)	217(22)	303(100)

No.	Name	CAS	RT	m/z	m/z	m/z	m/z
61	Aramite	140-57-8	25.60	319(37)	334(32)		185(100)
62	Tetrasul	2227-13-6	25.85	324(64)	254(68)		252(100)
63	Bupirimate	57839-19-1	26.00	316(41)	208(83)		273(100)
64	Flutolanil	66332-96-5	26.23	145(25)	323(14)		173(100)
65	Carboxin	5234-68-4	26.25	143(168)	87(52)		235(100)
66	p,p'-DDD	50-29-3	26.59	237(64)	199(12)	165(46)	235(100)
67	Ethion	562-12-2	26.69	384(13)	199(9)		231(100)
68	Etaconazole-1	1000290-09-5	26.81	173(85)	247(65)		245(100)
69	Sulprofos	35400-43-2	26.87	156(62)	280(11)		322(100)
70	Etaconazole-2	1000290-09-5	26.89	173(85)	247(65)		245(100)
71	Myclobutanil	88671-89-0	27.19	288(14)	150(45)		179(100)
72	Fensulfothion	115-90-2	27.94	308(22)	293(73)		292(100)
73	Diclofop-methyl	75736-33-3	28.08	281(50)	342(82)		253(100)
74	Propiconazole-1	60207-90-1	28.15	173(97)	261(65)		259(100)
75	propiconazole-2	60207-90-1	28.15	173(97)	261(65)		259(100)
76	Bifenthrin	82657-04-3	28.57	166(25)	165(23)		181(100)
77	Mirex	2385-85-5	28.72	237(49)	274(80)		272(100)
78	Carbosulfan	55285-14-8	28.80	118(95)	323(30)		160(100)
79	Nuarimol	63284-71-9	28.90	235(155)	203(108)		314(100)
80	Benodanil	15310-01-7	29.14	323(38)	203(22)		231(100)

Continued

TABLE 5.3 Retention Times, Quantifier and Qualifier Ions of 567 Pesticides and Chemical Pollutants Determined by GC-MS-MS—cont'd

No.	Compound	CAS	Retention Time (min)	Qualifier Ion/1	Qualifier Ion/2	Qualifier Ion/3	Quantifier Ion
81	Methoxychlor	72-43-5	29.38	228(16)	212(4)		227(100)
82	Oxadixyl	23135-22-0	29.50	233(18)	278(11)		163(100)
83	Tebuconazole	107534-96-3	29.51	163(55)	252(36)		250(100)
84	Tetramethrin	7696-12-0	29.59	135(3)	232(1)		164(100)
85	Norflurazon	27314-13-2	29.99	145(101)	102(47)		303(100)
86	Pyridaphenthion	119-12-0	30.17	199(48)	188(51)		340(100)
87	Phosmet	732-11-6	30.46	161(11)	317(4)		160(100)
88	Tetradifon	116-29-0	30.70	356(70)	159(196)		227(100)
89	Oxycarboxin	5259-88-1	31.00	267(52)	250(3)		175(100)
90	cis-Permethrin	74474-45-7	31.42	184(15)	255(2)		183(100)
91	Pyrazophos	13457-18-6	31.60	232(35)	373(19)		221(100)
92	trans-Permethrin	551877-74-8	31.68	184(15)	255(2)		183(100)
93	Cypermethrin	52315-07-8	33.19	152(23)	180(16)		181(100)
94	Fenvalerate-1	51630-58-1	34.45	225(53)	419(37)	181(41)	167(100)
95	fenvalerate-2	51630-58-1	34.79	225(54)	419(38)	181(42)	167(101)
96	Deltamethrin	52918-63-5	35.77	172(25)	174(25)		181(100)

Group B							
97	EPTC	759-94-4	8.54	189(30)	132(32)		128(100)
98	Butylate	2008-41-5	9.49	146(115)	217(27)		156(100)
99	Dichlobenil	1194-65-6	9.75	173(68)	136(15)		171(100)
100	Pebulate	1114-71-2	10.18	161(21)	203(20)		128(100)
101	Nitrapyrin	1929-82-4	10.89	196(97)	198(23)		194(100)
102	Mevinphos	7786-34-7	11.23	192(39)	164(29)		127(100)
103	Chloroneb	2675-77-6	11.85	193(67)	206(66)		191(100)
104	Tecnazene	117-18-0	13.54	203(135)	215(113)		261(100)
105	Heptanophos	23560-59-0	13.78	215(17)	250(14)		124(100)
106	Ethoprophos	13194-48-4	14.40	200(40)	242(23)	168(15)	158(100)
107	Hexachlorobenzene	118-74-1	14.69	286(81)	282(51)		284(100)
108	Propachlor	1918-16-7	14.73	176(45)	211(11)		120(100)
109	cis-Diallate	2303-16-4	14.75	236(37)	128(38)		234(100)
110	Trifluralin	1582-09-8	15.23	264(72)	335(7)		306(100)
111	trans-Diallate	2303-16-4	15.29	236(37)	128(38)		234(100)
112	Chlorpropham	5598-13-0	15.49	171(59)	153(24)		213(100)
113	Sulfotep	3689-24-5	15.55	202(43)	238(27)	266(24)	322(100)
114	Sulfallate	95-06-7	15.75	116(7)	148(4)		188(100)
115	Alpha-HCH	319-84-6	16.06	183(98)	221(47)	254(6)	219(100)

Continued

TABLE 5.3 Retention Times, Quantifier and Qualifier Ions of 567 Pesticides and Chemical Pollutants Determined by GC-MS-MS—cont'd

No.	Compound	CAS	Retention Time (min)	Qualifier Ion/1	Qualifier Ion/2	Qualifier Ion/3	Quantifier Ion
116	Terbufos	13071-79-9	16.83	153(25)	288(10)	186(13)	231(100)
117	Terbumeton	33693-04-8	17.20	169(66)	225(32)		210(100)
118	Profluralin	26399-36-0	17.36	304(47)	347(13)		318(100)
119	Dioxathion	78-34-2	17.51	197(43)	169(19)		270(100)
120	Propazine	139-40-2	17.67	229(67)	172(51)		214(100)
121	Chlorbufam	1967-16-4	17.85	153(53)	164(64)		223(100)
122	Dicloran	99-30-9	17.89	176(128)	160(52)		206(100)
123	Terbuthylazine	5915-41-3	18.07	229(33)	173(35)		214(100)
124	Monolinuron	1746-81-2	18.15	126(45)	214(51)		61(100)
125	Cyanophos	2636-26-2	18.73	180(8)	148(3)		243(100)
126	Flufenoxuron	101463-69-8	18.83	126(67)	307(32)		305(100)
127	Chlorpyrifos-methyl	2921-88-2	19.38	288(70)	197(5)		286(100)
128	Desmetryn	1014-69-3	19.64	198(60)	171(30)		213(100)
129	Dimethachlor	51218-45-2	19.80	197(47)	210(16)		134(100)
130	Alachlor	15972-60-8	20.03	237(35)	269(15)		188(100)

No.	Compound	CAS	RT				
131	Pirimiphos-methyl	29232-93-7	20.30	276(86)	305(74)		290(100)
132	Terbutryn	886-50-0	20.61	241(64)	185(73)		226(100)
133	Aspon	886-50-0	20.62	253(52)	378(14)		211(100)
134	Thiobencarb	28249-77-6	20.63	257(25)	259(9)		100(100)
135	Dicofol	115-32-2	21.33	141(72)	250(23)	251(4)	139(100)
136	Metolachlor	51218-45-2	21.34	162(159)	240(33)		238(100)
137	Pirimiphos-ethyl	23505-41-1	21.59	318(93)	304(69)		333(100)
138	Oxy-chlordane	27304-13-8	21.63	237(50)	185(68)		387(100)
139	Dichlofluanid	1085-98-9	21.68	226(74)	167(120)		224(100)
140	Methoprene	40596-69-8	21.71	191(29)	153(29)		73(100)
141	Bromofos	2104-96-3	21.75	329(75)	213(7)		331(100)
142	Ethofumesate	26225-79-6	21.84	161(54)	286(27)		207(100)
143	Isopropalin	33820-53-0	22.10	238(40)	222(4)		280(100)
144	Propanil	709-98-8	22.68	217(21)	163(62)		161(100)
145	Crufomate	299-86-5	22.93	182(154)	276(58)		256(100)
146	Isofenphos	25311-71-1	22.99	255(44)	185(45)		213(100)
147	Endosulfan-1	959-98-8	23.10	265(66)	339(46)		241(100)
148	Chlorfenvinphos	470-90-6	23.19	267(139)	269(92)		323(100)
149	Tolylfluanide	731-27-1	23.45	240(71)	137(210)		238(100)
150	cis-Chlordane	5103-71-9	23.55	375(96)	377(51)		373(100)

Continued

TABLE 5.3 Retention Times, Quantifier and Qualifier Ions of 567 Pesticides and Chemical Pollutants Determined by GC-MS-MS—cont'd

No.	Compound	CAS	Retention Time (min)	Qualifier Ion/1	Qualifier Ion/2	Qualifier Ion/3	Quantifier Ion
151	Butachlor	23184-66-9	23.82	160(75)	188(46)		176(100)
152	Chlozolinate	84332-86-5	23.83	188(83)	331(91)		259(100)
153	p,p'-DDE	72-54-8	23.92	316(80)	246(139)	248(70)	318(100)
154	Iodofenphos	18181-70-9	24.33	379(37)	250(6)		377(100)
155	Tetrachlorvinphos	22248-79-9	24.36	331(96)	333(31)		329(100)
156	Chlorbromuron	57160-47-1	24.37	294(17)	292(13)		61(100)
157	Profenofos	41198-08-7	24.65	374(39)	297(37)		339(100)
158	Buprofezin	69327-76-0	24.87	172(54)	305(24)		105(100)
159	Hexaconazole	79983-71-4	24.92	231(62)	256(26)		214(100)
160	o,p'-DDD	53-19-0	24.94	237(65)	165(39)	199(14)	235(100)
161	Chlorfenson	80-33-1	25.05	175(282)	177(103)		302(100)
162	Fluorochloridone	61213-25-0	25.14	313(64)	187(85)		311(100)
163	Endrin	72-20-8	25.15	317(30)	345(26)		263(100)
164	Paclobutrazol	76738-62-0	25.21	238(37)	167(39)		236(100)
165	o,p'-DDT	3424-82-6	25.56	237(63)	165(37)	199(14)	235(100)

No.	Compound	CAS	RT	m/z	m/z	m/z	m/z
166	Methoprotryne	40596-69-8	25.63	213(24)		271(17)	256(100)
167	Erbon	136-25-4	25.68	171(35)		223(30)	169(100)
168	Chloropropylate	5836-10-2	25.85	253(64)		141(18)	251(100)
169	Flamprop-methyl	52756-25-9	25.90	77(26)		276(11)	105(100)
170	Nitrofen	1836-75-5	26.12	253(90)	139(15)	202(48)	283(100)
171	Oxyfluorfen	42874-03-3	26.13	361(35)		300(35)	252(100)
172	Chlorthiophos	60238-56-4	26.52	360(52)		297(54)	325(100)
173	Flamprop-Isopropyl	52756-22-6	26.70	276(19)		363(3)	105(100)
174	Endosulfan-2	33213-65-9	26.72	265(66)		339(46)	241(100)
175	Carbofenothion	786-19-6	27.19	342(49)		199(28)	157(100)
176	p,p'-DDT	72-55-9	27.22	237(65)	165(34)	246(7)	235(100)
177	Benalaxyl	71626-11-4	27.54	206(32)		325(8)	148(100)
178	Edifenphos	17109-49-8	27.94	310(76)		201(37)	173(100)
179	Triazophos	24017-47-8	28.23	172(47)		257(38)	161(100)
180	Cyanofenphos	13067-93-1	28.43	169(56)		303(20)	157(100)
181	Chlorbenside sulfone	7082-99-7	28.88	99(14)		89(33)	127(100)
182	Endosulfan-Sulfate	1031-07-8	29.05	272(165)		389(64)	387(100)
183	Bromopropylate	18181-80-1	29.30	183(34)		339(49)	341(100)
184	Benzoylprop-ethyl	22212-55-1	29.40	365(36)		260(37)	292(100)

Continued

TABLE 5.3 Retention Times, Quantifier and Qualifier Ions of 567 Pesticides and Chemical Pollutants Determined by GC-MS-MS—cont'd

No.	Compound	CAS	Retention Time (min)	Qualifier Ion/1	Qualifier Ion/2	Qualifier Ion/3	Quantifier Ion
185	Fenpropathrin	39515-41-8	29.56	181(237)	349(25)		265(100)
186	EPN	2104-64-5	30.06	169(53)	323(14)		157(100)
187	Captafol	2425-06-1	30.11	183(5)	311(3)		79(100)
188	Hexazinone	51235-04-2	30.14	252(3)	128(12)		171(100)
189	Leptophos	21609-90-5	30.19	375(73)	379(28)		377(100)
190	Bifenox	42576-02-3	30.81	189(30)	310(27)		341(100)
191	Phosalone	2310-17-0	31.22	367(30)	154(20)		182(100)
192	Azinphos-methyl	86-50-0	31.41	132(71)	77(58)		160(100)
193	Fenarimol	60168-88-9	31.65	219(70)	330(42)		139(100)
194	Azinphos-ethyl	2642-71-9	32.01	132(103)	77(51)		160(100)
195	Cyfluthrin	68359-37-5	32.94	199(63)	226(72)		206(100)
196	Prochloraz	67747-09-5	33.07	308(59)	266(18)		180(100)
197	Coumaphos	56-72-4	33.22	226(56)	364(39)		362(100)
198	Fluvalinate	102851-06-9	34.94	252(38)	181(18)		250(100)

Group C

No.	Name	CAS	RT				
199	Dichlorvos	51338-27-3	7.80	185(34)	220(7)		109(100)
200	Biphenyl	92-52-4	9.00	153(40)	152(27)		154(100)
201	Propamocarb	24579-73-5	9.40	129(6)	188(5)		58(100)
202	Vernolate	1929-77-7	9.82	146(17)	203(9)		128(100)
203	3,5-Dichloroaniline	626-43-7	11.20	163(62)	126(10)		161(100)
204	Methacrifos	62610-77-9	11.86	208(74)	240(44)		125(100)
205	Molinate	2212-67-1	11.92	187(24)	158(2)		126(100)
206	2-Phenylphenol	90-43-7	12.47	169(72)	141(31)		170(100)
207	cis-1,2,3,6-Tetrahydrophthalimide	85-40-5	13.39	123(16)	122(16)		151(100)
208	Fenobucarb	3766-81-2	14.60	150(32)	107(8)		121(100)
209	Benfluralin	1861-40-1	15.23	264(20)	276(13)		292(100)
210	Hexaflumuron	86479-06-3	16.20	279(28)	277(43)		176(100)
211	Prometon	1610-18-0	16.66	225(91)	168(67)		210(100)
212	Triallate	2303-17-5	17.12	270(73)	143(19)		268(100)
213	Pyrimethanil	53112-28-0	17.28	199(45)	200(5)		198(100)
214	Gamma-HCH	77-06-5	17.48	219(93)	254(13)	221(40)	183(100)
215	Disulfoton	298-04-4	17.61	274(15)	186(18)		88(100)
216	Atrizine	1912-24-9	17.64	215(62)	173(29)		200(100)

Continued

TABLE 5.3 Retention Times, Quantifier and Qualifier Ions of 567 Pesticides and Chemical Pollutants Determined by GC-MS-MS—cont'd

No.	Compound	CAS	Retention Time (min)	Qualifier Ion/1	Qualifier Ion/2	Qualifier Ion/3	Quantifier Ion
217	Iprobenfos	26087-47-8	18.44	246(18)	288(17)		204(100)
218	Heptachlor	76-44-8	18.49	237(40)	337(27)		272(100)
219	Isazofos	42509-80-8	18.54	257(53)	285(39)	313(15)	161(100)
220	Plifenate	51366-25-7	18.87	175(96)	242(91)		217(100)
221	Fluchloralin	33245-39-5	18.89	326(87)	264(54)		306(100)
222	Transfluthrin	118712-89-3	19.04	165(23)	335(7)		163(100)
223	Fenpropimorph	67306-03-0	19.22	303(5)	129(9)		128(100)
224	Tolclofos-methyl	57018-04-9	19.69	267(36)	250(10)		265(100)
225	Propisochlor	86763-47-5	19.89	223(200)	146(17)		162(100)
226	Metobromuron	3060-89-7	20.07	258(11)	170(16)		61(100)
227	Ametryn	834-12-8	20.11	212(53)	185(17)		227(100)
228	Simetryn	1014-70-6	20.18	170(26)	198(16)		213(100)
229	Metribuzin	21087-64-9	20.33	199(21)	144(12)		198(100)
230	Dimethipin	55290-64-7	20.38	210(26)	103(20)		118(100)
231	HCH, epsilon	319-86-8	20.78	219(76)	254(15)	217(40)	181(100)

232	Dipropetryn	4147-51-7	20.82	240(42)	222(20)	255(100)
233	Formothion	2540-82-1	21.42	224(97)	257(63)	170(100)
234	Diethofencarb	87130-20-9	21.43	225(98)	151(31)	267(100)
235	Dimepiperate	61432-55-1	22.28	145(30)	263(8)	119(100)
236	Bioallethrin-1	584-79-2	22.29	136(24)	107(29)	123(100)
237	Bioallethrin-2	584-79-2	22.34	136(24)	107(29)	123(100)
238	Fenson	80-38-6	22.54	268(53)	77(104)	141(100)
239	o,p'-DDE	789-02-6	22.64	318(34)	176(26)	246(100)
240	Chlorthion	500-28-7	22.86	267(162)	299(45)	297(100)
241	Diphenamid	957-51-7	22.87	239(30)	165(43)	167(100)
242	Prallethrin	23031-36-9	23.11	105(17)	134(9)	123(100)
243	Penconazole	66246-88-6	23.17	250(33)	161(50)	248(100)
244	Tetraconazole	112281-77-3	23.35	338(33)	171(10)	336(100)
245	Mecarbam	2595-54-2	23.46	296(22)	329(40)	131(100)
246	Propaphos	7292-16-2	23.92	220(108)	262(34)	304(100)
247	Flumetralin	62924-70-3	24.10	157(25)	404(10)	143(100)
248	Triadimenol-1	55219-65-3	24.22	168(81)	130(15)	112(100)
249	Triadimenol-2	55219-65-3	24.94	168(71)	130(10)	112(100)
250	Pretilachlor	51218-49-6	24.67	238(26)	262(8)	162(100)
251	Kresoxim-methyl	143390-89-0	25.04	206(25)	131(66)	116(100)

Continued

239 o,p'-DDE row also contains: 248(70)

TABLE 5.3 Retention Times, Quantifier and Qualifier Ions of 567 Pesticides and Chemical Pollutants Determined by GC-MS-MS—cont'd

No.	Compound	CAS	Retention Time (min)	Qualifier Ion/1	Qualifier Ion/2	Qualifier Ion/3	Quantifier Ion
252	Fluazifop-butyl	79241-46-6	25.21	383(44)	254(49)		282(100)
253	Chlorfluazuron	71422-67-8	25.27	323(71)	356(8)		321(100)
254	Chlorobenzilate	510-15-6	25.90	253(65)	152(5)		251(100)
255	Flusilazole	85509-19-9	26.19	206(33)	315(9)		233(100)
256	Fluorodifen	15457-05-3	26.59	328(35)	162(34)		190(100)
257	Diniconazole	83657-24-3	27.03	270(65)	232(13)		268(100)
258	Piperonyl butoxide	51-03-6	27.46	177(33)	149(14)		176(100)
259	Dimefuron	34205-21-5	27.82	105(75)	267(36)		140(100)
260	Propargite	2312-35-8	27.87	350(7)	173(16)		135(100)
261	Mepronil	55814-41-0	27.91	269(26)	120(9)		119(100)
262	Diflufenican	83164-33-4	28.45	394(25)	267(14)		266(100)
263	Fludioxonil	131341-86-1	28.93	127(24)	154(21)		248(100)
264	Fenazaquin	120928-09-8	28.97	160(46)	117(10)		145(100)
265	Phenothrin	26002-80-2	29.08	183(74)	350(6)		123(100)
266	Fenoxycarb	79127-80-3	29.57	186(82)	116(93)		255(100)

267	Sethoxydim	74051-80-2	29.63	281(51)	219(36)	178(100)
268	Amitraz	33089-61-1	30.00	162(138)	132(168)	293(100)
269	Anilofos	64249-01-0	30.68	184(52)	334(10)	226(100)
270	Acrinathrin	101007-06-1	31.07	289(31)	247(12)	181(100)
271	Lambda-Cyhalothrin	91465-08-6	31.11	197(100)	141(20)	181(100)
272	Mefenacet	73250-68-7	31.29	120(35)	136(29)	192(100)
273	Permethrin	52645-53-1	31.57	184(14)	255(1)	183(100)
274	Pyridaben	96489-71-3	31.86	117(11)	364(7)	147(100)
275	Fluoroglycofen-ethyl	77502-90-7	32.01	428(20)	449(35)	447(100)
276	Bitertanol	55179-31-2	32.25	112(8)	141(6)	170(100)
277	Etofenprox	80844-07-1	32.75	376(4)	183(6)	163(100)
278	Cycloxydim	101205-02-1	33.05	279(7)	251(4)	178(100)
279	Alpha-Cypermethrin	67375-30-8	33.35	181(84)	165(63)	163(100)
280	Flucythrinate-1	70124-77-5	33.58	157(90)	451(22)	199(100)
281	Flucythrinate-2	70124-77-5	33.85	157(91)	451(23)	199(101)
282	Esfenvalerate	51630-58-1	34.65	225(158)	181(189)	419(100)
283	Difenconazole-1	1977-6-54	35.40	325(66)	265(83)	323(100)
284	Difenonazole-2	1977-6-54	35.49	325(69)	265(70)	323(100)
285	Flumioxazin	103361-09-7	35.50	287(24)	259(15)	354(100)
286	Flumiclorac-pentyl	87546-18-7	36.34	308(51)	318(29)	423(100)

Continued

TABLE 5.3 Retention Times, Quantifier and Qualifier Ions of 567 Pesticides and Chemical Pollutants Determined by GC-MS-MS—cont'd

No.	Compound	CAS	Retention Time (min)	Qualifier Ion/1	Qualifier Ion/2	Qualifier Ion/3	Quantifier Ion
Group D							
287	Dimefox	115-26-4	5.62	154(75)	153(17)		110(100)
288	Disulfoton-sulfoxide	2497-07-6	8.41	153(61)	184(20)		212(100)
289	Pentachlorobenzene	608-93-5	11.11	252(64)	215(24)		250(100)
290	Tri-iso-butyl phosphate	126-71-6	11.65	139(67)	211(24)		155(100)
291	Crimidine	535-89-7	13.13	156(90)	171(84)		142(100)
292	BDMC-1	672-99-1	13.25	202(104)	201(13)		200(100)
293	Chlorfenprop-methyl	14437-17-3	13.57	196(87)	197(49)		165(100)
294	Thionazin	297-97-2	14.04	192(39)	220(14)		143(100)
295	2,3,5,6-Tetrachloroaniline	3481-20-7	14.22	229(76)	158(25)		231(100)
296	Tri-n-butyl phosphate	126-73-8	14.33	211(61)	167(8)		155(100)
297	2,3,4,5-Tetrachloroanisole	938-86-3	14.66	203(70)	231(51)		246(100)
298	Pentachloroanisole	1825-21-4	15.19	265(100)	237(85)		280(100)
299	Tebutam	35256-85-0	15.30	106(38)	142(24)		190(100)

300	Dioxabenzofos	3811-49-2	16.14	201(26)	171(5)	216(100)
301	Methabenzthiazuron	18691-97-9	16.34	136(81)	108(27)	164(100)
302	Desisopropyl-atrazine	1007-28-9	16.69	158(84)	145(73)	173(100)
303	Simetone	673-04-1	16.69	196(40)	182(38)	197(100)
304	Atratone	1610-17-9	16.70	211(68)	197(105)	196(100)
305	Tefluthrin	79538-32-2	17.24	197(26)	161(5)	177(100)
306	Bromocylen	1715-40-8	17.43	357(99)	394(14)	359(100)
307	Trietazine	1912-26-1	17.53	229(51)	214(45)	200(100)
308	Etrimfos oxon	59399-24-5	17.83	277(35)	263(12)	292(100)
309	2,6-Dichlorobenzamide	2008-58-4	17.93	189(36)	175(62	173(100)
310	Cycluron	2163-69-1	17.95	198(36)	114(9)	89(100)
311	DE-PCB 28	2012-37-5	18.15	186(53)	258(97)	256(100)
312	DE-PCB 31	16606-02-3	18.19	186(53)	258(97)	256(100)
313	Desethyl-sebuthylazine	37019-18-4	18.32	174(32)	186(11)	172(100)
314	2,3,4,5-Tetrachloroaniline	634-83-3	18.55	229(76)	233(48)	231(100)
315	Musk ambrette	83-46-9	18.62	268(35)	223(18)	253(100)
316	Musk xylene	81-15-2	18.66	297(10)	128(20)	282(100)
317	Pentachloroaniline	527-20-8	18.91	263(63)	230(8)	265(100)

Continued

TABLE 5.3 Retention Times, Quantifier and Qualifier Ions of 567 Pesticides and Chemical Pollutants Determined by GC-MS-MS—cont'd

No.	Compound	CAS	Retention Time (min)	Qualifier Ion/1	Qualifier Ion/2	Qualifier Ion/3	Quantifier Ion
318	Aziprotryne	4658-28-0	19.11	184(83)	157(31)		199(100)
319	Isocarbamid	30979-48-7	19.24	185(2)	143(6)		142(100)
320	Sebutylazine	7286-69-3	19.26	214(14)	229(13)		200(100)
321	Musk moskene	116-66-5	19.46	278(12)	264(15)		263(100)
322	DE-PCB 52	35693-99-3	19.48	220(88)	255(32)		292(100)
323	Prosulfocarb	52888-80-9	19.51	252(14)	162(10)		251(100)
324	Dimethenamid	87674-68-8	19.55	230(43)	203(21)		154(100)
325	Fenchlorphos oxon	3983-45-7	19.72	287(70)	270(7)		285(100)
326	BDMC-2	672-99-1	19.74	202(101)	201(12)		200(100)
327	Monalide	7287-36-7	20.02	199(31)	239(45)		197(100)
328	Musk tibeten	145-39-1	20.40	266(25)	252(14)		251(100)
329	Isobenzan	297-78-9	20.55	375(31)	412(7)		311(100)
330	Pyrimitate	5221-49-8	20.59	153(116)	180(49)		305(100)
331	Octachlorostyrene	29082-74-4	20.60	343(94)	308(120)		380(100)
332	Isodrin	465-73-6	21.01	263(46)	195(83)		193(100)

#	Name	CAS	RT			
333	Isomethiozin	57052-04-7	21.06	198(86)	184(13)	225(100)
334	Trichloronat	327-98-0	21.10	269(86)	196(16)	297(100)
335	Dacthal	1861-32-1	21.25	332(31)	221(16)	301(100)
336	4,4-Dichlorobenzophenone	90-98-2	21.29	252(62)	215(26)	250(100)
337	Nitrothal-isopropyl	10552-74-6	21.69	254(54)	212(74)	236(100)
338	Musk ketone	541-91-3	21.70	294(28)	128(16)	279(100)
339	Rabenzazole	69899-24-7	21.73	170(26)	195(19)	212(100)
340	Cyprodinil	121552-61-2	21.94	225(62)	210(9)	224(100)
341	Isofenphos oxon	106848-93-5	22.04	201(2)	314(12)	229(100)
342	Fuberidazole	3878-19-1	22.10	155(21)	129(12)	184(100)
343	Dicapthon	2463-84-5	22.44	263(10)	216(10)	262(100)
344	MCPA-butoxyethyl ester	94-81-5	22.61	200(71)	182(41)	300(100)
345	DE-PCB 101	37680-73-2	22.62	254(66)	291(18)	326(100)
346	Methfuroxam	2873-17-8	22.71	212(4)	230(16)	229(100)
347	Isocarbophos	245-61-5	22.87	230(26)	289(22)	136(100)
348	Phorate sulfone	2588-04-7	23.15	171(30)	215(11)	199(100)
349	Chlorfenethol	80-06-8	23.29	253(66)	266(12)	251(100)
350	trans-Nonachlor	39765-80-5	23.62	407(89)	411(63)	409(100)

Continued

TABLE 5.3 Retention Times, Quantifier and Qualifier Ions of 567 Pesticides and Chemical Pollutants Determined by GC-MS-MS—cont'd

No.	Compound	CAS	Retention Time (min)	Qualifier Ion/1	Qualifier Ion/2	Qualifier Ion/3	Quantifier Ion
351	Dinobuton	973-21-7	23.88	240(15)	223(15)		211(100)
352	DEF	78-48-8	24.08	226(51)	258(55)		202(100)
353	Flurochloridone	61213-25-0	24.31	187(74)	313(66)		311(100)
354	Bromfenvinfos	33399-00-7	24.62	323(56)	295(18)		267(100)
355	Perthane	72-56-0	24.81	224(20)	178(9)		223(100)
356	Ditalimfos	5131-24-8	24.82	148(43)	299(34)		130(100)
357	DE-PCB 118	31508-00-6	25.08	254(38)	184(16)		326(100)
358	Mephosfolan	950-10-7	25.29	227(49)	168(60)		196(100)
359	4,4-Dibromobenzophenone	3988-03-2	25.30	259(30)	185(179)		340(100)
360	Flutriafol	76674-21-0	25.31	164(96)	201(7)		219(100)
361	Athidathion	19691-80-6	25.63	330(1)	129(12)		145(100)
362	DE-PCB 153	35065-27-1	25.64	290(62)	218(24)		360(100)
363	Diclobutrazole	62-73-7	25.95	272(68)	159(42)		270(100)
364	Disulfoton sulfone	2497-06-5	26.16	229(4)	185(11)		213(100)
365	Hexythiazox	78587-05-0	26.48	156(158)	184(93)		227(100)

366	DE-PCB 138	35065-28-2	26.84	290(68)	218(26)	360(100)
367	Triamiphos	1031-47-6	27.02	294(28)	251(16)	160(100)
368	Cyproconazole	113096-99-4	27.23	224(35)	223(11)	222(100)
369	Resmethrin-1	10453-86-8	27.26	143(83)	338(7)	171(100)
370	Resmethrin-2	10453-86-8	27.43	143(80)	338(7)	171(100)
371	Phthalic acid, benzyl butyl ester	85-68-7	27.56	312(4)	230(1)	206(100)
372	Clodinafop-propargyl	105512-06-9	27.74	238(96)	266(83)	349(100)
373	Fenthion sulfoxide	3761-41-9	28.06	279(290)	294(145)	278(100)
374	Fluotrimazole	31251-03-3	28.39	379((60)	233(36)	311(100)
375	Fluroxypr-1-methylheptyl ester	81406-37-3	28.45	254(67)	237(60)	366(100)
376	Fenthion sulfone	3761-42-0	28.55	136(25)	231(10)	310(100)
377	Metamitron	41394-05-2	28.63	174(52)	186(12)	202(100)
378	Triphenyl phosphate	115-86-6	28.65	233(16)	215(20)	326(100)
379	DE-PCB 180	35065-29-3	29.05	324(70)	359(20)	394(100)
380	Tebufenpyrad	119168-77-3	29.06	333(78)	276(44)	318(100)
381	Cloquintocet-mexyl	99607-70-2	29.32	194(32)	220(4)	192(100)
382	Lenacil	2164-08-1	29.70	136(6)	234(2)	153(100)
383	Bromuconazole-1	116255-48-2	29.90	175(65)	214(15)	173(100)

Continued

TABLE 5.3 Retention Times, Quantifier and Qualifier Ions of 567 Pesticides and Chemical Pollutants Determined by GC-MS-MS—cont'd

No.	Compound	CAS	Retention Time (min)	Qualifier Ion/1	Qualifier Ion/2	Qualifier Ion/3	Quantifier Ion
384	Desbrom-leptophos		30.15	171(97)	375(72)		377(100)
385	Bromuconazole-2	116255-48-2	30.72	175(67)	214(14)		173(100)
386	Nitralin	4726-14-1	30.92	274(58)	300(15)		316(100)
387	Fenamiphos sufoxide	31972-42-7	31.03	319(29)	196(22)		304(100)
388	Fenamiphos sulfone	31972-44-8	31.34	292(57)	335(7)		320(100)
389	Fenpiclonil	74738-17-3	32.37	238(66)	174(36)		236(100)
390	Fluquinconazole	136426-54-5	32.62	342(37)	341(20)		340(100)
391	Fenbuconazole	114369-43-6	34.02	198(51)	125(31)		129(100)
Group E							
392	Propoxur-1	158474-72-7	6.58	152(16)	111(9)		110(100)
393	XMC	2655-14-3	7.40	121(37)	107(114)		122(100)
394	Isoprocarb-1	2631-40-7	7.56	136(34)	103(20)		121(100)
395	Methamidophos	10265-92-6	9.37	95(112)	141(52)		94(100)
396	Acenaphthene	83-32-9	10.79	162(84)	160(38)		164(100)
397	Terbucarb-1	1918-11-2	10.89	220(51)	206(16)		205(100)

398	Dibutyl succinate	141-03-7	12.20	157(19)	175(5)	101(100)
399	Chlorethoxyfos	54593-83-8	13.43	125(67)	301(19)	153(100)
400	Isoprocarb-2	2631-40-7	13.69	136(34)	103(20)	121(100)
401	Tebuthiuron	34014-18-1	14.25	171(30)	157(9)	156(100)
402	Pencycuron	58810-48-3	14.30	180(65)	209(20)	125(100)
403	Demeton-S-methyl	301-12-2	15.19	142(43)	230(5)	109(100)
404	Propoxur-2	114-26-1	15.48	152(19)	111(8)	110(100)
405	Naled	133408-50-1	15.51	145(26)	185(15)	109(100)
406	Phenanthrene	66063-05-6	16.97	160(9)	189(16)	188(100)
407	Fenpyroximate	134098-61-6	17.49	142(21)	198(9)	213(100)
408	Tebupirimfos	96182-53-5	17.61	261(107)	234(100)	318(100)
409	Prohydrojasmon	23103-98-2	17.80	184(41)	254(7)	153(100)
410	Fenpropidin	67306-00-7	17.85	273(5)	145(5)	98(100)
411	Dichloran	99-30-9	18.10	206(87)	124(101)	176(100)
412	Pyroquilon	95737-68-1	18.28	130(69)	144(38)	173(100)
413	Propyzamide	114-26-1	19.01	255(23)	240(9)	173(100)
414	Pirimicarb	24151-93-7	19.08	238(23)	138(8)	166(100)
415	Benoxacor	98730-04-2	19.62	259(38)	176(19)	120(100)
416	Phosphamidon-1	65-01-8	19.66	138(62)	227(25)	264(100)
417	Bromobutide	74712-19-9	19.70	232(27)	296(6)	119(100)

Continued

TABLE 5.3 Retention Times, Quantifier and Qualifier Ions of 567 Pesticides and Chemical Pollutants Determined by GC-MS-MS—cont'd

No.	Compound	CAS	Retention Time (min)	Qualifier Ion/1	Qualifier Ion/2	Qualifier Ion/3	Quantifier Ion
418	Acetochlor	34256-82-1	19.84	162(59)	223(59)		146(100)
419	Tridiphane	58138-08-2	19.90	187(90)	219(46)		173(100)
420	Esprocarb	85785-20-2	20.01	265(10)	162(61)		222(100)
421	Terbucarb-2	1918-11-2	20.06	220(52)	206(16)		205(100)
422	Fenfuram	24691-80-3	20.35	201(29)	202(5)		109(100)
423	Acibenzolar-S-methyl	135158-54-2	20.42	135(64)	153(34)		182(100)
424	Benfuresate	68505-69-1	20.68	256(17)	121(18)		163(100)
425	Dithiopyr	97886-45-8	20.78	306(72)	286(74)		354(100)
426	Mefenoxam	70630-17-0	20.91	249(46)	279(11)		206(100)
427	Malaoxon	1364-78-2	21.17	268(11)	195(15)		127(100)
428	Phosphamidon-2	13171-21-6	21.36	138(54)	227(17)		264(100)
429	Chlorthal-dimethyl	1861-32-1	21.39	332(27)	221(17)		301(100)
430	Simeconazole	105024-66-6	21.41	278(14)	211(34)		121(100)
431	Terbacil	5902-51-2	21.50	160(70)	117(39)		161(100)
432	Thiazopyr	117718-60-2	21.91	363(73)	381(34)		327(100)

433	Dimethylvinphos	2274-67-1	22.21	297(56)	109(74)	295(100)
434	Zoxamide	156052-68-5	22.30	242(68)	299(9)	187(100)
435	Allethrin	584-79-2	22.60	107(24)	136(20)	123(100)
436	Quinoclamine	57369-32-1	22.89	172(259)	144(64)	207(100)
437	Methothrin-1	841-06-5	22.92	135(89)	104(41)	123(100)
438	Flufenacet	142459-58-3	23.09	211(61)	363(6)	151(100)
439	Methothrin-2	34388-29-9	23.19	135(73)	104(12)	123(100)
440	Fenoxanil	115852-48-7	23.58	189(14)	301(6)	140(100)
441	Furalaxyl	57646-30-7	23.97	301(24)	152(40)	242(100)
442	Mepanipyrim	110235-47-7	24.29	223(53)	221(9)	222(100)
443	Thiamethoxam	153719-23-4	24.38	212(92)	247(124)	182(100)
444	Captan	133-06-2	24.55	264(32)	236(10)	149(100)
445	Bromacil	314-40-9	24.73	207(46)	231(5)	205(100)
446	Picoxystrobin	137641-05-5	24.97	303(43)	367(9)	335(100)
447	Butamifos	8013-75-0	25.41	200(57)	232(37)	286(100)
448	Imazamethabenz-methyl	81405-85-8	25.50	187(117)	256(95)	144(100)
449	Methiocarb Sulfone	2178-25-1	25.56	185(40)	137(16)	200(100)
450	TCMTB	148477-71-8	25.59	238(108)	136(30)	180(100)
451	Metominostrobin	786-19-6	25.61	238(56)	196(75)	191(100)

Continued

TABLE 5.3 Retention Times, Quantifier and Qualifier Ions of 567 Pesticides and Chemical Pollutants Determined by GC-MS-MS—cont'd

No.	Compound	CAS	Retention Time (min)	Qualifier Ion/1	Qualifier Ion/2	Qualifier Ion/3	Quantifier Ion
452	Imazalil	35554-44-0	25.72	173(66)	296(5)		215(100)
453	Isoprothiolane	34123-59-6	25.87	231(82)	204(88)		290(100)
454	Cyflufenamid	180409-60-3	26.02	412(11)	294(11)		91(100)
455	Pyriminobac-Methyl	105779-78-0	26.34	330(107)	361(86)		302(100)
456	Methyl trithion	72-43-5	26.36	157(492)	125(247)		314(100)
457	Isoxathion	58769-20-3	26.51	105(341)	177(208)		313(100)
458	Quinoxyphen	2797-51-5	27.14	272(37)	307(29)		237(100)
459	Thifluzamide	130000-40-7	27.26	447(97)	194(308)		449(100)
460	Trifloxystrobin	141517-21-7	27.71	131(40)	222(30)		116(100)
461	Imibenconazole-des-benzyl	199338-48-2	27.86	270(35)	272(35)		235(100)
462	Imiprothrin-1	72693-72-5	28.31	151(55)	107(54)		123(100)
463	Fipronil	120068-37-3	28.34	369(69)	351(15)		367(100)
464	Imiprothrin-2	72693-72-5	28.50	151(21)	107(17)		123(100)
465	Epoxiconazole-1	106325-08-0	28.58	183(24)	138(35)		192(100)
466	Pyributicarb	129630-17-7	28.87	181(23)	108(64)		165(100)

467	Pyraflufen Ethyl	175013-18-0	28.91	349(41)	339(34)	412(100)
468	Thenylchlor	96491-05-3	29.12	288(25)	141(17)	127(100)
469	Clethodim	99129-21-2	29.21	205(50)	267(15)	164(100)
470	Chrysene	218-01-9	29.40	236(24)	120(16)	240(100)
471	Mefenpyr-diethyl	135590-91-9	29.55	299(131)	372(18)	227(100)
472	Etoxazole	153233-91-1	29.64	330(69)	359(65)	300(100)
473	Epoxiconazole-2	106325-08-0	29.73	183(13)	138(30)	192(100)
474	Famphur	52-85-7	29.80	125(27)	217(22)	218(100)
475	Pyriproxyfen	136191-64-5	30.06	226(8)	185(10)	136(100)
476	Iprodione	36734-19-7	30.24	244(65)	246(42)	187(100)
477	Picolinafen	13171-21-6	30.27	376(77)	266(11)	238(100)
478	Ofurace	52570-16-8	30.36	232(83)	204(35)	160(100)
479	Piperophos	117428-22-5	30.42	140(123)	122(114)	320(100)
480	Clomeprop	84496-56-0	30.48	288(279)	148(206)	290(100)
481	Fenamidone	161326-34-7	30.66	238(111)	206(32)	268(100)
482	Naproanilide	300-76-5	31.89	171(96)	144(100)	291(100)
483	Pyraclostrobin	77458-01-6	31.98	325(14)	283(21)	132(100)
484	Lactofen	77501-63-4	32.06	461(25)	346(12)	442(100)
485	Tralkoxydim	87820-88-0	32.14	226(7)	268(8)	283(100)
486	Pyraclofos	23950-58-5	32.18	194(79)	362(38)	360(100)

Continued

TABLE 5.3 Retention Times, Quantifier and Qualifier Ions of 567 Pesticides and Chemical Pollutants Determined by GC-MS-MS—cont'd

No.	Compound	CAS	Retention Time (min)	Qualifier Ion/1	Qualifier Ion/2	Qualifier Ion/3	Quantifier Ion
487	Dialifos	10311-84-9	32.27	357(143)	210(397)		186(100)
488	Spirodiclofen	149508-90-7	32.50	259(48)	277(28)		312(100)
489	Flurtamone	96525-23-4	32.78	199(63)	247(25)		333(100)
490	Pyriftalid	88678-67-5	32.94	274(71)	303(44)		318(100)
491	Silafluofen	124495-18-7	33.18	286(274)	258(289)		287(100)
492	Pyrimidifen	135186-78-6	33.63	186(32)	185(10)		184(100)
493	Butafenacil	134605-64-4	33.85	333(34)	180(35)		331(100)
494	Cafenstrole	125306-83-4	34.36	188(69)	119(25)		100(100)
495	Fluridone	59756-60-4	37.61	329(100)	330(100)		328(100)
Group F							
496	Tribenuron-methyl	106040-48-6	9.34	124(45)	110(18)		154(100)
497	Ethiofencarb	29973-13-5	11.00	168(34)	77(26)		107(100)
498	Dioxacarb	6988-21-2	11.10	166(44)	165(36)		121(100)
499	Dimethyl phthalate	131-11-3	11.54	194(7)	133(5)		163(100)
500	4-Chlorophenoxy acetic acid	122-88-3	11.84	141(93)	111(61)		200(100)

No.	Name	CAS	RT	m/z	m/z	m/z	m/z
501	Triflumizole	68694-11-1	12.16	221(61)	250(20)		182(100)
502	Phthalimide	5333-22-2	13.21	104(61)	103(35)		147(100)
503	Acephate	30560-19-1	13.89	94(50)	183(3)		136(100)
504	Diethyltoluamide	134-62-3	14.00	190(32)	191(31)		119(100)
505	2,4-D	94-75-7	14.35	234(63)	175(61)		199(100)
506	Carbaryl	63-25-2	14.42	115(100)	116(43)		144(100)
507	Cadusafos	95465-99-9	15.14	213(14)	270(13)		159(100)
508	Endothal	145-73-3	15.68	140(268)	68(745)		100(100)
509	Demetom-s	126-75-0	16.88	170(15)	143(11)		88(100)
510	Spiroxamine-1	118134-30-8	17.26	126(7)	198(5)		100(100)
511	Dicrotophos	141-66-2	17.31	237(11)	109(8)		127(100)
512	3.4.5-Trimethacarb	2686-99-9	17.70	193(32)	121(31)		136(100)
513	2,4,5-T	93-76-5	17.75	268(49)	209(36)		233(100)
514	3-Phenylphenol	580-51-8	18.11	141(23)	115(17)		170(100)
515	Furmecyclox	60568-05-0	18.22	251(6)	94(10)		123(100)
516	Spiroxamine-2	118134-30-8	18.23	126(5)	198(5)		100(100)
517	DMSA	1596-84-5	18.45	92(123)	121(8)		200(100)
518	Sobutylazine	7286-69-3	18.63	174(32)	186(11)		172(100)
519	Cinmethylin	87818-31-3	18.96	169(16)	154(14)		105(100)
520	Monocrotophos	6923-22-4	19.18	192(2)	223(4)	164(20)	127(100)

Continued

TABLE 5.3 Retention Times, Quantifier and Qualifier Ions of 567 Pesticides and Chemical Pollutants Determined by GC-MS-MS—cont'd

No.	Compound	CAS	Retention Time (min)	Qualifier Ion/1	Qualifier Ion/2	Qualifier Ion/3	Quantifier Ion
521	S421 (octachlorodipropyl ether)-1	127-90-2	19.31	132(96)	211(8)		130(100)
522	S421 (octachlorodipropyl ether)-2	127-90-2	19.57	132(97)	211(7)		130(100)
523	Dodemorph	1593-77-7	19.62	281(12)	238(10)		154(100)
524	Desmedipham	13684-56-5	19.76	109(75)	135(20)		181(100)
525	Fenchlorphos	3983-45-7	19.84	287(69)	270(6)		285(100)
526	Difenoxuron	14214-32-5	20.85	226(21)	242(15)		241(100)
527	Prothoate	2275-18-5	20.85	97(48)	285(14)		115(100)
528	Butralin	33629-47-9	22.18	224(16)	295(9)		266(100)
529	Pyrifenox-2	88283-41-4	22.47	294(16)	227(15)		262(100)
530	Dimethametryn	14214-32-5	22.75	255(9)	213(2)		212(100)
531	Pyrifenox-1	88283-41-4	23.46	294(18)	227(15)		262(100)
532	Phthalide	27355-22-2	23.51	272	215		243
533	Merphos	150-50-5	24.33	202(32)	258(31)		169(100)

534	Thiabendazole	148-79-8	24.97	174(87)	175(9)	201(100)
535	Iprovalicarb-1	140923-17-7	26.13	134(126)	158(62)	119(100)
536	Azaconazole	60207-31-0	26.50	173(59)	219(64)	217(100)
537	Iprovalicarb-2	140923-17-7	26.54	119(75)	158(48)	134(100)
538	Diofenolan-1	63837-33-2	26.76	300(60)	225(24)	186(100)
539	Diofenolan-2	63837-33-2	27.09	300(60)	225(29)	186(100)
540	Aclonifen	74070-46-5	27.24	212(65)	194(57)	264(100)
541	Chlorfenapyr	122453-73-0	27.47	328(54)	408(51)	247(100)
542	Bioresmethrin	28434-01-7	27.55	171(54)	143(31)	123(100)
543	Isoxadifen-Ethyl	141112-29-0	27.90	222(76)	294(44)	204(100)
544	Carfentrazone-Ethyl	128621-72-7	28.09	330(52)	290(53)	312(100)
545	Endrin aldehyde	7421-93-4	28.30	250(62)	279(36)	345(100)
546	Halosulfuran-methyl	100784-20-1	28.32	260(86)	295(33)	327(100)
547	Tricyclazole	41814-78-2	28.34	162(54)	161(40)	189(100)
548	Fenhexamid	126833-17-8	28.86	177(33)	301(13)	97(100)
549	Spiromesifen	283594-90-1	29.56	254(27)	370(14)	272(100)
550	Phenkapton	2275-14-1	29.62	153(79)	191(65)	121(100)
551	Fluazinam	79622-59-6	30.04	417(44)	371(29)	387(100)
552	Bifenazate	149877-41-8	30.38	258(99)	199(100)	300(100)
553	Endrin ketone	53494-70-5	30.40	250(28)	281(35)	317(100)

Continued

TABLE 5.3 Retention Times, Quantifier and Qualifier Ions of 567 Pesticides and Chemical Pollutants Determined by GC-MS-MS—cont'd

No.	Compound	CAS	Retention Time (min)	Qualifier Ion/1	Qualifier Ion/2	Qualifier Ion/3	Quantifier Ion
554	Norflurazon-desmethyl	23576-24-1	30.80	289(76)	88(35)		145(100)
555	Gamma-cyhaloterin-1	91465-08-6	31.10	197(84)	141(28)		181(100)
556	Metoconazole	125116-23-6	31.12	319(14)	250(17)		125(100)
557	Cyhalofop-butyl	122008-85-9	31.40	357(74)	229(79)		256(100)
558	Gamma-cyhalothrin-2	91465-08-6	31.40	197(77)	141(20)		181(100)
559	Cythioate	115-93-5	31.95	109(85)	125(71)		297(100)
560	Halfenprox	111872-58-3	32.81	237(5)	476(5)		263(100)
561	Acetamiprid	160430-64-8	33.67	152(114)	166(64)		126(100)
562	Boscalid	188425-85-6	34.16	140(229)	112(71)		342(100)
563	Tralomethrin-1	66841-25-6	35.51	253(147)	251(40)		181(100)
564	Tralomethrin-2	66841-25-6	35.97	253(140)	251(40)		181(100)
565	Tolfenpyrad	129558-76-5	36.00	197(72)	171(75)		383(100)
566	Dimethomorph	211867-47-9	37.40	387(32)	165(28)		301(100)
567	Azoxystrobin	131860-33-8	37.77	388(32)	404(31)		344(100)

5.1.2   Monitoring of Selected Ions for 567 Pesticides and Chemical Pollutants Determined by GC-MS

See Table 5.4.

TABLE 5.4 Monitoring of Selected Ions for 567 Pesticides and Chemical Pollutants Determined by GC-MS

No.	Retention Time (min)	Ion (amu)	Dwelling Time (ms)
Group A			
1	8.30	138, 158, 173	200
2	9.60	124, 140, 166, 172, 183, 211	90
3	10.50	121, 154, 234	200
4	10.75	120, 137, 179	200
5	11.70	154, 186, 215	200
6	14.40	167, 168, 169	200
7	14.90	121, 142, 143, 153, 183, 195, 196, 198, 230, 231, 260, 276, 292, 316	30
8	16.20	88, 125, 246	200
9	16.70	137, 138, 145, 172, 174, 179, 187, 202, 204, 205, 237, 246, 249, 295, 304	30
10	17.80	138, 173, 175, 181, 186, 194, 196, 201, 210, 225, 236, 255, 277, 292	30
11	18.80	150, 165, 173, 175, 222, 223, 251, 255, 279	50
12	19.20	125, 143, 229, 261, 263, 265, 293, 305, 307, 329	50
13	19.80	125, 261, 263, 265, 285, 287, 293, 305, 307, 329	50
14	20.10	170, 181, 184, 198, 200, 206, 212, 217, 219, 226, 227, 233, 234, 241, 246, 249, 254, 258, 263, 264, 266, 268, 285, 286, 314	10
15	21.40	143, 152, 153, 158, 169, 173, 180, 181, 208, 217, 219, 220, 247, 254, 256, 260, 275, 277, 278, 351, 353, 355	10
16	22.30	61, 143, 160, 162, 181, 186, 208, 210, 220, 235, 248, 252, 263, 268, 270, 291, 351, 353, 355	20
17	23.00	133, 143, 146, 157, 209, 211, 246, 268, 270, 274, 298, 303, 320, 357, 359, 373, 375, 377	20

Continued

TABLE 5.4 Monitoring of Selected Ions for 567 Pesticides and Chemical Pollutants Determined by GC-MS—cont'd

No.	Retention Time (min)	Ion (amu)	Dwelling Time (ms)
18	23.70	72, 104, 133, 145, 152, 157, 160, 162, 209, 211, 215, 253, 255, 260, 263, 267, 274, 277, 283, 285, 297, 302, 309, 345, 380	10
19	24.80	128, 145, 154, 157, 171, 175, 198, 217, 225, 240, 255, 258, 271, 283, 285, 288, 302, 303	20
20	25.50	154, 185, 217, 252, 253, 254, 288, 303, 319, 324, 334	50
21	26.00	87, 139, 143, 145, 165, 173, 199, 208, 231, 235, 237, 251, 253, 273, 316, 323, 384	20
22	26.80	145, 150, 156, 165, 173, 179, 199, 231, 235, 237, 245, 247, 280, 288, 322, 323, 384	20
23	27.90	165, 166, 173, 181, 253, 259, 261, 281, 292, 293, 308, 342	40
24	28.60	118, 160, 165, 166, 181, 203, 212, 227, 228, 231, 235, 237, 272, 274, 314, 323	30
25	29.30	135, 163, 164, 212, 227, 228, 232, 233, 250, 252, 278	40
26	30.00	102, 145, 159, 160, 161, 188, 199, 227, 303, 317, 340, 356	40
27	31.00	175, 183, 184, 220, 221, 223, 232, 250, 255, 267, 373	40
28	33.00	127, 180, 181	200
29	34.40	167, 181, 225, 419	150
30	35.70	172, 174, 181	200
Group B			
1	7.80	128, 132, 189	200
2	8.80	146, 156, 217	200
3	9.70	128, 136, 161, 171, 173, 203	90
4	10.70	127, 164, 192, 194, 196, 198	90
5	11.70	191, 193, 206	200
6	13.40	124, 203, 215, 250, 261	100
7	14.40	158, 168, 200, 242, 282, 284, 286	80

TABLE 5.4 Monitoring of Selected Ions for 567 Pesticides and Chemical Pollutants Determined by GC-MS—cont'd

No.	Retention Time (min)	Ion (amu)	Dwelling Time (ms)
8	14.70	116, 120, 128, 148, 153, 171, 176, 188, 202, 211, 213, 234, 236, 238, 264, 266, 282, 284, 286, 306, 322, 335	10
9	16.00	116, 148, 183, 188, 219, 221, 254	80
10	16.80	153, 186, 231, 288	150
11	17.10	153, 160, 164, 169, 172, 173, 176, 197, 206, 210, 214, 223, 225, 229, 270, 318, 330, 347	20
12	18.20	61, 126, 160, 173, 176, 206, 214, 229	60
13	18.70	126, 127, 134, 148, 164, 171, 172, 180, 192, 197, 198, 210, 213, 223, 243, 286, 288, 305, 307	20
14	19.90	134, 171, 188, 197, 198, 210, 213, 237, 269, 276, 290, 305	40
15	20.60	100, 185, 211, 226, 241, 253, 257, 259, 378	50
16	21.20	73, 139, 141, 153, 161, 162, 167, 185, 191, 207, 213, 224, 226, 237, 238, 240, 250, 251, 286, 304, 318, 329, 331, 333, 351, 353, 355, 387	10
17	22.00	161, 167, 207, 222, 224, 226, 238, 264, 280, 286, 351, 353, 355	40
18	22.70	161, 163, 170, 171, 182, 185, 205, 213, 217, 241, 255, 256, 265, 267, 269, 276, 323, 339	20
19	23.40	137, 160, 176, 188, 238, 240, 246, 248, 259, 267, 269, 316, 318, 323, 331, 373, 375, 377	20
20	23.90	61, 160, 166, 176, 188, 193, 194, 246, 248, 250, 259, 292, 294, 297, 316, 318, 329, 331, 333, 339, 374, 377, 379	20
21	24.90	61, 105, 165, 167, 172, 175, 177, 187, 199, 214, 231, 235, 236, 237, 238, 256, 263, 292, 294, 297, 302, 305, 311, 313, 317, 339, 345, 374	10
22	25.60	77, 105, 139, 141, 165, 169, 171, 199, 202, 213, 223, 235, 237, 251, 252, 253, 256, 271, 276, 283, 297, 300, 325, 360, 361	10
23	26.70	105, 157, 165, 195, 199, 235, 237, 246, 276, 297, 325, 339, 342, 360, 363	30

Continued

TABLE 5.4 Monitoring of Selected Ions for 567 Pesticides and Chemical Pollutants Determined by GC-MS—cont'd

No.	Retention Time (min)	Ion (amu)	Dwelling Time (ms)
24	27.60	148, 157, 161, 169, 172, 173, 201, 206, 257, 303, 310, 325	40
25	28.90	89, 99, 126, 127, 157, 161, 169, 172, 181, 183, 257, 260, 265, 272, 292, 303, 339, 341, 349, 365, 387, 389	10
26	29.80	79, 181, 183, 265, 311, 349	90
27	30.00	128, 157, 169, 171, 189, 252, 310, 323, 341, 375, 377, 379	40
28	31.20	132, 139, 154, 160, 161, 182, 189, 251, 310, 330, 341, 367	40
29	32.90	180, 199, 206, 226, 266, 308, 334, 362, 364	50
30	34.00	181, 250, 252	200
Group C			
1	7.30	109, 185, 220	200
2	8.70	152, 153, 154	200
3	9.30	58, 128, 129, 146, 188, 203	90
4	11.20	126, 161, 163	200
5	11.75	125, 126, 141, 158, 169, 170, 187, 208, 240	50
6	13.50	122, 123, 124, 151, 215, 250	90
7	14.70	107, 121, 150, 264, 276, 292	90
8	16.00	174, 202, 217	200
9	16.50	126, 141, 143, 156, 168, 176, 198, 199, 200, 210, 225, 268, 270, 277, 279	30
10	17.60	88, 173, 183, 186, 200, 215, 219, 254, 274	50
11	18.40	104, 130, 159, 161, 204, 237, 246, 257, 272, 285, 288, 313, 337	40
12	18.90	128, 129, 161, 163, 165, 175, 204, 217, 242, 246, 257, 264, 285, 288, 303, 306, 313, 326, 335	20
13	19.80	73, 89, 146, 162, 185, 212, 223, 227, 250, 265, 267	50

TABLE 5.4 Monitoring of Selected Ions for 567 Pesticides and Chemical Pollutants Determined by GC-MS—cont'd

No.	Retention Time (min)	Ion (amu)	Dwelling Time (ms)
14	20.30	61, 144, 146, 162, 170, 185, 198, 199, 212, 213, 223, 227, 258	40
15	20.70	61, 103, 118, 144, 170, 181, 198, 199, 210, 217, 219, 222, 240, 254, 255	30
16	21.35	108, 117, 151, 160, 161, 170, 219, 221, 224, 225, 257, 267, 351, 353, 355	30
17	22.20	107, 108, 119, 123, 136, 145, 176, 219, 221, 246, 248, 263, 318, 351, 353, 355	20
18	22.70	77, 141, 165, 167, 174, 176, 206, 234, 239, 246, 248, 267, 268, 297, 299, 318	20
19	23.20	105, 123, 134, 161, 248, 250, 267, 297, 299	50
20	23.50	131, 143, 157, 161, 171, 220, 248, 250, 262, 296, 304, 329, 336, 338, 404	30
21	24.30	112, 130, 162, 168, 238, 262	90
22	25.10	112, 116, 130, 131, 162, 168, 206, 233, 234, 235, 238, 262	40
23	25.30	254, 282, 321, 323, 356, 383	90
24	26.00	131, 152, 206, 233, 234, 236, 251, 253, 315	50
25	26.90	149, 162, 176, 177, 190, 232, 268, 270, 328	50
26	27.90	105, 119, 120, 135, 140, 173, 266, 267, 269, 350, 394	50
27	28.80	105, 117, 123, 140, 145, 160, 183, 266, 267, 350, 394	50
28	29.00	117, 123, 127, 145, 154, 160, 183, 248, 350	50
29	29.60	116, 178, 186, 191, 219, 255	90
30	30.30	132, 162, 178, 184, 219, 226, 281, 293, 334	50
31	31.10	120, 136, 141, 147, 181, 183, 184, 192, 197, 247, 255, 289, 309, 364	30
32	32.00	112, 141, 147, 170, 183, 184, 255, 309, 364, 428, 447, 449	40
33	32.60	112, 141, 163, 170, 183, 376, 428, 447, 449	50
34	33.10	163, 165, 178, 181, 251, 279	90
35	33.80	157, 199, 451	200

Continued

TABLE 5.4 Monitoring of Selected Ions for 567 Pesticides and Chemical Pollutants Determined by GC-MS—cont'd

No.	Retention Time (min)	Ion (amu)	Dwelling Time (ms)
36	34.70	181, 225, 250, 252, 419	100
37	35.40	259, 265, 287, 323, 325, 354	90
38	36.40	308, 318, 423	200
Group D			
1	5.50	110, 153, 154	200
2	8.00	153, 184, 212	200
3	11.00	139, 155, 211, 215, 250, 252	90
4	13.00	142, 156, 165, 171, 196, 197, 200, 201, 202	50
5	14.00	143, 155, 158, 167, 192, 203, 211, 220, 229, 231, 246	40
6	15.00	106, 142, 190, 237, 265, 280	90
7	16.00	108, 136, 145, 158, 164, 171, 173, 182, 186, 196, 197, 201, 211, 216, 213, 288	20
8	17.20	161, 174, 177, 197, 200, 202, 214, 229, 246, 357, 359, 394	40
9	17.90	89, 114, 128, 172, 173, 174, 175, 186, 189, 198, 223, 229, 230, 231, 233, 253, 256, 258, 263, 265, 268, 277, 282, 292, 297	10
10	19.20	142, 143, 154, 157, 162, 184, 185, 199, 200, 201, 202, 203, 214, 220, 229, 230, 247, 251, 252, 255, 263, 264, 270, 278, 285, 287, 292	10
11	20.00	153, 180, 197, 199, 200, 201, 202, 230, 239, 247, 251, 252, 266, 305, 308, 311, 343, 375, 380, 412	15
12	21.00	115, 184, 193, 195, 196, 198, 215, 221, 225, 250, 252, 263, 269, 276, 285, 297, 301, 332	20
13	21.60	128, 170, 194, 195, 210, 212, 224, 225, 236, 254, 279, 294	40
14	22.10	129, 155, 182, 184, 200, 201, 210, 212, 216, 224, 225, 229, 230, 254, 262, 263, 291, 300, 314, 326, 351, 353, 355	10
15	23.00	136, 171, 199, 215, 230, 251, 253, 266, 289, 407, 409, 411	40

TABLE 5.4 Monitoring of Selected Ions for 567 Pesticides and Chemical Pollutants Determined by GC-MS—cont'd

No.	Retention Time (min)	Ion (amu)	Dwelling Time (ms)
16	23.90	130, 148, 178, 187, 202, 211, 223, 224, 226, 240, 258, 267, 295, 299, 311, 313, 323	20
17	25.00	129, 130, 145, 148, 164, 168, 184, 185, 196, 201, 218, 219, 227, 254, 259, 290, 299, 326, 330, 340, 360	15
18	26.00	156, 159, 184, 185, 213, 218, 227, 229, 270, 272, 290, 360	40
19	27.10	143, 160, 171, 206, 222, 223, 224, 230, 238, 251, 266, 294, 312, 338, 349	30
20	28.00	136, 174, 186, 202, 215, 231, 233, 237, 254, 278, 279, 294, 310, 311, 326, 366, 379	20
21	29.00	136, 153, 192, 194, 220, 234, 276, 318, 324, 333, 359, 394	40
22	30.00	160, 161, 171, 173, 175, 214, 317, 375, 377	50
23	30.80	173, 175, 196, 213, 230, 274, 292, 300, 304, 316, 319, 320, 335, 373	30
24	32.40	147, 236, 238, 340, 341, 342	90
25	34.00	125, 129, 198	200
Group E			
1	6.10	110, 111, 152	200
2	7.00	103, 107, 121, 122, 136	100
3	9.00	94, 95, 141	200
4	10.40	160, 162, 164	200
5	12.00	101, 157, 175	200
6	12.90	103, 121, 125, 136, 153, 301	100
7	13.90	125, 156, 157, 171, 180, 209	100
8	14.80	109, 110, 111, 142, 145, 152, 185, 213, 230	40
9	16.80	98, 142, 145, 153, 160, 184, 189, 198, 213, 234, 254, 261, 273, 318	30
10	17.95	124, 130, 144, 173, 176, 187, 206	50
11	18.70	138, 166, 173, 238, 240, 255	90

Continued

TABLE 5.4 Monitoring of Selected Ions for 567 Pesticides and Chemical Pollutants Determined by GC-MS—cont'd

No.	Retention Time (min)	Ion (amu)	Dwelling Time (ms)
12	19.20	109, 119, 120, 135, 138, 146, 153, 162, 173, 176, 182, 187, 201, 202, 205, 206, 219, 220, 222, 223, 227, 232, 259, 264, 265, 296	15
13	20.30	109, 121, 127, 135, 153, 163, 182, 195, 201, 202, 206, 249, 256, 268, 279, 286, 306, 354	20
14	20.90	117, 121, 138, 160, 161, 211, 221, 227, 264, 278, 301, 327, 332, 363, 381	20
15	21.95	295, 297, 299	200
16	22.30	104, 107, 123, 135, 136, 144, 151, 172, 211, 363, 207	50
17	23.30	140, 152, 189, 242, 301	100
18	24.00	149, 182, 205, 207, 212, 221, 222, 223, 231, 236, 247, 264, 303, 335, 367	40
19	25.00	91, 136, 137, 144, 173, 180, 185, 187, 191, 196, 200, 204, 215, 231, 232, 238, 256, 286, 290, 294, 296, 412	15
20	26.10	105, 125, 157, 177, 302, 313, 314, 330, 361	50
21	26.90	116, 131, 194, 222, 235, 237, 270, 272, 307, 447, 449	50
22	28.00	107, 123, 138, 151, 183, 192, 351, 367, 369	50
23	28.60	108, 127, 141, 164, 165, 181, 205, 267, 288, 339, 349, 412	40
24	29.20	120, 125, 136, 138, 183, 185, 187, 192, 217, 218, 226, 227, 236, 240, 244, 246, 299, 300, 330, 359, 372	15
25	30.05	122, 140, 148, 160, 204, 206, 232, 238, 266, 268, 288, 290, 320, 376	15
26	31.60	132, 144, 171, 186, 194, 199, 210, 226, 247, 259, 268, 274, 277, 291, 303, 312, 318, 325, 333, 346, 357, 360, 362, 442, 461, 283	15
27	33.00	180, 184, 185, 186, 258, 286, 287, 331, 333	50
28	34.00	100, 119, 188	200
29	37.00	328, 329, 330	200

TABLE 5.4 Monitoring of Selected Ions for 567 Pesticides and Chemical Pollutants Determined by GC-MS—cont'd

No.	Retention Time (min)	Ion (amu)	Dwelling Time (ms)
Group F			
1	5.50	110, 124, 154	180
2	10.50	77, 107, 111, 121, 133, 141, 163, 165, 166, 168, 182, 194, 200, 221, 250	40
3	13.00	94, 103, 104, 115, 116, 136, 144, 147, 159, 175, 183, 199, 213, 234, 270	40
4	15.25	68, 100, 140	170
5	16.65	88, 109, 121, 127, 136, 143, 169, 170, 193, 209, 210, 225, 233, 237, 268	40
6	17.90	86, 92, 94, 101, 105, 115, 116, 121, 123, 138, 141, 154, 163, 166, 169, 170, 172, 174, 186, 200, 211, 238, 240, 251	20
7	19.30	122, 130, 132, 135, 154, 162, 181, 211, 222, 238, 265, 270, 281, 285, 287	35
8	20.30	97, 103, 115, 226, 241, 242, 285, 286, 306, 311, 354, 375	30
9	21.59	43, 109, 115, 142, 147, 163, 185, 212, 213, 224, 227, 240, 255, 262, 266, 294, 295, 297, 351, 353, 355	30
10	22.70	77, 115, 140, 141, 142, 151, 170, 185, 189, 211, 212, 213, 215, 227, 243, 255, 262, 267, 269, 272, 294, 301, 323, 363	30
11	24.00	112, 128, 135, 168, 169, 174, 175, 201, 237, 258, 272, 355, 378, 416	30
12	25.95	119, 134, 158, 173, 186, 194, 212, 217, 219, 225, 264, 300	40
13	27.35	123, 143, 161, 162, 171, 189, 247, 250, 253, 255, 260, 279, 290, 295, 312, 327, 328, 330, 342, 345, 408	40
14	28.30	97, 109, 118, 127, 128, 160, 161, 162, 163, 177, 189, 250, 260, 279, 290, 295, 301, 327, 345	30
15	29.30	88, 121, 125, 145, 153, 191, 199, 217, 218, 250, 254, 258, 272, 281, 289, 300, 317, 370, 371, 387, 417	30

Continued

TABLE 5.4 Monitoring of Selected Ions for 567 Pesticides and Chemical Pollutants Determined by GC-MS—cont'd

No.	Retention Time (min)	Ion (amu)	Dwelling Time (ms)
16	30.80	88, 125, 141, 145, 181, 197, 229, 250, 256, 289, 319, 357	30
17	31.75	109, 125, 237, 263, 274, 297, 303, 318, 476	50
18	33.50	112, 126, 140, 152, 166, 342	90
19	35.00	171, 181, 197, 251, 253, 383	80
20	36.80	165, 301, 344, 404, 387, 388	80

5.1.3 Retention Time, Quantifier and Qualifier Ions, and Collision Energies of 454 Pesticides and Chemical Pollutants Determined by GC-MS-MS

See Table 5.5.

TABLE 5.5 Retention Time, Quantifier and Qualifier Ions, Collision Energies of 454 Pesticides and Chemical Pollutants Determined by GC-MS-MS

No.	Compound	CAS	Retention Time (min)	Qualifier Ion	Quantifier Ion	Collision Energy (V)
Group A						
1	Allidochlor	93-71-0	8.9	138.0/96.0;138.0/110.0	138.0, 96.0	10:10
2	Dichlormid	37764-25-3	9.9	172.0/108.0;172.0/80.0	172.0, 108.0	10:15
3	Etridiazole	2593-15-9	10.3	211.0/183.0;211.0/140.0	211.0, 183.0	10:15
4	Chlormephos	24934-91-6	10.6	234.0/121.0;234.0/154.0	234.0, 121.0	10:10
5	Propham	122-42-9	11.5	179.0/93.0;179.0/137.0	179.0, 93.0	15:10
6	Thiometon	640-15-3	12.1	125.0/63.0;125.0/79.0	125.0, 63.0	25:15:00
7	Cycloate	1134-23-2	13.5	154.0/83.0;154.0/72.0	154.0, 83.0	10:10
8	Diphenylamin	122-39-4	14.6	169.0/168.0;169.0/167.0	169.0, 168.0	15:25
9	Ethalfluralin	55283-68-6	15.2	316.0/202.0;316.0/279.0	316.0, 202.0	25:10:00
10	Quintozene	82-68-8	16.6	295.0/237.0;295.0/265.0	295.0, 237.0	15:10
11	Atrazine-desethyl	6190-65-4	17.0	187.0/172.0;172.0/104.0	187.0, 172.0	5:12

Continued

TABLE 5.5 Retention Time, Quantifier and Qualifier Ions, Collision Energies of 454 Pesticides and Chemical Pollutants Determined by GC-MS-MS—cont'd

No.	Compound	CAS	Retention Time (min)	Qualifier Ion	Quantifier Ion	Collision Energy (V)
12	Clomazone	81777-89-1	17.1	204.0/107.0;204.0/78.0	204.0, 107.0	25:25:00
13	Diazinon	333-41-5	17.2	304.0/179.0;304.0/162.0	304.0, 179.0	8:08
14	Fonofos	944-22-9	17.4	246.0/109.0;246.0/137.0	246.0, 109.0	15:05
15	Etrimfos	38260-54-7	18.0	292.0/181.0;292.0/153.0	292.0, 181.0	5:25
16	Simazine	122-34-9	18.1	201.0/173.0;201.0/110.0	201.0, 173.0	5:25
17	Propetamphos	31218-83-4	18.3	194.0/166.0;194.0/94.0	194.0, 166.0	10:25
18	Secbumeton	26259-45-0	18.5	225.0/169.0;225.0/154.0	225.0, 169.0	5:15
19	Dichlofenthion	97-17-6	18.9	279.0/223.0;279.0/205.0	279.0, 223.0	10:25
20	Mexacarbate	315-18-4	19.0	165.0/134.0;165.0/150.0	165.0, 134.0	10:15
21	Pronamide	23950-58-5	19.0	173.0/145.0;173.0/109.0	173.0, 145.0	15:25
22	Dinitramine	29091-05-2	19.7	305.0/201.0;305.0/230.0	305.0, 201.0	15:15
23	Dimethoate	60-51-5	19.8	125.0/79.0;143.0/111.0	125.0, 79.0	8:12
24	Ronnel	299-84-3	19.8	285.0/270.0;285.0/240.0	285.0, 270.0	15:25
25	Prometryne	7287-19-6	20.3	241.0/199.0;241.0/184.0	241.0, 199.0	5:05
26	Vinclozolin	50471-44-8	20.6	285.0/212.0;285.0/178.0	285.0, 212.0	10:10

27	Beta-HCH	319-85-7	21.6	219.0/183.0;219.0/147.0	219.0, 183.0	10:20
28	Metalaxyl	57837-19-1	20.9	206.0/132.0;206.0/105.0	206.0, 132.0	15:15
29	Chlorpyrifos(ethyl)	2921-88-2	21.0	314.0/286.0;314.0/258.0	314.0, 286.0	5:05
30	Methyl-parathion	298-00-0	21.2	263.0/109.0;263.0/246.0	263.0, 109.0	12:05
31	Delta-HCH	319-86-8	21.6	219.0/183.0;219.0/147.0	219.0, 183.0	10:20
32	Anthraquinone	84-65-1	21.5	208.0/180.0;208.0/152.0	208.0, 180.0	15:15
33	Fenthion	55-38-9	21.7	278.0/109.0;278.0/169.0	278.0, 109.0	15:15
34	Malathion	121-75-5	21.9	173.0/99.0;173.0/127.0	173.0, 99.0	10:05
35	Fenitrothion	122-14-5	22.0	277.0/260.0;277.0/109.0	277.0, 260.0	5:15
36	Paraoxon-ethyl	311-45-5	22.1	275.0/99.0;275.0/149.0	275.0, 99.0	10:05
37	Triadimefon	43121-43-3	22.5	210.0/183.0;210.0/129.0	210.0, 183.0	5:10
38	Pendimethalin	40318-45-4	22.7	252.0/162.0;252.0/161.0	252.0, 162.0	10:25
39	Linuron	330-55-2	22.8	248.0/61.0;160.0/133.0	248.0, 61.0	15:12
40	Chlorbenside	103-17-3	23.0	270.0/125.0;270.0/127.0	270.0, 125.0	10:10
41	Bromophos-ethyl	4824-78-6	23.1	359.0/303.0;359.0/331.0	359.0, 303.0	10:10
42	Quinalphos	13593-03-8	23.3	157.0/102.0;157.0/129.0	157.0, 102.0	25:15:00
43	trans-Chlodane	5103-74-2	23.3	375.0/266.0;375.0/303.0	375.0, 266.0	15:10
44	Metazachlor	67129-08-2	23.6	209.0/132.0;209.0/133.0	209.0, 132.0	15:05
45	Prothiophos	34643-46-4	24.1	309.0/239.0;309.0/221.0	309.0, 239.0	15:25

Continued

TABLE 5.5 Retention Time, Quantifier and Qualifier Ions, Collision Energies of 454 Pesticides and Chemical Pollutants Determined by GC-MS-MS—cont'd

No.	Compound	CAS	Retention Time (min)	Qualifier Ion	Quantifier Ion	Collision Energy (V)
46	Chlorfurenol	2464-37-1	24.4	274.0/215.0;274.0/152.0	274.0,215.0	5:25
47	Procymidone	32809-16-8	24.7	283.0/96.0;283.0/255.0	283.0, 96.0	10:10
48	Methidathion	950-37-8	24.8	145.0/85.0;145.0/58.0	145.0, 85.0	10:25
49	Cyanazine	21725-46-2	25.1	225.0/189.0;225.0/68.0	225.0, 189.0	15:25
50	Napropamide	15299-99-7	25.1	271.0/72.0;271.0/128.0	271.0, 72.0	10:05
51	Oxadiazone	19666-30-9	25.3	258.0/175.0;258.0/112.0	258.0, 175.0	10:25
52	Fenamiphos	22224-92-6	25.6	303.0/195.0;303.0/288.0	303.0, 195.0	10:10
53	Tetrasul	2227-13-6	25.7	324.0/254.0;324.0/252.0	324.0, 254.0	15:15
54	Bupirimate	57839-19-1	26.3	273.0/108.0;273.0/193.0	273.0, 108.0	15:15
55	Carboxin	5234-68-4	26.6	235.0/143.0;235.0/87.0	235.0, 143.0	15:15
56	Flutolanil	66332-96-5	26.8	173.0/145.0;173.0/95.0	173.0, 145.0	10:25
57	p,p'-DDD	72-54-8	26.7	235.0/165.0;235.0/199.0	235.0, 165.0	15:15
58	Ethion	562-12-2	27.0	384.0/129.0;384.0/203.0	384.0, 129.0	25:05:00
59	Etaconazole-1	1000290-09-5	27.2	245.0/173.0;245.0/191.0	245.0, 173.0	10:10
60	Etaconazole-2	1000290-09-5	27.2	245.0/173.0;245.0/191.0	245.0, 173.0	10:10

61	Sulprofos	35400-43-2	27.0	322.0/97.0;322.0/156.0	322.0, 97.0	25:10:00
62	Myclobutanil	88671-89-0	27.8	179.0/125.0;179.0/90.0	179.0, 125.0	15:25
63	Fensulfothin	115-90-2	28.5	292.0/109.0;292.0/165.0	292.0, 109.0	15:15
64	Propiconazole	60207-90-1	28.4	259.0/69.0;259.0/173.0	259.0, 69.0	10:15
65	Bifenthrin	82657-04-3	28.6	181.0/165.0;181.0/166.0	181.0, 165.0	15:25
66	Mirex	2385-85-5	28.7	272.0/237.0;272.0/235.0	272.0, 237.0	10:10
67	Benodanil	15310-01-7	29.0	323.0/231.0;323.0/196.0	323.0, 231.0	10:05
68	Nuarimol	63284-71-9	29.1	314.0/139.0;314.0/111.0	314.0, 139.0	5:25
69	Methoxychlor	72-43-5	29.4	227.0/169.0;227.0/212.0	227.0, 169.0	15:15
70	Oxadixyl	77732-09-3	29.7	163.0/132.0;163.0/117.0	163.0, 132.0	10:25
71	Tetramethrin	7696-12-0	29.8	164.0/77.0;164.0/107.0	164.0, 77.0	25:10:00
72	Tebuconazole	107534-96-3	29.9	250.0/125.0;250.0/153.0	250.0, 125.0	15:10
73	Norflurazon	27314-13-2	30.4	303.0/145.0;303.0/302.0	303.0, 145.0	25:15:00
74	Pyridaphenthion	119-12-0	30.5	340.0/199.0;340.0/109.0	340.0, 199.0	5:15
75	Phosmet	732-11-6	30.8	160.0/77.0;160.0/105.0	160.0, 77.0	25:15:00
76	Tetradifon	116-29-0	30.9	356.0/159.0;356.0/229.0	356.0, 159.0	10:10
77	Pyrazophos	13457-18-6	31.8	221.0/193.0;221.0/149.0	221.0, 193.0	10:15
78	Cypermethrin	52315-07-8	36.1	181.0/152.0;181.0/87.0	181.0, 152.0	25:40:00

Continued

TABLE 5.5 Retention Time, Quantifier and Qualifier Ions, Collision Energies of 454 Pesticides and Chemical Pollutants Determined by GC-MS-MS—cont'd

No.	Compound	CAS	Retention Time (min)	Qualifier Ion	Quantifier Ion	Collision Energy (V)
Group B						
79	EPTC	759-94-4	8.5	132.0/90.0;132.0/62.0	132.0, 90.0	10:15
80	Butylate	2008-41-5	9.4	146.0/90.0;146.0/57.0	146.0, 90.0	5:10
81	Dichlobenil	1194-65-6	9.8	171.0/136.0;171.0/100.0	171.0, 136.0	15:15
82	Pebulate	1114-71-2	10.1	128.0/72.0;161.0/128.0	128.0, 72.0	7:05
83	Nitrapyrin	1929-82-4	10.9	194.0/133.0;194.0/158.0	194.0, 133.0	15:15
84	Chloroneb	2675-77-6	11.8	191.0/113.0;191.0/141.0	191.0, 113.0	10:10
85	Tecnazene	117-18-0	13.5	203.0/83.0;203.0/143.0	203.0, 83.0	10:10
86	Heptanophos	23560-59-0	14.0	124.0/89.0;124.0/63.0	124.0, 89.0	5:25
87	Ethoprophos	13194-48-4	14.5	158.0/97.0;158.0/114.0	158.0, 97.0	12:07
88	Hexachlorobenzene	118-74-1	14.3	284.0/249.0;284.0/214.0	284.0, 249.0	18:25
89	cis-Diallate	2303-16-4	14.7	234.0/150.0;234.0/192.0	234.0, 150.0	15:10
90	trans-Diallate	2303-16-4	14.7	234.0/150.0;234.0/192.0	234.0, 150.0	15:10
91	Propachlor	1918-16-7	15.0	176.0/77.0;176.0/120.0	176.0, 77.0	25:10:00
92	Trifluralin	1582-09-8	15.5	306.0/264.0;306.0/206.0	306.0, 264.0	12:15

Continued

93	Chlorpropham	101-21-3	15.7	213.0/171.0;213.0/127.0	213.0, 171.0	5:15
94	Sulfotep	3689-24-5	15.8	322.0/202.0;322.0/294.0	322.0, 202.0	10:05
95	Sulfallate	95-06-7	15.9	188.0/160.0;188.0/132.0	188.0, 160.0	10:10
96	Alpha-HCH	319-84-6	16.2	219.0/183.0;219.0/147.0	219.0, 183.0	5:15
97	Terbufos	13071-79-9	17.0	231.0/129.0;231.0/175.0	231.0, 129.0	25:15:00
98	Profluralin	26399-36-0	17.6	318.0/199.0;318.0/55.0	318.0, 199.0	10:10
99	Dioxathion	78-34-2	17.8	270.0/197.0;270.0/141.0	270.0, 197.0	5:15
100	Propazine	139-40-2	19.8	214.0/172.0;214.0/105.0	214.0, 172.0	5:10
101	Dicloran	99-30-9	18.2	206.0/176.0;206.0/124.0	206.0, 176.0	15:25
102	Terbuthylazine	5915-41-3	18.5	214.0/71.0;214.0/132.0	214.0, 71.0	15:10
103	Monolinuron	1746-81-2	18.5	126.0/99.0;214.0/61.0	126.0, 99.0	10:10
104	Chlorbufam	1967-16-4	18.7	164.0/111.0;164.0/75.0	164.0, 111.0	15:25
105	Flufenoxuron	101463-69-8	19.0	307.0/126.0;307.0/98.0	307.0, 126.0	25:25:00
106	Cyanophos	2636-26-2	19.2	243.0/109.0;243.0/79.0	243.0, 109.0	15:25
107	Chlorpyrifos-methyl	5598-13-0	19.5	286.0/93.0;286.0/271.0	286.0, 93.0	15:15
108	Desmetryn	1014-69-3	19.8	213.0/171.0;213.0/198.0	213.0, 171.0	5:10
109	Dimethachloro	51218-45-2	20.1	197.0/148.0;197.0/120.0	197.0, 148.0	10:15
110	Alachlor	15972-60-8	20.3	237.0/160.0;237.0/146.0	237.0, 160.0	8:20
111	Pirimiphos-methyl	29232-93-7	20.4	290.0/125.0;290.0/233.0	290.0, 125.0	15:05
112	Thiobencarb	28249-77-6	20.7	257.0/100.0;257.0/72.0	257.0, 100.0	5:25

TABLE 5.5 Retention Time, Quantifier and Qualifier Ions, Collision Energies of 454 Pesticides and Chemical Pollutants Determined by GC-MS-MS—cont'd

No.	Compound	CAS	Retention Time (min)	Qualifier Ion	Quantifier Ion	Collision Energy (V)
113	Terbutyrn	886-50-0	20.8	226.0/68.0;226.0/96.0	226.0, 68.0	25:15:00
114	Dicofol	115-32-2	21.4	250.0/139.0;250.0/215.0	250.0, 139.0	15:10
115	Metolachlor	51218-45-2	21.6	238.0/162.0;238.0/133.0	238.0, 162.0	15:25
116	oxy-Chlordane	27304-13-8	21.5	387.0/263.0;387.0/287.0	387.0, 263.0	10:25
117	Pirimiphos-ethyl	23505-41-1	21.7	333.0/168.0;333.0/180.0	333.0, 168.0	25:15:00
118	Methoprene	40596-69-8	21.7	153.0/111.0;153.0/83.0	153.0, 111.0	5;15
119	Bromofos	2104-96-3	21.8	331.0/286.0;331.0/316.0	331.0, 286.0	25:05:00
120	Dichlofluanid	1085-98-9	22.0	224.0/123.0;224.0/77.0	224.0, 123.0	10:25
121	Ethofumesate	26225-79-6	22.3	207.0/161.0;207.0/137.0	207.0, 161.0	5:15
122	Isopropalin	33820-53-0	22.5	280.0/238.0;280.0/180.0	280.0, 238.0	10:10
123	Propanil	709-98-8	23.3	163.0/90.0;163.0/99.0	163.0, 90.0	25:25:00
124	Endosulfan-1	959-98-8	23.0	241.0/206.0;241.0/170.0	241.0, 206.0	25:25:00
125	Crufomate	299-86-5	23.3	182.0/147.0;256.0/226.0	182.0, 147.0	12:25
126	Isofenphos	25311-71-1	23.3	255.0/121.0;255.0/213.0	255.0, 121.0	25:05:00
127	Chlorfenvinphos	470-90-6	23.5	323.0/267.0;323.0/159.0	323.0, 267.0	15:25

128	Chlorthiamid	1918-13-4	23.0	205.0/170.0;205.0/135.0	205.0, 170.0	5:25
129	cis-Chlordane	5103-71-9	23.6	373.0/301.0;373.0/266.0	373.0, 301.0	12:12
130	Tolylfluanid	731-27-1	24.0	238.0/91.0;238.0/137.0	238.0, 91.0	25:10:00
131	p,p'-DDE	72-55-9	23.9	318.0/248.0;318.0/246.0	318.0, 248.0	25:25:00
132	Butachlor	23184-66-9	23.9	176.0/150.0;176.0/126.0	176.0, 150.0	25:24:00
133	Chlozolinate	84332-86-5	24.3	259.0/188.0;259.0/153.0	259.0, 188.0	10:25
134	Tetrachlorvinphos	22248-79-9	24.6	331.0/109.0;331.0/127.0	331.0, 109.0	25:25:00
135	Chlorbromuron	13360-45-7	24.8	294.0/61.0;292.0/61.0	294.0, 61.0	15:12
136	Profenofos	41198-08-7	24.8	374.0/339.0;374.0/337.0	374.0, 339.0	5:10
137	Fluorochloridone	61213-25-0	25.0	187.0/159.0;187.0/109.0	187.0, 159.0	15:15
138	Buprofenzin	69327-76-0	25.0	105.0/77.0;172.0/116.0	105.0, 77.0	18:07
139	o,p'-DDD	53-19-0	25.1	235.0/165.0;235.0/199.0	235.0,165.0	15:15
140	Endrin	72-20-8	25.1	263.0/191.0;263.0/193.0	263.0, 191.0	20:12
141	Chlorfenson	80-33-1	25.4	302.0/111.0;302.0/175.0	302.0, 111.0	25:10:00
142	Paclobutrazol	76738-62-0	25.7	236.0/125.0;236.0/167.0	236.0, 125.0	15:10
143	Methoprotyne	841-06-5	25.9	256.0/212.0;256.0/170.0	256.0, 212.0	10:15
144	Chlorpropylate	5836-10-2	26.0	251.0/139.0;251.0/111.0	251.0, 139.0	15:25
145	Nitrofen	1836-75-5	26.4	283.0/162.0;283.0/253.0	283.0, 162.0	25:10:00
146	Oxyflurofen	42874-03-3	26.6	361.0/317.0;361.0/300.0	361.0, 317.0	5:10
147	Chlorthiophos	60238-56-4	26.7	360.0/325.0;360.0/297.0	360.0, 325.0	5:10

Continued

TABLE 5.5 Retention Time, Quantifier and Qualifier Ions, Collision Energies of 454 Pesticides and Chemical Pollutants Determined by GC-MS-MS—cont'd

No.	Compound	CAS	Retention Time (min)	Qualifier Ion	Quantifier Ion	Collision Energy (V)
148	Endosulfan-2	33213-65-9	26.9	241.0/206.0;241.0/170.0	241.0, 206.0	25:25:00
149	Flamprop-isopropyl	52756-22-6	27.1	276.0/105.0;276.0/77.0	276.0, 105.0	15:25
150	Flamprop-methyl	52756-25-9	27.1	276.0/105.0;276.0/77.0	276.0, 105.0	10:25
151	o,p′-DDT	789-02-6	26.7	235.0/165.0;235.0/199.0	235.0, 165.0	25:25:00
152	p,p′-DDT	50-29-3	26.7	235.0/165.0;235.0/199.0	235.0, 165.0	25:25:00
153	Carbofenothion	786-19-6	27.4	342.0/199.0;251.0/121.0	342.0, 199.0	25:25:00
154	Benalaxyl	71626-11-4	27.8	148.0/79.0;148.0/105.0	148.0, 79.0	25:15:00
155	Edifenphos	17109-49-8	28.2	173.0/109.0;310.0/201.0	173.0, 109.0	10:05
156	Cyanofenphos	13067-93-1	28.9	157.0/77.0;157.0/110.0	157.0, 77.0	25:15:00
157	Endosulfen sulfate	1031-07-8	29.4	387.0/289.0;387.0/253.0	387.0, 289.0	5:05
158	Bromopropylate	18181-80-1	29.5	341.0/183.0;341.0/185.0	341.0, 183.0	15:15
159	Benzoylprop-ethyl	22212-55-1	29.7	292.0/105.0;292.0/77.0	292.0, 105.0	5:25
160	Fenpropathrin	39515-41-8	29.8	265.0/210.0;265.0/89.0	265.0, 210.0	10:25
161	Leptophos	21609-90-5	30.2	377.0/362.0;377.0/296.0	377.0, 362.0	25:25:00
162	EPN	2104-64-5	30.4	157.0/63.0;157.0/110.0	157.0, 63.0	10:10

163	Hexazinone	51235-04-2	30.6	171.0/71.0;171.0/85.0	171.0, 71.0	15:15
164	Bifenox	42576-02-3	30.9	341.0/310.0;341.0/281.0	341.0, 310.0	5:10
165	Phosalone	2310-17-0	31.5	182.0/111.0;182.0/75.0	182.0, 111.0	15:25
166	Fenarimol	60168-88-9	31.8	330.0/139.0;330.0/251.0	330.0, 139.0	5:05
167	Azinphos-ethyl	2642-71-9	32.3	132.0/77.0;132.0/104.0	132.0, 77.0	15:05
168	Azinphos-methyl	86-50-0	32.3	132.0/77.0;132.0/104.0	132.0, 77.0	15:05
169	Prochloraz	67747-09-5	33.3	180.0/138.0;308.0/70.0	180.0, 138.0	15:15
170	Fluvalinate	102851-06-9	31.7	250.0/208.0;250.0/55.0	250.0,208.0	25:15:00
171	Cyfluthrin	68359-37-5	33.4	206.0/151.0;206.0/177.0	206.0, 151.0	25:25:00
Group C						
172	Propamocarb	24579-73-5	9.7	188.0/58.0;129.0/86.0	188.0, 58.0	2:03
173	Vernolate	1929-77-7	9.8	146.0/76.0;146.0/104.0	146.0, 76.0	10:05
174	3,5-Dichloroaniline	626-43-7	11.4	163.0/90.0;163.0/99.0	163.0, 90.0	15:25
175	Molinate	2212-67-1	12.0	126.0/55.0;126.0/83.0	126.0, 55.0	10:05
176	Methacrifos	62610-77-9	12.1	208.0/180.0;208.0/110.0	208.0, 180.0	5:15
177	2-Phenylphenol	90-43-7	12.6	169.0/141.0;141.0/115.0	169.0, 141.0	15:10
178	cis-1,2,3,6-Tetrahydrophthalimide	84-40-5	8.5	151.0/122.0;151.0/80.0	151.0, 122.0	10:05
179	Benfluralin	1861-40-1	15.6	292.0/264.0;292.0/160.0	292.0,264.0	10:15
180	Hexaflumuron	86479-06-3	16.7	176.0/148.0;176.0/121.0	176.0, 148.0	15:25

Continued

TABLE 5.5 Retention Time, Quantifier and Qualifier Ions, Collision Energies of 454 Pesticides and Chemical Pollutants Determined by GC-MS-MS—cont'd

No.	Compound	CAS	Retention Time (min)	Qualifier Ion	Quantifier Ion	Collision Energy (V)
181	Prometon	1610-18-0	16.9	225.0/183.0;225.0/168.0	225.0, 183.0	5:10
182	Triallate	2303-17-5	17.2	270.0/186.0;270.0/228.0	270.0, 186.0	15:10
183	Pyrimethanil	53112-28-0	17.4	200.0/199.0;183.0/102.0	200.0, 199.0	10:30
184	Gamma-HCH	58-89-9	17.8	219.0/183.0;219.0/147.0	219.0, 183.0	5:15
185	Disulfoton	298-04-4	17.9	88.0/60.0;88.0/59.0	88.0, 60.0	10:25
186	Atrizine	1912-24-9	18.0	200.0/122.0;200.0/94.0	200.0, 122.0	10:15
187	Heptachlor	76-44-8	18.4	272.0/237.0;272.0/235.0	272.0, 237.0	10:10
188	Iprobenfos	26087-47-8	18.8	204.0/91.0;204.0/122.0	204.0, 91.0	5:15
189	Isazofos	42509-80-8	19.0	257.0/119.0;257.0/162.0	257.0, 119.0	25:15:00
190	Plifenate	51366-25-7	19.0	175.0/147.0;175.0/111.0	175.0, 147.0	12:10
191	Transfluthrin	118712-89-3	19.3	163.0/91.0;163.0/143.0	163.0, 91.0	15:15
192	Fenpropimorph	67306-03-0	19.1	128.0/70.0;128.0/110.0	128.0, 70.0	15:15
193	Fluchloralin	33245-39-5	19.5	326.0/63.0;306.0/264.0	326.0, 63.0	15:10
194	Tolclofos-methyl	57018-04-9	19.9	267.0/252.0;267.0/93.0	267.0, 252.0	15:25
195	Ametryn	834-12-8	20.4	227.0/58.0;227.0/170.0	227.0, 58.0	25:10:00

196	Methobromuron	3060-89-7	20.6	258.0/61.0	258.0, 61.0	5
197	Metribuzin	21087-64-9	20.8	198.0/82.0;198.0/110.0	198.0, 82.0	15:15
198	Dimethipin	55290-64-7	21.0	210.0/76.0;210.0/124.0	210.0, 76.0	10:05
199	Dipropetryn	4147-51-7	21.1	255.0/222.0;255.0/138.0	255.0, 222.0	10:25
200	Diethofencarb	87130-20-9	21.8	225.0/96.0;225.0/168.0	225.0, 96.0	25:10:00
201	Dimepiperate	61432-55-1	22.5	119.0/91.0;119.0/65.0	119.0, 91.0	10:25
202	Bitertanol	55179-31-2	32.6	170.0/141.0;170.0/115.0	170.0, 141.0	25:25:00
203	Bioallethrin-1	584-79-2	22.6	123.0/81.0;123.0/69.0	123.0, 81.0	5:25
204	o,p'-DDE	3424-82-6	22.7	318.0/248.0;318.0/246.0	318.0, 248.0	15:15
205	Fenson	80-38-6	23.0	268.0/77.0;268.0/141.0	268.0, 77.0	25:05:00
206	Diphenamid	957-51-7	23.3	167.0/152.0;167.0/165.0	167.0, 152.0	15:15
207	Chlorthion	500-28-7	23.4	299.0/109.0;299.0/79.0	299.0, 109.0	15:25
208	Penconazole	66246-88-6	23.6	248.0/157.0;248.0/192.0	248.0, 157.0	25:15:00
209	Mecarbam	2595-54-2	24.0	296.0/196.0;296.0/168.0	296.0, 196.0	10:25
210	Triadimenol	55219-65-3	24.8	168.0/70.0;128.0/100.0	168.0, 70.0	10:15
211	Tetraconazole	112281-77-3	23.7	336.0/204.0;336.0/156.0	336.0, 204.0	25:25:00
212	Flumetrialin	62924-70-3	24.5	143.0/107.0;143.0/108.0	143.0, 107.0	25:25:00
213	Pretilachlor	51218-49-6	25.0	162.0/147.0;162.0/132.0	162.0, 147.0	10:15
214	Kresoxim-methyl	143390-89-0	25.3	131.0/89.0;131.0/130.0	131.0, 89.0	25:10:00
215	Fluazifop-butyl	79241-46-6	23.9	383.0/282.0;383.0/254.0	383.0, 282.0	10:25

Continued

TABLE 5.5 Retention Time, Quantifier and Qualifier Ions, Collision Energies of 454 Pesticides and Chemical Pollutants Determined by GC-MS-MS—cont'd

No.	Compound	CAS	Retention Time (min)	Qualifier Ion	Quantifier Ion	Collision Energy (V)
216	Chlorfluazuron	71422-67-8	25.7	321.0/304.0;323.0/306.0	321.0, 304.0	25:25:00
217	Chlorobenzilate	510-15-6	26.2	251.0/139.0;251.0/111.0	251.0, 139.0	15:25
218	Flusilazole	85509-19-9	26.7	233.0/165.0;233.0/152.0	233.0, 165.0	15:15
219	Fluorodifen	15457-05-3	27.4	190.0/126.0;190.0/75.0	190.0, 126.0	10:25
220	Diniconazole	83657-24-3	27.5	268.0/232.0;268.0/136.0	268.0, 232.0	10:25
221	Piperonyl butoxide	51-03-6	27.7	176.0/103.0;176.0/131.0	176.0, 103.0	25:15:00
222	Mepronil	55814-41-0	28.4	119.0/91.0;119.0/65.0	119.0, 91.0	10:25
223	Diflufenican	83164-33-4	28.8	266.0/218.0;266.0/246.0	266.0, 218.0	25:10:00
224	Fenazaquin	120928-09-8	29.1	145.0/117.0;145.0/91.0	145.0, 117.0	10:25
225	Phenothrin	26002-80-2	29.4	123.0/81.0;123.0/79.0	123.0, 81.0	10:12
226	Fludioxonil	131341-86-1	29.6	248.0/154.0;248.0/182.0	248.0, 154.0	15:08
227	Fenoxycarb	79127-80-3	29.9	255.0/186.0;255.0/129.0	255.0, 186.0	10:25
228	Amitraz	33089-61-1	30.4	293.0/162.0;293.0/132.0	293.0, 162.0	5:15
229	Anilofos	64249-01-0	31.1	226.0/184.0;226.0/157.0	226.0, 184.0	5:10
230	Acrinathrin	101007-06-1	31.4	289.0/93.0;289.0/77.0	289.0, 93.0	5:25

231	Permethrin	52645-53-1	31.7	183.0/168.0;183.0/153.0	183.0, 168.0	15:15
232	Mefenacet	73250-68-7	31.6	192.0/136.0;192.0/109.0	192.0, 136.0	15:25
233	Lambda-cyhalothrin	91465-08-6	31.6	197.0/141.0;197.0/91.0	197.0, 141.0	10:25
234	Pyridaben	96489-71-3	32.1	147.0/117.0;147.0/132.0	147.0, 117.0	25:15:00
235	Flucythrinate	70124-77-5	32.3	199.0/107.0;199.0/157.0	199.0, 107.0	25:05:00
236	Fluoroglycofen-ethyl	77502-90-7	32.4	428.0/252.0;449.0/347.0	428.0, 252.0	15:05
237	Bioallethrin-2	584-79-2	22.6	123.0/81.0;123.0/79.0	123.0, 81.0	10:25
238	Etofenprox	80844-07-1	32.9	163.0/107.0;163.0/135.0	163.0, 107.0	15:10
239	Alpha-cypermethrin	67375-30-8	33.7	163.0/91.0;163.0/127.0	163.0, 91.0	10:05
240	Cycloxydim	101205-02-1	33.5	178.0/81.0;178.0/108.0	178.0, 81.0	25:10:00
241	Esfenvalerate	66230-04-4	35.0	419.0/167.0;419.0/225.0	419.0,167.0	10:05
242	Difenconazole	119446-68-3	35.8	323.0/265.0;323.0/202.0	323.0, 265.0	15:25
243	Flumioxazin	103361-09-7	36.1	354.0/176.0;354.0/326.0	354.0, 176.0	15:10
244	Flumiclorac-pentyl	87546-18-7	37.1	423.0/318.0;423.0/308.0	423.0, 318.0	10:15
Group D						
245	Dimefox	115-26-4	5.7	154.0/111.0;154.0/121.0	154.0, 111.0	15:15
246	Disulfoton-sulfoxide	2497-07-6	8.4	212.0/97.0;212.0/174.0	212.0, 97.0	15:25
247	Pentachlorobenzen	608-93-5	10.8	250.0/215.0;250.0/177.0	250.0, 215.0	15:25
248	Crimidine	535-89-7	13.4	142.0/106.0;142.0/67.0	142.0, 106.0	10:25
249	BDMC-1	672-99-1	13.5	202.0/121.0;202.0/77.0	202.0, 121.0	5:15

Continued

TABLE 5.5 Retention Time, Quantifier and Qualifier Ions, Collision Energies of 454 Pesticides and Chemical Pollutants Determined by GC-MS-MS—cont'd

No.	Compound	CAS	Retention Time (min)	Qualifier Ion	Quantifier Ion	Collision Energy (V)
250	Chlorfenprop-methyl	14437-17-3	13.7	196.0/165.0;196.0/137.0	196.0, 165.0	10:20
251	Thionazin	297-97-2	14.3	143.0/79.0;143.0/52.0	143.0, 79.0	15:25
252	2,3,5,6-Tetrachloroaniline	3481-20-7	14.2	231.0/158.0;231.0/160.0	231.0, 158.0	25:25:00
253	Tri-n-butyl-phosphate	126-73-8	14.5	211.0/99.0;211.0/155.0	211.0, 99.0	10:05
254	Pentachloroanisole	1825-21-4	15.0	280.0/265.0;280.0/237.0	280.0, 265.0	10:25
255	Tebutam	35256-85-0	15.5	190.0/57.0;190.0/106.0	190.0, 57.0	10:10
256	Methabenzthiazuron	18691-97-9	16.6	164.0/136.0;164.0/108.0	164.0, 136.0	10:25
257	Simeton	673-04-1	16.9	197.0/169.0;197.0/111.0	197.0, 169.0	5:25
258	Atratone	1610-17-9	16.9	211.0/169.0;211.0/196.0	211.0, 169.0	5:10
259	Desisopropyl-atrazine	1007-28-9	17.1	173.0/145.0;173.0/158.0	173.0,145.0	10:10
260	Tefluthrin	79538-32-2	17.4	177.0/127.0;177.0/161.0	177.0, 127.0	13:25
261	Bromocylen	1715-40-8	17.3	359.0/243.0;359.0/242.0	359.0, 243.0	15:15
262	Trietazine	1912-26-1	17.8	229.0/200.0;229.0/186.0	229.0, 200.0	5:05
263	DE-PCB 28	2012-37-5	18.1	256.0/186.0;256.0/151.0	256.0, 186.0	25:25:00

No.	Name	CAS	RT	Ions	Quant	Ratio
264	DE-PCB 31	16606-02-3	18.1	256.0/186.0;258.0/186.0	256.0, 186.0	25:15:00
265	Cycluron	2163-69-1	18.3	198.0/89.0;198.0/72.0	198.0, 89.0	5:15
266	2,6-Dichlorodenzamide	2008-58-4	18.5	173.0/109.0;173.0/145.0	173.0, 109.0	25:15:00
267	Desethyl-sebuthylazine	37019-18-4	18.7	172.0/94.0;172.0/169.0	172.0, 94.0	25:15:00
268	2,3,4,5-Tetrachloroaniline	634-83-3	18.7	231.0/158.0;231.0/160.0	231.0, 158.0	20:15
269	Musk ambrette	83-46-9	18.9	253.0/106.0;253.0/91.0	253.0, 106.0	10:15
270	Pentachloroaniline	527-20-8	19.0	263.0/192.0;263.0/156.0	263.0, 192.0	15:25
271	Aziprotryne	4658-28-0	19.4	199.0/184.0;199.0/157.0	199.0, 184.0	10:10
272	DE-PCB 52	35693-99-3	19.4	292.0/257.0;292.0/222.0	292.0, 257.0	10:15
273	Sebutylazine	7286-69-3	19.6	200.0/104.0;200.0/122.0	200.0, 104.0	15:10
274	Isocarbamid	30979-48-7	19.8	142.0/70.0;142.0/113.0	142.0, 70.0	10:15
275	Prosulfocarb	52888-80-9	19.6	251.0/128.0;251.0/86.0	251.0, 128.0	5:10
276	Dimethenamid	87674-68-8	19.9	230.0/154.0;230.0/111.0	230.0, 154.0	8:25
277	Monalide	7287-36-7	20.5	239.0/197.0;239.0/85.0	239.0, 197.0	5:15
278	Octachlorostyrene	29082-74-4	20.3	380.0/310.0;380.0/307.0	380.0,310.0	25:25:00
279	Paraoxon-methyl	950-35-6	20.5	230.0/193.0;230.0/195.0	230.0, 193.0	10:10
280	Isobenzan	297-78-9	20.5	311.0/275.0;311.0/240.0	311.0, 275.0	10:15
281	Isodrin	465-73-6	20.9	193.0/123.0;193.0/157.0	193.0, 123.0	25:15:00

Continued

TABLE 5.5 Retention Time, Quantifier and Qualifier Ions, Collision Energies of 454 Pesticides and Chemical Pollutants Determined by GC-MS-MS—cont'd

No.	Compound	CAS	Retention Time (min)	Qualifier Ion	Quantifier Ion	Collision Energy (V)
282	Trichloronat	327-98-0	21.2	297.0/269.0;297.0/223.0	297.0, 269.0	15:25
283	Isomethiozin	57052-04-7	21.3	198.0/82.0;198.0/110.0	198.0, 82.0	10:10
284	Dacthal	1861-32-1	21.4	301.0/223.0;301.0/273.0	301.0, 223.0	25:15:00
285	4,4-Dichlorobenzophenone	90-98-2	21.4	250.0/139.0;250.0/215.0	250.0, 139.0	10:05
286	Rabenzazole	69899-24-7	21.9	212.0/195.0;212.0/188.0	212.0, 195.0	15:25
287	Nitrothal-isopropyl	10552-74-6	22.0	254.0/212.0;254.0/165.0	254.0, 212.0	10:25
288	Cyprodinil	121552-61-2	22.0	224.0/208.0;224.0/222.0	224.0, 208.0	15:15
289	Fuberidazole	3878-19-1	22.6	184.0/156.0;184.0/129.0	184.0, 156.0	10:15
290	DE-PCB 101	37680-73-2	22.5	326.0/256.0;326.0/254.0	326.0, 256.0	25:25:00
291	Methfuroxam	2873-17-8	22.8	229.0/137.0;229.0/67.0	229.0, 137.0	5:25
292	Dicapthon	2463-84-5	22.9	262.0/216.0;262.0/123.0	262.0, 216.0	15:25
293	MCPA-butoxyethyl ester	19480-43-4	22.8	300.0/182.0;300.0/200.0	300.0, 182.0	5:10
294	Isocarbophos	245-61-5	23.4	136.0/108.0;136.0/69.0	136.0, 108.0	10:25
295	trans-Nonachlor	39765-80-5	23.6	409.0/300.0;409.0/302.0	409.0, 300.0	25:25:00

No.	Compound	CAS	RT	Ions	Confirming ions	Time
296	DEF	78-48-8	24.3	202.0/113.0;202.0/147.0	202.0, 113.0	15:05
297	Flurochloridone	61213-25-0	25.0	311.0/174.0;311.0/311/103.0	311.0,174.0	10:10
298	Bromfenvinfos	33399-00-7	25.0	323.0/267.0;323.0/159.0	323.0, 267.0	15:25
299	Perthane	72-56-0	24.9	223.0/179.0;223.0/193.0	223.0, 179.0	25:25:00
300	Ditalimfos	5131-24-8	25.2	130.0/102.0;130.0/75.0	130.0, 102.0	10:25
301	DE-PCB 118	31508-00-6	25.0	326.0/256.0;326.0/254.0	326.0, 256.0	25:25:00
302	4,4-Dibromobenzophenone	3988-03-2	25.4	340.0/183.0;340.0/185.0	340.0, 183.0	15:15
303	Flutriafol	76674-21-0	25.8	219.0/123.0;219.0/95.0	219.0, 123.0	15:25
304	Mephosfolan	950-10-7	26.0	196.0/140.0;196.0/168.0	196.0, 140.0	10:05
305	DE-PCB 153	35065-27-1	25.6	360.0/290.0;360.0/325.0	360.0, 290.0	20:15
306	Diclobutrazole	75736-33-3	26.4	272.0/161.0;272.0/102.0	272.0, 161.0	10:25
307	Disulfoton sulfone	2497-06-5	27.0	213.0/153.0;213.0/125.0	213.0, 153.0	5:10
308	DE-PCB 138	35065-28-2	26.8	360.0/325.0;360.0/290.0	360.0, 325.0	15:15
309	Resmethrin-1	10453-86-8	27.5	171.0/143.0;171.0/128.0	171.0, 143.0	5:10
310	Resmethrin-2	10453-86-8	27.5	171.0/143.0;171.0/128.0	171.0, 143.0	5:10
311	Cyproconazole	113096-99-4	27.8	222.0/125.0;222.0/82.0	222.0, 125.0	15:10
312	Phthalic acid, benzyl butyl ester	85-68-7	27.9	206.0/149.0;206.0/93.0	206.0, 149.0	5:25
313	Clodinafop-propargyl	105512-06-9	28.1	349.0/266.0;349.0/238.0	349.0, 266.0	10:15

Continued

TABLE 5.5 Retention Time, Quantifier and Qualifier Ions, Collision Energies of 454 Pesticides and Chemical Pollutants Determined by GC-MS-MS—cont'd

No.	Compound	CAS	Retention Time (min)	Qualifier Ion	Quantifier Ion	Collision Energy (V)
314	Fluotrimazole	31251-03-3	28.6	379.0/276.0;379.0/262.0	379.0, 276.0	10:15
315	Fluroxypr-1-methylheptyl ester	81406-37-3	28.8	366.0/181.0;366.0/209.0	366.0, 181.0	15:15
316	Triphenyl phosphate	115-86-6	28.9	326.0/170.0;326.0/215.0	326.0, 170.0	15:15
317	Metamitron	41394-05-2	29.2	202.0/174.0;202.0/104.0	202.0, 174.0	5:25
318	DE-PCB 180	35065-29-3	29.0	394.0/324.0;394.0/359.0	394.0, 324.0	25:15:00
319	Tebufenpyrad	119168-77-3	29.2	333.0/171.0;333.0/276.0	333.0, 171.0	15:05
320	Cloquintocet-mexyl	99607-70-2	29.5	192.0/190.0;192.0/162.0	192.0, 190.0	15:25
321	Lenacil	2164-08-1	31.5	153.0/110.0;153.0/136.0	153.0, 110.0	15:15
322	Bromuconazole-1	116255-48-2	31.0	173.0/109.0;173.0/145.0	173.0, 109.0	25:15:00
323	Bromuconazole-2	116255-48-2	31.0	173.0/109.0;173.0/145.0	173.0, 109.0	25:15:00
324	Nitralin	4726-14-1	31.5	274.0/169.0;274.0/216.0	274.0, 169.0	10:08
325	Fenamiphos sulfone	31972-44-8	31.9	320.0/292.0;320.0/79.0	320.0, 292.0	10:25
326	Fenpiclonil	74738-17-3	32.9	236.0/174.0;236.0/201.0	236.0, 174.0	25:15:00
327	Fluquinconazole	607-68-1	32.8	340.0/298.0;340.0/286.0	340.0, 298.0	15:25
328	Fenbuconazole	114369-43-6	34.5	198.0/129.0;198.0/102.0	198.0, 129.0	15:25

Group E

329	XMC	2655-14-3	5.5	107.0/79.0;107.0/77.0	107.0, 79.0	10:15
330	Dibutyl succinate	141-03-7	12.2	101.0/73.0;101.0/100.0	101.0, 73.0	10:25
331	Propoxur-1	114-26-1	6.5	110.0/63.0;110.0/64.0	110.0, 63.0	25:15:00
332	Isoprocarb-1	2631-40-7	7.5	121.0/77.0;121.0/103.0	121.0, 77.0	15:10
333	Terbucarb-1	1918-11-2	10.6	220.0/205.0;220.0/145.0	220.0, 205.0	10:25
334	Chlorethoxyfos	54593-83-8	13.4	301.0/97.0;301.0/153.0	301.0, 97.0	15:05
335	Isoprocarb-2	2631-40-5	13.7	121.0/77.0;121.0/103.0	121.0, 77.0	15:10
336	Tebuthiuron	34014-18-1	14.2	156.0/74;156.0/89	156.0, 74	15:10
337	Pencycuron	66063-05-6	14.3	209.0/180.0;209.0/125.0	209.0, 180.0	5:25
338	Demeton-s-methyl	919-86-8	15.2	142.0/79.0;142.0/112.0	142.0, 79.0	10:05
339	Propoxur-2	114-26-1	15.5	110.0/63.0;110.0/64.0	110.0, 63.0	25:15:00
340	Phenanthrene	65-01-8	17.0	189.0/161.0;189.0/185.0	189.0, 161.0	25:25:00
341	Fenpyroximate	134098-61-6	17.5	213.0/77.0;213.0/212.0	213.0, 77.0	25:10:00
342	Tebupirimfos	96182-53-5	17.5	318.0/152.0;318.0/276.0	318.0, 152.0	10:05
343	Fenpropidin	67306-00-7	17.7	98.0/70.0;98.0/69.0	98.0, 70.0	10:15
344	Fluometuron	2164-17-2	18.2	232.0/72.0;232.0/187.0	232.0, 72.0	10:05
345	Dichloran	99-30-9	18.2	206.0/176.0;206.0/123.0	206.0, 176.0	10:25
346	Pyroquilon	57369-32-1	18.4	173.0/130.0;173.0/144.0	173.0, 130.0	15:15

Continued

TABLE 5.5 Retention Time, Quantifier and Qualifier Ions, Collision Energies of 454 Pesticides and Chemical Pollutants Determined by GC-MS-MS—cont'd

No.	Compound	CAS	Retention Time (min)	Qualifier Ion	Quantifier Ion	Collision Energy (V)
347	Phosphamidon-1	13171-21-6	19.7	264.0/127.0;264.0/193.0	264.0,127.0	15:05
348	Benoxacor	98730-04-2	19.7	259.0/120.0;259.0/176.0	259.0, 120.0	25:10:00
349	Acetochlor	34256-82-1	19.8	146.0/131.0;146.0/118.0	146.0, 131.0	10:10
350	Tridiphane	58138-08-2	19.9	187.0/159.0;187.0/123.0	187.0, 159.0	10:25
351	Propyzamide	23950-58-5	19.0	173.0/145.0;173.0/109.0	173.0, 145.0	15:25
352	Terbucarb-2	1918-11-2	20.0	220.0/205.0;220.0/145.0	220.0, 205.0	10:25
353	Fenfuram	24691-80-3	20.4	201.0/109.0;201.0/184.0	201.0, 109.0	10:05
354	Acibenzolar-S-methyl	135158-54-2	20.5	182.0/153.0;182.0/107.0	182.0, 153.0	25:25:00
355	Benfuresate	68505-69-1	20.7	163.0/91.0;163.0/107.0	163.0, 91.0	25:05:00
356	Mefenoxam	70630-17-0	20.9	206.0/162.0;206.0/132.0	206.0, 162.0	5:10
357	Phosphamidon-2	13171-21-6	21.4	264.0/127.0;264.0/193.0	264.0, 127.0	15:05
358	Simeconazole	149508-90-7	21.4	121.0/101.0;121.0/75.0	121.0, 101.0	10:25
359	Chlorthal-dimethyl	1861-32-1	21.4	301.0/223.0;301.0/273.0	301.0, 223.0	25:15:00
360	Terbacil	5902-51-2	21.7	161.0/88.0;161.0/144.0	161.0, 88.0	25:15:00
361	Thiazopyr	117718-60-2	21.9	363.0/300.0;363.0/272.0	363.0, 300.0	15:25

No.	Name	CAS	RT	Transitions	Ions	Ratio
362	Dimethylvinphos	2274-67-1	22.2	295.0/109.0;295.0/127.0	295.0, 109.0	15:10
363	Zoxamide	156052-68-5	22.2	242.0/214.0;242.0/187.0	242.0, 214.0	10:15
364	Allethrin	584-79-2	23.1	123.0/81.0;123.0/79.0	123.0, 81.0	5:15
365	Quinoclamine	2797-51-5	23.0	172.0/89.0;172.0/128.0	172.0, 89.0	25:10:00
366	Fenoxanil	115852-48-7	23.6	140.0/85.0;140.0/71.0	140.0, 85.0	10:25
367	Furalaxyl	57646-30-7	24.0	242.0/95.0;301.0/224.0	242.0, 95.0	10:15
368	Thiamethoxam	153719-23-4	24.5	247.0/182.0;247.0/212.0	247.0, 182.0	10:05
369	Thiabendazole	148-79-8	24.7	201.0/174.0;201.0/130.0	201.0, 174.0	10:25
370	Bromacil	314-40-9	24.8	205.0/188.0;205.0/162.0	205.0, 188.0	10:15
371	Picoxystrobin	117428-22-5	24.9	335.0/173.0;335.0/303.0	335.0, 173.0	10:10
372	Methiocarb sulfone	2178-25-1	25.6	185.0/77.0;185.0/121.0	185.0,77.0	25:10:00
373	Butamifos	36335-67-8	25.4	286.0/202.0;286.0/185.0	286.0,202.0	15:10
374	TCMTB	21564-17-0	25.7	180.0/136.0;180.0/109.0	180.0,136.0	15:25
375	Imazalil		25.7	215.0/173.0;215.0/145.0	215.0,173.0	5:25
376	Cyflufenamid	180409-60-3	26.0	412.0/295.0;412.0/118.0	412.0, 295.0	5:15
377	Isoxathion	18854-01-8	26.5	313.0/177.0;313.0/130.0	313.0, 177.0	5:15
378	Quinoxyphen	124495-18-7	27.1	273.0/208.0;273.0/182.0	273.0, 208.0	25:25:00
379	Imibenconazole-des-benzyl	199338-48-2	27.1	272.0/237.0;272.0/235.0	272.0, 237.0	5:05
380	Trifloxystrobin	141517-21-7	27.7	222.0/162.0;222.0/190.0	222.0, 162.0	5:10

Continued

TABLE 5.5 Retention Time, Quantifier and Qualifier Ions, Collision Energies of 454 Pesticides and Chemical Pollutants Determined by GC-MS-MS—cont'd

No.	Compound	CAS	Retention Time (min)	Qualifier Ion	Quantifier Ion	Collision Energy (V)
381	Fipronil	120068-37-3	28.4	367.0/213.0;367.0/255.0	367.0, 213.0	25:15:00
382	Pyraflufen ethyl	129630-17-7	28.9	412.0/349.0;412.0/307.0	412.0, 349.0	15:25
383	Thenylchlor	96491-05-3	29.1	288.0/141.0;288.0/174.0	288.0, 141.0	10:05
384	Clethobim	136191-64-5	29.2	205.0/176.0;205.0/148.0	205.0, 176.0	15:25
385	Mefenpyr-diethyl	135590-91-9	29.5	299.0/253.0;299.0/190.0	299.0, 253.0	5:25
386	Chrysene	218-01-9	29.3	228.0/226.0;229.0/227.0	228.0, 226.0	25:25:00
387	Epoxiconazole-1	106325-08-0	29.7	192.0/138.0;192.0/111.0	192.0, 138.0	10:25
388	Epoxiconazole-2	106325-08-0	29.7	192.0/138.0;192.0/111.0	192.0, 138.0	10:25
389	Pyriproxyfen	95737-68-1	30.0	136.0/78.0;136.0/96.0	136.0, 78.0	25:15:00
390	Piperophos	24151-93-7	30.4	321.0/122.0;321.0/123.0	321.0, 122.0	15:05
391	Fenamidone	161326-34-7	30.7	268.0/180.0;268.0/77.0	268.0, 180.0	15:25
392	Pyraclostrobin	175013-18-0	31.9	132.0/77.0;132.0/104.0	132.0, 77.0	15:10
393	Tralkoxydim	87820-88-0	32.1	283.0/227.0;283.0/137.0	283.0, 227.0	10:25
394	Pyraclofos	77458-01-6	32.1	360.0/97.0;360.0/194.0	360.0, 97.0	25:15:00
395	Dialifos	10311-84-9	32.1	210.0/148.0;210.0/151.0	210.0, 148.0	15:15

396	Spirodiclofen	148477-71-8	32.4	312.0/259.0;312.0/294.0	312.0, 259.0	10:05
397	Flurtamone	96525-23-4	32.8	333.0/120.0;199.0/157.0	333.0, 120.0	10:15
398	Silafluofen	105024-66-6	33.1	287.0/259.0;287.0/179.0	287.0, 259.0	5:25
399	Pyrimidifen	105779-78-0	33.5	184.0/169.0;184.0/141.0	184.0, 169.0	15:25
400	Butafenacil	134605-64-4	33.8	331.0/180.0;331.0/152.0	331.0, 180.0	12:25
401	Acetamiprid	160430-64-8	33.9	152.0/116.0;166.0/139.0	152.0, 116.0	20:08
402	Fluridone	59756-60-4	37.6	328.0/259.0;328.0/189.0	328.0, 259.0	30:38:00
Group F						
403	Tribenron-methyl	106040-48-6	9.4	154.0/124.0;124.0/83.0	154.0, 124.0	20:20
404	Dioxacarb	6988-21-2	11.0	166.0/165.0;166.0/121.0	166.0, 165.0	20:20
405	Ethiofencarb	29973-13-5	11.0	168.0/107.0;168.0/77.0	168.0, 107.0	20:20
406	Dimethyl phthalate	131-11-3	11.5	163.0/77.0;163.0/133.0	163.0, 77.0	20:20
407	4-Chlorophenoxy acetic acid	122-88-3	11.9	200.0/141.0;200.0/111.0	200.0, 141.0	20:20
408	Phthalimide	85-41-6	13.3	147.0/103.0;147.0/76.0	147.0, 103.0	20:20
409	Carbaryl	63-25-2	14.5	144.0/115.0;144.0/116.0	144.0, 115.0	20:20
410	2,4-D	94-75-7	14.9	234.0/199.0;234.0/73.0	234.0, 199.0	20:20
411	Cadusafos	95465-99-9	15.1	159.0/97.0;159.0/131.0	159.0, 97.0	20:20
412	Demetom-s	126-75-0	17.0	170.0/114.0;170.0/97.0	170.0, 114.0	20:20
413	Dicrotophos	141-66-2	17.5	127.0/109.0;127.0/95.0	127.0, 109.0	20:20

Continued

TABLE 5.5 Retention Time, Quantifier and Qualifier Ions, Collision Energies of 454 Pesticides and Chemical Pollutants Determined by GC-MS-MS—cont'd

No.	Compound	CAS	Retention Time (min)	Qualifier Ion	Quantifier Ion	Collision Energy (V)
414	3,4,5-Trimethacarb	2686-99-9	17.7	193.0/136.0;193.0/121.0	193.0, 136.0	20:20
415	2,4,5-T	93-76-5	17.8	233.0/190.0;233.0/159.0	233.0, 190.0	20:20
416	DMSA	1596-84-5	18.7	200.0/108.0;200.0/92.0	200.0, 108.0	20:20
417	Pirimicarb	23103-98-2	19.1	238.0/166.0;238.0/96.0	238.0, 166.0	10:10
418	Dodemorph-1	1593-77-7	19.4	154.0/82.0;154.0/96.0	154.0, 82.0	10:10
419	Dodemorph-2	1593-77-7	19.4	154.0/82.0;154.0/96.0	154.0, 82.0	10:10
420	Desmedipham	13684-56-5	19.9	181.0/109.0;181.0/80.0	181.0, 109.0	10:10
421	Fenchlorphos	3983-45-7	19.8	287.0/272.0;287.0/242.0	287.0, 272.0	10:10
422	S421 (octachlorodipropyl ester)-1	127-90-2	19.5	132.0/97.0;132.0/60.0	132.0,97.0	10:10
423	S421 (octachlorodipropyl ester)-2	127-90-2	19.5	132.0/97.0;132.0/60.0	132.0,97.0	10:10
424	Esprocarb	85785-20-2	20.0	222.0/91.0;222.0/162.0	222.0, 91.0	10:10
425	Telodrin	297-78-9	20.5	311.0/240.0;311.0/241.0	311.0, 240.0	10:10

426	Difenoxuron	14214-32-5	20.9	241.0/226.0;241.0/170.0	241.0, 226.0	10:10
427	Butralin	33629-47-9	22.2	266.0/190.0;266.0/174.0	266.0, 190.0	12:12
428	Pyrifenox-1	88283-41-4	23.5	262.0/200.0;262.0/192.0	262.0, 200.0	12:12
429	Dimethametryn	22936-75-0	22.8	212.0/122.0;212.0/94.0	212.0, 122.0	12:12
430	Flufenacet	142459-58-3	23.1	211.0/96.0;211.0/123.0	211.0, 96.0	12:12
431	Aclonifen	74070-46-5	22.1	264.0/194.0;264.0/194.0	264.0, 194.0	12:12
432	Pyrifenox-2	88283-41-4	23.5	262.0/200.0;262.0/192.0	262.0, 200.0	12:12
433	Flubenzimine	37893-02-0	25.8	416.0/186.0;416.0/212.0	416.0,186.0	12:12
434	Azaconazole	60207-31-0	26.6	217.0/173.0;217.0/145.0	217.0, 173.0	14:14
435	Iprovalicarb-1	140923-17-7	26.7	134.0/93.0;134.0/91.0	134.0, 93.0	14:14
436	Irpovalicarb-2	140923-17-7	26.7	134.0/93.0;134.0/91.0	134.0, 93.0	14:14
437	Diofenolan-1	63837-33-2	26.8	186.0/109.0;186.0/158.0	186.0, 109.0	14:14
438	Diofenolan-2	63837-33-2	26.8	186.0/109.0;186.0/158.0	186.0, 109.0	14:14
439	Chlorfenapyr	122453-73-0	27.6	408.0/59.0;408.0/363.0	408.0, 59.0	14:14
440	Bioresmethrin	28434-01-7	27.5	171.0/143.0;171.0/128.0	171.0, 143.0	14:14
441	Carfentrazone-ethyl	128621-72-7	28.3	330.0/310.0;330.0/241.0	330.0, 310.0	18:18
442	Diclofop-methyl	51338-27-3	28.2	342.0/255.0;342.0/184.0	342.0, 255.0	18:18
443	Endrin aldehyde	7421-93-4	28.4	345.0/281.0;345.0/317.0	345.0, 281.0	18:18
444	Halosulfuran-methyl	100784-20-1	28.4	327.0/295.0;327.0/260.0	327.0, 295.0	18:18
445	Carbosulfan	55285-14-8	28.4	160.0/104.0;160.0/62.0	160.0, 104.0	18:18

Continued

TABLE 5.5 Retention Time, Quantifier and Qualifier Ions, Collision Energies of 454 Pesticides and Chemical Pollutants Determined by GC-MS-MS—cont'd

No.	Compound	CAS	Retention Time (min)	Qualifier Ion	Quantifier Ion	Collision Energy (V)
446	Fenhexamid	126833-17-8	28.4	177.0/113.0;177.0/78.0	177.0,113.0	18:18
447	Spiromesifen	283594-90-1	29.7	272.0/254.0;272.0/209.0	272.0, 254.0	25:25:00
448	Famphur	52-85-7	29.8	218.0/109.0;218.0/79.0	218.0, 109.0	25:25:00
449	Endrin ketone	53494-70-5	30.5	317.0/245.0;317.0/209.0	317.0, 245.0	25:25:00
450	Metconazole	125116-23-6	31.2	125.0/89.0;125.0/99.0	125.0, 89.0	18:18
451	Cyhalofop-butyl	122008-85-9	31.5	357.0/256.0;357.0/229.0	357.0, 256.0	18:18
452	Boscalid	188425-85-6	34.3	342.0/140.0;342.0/112.0	342.0, 140.0	18:18
453	Tolfenpyrad	129558-76-5	35.9	383.0/171.0;383.0/145.0	383.0,171.0	18:18
454	Dimethomorph	211867-47-9	37.6	301.0/165.0;301.0/139.0	301.0, 165.0	100:100

5.1.4   Monitoring of Selected Ions for 454 Pesticides and Chemical Pollutants Determined by GC-MS-MS

See Table 5.6.

TABLE 5.6 Monitoring of Selected Ions for 454 Pesticides and Chemical Pollutants Determined by GC-MS-MS

No.	Time (min)	Ion (amu)	Dwell Time (ms)
Group A			
1	6.0	138.0/96.0, 138.0/110.0, 172.0/108.0, 172.0/80.0, 211.0/183.0, 211.0/140.0, 234.0/121.0, 234.0/154.0, 179.0/93.0, 179.0/137.0, 125.0/63.0, 125.0/79.0, 154.0/83.0, 154.0/72.0, 169.0/168.0, 169.0/167.0, 316.0/202.0, 316.0/279.0, 295.0/237.0, 295.0/265.0	10
2	16.9	187.0/172.0, 172.0/104.0, 204.0/107.0, 204.0/78.0, 304.0/179.0, 304.0/162.0, 246.0/109.0, 246.0/137.0, 292.0/181.0, 292.0/153.0, 201.0/173.0, 201.0/110.0, 194.0/166.0, 194.0/94.0, 225.0/169.0, 225.0/154.0, 279.0/223.0, 279.0/205.0, 165.0/134.0, 165.0/150.0, 173.0/145.0, 173.0/109.0, 305.0/201.0, 305.0/230.0, 285.0/270.0, 285.0/240.0, 125.0/79.0, 143.0/111.0	7
3	20.0	241.0/199.0, 241.0/184.0, 285.0/212.0, 285.0/178.0, 206.0/132.0, 206.0/105.0, 314.0/286.0, 314.0/258.0, 263.0/109.0, 263.0/246.0	18
4	21.32	208.0/180.0, 208.0/152.0, 219.0/183.0, 219.0/147.0, 219.0/183.0, 219.0/147.0, 278.0/109.0, 278.0/169.0, 173.0/99.0, 173.0/127.0, 277.0/260.0, 277.0/109.0, 275.0/99.0, 275.0/149.0, 210.0/183.0, 210.0/129.0, 252.0/162.0, 252.0/161.0, 248.0/61.0, 160.0/133.0, 270.0/125.0, 270.0/127.0, 359.0/303.0, 359.0/331.0, 157.0/102.0, 157.0/129.0, 375.0/266.0, 375.0/303.0, 209.0/132.0, 209.0/133.0, 353/282, 353/263	6
5	23.8	309.0/239.0, 309.0/221.0, 274.0/215.0, 274.0/152.0, 283.0/96.0, 283.0/255.0, 145.0/85.0, 145.0/58.0, 225.0/189.0, 225.0/68.0, 271.0/72.0, 271.0/128.0, 258.0/175.0, 258.0/112.0, 303.0/195.0, 303.0/288.0, 324.0/254.0, 324.0/252.0	10
6	26.0	273.0/108.0, 273.0/193.0, 235.0/143.0, 235.0/87.0, 235.0/165.0, 235.0/199.0, 173.0/145.0, 173.0/95.0, 384.0/129.0, 384.0/203.0, 322.0/97.0, 322.0/156.0, 245.0/173.0, 245.0/191.0, 245.0/173.0, 245.0/191.0, 179.0/125.0, 179.0/90.0	10

Continued

TABLE 5.6 Monitoring of Selected Ions for 454 Pesticides and Chemical Pollutants Determined by GC-MS-MS—cont'd

No.	Time (min)	Ion (amu)	Dwell Time (ms)
7	27.9	259.0/69.0, 259.0/173.0, 292.0/109.0, 292.0/165.0, 181.0/165.0, 181.0/166.0, 272.0/237.0, 272.0/235.0, 323.0/231.0, 323.0/196.0, 314.0/139.0, 314.0/111.0, 227.0/169.0, 227.0/212.0, 163.0/132.0, 163.0/117.0, 164.0/77.0, 164.0/107.0, 250.0/125.0, 250.0/153.0	9
8	30.0	303.0/145.0, 303.0/302.0, 340.0/199.0, 340.0/109.0, 160.0/77.0, 160.0/105.0, 356.0/159.0, 356.0/229.0, 221.0/193.0, 221.0/149.0, 181.0/152.0, 181.0/87.0	15
Group B			
1	6.0	132.0/90.0, 132.0/62.0, 146.0/90.0, 146.0/57.0, 171.0/136.0, 171.0/100.0, 128.0/72.0, 161.0/128.0, 194.0/133.0, 194.0/158.0, 191.0/113.0, 191.0/141.0	12
2	12.4	203.0/83.0, 203.0/143.0, 124.0/89.0, 124.0/63.0, 284.0/249.0, 284.0/214.0, 158.0/97.0, 158.0/114.0, 234.0/150.0, 234.0/192.0, 234.0/150.0, 234.0/192.0, 176.0/77.0, 176.0/120.0, 306.0/264.0, 306.0/206.0, 213.0/171.0, 213.0/127.0, 322.0/202.0, 322.0/294.0, 188.0/160.0, 188.0/132.0, 219.0/183.0, 219.0/147.0, 231.0/129.0, 231.0/175.0	9
3	17.3	318.0/199.0, 318.0/55.0, 270.0/197.0, 270.0/141.0, 206.0/176.0, 206.0/124.0, 214.0/71.0, 214.0/132.0, 126.0/99.0, 214.0/61.0, 164.0/111.0, 164.0/75.0, 307.0/126.0, 307.0/98.0, 243.0/109.0, 243.0/79.0, 286.0/93.0, 286.0/271.0, 214.0/172.0, 214.0/105.0, 213.0/171.0, 213.0/198.0, 197.0/148.0, 197.0/120.0, 237.0/160.0, 237.0/146.0, 290.0/125.0, 290.0/233.0, 257.0/100.0, 257.0/72.0, 226.0/68.0, 226.0/96.0	6
4	21.0	250.0/139.0, 250.0/215.0, 387.0/263.0, 387.0/287.0, 238.0/162.0, 238.0/133.0, 153.0/111.0, 153.0/83.0, 333.0/168.0, 333.0/180.0, 331.0/286.0, 331.0/316.0, 224.0/123.0, 224.0/77.0, 207.0/161.0, 207.0/137.0, 280.0/238.0, 280.0/180.0	10
5	22.7	241.0/206.0, 241.0/170.0, 205.0/170.0, 205.0/135.0, 163.0/90.0, 163.0/99.0, 255.0/121.0, 255.0/213.0, 182.0/147.0, 256.0/226.0, 323.0/267.0, 323.0/159.0, 373.0/301.0, 373.0/266.0, 176.0/150.0, 176.0/126.0, 318.0/248.0, 318.0/246.0, 238.0/91.0, 238.0/137.0, 259.0/188.0, 259.0/153.0, 331.0/109.0, 331.0/127.0, 353/282, 353/263	10

TABLE 5.6 Monitoring of Selected Ions for 454 Pesticides and Chemical Pollutants Determined by GC-MS-MS—cont'd

No.	Time (min)	Ion (amu)	Dwell Time (ms)
6	24.07	331.0/109.0, 331.0/127.0, 294.0/61.0,292.0/61.0,374.0/339.0, 374.0/337.0, 105.0/77.0, 172.0/116.0, 187.0/159.0, 187.0/109.0, 235.0/165.0, 235.0/199.0, 263.0/191.0, 263.0/193.0, 302.0/111.0, 302.0/175.0	10
7	25.55	236.0/125.0, 236.0/167.0, 256.0/212.0, 256.0/170.0, 251.0/139.0, 251.0/111.0, 283.0/162.0, 283.0/253.0, 361.0/317.0, 361.0/300.0, 360.0/325.0, 360.0/297.0, 235.0/165.0, 235.0/199.0, 235.0/165.0, 235.0/199.0, 241.0/206.0, 241.0/170.0, 276.0/105.0, 276.0/77.0, 276.0/105.0, 276.0/77.0, 342.0/199.0, 251.0/121.0	8
8	27.5	148.0/79.0, 148.0/105.0, 173.0/109.0, 310.0/201.0, 157.0/77.0, 157.0/110.0, 387.0/289.0, 387.0/253.0, 341.0/183.0, 341.0/185.0, 292.0/105.0, 292.0/77.0, 265.0/210.0, 265.0/89.0	14
9	30.1	377.0/362.0, 377.0/296.0, 157.0/63.0, 157.0/110.0, 171.0/71.0, 171.0/85.0, 341.0/310.0, 341.0/281.0, 182.0/111.0, 182.0/75.0, 250.0/208.0, 250.0/55.0, 330.0/139.0, 330.0/251.0, 132.0/77.0, 132.0/104.0, 132.0/77.0, 132.0/104.0, 180.0/138.0, 308.0/70.0, 206.0/151.0, 206.0/177.0	9
Group C			
1	6.0	151.0/122.0, 151.0/80.0, 188.0/58.0, 129.0/86.0, 146.0/76.0, 146.0/104.0, 163.0/90.0, 163.0/99.0, 126.0/55.0, 126.0/83.0, 208.0/180.0, 208.0/110.0, 169.0/141.0, 141.0/115.0	15
2	14.0	292.0/264.0, 292.0/160.0, 176.0/148.0, 176.0/121.0, 225.0/183.0, 225.0/168.0, 270.0/186.0, 270.0/228.0, 200.0/199.0, 183.0/102.0, 219.0/183.0, 219.0/147.0, 88.0/60.0, 88.0/59.0, 200.0/122.0, 200.0/94.0	12
3	18.2	272.0/237.0, 272.0/235.0, 204.0/91.0, 204.0/122.0, 257.0/119.0, 257.0/162.0, 175.0/147.0, 175.0/111.0, 128.0/70.0, 128.0/110.0, 163.0/91.0, 163.0/143.0, 326.0/63.0, 306.0/264.0, 267.0/252.0, 267.0/93.0	12
4	20.2	227.0/58.0, 227.0/170.0, 258.0/61.0, 198.0/82.0, 198.0/110.0, 210.0/76.0, 210.0/124.0, 255.0/222.0, 255.0/138.0, 225.0/96.0, 225.0/168.0	15

Continued

TABLE 5.6 Monitoring of Selected Ions for 454 Pesticides and Chemical Pollutants Determined by GC-MS-MS—cont'd

No.	Time (min)	Ion (amu)	Dwell Time (ms)
5	22.0	119.0/91.0, 119.0/65.0, 123.0/81.0, 123.0/69.0, 123.0/81.0, 123.0/79.0, 318.0/248.0, 318.0/246.0, 268.0/77.0, 268.0/141.0, 353/282, 353/263	16
6	23.15	167.0/152.0, 167.0/165.0, 299.0/109.0, 299.0/79.0, 248.0/157.0, 248.0/192.0, 336.0/204.0, 336.0/156.0, 383.0/282.0, 383.0/254.0, 296.0/196.0, 296.0/168.0, 143.0/107.0, 143.0/108.0, 168.0/70.0, 128.0/100.0, 162.0/147.0, 162.0/132.0, 131.0/89.0, 131.0/130.0, 321.0/304.0, 323.0/306.0	9
7	25.9	251.0/139.0, 251.0/111.0, 233.0/165.0, 233.0/152.0, 190.0/126.0, 190.0/75.0, 268.0/232.0, 268.0/136.0, 176.0/103.0, 176.0/131.0	19
8	27.9	119.0/91.0, 119.0/65.0, 266.0/218.0, 266.0/246.0, 145.0/117.0, 145.0/91.0, 123.0/81.0, 123.0/79.0, 248.0/154.0, 248.0/182.0, 255.0/186.0, 255.0/129.0	18
9	30.1	293.0/162.0, 293.0/132.0, 226.0/184.0, 226.0/157.0, 289.0/93.0, 289.0/77.0, 192.0/136.0, 192.0/109.0, 197.0/141.0, 197.0/91.0, 183.0/168.0, 183.0/153.0	17
10	31.9	147.0/117.0, 147.0/132.0, 199.0/107.0, 199.0/157.0, 428.0/252.0, 449.0/347.0, 170.0/141.0, 170.0/115.0, 163.0/107.0, 163.0/135.0, 178.0/81.0, 178.0/108.0, 163.0/91.0, 163.0/127.0	14
11	34.0	419.0/167.0, 419.0/225.0, 323.0/265.0, 323.0/202.0, 354.0/176.0, 354.0/326.0, 423.0/318.0, 423.0/308.0	25
Group D			
1	5.0	154.0/111.0, 154.0/121.0, 212.0/97.0, 212.0/174.0, 250.0/215.0, 250.0/177.0, 142.0/106.0, 142.0/67.0, 202.0/121.0, 202.0/77.0, 196.0/165.0, 196.0/137.0	15
2	13.9	231.0/158.0, 231.0/160.0, 143.0/79.0, 143.0/52.0, 211.0/99.0, 211.0/155.0, 280.0/265.0, 280.0/237.0, 190.0/57.0, 190.0/106.0, 164.0/136.0, 164.0/108.0, 197.0/169.0, 197.0/111.0, 211.0/169.0, 211.0/196.0, 173.0/145.0, 173.0/158.0, 359.0/243.0, 359.0/242.0, 177.0/127.0, 177.0/161.0, 229.0/200.0, 229.0/186.0	8
3	17.9	256.0/186.0, 256.0/151.0, 256.0/186.0, 258.0/186.0, 198.0/89.0, 198.0/72.0, 173.0/109.0, 173.0/145.0, 172.0/94.0, 172.0/169.0, 231.0/158.0, 231.0/160.0, 253.0/106.0, 253.0/91.0, 263.0/192.0, 263.0/156.0	12

TABLE 5.6 Monitoring of Selected Ions for 454 Pesticides and Chemical Pollutants Determined by GC-MS-MS—cont'd

No.	Time (min)	Ion (amu)	Dwell Time (ms)
4	19.2	199.0/184.0, 199.0/157.0, 292.0/257.0, 292.0/222.0, 200.0/104.0, 200.0/122.0, 251.0/128.0, 251.0/86.0, 142.0/70.0, 142.0/113.0, 230.0/154.0, 230.0/111.0, 380.0/310.0, 380.0/307.0, 230.0/193.0, 230.0/195.0, 311.0/275.0, 311.0/240.0, 239.0/197.0, 239.0/85.0	10
5	20.6	193.0/123.0, 193.0/157.0, 297.0/269.0, 297.0/223.0, 198.0/82.0, 198.0/110.0, 301.0/223.0, 301.0/273.0, 250.0/139.0, 250.0/215.0	20
6	21.7	212.0/195.0, 212.0/188.0, 224.0/208.0, 224.0/222.0, 254.0/212.0, 254.0/165.0, 326.0/256.0, 326.0/254.0, 184.0/156.0, 184.0/129.0, 229.0/137.0, 229.0/67.0, 300.0/182.0, 300.0/200.0, 262.0/216.0, 262.0/123.0, 136.0/108.0, 136.0/69.0, 353/282, 353/263	9
7	23.5	409.0/300.0, 409.0/302.0, 202.0/113.0, 202.0/147.0, 223.0/179.0, 223.0/193.0, 323.0/267.0, 323.0/159.0, 311.0/174.0, 311.0/311/103.0, 326.0/256.0, 326.0/254.0, 130.0/102.0, 130.0/75.0, 340.0/183.0, 340.0/185.0, 360.0/290.0, 360.0/325.0, 219.0/123.0, 219.0/95.0, 196.0/140.0, 196.0/168.0	9
8	26.15	272.0/161.0, 272.0/102.0, 360.0/325.0, 360.0/290.0, 213.0/153.0, 213.0/125.0, 171.0/143.0, 171.0/128.0, 171.0/143.0, 171.0/128.0, 222.0/125.0, 222.0/82.0, 206.0/149.0, 206.0/93.0, 349.0/266.0, 349.0/238.0	12
9	28.2	379.0/276.0, 379.0/262.0, 366.0/181.0, 366.0/209.0, 326.0/170.0, 326.0/215.0, 394.0/324.0, 394.0/359.0, 333.0/171.0, 333.0/276.0, 202.0/174.0, 202.0/104.0, 192.0/190.0, 192.0/162.0	15
10	29.7	173.0/109.0, 173.0/145.0, 173.0/109.0, 173.0/145.0, 274.0/169.0, 274.0/216.0, 153.0/110.0, 153.0/136.0, 320.0/292.0, 320.0/79.0, 340.0/298.0, 340.0/286.0, 236.0/174.0, 236.0/201.0, 198.0/129.0, 198.0/102.0	12
Group E			
1	5.0	107.0/79.0, 107.0/77.0, 110.0/63.0, 110.0/64.0, 121.0/77.0, 121.0/103.0, 220.0/205.0, 220.0/145.0, 101.0/73.0, 101.0/100.0	20
2	12.5	301.0/97.0, 301.0/153.0, 121.0/77.0, 121.0/103.0, 156.0/74, 156.0/89, 209.0/180.0, 209.0/125.0, 142.0/79.0, 142.0/112.0, 110.0/63.0, 110.0/64.0	16

Continued

TABLE 5.6 Monitoring of Selected Ions for 454 Pesticides and Chemical Pollutants Determined by GC-MS-MS—cont'd

No.	Time (min)	Ion (amu)	Dwell Time (ms)
3	16.0	189.0/161.0, 189.0/185.0, 213.0/77.0, 213.0/212.0, 318.0/152.0, 318.0/276.0, 98.0/70.0, 98.0/69.0, 206.0/176.0, 206.0/123.0, 232.0/72.0, 232.0/187.0, 173.0/130.0, 173.0/144.0	14
4	18.7	173.0/145.0, 173.0/109.0, 259.0/120.0, 259.0/176.0, 264.0/127.0, 264.0/193.0, 146.0/131.0, 146.0/118.0, 187.0/159.0, 187.0/123.0, 220.0/205.0, 220.0/145.0, 201.0/109.0, 201.0/184.0, 182.0/153.0, 182.0/107.0, 163.0/91.0, 163.0/107.0, 206.0/162.0, 206.0/132.0	10
5	21.1	301.0/223.0, 301.0/273.0, 264.0/127.0, 264.0/193.0, 121.0/101.0, 121.0/75.0, 161.0/88.0, 161.0/144.0, 363.0/300.0, 363.0/272.0, 295.0/109.0, 295.0/127.0, 242.0/214.0, 242.0/187.0, 172.0/89.0, 172.0/128.0, 123.0/81.0, 123.0/79.0, 353/282, 353/263	9
6	23.2	140.0/85.0, 140.0/71.0, 242.0/95.0, 301.0/224.0, 247.0/182.0, 247.0/212.0, 201.0/174.0, 201.0/130.0, 205.0/188.0, 205.0/162.0, 335.0/173.0, 335.0/303.0	17
7	25.1	286.0/202.0, 286.0/185.0, 185.0/77.0, 185.0/121.0, 180.0/136.0, 180.0/109.0, 215.0/173.0, 215.0/145.0	20
8	25.8	412.0/295.0, 412.0/118.0, 313.0/177.0, 313.0/130.0, 272.0/237.0, 272.0/235.0, 273.0/208.0, 273.0/182.0, 222.0/162.0, 222.0/190.0, 367.0/213.0, 367.0/255.0	17
9	28.8	412.0/349.0, 412.0/307.0, 288.0/141.0, 288.0/174.0, 205.0/176.0, 205.0/148.0, 228.0/226.0, 229.0/227.0, 299.0/253.0, 299.0/190.0, 192.0/138.0, 192.0/111.0, 192.0/138.0, 192.0/111.0, 136.0/78.0, 136.0/96.0, 321.0/122.0, 321.0/123.0, 268.0/180.0, 268.0/77.0	10
10	31.3	132.0/77.0, 132.0/104.0, 283.0/227.0, 283.0/137.0, 360.0/97.0, 360.0/194.0, 210.0/148.0, 210.0/151.0, 312.0/259.0, 312.0/294.0, 333.0/120.0, 199.0/157.0	17
11	32.9	287.0/259.0, 287.0/179.0, 184.0/169.0, 184.0/141.0, 331.0/180.0, 331.0/152.0, 152.0/116.0, 166.0/139.0, 328.0/259.0, 328.0/189.0	20
Group F			
1	7.0	154.0/124.0, 124.0/83.0, 166.0/165.0, 166.0/121.0, 168.0/107.0, 168.0/77.0, 163.0/77.0, 163.0/133.0, 200.0/141.0, 200.0/111.0	20
2	12.5	147.0/103.0, 147.0/76.0, 144.0/115.0, 144.0/116.0, 234.0/199.0, 234.0/73.0, 159.0/97.0, 159.0/131.0	20

TABLE 5.6 Monitoring of Selected Ions for 454 Pesticides and Chemical Pollutants Determined by GC-MS-MS—cont'd

No.	Time (min)	Ion (amu)	Dwell Time (ms)
3	16.4	170.0/114.0, 170.0/97.0, 127.0/109.0, 127.0/95.0, 193.0/136.0, 193.0/121.0, 233.0/190.0, 233.0/159.0, 200.0/108.0, 200.0/92.0	20
4	18.8	238.0/166.0, 238.0/96.0, 154.0/82.0, 154.0/96.0, 154.0/82.0, 154.0/96.0, 132.0/97.0, 132.0/60.0, 132.0/97.0, 132.0/60.0, 287.0/272.0, 287.0/242.0, 181.0/109.0, 181.0/80.0, 222.0/91.0, 222.0/162.0, 311.0/240.0, 311.0/241.0, 241.0/226.0, 241.0/170.0	10
5	21.4	264.0/194.0, 264.0/194.0, 266.0/190.0, 266.0/174.0, 212.0/122.0, 212.0/94.0, 211.0/96.0, 211.0/123.0, 262.0/200.0, 262.0/192.0, 262.0/200.0, 262.0/192.0, 416.0/186.0, 416.0/212.0, 353/282, 353/263	12
6	26.0	217.0/173.0, 217.0/145.0, 134.0/93.0, 134.0/91.0, 134.0/93.0, 134.0/91.0, 186.0/109.0, 186.0/158.0, 186.0/109.0, 186.0/158.0, 171.0/143.0, 171.0/128.0, 408.0/59.0, 408.0/363.0	14
7	27.9	342.0/255.0, 342.0/184.0330.0/310.0, 330.0/241.0, 177.0/113.0, 177.0/78.0, 345.0/281.0, 345.0/317.0, 327.0/295.0, 327.0/260.0, 160.0/104.0, 160.0/62.0	18
8	29.1	272.0/254.0, 272.0/209.0, 218.0/109.0, 218.0/79.0, 317.0/245.0, 317.0/209.0	25
9	30.7	125.0/89.0, 125.0/99.0, 357.0/256.0, 357.0/229.0, 342.0/140.0, 342.0/112.0, 383.0/171.0, 383.0/145.0	18
10	36.5	301.0/165.0, 301.0/139.0	100

5.1.5   Retention Time, Quantifier and Qualifier Ions, and Collision Energies of 284 Environmental Pollutants by GC-MS-MS

See Table 5.7.

TABLE 5.7 Retention Time, Quantifier and Qualifier Ions, and Collision Energies of 284 Environmental Pollutants by GC-MS-MS

No.	Compound	Retention Time (min)	Qualifier Ion	Quantifier Ion	Collision Energy (V)
	Heptachlor (ISTD)	22.04	353/263;353/282	353/263	17:17
Group A					
1	2-Chlorobiphenyl	11.04	188/152;188/153	188/152	20:10
2	2,2'-Dichlorobiphenyl	13.39	152/151;152/150	152/151	20:40
3	2,4'-Dichlorobiphenyl	14.91	224/152;224/151	224/152	30:50
4	2,2',6-Trichlorobiphenyl	15.77	256/221;256/186	256/221	10:20
5	3,4-Dichlorobiphenyl	16.55	222/152;222/151	222/152	30:50
6	2,3',6-Trichlorobiphenyl	16.94	186/151;186/150	186/151	20:30
7	2,2',3-Trichlorobiphenyl	17.40	256/186;256/221	256/186	20:10
8	2,3',4-Trichlorobiphenyl	17.93	256/186;256/151	256/186	30:40
9	2,3,4-Trichlorobiphenyl	18.60	256/186;186/151	256/186	20:20

10	2,3,3'-Trichlorobiphenyl	18.81	186/151;186/150	186/151	20:30
11	3,3',5-Trichlorobiphenyl	19.24	186/151;186/150	186/151	20:30
12	2,2',3,5-Tetrachlorobiphenyl	19.57	294/222;294/150	294/222	30:50
13	2,3,5,6-Tetrachlorobiphenyl	19.67	292/222;292/220	292/222	20:20
14	2,2',4,6,6'-Pentachlorobiphenyl	19.84	254/184;254/219	254/184	30:20
15	2,3',5,5'-Tetrachlorobiphenyl	20.52	292/220;292/150	292/220	30:50
16	2,2',4,5',6-Pentachlorobiphenyl	20.73	326/256;326/184	326/256	40:50
17	2,2',3,4-Tetrachlorobiphenyl	20.93	292/220;292/150	292/220	30:50
18	2,3',4,5-Tetrachlorobiphenyl	21.22	292/220;292/185	292/220	30:40
19	2,2',3,3'-Tetrachlorobiphenyl	21.43	292/220;292/150	292/220	30:50
20	2,4,4',5-Tetrachlorobiphenyl	21.57	290/220;290/150	290/220	20:50
21	2,2',4,5,6'-Pentachlorobiphenyl	21.72	254/184;254/219	254/184	30:20
22	2,2',3,5',6-Pentachlorobiphenyl	21.97	254/184;254/219	254/184	30:20
23	2,2',3,5,5'-Pentachlorobiphenyl	22.46	184/149;328/256	184/149	20:40
24	2,2',4,4',5-Pentachlorobiphenyl	22.77	326/184;326/256	326/184	50:50
25	2,2',3,3',6-Pentachlorobiphenyl	22.92	254/184;254/219	254/184	30:20
26	2,3,3',4,6-Pentachlorobiphenyl	23.24	326/184;326/256	326/184	50:40
27	2,2',3,3',5-Pentachlorobiphenyl	23.42	184/149;184/123	184/149	20:30
28	2,2',3,4,5-Pentachlorobiphenyl	23.53	326/291;326/256	326/291	10:20
29	2',3,4,5,6'-Pentachlorobiphenyl	23.66	254/184;254/219	254/184	20:20

Continued

TABLE 5.7 Retention Time, Quantifier and Qualifier Ions, and Collision Energies of 284 Environmental Pollutants by GC-MS-MS—cont'd

No.	Compound	Retention Time (min)	Qualifier Ion	Quantifier Ion	Collision Energy (V)
30	2,2′,3,4,5′-Pentachlorobiphenyl	23.87	328/256;328/258	328/256	30:30
31	2,3,3′,4′,6-Pentachlorobiphenyl	24.29	324/254;324/184	324/254	30:50
32	2,2′,3,3′,5,6′-Hexachlorobiphenyl	24.58	325/290;325/288	325/290	10:10
33	2′,3,4,5,5′-Pentachlorobiphenyl	24.83	326/256;326/254	326/256	30:30
34	2′,3,4,4′,5-Pentachlorobiphenyl	24.97	328/256;328/258	328/256	35:35
35	2,3′,4,4′,5-Pentachlorobiphenyl	25.13	326/256;326/254	326/256	30:30
36	2,2′,3,3′,5,6-Hexachlorobiphenyl	25.31	325/290;325/288	325/290	10:10
37	2,3,4,4′,5-Pentachlorobiphenyl	25.45	254/184;254/219	254/184	20:20
38	2,3′,4,4′,5′,6-Hexachlorobiphenyl	25.69	358/218;358/288	358/218	50:40
39	3,3′,4,5,5′-Pentachlorobiphenyl	26.27	326/256;326/254	326/256	20:20
40	2,2′,3,4,4′,5-Hexachlorobiphenyl	26.43	362/290;362/292	362/290	25:25
41	2,3,3′,4′,5,6-Hexachlorobiphenyl	26.86	360/290;360/288	360/290	30:30
42	2,2′,3,3′,5,5′,6-Heptachlorobiphenyl	26.93	396/326;396/324	396/326	30:30
43	2,2′,3,4′,5,5′,6-Heptachlorobiphenyl	27.25	396/361;396/359	396/361	10:10
44	2,3,3′,4′,5,5′-Hexachlorobiphenyl	27.70	358/288;358/218	358/288	30:50

45	2,2',3,3',5,5',6,6'-Octachlorobiphenyl	28.07	432/360;432/362	432/360	30:30
46	2,2',3,4,4',5,6,6'-Octachlorobiphenyl	28.30	432/360;432/362	432/360	30:30
47	2,2',3,3',4,4',6,6'-Octachlorobiphenyl	28.58	428/358;428/356	428/358	30:30
48	2,3,3',4,5,5',6-Heptachlorobiphenyl	28.96	396/324;396/326	396/324	40:40
49	2,3,3',4',5,5',6-Heptachlorobiphenyl	29.31	324/254;324/252	324/254	30:30
50	2,3,3',4,4',5,6-Heptachlorobiphenyl	30.19	394/324;394/322	394/324	20:20
51	3,3',4,4',5,5'-Hexachlorobiphenyl	30.48	358/288;362/290	358/288	20:20
52	2,2',3,3',4,4',5,6-Octachlorobiphenyl	31.18	428/358;428/356	428/358	30:30
53	2,2',3,3',4,4',5,5',6-Nonachlorobiphenyl	32.44	466/394;466/396	466/394	40:40
54	2,2',3,3',4,4',5,5',6,6'-Decachlorobiphenyl	32.78	500/429;500/428	500/429	30:30
	Group B				
55	3-Chlorobiphenyl	12.46	188/152;188/151	188/152	30:50
56	2,4-Dichlorobiphenyl	14.07	224/152;152/151	224/152	10:20
57	2,3-Dichlorobiphenyl	14.97	222/152;152/151	222/152	10:20
58	3,3'-Dichlorobiphenyl	16.34	224/152;224/151	224/152	20:50
59	3,4-Dichlorobiphenyl	16.63	152/151;152/150	152/151	20:40
60	2,4',6-Trichlorobiphenyl	17.23	256/186;256/151	256/186	30:40
61	2,4,5-Trichlorobiphenyl	17.42	256/151;256/150	256/151	40:50

Continued

TABLE 5.7 Retention Time, Quantifier and Qualifier Ions, and Collision Energies of 284 Environmental Pollutants by GC-MS-MS—cont'd

No.	Compound	Retention Time (min)	Qualifier Ion	Quantifier Ion	Collision Energy (V)
62	2,2′,4,6-Tetrachlorobiphenyl	17.94	292/220;292/222	292/220	40:40
63	2,2′,5,6′-Tetrachlorobiphenyl	18.68	292/150;292/220	292/150	50:50
64	2,3,4′-Trichlorobiphenyl	19.11	256/186;256/151	256/186	20:40
65	2,3′,5′,6-Tetrachlorobiphenyl	19.44	290/220;290/150	290/220	30:50
66	3,4′,5-Trichlorobiphenyl	19.61	258/151;258/186	258/151	50:40
67	2,3,4,6-Tetrachlorobiphenyl	19.68	292/222;292/150	292/222	20:50
68	3,4,5-Trichlorobiphenyl	19.93	258/151;258/186	258/151	30:50
69	3,3′,4-Trichlorobiphenyl	20.55	186/151;186/150	186/151	20:30
70	2,3,4′,6-Tetrachlorobiphenyl	20.83	220/150;294/222	220/150	30:20
71	3,4,4′-Trichlorobiphenyl	20.97	186/151;186/150	186/151	20:30
72	3,3′,5,5′-Tetrachlorobiphenyl	21.39	292/220;292/150	292/220	30:50
73	2,3,3′,5′-Tetrachlorobiphenyl	21.45	292/222;292/220	292/222	20:20
74	2,3′,4,5′,6-Pentachlorobiphenyl	21.57	328/256;328/258	328/256	40:40
75	2,2′,3,5,6-Pentachlorobiphenyl	21.76	326/291;326/289	326/291	10:10
76	2,3′,4,4′-Tetrachlorobiphenyl	22.00	292/220;292/222	292/220	20:20

No.	Name	RT	Ions	Ion	Ratio
77	2,2',3,4',5-Pentachlorobiphenyl	22.58	324/254;326/291	324/254	30:10
78	2,3,3',5',6-Pentachlorobiphenyl	22.82	324/254;326/256	324/254	30:20
79	2,2',3,4,6'-Pentachlorobiphenyl	22.93	254/184;254/219	254/184	30:20
80	2,2',3,5,6,6'-Hexachlorobiphenyl	23.26	358/288;358/218	358/288	30:50
81	2,2',3,4,6,6'-Hexachlorobiphenyl	23.45	290/218;290/220	290/218	30:30
82	2,3,4,4',6-Pentachlorobiphenyl	23.59	326/256;326/254	326/256	35:35
83	2,2',4,4',5,6'-Hexachlorobiphenyl	23.66	358/288;358/218	358/288	40:50
84	2,2',3,4,4'-Pentachlorobiphenyl	23.97	326/256;326/254	326/256	40:40
85	2,2',3,5,5',6-Hexachlorobiphenyl	24.36	358/288;358/323	358/288	30:10
86	2,2',3,4,4',6-Hexachlorobiphenyl	24.67	358/288;358/218	358/288	30:50
87	2,2',3,4,4',6'-Hexachlorobiphenyl	24.86	360/325;360/290	360/325	10:20
88	2,3,3',4',5-Pentachlorobiphenyl	25.04	254/184;254/219	254/184	30:20
89	2,2',3,4,5,6'-Hexachlorobiphenyl	25.20	290/218;290/220	290/218	30:30
90	2,2',3,4,5,6-Hexachlorobiphenyl	25.32	362/290;362/237	362/237	10:20
91	2,2',3,4',5,5'-Hexachlorobiphenyl	25.45	358/288;358/323	358/288	20:10
92	2',3,3',4,5-Pentachlorobiphenyl	25.76	326/256;326/254	326/256	40:41
93	2,2',3,4,5,5'-Hexachlorobiphenyl	26.28	290/218;290/220	290/218	30:30
94	2,2',3,3',4,5'-Hexachlorobiphenyl	26.67	358/288;358/218	358/288	50:50
95	2,2',3,4,4',5'-Hexachlorobiphenyl	26.87	360/290;360/288	360/290	30:30
96	2,2',3,3',4,5',6-Heptachlorobiphenyl	26.97	359/324;359/322	359/324	10:10

Continued

TABLE 5.7 Retention Time, Quantifier and Qualifier Ions, and Collision Energies of 284 Environmental Pollutants by GC-MS-MS—cont'd

No.	Compound	Retention Time (min)	Qualifier Ion	Quantifier Ion	Collision Energy (V)
97	2,2′,3,4,4′,5′,6-Heptachlorobiphenyl	27.42	396/361;396/359	396/361	10:10
98	2,3,4,4′,5,6-Hexachlorobiphenyl	27.80	362/290;362/292	362/290	20:20
99	2,2′,3,3′,4,4′-Hexachlorobiphenyl	28.09	325/290;325/288	325/290	10:10
100	2,2′,3,3′,4,5′,6,6′-Octachlorobiphenyl	28.34	428/358;428/356	428/358	30:30
101	2,2′,3,3′,4,5,6-Heptachlorobiphenyl	28.74	396/361;396/359	396/361	10:10
102	2,3,3′,4,4′,5′-Hexachlorobiphenyl	29.05	360/290;362/290	360/290	30:30
103	2,3,3′,4,4′,6-Heptachlorobiphenyl	29.40	396/326;396/324	396/326	30:30
104	2,2′,3,4,4′,5,5′,6-Octachlorobiphenyl	30.26	358/288;358/286	358/288	40:40
105	2,2′,3,3′,4,5,5′,6,6′-Nonachlorobiphenyl	30.73	462/392;462/390	462/392	30:30
106	2,2′,3,3′,4,4′,5,5′-Octachlorobiphenyl	31.77	358/288;358/286	358/288	30:30
	Group C				
107	4-Chlorobiphenyl	12.66	188/152;188/153	188/152	20:10
108	2,5-Dichlorobiphenyl	14.09	224/152;224/151	224/152	20:40
109	3,5-Dichlorobiphenyl	15.27	222/152;222/151	222/152	20:50
110	2,2′,5-Trichlorobiphenyl	16.47	186/151;186/150	186/151	20:30

111	2,3,6-Trichlorobiphenyl	16.82	258/151;258/150	258/151	50:50
112	2,3,5-Trichlorobiphenyl	17.29	186/151,186/150	186/151	20:30
113	2,2',6,6'-Tetrachlorobiphenyl	17.87	292/222;292/220	292/222	30:30
114	2,4',5-Trichlorobiphenyl	18.19	258/151;258/166	258/151	50:50
115	2',3,4-Trichlorobiphenyl	18.73	258/186;258/188	258/186	20:20
116	2,3',4,6-Tetrachlorobiphenyl	19.18	294/222;220/150	294/222	20:40
117	2,4',6-Tetrachlorobiphenyl	19.52	292/220;292/150	292/220	30:50
118	2,2',3,6'-Tetrachlorobiphenyl	19.62	292/220;292/222	292/220	30:30
119	2,2',4,4'-Tetrachlorobiphenyl	19.70	290/220;290/255	290/220	20:20
120	2,2',3,5'-Tetrachlorobiphenyl	20.48	292/150;292/220	292/150	50:40
121	2,2',3,4'-Tetrachlorobiphenyl	20.56	294/222;294/150	294/222	30:50
122	2,3',4',6-Tetrachlorobiphenyl	20.84	294/220;220/150	294/220	20:40
123	2,2',3,6,6'-Pentachlorobiphenyl	21.02	324/254;328/256	324/254	20:20
124	2,2',3,4,6-Pentachlorobiphenyl	21.39	328/256;328/258	328/256	40:40
125	2,2',3,5,6'-Pentachlorobiphenyl	21.46	254/184;254/219	254/184	30:20
126	2,2',3',4,6-Pentachlorobiphenyl	21.67	254/184;254/219	254/184	20:20
127	2',3,4,5-Tetrachlorobiphenyl	21.85	220/150;294/222	220/150	30:20
128	2,2',3,4',6-Pentachlorobiphenyl	22.13	328/256;328/258	328/256	30:30
129	2,2',4,5,5'-Pentachlorobiphenyl	22.66	328/256;328/293	328/256	30:10
130	2,3,3',4'-Tetrachlorobiphenyl	22.84	290/220;290/150	290/220	30:50

Continued

TABLE 5.7 Retention Time, Quantifier and Qualifier Ions, and Collision Energies of 284 Environmental Pollutants by GC-MS-MS—cont'd

No.	Compound	Retention Time (min)	Qualifier Ion	Quantifier Ion	Collision Energy (V)
131	2,3′,4,4′,6-Pentachlorobiphenyl	23.01	326/256;326/254	326/256	40:40
132	3,3′,4,5′-Tetrachlorobiphenyl	23.31	220/150;294/222	220/150	30:20
133	2,3,4,5,6-Pentachlorobiphenyl	23.48	328/256;328/258	328/256	30:30
134	2,3,4′,5,6-Pentachlorobiphenyl	23.59	328/256;328/258	328/256	40:40
135	3,3′,4,5-Tetrachlorobiphenyl	23.72	294/222;294/150	294/222	30:50
136	2,2′,3,3′,6,6′-Hexachlorobiphenyl	23.99	362/290;362/292	362/290	20:20
137	2,2′,3,4,5′,6-Hexachlorobiphenyl	24.48	360/290;360/288	360/290	30:30
138	3,3′,4,4′-Tetrachlorobiphenyl	24.77	290/220;290/150	290/220	30:50
139	2,2′,3,4′,5′,6-Hexachlorobiphenyl	24.86	360/325;360/290	360/325	10:20
140	2,2′,3,4′,5,6,6′-Heptachlorobiphenyl	25.06	324/254;324/252	324/254	30:30
141	2,2′,3,3′,5,5′-Hexachlorobiphenyl	25.22	358/288;358/323	358/288	30:20
142	2,3,3′,5,5′,6-Hexachlorobiphenyl	25.39	358/218;358/288	358/218	50:40
143	2,3,3′,4,5′,6-Hexachlorobiphenyl	25.47	362/290;362/292	362/290	20:20
144	2,2′,3,3′,4,6′-Hexachlorobiphenyl	26.08	360/290;360/288	360/290	25:25
145	2,3,3′,4,4′-Pentachlorobiphenyl	26.34	254/184;254/219	254/184	30:20

No.	Compound	RT			
146	2,2',3,4,5,6,6'-Heptachlorobiphenyl	26.69	324/254;324/252	324/254	30:30
147	2,3,3',4,4',6-Hexachlorobiphenyl	26.92	290/218;290/220	290/218	30:30
148	2,2',3,4,4',5,6'-Heptachlorobiphenyl	27.16	398/326;398/328	398/326	30:30
149	2,3,3',4,5,5'-Hexachlorobiphenyl	27.44	358/288;362/290	358/288	20:20
150	2,3',4,4',5,5'-Hexachlorobiphenyl	27.91	358/218;358/288	358/218	50:40
151	2,2',3,3',4,5,6'-Heptachlorobiphenyl	28.17	324/254;324/252	324/254	30:30
152	2,2',3,3',4',5,6-Heptachlorobiphenyl	28.44	394/324;394/322	394/324	30:30
153	2,3,3',4,4',5-Hexachlorobiphenyl	28.82	362/290;360/290	362/290	30:40
154	2,2',3,4,4',5,5'-Heptachlorobiphenyl	29.23	396/324;396/326	396/324	30:30
155	2,2',3,3',4,5,5',6-Octachlorobiphenyl	30.01	430/360;430/358	430/360	30:30
156	2,2',3,3',4,4',5,6'-Octachlorobiphenyl	30.27	358/288;358/286	358/288	30:30
157	2,2',3,3',4,4',5,6,6'-Nonachlorobiphenyl	30.94	464/463;464/394	464/463	10:30
158	2,3,3',4,4',5,5',6-Octachlorobiphenyl	31.90	430/360;430/358	430/360	20:20
	Group D				
159	2,6-Dichlorobiphenyl	13.27	152/151;152/150	152/151	20:40
160	2,3'-Dichlorobiphenyl	14.67	152/151;152/150	152/151	20:40
161	2,4,6-Trichlorobiphenyl	15.50	186/151;186/150	186/151	20:30

Continued

TABLE 5.7 Retention Time, Quantifier and Qualifier Ions, and Collision Energies of 284 Environmental Pollutants by GC-MS-MS—cont'd

No.	Compound	Retention Time (min)	Qualifier Ion	Quantifier Ion	Collision Energy (V)
162	2,2',4-Trichlorobiphenyl	16.48	221/186;221/151	221/186	20:40
163	4,4'-Dichlorobiphenyl	16.91	222/152;222/151	222/152	20:40
164	2',3,5-Trichlorobiphenyl	17.39	258/186;258/188	258/186	20:20
165	2,3',5-Trichlorobiphenyl	17.89	258/186;258/151	258/186	20:40
166	2,4,4'-Trichlorobiphenyl	18.22	256/151;256/150	256/151	50:50
167	2,2',4,6'-Tetrachlorobiphenyl	18.80	294/222;294/224	294/222	30:30
168	2,2',3,6-Tetrachlorobiphenyl	19.20	220/150;220/185	220/150	30:20
169	2,2',5,5'-Tetrachlorobiphenyl	19.56	220/150;220/185	220/150	30:20
170	2,2',4,5'-Tetrachlorobiphenyl	19.64	290/220;290/185	290/220	40:40
171	2,2',4,5-Tetrachlorobiphenyl	19.75	220/150;220/185	220/150	10:10
172	2,3,3',6-Tetrachlorobiphenyl	20.49	220/150;220/185	220/150	30:20
173	2,3',4,5'-Tetrachlorobiphenyl	20.62	294/222;294/220	294/222	30:30
174	2,2',4,4',6-Pentachlorobiphenyl	20.87	328/256;328/184	328/256	30:50
175	2,3,3',5-Tetrachlorobiphenyl	21.05	220/150;220/185	220/150	30:20
176	2,3,4',5-Tetrachlorobiphenyl	21.41	292/220;292/222	292/220	30:30

177	2,3,4,5-Tetrachlorobiphenyl	21.47	294/222;294/150	294/222	30:50
178	2,2',4,4',6,6'-Hexachlorobiphenyl	21.70	360/290;360/288	360/290	30:30
179	2,3',4',5-Tetrachlorobiphenyl	21.91	294/222;220/150	294/222	20:40
180	2,3,3',4-Tetrachlorobiphenyl	22.42	292/222;292/220	292/222	20:20
181	2,3,4,4'-Tetrachlorobiphenyl	22.76	294/222;294/224	294/222	20:20
182	2,2',3,4',6,6'-Hexachlorobiphenyl	22.85	325/290;325/288	325/290	10:10
183	2,3,3',5,6-Pentachlorobiphenyl	23.15	326/256;326/254	326/256	30:30
184	2,2',3,4',5,6'-Hexachlorobiphenyl	23.39	362/327;362/290	362/327	10:20
185	2,3,3',5,5'-Pentachlorobiphenyl	23.49	324/254;328/256	324/254	20:20
186	2,2',3',4,5-Pentachlorobiphenyl	23.65	328/256;328/293	328/256	20:20
187	2,3',4,5,5'-Pentachlorobiphenyl	23.73	254/184;254/219	254/184	20:20
188	3,4,4',5-Tetrachlorobiphenyl	24.21	290/220;290/150	290/220	30:50
189	2,2',3,4',5,6-Hexachlorobiphenyl	24.57	290/218;290/220	290/218	30:30
190	2,2',3,3',4-Pentachlorobiphenyl	24.82	328/256;328/258	328/256	40:40
191	2,3,3',4,5'-Pentachlorobiphenyl	24.95	254/184;254/219	254/184	30:20
192	2,3,3',4,5-Pentachlorobiphenyl	25.10	328/256;328/184	328/256	40:50
193	2,2',3,4,4',6,6'-Heptachlorobiphenyl	25.31	396/326;396/324	396/326	30:30
194	2,2',3,3',4,6-Hexachlorobiphenyl	25.44	360/290;360/288	360/290	30:30
195	2,2',4,4',5,5'-Hexachlorobiphenyl	25.67	290/218;290/220	290/218	20:20

Continued

TABLE 5.7 Retention Time, Quantifier and Qualifier Ions, and Collision Energies of 284 Environmental Pollutants by GC-MS-MS—cont'd

No.	Compound	Retention Time (min)	Qualifier Ion	Quantifier Ion	Collision Energy (V)
196	2,2′,3,3′,5,6,6′-Heptachlorobiphenyl	26.14	398/326;398/328	398/326	20:20
197	2,2′,3,3′,4,6,6′-Heptachlorobiphenyl	26.39	324/254;324/252	324/254	40:40
198	2,3,3′,4,5,6-Hexachlorobiphenyl	26.85	360/290;360/288	360/290	20:20
199	2,3,3′,4′,5′,6-Hexachlorobiphenyl	26.92	360/290;360/288	360/290	20:20
200	2,2′,3,3′,4,5-Hexachlorobiphenyl	27.23	325/290;325/218	325/290	10:40
201	3,3′,4,4′,5-Pentachlorobiphenyl	27.69	254/184;254/220	254/184	30:20
202	2,2′,3,4,5,5′,6-Heptachlorobiphenyl	27.92	394/322;394/320	394/320	30:30
203	2,2′,3,4,4′,5,6-Heptachlorobiphenyl	28.17	394/324;394/322	394/324	30:30
204	2,2′,3,3′,4,4′,6-Heptachlorobiphenyl	28.49	398/326;398/328	398/326	30:30
205	2,2′,3,3′,4,5,5′-Heptachlorobiphenyl	28.93	394/324;394/322	394/324	30:30
206	2,2′,3,3′,4,5,6,6′-Octachlorobiphenyl	29.27	358/288;358/286	358/288	40:40
207	2,2′,3,3′,4,5,5′,6′-Octachlorobiphenyl	30.13	432/360;432/361	432/360	30:30
208	2,2′,3,3′,4,4′,5-Heptachlorobiphenyl	30.36	359/324;359/322	359/324	20:20
209	2,3,3′,4,4′,5,5′-Heptachlorobiphenyl	31.02	394/324;394/322	394/324	20:20

	Group E			
210	Naphthalene	6.41	128/101;128/77	128/101	15:15
211	Isoprotuton	6.58	146/128;146/91	146/128	15:15
212	Dichlorvos	7.88	185/93;185/109	185/93	15:10
213	Carbofuran	8.36	164/149;164/103	164/149	15:25
214	Methamidophos	9.35	141/95;141/80	141/95	10:15
215	Acenaphthylene	10.55	152/126;151/99	152/126	15:25
216	Acenaphthene	10.85	152/126;151/99	152/126	15:25
217	Fluorene	12.94	165/164;165/163	165/164	25:25
218	Hexachlorobenzene	14.36	284/249;284/214	284/249	18:25
219	Ethoprophos	14.40	158/97;158/114	158/97	12:7
220	Chlordimeform	14.91	196/181;196/152	196/181	5:25
221	Trifluralin	15.37	306/264;306/206	306/264	12:15
222	α-HCH	16.14	219/183;219/147	219/183	5:15
223	Omethoate	16.82	156/110;156/80	156/110	5:10
224	Anthracene	17.03	176/150;178/152	176/150	20:12
225	Clomazone	17.04	204/107;204/78	204/107	25:25
226	Diazinon	17.09	304/179;304/162	304/179	8:8
227	Phenanthrene	17.13	178/150;178/151	178/150	45:40

Continued

TABLE 5.7 Retention Time, Quantifier and Qualifier Ions, and Collision Energies of 284 Environmental Pollutants by GC-MS-MS—cont'd

No.	Compound	Retention Time (min)	Qualifier Ion	Quantifier Ion	Collision Energy (V)
228	γ-HCH	17.72	219/183;219/147	219/183	5:15
229	Atrazine	17.95	215/173;215/200	215/173	5:5
230	Simazine	18.03	201/173;201/138	201/173	5:15
231	Heptachlor	18.40	272/237;272/235	272/237	25:25
232	Pirimicarb	18.98	238/166;238/96	238/166	15:25
233	Dimethoate	19.32	125/79;143/111	125/79	8:12
234	Aldrin	19.41	263/193;263/191	263/193	25:35
235	Alachlor	20.16	237/160;237/146	237/160	8:20
236	Prometryne	20.19	241/199;241/184	241/199	5:5
237	Chlorothalonil	20.35	266/231;266/170	266/231	20:35
238	Phthalic acid bis-butyl ester	20.69	149/121;149/93	149/121	10:10
239	β-HCH	20.72	219/183;219/147	219/183	10:20
240	Chlorpyrifos	20.92	314/286;314/258	314/286	5:5
241	Parathion-methyl	21.05	263/109;263/246	263/109	12:5
242	Dicofol	21.34	250/139;250/215	250/139	15:10

243	Metolachlor	21.44	238/162;238/133	238/162	15:25
244	δ-HCH	21.50	219/183;219/147	219/183	10:20
245	Triadimefon	22.42	210/183;210/129	210/183	5:10
246	Fluoranthene	22.58	202/152;202/176	202/152	30:30
247	2,4'-DDE	22.70	246/176;246/211	246/176	25:25
248	cis-Chlordane	23.21	373/266;373/301	373/266	12:12
249	Phenthoate	23.38	274/246;274/121	274/246	5:25
250	trans-Chlordane	23.50	373/266;373/301	373/266	12:12
251	Pyrene	23.62	202/199;202/200	202/199	45:40
252	4,4'-DDE	23.90	246/176;246/211	246/176	25:25
253	Butachlor	23.98	176/150;176/126	176/150	25:25
254	Dieldrin	24.47	277/241;277/207	277/241	12:12
255	2,4'-DDD	25.04	235/165;235/199	235/165	15:15
256	Buprofezin	25.05	105/77;172/116	105/77	18:7
257	Endrin	25.06	263/191;263/193	263/191	20:12
258	2,4'-DDT	25.47	235/165;235/199	235/165	25:25
259	Nithophen	26.27	283/162;283/202	283/162	25:25
260	Oxyfluorfen	26.45	300/223;188/144	300/223	18:17
261	4,4'-DDD	26.73	235/165;235/199	235/165	15:15

Continued

TABLE 5.7 Retention Time, Quantifier and Qualifier Ions, and Collision Energies of 284 Environmental Pollutants by GC-MS-MS—cont'd

No.	Compound	Retention Time (min)	Qualifier Ion	Quantifier Ion	Collision Energy (V)
262	4,4'-DDT	27.23	235/199;235/165	235/199	25:25
263	Phthalic acid benzyl butyl ester	27.84	206/149;149/65	206/149	5:25
264	Propargite	28.08	173/117;173/145	173/117	10:10
265	Tricyclazole	28.39	189/162;189/135	189/162	10:15
266	Triazophos	28.54	161/134;161/106	161/134	8:15
267	Mirex	28.70	272/237;272/235	272/237	15:15
268	Benzo(a)anthrancene	29.27	228/226;228/202	228/226	30:30
269	Phthalic acid bis-2-ethylhexyl ester	29.47	167/149;167/65	167/149	10:25
270	Amitraz	30.37	293/162;293/132	293/162	5:15
271	Lamba-cyhalothrin	31.41	197/141;197/161	197/141	15:5
272	Pyridaben	32.07	147/117;147/132	147/117	25:15
273	Benzo(b)fluoranthene	32.94	252/250;252/224	252/250	40:50
274	Benzo(k)fluoranthene	32.94	252/250;252/224	252/250	40:50
275	Cyfluthrin	33.33	206/151;206/177	206/151	15:20
276	Cypermethrin	33.53	163/127;163/91	163/127	5:10

277	Benzo(a)pyrene	33.70	252/250;252/226	252/250	25
278	Acetamiprid	33.78	152/116;166/139	152/116	20:8
279	Fenvalerate-1	34.61	419/225;419/167	419/225	5:5
280	Fenvalerate-2	34.96	419/225;419/167	419/225	5:5
281	Deltamethrin	35.98	181/152;181/127	181/152	25:25
282	Indeno(1,2,3-cd)pyrene	37.63	276/274;276/248	276/274	40:50
283	Dibenzo[a,h]anthracene	37.83	278/276;278/274	278/276	40:55
284	Benzo(g,h,i)peryene	38.64	274/272;274/248	274/272	25:25

5.1.6   Retention Times, Selective Ions, and Relative Abundances of Endosulfans Determined by GC-NCI-MS

See Table 5.8.

TABLE 5.8 Retention Times, Selective Ions, and Relative Abundances of Endosulfans Determined by GC-NCI-MS

No.	Compound	Retention Time (min)	Selective Ion	Relative Abundance
1	Endosulfan I	11.14	406, 408, 372	100: 55: 44.6
2	Endosulfan II	12.56	406, 408, 372	100: 78: 17.5

5.1.7 Retention Times, Quantifying and Qualifying Ions, Declustering Potentials, and Collision Energies of 9 Environmental Pollutants Determined by LC-MS-MS

See Table 5.9.

TABLE 5.9 Retention Times, Quantifying and Qualifying Ions, Declustering Potentials, and Collision Energies of 9 Environmental Pollutants Determined by LC-MS-MS

No.	Compound	Retention Time (min)	Qualifying Ion	Quantifying Ion	Declustering Potential (V)	COLLISION energy (V)	Collision Chamber Outlet Voltage (V)
1.	Trichlorphon	9.25	257/109;257/127.1	257/109	28	25:23	2;2
2.	Metsulfuron-methyl	11.18	382/167.1;382/199.1	382/167.1	31	25:30	3;3
3.	Chlorolurons	11.74	213.1/72.0;213.1/140.1	213.1/72.0	29	38:33	3;3
4.	2,4-D	12.12	219/124.8;219/89	219/160.8	−35	−21:−38:−50	−2;−2;−2
5.	Bensulfuron-methyl	12.57	411.1/149.1;411.1/182.1	411.1/149.1	29	31:30	2;2
6.	Propanil	13.25	218/162.1;218/127	218/162.1	45	23:41	2;2
7.	Fipronil	13.59	436.9/368;436.9/290	436.9/368	44	23:35	4;4
8.	Phoxim	18.53	299/129;299/97	299/129	36	18:35	3;2
9.	Hexythiazox	22.05	353.1/228.1;353.1/168.1	353.1/228.1	50	21:35	2.5;1.8

5.1.8 Retention Times, Quantifying and Qualifying Ions, Fragmentor, and Collision Energies of 569 Pesticides and Chemical Pollutants Determined by LC-MS-MS

See Table 5.10.

TABLE 5.10 Retention Times, Quantifying and Qualifying Ions, Fragmentor, and Collision Energies of 569 Pesticides and Chemical Pollutants Determined by LC-MS-MS

No.	Compound	CAS	Retention Time (min)	Qualifying Ion	Quantifying Ion	Fragmentor (V)	Collision Energy (V)
1	Propham	122-42-9	8.80	180.1/138.0;180.1/120	180.1/138.0	80	5:15
2	Isoprocarb	2631-40-7	8.38	194.1/95.0;194.1/137.1	194.1/95.0	80	20:5
3	3,4,5-Trimethacarb	2686-99-9	8.38	194.2/137.2;194.2/122.2	194.2/137.2	80	5:20
4	Cycluron	2163-69-1	7.73	199.4/72.0;199.4/89.0	199.4/72.0	120	25:15
5	Carbaryl	63-25-2	7.45	202.1/145.1;202.1/127.1	202.1/145.1	80	10:5
6	Propachlor	1918-16-7	8.75	212.1/170.1;212.1/94.1	212.1/170.1	100	10:30
7	Rabenzazole	69899-24-7	7.54	213.2/172;213.2/118.0	213.2/172	120	25:25
8	Simetryn	1014-70-6	5.32	214.2/124.1;214.2/96.1	214.2/124.1	120	20:25
9	Monolinuron	1746-81-2	7.82	215.1/126.0;215.1/148.1	215.1/126.0	100	15:10
10	Mevinphos	7786-34-7	5.17	225.0/127.0;225.0/193	225.0/127.0	80	15:1
11	Aziprotryne	4658-28-0	10.40	226.1/156.1;226.1/198.1	226.1/156.1	100	10:10

No.	Name	CAS					
12	Secbumeton	26259-45-0	5.56	226.2/170.1;226.2/142.1	226.2/170.1	120	20:25
13	Cyprodinil	121552-61-2	9.24	226.0/93.0;226.0/108	226.0/93.0	120	40:30
14	Buturon	3766-60-7	9.38	237.1/84.1;237.1/126.1	237.1/84.1	120	30:15
15	Carbetamide	16118-49-3	5.80	237.1/192.1;237.1/118.1	237.1/192.1	80	5:10
16	Pirimicarb	23103-98-2	4.20	239.2/72.0;239.2/182.2	239.2/72.0	120	20:15
17	Clomazone	81777-89-1	9.36	240.1/125.0;240.1/89.1	240.1/125.0	100	20:50
18	Cyanazine	21725-46-2	6.38	241.1/214.1;241.1/174	241.1/214.1	120	15:15
19	Prometryne	7287-19-6	7.66	242.2/158.1;242.2/200.2	242.2/158.1	120	20:20
20	Paraoxon methyl	950-35-6	6.20	248.0/202.1;248.0/90	248.0/202.1	120	20:30
21	4,4-Dichlorobenzophenone	90-98-2	12.00	251.1/111.1;251.1/139.0	251.1/111.1	100	35:20
22	Thiacloprid	111988-49-9	5.65	253.1/126.1;253.1/186.1	253.1/126.1	120	20:10
23	Imidacloprid	138261-41-3	4.73	256.1/209.1;256.1/175.1	256.1/209.1	80	10:10
24	Ethidimuron	30043-49-3	4.62	265.1/208.1;265.1/162.1	265.1/208.1	80	10:25
25	Isomethiozin	57052-04-7	14.20	269.1/200.0;269.1/172.1	269.1/200.0	120	15:25
26	cis- and trans- diallate	2303-16-4	17.40	270.0/86;270.0/109.0	270.0/86	100	15:35
27	Acetochlor	34256-82-1	13.70	270.2/224;270.2/148.2	270.2/224	80	5:20
28	Nitenpyram	150824-47-8	3.87	271.1/224.1;271.1/237.1	271.1/224.1	100	15:15
29	Methoprotryne	841-06-5	6.47	272.2/198.2;272.2/170.1	272.2/198.2	140	25:30
30	Dimethenamid	87674-68-8	10.50	276.1/244.1;276.1/168.1	276.1/244.1	120	10:15
31	Terrbucarb	1918-11-2	16.50	278.2/166.1;278.2/109	278.2/166.1	80	15:30

Continued

TABLE 5.10 Retention Times, Quantifying and Qualifying Ions, Fragmentor, and Collision Energies of 569 Pesticides and Chemical Pollutants Determined by LC-MS-MS—cont'd

No.	Compound	CAS	Retention Time (min)	Qualifying Ion	Quantifying Ion	Fragmentor (V)	Collision Energy (V)
32	Penconazole	66246-88-6	13.70	284.1/70;284.1/159.0	284.1/70	120	15:20
33	Myclobutanil	88671-89-0	12.10	289.1/125;289.1/70.0	289.1/125	120	20:15
34	Imazethapyr	81385-77-5	5.60	290.2/177.1;290.2/245.2	290.2/177.1	120	25:20
35	Paclobutrazol	76738-62-0	10.32	294.2/70.0;294.2/125	294.2/70.0	100	15:25
36	Fenthion sulfoxide	3761-41-9	7.31	295.1/109;295.1/280.0	295.1/109	140	35:20
37	Triadimenol	55219-65-3	10.15	296.1/70.0;296.1/99.1	296.1/70.0	80	10:10
38	Butralin	33629-47-9	18.60	296.1/240.1;296.1/222.1	296.1/240.1	100	10:20
39	Spiroxamine	118134-30-8	9.90	298.2/144.2;298.2/100.1	298.2/144.2	120	20:35
40	Tolclofos methyl	57018-04-9	16.60	301.2/269;301.2/125.2	301.2/269	120	15:20
41	Desmedipham	13684-56-5	10.65	301.2/182.1;301.2/136.1	301.2/182.1	80	5:20
42	Methidathion	950-37-8	10.69	303.0/145.1;303.0/85	303.0/145.1	80	5:10
43	Allethrin	584-79-2	18.10	303.2/135.1;303.2/123.2	303.2/135.1	60	10:20
44	Diazinon	333-41-5	15.95	305.0/169.1;305.0/153.2	305.0/169.1	160	20:20
45	Edifenphos	17109-49-8	3.00	311.1/283.0;311.1/109	311.1/283.0	100	10:35
46	Pretilachlor	51218-49-6	17.15	312.1/252.1;312.1/176.2	312.1/252.1	100	15:30

47	Flusilazole	85509-19-9	13.60	316.1/247.1;316.1/165.1	316.1/247.1	120	15:20
48	Iprovalicarb	140923-17-7	12.00	321.1/119.0;321.1/203.2	321.1/119.0	100	25:5
49	Benodanil	15310-01-7	9.80	324.1/203;324.1/231.0	324.1/203	120	25:40
50	Flutolanil	66332-96-5	14.00	324.2/262.1;324.2/282.1	324.2/262.1	120	20:10
51	Famphur	52-85-7	10.30	326.0/217;326.0/281	326.0/217	100	20:10
52	Benalaxyl	71626-11-4	15.19	326.2/148.1;326.2/294	326.2/148.1	120	1:5
53	Diclobutrazole	75736-33-3	12.20	328.0/159;328.0/70.0	328.0/159	120	35:30
54	Etaconazole	1000290-09-5	11.75	328.1/159.1;328.1/205.1	328.1/159.1	80	25:20
55	Fenarimol	60168-88-9	12.20	331.0/268.1;331.0/81	331.0/268.1	120	25:30
56	Phthalic acid, dicyclobexyl ester	84-61-7	4.35	313.2/149.1;313.2/205.0	313.2/149.1	100	5:1
57	Tetramethirn	7696-12-0	17.85	332.2/164.1;332.2/135.1	332.2/164.1	100	15:15
58	Dichlofluanid	1085-98-9	15.16	333/123;333/224.0	333/123	80	20:10
59	Cloquintocet mexyl	99607-70-2	17.36	336.1/238.1;336.1/192.1	336.1/238.1	120	15:20
60	Bitertanol	55179-31-2	13.90	338.2/70;338.2/269.2	338.2/70	60	5:1
61	Chlorpirifos methyl	5598-13-0	16.72	322/125;322/290	322/125	80	15:15
62	Tepraloxydim	149979-41-9	12.73	342.2/250.2;342.2/166.1	342.2/250.2	120	10:25
63	Thiophanate methyl	23564-05-8	6.28	343.1/151.1;343.1/311.1	343.1/151.1	120	20:10
64	Azinphos ethyl	2642-71-9	14.00	346.0/233;346.0/261.1	346.0/233	120	10:5
65	Clodinafop propargyl	105512-06-9	16.09	350.1/266.1;350.1/238.1	350.1/266.1	120	15:20

Continued

TABLE 5.10 Retention Times, Quantifying and Qualifying Ions, Fragmentor, and Collision Energies of 569 Pesticides and Chemical Pollutants Determined by LC-MS-MS—cont'd

No.	Compound	CAS	Retention Time (min)	Qualifying Ion	Quantifying Ion	Fragmentor (V)	Collision Energy (V)
66	Triflumuron	64628-44-0	15.59	359.0/156.1;359.0/139	359.0/156.1	120	15:30
67	Isoxaflutole	141112-29-0	12.00	360.0/251.1;360.0/220.1	360.0/251.1	120	10:45
68	Anilofos	64249-01-0	17.35	367.9/145.2;367.9/205	367.9/145.2	120	20:5
69	Thiophanat ethyl	23564-06-9	9.32	371.1/151.1;371.1/325	371.1/151.1	120	15:10
70	Quizalofop-ethyl	76578-14-8	17.40	373.0/299.1;373.0/91	373.0/299.1	140	15:30
71	Haloxyfop-methyl	69806-40-2	17.11	376.0/316.0;376.0/288	376.0/316.0	120	15:20
72	Fluazifop butyl	79241-46-6	18.24	384.1/282.1;384.1/328.1	384.1/282.1	120	20:15
73	Bromophos-ethyl	4824-78-6	19.15	393.0/337.0;393.0/162.1	393.0/337.0	100	20:30
74	Bensulide	741-58-2	16.18	398.0/158.1;398.0/314.0	398.0/158.1	80	20:5
75	Triasulfuron	82097-50-5	7.27	402.1/167.1;402.1/141.1	402.1/167.1	120	15:20
76	Bromfenvinfos	33399-00-7	15.22	402.9/170.0;402.9/127	402.9/170.0	100	35:20
77	Azoxystrobin	131860-33-8	12.50	404.0/372.0;404.0/344.1	404.0/372.0	120	10:15
78	Pyrazophos	13457-18-6	16.20	374/222;374/194	374/222	120	20:30
79	Bensultap	17606-31-4	9.23	432.0/290.2;432.0/104	432.0/290.2	140	15:30
80	Flufenoxuron	101463-69-8	18.30	489.0/158.1;489.0/141.1	489.0/158.1	80	10:15

81	Indoxacarb	12124-97-9	17.43	528.0/150;528.0/218	528.0/150	120	20:20
82	Emamectin benzoate	15569-91-8	17.00	886.7/158.2;886.7/126.1	886.7/158.2	150	40:40
83	Ethylene thiourea	96-45-7	0.74	103/60;103.0/86.0	103/60	100	35:10
84	Daminozide	1596-84-	0.74	161.1/143.1;161.1/102.2	161.1/143.1	80	15:15
85	Dazomet	533-74-4	3.80	163.1/120.0;163.1/77	163.1/120.0	80	10:35
86	Nicotine	54-11-5	0.74	163.2/130.1;163.2/117.1	163.2/130.1	100	25:30
87	Fenuron	101-42-8	4.50	165.1/72.0;165.1/120	165.1/72.0	120	15:15
88	Cyromazine	66215-27-8	0.74	167/85;167.0/125.0	167/85	120	25:20
89	Crimidine	535-89-7	4.47	172.1/107.1;172.1/136.2	172.1/107.1	120	30:25
90	Acephate	30560-19-1	0.74	184.1/143.0;184.1/95	184.1/143.0	60	5:20
91	Molinate	2212-67-1	11.30	188.1/126.1;188.1/83	188.1/126.1	120	10:15
92	Carbendazim	10605-21-7	3.30	192.1/160.1;192.1/132.1	192.1/160.1	80	15:20
93	6-Chloro-4-hydroxy-3-phenyl-pyridazin	40020-01-7	12.86	207.1/77;207.1/104.0	207.1/77	120	25:35
94	Propoxur	114-26-1	6.79	210.1/111;210.1/168.1	210.1/111	80	10:5
95	Isouron	55861-78-4	6.11	212.2/167.1;212.2/72.0	212.2/167.1	120	15:25
96	Chlorotoluron	15545-48-9	7.23	213.1/72.0;213.1/140.1	213.1/72.0	80	25:25
97	Thiofanox	39196-18-4	1.00	241/184;241/57.1	241/184	120	15:5
98	Chlorbufam	1967-16-4	11.67	224.1/172.1;224.1/154.1	224.1/172.1	120	5:15
99	Bendiocarb	22781-23-3	6.87	224.1/109;224.1/167.1	224.1/109	80	5:10

Continued

TABLE 5.10 Retention Times, Quantifying and Qualifying Ions, Fragmentor, and Collision Energies of 569 Pesticides and Chemical Pollutants Determined by LC-MS-MS—cont'd

No.	Compound	CAS	Retention Time (min)	Qualifying Ion	Quantifying Ion	Fragmentor (V)	Collision Energy (V)
100	Propazine	139-40-2	9.37	229.9/146.1;229.9/188.1	229.9/146.1	120	20:15
101	Terbuthylazine	5915-41-3	10.15	230.1/174.1;230.1/132	230.1/174.1	120	15:20
102	Diuron	330-54-1	7.82	233.1/72.0;233.1/160.1	233.1/72.0	120	20:20
103	Chlormephos	24934-91-6	13.70	235/125;235.0/75	235/125	100	10:10
104	Carboxin	5234-68-4	7.67	236.1/143.1;236.1/87	236.1/143.1	120	15:20
105	Difenzoquat-methyl sulfate	43222-48-6	5.51	249.1/130.0;249.1/193.1	249.1/130.0	140	40:30
106	Clothianidin	210880-92-5	4.40	250.2/169.1;250.2/132.0	250.2/169.1	80	10:15
107	Pronamide	23950-58-5	11.81	256.1/190.1;256.1/173	256.1/190.1	80	10:20
108	Dimethachlor	50563-36-5	8.96	256.1/224.2;256.1/148.2	256.1/224.2	120	10:20
109	Methobromuron	40596-69-8	8.25	259.0/170.1;259/148	259.0/170.1	80	15:15
110	Phorate	298-02-2	16.55	261.0/75.0;261/199	261.0/75.0	80	10:5
111	Aclonifen	74070-46-5	14.70	265.1/248.0;265.1/193	265.1/248.0	120	15:15
112	Mephosfolan	950-10-7	5.97	270.1/140.1;270.1/168.1	270.1/140.1	100	25:15
113	Imibenzonazole-des-benzyl	199338-48-2	5.96	271.0/174.0;271/70	271.0/174.0	120	25:25
114	Neburon	555-37-3	14.17	275.1/57;275.1/88.1	275.1/57	120	20:15

115	Mefenoxam	70630-17-0	7.92	280.1/192.1;280.1/220.0	280.1/192.1	100	15:10
116	Prothoate	2275-18-5	4.78	286.1/227.1;286.1/199	286.1/227.1	100	5:15
117	Ethofumesate	26225-79-6	12.86	287/121;287.0/161.0	287/121	80	10:20
118	Iprobenfos	26087-47-8	13.50	289.1/91;289.1/205.1	289.1/91	80	25:5
119	TEPP	107-49-3	5.64	291.1/179.0;291.1/99	291.1/179.0	100	20:35
120	Cyproconazole	113096-99-4	10.59	292.1/70.0;292.1/125	292.1/70.0	120	15:15
121	Thiamethoxam	153719-23-4	4.05	292.1/211.2;292.1/181.1	292.1/211.2	80	10:20
122	Crufomate	299-86-5	11.56	292.1/236.0;292.1/108.1	292.1/236.0	120	20:30
123	Etrimfos	38260-54-7	6.16	293.1/125.0;293.1/265.1	293.1/125.0	80	20:15
124	Coumatetralyl	5836-29-3	4.68	293.2/107;293.2/175.1	293.2/107	140	35:25
125	Cythioate	115-93-5	6.59	298/217.1;298.0/125.0	298/217.1	100	15:25
126	Phosphamidon	13171-21-6	5.77	300.1/174.1;300.1/127.0	300.1/174.1	120	10:20
127	Phenmedipham	13864-63-4	10.69	301.1/168.1;301.1/136	301.1/168.1	80	5:20
128	Bifenazate	149877-41-8	13.28	301.2/198.1;301.2/170.1	301.2/198.1	60	5:20
129	Fenhexamid	126833-17-8	12.33	302.0/97.1;302/55	302.0/97.1	80	30:25
130	Flutriafol	76674-21-0	7.55	302.1/70;302.1/123.0	302.1/70	120	15:20
131	Furalaxyl	57646-30-7	10.77	302.2/242.2;302.2/270.2	302.2/242.2	100	15:5
132	Bioallethrin	584-79-2	18.00	303.1/135.1;303.1/107	303.1/135.1	80	10:20
133	Cyanofenphos	13067-93-1	16.44	304.0/157.0;304/276	304.0/157.0	100	20:10
134	Pirimiphos methyl	29232-93-7	15.50	306.2/164;306.2/108.1	306.2/164	120	20:30

Continued

TABLE 5.10 Retention Times, Quantifying and Qualifying Ions, Fragmentor, and Collision Energies of 569 Pesticides and Chemical Pollutants Determined by LC-MS-MS—cont'd

No.	Compound	CAS	Retention Time (min)	Qualifying Ion	Quantifying Ion	Fragmentor (V)	Collision Energy (V)
135	Buprofezin	69327-76-0	13.34	306.2/201;306.2/116.1	306.2/201	120	15:10
136	Disulfoton sulfone	2497-06-5	9.79	307.0/97.0;307/125	307.0/97.0	100	30:10
137	Fenazaquin	120928-09-8	18.80	307.2/57.1;307.2/161.2	307.2/57.1	120	20:15
138	Triazophos	24017-47-8	13.80	314.1/162.1;314.1/286	314.1/162.1	120	20:10
139	DEF	78-48-8	19.21	315.1/169.0;315.1/113	315.1/169.0	100	10:20
140	Pyriftalid	135186-78-6	12.00	319.0/139.1;319/179	319.0/139.1	140	35:35
141	Metconazole	125116-23-6	13.77	320.2/70.0;320.2/125	320.2/70.0	140	35:55
142	Pyriproxyfen	95737-68-1	18.00	322.1/96.0;322.1/227.1	322.1/96.0	120	15:10
143	Cycloxydim	101205-02-1	17.00	326.2/280.2;326.2/180.2	326.2/280.2	120	10:15
144	Isoxaben	82558-50-7	13.21	333.1/165.0;333.1/150.1	333.1/165.0	120	15:50
145	Flurtamone	96525-23-4	11.25	334.1/247.1;334.1/303	334.1/247.1	120	30:20
146	Trifluralin	1582-09-8	12.86	336/138.9;336/103	336/138.9	120	20:45
147	Flamprop methyl	52756-25-9	13.20	336.1/105.1;336.1/304.0	336.1/105.1	80	20:5
148	Bioresmethrin	28434-01-7	19.39	339.2/171.1;339.2/143.1	339.2/171.1	100	15:25
149	Propiconazole	60207-90-1	14.29	342.1/159.1;342.1/69	342.1/159.1	120	20:20

150	Chlorpyrifos	2921-88-2	18.29	350/198;350.0/79.0	350/198	100	20:35
151	Fluchloralin	33245-39-5	17.68	356.0/314.1;356/63	356.0/186	80	15:30
152	Chlorsulfuron	64902-72-3	6.96	358.0/141.1;358/167	358.0/141.1	120	15:15
153	Clethodim	99129-21-2	17.60	360.1/164.1;360.1/268.0	360.1/164.1	120	20:10
154	Flamprop isopropyl	52756-22-6	16.00	364.1/105.1;364.1/304.1	364.1/105.1	80	20:5
155	Tetrachlorvinphos	22248-79-9	13.70	365.0/127.0;365/239	365.0/127.0	120	15:15
156	Propargite	2312-35-8	18.77	368.1/231;368.1/175.1	368.1/231	100	5:15
157	Bromuconazole	116255-48-2	12.70	376.0/159.0;376/70	376.0/159.0	80	20:20
158	Picolinafen	137641-05-5	17.74	377.0/238.0;377/359	377.0/238.0	120	20:20
159	Fluthiacet methyl	117337-19-6	14.80	404/215;404.0/274.0	404/215	180	50:10
160	Trifloxystrobin	141517-21-7	17.44	409.3/186.1;409.3/206.2	409.3/186.1	120	15:10
161	Chlorimuron ethyl	90982-32-4	11.59	415.0/186.1;415/213.1	415.0/186.1	120	10:10
162	Hexaflumuron	86479-06-3	16.90	461.1/141.1;461.0/158.1	461/141.1	120	35:35
163	Novaluron	116714-46-6	17.39	493.0/158.0;493/141.1	493.0/158.0	80	15:55
164	Hydramethylnon	67485-29-4	17.58	495.2/323.2;495.2/171	495.2/323.2	100	35:50
165	Flurazuron	86811-58-7	18.10	506/158.1;506.0/141.1	506/158.1	120	15:50
166	Maleic hydrazide	123-33-1	0.73	113.1/67.1;113.1/85	113.1/67.1	100	20:20
167	Methamidophos	10265-92-6	0.74	142.1/94.0;142.1/125	142.1/94.0	80	15:10
168	EPTC	759-94-4	14.00	190.2/86;190.2/128.1	190.2/86	100	10:10
169	Diethyltoluamide	134-62-3	7.70	192.2/119.0;192.2/91	192.2/119.0	100	15:30

Continued

TABLE 5.10 Retention Times, Quantifying and Qualifying Ions, Fragmentor, and Collision Energies of 569 Pesticides and Chemical Pollutants Determined by LC-MS-MS—cont'd

No.	Compound	CAS	Retention Time (min)	Qualifying Ion	Quantifying Ion	Fragmentor (V)	Collision Energy (V)
170	Monuron	150-68-5	5.94	199/72;199.0/126.0	199/72	120	15:15
171	Pyrimethanil	53112-28-0	6.70	200.2/107.0;200.2/183.1	200.2/107.0	120	25:25
172	Fenfuram	24691-80-3	7.48	202.1/109.0;202.1/83	202.1/109.0	120	20:20
173	Quinoclamine	2797-51-5	6.09	208.1/105.0;208.1/154.1	208.1/105.0	120	30:20
174	Fenobucarb	3766-81-2	9.92	208.2/95.0;208.2/152.1	208.2/95.0	80	10:5
175	Ethirimol	23947-60-6	4.29	210.2/140.1;210.2/98	210.2/140.1	120	25:30
176	Propanil	709-98-8	9.09	218.0/162.1;218/127	218.0/162.1	120	15:20
177	Carbofuran	1563-66-2	6.81	222.3/165.1;222.3/123.1	222.3/165.1	120	5:20
178	Acetamiprid	160430-64-8	4.86	223.2/126.0;223.2/56	223.2/126.0	120	15:15
179	Mepanipyrim	110235-47-7	12.23	224.2/77.0;224.2/106	224.2/77.0	120	30:25
180	Prometon	1610-18-0	5.40	226.2/142;226.2/184.1	226.2/142	120	20:20
181	Methiocarb	2032-65-7	4.51	226.2/121.1;226.2/169.1	226.2/121.1	80	10:5
182	Metoxuron	19937-59-8	5.59	229.1/72.0;229.1/156.1	229.1/72.0	120	20:20
183	Dimethoate	60-51-5	4.88	230.0/199.0;230/171	230.0/199.0	80	5:10
184	Methfuroxam	2873-17-8	10.42	230.2/137.1;230.2/111.1	230.2/137.1	120	20:15

185	Fluometuron	2164-17-2	7.27	233.1/72.0;233.1/160	233.1/72.0	120	20:20
186	Dicrotophos	141-66-2	3.97	238.1/112.1;238.1/193	238.1/112.1	80	10:5
187	Monalide	7287-36-7	14.50	240.1/85.1;240.1/57	240.1/85.1	120	15:35
188	Diphenamid	957-51-7	9.00	240.1/134.1;240.1/167.1	240.1/134.1	120	20:25
189	Ethoprophos	13194-48-4	11.98	243.1/173;243.1/215.0	243.1/173	120	10:10
190	Fonofos	944-22-9	16.10	247.1/109.0;247.1/137.1	247.1/109.0	80	15:5
191	Etridiazol	2593-15-9	17.20	247.1/183.1;247.1/132.0	247.1/183.1	120	15:15
192	Furmecyclox	60568-05-0	14.00	252.2/170.1;252.2/110.1	252.2/170.1	100	10:25
193	Hexazinone	51235-04-2	5.66	253.2/171.1;253.2/71	253.2/171.1	120	15:20
194	Dimethametryn	14214-32-5	8.79	256.2/186.1;256.2/96.1	256.2/186.1	140	20:35
195	Trichlorphon	52-68-6	4.21	257/221;257.0/109.0	257/221	120	10:20
196	Demeton(O+S)	126-75-0	8.59	259.1/89.0;259.1/61	259.1/89.0	60	10:35
197	Benoxacor	98730-04-2	10.83	260.0/149.2;260/134.1	260.0/149.2	120	15:20
198	Bromacil	314-40-9	5.78	261.0/205.0;261/188	261.0/205.0	80	10:20
199	Phorate sulfoxide	2588-03-6	7.34	277/143;277/199	277/143	100	15:5
200	Brompyrazon	3042-84-0	4.69	266.0/92.0;266/104	266.0/92.0	120	30:30
201	Oxycarboxin	5259-88-1	5.38	268.0/175.0;268/147.1	268.0/175.0	100	10:20
202	Mepronil	55814-41-0	13.15	270.2/119.1;270.2/228.2	270.2/119.1	100	30:15
203	Disulfoton	298-04-4	16.80	275.0/89.0;275/61	275.0/89.0	80	5:20
204	Fenthion	55-38-9	15.54	279.0/169.1;279/247	279.0/169.1	120	15:10

Continued

TABLE 5.10 Retention Times, Quantifying and Qualifying Ions, Fragmentor, and Collision Energies of 569 Pesticides and Chemical Pollutants Determined by LC-MS-MS—cont'd

No.	Compound	CAS	Retention Time (min)	Qualifying Ion	Quantifying Ion	Fragmentor (V)	Collision Energy (V)
205	Metalaxyl	57837-19-1	7.75	280.1/192.2;280.1/220.2	280.1/192.2	120	15:20
206	Ofurace	58810-48-3	7.65	282.1/160.2;282.1/254.2	282.1/160.2	120	20.1
207	Dodemorph	1593-77-7	8.45	282.3/116.1;282.3/98.1	282.3/116.1	120	20:30
208	Fosthiazate	98886-44-3	4.38	284.1/228.1;284.1/104	284.1/228.1	80	5:20
209	Imazamethabenz-methyl	81405-85-8	5.33	289.1/229;289.1/86.0	289.1/229	120	15:25
210	Disulfoton-sulfoxide	2497-07-6	7.38	291.0/185;291/157	291.0/185	80	10:20
211	Isoprothiolane	50512-35-1	13.17	291.1/189.1;291.1/231.1	291.1/189.1	80	20:5
212	Imazalil	35554-44-0	6.86	297.0/159.0;297/255	297.0/159.0	120	20:20
213	Phoxim	14816-18-3	16.80	299.0/77.0;299/129	299.0/77.0	80	20:10
214	Quinalphos	13593-03-8	14.80	299.1/147.1;299.1/163.1	299.1/147.1	120	20:20
215	Ditalimfos	5131-24-8	13.53	300/148.1;300.0/244	300/148.1	80	15:10
216	Fenoxycarb	79127-80-3	18.10	362.1/288.0;362.1/244	362.1/288.0	120	20:20
217	Pyrimitate	5221-49-8	14.00	306.1/170.2;306.1/154.2	306.1/170.2	120	20:20
218	Fensulfothin	115-90-2	8.55	309.0/157.1;309/253	309.0/157.1	120	25:15
219	Fluorochloridone	61213-25-0	13.80	312.1/292.1;312.1/89	312.1/292.1	100	25:25

220	Butachlor	23184-66-9	18.00	312.2/238.1;312.2/162	312.2/238.1	80	10:20
221	Imazaquin	81335-37-7	6.27	312.2/199.1;312.2/267	312.2/199.1	160	25:20
222	Kresoxim-methyl	143390-89-0	15.20	314.1/267;314.1/206	314.1/267	80	5:5
223	Triticonazole	131983-72-7	10.55	318.2/70.0;318.2/125.1	318.2/70.0	120	15:35
224	Fenamiphos sulfoxide	31972-42-7	5.87	320.1/171.1;320.1/292.1	320.1/171.1	140	25:15
225	Thenylchlor	96491-05-3	14.00	324.1/127.0;324.1/59	324.1/127.0	80	10:45
226	Fenoxanil	115852-48-7	18.81	329.1/302.0;329.1/189.1	329.1/302.0	80	5:30
227	Fluridone	59756-60-4	10.30	330.1/309.1;330.1/259.2	330.1/309.1	160	40:55
228	Epoxiconazole	106325-08-0	18.81	330.1/141.1;330.1/121.1	330.1/141.1	120	20:20
229	Chlorphoxim	14816-20-7	17.15	333.0/125.0;333/163.1	333.0/125.0	80	5:5
230	Fenamiphos sulfone	31972-44-8	6.63	336.1/188.2;336.1/266.2	336.1/188.2	120	30:20
231	Fenbuconazole	114369-43-6	13.40	337.1/70;337.1/125.0	337.1/70	120	20:20
232	Isofenphos	25311-71-1	17.25	346.1/217.0;346.1/245	346.1/217.0	80	20:10
233	Phenothrin	26002-80-2	19.70	351.1/183.2;351.1/237	351.1/183.2	100	15:5
234	Fentin-chloride	6369-58-7	7.00	351.1/120;351.1/170	351.1/120	180	40:30
235	Piperophos	24151-93-7	17.00	354.1/171;354.1/143.0	354.1/171	100	20:30
236	Piperonyl butoxide	51-03-6	17.75	356.2/177.1;356.2/119	356.2/177.1	100	10:35
237	Oxyfluorfen	42874-03-3	18.00	362.0/316.1;362/237.1	362.0/316.1	120	10:25
238	Coumaphos	56-72-4	16.42	363.1/227.2;363.1/307.1	363.1/227.2	120	20:15
239	Flufenacet	142459-58-3	14.00	364.0/194.0;364/152	364.0/194.0	80	5:10

Continued

TABLE 5.10 Retention Times, Quantifying and Qualifying Ions, Fragmentor, and Collision Energies of 569 Pesticides and Chemical Pollutants Determined by LC-MS-MS—cont'd

No.	Compound	CAS	Retention Time (min)	Qualifying Ion	Quantifying Ion	Fragmentor (V)	Collision Energy (V)
240	Phosalone	2310-17-0	16.79	368.1/182.0;368.1/322	368.1/182.0	80	10:5
241	Methoxyfenozide	161050-58-4	13.41	313/149;313/91	313/149	100	10:35
242	Prochloraz	67747-09-5	11.79	376.1/308.0;376.1/266	376.1/308.0	80	10:10
243	Aspon	3244-90-4	19.22	379.1/115.0;379.1/210	379.1/115.0	80	30:15
244	Ethion	562-12-2	18.46	385.0/199.1;385/171	385.0/199.1	80	5:15
245	Diafenthiuron	80060-09-9	18.90	385.0/329.2;385.0/278.2	385.0/329.2	140	15:35
246	Thifensulfuron-methyl	79277-27-3	6.40	388.1/167;388.1/141.1	388.1/167	120	10:10
247	Ethoxysulfuron	126801-58-9	11.86	399.2/261.1;399.2/218.1	399.2/261.1	120	25:25
248	Dithiopyr	97886-45-8	17.81	402.0/354.0;402/272	402.0/354.0	120	20:30
249	Spirodiclofen	148477-71-8	19.28	411.1/71.0;411.1/313.1	411.1/71.0	100	10:5
250	Fenpyroximate	134098-61-6	18.66	422.2/366.2;422.2/135	422.2/366.2	120	10:35
251	Flumiclorac-pentyl	87546-18-7	18.00	441.1/308.0;441.1/354	441.1/308.0	100	25:10
252	Temephos	3383-96-8	18.30	467.0/125.0;467.0/155.0	467.0/125.0	100	30:30
253	Butafenacil	134605-64-4	15.00	492.0/180.0;492.0/331.0	492.0/180.0	120	35:25
254	Spinosad	131929-63-0	14.30	732.4/142.2;732.4/98.1	732.4/142.2	180	30:75

255	Mepiquat chloride	24307-26-4	0.71	114.1/98.1;114.1/58	114.1/98.1	140	30:30
256	Allidochlor	93-71-0	5.78	174.1/98.1;174.1/81	174.1/98.1	100	10:15
257	Propamocarb	24579-73-5	2.84	190.1/102.1;190.1/74.1	190.1/102.1	110	20:30
258	Tricyclazole	41814-78-2	5.06	190.1/136.1;190.1/163.1	190.1/136.1	120	30:25
259	Thiabendazole	148-79-8	3.32	202.1/175.1;202.1/131.1	202.1/175.1	120	30:30
260	Metamitron	41394-05-2	4.18	203.1/175.1;203.1/104	203.1/175.1	120	15:20
261	Isoproturon	34123-59-6	7.44	207.2/72.0;207.2/165.1	207.2/72.0	120	15:15
262	Atratone	1610-17-9	4.46	212.2/170.2;212.2/100.1	212.2/170.2	120	15:30
263	Desmetryn	1014-69-3	4.92	214.1/172.1;214.1/82.1	214.1/172.1	120	15:25
264	Metribuzin	21087-64-9	7.16	215.1/187.2;215.1/131.1	215.1/187.2	120	15:20
265	DMST	66840-71-9	7.06	215.3/106.1;215.3/151.2	215.3/106.1	80	10:5
266	Cycloate	1134-23-2	15.95	216.2/83.0;216.2/154.1	216.2/83.0	120	15:10
267	Atrazine	1912-24-9	7.20	216.0/174.2;216.0/132	216.0/174.2	120	15:20
268	Butylate	2008-41-5	17.20	218.1/57.0;218.1/156.2	218.1/57.0	80	10:5
269	Pymetrozin	123312-89-0	0.73	218.1/105.1;218.1/78	218.1/105.1	100	20:40
270	Chloridazon	1968-60-8	4.35	222.1/104.0;222.1/92	222.1/104.0	120	25:35
271	Sulfallate	95-06-7	15.25	224.1/116.1;224.1/88.2	224.1/116.1	100	10:20
272	Ethiofencarb	29973-13-5	4.48	227/107;227/164	227/107	80	5:5
273	Terbumeton	33693-04-8	5.25	226.2/170.1;226.2/114	226.2/170.1	120	15:20
274	Cyprazine	22936-86-3	7.15	228.2/186.1;228.2/108.1	228.2/186.1	120	15:25

Continued

TABLE 5.10 Retention Times, Quantifying and Qualifying Ions, Fragmentor, and Collision Energies of 569 Pesticides and Chemical Pollutants Determined by LC-MS-MS—cont'd

No.	Compound	CAS	Retention Time (min)	Qualifying Ion	Quantifying Ion	Fragmentor (V)	Collision Energy (V)
275	Ametryn	834-12-8	5.85	228.2/186;228.2/68	228.2/186	120	20:35
276	Tebuthiuron	34014-18-1	5.30	229.2/172.2;229.2/116	229.2/172.2	120	15:20
277	Trietazine	1912-26-1	12.00	230.1/202;230.1/132.1	230.1/202	160	20:20
278	Sebutylazine	7286-69-3	8.65	230.1/174.1;230.1/104	230.1/174.1	12	15:30
279	Dibutyl succinate	141-03-7	14.80	231.1/101;231.1/157.1	231.1/101	60	1:10
280	Tebutam	35256-85-0	13.04	234.2/91.1;234.2/192.2	234.2/91.1	120	20:15
281	Thiofanox-sulfoxide	39184-27-5	4.08	235.1/104.0;235.1/57	235.1/104.0	60	5:20
282	Cartap hydrochloride	15263-52-2	5.90	238/73;238.0/150	238/73	100	30:10
283	Methacrifos	62610-77-9	10.03	241/209;241.0/125.0	241/209	60	5:20
284	Terbutryn	886-50-0	7.44	242.2/186.1;242.2/71	242.2/186.1	120	15:20
285	Triazoxide	72459-58-6	5.66	248/68;248.0/95	248/68	100	35:25
286	Thionazin	297-97-2	8.84	249.1/97.0;249.1/193	249.1/97.0	80	30:10
287	Linuron	330-55-2	9.84	249.0/160.1;249/182.1	249.0/160.1	100	15:15
288	Heptanophos	23560-59-0	7.85	251.0/127.0;251/109	251.0/127.0	80	10:30
289	Prosulfocarb	52888-80-9	17.10	252.1/91.0;252.1/128.1	252.1/91.0	120	15:10

Continued

No.	Name	CAS	RT	Transitions			
290	Dipropetryn	4147-51-7	8.58	256.1/144.1;256.1/214	256.1/144.1	140	30:20
291	Thiobencarb	28249-77-6	15.80	258.1/125.0;258.1/89	258.1/125.0	80	20:55
292	Tri-iso-butyl phosphate	126-71-6	15.45	267.1/99;267.1/155.1	267.1/99	80	20:5
293	Tri-n-butyl phosphate	126-73-8	15.45	267.2/99.0;267.2/155.1	267.2/99.0	80	5:15
294	Diethofencarb	87130-20-9	10.40	268.1/226.2;268.1/152.1	268.1/226.2	80	5:20
295	Alachlor	15972-60-8	13.15	270.2/238.2;270.2/162.2	270.2/238.2	80	10:20
296	Cadusafos	95465-99-9	15.27	271.1/159.1;271.1/131	271.1/159.1	80	10:20
297	Metazachlor	67129-08-2	8.36	278.1/134.1;278.1/210.1	278.1/134.1	80	20:5
298	Propetamphos	31218-83-4	13.60	282.1/138;282.1/156.1	282.1/138	80	15:10
299	Terbufos	13071-79-9	13.70	289/57;289.0/103.1	289/57	80	20:5
300	Simeconazole	149508-90-7	11.00	294.2/70.1;294.2/135.1	294.2/70.1	120	15:15
301	Triadimefon	43121-43-3	11.88	294.2/69;294.2/197.1	294.2/69	100	20:15
302	Phorate sulfone	2588-04-7	9.34	293.0/171.0;293/143.1	293.0/171.0	60	5:15
303	Tridemorph	24602-86-6	14.00	298.3/130.1;298.3/57.1	298.3/130.1	160	25:35
304	Mefenacet	73250-68-7	11.60	299.1/148.1;299.1/120.1	299.1/148.1	100	15:25
305	Azaconazole	60207-31-0	12.37	300.1/231.1;300.1/159.0	300.1/231.1	100	15:30
306	Fenamiphos	22224-92-6	8.97	304.0/216.9;304.0/202	304.0/216.9	100	20:35
307	Fenpropimorph	67564-91-4	9.10	304.0/147.2;304.0/130	304.0/147.2	120	30:30
308	Tebuconazole	107534-96-3	12.44	308.2/70.0;308.2/125	308.2/70.0	100	25:25
309	Isopropalin	33820-53-0	19.05	310.2/225.7;310.2/207.7	310.2/225.7	120	15:20

TABLE 5.10 Retention Times, Quantifying and Qualifying Ions, Fragmentor, and Collision Energies of 569 Pesticides and Chemical Pollutants Determined by LC-MS-MS—cont'd

No.	Compound	CAS	Retention Time (min)	Qualifying Ion	Quantifying Ion	Fragmentor (V)	Collision Energy (V)
310	Nuarimol	63284-71-9	9.20	315.1/252.1;315.1/81	315.1/252.1	120	25:30
311	Bupirimate	41483-43-6	9.52	317.2/166;317.2/272	317.2/166	120	25:20
312	Azinphos-methyl	86-50-0	10.45	318.1/125;318.1/160	318.1/125	80	15:10
313	Tebupirimfos	96182-53-5	18.15	319.1/277.1;319.1/153.2	319.1/277.1	120	10:30
314	Phenthoate	2597-03-7	15.57	321.1/247;321.1/163.1	321.1/247	80	5:10
315	Sulfotep	3689-24-5	16.35	323/171.1;323.0/143	323/171.1	120	10:20
316	Sulprofos	35400-43-2	18.40	323.0/219.1;323/247	323.0/219.1	120	15:10
317	EPN	2104-64-5	17.10	324/296;324.0/157.1	324/296	120	10:20
318	Azamethiphos	35575-96-3	6.05	325.0/183.0;325/139	325.0/183.0	80	15:25
319	Diniconazole	83657-24-3	13.67	326.1/70.0;326.1/159	326.1/70.0	120	25:30
320	Flumetsulam	98967-40-9	4.95	326.1/129.0;326.1/262.1	326.1/129.0	120	30:20
321	Sethoxydim	74051-80-2	5.36	328.2/282.2;328.2/178.1	328.2/282.2	100	10:15
322	Pencycuron	66063-05-6	16.33	329.2/125.0;329.2/218.1	329.2/125.0	120	20:15
323	Mecarbam	2595-54-2	14.46	330/227;330.0/199.0	330/227	80	5:10
324	Tralkoxydim	87820-88-0	18.09	330.2/284.2;330.2/138.1	330.2/284.2	100	10:20

325	Malathion	121-75-5	13.20	331.0/127.1;331/99	331.0/127.1	80	5:10
326	Pyributicarb	88678-67-5	18.26	331.1/181.1;331.1/108	331.1/181.1	120	10:20
327	Pyridaphenthion	119-12-0	12.32	341.1/189.2;341.1/205.2	341.1/189.2	120	20:20
328	Pirimiphos-ethyl	23505-41-1	17.75	334.2/198.2;334.2/182.2	334.2/198.2	120	20:25
329	Thiodicarb	59669-26-0	6.55	355.1/88.0;355.1/163	355.1/88.0	80	15:5
330	Pyraclofos	77458-01-6	15.34	361.1/257.0;361.1/138	361.1/257.0	120	25:35
331	Picoxystrobin	117428-22-5	15.40	368.1/145.0;368.1/205.0	368.1/145.0	80	20:5
332	Tetraconazole	112281-77-3	12.54	372.0/159.0;372/70	372.0/159.0	120	35:35
333	Mefenpyr-diethyl	135590-91-9	16.80	373/327;373.0/160.0	373/327	80	15:35
334	Profenefos	41198-08-7	16.74	373.0/302.9;373/345	373.0/302.9	120	15:10
335	Pyraclostrobin	175013-18-0	16.04	388/163;388.0/194.0	388/163	120	20:10
336	Dimethomorph	211867-47-9	16.04	388.1/165.1;388.1/301.1	388.1/165.1	120	25:20
337	Kadethrin	58769-20-3	17.95	397.1/171.1;397.1/128	397.1/171.1	100	15:55
338	Thiazopyr	117718-60-2	16.15	397.1/377;397.1/335.1	397.1/377	140	20:30
339	Benfuracarb-methyl	82560-54-1	8.60	411.1/149.1;411.1/182.1	411.1/149.1	100	20:20
340	Cinosulfuron	94593-91-6	6.53	414.1/183.1;414.1/157.1	414.1/183.1	120	10:20
341	Pyrazosulfuron-ethyl	93699-74-6	17.20	415.1/182.1415.1/369.1	415.1/182.1	120	15:10
342	Metosulam	139528-85-1	7.60	418.0/175.1;418/354	418.0/175.1	120	25:20
343	Chlorfluazuron	71422-67-8	18.53	540.0/383.0;540/158.2	540.0/383.0	120	15:15
344	4-Aminopyridine	504-24-5	0.72	95.1/52.1;95.1/78.1	95.1/52.1	120	25:5

Continued

TABLE 5.10 Retention Times, Quantifying and Qualifying Ions, Fragmentor, and Collision Energies of 569 Pesticides and Chemical Pollutants Determined by LC-MS-MS—cont'd

No.	Compound	CAS	Retention Time (min)	Qualifying Ion	Quantifying Ion	Fragmentor (V)	Collision Energy (V)
345	Chlormequat	999-81-5	0.72	122.1/58.1;122.1/63.1	122.1/58.1	100	35:20
346	Methomyl	16752-77-5	3.76	163.2/88.1;163.2/106.1	163.2/88.1	80	5:10
347	Pyroquilon	57369-32-1	5.87	174.1/117.1;174.1/132.2	174.1/117.1	140	35:25
348	Fuberidazole	3878-19-1	3.66	185.2/157.2;185.2/92.1	185.2/157.2	120	20:25
349	Isocarbamid	30979-48-7	4.35	186.2/87.1;186.2/130.1	186.2/87.1	80	20:5
350	Butocarboxim	34681-10-2	5.30	213/75.1;213/156.1	213/75.1	100	15:5
351	Chlordimeform	6164-98-3	4.13	197.2/117.1;197.2/89.1	197.2/117.1	120	25:50
352	Cymoxanil	57966-95-7	4.95	199.1/111.1;199.1/128.1	199.1/111.1	80	20:15
353	Vernolate	1929-77-7	3.47	204.2/128.2;204.2/175.5	204.2/128.2	100	10:10
354	Chlorthiamid	1918-13-4	5.80	206.0/189.0;206.0/119	206.0/189.0	80	15:50
355	Aminocarb	2032-59-9	0.75	209.3/137.1;209.3/152.1	209.3/137.1	100	20:10
356	Dimethirimol	5221-53-4	4.20	210.2/71.1;210.2/140	210.2/71.1	120	25:20
357	Omethoate	1113-02-6	0.75	214.1/125.0;214.1/183.0	214.1/125.0	80	20:5
358	Ethoxyquin	91-53-2	7.19	218.2/174.2;218.2/160.1	218.2/174.2	120	30:35
359	Dichlorvos	62-73-7	4.20	222.9.0/109.0;222.9/79	222.9.0/109.0	120	15:30

360	Aldicarb sulfone	1646-88-4	3.50	223.1/76;223.1/148.0	223.1/76	80	5:5
361	Dioxacarb	6988-21-2	4.70	224.1/123.1;224.1/167.1	224.1/123.1	80	15:5
362	Benzyladenine	1214-39-7	4.16	226.1/91.1;226.1/148.0	226.1/91.1	140	20:15
363	Demeton-s-methyl	919-86-8	6.25	253/89.0;253/61	253/89.0	80	10:35
364	Ethiofencarb-sulfoxide	53380-22-6	3.95	242.2/107.1;242.2/185.1	242.2/107.1	80	15:5
365	Cyanohos	2636-26-2	6.89	244.2/180.0;244.2/125.0	244.2/180.0	120	20:15
366	Thiometon	640-15-3	7.16	247.1/171.0;247.1/89.1	247.1/171.0	100	10:10
367	Folpet	133-07-3	12.82	260.0/130.0;260.0/102.3	260.0/130.0	100	10:40
368	Demeton-s-methyl sulfone	17040-19-6	3.96	263.1/169.1;263.1/125	263.1/169.1	80	15:20
369	Dimepiperate	61432-55-1	16.82	286.1/168;286.1/119.1	286.1/168	80	10:10
370	Fenpropidin	67306-00-7	8.96	274.0/147.1;274.0/86.1	274.0/147.1	160	25:25
371	Amidithion	919-76-6	14.25	274.1/97;274.1/122	274.1/97	140	20:15
372	Imazapic	104098-48-8	4.80	276.2/163.2;276.2/216.2;276.2/86.1	276.2/163.2	120	20:20:25
373	Paraoxon-ethyl	311-45-5	8.00	276.2/220.1;276.2/94.1	276.2/220.1	100	10:40
374	Aldimorph	174-28-5	14.10	284.4/57.2;284.4/98.1	284.4/57.2	160	30:30
375	Vinclozolin	50471-44-8	14.66	286.1/242;286.1/145.1	286.1/242	100	5:45
376	Uniconazole	83657-22-1	11.69	292.1/70.1;292.1/125.1	292.1/70.1	120	30:30
377	Pyrifenox	88283-41-4	7.42	295.0/93.1;295.0/163.0	295.0/93.1	120	15:15
378	Chlorthion	500-28-7	14.45	298.0/125.0;298/109	298.0/125.0	100	15:20

Continued

TABLE 5.10 Retention Times, Quantifying and Qualifying Ions, Fragmentor, and Collision Energies of 569 Pesticides and Chemical Pollutants Determined by LC-MS-MS—cont'd

No.	Compound	CAS	Retention Time (min)	Qualifying Ion	Quantifying Ion	Fragmentor (V)	Collision Energy (V)
379	Dicapthon	2463-84-5	14.47	298/125;298.0/266.1	298/125	80	10:10
380	Clofentezine	74115-24-5	16.18	303.0/138.0;303.0/156.0	303.0/138.0	100	25:25
381	Norflurazon	27314-13-2	8.08	304.0/284;304.0/160.1	304.0/284	140	25:35
382	Triallate	2303-17-5	18.52	304.0/143.0;304.0/86.1	304.0/143.0	120	25:15
383	Ziram	137-30-4	7.32	305/185.9;305/88.1	305/185.9	120	10:20
384	Quinoxyphen	124495-18-7	17.05	308.0/197.0;308.0/272.0	308.0/197.0	180	35:35
385	Fenthion sulfone	3761-42-0	8.71	311.1/125.0;311.1/109	311.1/125.0	140	15:20
386	Flurochloridone	61213-25-0	13.34	312.2/292.2;312.2/53.1	312.2/292.2	140	25:30
387	Phthalic acid, benzyl butyl ester	85-68-7	17.34	313.2/91.1;313.2/149.0; 313.2/205.1	313.2/91.1	80	10:10:5
388	Isazofos	42509-80-8	13.67	314.1/162.1;314.1/120.0	314.1/162.1	100	10:35
389	Dichlofenthion	97-17-6	18.15	315.0/259.0;315/287	315.0/259.0	100	10:5
390	Vamidothion sulfone	70898-34-9	2.45	178/87;178/60	178/87	100	15:10
391	Terbufos sulfone	56070-16-7	12.57	321.2/171.1;321.2/143.0	321.2/171.1	80	5:15
392	Dinitramine	29091-05-2	15.80	323.1/305.0;323.1/247.0	323.1/305.0	120	10:15

393	Cyazofamid	120116-88-3	5.10	325.2/261.3;325.2/108.0	325.2/261.3	80	5:15
394	Trichloronat	327-98-0	18.98	333.1/304.9;333.1/161.8	333.1/304.9	100	10:45
395	Resmethrin-2	10453-86-8	12.35	339.2/171.1;339.2/143.1	339.2/171.1	80	10:25
396	Boscalid	188425-85-6	12.20	343.2/307.2;343.2/271	343.2/307.2	140	20:35
397	Nitralin	4726-14-1	15.15	346.1/304.1;346.1/262.1	346.1/304.1	100	10:20
398	Fenpropathrin	395151-41-8	19.00	350.2/125.2;350.2/97	350.2/125.2	120	5:20
399	Hexythiazox	78587-05-0	18.23	353.1/168.1;353.1/228.1	353.1/168.1	120	20:10
400	Florasulam	145701-23-1	6.80	360.2/129.1;360.2/192	360.2/129.1	120	30:15
401	Benzoximate	29104-30-1	17.00	386.1/197;386.1/199.2	386.1/197	140	30:30
402	Benzoylprop-ethyl	22212-55-1	16.00	366.1/105.0;366.1/77	366.1/105.0	80	15:35
403	Pyrimidifen	105779-78-0	13.69	378.2/184.1;378.2/150.2	378.2/184.1	140	15:40
404	Furathiocarb	65907-30-4	17.85	383.3/195.1;383.3/252.1; 383.3/167	383.3/195.1	100	10:5:25
405	trans-Permethrin	551877-74-8	21.00	391.3/149.1;391.3/167.1	391.3/149.1	100	10:10
406	Etofenprox	80844-07-1	19.73	394/177;394/359	394/177	100	15:5
407	Pyrazoxyfen	71561-11-0	14.30	403.2/91.1;403.2/105.1; 403.2/139.1	403.2/91.1	140	25:20:20
408	Flubenzimine	37893-02-0	14.48	417.0/397;417.0/167.1	417.0/397	100	10:25
409	Zeta cypermethrin	52315-07-8	20.45	433.3/416.2;433.3/191.2	433.3/416.2	100	5:10
410	Haloxyfop-2-ethoxyethyl	87237-48-7	17.65	434.1/316.0;434.1/288.0; 434.1/91.2	434.1/316.0	120	15:20:45

Continued

TABLE 5.10 Retention Times, Quantifying and Qualifying Ions, Fragmentor, and Collision Energies of 569 Pesticides and Chemical Pollutants Determined by LC-MS-MS—cont'd

No.	Compound	CAS	Retention Time (min)	Qualifying Ion	Quantifying Ion	Fragmentor (V)	Collision Energy (V)
411	Esfenvalerate	66230-04-4	8.23	437.2/206.9;437.2/154.2	437.2/206.9	80	35:20
412	Fluoroglycofen-ethyl	77501-90-7	17.70	344/300;344/233	344/300	120	15:20
413	Tau-fluvalinate	102851-06-9	19.58	503.2/181.2;503.2/208.1	503.2/181.2	80	25:15
414	Acrylamide	79-06-1	0.73	72.0/55;72.0/27	72.0/55	100	10:10
415	Tert-butylamine	75-64-9	0.65	74.1/46;74.1/56.8	74.1/46	120	5:5
416	Hymexazol	10004-44-1	2.65	100.1/54.1;100.1/44.2; 100.1/28	100.1/54.1	100	10:15:15
417	Chlormequat chloride	999-81-5	0.69	122.1/58.1;122.1/63	122.1/58.1	120	30:20
418	Phthalimide	115-26-4	0.74	148.0/130.1;148.0/102	148.0/130.1	100	10:25
419	Dimefox	1129-41-5	3.88	155.1/110.1;155.1/135	155.1/110.1	120	20:10
420	Metolcarb	122-39-4	6.50	166.2/109.0;166.2/97.1	166.2/109.0	80	15:50
421	Diphenylamin	86-86-2	13.06	170.2/93.1;170.2/152	170.2/93.1	120	30:30
422	1-Naphthy acetamide	6190-65-4	5.30	186.2/141.1;186.2/115.1	186.2/141.1	100	15:45
423	Atrazine-desethyl	2008-58-4	4.43	188.2/146.1;188.2/104.1	188.2/146.1	120	10:20
424	2,6-Dichlorobenzamide	116-06-3	3.85	190.1/173.0;190.1/145	190.1/173.0	100	20:30

425	Aldicarb	116-06-3	5.42	213/89;213/116	213/89	100	30:10
426	Dimethyl phthalate	131-11-3	3.50	217/86;217/156	217/86	100	15:20
427	Chlordimeform hydrochloride	19750-95-9	4.00	197.2/117.1;197.2/89.1	197.2/117.1	120	25:50
428	Simeton	673-04-1	3.94	198.2/100.1;198.2/128.2	198.2/100.1	120	25:20
429	Dinotefuran	165252-70-0	3.06	203.3/129.2;203.3/87.1	203.3/129.2	80	5:10
430	Pebulate	1114-71-2	16.05	204.2/72.1;204.2/128	204.2/72.1	100	10:10
431	Acibenzolar-S-methyl	135158-54-2	10.00	211.1/91;211.1/136.0	211.1/91	120	20:30
432	Dioxabenzofos	3811-49-2	10.15	217/77.1;217.0/107.1	217/77.1	100	40:30
433	Oxamyl	23135-22-0	3.46	241/72;242/121	241/72	120	15:10
434	Thidiazuron	51707-55-2	5.60	221.1/102.0;221.1/128	221.1/102.0	100	15:5
435	Methabenzthiazuron	18691-97-9	6.80	222.2/165.1;222.2/149.9	222.2/165.1	100	15:35
436	Butoxycarboxim	34681-23-7	3.30	223.2/63;223.2/106.1	223.2/63	80	10:5
437	Mexacarbate	315-18-4	4.00	233.2/151.2;233.2/166.2	233.2/151.2	100	15:10
438	Demeton-S-methyl sulfoxide	301-12-2	3.42	247.1/109;247.1/169.1	247.1/109	80	20:10
439	Thiofanox sulfone	39184-59-3	7.30	251.1/57.2;251.1/76.1	251.1/57.2	80	5:5
440	Phosfolan	947-02-4	4.95	256.2/140.0;256.2/228	256.2/140.0	100	25:10
441	Triclopyr	55335-06-3	17.70	256.2/146;256.2/212	256.2/146	100	20:5
442	Demeton-S	126-75-0	5.44	259.1/89.1;259.1/61	259.1/89.1	60	10:35
443	Imazapyr	81334-34-1	5.00	260/173;260/216	260/173	80	15:5

Continued

TABLE 5.10 Retention Times, Quantifying and Qualifying Ions, Fragmentor, and Collision Energies of 569 Pesticides and Chemical Pollutants Determined by LC-MS-MS—cont'd

No.	Compound	CAS	Retention Time (min)	Qualifying Ion	Quantifying Ion	Fragmentor (V)	Collision Energy (V)
444	Fenthion oxon	6552-12-1	8.15	263.2/230;263.2/216	263.2/230	100	10;20
445	Napropamide	15299-99-7	12.45	272.2/171.1;272.2/129.2	272.2/171.1	120	15;15
446	Fenitrothion	122-14-5	13.60	278.1/125.0;278.1/246	278.1/125.0	140	15;15
447	Phthalic acid, dibutyl ester	84-74-2	17.50	279.2/149.0;279.2/121.1	279.2/149.0	80	10;45
448	Metolachlor	51218-45-2	13.15	284.1/252.2;284.1/176.2	284.1/252.2	120	10;15
449	Procymidone	32809-16-8	13.33	284/256;284/145	284/256	140	10;45
450	Vamidothion	2275-23-2	4.18	288.2/146.1;288.2/118.1	288.2/146.1	80	10;20
451	Chloroxuron	1982-47-4	9.00	291.2/72.1;291.2/218.1	291.2/72.1	120	20;30
452	Triamiphos	1031-47-6	6.58	295.2/135.1;295.2/92	295.2/135.1	100	25;35
453	Prallethrin	23031-36-9	7.25	301/105;301/169	301/105	80	5;20
454	Cumyluron	99485-76-4	11.70	303.3/185.1;303.3/125	303.3/185.1	100	5;45
455	Imazamox	114311-32-9	3.00	304.2/260;304.2/186	304.2/260	100	5;40
456	Warfarin	81-81-2	10.30	309.2/163.1;309.2/251.2	309.2/163.1	100	20;15
457	Phosmet	732-11-6	11.14	318.0/160.1;318.0/133	318.0/160.1	80	10;35
458	Ronnel	299-84-3	17.70	320.9/125.0;320.9/288.8	320.9/125.0	120	10;10

459	Pyrethrin	121-29-9	18.78	329.2/161.1;329.2/133.1	329.2/161.1	100	5:15
460	Phthalic acid, biscyclohexyl ester	84-61-7	19.10	331.3/149.1;331.3/167.1; 331.3/249	331.3/149.1	80	10:5:5
461	Carpropamid	104030-54-8	15.36	334.2/196.1;334.2/139.1	334.2/196.1	120	10:15
462	Tebufenpyrad	119168-77-3	17.32	334.3/147;334.3/117.1	334.3/147	160	25:40
463	Tebufenozide	112410-23-8	14.70	297/133;97/105	297/133	80	15:35
464	Iminoctadine triacetate	39202-40-9	3.18	356.3/314.3;356.3/297.4; 356.3/322.4	356.3/314.3	160	20:25:25
465	Chlorthiophos	60238-56-4	18.58	361.0/305.0;361/225	361.0/305.0	100	10:15
466	Naled	300-76-5	17.38	378.8/108.9;378.8/127.1	378.8/108.9	100	40:15
467	Dialifos	142891-20-1	17.15	394.0/208;394.0/187	394.0/208	100	5:20
468	Cinidon-ethyl	83-79-4	17.63	394.2/348.1;394.2/107.1	394.2/348.1	120	15:45
469	Rotenone	86598-92-7	14.00	395.3/213.2;395.3/192.2	395.3/213.2	160	20:20
470	Imibenconazole	86598-92-7	17.16	411.0/125.1;411.0/171.1; 411/342	411.0/125.1	120	25:15:10
471	Propaquiafop	111479-05-1	17.56	444.2/100.1;444.2/299.1	444.2/100.1	140	15:25
472	Lactofen	77501-63-4	18.23	479.1/344.0;479.1/223	479.1/344.0	120	15:35
473	2,3,4,5-Tetrachloroaniline	634-83-3	0.72	229.7/194;229.7/159	229.7/194	100	20:35
474	Benzofenap	85692-44-2	16.95	431/105;431/119	431/105	140	30:20
475	Dinoseb acetate	2813-95-8	0.75	283.1/89.2;283.1/133.1; 283.1/177.2	283.1/89.2	120	10:10:10

Continued

TABLE 5.10 Retention Times, Quantifying and Qualifying Ions, Fragmentor, and Collision Energies of 569 Pesticides and Chemical Pollutants Determined by LC-MS-MS—cont'd

No.	Compound	CAS	Retention Time (min)	Qualifying Ion	Quantifying Ion	Fragmentor (V)	Collision Energy (V)
476	Pyriminobac-methyl	147411-69-6	5.10	362.3/330;362.3/256	362.3/330	100	10:25
477	Propisochlor	86763-47-5	15.00	284/224;284/212	284/224	80	5:15
478	Silafluofen	105024-66-6	20.80	412/91;412/72.1	412/91	100	40:30
479	Triphenyl phosphate	115-86-6	4.90	317.1/166;317.1/210	317.1/166	120	25:25
480	Etobenzanid	79540-50-4	15.65	340/149;340/121.1	340/149	120	20:30
481	Fentrazamide	158237-07-1	16.00	372.1/219;372.1/83.2	372.1/219	200	5:35
482	Pentachloroaniline	527-20-8	14.30	285/99.1;285/127	285/99.1	100	15:5
483	Carbosulfan	55285-14-8	19.50	381.2/118.1;381.2/160.2	381.2/118.1	100	10:10
484	Cyphenothrin	39515-40-7	19.40	376.2/151.2;376.2/123.2	376.2/151.2	100	5:15
485	Dieldrin	60-57-1	3.91	377/333;377/221.2	377/333	100	5:35
486	Dimefuron	34205-21-5	10.30	339.1/167;339.1/72.1	339.1/167	140	20:30
487	Etoxazole	153233-91-1	3.36	360.2/141.1;360.2/304	360.2/141.1	100	30:15
488	Malaoxon	1364-78-2	13.80	331/99;331/127	331/99	120	20:5
489	Chlorbenside sulfone	7082-99-7	9.86	299/235;299/125	299/235	100	5:25
490	Dodine	2439-10-3	7.46	228.2/57.3;228.2/60.1	228.2/57.3	160	25:20

491	Propylene thiourea	2122-19-2	0.73	117/60.1;117/58	117/60.1	100	35:15
492	Dalapon	17040-19-6	0.60	140.8/58.8;140.8/62.9	140.8/58.8	100	10:15
493	Ethephon	16672-87-0	0.67	142.8/106.9;142.8/79.2	142.8/106.9	80	5:30
494	Flupropanate	756-09-2	0.97	144.9/81;144.9/101.5	144.9/81	100	15:5
495	2,6-Difluorobenzoic acid	385-00-2	0.73	153/93.1;158/113	153/93.1	60	20:5
496	Trichloroacetic acid sodium salt	650-51-1	0.74	160.9/116.9;160.9/95.9	160.9/116.9	120	7:29
497	Tert-butyl-4-hydroxyanisole	25013-16-5	1.96	163.2/147;163.2/107	163.2/147	120	19:19
498	2-Phenylphenol	90-43-7	9.78	169/115;169/93	169/115	140	35:20
499	3-Phenylphenol	580-51-8	9.78	169/115;169/141.1	169/115	140	35:35
500	Clopyralld	1702-17-6	2.14	190/146;190/74	190/146	60	5:45
501	DNOC	534-52-1	4.19	197.1/180;197.1/108.9	197.1/180	120	15:20
502	Cloprop	101-10-0	3.38	199/127;199/71	199/127	80	5:5
503	Dicloran	99-30-9	8.82	205.1/169.3;205.1/123.2	205.1/169.3	120	15:30
504	Aminopyralid	150114-71-9	4.29	205/160.7;205/125	205/160.7	80	5:10
505	Chlorpropham	101-21-3	12.55	212/152;212/57	212/152	80	5:20
506	Mecoprop	5902-51-2	4.46	213.1/141;213.1/71	213.1/141	80	5:5
507	Terbacil	5902-51-2	5.94	215.1/159;215.1/73	215.1/159	120	10:40
508	2,4-D	1918-00-9	4.28	218.9/161;218.9/125	218.9/161	80	5:20
509	Dicamba	94-81-5	0.75	219/175;219/145	219/175	60	5:5

Continued

TABLE 5.10 Retention Times, Quantifying and Qualifying Ions, Fragmentor, and Collision Energies of 569 Pesticides and Chemical Pollutants Determined by LC-MS-MS—cont'd

No.	Compound	CAS	Retention Time (min)	Qualifying Ion	Quantifying Ion	Fragmentor (V)	Collision Energy (V)
510	MCPB	140-56-7	5.53	227/141;227/105	227/141	80	10:25
511	Fenaminosulf	15165-67-0	0.76	228/183.5;228/79.8	228/183.5	80	5:5
512	Dichlorprop	1918-02-1	13.00	232.9/161.1;232.9/125	232.9/161.1	80	5:10
513	Picloram	25057-89-0	0.74	238.9/194.8;238.9/122.9	238.9/194.8	80	5:20
514	Bentazone	88-85-7	3.69	239/132;239/197	239/132	140	20:15
515	Dinoseb	1420-07-1	6.13	239/193;239/163	239/193	120	22:25
516	Dinoterb	68157-60-8	6.13	239/207;239/176.1	239/207	140	25:35
517	Forchlorfenuron	68157-60-8	7.35	246.1/127;246.1/91.1	246.1/127	80	5:25
518	2,4-DB	131341-86-1	21.79	246.9/160.9;246.9/125	246.9/160.9	80	15:25
519	Fludioxonil	1689-84-5	11.10	247/180;247/126	247/180	140	10:10
520	Trinexapac-ethyl	93-76-5	5.73	251.1/177.1;251.1/137.1	251.1/177.1	120	15:15
521	2,4,5-T	69377-81-7	2.63	253/195;253/158.9	253/195	80	5:25
522	Fluroxypyr	80-06-8	2.63	252.9/195;252.9/232.5	252.9/195	80	5:5
523	Chlorfenethol	93-72-1	11.81	265/96.7;265/152.7	265/96.7	120	15:5
524	Fenoprop	113136-77-9	1.83	267/194.5;267/77.1	267/194.5	120	15:20

525	Cyclanilide	1689-84-5	3.44	272/159.9;272/228	272/159.9	120	15:5
526	Bromoxynil	87-86-5	4.07	274/79.1;274/167	274/79.1	120	40:40
527	Pentachlorophenol	245-61-5	1.85	282/254;282/160	282/254	120	15:17
528	Isocarbophos	132-66-1	0.75	288.1/228;288.1/214	288.1/228	120	10:12
529	Naptalam	132-66-1	4.30	290/246;290/168.3	290/246	100	10:30
530	Chlorobenzuron	57160-47-1	14.05	306.9/154;306.9/125.9	306.9/154	100	5:20
531	Chloramphenicolum	66003-55-2	5.07	321/152;321/257	321/152	100	15:10
532	Alloxydim-sodium	123343-16-8	3.49	322.2/222;322.2/190	322.2/222	120	20:35
533	Pyrithlobac sodium	973-21-7	7.19	325.1/183.1;325.1/118.9	325.1/183.1	160	35:55
534	Dinobuton	15457-05-3	6.90	325.2/78.7;325.2/182.9	325.2/78.7	120	40:30
535	Fluorodifen	15457-05-3	0.85	326.9/263.9;326.9/124	326.9/263.9	100	15:30
536	Dimehypo	7772-98-7	0.71	331.9/118.8;331.9/79.9	331.9/118.8	140	20:45
537	Fempxaprop-ethyl	66441-23-4	5.82	332/151.9;332/260	332/151.9	100	15:5
538	Diflufenzopyr-sodium	122-16-7	0.72	333.1/160;333.1/134; 333.1/128	333.1/160	80	5:10:10
539	Sulfanitran	104206-82-8	5.77	334/137;334/197	334/137	120	28:29
540	Mesotrion	19044-88-3	0.74	338.1/291;338.1/212.2	338.1/291	80	5:40
541	Oryzalin	77-06-5	14.04	345.1/281.1;345/146.9; 345/78.1	345/281.1	120	10:10:5
542	Gibberellic acid	50594-66-6	0.74	345.1/143	345.1/143	120	15:10:15

Continued

TABLE 5.10 Retention Times, Quantifying and Qualifying Ions, Fragmentor, and Collision Energies of 569 Pesticides and Chemical Pollutants Determined by LC-MS-MS—cont'd

No.	Compound	CAS	Retention Time (min)	Qualifying Ion	Quantifying Ion	Fragmentor (V)	Collision Energy (V)
543	Acifluorfen	76-44-8	6.40	345.1/143;345.1/221.1; 345.1/240	360/316	80	5:25
544	Heptachlor	51366-25-7	0.55	360/316;360/194.9	369.2/233.1	100	10:5
545	Plifenate	689-83-4	3.14	369.2/233.1;369.2/301	368.9/147.9	120	20:15
546	Ioxynil	131807-57-3	4.63	368.9/147.9;368.9/326	369.8/126.9	120	35:35
547	Famoxadone	74223-64-6	16.52	369.8/126.9;369.8/215	373/282	120	20:15
548	Metsulfuron-methyl	122836-35-5	4.35	373/282;373/328.9	380/138.9	100	20:50
549	Sulfentrazone	83164-33-4	6.54	380/138.9;380/107.1	385/307	100	25:40
550	Diflufenican	83164-33-4	17.30	385/307;385/199.3	393.1/329.1	100	10:10
551	Ethiprole	181274-15-7	10.74	393.1/329.1;393.1/272	394.9/331	100	5:25
552	Propoxycarbzone-sodium	104040-78-0	5.05	394.9/331;394.9/250	397.1/156.1	80	5:10
553	Flazasulfuron	106917-52-6	4.84	397.1/156.1;397.1/113	406.1/250.9	100	10:20
554	Flusulfamide	1031-07-8	11.15	406.1/250.9;406.1/153.8	413/171	160	40:40
555	Endosulfan-sulfate	136849-15-5	4.08	413/171;413/179	418.8/113	80	20:25

No.	Name	CAS					
556	Cyclosulfamuron	26644-46-2	7.60	420.2/238.8;420.2/420.2/265.4	420.2/238.8	100	10:5
557	Triforine	100784-20-1	0.59	431/231.1;431/116.9	431/231.1	120	12:17
558	Halosulfuron-methyl	72178-02-0	4.70	432.9/252;432.9/153.7	432.9/252	120	15:30
559	Fomesafen	34014-18-1	7.13	437/195.1;437/222.1	437/195.1	140	40:40
560	Tecloftalam	79622-59-6	0.99	444/400;444/212.9	444/400	120	8:17
561	Fluazinam	86811-58-7	17.25	462.9/415.9;462.9/398	462.9/415.9	120	20:15
562	Fluazuron	144550-36-7	18.19	504.2/305.1;504.2/156	504.2/305.1	120	11:13
563	Iodosulfuron-methyl sodium	103055-07-8	4.48	505.9/139;505.9/308	505.9/139	120	25:15
564	Lufenuron	130000-40-7	18.15	508.9/339.1;508.9/326;508.9/174.8	508.9/339.1	100	5:5:5
565	Thifluzamide	4234-79-1	0.74	525/166;525/125	525/166	120	20:40
566	Kelevan	101007-06-1	19.50	628.1/169;628.1/422.6	628.1/169	120	24:22
567	Acrinathrin	1225-60-61	19.60	540/345;540/372	540/345	120	15:5
568	Iodosulfuron-methyl	29082-74-4	4.48	506/139;506/307.9	506/139	120	25:15
569	Octachlorostyrene	29082-74-4	19.85	374.9/290.2;374.9/246	374.9/290.2	80	10:10

5.1.9 Accurate Masses, Retention Times, Parent Ions, and Collision Energies of 492 Pesticides and Chemical Pollutants Determined by LC-TOF-MS

See Table 5.11.

TABLE 5.11 Accurate Masses, Retention Times, Parent Ions, and Collision Energies of 492 Pesticides and Chemical Pollutants Determined by LC-TOF-MS

No.	Compound	CAS	Molecular Formula	Accurate Mass	Score	Accurate Mass Error	ESI –	ESI+	Retention Time (min)	Parent Ion	Ionic Speciation	Colli. Energy	Colli. Energy Collect Range
1	1-Naphthyl acetamide	86-86-2	C12H11NO	185.0841	99.8	−1.0		X	4.57	186.0912	[M+H]+	10	5–20
2	2,4-D	94-75-7	C8H6Cl2O3	219.9694	98.6	1.6	X		3.25	218.9625	[M−H]−	5	5–20
3	2,6-Dichloro benzamide	2008-58-4	C7H5Cl2NO	188.9748	93.3	−0.8		X	3.24	189.9819	[M+H]+	20	5–20
4	3,4,5-Trimethacarb	2686-99-9	C11H15NO2	193.1103	90.5	−6.4		X	7.37	194.1163	[M+H]+	5	5–20
5	6-Chloro-4-hydroxy-3-phenyl-pyridazin	40020-01-7	C10H7ClN2O	206.0247	97.8	3.2		X	4.03	207.0326	[M+H]+	30	20–35
6	Acetamiprid	135410-20-7	C10H11ClN4	222.0672	97.8	−1.9		X	4.05	223.0743	[M+H]+	15	10–25

Continued

No.	Name	CAS	Formula							Adduct		
7	Acetochlor	34256-82-1	$C_{14}H_{20}ClNO_2$	269.1183	98.4	1.8	X	12.76	270.1260	[M+H]+	10	5–20
8	Aclonifen	74070-46-5	$C_{12}H_9ClN_2O_3$	264.0302	94.6	−0.2	X	13.91	265.0373	[M+H]+	15	5–20
9	Albendazole	54965-21-8	$C_{12}H_{15}N_3O_2S$	265.0885	96.8	−2.2	X	6.37	266.0952	[M+H]+	15	5–20
10	Aldicarb+	116-06-3	$C_7H_{14}N_2O_2S$	190.0776	90.7	−4.0	X	4.75		[M+Na]+		
11	Aldicarb sulfone	1646-88-4	$C_7H_{14}N_2O_4S$	222.0674	99.8	−0.5	X	2.66	223.0739	[M+H]+	5	5–20
12	Aldimorph※	1704-28-5	$C_{18}H_{37}NO$	283.2875	97.5	−2.7	X	11.70	284.2941	[M+H]+	35	25–40
13	Allidochlor	93-71-0	$C_8H_{12}ClNO$	173.0607	90.4	−0.4	X	5.07	174.0679	[M+H]+	10	5–20
14	Ametryn※	834-12-8	$C_9H_{17}N_5S$	227.1205	94.2	−3.5	X	6.87	228.1273	[M+H]+	25	10–25
15	Amidithion	919-76-6	$C_7H_{16}NO_4PS_2$	273.0258	97.8	0.8	X	4.40	274.0333	[M+H]+	10	5–20
16	Amidosulfuron	120923-37-7	$C_9H_{15}N_5O_7S_2$	369.0413	97.9	−1.3	X	6.17	370.0480	[M+H]+	10	5–20
17	Aminocarb	2032-59-9	$C_{11}H_{16}N_2O_2$	208.1212	94.8	−3.9	X	2.01	209.1276	[M+H]+	10	5–20
18	Aminopyralid	150114-71-9	$C_6H_4Cl_2N_2O_2$	205.9650	98.0	1.4	X	1.62	206.9726	[M+H]+	10	5–20
19	Anilofos	64249-01-0	$C_{13}H_{19}ClNO_3PS_2$	367.0233	99.0	−1.5	X	14.94	368.0299	[M+H]+	10	5–20

TABLE 5.11 Accurate Masses, Retention Times, Parent Ions, and Collision Energies of 492 Pesticides and Chemical Pollutants Determined by LC-TOF-MS—cont'd

No.	Compound	CAS	Molecular Formula	Accurate Mass	Score	Accurate Mass Error	ESI −	ESI+	Retention Time (min)	Parent Ion	Ionic Speciation	Colli. Energy	Colli. Energy Collect Range
20	Aspon	3244-90-4	$C_{12}H_{28}O_5P_2S_2$	378.0853	99.0	1.0		X	19.03	379.0927	[M+H]+	10	5–20
21	Asulam	3337-71-1	C8H10N2O4S	230.0361	95.6	−2.9		X	2.77	231.0430	[M+H]+	5	5–20
22	Athidathion	19691-80-6	$C_8H_{15}N_2O_4PS_3$	329.9932	98.4	2.1		X	13.46	331.0012	[M+H]+	5	5–20
23	Atratone	1610-17-9	C9H17N5O	211.1433	94.7	−4.3		X	3.74	212.1497	[M+H]+	25	15–30
24	Atrazine*	1912-24-9	C8H14ClN5	215.0938	90.1	−4.5		X	6.50	216.1001	[M+H]+	25	10–25
25	Atrazine-Desethyl	6190-65-4	C6H10ClN5	187.0625	96.9	−1.9		X	3.78	188.0694	[M+H]+	20	10–25
26	Azaconazole*	60207-31-0	$C_{12}H_{11}Cl_{2}N_3O_2$	299.0228	99.8	6.9		X	6.91	300.0303	[M+H]+	15	10–25
27	Azamethiphos*	35575-96-3	$C_9H_{10}ClN_2O_5PS$	323.9737	99.1	−0.4		X	5.41	324.9808	[M+H]+	10	5–20
28	Azinphos-ethyl	2642-71-9	C12H16N3O3PS2	345.0371	98.5	−1.3		X	13.40	346.0439	[M+H]+	15	5–20

#	Name	CAS	Formula						Adduct		
29	Azinphos-methyl	86-50-0	C10H12N3O3PS2	317.0058	96.7	-0.6	X	9.63	318.0138	[M+H]+	5	5-20
30	Aziprotryne※	4658-28-0	C7H11N7S	225.0797	91.1	-1.1	X	9.80	226.0867	[M+H]+	10	5-20
31	Azoxystrobin	131860-33-8	C22H17N3O5	403.1168	96.2	1.0	X	11.30	404.1245	[M+H]+	10	5-20
32	Benalaxyl	71626-11-4	C20H23NO3	325.1678	84.6	-0.9	X	14.23	326.1748	[M+H]+	10	5-20
33	Bendiocarb※	22781-23-3	C11H13NO4	223.0845	93.0	2.7	X	5.88	224.0923	[M+H]+	5	5-20
34	Benodanil※	15310-01-7	C13H10INO	322.9807	99.7	-0.7	X	8.51	323.9878	[M+H]+	20	10-25
35	Benoxacor	98730-04-2	C11H11Cl2NO2	259.0167	98.9	2.0	X	9.98	260.0245	[M+H]+	20	15-30
36	Bensulfuron-methyl	83055-99-6	C16H18N4O7S	410.0896	95.6	-2.4	X	7.96	411.0960	[M+H]+	15	5-20
37	Bensulide※	741-58-2	C14H24NO4PS3	397.0605	99.5	0.5	X	15.32	398.0679	[M+H]+	5	5-20
38	Bensultap※	17606-31-4	C17H21NO4S4	431.0353	94.0	-1.7	X	7.71	432.0419	[M+H]+	25	15-30
39	Benzofenap	82692-44-2	C22H20Cl2N2O3	430.0851	98.2	2.0	X	16.34	431.0931	[M+H]+	15	10-25
40	Benzoximate※	29104-30-1	C18H18ClNO5	363.0874	95.1	-1.1	X	16.47	364.0951	[M+H]+	5	5-20
41	Benzoylprop-ethyl	22212-55-1	C18H17Cl2NO3	365.0586	98.2	-0.8	X	15.36	366.0656	[M+H]+	5	5-20

Continued

TABLE 5.11 Accurate Masses, Retention Times, Parent Ions, and Collision Energies of 492 Pesticides and Chemical Pollutants Determined by LC-TOF-MS—cont'd

No.	Compound	CAS	Molecular Formula	Accurate Mass	Score	Accurate Mass Error	ESI −	ESI+	Retention Time (min)	Parent Ion	Ionic Speciation	Colli. Energy	Colli. Energy Collect Range
42	Benzyladenine	1214-39-7	C12H11N5	225.1015	93.3	−4.3		X	3.30	226.1078	[M+H]+	15	10–25
43	Bifenazate	149877-41-8	C17H20N2O3	300.1474	95.5	−2.1		X	12.39	301.1553	[M+H]+	5	5–20
44	Bioallethrin*	584-79-2	C19H26O3	302.1882	98.7	0.3		X	17.59	303.1951	[M+H]+	5	5–20
45	Bioresmethrin*	28434-01-7	C22H26O3	338.1882	99.6	0.3		X	19.22	339.1956	[M+H]+	15	5–20
46	Bitertanol	55179-31-2	$C_{20}H_{23}N_3O_2$	337.1790	95.9	−2.6		X	12.88	338.1854	[M+H]+	5	5–20
47	Bromacil*	314-40-9	$C_9H_{13}BrN_2O_2$	260.0160	99.3	0.0		X	4.93	261.0232	[M+H]+	5	5–20
48	Bromfenvinfos	33399-00-7	$C_{12}H_{14}BrCl_2O_4P$	401.9190	99.3	−0.8		X	14.25	402.9265	[M+H]+	5	5–20
49	Bromobutide	74712-19-9	C15H22BrNO	311.0885	99.7	0.4		X	13.92	312.0959	[M+H]+	5	5–20
50	Bromophos-ethyl*	4824-78-6	C10H12BrCl2O3PS	391.8805	98.9	0.9		X	18.88	392.8885	[M+H]+	5	5–20

No.	Name	CAS	Formula									
51	Brompyrazon	3042-84-0	C10H8BrN3O	264.9851	97.9	-2.3	X	3.89	265.9923	[M+H]+	35	20-35
52	Bromuconazole	116255-48-2	C13H12BrCl2N3O	374.9541	99.5	-0.7	X	10.50	375.9609	[M+H]+	20	10-25
53	Bupirimate	41483-43-6	C13H24N4O3S	316.1569	95.8	-2.1	X	12.96	317.1635	[M+H]+	25	15-30
54	Buprofezin	69327-76-0	C16H23N3OS	305.1562	92.8	-3.2	X	13.56	306.1625	[M+H]+	10	5-20
55	Butachlor※	23184-66-9	C17H26ClNO2	311.1652	95.5	-1.2	X	17.61	312.1727	[M+H]+	10	5-20
56	Butafenacil	134605-64-4	C20H18ClF3N2O6	474.0806	95.1	2.9	X	14.33	492.1157	[M+NH4]+	10	5-20
57	Butamifos	36335-67-8	C13H21N2O4PS	332.0960	99.3	0.6	X	16.62	333.1035	[M+H]+	5	5-20
58	Butocarboxim※	34681-10-2	C7H14N2O2S	190.0776	96.9	-1.6	X	4.48		[M+Na]+		
59	Butocarboxim-sulfoxide	34681-24-8	C7H14N2O3S	206.0725	96.0	-3.7	X	2.24	207.0789	[M+H]+	5	5-20
60	Butoxycarboxim※	34681-23-7	C7H14N2O4S	222.0674	99.8	-0.5	X	2.66	223.0745	[M+H]+	5	5-20
61	Butralin	33629-47-9	C14H21N3O4	295.1532	99.3	0.2	X	18.28	296.1605	[M+H]+	10	5-20
62	Butylate※	2008-41-5	C11H23NOS	217.1500	97.6	3.6	X	16.77	218.1581	[M+H]+	10	5-20

Continued

TABLE 5.11 Accurate Masses, Retention Times, Parent Ions, and Collision Energies of 492 Pesticides and Chemical Pollutants Determined by LC-TOF-MS—cont'd

No.	Compound	CAS	Molecular Formula	Accurate Mass	Score	Accurate Mass Error	ESI −	ESI+	Retention Time (min)	Parent Ion	Ionic Speciation	Colli. Energy	Colli. Energy Collect Range
63	Cadusafos	95465-99-9	$C_{10}H_{23}O_2PS_2$	270.0877	94.6	−3.5		X	14.78	271.0940	[M+H]+	10	5–20
64	Cafenstrole	125306-83-4	$C_{16}H_{22}N_4O_3S$	350.1413	98.9	1.3		X	12.94	351.1491	[M+H]+	5	5–20
65	Carbaryl	63-25-2	C12H11NO2	201.0790	99.8	−0.2		X	6.40	202.0862	[M+H]+	5	5–20
66	Carbendazim	10605-21-7	C9H9N3O2	191.0695	90.3	−3.3		X	2.42	192.0761	[M+H]+	15	5–20
67	Carbetamide	16118-49-3	C12H16N2O3	236.1161	97.3	−1.7		X	4.75	237.1229	[M+H]+	5	5–20
68	Carbofuran+	1563-66-2	C12H15NO3	221.1052	92.0	−3.6		X	5.96	222.1117	[M+H]+	10	5–20
69	Carbofuran-3-hydroxy	16655-82-6	C12H15NO4	237.1001	96.8	0.1		X	3.67	238.1074	[M+H]+	5	5–20
70	Carbophenothion*	786-19-6	C11H16ClO2PS3	341.9739	98.8	1.6		X	18.27	342.9817	[M+H]+	5	5–20
71	Carboxin	5234-68-4	C12H13NO2S	235.0667	94.8	−4.5		X	6.63	236.0729	[M+H]+	15	5–20

72	Carfentrazone-ethyl	128639-02-1	C15H14Cl2F3N3O3	411.0364	99.4	-0.1		X	14.39	412.0436	[M+H]+	15	5–20
73	Carpropamid	104030-54-8	C15H18Cl3NO	333.0454	98.8	-0.8		X	14.77	334.0524	[M+H]+	5	5–20
74	Cartap※	15263-53-3	C7H15N3O2S2	237.0606	94.5	-2.5		X	0.85	238.0676	[M+H]+	15	5–20
75	Chlordimeform△	19750-95-9	C10H13ClN2	196.0767	96.6	-2.6		X	3.43	197.0835	[M+H]+	20	15–30
76	Chlorfenvinphos※	470-90-6	C12H14Cl3O4P	357.9695	99.3	0.3		X	13.85	358.9762	[M+H]+	5	5–20
77	Chloridazon	1698-60-8	C10H8ClN3O	221.0356	94.8	-3.7		X	3.74	222.0427	[M+H]+	30	20–35
78	Chlorimuron-ethyl	90982-32-4	C15H15ClN4O6S	414.0401	99.3	-0.6		X	10.92	415.0474	[M+H]+	10	5–20
79	Chlorotoluron	15545-48-9	C10H13ClN2O	212.0716	94.8	-3.7		X	6.24	213.0781	[M+H]+	15	5–20
80	Chloroxuron※	1982-47-4	C15H15ClN2O2	290.0822	94.9	-2.4		X	10.26	291.0887	[M+H]+	15	10–25
81	Chlorphoxim	14816-20-7	C12H14ClN2O3PS	332.0151	97.0	0.4		X	16.55	333.0225	[M+H]+	5	5–20
82	Chlorpyrifos-ethyl	2921-88-2	C9H11Cl3NO3PS	348.9263	98.0	-0.4		X	17.83	349.9334	[M+H]+	10	5–20
83	Chlorpyrifos-methyl	5598-13-0	C7H7Cl3NO3PS	320.8950	99.0	1.7		X	7.88	321.9028	[M+H]+	15	5–20

Continued

TABLE 5.11 Accurate Masses, Retention Times, Parent Ions, and Collision Energies of 492 Pesticides and Chemical Pollutants Determined by LC-TOF-MS—cont'd

No.	Compound	CAS	Molecular Formula	Accurate Mass	Score	Accurate Mass Error	ESI−	ESI+	Retention Time (min)	Parent Ion	Ionic Speciation	Colli. Energy	Colli. Energy Collect Range
84	Chlorsulfuron	64902-72-3	C12H12ClN5O4S	357.0299	99.6	−0.1		X	6.27	358.0372	[M+H]+	10	5–20
85	Chlorthiophos*	60238-56-4	C11H15Cl2O3PS2	359.9577	99.7	0.4		X	18.27	360.9648	[M+H]+	10	5–20
86	Chromafenozide	143807-66-3	C24H30N2O3	394.2256	97.2	−1.5		X	13.12	395.2326	[M+H]+	5	5–20
87	Cinmethylin	87818-31-3	C18H26O2	274.1933	99.9	0.0		X	17.33	275.2006	[M+H]+	5	5–20
88	Cinosulfuron*	94593-91-6	C15H19N5O7S	413.1005	96.5	−2.2		X	5.79	414.1070	[M+H]+	10	5–20
89	Clethodim	99129-21-2	C17H26ClNO3S	359.1322	90.2	0.5		X	17.01	360.1396	[M+H]+	10	5–20
90	Clodinafop free acid	114420-56-3	C14H11ClFNO4	311.0361	99.9	−0.4		X	8.46	312.0432	[M+H]+	15	5–20
91	Clodinafop-propargyl	105512-06-9	C17H13ClFNO4	349.0517	96.8	−1.4		X	15.27	350.0586	[M+H]+	15	5–20
92	Clofentezine	74115-24-5	C14H8Cl2N4	302.0126	93.0	−0.3		X	15.52	303.0200	[M+H]+	5	5–20

No.	Name	CAS	Formula	Mass	%	Δ		RT	m/z	Adduct		
93	Clomazone	81777-89-1	C12H14ClNO2	239.0713	99.6	1.6	X	8.08	240.0791	[M+H]+	15	5–20
94	Clomeprop	84496-56-0	C16H15Cl2NO2	323.04798	97.4	2.7	X	16.71	324.0562	[M+H]+	15	5–20
95	Cloquintocet-mexyl	99607-70-2	C18H22ClNO3	335.1288	93.3	−2.5	X	16.62	336.1352	[M+H]+	15	5–20
96	Cloransulam-methyl	147150-35-4	C15H13ClFN5O5S	429.0310	96.3	−1.2	X	7.87	430.0377	[M+H]+	10	5–20
97	Clothianidin	210880-92-5	C6H8ClN5O2S	249.0087	99.4	−0.1		3.61	250.0160	[M+H]+	5	5–20
98	Crotoxyphos	7700-17-6	C14H19O6P	314.0919	99.7	0.7	X	9.82	332.1259	[M+NH4]+	5	5–20
99	Crufomate	299-86-5	C12H19ClNO3P	291.0791	99.7	−0.6	X	10.92	292.0862	[M+H]+	20	10–25
100	Cumyluron	99485-76-4	C17H19ClN2O	302.1186	95.5	−2.7	X	10.96	303.1265	[M+H]+	10	5–20
101	Cyanazine※	21725-46-2	C9H13ClN6	240.0890	95.9	−1.1	X	5.32	241.0960	[M+H]+	20	10–25
102	Cycloate※	1134-23-2	C11H21NOS	215.1344	92.5	0.6	X	15.51	216.1418	[M+H]+	15	5–20
103	cyclosulfamuron	136849-15-5	C17H19N5O6S	421.1056	99.8	0.3	X	12.39	422.1130	[M+H]+	10	5–20
104	Cycluron※	8015-55-2	C11H22N2O	198.1732	94.1	−4.9	X	6.55	199.1795	[M+H]+	20	10–25

Continued

TABLE 5.11 Accurate Masses, Retention Times, Parent Ions, and Collision Energies of 492 Pesticides and Chemical Pollutants Determined by LC-TOF-MS—cont'd

No.	Compound	CAS	Molecular Formula	Accurate Mass	Score	Accurate Mass Error	ESI –	ESI+	Retention Time (min)	Parent Ion	Ionic Speciation	Colli. Energy	Colli. Energy Collect Range
105	Cyflufenamid	136489-15-5	C20H17F5N2O2	412.1210	99.9	0.1		X	16.71	413.1283	[M+H]+	10	5–20
106	Cyprazine	22936-86-3	C9H14ClN5	227.0938	94.9	−1.7		X	6.52	228.1007	[M+H]+	20	10–25
107	Cyproconazole	94361-06-5	C15H18ClN3O	291.1138	96.0	−1.5		X	9.41	292.1206	[M+H]+	15	5–20
108	Cyprodinil	121552-61-2	C14H15N3	225.1266	96.0	−3.5		X	8.14	226.1331	[M+H]+	40	25–40
109	Cyromazine	66215-27-8	C6H10N6	166.0967	92.6	−5.2		X	1.64	167.1037	[M+H]+	25	15–30
110	Daminozide	1596-84-5	C6H12N2O3	160.0848	95.9	−4.8		X	0.87	161.0913	[M+H]+	10	5–20
111	Dazomet	533-74-4	C5H10N2S2	162.0285	98.6	2.6		X	3.03	163.0362	[M+H]+	15	5–20
112	Demeton-S	126-75-0	C8H19O3PS2	258.0513	95.6	−2.9		X	7.66	259.0586	[M+H]+	5	5–20
113	Demeton-S sulfoxide	2496-92-6	C8H19O4PS2	274.0462	95.4	−2.4		X	3.65	275.0529	[M+H]+	10	5–20

No.	Name	CAS	Formula	Mass	Purity (%)	Error		RT	m/z	Ion	CE	Voltage
114	Demeton-S-methyl※	919-86-8	C6H15O3PS2	230.0200	99.7	0.8	x	5.40		[M+Na]+		5–20
115	Demeton-S-methyl sulfone	17040-19-6	C6H15O5PS2	262.0099	98.4	−0.6	x	3.16	263.0174	[M+H]+	15	5–20
116	Demeton-S-methyl sulfoxide	301-12-2	C6H15O4PS2	246.0149	97.1	−2.4	x	2.74	247.0215	[M+H]+	10	10–25
117	Desamino-metamitron	36993-94-9	C10H9N3O	187.0746	99.3	−4.8	x	3.28	188.0809	[M+H]+	25	5–20
118	Desethyl-sebuthylazine	37019-18-4	C7H12ClN5	201.0781	99.7	−0.2	x	4.59	202.0853	[M+H]+	20	10–25
119	Desisopropyl-atrazine	1007-28-9	C5H8ClN5	173.0468	95.8	−0.2	x	3.05	174.0540	[M+H]+	25	5–20
120	Desmedipham	13684-56-5	C16H16N2O4	300.1110	99.9	0.5	x	9.46	301.1183	[M+H]+	5	15–30
121	Desmethyl-norflurazon	23576-24-1	C11H7ClF3N3O	289.0230	98.9	2.0	x	6.13	290.0308	[M+H]+	30	5–20
122	Desmethyl-pirimicarb	30614-22-3	C10H16N4O2	224.1273	95.9	−3.2	x	3.31	225.1339	[M+H]+	10	10–25
123	Desmetryn	1014-69-3	C8H15N5S	213.1048	89.4	−3.8	x	4.24	214.1112	[M+H]+	25	10–25
124	Diafenthiuron※	80060-09-9	C23H32N2OS	384.2235	97.7	−0.6	x	18.64	385.2305	[M+H]+	20	10–25
125	Dialifos※	10311-84-9	C14H17ClNO4PS2	393.0025	99.1	1.1	x	16.61	394.0103	[M+H]+	5	5–20

Continued

TABLE 5.11 Accurate Masses, Retention Times, Parent Ions, and Collision Energies of 492 Pesticides and Chemical Pollutants Determined by LC-TOF-MS—cont'd

No.	Compound	CAS	Molecular Formula	Accurate Mass	Score	Accurate Mass Error	ESI−	ESI+	Retention Time (min)	Parent Ion	Ionic Speciation	Colli. Energy	Colli. Energy Collect Range
126	Diallate	2303-16-4	C10H17Cl2NOS	269.0408	96.6	2.0		X	16.83	270.0485	[M+H]+	15	5–20
127	Diazinon	333-41-5	C12H21N2O3PS	304.1011	96.6	−1.8		X	15.05	305.1089	[M+H]+	20	10–25
128	Dibutyl succinate	141-03-7	C12H22O4	230.1518	98.9	−0.6		X	14.33	231.1593	[M+H]+	5	5–20
129	Dichlofenthion*	97-17-6	C10H13Cl2O3PS	313.9700	99.1	1.0		X	17.75	314.9776	[M+H]+	15	5–20
130	Diclobutrazole*	75736-33-3	C15H19Cl2N3O	327.0905	99.9	0.0		X	11.81	328.0978	[M+H]+	15	5–20
131	Dicloran	99-30-9	C6H4Cl2N2O2	205.9650	97.9	−1.4	X		7.89	204.9574	[M−H]−	35	25–40
132	Diclosulam	145701-21-9	C13H10Cl2FN5O3S	404.9865	99.8	0.2		X	8.35	405.9939	[M+H]+	10	5–20
133	Dicrotophos*	141-66-2	C8H16NO5P	237.0766	96.2	−2.6		X	3.16	238.0832	[M+H]+	10	5–20
134	Diethofencarb	87130-20-9	C14H21NO4	267.1471	99.1	−0.8		X	9.73	268.1541	[M+H]+	5	5–20

No.	Name	CAS	Formula	Mass	%	Error			RT	Ion Mass	Adduct		Range
135	Diethyltoluamide	134-62-3	C12H17NO	191.1310	96.7	-5.5		X	6.81	192.1373	[M+H]+	20	10-25
136	Difenoconazole	119446-68-3	C19H17Cl2N3O3	405.0647	99.6	0.3		X	14.73	406.0721	[M+H]+	15	5-20
137	Difenoxuron※	14214-32-5	C16H18N2O3	286.1317	95.3	-2.8		X	7.15	287.1381	[M+H]+	15	10-25
138	Diflufenican	83164-33-4	C19H11F5N2O2	394.0741	98.2	1.8	X		16.41	393.0675	[M-H]-	15	5-20
139	Dimefox※	115-26-4	C4H12FN2OP	154.0671	97.0	-4.3		X	3.21	155.0737	[M+H]+	20	10-25
140	Dimefuron※	34205-21-5	C15H19ClN4O3	338.1146	98.6	1.0		X	8.23	339.1218	[M+H]+	20	10-25
141	Dimepiperate※	61432-55-1	C15H21NOS	263.1344	95.4	1.1		X	16.14	264.1419	[M+H]+	5	5-20
142	Dimethachlor	50563-36-5	C13H18ClNO2	255.1026	94.3	1.1		X	7.84	256.1101	[M+H]+	10	5-20
143	Dimethametryn	22936-75-0	C11H21N5S	255.1518	93.8	-3.8		X	7.69	256.1580	[M+H]+	25	10-25
144	Dimethenamid	87674-68-8	C12H18ClNO2S	275.0747	99.7	-0.5		X	9.81	276.0819	[M+H]+	10	5-20
145	Dimethirimol※	5221-53-4	C11H19N3O	209.1528	92.2	-5.2		X	3.72	210.1590	[M+H]+	30	20-35
146	Dimethoate	60-51-5	C5H12NO3PS2	228.9996	97.3	-2.0		X	3.90	230.0066	[M+H]+	10	5-20
147	Dimethomorph	110488-70-5	C21H22ClNO4	387.1237	96.4	-1.2		X	8.97	388.1305	[M+H]+	20	10-25

Continued

TABLE 5.11 Accurate Masses, Retention Times, Parent Ions, and Collision Energies of 492 Pesticides and Chemical Pollutants Determined by LC-TOF-MS—cont'd

No.	Compound	CAS	Molecular Formula	Accurate Mass	Score	Accurate Mass Error	ESI −	ESI+	Retention Time (min)	Parent Ion	Ionic Speciation	Colli. Energy	Colli. Energy Collect Range
148	Diniconazole	83657-24-3	C15H17Cl2N3O	325.0749	99.6	−0.2		X	13.14	326.0821	[M+H]+	20	10–25
149	Dinitramine*	29091-05-2	C11H13F3N4O4	322.0889	93.1	−3.1		X	15.10	323.0973	[M+H]+	20	10–25
150	Dinotefuran	165252-70-0	C7H14N4O3	202.1066	96.2	0.1		X	2.41	203.1140	[M+H]+	10	5–20
151	Dioxabenzofos	3811-49-2	C8H9O3PS	216.0010	98.5	2.4		X	9.40	217.0088	[M+H]+	5	5–20
152	Dioxacarb*	6988-21-2	C11H13NO4	223.0845	95.4	−3.5		X	3.93	224.0917	[M+H]+	5	5–20
153	Diphenamid*	957-51-7	C16H17NO	239.1310	95.0	−3.4		X	8.75	240.1374	[M+H]+	20	10–25
154	Diphenylamine	122-39-4	C12H11N	169.0892	91.2	−1.8		X	12.48	170.0962	[M+H]+	25	10–25
155	Dipropetryn	4147-51-7	C11H21N5S	255.1518	93.8	−3.8		X	7.69	256.1583	[M+H]+	25	15–30
156	Disulfoton sulfone	2497-06-5	C8H19O4PS3	306.0183	99.5	−0.6		X	8.67	307.0254	[M+H]+	10	5–20

No.	Name	CAS	Formula	Mass	Purity	Error		RT	m/z	Adduct	CE	Range
157	Disulfoton sulfoxide	2497-07-6	C8H19O3PS3	290.0234	95.9	-2.5	X	6.48	291.0304	[M+H]+	10	5-20
158	Ditalimfos※	5131-24-8	C12H14NO4PS	299.0381	99.8	0.2	X	12.74	300.0455	[M+H]+	10	5-20
159	Dithiopyr	97886-45-8	C15H16F5NO2S2	401.0543	90.5	2.3	X	17.32	402.0618	[M+H]+	25	10-25
160	Diuron	330-54-1	C9H10Cl2N2O	232.0170	98.6	-0.7	X	6.82	233.0242	[M+H]+	15	5-20
161	Dodemorph	1593-77-7	C18H35NO	281.2719	97.0	-2.8	X	6.26	282.2784	[M+H]+	25	15-30
162	Edifenphos	17109-49-8	C14H15O2PS2	310.0251	99.6	0.1	X	13.66	311.0324	[M+H]+	15	5-20
163	Emamectin-benzoate	155569-91-8	C49H75NO13	885.5238	97.8	-0.6	X	17.10	886.5305	[M+H]+	30	20-35
164	Epoxiconazole	106325-08-0	C17H13ClFN3O	329.0731	99.6	0.6	X	11.36	330.0806	[M+H]+	15	5-20
165	esprocarb	85785-20-2	C15H23NOS	265.1500	98.5	0.9	X	17.30	266.1575	[M+H]+	10	5-20
166	Etaconazole	71245-23-3	C14H15Cl2N3O2	327.0541	96.2	0.2	X	11.12	328.0617	[M+H]+	20	10-25
167	Ethametsulfuron-methyl	97780-06-8	C15H18N6O6S	410.1009	95.9	-2.0	X	6.53	411.1073	[M+H]+	15	5-20
168	Ethidimuron※	30043-49-3	C7H12N4O3S2	264.0351	99.3	0.6	X	3.68	265.0425	[M+H]+	5	5-20

Continued

TABLE 5.11 Accurate Masses, Retention Times, Parent Ions, and Collision Energies of 492 Pesticides and Chemical Pollutants Determined by LC-TOF-MS—cont'd

No.	Compound	CAS	Molecular Formula	Accurate Mass	Score	Accurate Mass Error	ESI −	ESI +	Retention Time (min)	Parent Ion	Ionic Speciation	Colli. Energy	Colli. Energy Collect Range
169	Ethiofencarb*	29973-13-5	C11H15NO2S	225.0824	97.0	−1.9		X	6.71	226.0892	[M+H]+	5	5–20
170	Ethiofencarb-sulfone	53380-23-7	C11H15NO4S	257.0722	98.4	−0.8		X	3.64	258.0791	[M+H]+	5	5–20
171	Ethiofencarb-sulfoxide	53380-22-6	C11H15NO3S	241.0773	96.7	−2.8		X	3.28	242.0839	[M+H]+	5	5–20
172	Ethion*	563-12-2	C9H22O4P2S4	383.9876	91.6	−3.5		X	18.10	384.9943	[M+H]+	5	5–20
173	Ethiprole	181587-01-9	C13H9Cl2F3N4OS	395.9826	99.5	−0.6		X	9.50	396.9897	[M+H]+	20	10–25
174	Ethirimol*	23947-60-6	C11H19N3O	209.1528	93.3	−4.4		X	3.75	210.1592	[M+H]+	25	15–30
175	Ethoprophos+	13194-48-4	C8H19O2PS2	242.0564	99.6	−0.1		X	10.70	243.0641	[M+H]+	15	5–20
176	Ethoxyquin	91-53-2	C14H19NO	217.1467	96.9	−2.6		X	6.41	218.1534	[M+H]+	30	20–35
177	Ethoxysulfuron	126801-58-9	C15H18N4O7S	398.0896	99.3	−0.5		X	10.80	399.0967	[M+H]+	10	5–20

178	Etobenzanid	79540-50-4	C16H15Cl2NO3	339.0429	97.6	2.8		X	15.00	340.0512	[M+H]+	20	10–25
179	Etoxazole	153233-91-1	C21H23F2NO2	359.1697	96.5	−1.7		X	18.28	360.1764	[M+H]+	20	10–25
180	Etrimfos※	38260-54-7	C10H17N2O4PS	292.0647	95.9	−2.3		X	14.74	293.0713	[M+H]+	20	10–25
181	Famoxadone	131807-57-3	C22H18N2O4	374.1267	98.4	0.5	X		15.63	373.1196	[M−H]−	20	10–25
182	Famphur	52-85-7	C10H16NO5PS2	325.0208	98.8	−0.3		X	7.25	326.0279	[M+H]+	15	5–20
183	Fenamidone	161326-34-7	C17H17N3OS	311.1092	99.3	−0.2		X	11.04	312.1164	[M+H]+	10	5–20
184	Fenamiphos+	22224-92-6	C13H22NO3PS	303.1058	93.0	−2.1		X	10.71	304.1133	[M+H]+	15	5–20
185	Fenamiphos sulfone	31972-44-8	C13H22NO5PS	335.0956	99.4	0.7		X	5.74	336.1031	[M+H]+	15	5–20
186	Fenamiphos sulfoxide	31972-43-7	C13H22NO4PS	319.1007	98.0	−1.9		X	4.72	320.1074	[M+H]+	15	10–25
187	Fenarimol	60168-88-9	C17H12Cl2N2O	330.0327	99.9	0.3		X	10.78	331.0399	[M+H]+	25	15–30
188	Fenazaquin	120928-09-8	C20H22N2O	306.1732	97.5	−1.7		X	18.38	307.1798	[M+H]+	20	10–25
189	Fenbuconazole	114369-43-6	C19H17ClN4	336.1142	94.1	1.4		X	12.61	337.1219	[M+H]+	20	10–25
190	Fenfuram※	24691-80-3	C12H11NO2	201.0790	95.0	−4.1		X	6.85	202.0854	[M+H]+	15	5–20

Continued

TABLE 5.11 Accurate Masses, Retention Times, Parent Ions, and Collision Energies of 492 Pesticides and Chemical Pollutants Determined by LC-TOF-MS—cont'd

No.	Compound	CAS	Molecular Formula	Accurate Mass	Score	Accurate Mass Error	ESI–	ESI+	Retention Time (min)	Parent Ion	Ionic Speciation	Colli. Energy	Colli. Energy Collect Range
191	Fenhexamid	126833-17-8	C14H17Cl2NO2	301.0636	99.9	0.1		X	11.27	302.0709	[M+H]+	25	15–30
192	Fenobucarb	3766-81-2	C12H17NO2	207.1259	96.3	3.9		X	9.03	208.1340	[M+H]+	5	5–20
193	Fenoxanil	115852-48-7	C15H18Cl2N2O2	328.0745	96.7	–2.0		X	14.13	329.0811	[M+H]+	5	5–20
194	Fenoxaprop-ethyl	66441-23-4	C18H16ClNO5	361.0717	99.8	0.1		X	16.77	362.0790	[M+H]+	20	10–25
195	Fenoxycarb	72490-01-8	C17H19NO4	301.1314	91.5	0.6		X	13.13	302.1388	[M+H]+	10	5–20
196	Fenpropidin	67306-00-7	C19H31N	273.2457	94.0	–4.1		X	7.10	274.2519	[M+H]+	35	25–40
197	Fenpropimorph	67564-91-4	C20H33NO	303.2562	95.6	–3.1		X	7.26	304.2625	[M+H]+	35	25–40
198	Fenpyroximate	134098-61-6	C24H27N3O4	421.2002	95.8	–1.9		X	18.29	422.2066	[M+H]+	15	5–20
199	Fensulfothion	115-90-2	C11H17O4PS2	308.0306	99.7	–0.2		X	7.56	309.0378	[M+H]+	20	10–25

	Name	CAS	Formula									
200	Fenthion	55-38-9	C10H15O3PS2	278.0200	99.4	0.5	X	8.93	279.0267	[M+H]+	15	5–20
201	Fenthion oxon	6552-12-1	C10H15O4PS	262.0429	97.9	−1.3	X	7.36	263.0498	[M+H]+	15	5–20
202	Fenthion oxon sulfone	14086-35-2	C10H15O6PS	294.0327	99.4	1.2	X	4.19	295.0403	[M+H]+	20	10–25
203	Fenthion oxon sulfoxide	6552-13-2	C10H15O5PS	278.0378	97.5	−1.6	X	3.62	279.0446	[M+H]+	20	10–25
204	Fenthion sulfone	3761-42-0	C10H15O5PS2	310.0099	94.7	0.3	X	7.88	311.0176	[M+H]+	20	10–25
205	Fenthion sulfoxide	3761-41-9	C10H15O4PS2	294.0149	96.7	−1.6	X	6.15	295.0217	[M+H]+	20	10–25
206	Fenuron*	101-42-8	C9H12N2O	164.0950	93.6	−4.4	X	3.71	165.1016	[M+H]+	10	5–20
207	Flamprop-isopropyl	52756-22-6	C19H19ClFNO3	363.1038	98.3	−2.1	X	15.29	364.1103	[M+H]+	5	5–20
208	Flamprop-methyl	52756-25-9	C17H15ClFNO3	335.0725	99.4	−1.1	X	12.32	336.0794	[M+H]+	5	5–20
209	Flazasulfuron	104040-78-0	C13H12F3N5O5S	407.0511	98.8	−0.6	X	8.04	408.0581	[M+H]+	10	5–20
210	Florasulam	145701-23-1	C12H8F3N5O3S	359.0300	99.8	0.3	X	5.99	360.0374	[M+H]+	15	5–20
211	Fluazifop-butyl	69806-50-4	C19H20F3NO4	383.1344	91.1	−4.8	X	17.77	384.1398	[M+H]+	20	10–25
212	Flubenzimine*	37893-02-0	C17H10F6N4S	416.0530	94.9	3.5	X	17.72	417.0618	[M+H]+	15	5–20

Continued

TABLE 5.11 Accurate Masses, Retention Times, Parent Ions, and Collision Energies of 492 Pesticides and Chemical Pollutants Determined by LC-TOF-MS—cont'd

No.	Compound	CAS	Molecular Formula	Accurate Mass	Score	Accurate Mass Error	ESI −	ESI+	Retention Time (min)	Parent Ion	Ionic Speciation	Colli. Energy	Colli. Energy Collect Range
213	Fludioxonil	131341-86-1	C12H6F2N2O2	248.0397	96.7	−2.2	X		9.85	247.0317	[M−H]−	40	30–45
214	Flufenacet	142459-58-3	C14H13F4N3O2S	363.0665	97.6	−1.6		X	12.23	364.0732	[M+H]+	10	5–20
215	Flufenoxuron	101463-69-8	C21H11ClF6N2O3	488.0362	99.6	0.1		X	17.90	489.0436	[M+H]+	10	5–20
216	Flumetsulam	98967-40-9	C12H9F2N5O2S	325.0445	98.4	−1.3		X	4.23	326.0513	[M+H]+	20	5–20
217	Flumiclorac-pentyl	87546-18-7	C21H23ClFNO5	423.1249	98.5	−0.5		X	17.62	441.1584	[M+NH4]+	10	5–20
218	Fluometuron	2164-17-2	C10H11F3N2O	232.0824	97.8	−1.0		X	6.41	233.0894	[M+H]+	20	5–20
219	Fluoroglycofen-ethyl	77501-90-7	C18H13ClF3NO7	447.0333	99.8	0.6		X	17.26	465.0674	[M+NH4]+	5	5–20
220	Fluquinconazole	136426-54-5	C16H8Cl2FN5O	375.0090	99.6	−0.3		X	11.62	376.0162	[M+H]+	20	5–20
221	Fluridone*	59756-60-4	C19H14F3NO	329.1028	95.2	−3.2		X	9.30	330.1090	[M+H]+	40	25–40

No.	Compound	CAS	Formula	Mass	%	Error		RT	m/z	Adduct	CV	CE
222	Flurochloridone	61213-25-0	C12H10Cl2F3NO	311.0092	93.1	−1.7	X	13.15	312.0169	[M+H]+	25	15–30
223	Flurtamone	96525-23-4	C18H14F3NO2	333.0977	99.8	−0.5	X	10.04	334.1048	[M+H]+	25	15–30
224	Flusilazole	85509-19-9	C16H15F2N3Si	315.1003	99.4	−0.2	X	12.48	316.1075	[M+H]+	20	10–25
225	Fluthiacet-Methyl	117337-19-6	C15H15ClFN3O3S2	403.0227	99.1	−0.7	X	13.97	404.0297	[M+H]+	30	25–40
226	Flutolanil	66332-96-5	C17H16F3NO2	323.1133	97.4	0.7	X	13.08	324.1208	[M+H]+	10	5–20
227	Flutriafol	76674-21-0	C16H13F2N3O	301.1027	98.5	−1.6	X	6.53	302.1095	[M+H]+	10	5–20
228	Fonofos+,※	994-22-9	C10H15OPS2	246.0302	98.7	−1.5	X	15.41	247.0371	[M+H]+	5	5–20
229	Foramsulfuron	173159-57-4	C17H20N6O7S	452.1114	96.5	−1.5	X	5.12	453.1180	[M+H]+	10	5–20
230	Forchlorfenuron	68157-60-8	C12H10ClN3O	247.0512	99.3	−0.1	X	6.47	248.0585	[M+H]+	10	5–20
231	Fosthiazate	98886-44-3	C9H18NO3PS2	283.0466	95.2	−3.1	X	6.53	284.0530	[M+H]+	10	5–20
232	Fuberidazole	3878-19-1	C11H8N2O	184.0637	91.6	−5.8	X	2.79	185.0699	[M+H]+	30	15–30
233	Furalaxyl※	57646-30-7	C17H19NO4	301.1314	96.7	−1.4	X	9.51	302.1382	[M+H]+	10	5–20

Continued

TABLE 5.11 Accurate Masses, Retention Times, Parent Ions, and Collision Energies of 492 Pesticides and Chemical Pollutants Determined by LC-TOF-MS—cont'd

No.	Compound	CAS	Molecular Formula	Accurate Mass	Score	Accurate Mass Error	ESI −	ESI+	Retention Time (min)	Parent Ion	Ionic Speciation	Colli. Energy	Colli. Energy Collect Range
234	Furathiocarb※	65907-30-4	C18H26N2O5S	382.1562	95.9	−2.9		X	17.43	383.1627	[M+H]+	10	5–20
235	Furmecyclox※	60568-05-3	C14H21NO3	251.1521	94.2	−3.1		X	13.26	252.1586	[M+H]+	15	5–20
236	Halosulfuran-methyl	100784-20-1	C13H15ClN6O7S	434.0412	99.7	−0.3		X	10.01	435.0484	[M+H]+	10	5–20
237	Haloxyfop-ehyoxyethyl	87237-48-7	C19H19ClF3NO5	433.0904	99.6	−0.8		X	17.21	434.0973	[M+H]+	10	5–20
238	Haloxyfop-methyl	69806-40-2	C16H13ClF3NO4	375.0485	96.2	−3.2		X	16.41	376.0546	[M+H]+	15	5–20
239	Hexaconazole	79983-71-4	C14H17Cl2N3O	313.0749	99.4	0.8		X	12.39	314.0825	[M+H]+	15	5–20
240	Hexazinone※	51235-04-2	C12H20N4O2	252.1586	99.0	−1.0		X	4.78	253.1656	[M+H]+	10	5–20
241	Hexythiazox	78587-05-0	C17H21ClN2O2S	352.1012	99.4	0.2		X	17.88	353.1079	[M+H]+	10	5–20
242	Hydramethylnon	67485-29-4	C25H24F6N4	494.19052	96.1	−0.4		X	18.00	495.1976	[M+H]+	30	30–45

No.	Name	CAS	Formula	Mass	%	Error		RT	[M+H]	Adduct		Range
243	Imazalil	35554-44-0	C14H14Cl2N2O	296.0483	97.7	−1.7	X	6.36	297.0550	[M+H]+	25	15–30
244	Imazamethabenz-methyl	81405-85-8	C16H20N2O3	288.1474	95.4	−2.5	X	4.43	289.1540	[M+H]+	20	15–30
245	Imazamox	114311-32-9	C15H19N3O4	305.1376	89.6	−2.6	X	3.80	306.1442	[M+H]+	25	15–30
246	Imazapic	104098-48-8	C14H17N3O3	275.1270	92.2	−3.9	X	3.85	276.1332	[M+H]+	25	15–30
247	Imazapyr*	81334-34-1	C13H15N3O3	261.1113	92.1	−4.4	X	3.16	262.1174	[M+H]+	25	15–30
248	Imazaquin	81335-37-7	C17H17N3O3	311.1270	96.7	−2.4	X	5.25	312.1335	[M+H]+	25	15–30
249	Imazethapyr	81335-77-5	C15H19N3O3	289.1426	96.3	−2.2	X	4.56	290.1493	[M+H]+	25	15–30
250	Imazosulfuron	122548-33-8	C14H13ClN6O5S	412.0357	99.9	−0.2	X	7.97	413.0429	[M+H]+	10	5–20
251	Imibenconazole	86598-92-7	C17H13Cl3N4S	409.9927	99.6	−0.2	X	16.59	410.9999	[M+H]+	15	10–25
252	Imidacloprid	138261-41-3	C9H10ClN5O2	255.0523	96.0	0.7	X	3.79	256.0597	[M+H]+	10	5–20
253	inabenfide	82211-24-3	C19H15ClN2O2	338.0822	99.4	0.1	X	7.99	339.0895	[M+H]+	15	5–20
254	Indoxacarb	144171-61-9	C22H17ClF3N3O7	527.0707	99.7	−0.4	X	16.77	528.0778	[M+H]+	10	10–25

Continued

TABLE 5.11 Accurate Masses, Retention Times, Parent Ions, and Collision Energies of 492 Pesticides and Chemical Pollutants Determined by LC-TOF-MS—cont'd

No.	Compound	CAS	Molecular Formula	Accurate Mass	Score	Accurate Mass Error	ESI−	ESI+	Retention Time (min)	Parent Ion	Ionic Speciation	Colli. Energy	Colli. Energy Collect Range
255	Iodosulfuron-methyl	144550-36-7	C14H14IN5O6S	506.9710	99.7	−0.3		X	7.96	507.9293	[M+H]+	10	5–20
256	Iprobenfos	26087-47-8	C13H21O3PS	288.0949	97.2	−1.9		X	12.53	289.1018	[M+H]+	5	5–20
257	Iprovalicarb	140923-17-7	C18H28N2O3	320.2100	94.2	−3.3		X	10.67	321.2162	[M+H]+	5	5–20
258	Isazofos+,※	42509-80-8	C9H17ClN3O3PS	313.0417	94.5	−3.8		X	13.81	314.0477	[M+H]+	15	5–20
259	Isocarbamid※	30979-48-7	C8H15N3O2	185.1164	96.7	−3.2		X	3.68	186.1231	[M+H]+	10	5–20
260	Isocarbophos	24353-61-5	C11H16NO4PS	289.0538	99.9	−0.7		X	8.84		[M+Na]+		
261	Isofenphos※	25311-71-1	C15H24NO4PS	345.1164	93.9	−4.1		X	16.65		[M+Na]+		
262	Isofenphos oxon	31120-85-1	C15H24NO5P	329.1392	99.7	−0.3		X	9.80	330.1464	[M+H]+	5	5–20
263	Isomethiozin	57052-04-7	C12H20N4OS	268.1358	84.3	−2.0		X	13.53	269.1433	[M+H]+	15	5–20

264	Isoprocarb	2631-40-5	C11H15NO2	193.1103	94.3	-4.6		X	7.21	194.1167	[M+H]+	20	10-25
265	Isopropalin※	33820-53-0	C15H23N3O4	309.1689	99.0	-0.5		X	18.91	310.1756	[M+H]+	20	10-25
266	Isoprothiolane※	50512-35-1	C12H18O4S2	290.0647	98.4	-1.8		X	12.42	291.0714	[M+H]+	5	5-20
267	Isoproturon	34123-59-6	C12H18N2O	206.1419	90.8	-6.3		X	6.81	207.1479	[M+H]+	15	10-25
268	Isouron	55861-78-4	C10H17N3O2	211.1321	96.2	-3.4		X	5.14	212.1387	[M+H]+	15	5-20
269	Isoxaben	82558-50-7	C18H24N2O4	332.1736	95.5	-1.7		X	12.28	333.1803	[M+H]+	10	5-20
270	Isoxadifen-ethyl	163520-33-0	C18H17NO3	295.1208	95.3	0.5		X	14.61	296.1282	[M+H]+	10	5-20
271	Isoxaflutole	141112-29-0	C15H12F3NO4S	359.0439	95.1	0.8	X	X	4.47	360.0515	[M+H]+	5	5-20
272	Isoxathion※	18854-01-8	C13H16NO4PS	313.0538	99.8	-0.2		X	16.37	314.0610	[M+H]+	10	5-20
273	Kadethrin	58769-20-3	C23H24O4S	396.1395	99.3	1.2		X	17.55	414.1738	[M+NH4]+	5	5-20
274	Kelevan	4234-79-1	C17H12Cl10O4	629.7621	95.2	2.5			19.10	628.7559	[M-H]-	25	15-30
275	Kresoxim-methyl	143390-89-0	C18H19NO4	313.1314	97.1	-2.4		X	14.43	314.1391	[M+H]+	5	5-20

Continued

TABLE 5.11 Accurate Masses, Retention Times, Parent Ions, and Collision Energies of 492 Pesticides and Chemical Pollutants Determined by LC-TOF-MS—cont'd

No.	Compound	CAS	Molecular Formula	Accurate Mass	Score	Accurate Mass Error	ESI −	ESI+	Retention Time (min)	Parent Ion	Ionic Speciation	Colli. Energy	Colli. Energy Collect Range
276	Lactofen	77501-63-4	C19H15ClF3NO7	461.0489	96.4	−2.0		X	17.66	479.0821	[M+NH4]+	5	5–20
277	Linuron	330-55-2	C9H10Cl2N2O2	248.0119	97.2	1.1		X	9.29	249.0196	[M+H]+	15	5–20
278	Malaoxon	1634-78-2	C10H19O7PS	314.0589	95.5	−2.6		X	5.87	315.0653	[M+H]+	5	5–20
279	Malathion	121-75-5	C10H19O6PS2	330.0361	99.6	−0.8		X	12.74	331.0431	[M+H]+	5	5–20
280	Mecarbam※	2595-54-2	C10H20NO5PS2	329.0521	99.5	−1.3		X	13.91	330.0590	[M+H]+	5	5–20
281	Mefenacet※	73250-68-7	C16H14N2O2S	298.0776	95.9	−3.0		X	11.09	299.0840	[M+H]+	10	5–20
282	Mefenpyr-diethyl	135590-91-9	C16H18Cl2N2O4	372.0644	97.2	−1.1		X	15.75	373.0711	[M+H]+	5	5–20
283	Mepanipyrim	110235-47-7	C14H13N3	223.1110	92.1	−4.6		X	11.43	224.1172	[M+H]+	35	25–40
284	Mephosfolan	950-10-7	C8H16NO3PS2	269.0309	95.4	−2.9		X	5.00	270.0374	[M+H]+	15	5–20

285	Mepiquat chloride	7003-32-9	C7H15N	113.1205	96.9	-5.5		X	0.81	114.1271	[M+H]+	30	15-30
286	Mepronil※	55814-41-0	C17H19NO2	269.1416	96.9	-5.0		X	12.41	270.1474	[M+H]+	15	5-20
287	Mesosulfuron-methyl	208465-21-8	C17H21N5O9S2	503.0781	99.6	0.2		X	6.86	504.0855	[M+H]+	15	5-20
288	Metalaxyl	57837-19-1	C15H21NO4	279.1471	96.4	-2.3		X	6.84	280.1537	[M+H]+	10	5-20
289	Metalaxyl-M	70630-17-0	C15H21NO4	279.1471	95.0	-2.9		X	6.88	280.1540	[M+H]+	10	5-20
290	Metamitron	41394-05-2	C10H10N4O	202.0855	93.7	-4.2		X	3.57	203.0919	[M+H]+	25	15-30
291	Metazachlor	67129-08-2	C14H16ClN3O	277.0982	96.8	-2.4		X	7.63	278.1048	[M+H]+	5	5-20
292	Metconazole	125116-23-6	C17H22ClN3O	319.1451	96.9	-3.4		X	12.77	320.1513	[M+H]+	20	10-25
293	Methabenzthiazuron	18691-97-9	C10H11N3OS	221.0623	94.8	-4.0		X	6.07	222.0687	[M+H]+	10	5-20
294	Methamidophos+	10265-92-6	C2H8NO2PS	141.0013	99.3	-2.5		X	1.68	142.0082	[M+H]+	10	5-20
295	Methiocarb	2032-65-7	C11H15NO2S	225.0824	90.0	-6.8		X	8.94	226.0881	[M+H]+	5	5-20
296	Methiocarb Sulfone	2179-25-1	C11H15NO4S	257.0722	93.7	-1.3		X	4.31	258.0793	[M+H]+	5	5-20

Continued

TABLE 5.11 Accurate Masses, Retention Times, Parent Ions, and Collision Energies of 492 Pesticides and Chemical Pollutants Determined by LC-TOF-MS—cont'd

No.	Compound	CAS	Molecular Formula	Accurate Mass	Score	Accurate Mass Error	ESI −	ESI+	Retention Time (min)	Parent Ion	Ionic Speciation	Colli. Energy	Colli. Energy Collect Range
297	Methiocarb sulfoxide	2635-10-1	C11H15NO3S	241.0773	94.5	−3.6		X	3.51	242.0837	[M+H]+	5	5–20
298	Methomyl	16752-77-5	C5H10N2O2S	162.0463	99.5	−1.3		X	2.95	163.0533	[M+H]+	5	5–20
299	Methoprotryne※	841-06-5	C11H21N5OS	271.1467	91.4	−3.9		X	5.25	272.1531	[M+H]+	25	15–30
300	Methoxyfenozide	161050-58-4	C22H28N2O3	368.2100	99.3	0.4		X	12.57	369.2171	[M+H]+	5	5–20
301	Metobromuron※	3060-89-7	C9H11BrN2O2	258.0004	99.9	0.0		X	7.20	259.0078	[M+H]+	15	5–20
302	Metolachlor※	51218-45-2	C15H22ClNO2	283.1339	95.2	−1.4		X	12.51	284.1408	[M+H]+	10	5–20
303	Metolcarb	1129-41-5	C9H11NO2	165.0790	90.5	0.9		X	5.15	166.0864	[M+H]+	5	5–20
304	Metominostrobin-(E)	133408-50-1	C16H16N2O3	284.1161	96.5	−1.7		X	8.05	285.1229	[M+H]+	10	5–20
305	Metominostrobin-(Z)	133408-51-2	C16H16N2O3	284.1161	96.6	−1.8		X	7.25	285.1228	[M+H]+	5	5–20

Continued

No.	Name	CAS	Formula									
306	Metosulam	139528-85-1	C14H13Cl2N5O4S	417.0065	99.7	0.7	X	6.86	418.0141	[M+H]+	20	10–25
307	Metoxuron*	19937-59-8	C10H13ClN2O2	228.0666	95.9	−2.8	X	4.70	229.0732	[M+H]+	15	5–20
308	Metribuzin	21087-64-9	C8H14N4OS	214.0888	93.8	−2.7	X	5.40	215.0955	[M+H]+	25	15–30
309	Metsulfuron-methyl	74223-64-6	C14H15N5O6S	381.0743	99.6	−0.2	X	5.66	382.0815	[M+H]+	10	5–20
310	Mevinphos*	7786-34-7	C7H13O6P	224.0450	96.7	−3.4	X	3.67	225.0515	[M+H]+	5	5–20
311	Mexacarbate	315-18-4	C12H18N2O2	222.1368	81.7	2.7	X	4.17	223.1439	[M+H]+	15	5–20
312	Molinate	2212-67-1	C9H17NOS	187.1031	91.5	−7.0	X	10.14	188.1096	[M+H]+	15	5–20
313	Monocrotophos+,*	6923-22-4	C7H14NO5P	223.0610	94.2	−4.4	X	2.88	224.0673	[M+H]+	5	5–20
314	Monolinuron	1746-81-2	C9H11ClN2O2	214.0509	96.6	3.5	X	6.74	215.0590	[M+H]+	15	5–20
315	Monuron*	150-68-5	C9H11ClN2O	198.0560	95.5	−4.3	X	5.07	199.0624	[M+H]+	15	5–20
316	Myclobutanil	88761-89-0	C15H17ClN4	288.1142	98.0	−1.1	X	10.75	289.1211	[M+H]+	15	5–20
317	Napropamide	52570-16-8	C19H17NO2	291.1259	97.8	0.4	X	13.69	292.1334	[M+H]+	10	5–20

TABLE 5.11 Accurate Masses, Retention Times, Parent Ions, and Collision Energies of 492 Pesticides and Chemical Pollutants Determined by LC-TOF-MS—cont'd

No.	Compound	CAS	Molecular Formula	Accurate Mass	Score	Accurate Mass Error	ESI −	ESI+	Retention Time (min)	Parent Ion	Ionic Speciation	Colli. Energy	Colli. Energy Collect Range
318	Napropamide	15299-99-7	C17H21NO2	271.1572	96.6	−2.5		X	11.77	272.1638	[M+H]+	15	5–20
319	Naptalam*	132-66-1	C18H13NO3	291.0895	99.9	0.3		X	5.59	292.0969	[M+H]+	5	5–20
320	Neburon*	555-37-3	C12H16Cl2N2O	274.0640	99.6	−0.6		X	13.32	275.0711	[M+H]+	20	10–25
321	Nitenpyram	120738-89-8	C11H15ClN4O2	270.0884	96.4	−2.5		X	2.87	271.0950	[M+H]+	15	5–20
322	Nitralin*	4726-14-1	C13H19N3O6S	345.0995	98.4	−0.5		X	14.42	346.1066	[M+H]+	15	5–20
323	Norflurazon*	27314-13-2	C12H9ClF3N3O	303.0386	99.7	0.3		X	7.24	304.0459	[M+H]+	35	20–35
324	Nuarimol	63284-71-9	C17H12ClFN2O	314.0622	99.8	−0.4		X	8.27	315.0694	[M+H]+	25	15–30
325	Ofurace*	58810-48-3	C14H16ClNO3	281.0819	95.4	−2.8		X	3.79	282.0883	[M+H]+	10	5–20
326	Omethoate*	1113-02-6	C5H12NO4PS	213.0225	94.0	−4.8		X	2.17	214.0287	[M+H]+	10	5–20

No.	Name	CAS	Formula	Mass	Purity	Error		RT	[M+H]+ mass	Ion		Range
327	Oxadixyl※	77732-09-3	C14H18N2O4	278.1267	98.8	-1.2	X	5.13	279.1336	[M+H]+	5	5–20
328	Oxamyl	23135-22-0	C7H13N3O3S	219.0678	99.5	0.5	X	2.79	220.0748	[M+H]+	5	5–20
329	Oxamyl-oxime	30558-43-1	C5H10N2O2S	162.0463	99.2	-1.5	X	2.27	163.0533	[M+H]+	5	5–20
330	Oxycarboxin※	5259-88-1	C12H13NO4S	267.0565	99.7	0.1	X	4.54	268.0638	[M+H]+	10	5–20
331	Oxyfluorfen	42874-03-3	C15H11ClF3NO4	361.0329	98.1	0.6	X	17.59	362.0403	[M+H]+	15	5–20
332	Paclobutrazol	76738-62-0	C15H20ClN3O	293.1295	99.5	-0.1	X	8.85	294.1367	[M+H]+	15	5–20
333	Paraoxon-ethyl	311-45-5	C10H14NO6P	275.0559	99.7	-0.7	X	7.21	276.0623	[M+H]+	10	5–20
334	Paraoxon-methyl	950-35-6	C8H10NO6P	247.0246	99.1	-1.2	X	5.14	248.0316	[M+H]+	20	15–30
335	Pebulate※	1114-71-2	C10H21NOS	203.1344	96.2	-3.9	X	15.47	204.1405	[M+H]+	10	10–20
336	Penconazole	66246-88-6	C13H15Cl2N3	283.0643	99.8	0.2	X	12.50	284.0716	[M+H]+	10	5–20
337	Pencycuron	66063-05-6	C19H21ClN2O	328.1342	97.8	-1.4	X	15.84	329.1410	[M+H]+	15	10–25
338	Phenmedipham	13684-63-4	C16H16N2O4	300.1110	99.9	-0.4	X	9.40	301.1183	[M+H]+	5	5–10

Continued

TABLE 5.11 Accurate Masses, Retention Times, Parent Ions, and Collision Energies of 492 Pesticides and Chemical Pollutants Determined by LC-TOF-MS—cont'd

No.	Compound	CAS	Molecular Formula	Accurate Mass	Score	Accurate Mass Error	ESI –	ESI+	Retention Time (min)	Parent Ion	Ionic Speciation	Colli. Energy	Colli. Energy Collect Range
339	Phenthoate※	2597-03-7	C12H17O4PS2	320.0306	97.9	−0.2		X	15.11	321.0378	[M+H]+	5	5–20
340	Phorate+,※	298-02-2	C7H17O2PS3	260.0128	97.7	−1.6		X	15.79	261.0206	[M+H]+	5	5–20
341	Phorate sulfone	2588-04-7	C7H17O4PS3	292.0027	99.4	−0.9		X	8.72	293.0097	[M+H]+	5	5–20
342	Phorate sulfoxide	2588-03-6	C7H17O3PS3	276.0077	95.1	−2.7		X	6.44	277.0141	[M+H]+	5	5–20
343	Phosalone	2310-17-0	C12H15ClNO4PS2	366.9869	97.8	−0.5		X	16.15	367.9945	[M+H]+	5	5–20
344	Phosfolan+	947-02-4	C7H14NO3PS2	255.0153	96.2	−3.3		X	4.25	256.0217	[M+H]+	15	5–20
345	Phosmet	732-11-6	C11H12NO4PS2	316.9945	98.9	1.8		X	10.41	318.0025	[M+H]+	5	5–20
346	Phosmet-oxon	3735-33-9	C11H12NO5PS	301.0174	97.1	−1.5		X	4.81	302.0243	[M+H]+	5	5–20
347	Phosphamidon+,※	13171-21-6	C10H19ClNO5P	299.0689	97.0	−1.8		X	4.79	300.0756	[M+H]+	10	5–20

No.	Name	CAS	Formula	Mass	Purity	Error		X	RT	m/z	Adduct		Range
348	Phoxim	14816-18-3	C12H15N2O3PS	298.0541	99.4	0.3		X	16.14	299.0615	[M+H]+	5	5–20
349	Phthalic acid, benzyl butyl ester	85-68-7	C19H20O4	312.1362	99.1	0.9		X	16.78	313.1437	[M+H]+	5	5–20
350	Phthalic acid, dicyclohexyl ester	84-61-7	C20H26O4	330.1831	99.8	0.6		X	18.87	331.1906	[M+H]+	5	5–20
351	Phthalic acid, bis-butyl	84-74-2	C16H22O4	278.1518	98.4	−1.8		X	16.89	279.1586	[M+H]+	5	5–20
352	Picloram	1918-02-1	C6H3Cl3N2O2	239.9260	99.1	−0.5		X	2.62	240.9333	[M+H]+	10	5–20
353	Picolinafen	137641-05-5	C19H12F4N2O2	376.0835	99.1	−0.3		X	17.18	377.0906	[M+H]+	20	10–25
354	Picoxystrobin	117428-22-5	C18H16F3NO4	367.1031	98.3	−1.4		X	14.81	368.1098	[M+H]+	5	5–20
355	Piperonyl butoxide	51-03-6	C19H30O5	338.2093	96.6	−2.1		X	17.20	356.2423	[M+NH4]+	5	5–20
356	Piperophos	24151-93-7	C14H28NO3PS2	353.1248	95.1	−2.2		X	16.35	354.1312	[M+H]+	15	5–20
357	Pirimicarb	23103-98-2	C11H18N4O2	238.1430	95.0	−3.6		X	4.65	239.1494	[M+H]+	15	5–20
358	Pirimicarb-desmethyl-formamido	27218-04-8	C11H16N4O3	252.1222	93.3	−1.8		X	5.20	253.1290	[M+H]+	10	5–20
359	Pirimiphos-ethyl*	23505-41-1	C13H24N3O3PS	333.1276	92.2	−4.2		X	17.40	334.1335	[M+H]+	20	10–25

Continued

TABLE 5.11 Accurate Masses, Retention Times, Parent Ions, and Collision Energies of 492 Pesticides and Chemical Pollutants Determined by LC-TOF-MS—cont'd

No.	Compound	CAS	Molecular Formula	Accurate Mass	Score	Accurate Mass Error	ESI –	ESI+	Retention Time (min)	Parent Ion	Ionic Speciation	Colli. Energy	Colli. Energy Collect Range
360	Pirimiphos-methyl	29232-93-7	C11H20N3O3PS	305.0963	96.4	−2.4		X	16.08	306.1028	[M+H]+	25	15–30
361	Prallethrin	23031-36-9	C19H24O3	300.1725	93.5	2.0		X	16.41	301.1793	[M+H]+	10	5–20
362	Pretlachlor*	51218-49-6	C17H26ClNO2	311.1652	93.6	0.5		X	16.34	312.1726	[M+H]+	10	5–20
363	Primisulfuron-methyl	86209-51-0	C15H12F4N4O7S	468.0363	99.6	−0.6		X	12.07	469.0433	[M+H]+	10	5–20
364	Prochloraz	67747-09-5	C15H16Cl3N3O2	375.0308	99.5	−0.6		X	13.37	376.0379	[M+H]+	5	5–20
365	Profenofos*	41198-08-7	C11H15BrClO3PS	371.9351	99.4	1.0		X	16.29	372.9429	[M+H]+	10	5–20
366	Promecarb*	2631-37-0	C12H17NO2	207.1259	96.0	−1.1		X	2.29	208.1330	[M+H]+	5	5–20
367	Prometon	1610-18-0	C10H19N5O	225.1590	91.1	−5.8		X	4.34	226.1650	[M+H]+	25	15–30
368	Prometryne*	7287-19-6	C10H19N5S	241.1361	94.9	−3.1		X	9.00	242.1426	[M+H]+	20	10–25

369	Pronamide	23950-58-5	C12H11Cl2NO	255.0218	99.6	-0.7	X	11.22	256.0289	[M+H]+	10	5-20
370	Propachlor	1918-16-7	C11H14ClNO	211.0764	93.6	-3.2	X	7.52	212.0830	[M+H]+	15	5-20
371	Propamocarb	24579-73-5	C9H20N2O2	188.1525	94.8	-4.1	X	2.30	189.1588	[M+H]+	15	5-20
372	Propanil	709-98-8	C9H9Cl2NO	217.0061	96.8	1.0	X	8.17	218.0136	[M+H]+	20	10-25
373	Propaphos	7292-16-2	C13H21O4PS	304.0898	95.1	-1.0	X	13.29	305.0968	[M+H]+	5	5-20
374	propaquizafop	111479-05-1	C22H22ClN3O5	443.1248	96.2	-0.6	X	17.06	444.1318	[M+H]+	15	5-20
375	Propargite	2312-35-8	C19H26O4S	350.1552	97.9	-1.2	X	18.43	368.1886	[M+NH4]+	5	5-20
376	Propazine*	139-40-2	C9H16ClN5	229.1094	93.7	-3.2	X	8.24	230.1159	[M+H]+	20	10-25
377	Propetamphos	31218-83-4	C10H20NO4PS	281.0851	97.3	2.6	X	13.13	282.0931	[M+H]+	5	5-20
378	Propiconazole	60207-90-1	C15H17Cl2N3O2	341.0698	99.6	-0.5	X	13.27	342.0768	[M+H]+	20	10-25
379	Propisochlor	86763-47-5	C15H22ClNO2	283.1339	99.2	-1.3	X	14.47	284.1411	[M+H]+	10	5-20
380	Propoxur*	114-26-1	C11H15NO3	209.1052	98.9	0.5	X	5.80	210.1126	[M+H]+	5	5-20

Continued

TABLE 5.11 Accurate Masses, Retention Times, Parent Ions, and Collision Energies of 492 Pesticides and Chemical Pollutants Determined by LC-TOF-MS—cont'd

No.	Compound	CAS	Molecular Formula	Accurate Mass	Score	Accurate Mass Error	ESI −	ESI+	Retention Time (min)	Parent Ion	Ionic Speciation	Colli. Energy	Colli. Energy Collect Range
381	Propoxycarbazone	181274-15-7	C15H18N4O7S	398.0896	99.7	0.0		X	5.88	399.0951	[M+H]+	5	5–20
382	Prosulfocarb	52888-80-9	C14H21NOS	251.1344	93.6	−3.1		X	16.67	252.1409	[M+H]+	10	5–20
383	Prothoate※	2275-18-5	C9H20NO3PS2	285.0622	96.0	−1.7		X	7.97	286.0690	[M+H]+	10	5–20
384	Pymetrozine	123312-89-0	C10H11N5O	217.0964	95.9	−3.7		X	1.85	218.1029	[M+H]+	15	10–25
385	Pyraclofos※	77458-01-6	C14H18ClN2O3PS	360.0464	99.6	−0.6		X	14.83	361.0535	[M+H]+	20	10–25
386	Pyraclostrobin	175013-18-0	C19H18ClN3O4	387.0986	97.7	−1.7		X	15.55	388.1052	[M+H]+	10	5–20
387	Pyraflufen Ethyl	129630-17-7	C15H13Cl2F3N2O4	412.0205	99.8	−0.5		X	15.15	413.0276	[M+H]+	20	10–25
388	Pyrazolynate (Pyrazolate)	58011-68-0	C19H16Cl2N2O4S	438.0208	99.8	0.1		X	16.01	439.0281	[M+H]+	15	5–20
389	Pyrazophos	13457-18-6	C14H20N3O5PS	373.0861	99.2	−0.7		X	15.28	374.0931	[M+H]+	20	10–25

	Name	CAS	Formula									
390	Pyrazosulfuron-ethyl	93697-74-6	C14H18N6O7S	414.0958	99.8	0.1	X	9.83	415.1031	[M+H]+	10	5–20
391	Pyrazoxyfen※	71567-11-0	C20H16Cl2N2O3	402.0538	99.8	0.4	X	14.06	403.0612	[M+H]+	20	10–25
392	Pyributicarb	88678-67-5	C18H22N2O2S	330.1402	99.4	0.0	X	17.96	331.1475	[M+H]+	15	5–20
393	Pyridaben	96489-71-3	C19H25ClN2OS	364.1376	91.5	−0.9	X	18.95	365.1447	[M+H]+	5	5–20
394	Pyridalyl	179101-81-6	C18H14Cl4F3NO3	488.9680	99.6	−0.6	X	20.31	489.9751	[M+H]+	10	10–25
395	Pyridaphenthion	119-12-0	C14H17N2O4PS	340.0647	98.6	−1.6	X	11.76	341.0734	[M+H]+	15	10–25
396	Pyridate	55512-33-9	C19H23ClN2O2S	378.1169	96.2	−0.9	X	19.74	379.1238	[M+H]+	5	5–20
397	Pyrifenox※	88283-41-4	C14H12Cl2N2O	294.0327	96.7	−2.8	X	6.42	295.0391	[M+H]+	15	5–20
398	Pyrimethanil	53112-28-0	C12H13N3	199.1110	91.3	−6.2	X	5.89	200.1170	[M+H]+	35	25–40
399	Pyrimidifen	105779-78-0	C20H28ClN3O2	377.1870	96.1	−2.2	X	16.30	378.1934	[M+H]+	20	10–25
400	Pyriminobac-methyl(z)	147411-70-9	C17H19N3O6	361.1274	95.9	−2.4	X	9.44	362.1338	[M+H]+	10	5–20
401	Pyrimitate	5221-49-8	C11H20N3O3PS	305.0963	97.7	−0.3	X	14.90	306.1034	[M+H]+	15	10–25

Continued

TABLE 5.11 Accurate Masses, Retention Times, Parent Ions, and Collision Energies of 492 Pesticides and Chemical Pollutants Determined by LC-TOF-MS—cont'd

No.	Compound	CAS	Molecular Formula	Accurate Mass	Score	Accurate Mass Error	ESI−	ESI+	Retention Time (min)	Parent Ion	Ionic Speciation	Colli. Energy	Colli. Energy Collect Range
402	Pyriproxyfen	95737-68-1	C20H19NO3	321.1365	94.1	−2.0		X	17.59	322.1431	[M+H]+	10	5–20
403	Pyroquilon*	57369-32-1	C11H11NO	173.0841	92.9	−5.0		X	4.99	174.0905	[M+H]+	30	20–35
404	Quinalphos*	13593-03-8	C12H15N2O3PS	298.0541	95.5	−2.9		X	14.13	299.0605	[M+H]+	15	10–25
405	Quinclorac*	84087-01-4	C10H5Cl2NO2	240.9697	99.8	0.5		X	4.15	241.9771	[M+H]+	10	5–20
406	Quinmerac	90717-03-6	C11H8ClNO2	221.0244	98.6	−2.0		X	3.66	222.0312	[M+H]+	10	5–20
407	Quinoclamine	2797-51-5	C10H6ClNO2	207.0087	99.1	0.3		X	5.24	208.0160	[M+H]+	25	10–25
408	Quinoxyphen	124495-18-7	C15H8Cl2FNO	306.9967	99.7	−0.8		X	16.47	308.0037	[M+H]+	35	25–40
409	Quizalofop-ethyl	76578-14-8	C19H17ClN2O4	372.0877	97.1	−1.1		X	16.76	373.0945	[M+H]+	20	10–25
410	Rabenzazole	40341-04-6	C12H12N4	212.1062	91.8	−5.1		X	6.46	213.1124	[M+H]+	30	20–35

411	Resmethrin※	10453-86-8	C22H26O3	338.1882	99.5	0.5	X	19.20	339.1956	[M+H]+	15	5–20
412	Rimsulfuron	122931-48-0	C14H17N5O7S2	431.0569	99.6	−0.2	X	6.23	432.0641	[M+H]+	10	5–20
413	Rotenone	83-79-4	C23H22O6	394.1416	99.2	−0.5	X	13.34	395.1478	[M+H]+	25	15–30
414	Sebutylazine	7286-69-3	C9H16ClN5	229.1094	93.1	−3.5	X	8.02	230.1159	[M+H]+	20	10–25
415	Secbumeton※	26259-45-0	C10H19N5O	225.1590	92.8	−4.7	X	4.43	226.1650	[M+H]+	25	15–30
416	Sethoxydim※	74051-80-2	C17H29NO3S	327.1868	97.0	−1.6	X	17.32	328.1936	[M+H]+	15	5–20
417	Simazine	122-34-9	C7H12ClN5	201.0781	97.9	−1.7	X	5.11	202.0850	[M+H]+	25	15–30
418	Simeconazole	149508-90-7	C14H20FN3OSi	293.1360	95.6	−1.6	X	10.42	294.1428	[M+H]+	15	5–20
419	Simeton	673-04-1	C8H15N5O	197.1277	96.1	−2.9	X	3.69	198.1344	[M+H]+	30	20–35
420	Simetryn	1014-70-6	C8H15N5S	213.1048	94.7	−4.9	X	4.23	214.1110	[M+H]+	25	15–30
421	Sobutylazine	7286-69-3	C9H16ClN5	229.1094	99.3	−0.8	X	8.07	230.1165	[M+H]+	20	10–25
422	Spinosad	168316-95-8	C41H65NO10	731.4609	97.4	−1.4	X	11.92	732.4670	[M+H]+	25	10–25
423	Spirodiclofen	148477-71-8	C21H24Cl2O4	410.1052	95.3	−2.7	X	19.08	411.1112	[M+H]+	5	5–20

Continued

TABLE 5.11 Accurate Masses, Retention Times, Parent Ions, and Collision Energies of 492 Pesticides and Chemical Pollutants Determined by LC-TOF-MS—cont'd

No.	Compound	CAS	Molecular Formula	Accurate Mass	Score	Accurate Mass Error	ESI –	ESI+	Retention Time (min)	Parent Ion	Ionic Speciation	Colli. Energy	Colli. Energy Collect Range
424	Spiroxamine	118134-30-8	C18H35NO2	297.2668	95.1	−2.9		X	8.99	298.2732	[M+H]+	20	10–25
425	Sulfallate	95-06-7	C8H14ClNS2	223.0256	98.1	−0.3		X	14.77	224.0325	[M+H]+	5	5–20
426	Sulfentrazone	122836-35-5	C11H10Cl2F2N4O3S	385.9819	97.8	2.5		X	6.51	404.0165	[M+NH4]+	5	5–20
427	Sulfotep+,※	3689-24-5	C8H20O5P2S2	322.0227	98.9	0.4		X	15.87	323.0301	[M+H]+	10	5–20
428	Sulprofos※	35400-43-2	C12H19O2PS3	322.0285	98.7	1.6		X	18.11	323.0363	[M+H]+	10	5–20
429	Tebuconazole	107534-96-3	C16H22ClN3O	307.1451	96.8	1.0		X	11.86	308.1527	[M+H]+	20	10–25
430	Tebufenozide	112410-23-8	C22H28N2O2	352.2151	98.0	−0.3		X	14.09	353.2222	[M+H]+	5	5–20
431	Tebufenpyrad	119168-77-3	C18H24ClN3O	333.1608	99.4	0.2		X	16.79	334.1682	[M+H]+	30	20–35
432	Tebupirimfos	96182-53-5	C13H23N2O3PS	318.1167	97.0	−1.6		X	17.71	319.1235	[M+H]+	10	5–20
433	Tebutam※	35256-85-0	C15H23NO	233.1780	96.7	−3.1		X	12.54	234.1845	[M+H]+	15	5–20

No.	Name	CAS	Formula	Mass	%	ppm			RT	m/z	Adduct		Range
434	Tebuthiuron*	34014-18-1	C9H16N4OS	228.1045	95.0	-3.5	X		4.66	229.1110	[M+H]+	15	5-20
435	Temephos*	3383-96-8	C16H20O6P2S3	465.9897	98.5	0.5	X		17.85	484.0224	[M+NH4]+	5	5-20
436	TEPP	107-49-3	C8H20O7P2	290.0684	99.6	-0.9	X		4.70	291.0754	[M+H]+	10	5-20
437	Tepraloxydim	149979-41-9	C17H24ClNO4	341.1394	99.7	0.2	X		11.45	342.1467	[M+H]+	10	5-20
438	Terbacil*	5902-51-2	C9H13ClN2O2	216.0666	98.4	-0.2		X	5.12	215.0592	[M-H]-	15	5-20
439	Terbucarb	1918-11-2	C17H27NO2	277.2042	99.7	0.5	X		15.93	278.2116	[M+H]+	5	5-20
440	Terbufos+,*	13071-79-9	C9H21O2PS3	288.0441	98.3	0.6	X		17.56	289.0519	[M+H]+	5	5-20
441	Terbufos sulfone O-analogue	56070-15-6	C9H21O5PS2	304.0568	95.3	-1.8	X		5.29	305.0635	[M+H]+	5	5-20
442	Terbumeton	33693-04-8	C10H19N5O	225.1590	95.4	-3.8	X		4.44	226.1654	[M+H]+	15	10-25
443	Terbuthylazine	5915-41-3	C9H16ClN5	229.1094	93.5	-3.2	X		8.94	230.1159	[M+H]+	15	5-20
444	Terbutryne	886-50-0	C10H19N5S	241.1361	95.3	-4.3	X		9.53	242.1428	[M+H]+	15	5-20
445	Tetrachlorvinphos*	22248-79-9	C10H9Cl4O4P	363.8993	99.9	-0.2	X		12.81	364.9065	[M+H]+	5	5-20

Continued

TABLE 5.11 Accurate Masses, Retention Times, Parent Ions, and Collision Energies of 492 Pesticides and Chemical Pollutants Determined by LC-TOF-MS—cont'd

No.	Compound	CAS	Molecular Formula	Accurate Mass	Score	Accurate Mass Error	ESI−	ESI+	Retention Time (min)	Parent Ion	Ionic Speciation	Colli. Energy	Colli. Energy Collect Range
446	Tetraconazole	112281-77-3	C13H11Cl2F4N3O	371.0215	99.5	0.4		X	11.97	372.0290	[M+H]+	20	15–30
447	Tetramethrin※	7696-12-0	C19H25NO4	331.1784	97.6	−1.6		X	17.37	332.1853	[M+H]+	10	5–20
448	Thenylchlor	96491-05-3	C16H18ClNO2S	323.0747	99.7	−0.5		X	13.11	324.0819	[M+H]+	5	5–20
449	Thiabendazole	148-79-8	C10H7N3S	201.0361	97.2	−4.3		X	3.04	202.0423	[M+H]+	35	20–35
450	Thiacloprid	111988-49-9	C10H9ClN4S	252.0236	99.8	0.0		X	4.61	253.0309	[M+H]+	15	10–25
451	Thiamethoxam	153719-23-4	C8H10ClN5O3S	291.0193	99.7	0.0		X	3.24	292.0266	[M+H]+	5	5–20
452	Thiazopyr※	117718-60-2	C16H17F5N2O2S	396.0931	97.2	−1.7		X	15.58	397.0997	[M+H]+	35	25–40
453	Thidiazuron	51707-55-2	C9H8N4OS	220.0419	99.7	−0.2		X	4.92	221.0491	[M+H]+	15	5–20
454	Thifensulfuron-methyl	79277-27-3	C12H13N5O6S2	387.0307	99.7	0.3		X	5.41	388.0380	[M+H]+	10	5–20

455	Thiobencarb	28249-77-6	C12H16ClNOS	257.0641	98.8	0.1	X	15.32	258.0714	[M+H]+	10	5–20
456	Thiodicarb	59669-26-0	C10H18N4O4S3	354.0490	96.0	2.5	X	5.89	355.0574	[M+H]+	5	5–20
457	Thiofanox sulfone	39184-59-3	C9H18N2O4S	250.0987	98.1	0.1	X	3.97	251.1060	[M+H]+	5	5–20
458	Thiofanox-sulfon	39196-18-4	C9H18N2O4S	250.0987	96.9	1.2	X	3.97	251.1063	[M+H]+	5	5–20
459	Thiofanox-sulfoxide	39184-27-5	C9H18N2O3S	234.1038	96.7	−2.2	X	3.36	235.1106	[M+H]+	5	5–20
460	Thionazin*	297-97-2	C8H13N2O3PS	248.0385	98.1	−0.8	X	8.24	249.0456	[M+H]+	10	5–20
461	Thiophanate-methyl	23564-05-8	C12H14N4O4S2	342.0457	97.6	−2.4	X	5.57	343.0525	[M+H]+	5	5–20
462	Thiophanat-ethyl	23564-06-9	C14H18N4O4S2	370.0770	99.5	−0.6	X	8.02	371.0843	[M+H]+	10	5–20
463	Thiram	137-26-8	C6H12N2S4	239.9883	99.6	0.3		6.54	240.9956	[M+H]+	5	5–20
464	Tolclofos-methyl	57018-04-9	C9H11Cl2O3PS	299.9544	99.6	1.0	X	15.79	300.9620	[M+H]+	15	10–25
465	Tolfenpyrad	129558-76-5	C21H22ClN3O2	383.1401	99.7	0.2	X	17.04	384.1477	[M+H]+	25	15–30
466	Tralkoxydim	87820-88-0	C20H27NO3	329.1991	97.9	−0.6	X	14.75	330.2062	[M+H]+	10	5–20

Continued

TABLE 5.11 Accurate Masses, Retention Times, Parent Ions, and Collision Energies of 492 Pesticides and Chemical Pollutants Determined by LC-TOF-MS—cont'd

No.	Compound	CAS	Molecular Formula	Accurate Mass	Score	Accurate Mass Error	ESI−	ESI+	Retention Time (min)	Parent Ion	Ionic Speciation	Colli. Energy	Colli. Energy Collect Range
467	Triadimefon	43121-43-3	C14H16ClN3O2	293.0931	99.9	−0.7		X	11.33	294.0996	[M+H]+	15	5–20
468	Triadimenol	55219-65-3	C14H18ClN3O2	295.1088	96.7	−0.9		X	8.65	296.1158	[M+H]+	5	5–20
469	Tri-allate	2303-17-5	C10H16Cl3NOS	303.0018	93.0	−1.9		X	18.19	304.0091	[M+H]+	15	5–20
470	Triasulfuron	82097-50-5	C14H16ClN5O5S	401.0561	99.5	−0.3		X	6.16	402.0637	[M+H]+	15	5–20
471	Triazophos※	24017-47-8	C12H16N3O3PS	313.0650	99.7	0.0		X	12.90	314.0723	[M+H]+	15	5–20
472	Triazoxide	72459-58-6	C10H6ClN5O	247.0261	97.1	−2.5		X	5.92	248.0327	[M+H]+	35	25–40
473	Tribenuron-methyl	101200-48-0	C15H17N5O6S	395.0900	99.2	−0.3		X	8.05	396.0974	[M+H]+	5	5–20
474	Tributos(DEF)※	78-48-8	C12H27OPS3	314.0962	96.6	−1.9		X	18.98	315.1029	[M+H]+	15	5–20
475	Trichlorfon	52-68-6	C4H8Cl3O4P	255.9226	96.8	3.5		X	3.43	256.9308	[M+H]+	10	5–20
476	Tricyclazole	41814-78-2	C9H7N3S	189.0361	95.2	−4.5		X	4.34	190.0425	[M+H]+	30	20–35

Continued

477	Tridemorph	24602-86-6	C19H39NO	297.3032	94.2	−3.2	X	14.72	298.3095	[M+H]+	35	20–35
478	Trietazine*	1912-26-1	C9H16ClN5	229.1094	97.0	−2.5	X	11.51	230.1159	[M+H]+	30	15–30
479	Trifloxystrobin	141517-21-7	C20H19F3N2O4	408.1297	95.3	−1.8	X	16.83	409.1362	[M+H]+	10	5–20
480	Triflumizole	99387-89-0	C15H15ClF3N3O	345.0856	95.5	−2.4	X	15.17	346.0920	[M+H]+	5	5–20
481	Triflumuron	64628-44-0	C15H10ClF3N2O3	358.0332	97.8	1.9	X	14.65	359.0411	[M+H]+	10	5–20
482	Triflusulfuron-methyl	126535-15-7	C17H19F3N6O6S	492.1039	99.1	−0.6	X	12.08	493.1109	[M+H]+	10	5–20
483	Tri-n-butyl phosphate	126-73-8	C12H27O4P	266.1647	95.4	−3.8	X	14.94	267.1710	[M+H]+	5	5–20
484	Trinexapac-ethyl	95266-40-3	C13H16O5	252.0998	97.7	3.0	X	7.68	253.1078	[M+H]+	10	5–20
485	Triphenyl phosphate	115-86-6	C18H15O4P	326.0708	99.4	1.1	X	15.11	327.0784	[M+H]+	30	20–35
486	Triticonazole	131983-72-7	C17H20ClN3O	317.1295	99.2	−0.4	X	9.52	318.1366	[M+H]+	10	5–20
487	Uniconazole	83657-22-1	C15H18ClN3O	291.1138	93.5	0.8	X	10.73	292.1213	[M+H]+	20	15–30
488	Validamycin	37248-47-8	C20H35NO13	497.21084	99.2	−1.0	X	0.69		[M+Na]+		

TABLE 5.11 Accurate Masses, Retention Times, Parent Ions, and Collision Energies of 492 Pesticides and Chemical Pollutants Determined by LC-TOF-MS—cont'd

No.	Compound	CAS	Molecular Formula	Accurate Mass	Score	Accurate Mass Error	ESI –	ESI+	Retention Time (min)	Parent Ion	Ionic Speciation	Colli. Energy	Colli. Energy Collect Range
489	Vamidothion*	2275-23-2	$C_8H_{18}NO_4PS_2$	287.0415	93.5	−3.3		X	3.49	288.0479	[M+H]+	5	5–20
490	Vamidothion sulfone	70898-34-9	$C_8H_{18}NO_6PS_2$	319.0313	97.8	−0.9		X	2.97	320.0384	[M+H]+	10	5–20
491	Vamidothion sulfoxide	2300-00-9	$C_8H_{18}NO_5PS_2$	303.0364	95.9	−2.5		X	2.54	304.0429	[M+H]+	10	5–20
492	Zoxamide	156052-68-5	$C_{14}H_{16}Cl_3NO_2$	335.0247	99.6	−0.3		X	15.09	336.0319	[M+H]+	15	5–20

Note: * The pesticides forbidden by the European Union (totaling 118 kinds); + The restricted pesticides in vegetables, fruit trees, teas, and Chinese traditional materials (including 13 kinds); △ The prohibited pesticides by the country (1 kind).

5.2   LINEAR EQUATION PARAMETERS OF 1200 PESTICIDES AND CHEMICAL POLLUTANTS DETERMINED BY GC-MS, GC-MS-MS, AND LC-MS-MS

5.2.1   Linear Equations, Linear Ranges, and Correlation Coefficients of 567 Pesticides and Chemical Pollutants Determined by GC-MS

See Table 5.12.

TABLE 5.12 Linear Equations, Linear Ranges, and Correlation Coefficients of 567 Pesticides and Chemical Pollutants Determined by GC-MS

No.	Compound	CAS	Linear Equation	Linear Range	Correlation Coeffient
1	Aldrin	309-00-2	$Y = 6.16 \times 10^5 X - 6.01 \times 10^4$	0.1250–5.000	0.9990
2	Allidochlor	93-71-0	$Y = 4.83 \times 10^5 X - 5.48 \times 10^4$	0.1250–5.000	0.9986
3	Anthraquinone	84-65-1	$Y = 2.27 \times 10^6 X - 3.41 \times 10^5$	0.0625–2.500	0.9862
4	Aramite	140-57-8	$Y = 3.89 \times 10^5 X - 3.66 \times 10^4$	0.0625–2.500	0.9927
5	Atrazine-Desethyl	6190-65-4	$Y = 2.26 \times 10^6 X - 1.93 \times 10^5$	0.0625–2.500	0.9956
6	Benodanil	15310-01-7	$Y = 5.34 \times 10^6 X - 2.7 \times 10^6$	0.1875–7.500	0.9899
7	Beta-HCH	319-85-7	$Y = 6.87 \times 10^5 X - 3.02 \times 10^4$	0.0625–2.500	0.9992
8	Bifenthrin	82657-04-3	$Y = 7.25 \times 10^6 X - 7.29 \times 10^5$	0.0625–2.500	0.9936

Continued

TABLE 5.12 Linear Equations, Linear Ranges, and Correlation Coefficients of 567 Pesticides and Chemical Pollutants Determined by GC-MS—cont'd

No.	Compound	CAS	Linear Equation	Linear Range	Correlation Coeffient
9	Bromophos-ethyl	4824-78-6	$Y = 8.39 \times 10^5 X - 6.63 \times 10^4$	0.0625–2.500	0.9973
10	Bupirimate	57839-19-1	$Y = 1.73 \times 10^6 X - 1.57 \times 10^5$	0.0625–2.500	0.9954
11	Carbosulfan	55285-14-8	$Y = 1.06 \times 10^6 X - 2.53 \times 10^5$	0.1875–7.500	0.9987
12	Carboxin	5234-68-4	$Y = 1.49 \times 10^6 X - 5.15 \times 10^5$	0.1875–7.500	0.9930
13	Chlorbenside	103-17-3	$Y = 6.55 \times 10^5 X - 1.37 \times 10^5$	0.1250–5.00	0.9946
14	Chlordimeform	6164-98-3	$Y = 3.39 \times 10^5 X - 4.28 \times 10^4$	0.1250–2.500	0.9966
15	Chlorfurenol	2464-37-1	$Y = 4.48 \times 10^6 X - 1.29 \times 10^6$	0.1875–7.500	0.9960
16	Chlormephos	24934-91-6	$Y = 1.19 \times 10^6 X + 1.14 \times 10^5$	0.1250–5.000	0.9928
17	Chlorpyifos(Ethyl)	2921-88-2	$Y = 6.84 \times 10^5 X - 4.74 \times 10^4$	0.0625–2.500	0.9985
18	cis-permethrin	74774-45-7	$Y = 4.92 \times 10^6 X - 4.33 \times 10^5$	0.0625–2.500	0.9935
19	Clomazone	81777-89-1	$Y = 3.04 \times 10^6 X - 2.30 \times 10^5$	0.0625–2.500	0.9976
20	Cyanazine	21725-46-2	$Y = 1.26 \times 10^6 X - 4.10 \times 10^5$	0.1875–7.500	0.9952
21	Cycloate	1134-23-2	$Y = 2.26 \times 10^6 X - 1.16 \times 10^5$	0.0625–2.500	0.9986
22	Cypermethrin	52315-07-8	$Y = 6.63 \times 10^5 X - 6.98 \times 10^5$	0.1875–7.500	0.9190
23	Cyprazine	22936-86-3	$Y = 1.81 \times 10^6 X - 1.38 \times 10^5$	0.0625–2.500	0.9978
24	Delta-HCH	319-86-8	$Y = 6.21 \times 10^5 X - 7.51 \times 10^4$	0.1250–5.000	0.9986

No.	Name	CAS	Equation	Range	R
25	Deltamethrin	52918-63-5	$Y=9.80\times10^5X-6.51\times10^5$	0.3750–15.00	0.9921
26	Diazinon	333-41-5	$Y=6.86\times10^5X-4.99\times10^4$	0.0625–2.500	0.9980
27	Dichlofenthion	97-17-6	$Y=1.45\times10^6X-8.71\times10^4$	0.0625–2.500	0.9986
28	Dichlormid	37764-25-3	$Y=3.26\times10^7X+1.03\times10^6$	0.0063–2.500	0.9981
29	Dichlorofop-methyl	51338-27-3	$Y=1.23\times10^6X-4.28\times10^4$	0.0625–2.500	0.9958
30	Dieldrin	60-57-1	$Y=3.00\times10^5X-2.58\times10^4$	0.1250–5.000	0.9994
31	Dimethoate	60-51-5			
32	Dinitramine	29091-05-2	$Y=8.42\times10^5X-3.98\times10^5$	0.2500–10.00	0.9956
33	Diphenylamin	122-39-4	$Y=5.81\times10^6X-3.97\times10^5$	0.0625–2.500	0.9975
34	Etaconazole-1	1000290-09-5	$Y=5.57\times10^5X-1.69\times10^5$	0.1875–7.500	0.9948
35	Etaconazole-2	1000290-09-5	$Y=8.08\times10^5X-1.78\times10^5$	0.1875–7.500	0.9976
36	Ethalfluralin	55283-68-6	$Y=5.28\times10^5X-2.60\times10^5$	0.2500–10.00	0.9926
37	Ethion	562-12-2	$Y=1.91\times10^6X-4.27\times10^5$	0.1250–5.000	0.9944
38	Etridiazol	2593-15-9	$Y=8.12\times10^5X-2.65\times10^5$	0.1875–7.500	0.9945
39	Etrimfos	38260-54-7	$Y=8.55\times10^5X-6.25\times10^4$	0.0625–2.500	0.9976
40	Fenamiphos	22224-92-6	$Y=1.80\times10^6X-8.73\times10^5$	0.1875–7.500	0.9864
41	Fenitrothion	122-14-5	$Y=1.13\times10^6X-2.61\times10^5$	0.1250–5.000	0.9957
42	Fenothiocarb	62850-32-2	$Y=2.27\times10^4X-1.21\times10^4$	1.0000–5.000	0.9955
43	Fensulfothion	115-90-2	$Y=3.84\times10^5X-5.31\times10^4$	0.1250–5.000	0.9900
44	Fenthion	55-38-9	$Y=2.91\times10^6X-2.33\times10^5$	0.0625–2.500	0.9973

Continued

TABLE 5.12 Linear Equations, Linear Ranges, and Correlation Coefficients of 567 Pesticides and Chemical Pollutants Determined by GC-MS—cont'd

No.	Compound	CAS	Linear Equation	Linear Range	Correlation Coeffient
45	Fenvalerate-1	51630-58-1	$Y = 1.54 \times 10^6 X - 4.39 \times 10^5$	0.2500–10.00	0.9928
46	Fenvalerate-2	51630-58-1	$Y = 1.54 \times 10^6 X - 4.39 \times 10^5$	0.2500–10.00	0.9928
47	Flutolanil	66332-96-5	$Y = 6.85 \times 10^6 X - 7.11 \times 10^5$	0.0625–2.500	0.9951
48	Folpet	133-07-3	$Y = 4.67 \times 10^5 X - 1.14 \times 10^6$	0.7500–30.00	0.9739
49	Fonofos	944-22-9	$Y = 1.43 \times 10^6 X - 1.10 \times 10^5$	0.0625–2.500	0.9975
50	Linuron	330-55-2	$Y = 4.30 \times 10^5 X - 3.08 \times 10^5$	0.2500–10.00	0.9730
51	Malathion	121-75-5	$Y = 1.43 \times 10^6 X - 5.23 \times 10^5$	0.2500–10.00	0.9974
52	Metalaxyl	57837-19-1	$Y = 1.23 \times 10^6 X - 2.59 \times 10^5$	0.1875–7.500	0.9982
53	Metazachlor	67129-08-2	$Y = 1.18 \times 10^6 X - 2.26 \times 10^5$	0.1875–7.500	0.9984
54	Methidathion	950-37-8	$Y = 2.58 \times 10^5 X - 5.55 \times 10^5$	0.1250–5.000	0.9948
55	Methoxychlor	72-43-5	$Y = 5.40 \times 10^6 X - 6.40 \times 10^5$	0.0625–2.500	0.9921
56	Methyl-Parathion	3060-89-7	$Y = 1.06 \times 10^6 X - 6.40 \times 10^5$	0.2500–10.00	0.9892
57	Mexacarbate	315-18-4	$Y = 2.74 \times 10^6 X - 9.75 \times 10^5$	0.1875–7.500	0.9924
58	Mirex	2385-85-5	$Y = 1.29 \times 10^6 X - 6.90 \times 10^4$	0.0625–2.500	0.9974
59	Myclobutanil	88671-89-0	$Y = 1.90 \times 10^6 X - 1.74 \times 10^5$	0.0625–2.500	0.9958

Continued

60	Napropamide	15299-99-7	$Y = 9.05 \times 10^5 X - 2.34 \times 10^5$	0.1875–7.500	0.9963
61	Norflurazon	27314-13-2	$Y = 1.16 \times 10^6 X - 1.33 \times 10^5$	0.0625–2.500	0.9879
62	Nuarimol	63284-71-9	$Y = 6.42 \times 10^5 X - 1.13 \times 10^5$	0.1250–5.000	0.9962
63	Oxadiazone	19666-30-9	$Y = 1.56 \times 10^6 X - 1.01 \times 10^5$	0.0625–2.500	0.9980
64	Oxadxyl	23135-22-0	$Y = 7.29 \times 10^5 X - 6.65 \times 10^4$	0.0625–2.500	0.9897
65	Oxycarboxin	5259-88-1	$Y = 1.34 \times 10^6 X - 1.23 \times 10^6$	0.3750–15.00	0.9850
66	p,p'-DDD	72-54-8	$Y = 4.46 \times 10^6 X - 5.15 \times 10^5$	0.0625–2.500	0.9930
67	Paraoxon-ethyl	311-45-5	$Y = 3.39 \times 10^5 X - 2.09 \times 10^5$	0.2500–10.00	0.9892
68	Parathion	56-38-2	$Y = 1.08 \times 10^6 X - 5.62 \times 10^5$	0.2500–10.000	0.9935
69	Pendimethalin	40318-45-4	$Y = 2.46 \times 10^6 X - 1.12 \times 10^6$	0.2500–10.00	0.9960
70	Phenthoate	2597-03-7	$Y = 1.37 \times 10^6 X - 2.52 \times 10^5$	0.1250–5.000	0.9970
71	Phorate	298-02-2	$Y = 5.32 \times 10^5 X - 4.50 \times 10^4$	0.0625–2.500	0.9969
72	Phosmet	732-11-6	$Y = 2.04 \times 10^6 X - 6.08 \times 10^5$	0.1250–5.000	0.9861
73	Procymidone	32809-16-8	$Y = 1.18 \times 10^6 X - 6.98 \times 10^4$	0.0625–2.500	0.9979
74	Prometrye	7287-19-6	$Y = 1.97 \times 10^6 X - 1.82 \times 10^5$	0.0625–2.500	0.9961
75	Pronamide	23950-58-5	$Y = 2.33 \times 10^6 X - 2.70 \times 10^5$	0.0625–2.500	0.9905
76	Propetamphos	31218-83-4	$Y = 2.44 \times 10^6 X - 2.24 \times 10^5$	0.0625–2.500	0.9968
77	Propham	122-42-9	$Y = 1.01 \times 10^6 X - 5.76 \times 10^4$	0.0625–2.500	0.9975
78	Propiconazole-1	60207-90-1	$Y = 6.40 \times 10^5 X - 1.94 \times 10^5$	0.1875–7.500	0.9942

TABLE 5.12 Linear Equations, Linear Ranges, and Correlation Coefficients of 567 Pesticides and Chemical Pollutants Determined by GC-MS—cont'd

No.	Compound	CAS	Linear Equation	Linear Range	Correlation Coeffient
79	Propiconazole-2	60207-90-1	$Y = 6.40 \times 10^5 X - 1.94 \times 10^5$	0.1875–7.500	0.9942
80	Prothiophos	34643-46-4	$Y = 9.17 \times 105 X - 9.14 \times 104$	0.0625–2.500	0.9957
81	Pyrazophos	13457-18-6	$Y = 2.23 \times 10^6 X - 5.32 \times 10^5$	0.1250–5.000	0.9904
82	Pyridaphenthion	119-12-0	$Y = 1.04 \times 10^6 X - 1.47 \times 10^5$	0.0625–2.500	0.9841
83	Quinalphos	84087-01-4	$Y = 1.93 \times 10^6 X - 1.91 \times 10^5$	0.0625–2.500	0.9959
84	Quintozene	82-68-8	$Y = 2.98 \times 10^5 X - 5.32 \times 10^4$	0.1250–5.000	0.9958
85	Ronnel	83-79-4	$Y = 2.37 \times 10^6 X - 3.41 \times 10^5$	0.1250–5.000	0.9978
86	Secbumeton	26259-45-0	$Y = 3.16 \times 10^6 X - 3.21 \times 10^5$	0.0625–2.500	0.9952
87	Sulprofos	35400-43-2	$Y = 1.45 \times 10^6 X - 2.67 \times 10^5$	0.1250–5.000	0.9959
88	Tebuconazole	107534-96-3	$Y = 1.39 \times 10^6 X - 4.05 \times 10^5$	0.1875–7.500	0.9941
89	Tetradifon	116-29-0	$Y = 4.97 \times 10^5 X - 2.54 \times 10^4$	0.0625–2.500	0.9966
90	Tetramethrin	7696-12-0	$Y = 4.35 \times 10^6 X - 9.91 \times 10^5$	0.1250–5.000	0.9928
91	Tetrasul	2227-13-6	$Y = 1.60 \times 10^6 X - 1.21 \times 10^5$	0.0625–2.500	0.9958
92	Thiometon	640-15-3	$Y = 3.00 \times 10^6 X - 2.40 \times 10^5$	0.0625–2.500	0.9967
93	trans-Chlodane	5103-74-2	$Y = 1.09 \times 10^6 X - 5.33 \times 10^4$	0.0625–2.500	0.9989

No.	Name	CAS	Equation	Range	R^2
94	trans-Permethrin	551877-74-8	$Y=4.19\times10^6 X-4.17\times10^5$	0.0625–2.500	0.9911
95	Triadimefon	43121-43-3	$Y=1.20\times10^6 X-2.03\times10^5$	0.1250–5.000	0.9976
96	Vinclozolin	50471-44-8	$Y=4.84\times10^5 X-3.88\times10^4$	0.0625–2.500	0.9973
97	Alachlor	15972-60-8	$Y=1.27\times10^6 X-2.26\times10^5$	0.1875–7.500	0.9990
98	Alpha-HCH	319-84-6	$Y=6.51\times10^5 X-2.25\times10^4$	0.0625–2.500	0.9994
99	Aspon	3244-90-4	$Y=3.58\times10^6 X-1.49\times10^5$	0.1250–5.000	0.9994
100	Azinphos-ethyl	2642-71-9	$Y=1.15\times10^6 X-2.82\times10^5$	0.1250–5.000	0.9930
101	Azinphos-methyl	86-50-0	$Y=1.82\times10^5 X-1.18\times10^5$	0.7500–15.00	0.9971
102	Benalaxyl	71626-11-4	$Y=3.38\times10^6 X-2.55\times10^5$	0.0625–2.500	0.9972
103	Benzoylprop-ethyl	22212-55-1	$Y=5.20\times10^5 X-1.00\times10^5$	0.1875–7.500	0.9983
104	Bifenox	42576-02-3	$Y=1.24\times10^6 X-5.52\times10^5$	0.2500–10.00	0.9826
105	Bromofos	2104-96-3	$Y=1.50\times10^6 X-1.98\times10^5$	0.1250–5.000	0.9987
106	Bromopropylate	1818-80-1	$Y=3.62\times10^6 X-6.59\times10^5$	0.1250–5.000	0.9963
107	Buprofezin	69327-76-0	$Y=7.95\times10^6 X-8.95\times10^5$	0.1250–5.000	0.9987
108	Butachlor	23184-66-9	$Y=1.84\times10^6 X-3.26\times10^5$	0.1250–5.000	0.9970
109	Butylate	2008-41-5	$Y=1.10\times10^6 X-1.13\times10^5$	0.1875–7.500	0.9994
110	Captafol	2425-06-1	$Y=2.33\times10^5 X-9.85\times10^5$	1.1250–45.00	0.9564
111	Carbofenothion	786-19-6	$Y=1.32\times10^6 X-2.79\times10^5$	0.1250–5.000	0.9955
112	Chlorbenside sulfone	7082-99-7	$Y=1.89\times10^6 X-2.89\times10^5$	0.1250–5.000	0.9966

Continued

TABLE 5.12 Linear Equations, Linear Ranges, and Correlation Coefficients of 567 Pesticides and Chemical Pollutants Determined by GC-MS—cont'd

No.	Compound	CAS	Linear Equation	Linear Range	Correlation Coeffient
113	Chlorbromuron	57160-47-1	$Y = 1.63 \times 10^5 X - 6.33 \times 10^5$	1.5000–60.00	0.9843
114	Chlorbufam	1967-16-4	$Y = 5.75 \times 10^5 X - 2.01 \times 10^5$	0.2500–10.00	0.9921
115	Chlorfenson	80-33-1	$Y = 5.95 \times 10^5 X - 8.13 \times 10^4$	0.1250–5.000	0.9980
116	Chlorfenvinphos	470-90-6	$Y = 8.38 \times 10^5 X - 2.30 \times 10^5$	0.1875–7.500	0.9976
117	Chloroneb	2675-77-6	$Y = 2.04 \times 10^6 X - 9.36 \times 10^4$	0.0625–2.500	0.9990
118	Chloropropylate	5836-10-2	$Y = 2.99 \times 10^6 X - 2.62 \times 10^5$	0.0625–2.500	0.9970
119	Chlorprifos-methyl	5598-13-0	$Y = 1.55 \times 10^6 X - 1.25 \times 10^5$	0.0625–2.500	0.9972
120	Chlorpropham	2921-88-2	$Y = 7.24 \times 10^5 X - 1.22 \times 10^5$	0.1250–5.000	0.9971
121	Chlorthiophos	60238-56-4	$Y = 6.89 \times 10^5 X - 1.58 \times 10^5$	0.1875–7.500	0.9978
122	Chlozolinate	84332-86-5	$Y = 6.77 \times 10^5 X - 6.41 \times 10^4$	0.1250–5.000	0.9994
123	cis-chlordane	5103-71-9	$Y = 9.13 \times 10^5 X - 7.13 \times 10^4$	0.1250–5.000	0.9994
124	cis-diallate	2303-16-4	$Y = 2.70 \times 10^5 X - 1.65 \times 10^4$	0.1250–5.000	0.9998
125	Coumaphos	56-72-4	$Y = 7.20 \times 10^5 X - 4.57 \times 10^5$	0.3750–15.00	0.9955
126	Crufomate	299-86-5	$Y = 1.85 \times 10^6 X - 1.63 \times 10^6$	0.3750–15.00	0.9914
127	Cyanofenphos	13067-93-1	$Y = 2.34 \times 10^6 X - 1.98 \times 10^5$	0.0625–2.500	0.9967

Continued

128	Cyanohos	2636-26-2	$Y = 2.01 \times 10^6 X - 2.99 \times 10^5$	0.1250–5.000	0.9980
129	Cyfluthrin	68359-37-5	$Y = 2.38 \times 10^5 X - 1.95 \times 10^5$	0.7500–30.00	0.9966
130	Desmetryn	1014-69-3	$Y = 1.92 \times 10^6 X - 1.86 \times 10^5$	0.0625–2.500	0.9964
131	Dichlobenil	1194-65-6	$Y = 2.99 \times 10^6 X - 1.58 \times 10^4$	0.0250–0.500	0.9998
132	Dichlofluanid	1085-98-9	$Y = 5.53 \times 10^5 X - 3.40 \times 10^5$	0.3750–15.00	0.9977
133	Dicloran	99-30-9	$Y = 6.59 \times 10^5 X - 1.86 \times 10^5$	0.2500–10.00	0.9963
134	Dicofol	115-32-2	$Y = 1.00 \times 10^6 X - 1.25 \times 10^5$	0.1250–5.000	0.9980
135	Dimethachlor	51218-45-2	$Y = 3.98 \times 10^6 X - 7.53 \times 10^5$	0.1875–7.500	0.9989
136	Dioxathion	78-34-2	$Y = 1.58 \times 10^5 X - 2.19 \times 10^4$	0.2500–10.00	0.9993
137	Edifenphos	17109-49-8	$Y = 1.25 \times 10^6 X - 3.66 \times 10^5$	0.1250–5.000	0.9912
138	Endosulfan-1	959-98-8	$Y = 1.47 \times 10^5 X - 3.22 \times 10^4$	0.3750–15.00	0.9991
139	Endosulfan-2	33213-65-9	$Y = 2.45 \times 10^4 X - 1.10 \times 10^4$	1.5000–15.00	0.9994
140	Endosulfan-sulfate	1031-07-8	$Y = 2.46 \times 10^5 X - 3.43 \times 10^4$	0.1875–7.500	0.9985
141	Endrin	72-20-8	$Y = 3.08 \times 10^5 X - 2.37 \times 10^5$	0.7500–30.00	0.9982
142	EPN	2104-64-5	$Y = 2.08 \times 10^6 X - 1.17 \times 10^6$	0.2500–10.00	0.9913
143	EPTC	759-94-4	$Y = 1.57 \times 10^6 X - 1.31 \times 10^5$	0.1875–7.500	0.9993
144	Erbon	136-25-4	$Y = 4.86 \times 10^5 X - 4.10 \times 10^4$	0.1250–2.500	0.9994
145	Ethofumesate	26225-79-6	$Y = 2.00 \times 10^6 X - 2.03 \times 10^5$	0.1250–5.000	0.9992
146	Ethoprophos	13194-48-4	$Y = 1.04 \times 10^6 X - 2.32 \times 10^5$	0.1875–7.500	0.9982

TABLE 5.12 Linear Equations, Linear Ranges, and Correlation Coefficients of 567 Pesticides and Chemical Pollutants Determined by GC-MS—cont'd

No.	Compound	CAS	Linear Equation	Linear Range	Correlation Coeffient
147	Fenarimol	60168-88-9	$Y = 1.19 \times 10^6 X - 1.33 \times 10^5$	0.1250–5.000	0.9993
148	Fenpropathrin	39515-41-8	$Y = 4.35 \times 10^5 X - 7.38 \times 10^4$	0.1250–5.000	0.9966
149	Flamprop-isopropyl	52756-22-6	$Y = 7.37 \times 10^6 X - 5.64 \times 10^5$	0.0625–2.500	0.9974
150	Flamprop-methyl	52756-25-9	$Y = 6.68 \times 10^6 X - 4.53 \times 10^5$	0.0625–2.500	0.9983
151	Flufenoxuron	101463-69-8	$Y = 1.40 \times 10^5 X - 5.53 \times 10^4$	0.1875–7.500	0.9929
152	Fluorochloridone	61213-25-0	$Y = 1.14 \times 10^5 X - 1.75 \times 10^4$	0.5000–5.000	0.9994
153	Fluvalinate	102851-06-9	$Y = 3.44 \times 10^6 X - 3.89 \times 10^6$	0.7500–30.00	0.9954
154	Heptanophos	23560-59-0	$Y = 1.71 \times 10^6 X - 4.01 \times 10^5$	0.1875–7.500	0.9979
155	Hexachlorobenzene	118-74-1	$Y = 1.80 \times 10^6 X - 5.09 \times 10^4$	0.0625–2.500	0.9996
156	Hexaconazole	79983-71-4	$Y = 5.28 \times 10^4 X - 2.04 \times 10^4$	0.7500–15.00	0.9989
157	Hexazinone	51235-04-2	$Y = 5.21 \times 10^6 X - 1.27 \times 10^6$	0.1875–7.500	0.9970
158	Iodofenphos	18181-70-9	$Y = 1.32 \times 10^6 X - 3.31 \times 10^5$	0.1250–5.000	0.9940
159	Isofenphos	25311-71-1	$Y = 1.54 \times 10^6 X - 2.40 \times 10^5$	0.1250–5.000	0.9982
160	Isopropalin	33820-53-0			
161	Leptophos	21609-90-5			

162	Methoprene	40596-69-8	$Y=3.30\times10^{6}X-1.31\times10^{6}$	0.2500–10.00	0.9975
163	Methoprotryne	841-06-5	$Y=1.80\times10^{6}X-5.13\times10^{5}$	0.1875–7.500	0.9973
164	Metolachlor	51218-45-2	$Y=2.13\times10^{6}X-1.69\times10^{5}$	0.0625–2.500	0.9977
165	Mevinphos	7786-34-7	$Y=1.77\times10^{6}X-3.53\times10^{5}$	0.1250–5.000	0.9957
166	Monolinuron	1746-81-2	$Y=1.78\times10^{6}X-9.80\times10^{5}$	0.2500–10.00	0.9888
167	Nitrapyrin	1929-82-4	$Y=1.41\times10^{6}X-4.43\times10^{5}$	0.1875–7.500	0.9957
168	Nitrofen	1836-75-5	$Y=1.20\times10^{6}X-1.04\times10^{6}$	0.3750–15.00	0.9905
169	o,p′-DDD	53-19-0	$Y=1.16\times10^{5}X-2.92\times10^{4}$	0.1250–2.500	0.9997
170	o,p′-DDT	789-02-6	$Y=2.71\times10^{6}X-4.24\times10^{5}$	0.1250–5.000	0.9978
171	Oxychlordane	27304-13-8	$Y=1.97\times10^{5}X-1.23\times10^{4}$	0.2500–2.500	0.9997
172	Oxyflurofen	42874-03-3	$Y=1.54\times10^{6}X-7.81\times10^{5}$	0.2500–10.00	0.9945
173	p,p′-DDE	72-55-9	$Y=1.41\times10^{6}X-5.29\times10^{4}$	0.0625–2.500	0.9994
174	p,p′-DDT	50-29-3	$Y=2.76\times10^{6}X-5.05\times10^{5}$	0.1250–5.000	0.9971
175	Paclobutrazol	76738-62-0	$Y=1.76\times10^{6}X-6.51\times10^{5}$	0.1875–7.500	0.9941
176	Pebulate	1114-71-2	$Y=2.11\times10^{6}X-2.26\times10^{5}$	0.1875–7.500	0.9994
177	Phosalone	2310-17-0	$Y=1.19\times10^{6}X-2.42\times10^{5}$	0.1250–5.000	0.9955
178	Pirimiphos-ethyl	23505-41-1	$Y=1.16\times10^{6}X-1.83\times10^{5}$	0.1250–5.000	0.9980
179	Pirimiphos-methyl	29232-93-7	$Y=1.36\times10^{6}X-1.07\times10^{5}$	0.0625–2.500	0.9977
180	Prochloraz	67747-09-5	$Y=5.01\times10^{5}X-5.72\times10^{4}$	0.3750–15.00	0.9919

Continued

TABLE 5.12 Linear Equations, Linear Ranges, and Correlation Coefficients of 567 Pesticides and Chemical Pollutants Determined by GC-MS—cont'd

No.	Compound	CAS	Linear Equation	Linear Range	Correlation Coeffient
181	Profenofos	41198-08-7	$Y = 4.53 \times 10^5 X - 2.76 \times 10^5$	0.3750–15.00	0.9969
182	Profluralin	26399-36-0	$Y = 9.72 \times 10^5 X - 4.42 \times 10^5$	0.2500–10.00	0.9955
183	Propachlor	1918-16-7	$Y = 2.89 \times 10^6 X - 3.00 \times 10^5$	0.1875–7.500	0.9998
184	Propanil	709-98-8	$Y = 1.99 \times 10^6 X - 4.75 \times 10^5$	0.1250–5.000	0.9930
185	Propazine	139-40-2	$Y = 1.63 \times 10^6 X - 9.61 \times 10^4$	0.0625–2.500	0.9988
186	Sulfallate	95-06-7	$Y = 2.84 \times 10^6 X - 6.05 \times 10^5$	0.1250–5.000	0.9958
187	Sulfotep	3689-24-5	$Y = 1.61 \times 10^6 X - 8.57 \times 10^4$	0.0625–2.500	0.9991
188	Tecnazene	117-18-0	$Y = 3.92 \times 10^5 X - 6.99 \times 10^4$	0.2500–5.000	0.9985
189	Terbufos	13071-79-9	$Y = 1.77 \times 10^6 X - 2.82 \times 10^5$	0.1250–5.000	0.9977
190	Terbumeton	33693-04-8	$Y = 2.47 \times 10^6 X - 5.48 \times 10^5$	0.1875–7.500	0.9983
191	Terbuthylazine	5915-41-3	$Y = 9.62 \times 10^5 X - 1.21 \times 10^5$	0.0625–2.500	0.9886
192	Terbutryn	886-50-0	$Y = 1.95 \times 10^6 X - 3.07 \times 10^5$	0.1250–5.000	0.9980
193	Tetrachlorvinphos	22248-79-9	$Y = 1.61 \times 10^6 X - 4.88 \times 10^5$	0.1875–7.500	0.9975
194	Thiobencarb	28249-77-6	$Y = 4.00 \times 10^6 X - 5.32 \times 10^5$	0.1250–5.000	0.9987
195	Tolyfluanide	731-27-1	$Y = 5.19 \times 10^5 X - 1.68 \times 10^5$	0.1875–7.500	0.9981

No.	Name	CAS	Equation	Range	r
196	trans-Diallate	2303-16-4	$Y = 1.02 \times 10^6 X - 9.85 \times 10^4$	0.1250–5.000	0.9994
197	Triazophos	24017-47-8	$Y = 1.05 \times 10^6 X - 3.63 \times 10^5$	0.1875–7.500	0.9944
198	Trifluralin	1582-09-8	$Y = 1.83 \times 10^6 X - 4.47 \times 10^5$	0.1250–5.000	0.9930
199	2-Phenylphenol	90-43-7	$Y = 1.70 \times 10^6 X - 3.54 \times 10^4$	0.0625–1.2500	0.9946
200	3,5-Dichloroaniline	626-43-7	$Y = 9.29 \times 10^5 X - 1.10 \times 10^5$	0.0625–2.500	0.9905
201	Acrinathrin	101007-06-1	$Y = 1.21 \times 10^6 X - 2.74 \times 10^5$	0.1250–5.000	0.9939
202	Alpha-cypermethrin	67375-30-8	$Y = 1.42 \times 10^6 X - 2.51 \times 10^5$	0.1250–5.000	0.9974
203	Ametryn	834-12-8	$Y = 1.96 \times 10^6 X - 4.43 \times 10^5$	0.1875–7.500	0.9989
204	Amitraz	33089-61-1	$Y = 4.35 \times 10^5 X - 8.43 \times 10^4$	0.1875–7.500	0.9993
205	Anilofos	64249-01-0	$Y = 5.65 \times 10^5 X - 1.83 \times 10^5$	0.2500–5.000	0.9954
206	Atrizine	1912-24-9	$Y = 1.60 \times 10^6 X - 1.05 \times 10^5$	0.0625–2.500	0.9991
207	Benfluralin	1861-40-1	$Y = 2.07 \times 10^6 X - 2.74 \times 10^5$	0.0625–2.500	0.9888
208	Bioallethrin-1	584-79-2	$Y = 1.37 \times 10^6 X - 7.09 \times 10^5$	0.2500–10.00	0.9941
209	Bioallethrin-2	584-79-2	$Y = 1.49 \times 10^6 X - 5.77 \times 10^5$	0.2500–10.00	0.9983
210	Biphenyl	92-52-4	$Y = 6.61 \times 10^6 X - 1.66 \times 10^5$	0.0625–2.500	0.9998
211	Bitertanol	55179-31-2	$Y = 4.06 \times 10^6 X - 1.42 \times 10^6$	0.1875–7.500	0.9931
212	Chlorfluazuron	71422-67-8	$Y = 1.55 \times 10^5 X - 9.15 \times 10^3$	0.3750–7.500	0.9980
213	Chlorobenzilate	510-15-6	$Y = 2.62 \times 10^6 X - 2.29 \times 10^5$	0.0625–2.500	0.9977
214	Chlorthion	500-28-7	$Y = 3.72 \times 10^5 X - 2.23 \times 10^5$	0.5000~5.000	0.9857

Continued

TABLE 5.12 Linear Equations, Linear Ranges, and Correlation Coefficients of 567 Pesticides and Chemical Pollutants Determined by GC-MS—cont'd

No.	Compound	CAS	Linear Equation	Linear Range	Correlation Coeffient
215	cis-1,2,3,6tetrahydrophthalimide	85-40-5	$Y = 1.06 \times 10^6 X - 2.31 \times 10^5$	0.1875–7.500	0.9985
216	Cycloxydim	101205-02-1	$Y = 7.64 \times 10^5 X + 8.13 \times 10^5$	0.7500–30.00	0.9907
217	Dichlorvos	62-73-7	$Y = 2.58 \times 10^6 X - 9.22 \times 10^5$	0.3750–15.00	0.9990
218	Diethofencarb	87130-20-9	$Y = 1.15 \times 10^6 X - 6.62 \times 10^5$	0.3750~15.00	0.9977
219	Difenonazole-1	1977-6-54	$Y = 1.08 \times 10^6 X - 3.85 \times 10^5$	0.3750–15.00	0.9860
220	Difenonazole-2	1977-6-54	$Y = 1.08 \times 10^6 X - 3.85 \times 10^5$	0.3750–15.00	0.9860
221	Diflufenican	83164-33-4	$Y = 3.75 \times 10^6 X - 4.25 \times 10^5$	0.0625–2.500	0.9950
222	Dimefuron	34205-21-5	$Y = 2.79 \times 10^5 X - 1.10 \times 10^4$	0.2500–10.00	0.9885
223	Dimepiperate	61432-55-1	$Y = 2.12 \times 10^4 X + 4.02 \times 10^4$	2.500~5.000	1.0000
224	Dimethipin	55290-64-7			
225	Diniconazole	83657-24-3	$Y = 1.62 \times 10^6 X - 6.37 \times 10^5$	0.1875–7.500	0.9936
226	Diphenamid	957-51-7	$Y = 4.01 \times 10^6 X - 2.59 \times 10^5$	0.0625~2.500	0.9989
227	Dipropetryn	4147-51-7	$Y = 1.67 \times 10^6 X - 1.52 \times 10^5$	0.0625~2.500	0.9976
228	Disulfoton	298-04-4	$Y = 2.37 \times 10^6 X - 1.88 \times 10^5$	0.0625–2.500	0.9983
229	Esfenvalerate	51630-58-1	$Y = 3.68 \times 10^5 X - 7.15 \times 10^4$	0.2500–10.00	0.9983

Continued

No.	Name	CAS	Equation	Range	R
230	Etofenprox	80844-07-1	$Y=6.57\times10^6X-3.60\times10^5$	0.0625–2.500	0.9961
231	Fenazaquin	120928-09-8	$Y=5.25\times10^6X-5.73\times10^5$	0.0625–2.500	0.9950
232	Fenobucarb	3766-81-2	$Y=5.44\times10^6X-7.85\times10^5$	0.1250–5.000	0.9982
233	Fenoxycarb	79127-80-3	$Y=1.81\times10^5X+1.11\times10^5$	0.3750–15.00	0.9902
234	Fenpropimorph	67306-03-0	$Y=8.00\times10^6X-5.62\times10^5$	0.0625–2.500	0.9989
235	Fenson	80-38-6	$Y=2.26\times10^6X-1.16\times10^5$	0.0625~2.500	0.9993
236	Fluazifop-butyl	79241-46-6	$Y=2.18\times10^6X-2.27\times10^5$	0.0625–2.500	0.9961
237	Fluchloralin	33245-39-5	$Y=9.56\times10^5X-4.69\times10^5$	0.2500–10.00	0.9952
238	Flucythrinate-1	70124-77-5			
239	Flucythrinate-2	70124-77-5			
240	Fludioxonil	131341-86-1	$Y=2.12\times10^6X-2.08\times10^5$	0.0625–2.500	0.9963
241	Flumetralin	62924-70-3	$Y=2.51\times10^6X-7.70\times10^5$	0.1250~5.000	0.9873
242	Flumiclorac-pentyl	87546-18-7	$Y=9.93\times10^5X-2.33\times10^5$	0.1250–5.000	0.9903
243	Flumioxazin	103361-09-7			
244	Fluorodifen	15457-05-3			
245	Fluoroglycofen-ethyl	77502-90-7	$Y=4.13\times10^5X-8.77\times10^5$	0.7500–30.00	0.9826
246	Flusilazole	85509-19-9	$Y=4.23\times10^6X-1.07\times10^6$	0.1875–7.500	0.9983
247	Formothion	2540-82-1	$Y=1.12\times10^5X-3.13\times10^4$	0.5000~5.000	0.9993
248	Gamma-HCH	77-06-5	$Y=7.39\times10^5X-3.27\times10^4$	0.1250–5.000	0.9998
249	HCH, epsilon-	58-89-9	$Y=1.443X-1.733\times10^{-1}$	0.125–5	0.9960

TABLE 5.12 Linear Equations, Linear Ranges, and Correlation Coefficients of 567 Pesticides and Chemical Pollutants Determined by GC-MS—cont'd

No.	Compound	CAS	Linear Equation	Linear Range	Correlation Coeffient
250	Heptachlor	76-44-8	$Y = 6.01 \times 10^5 X - 1.16 \times 10^5$	0.1875~7.500	0.9990
251	Hexaflumuron	86479-06-3			
252	Iprobenfos	26087-47-8	$Y = 1.47 \times 10^6 X - 6.69 \times 10^5$	0.1875~7.500	0.9895
253	Isazofos	42509-80-8	$Y = 8.48 \times 10^5 X - 9.19 \times 10^4$	0.1250~5.000	0.9994
254	Kresoxim-methyl	143390-89-0	$Y = 2.45 \times 10^6 X - 2.13 \times 10^5$	0.0625~2.500	0.9976
255	Lambda-cyhalothrin	91465-08-6	$Y = 1.83 \times 10^6 X - 1.43 \times 10^5$	0.0625~2.500	0.9982
256	Mecarbam	2595-54-2	$Y = 2.87 \times 10^5 X - 2.80 \times 10^5$	0.2500~10.00	0.9986
257	Mefenacet	73250-68-7	$Y = 2.02 \times 10^6 X - 7.14 \times 10^5$	0.1875~7.500	0.9928
258	Mepronil	55814-41-0	$Y = 5.96 \times 10^6 X - 6.83 \times 10^5$	0.0625~2.500	0.9941
259	Methacrifos	62610-77-9	$Y = 1.10 \times 10^6 X - 5.10 \times 10^4$	0.0625~2.500	0.9994
260	Methobromuron	40596-69-8	$Y = 1.67 \times 10^5 X - 6.17 \times 10^4$	0.3750~15.00	0.9986
261	Metribuzin	21087-64-9	$Y = 1.61 \times 10^6 X - 3.48 \times 10^5$	0.1875~7.500	0.9986
262	Molinate	2212-67-1	$Y = 2.55 \times 10^6 X - 8.80 \times 10^4$	0.0625~2.500	0.9996
263	o,p'-DDE	3424-82-6	$Y = 2.54 \times 10^6 X - 7.65 \times 10^4$	0.0625~2.500	0.9998
264	Penconazole	66246-88-6	$Y = 2.32 \times 10^6 X - 4.76 \times 10^5$	0.1875~7.500	0.9992

Continued

265	Permethrin	52645-53-1	$Y = 2.92 \times 10^6 X - 5.21 \times 10^5$	0.1250–5.000	0.9967
266	Phenothrin	26002-80-2	$Y = 2.13 \times 10^6 X - 2.47 \times 10^5$	0.0625–2.500	0.9950
267	Piperonyl butoxide	51-03-6	$Y = 3.50 \times 10^6 X - 4.69 \times 10^5$	0.0625–2.500	0.9919
268	Plifenate	51366-25-7	$Y = 4.07 \times 10^5 X - 7.16 \times 10^4$	0.2500–5.000	0.9992
269	Prallethrin	23031-36-9	$Y = 3.52 \times 10^6 X - 1.22 \times 10^6$	0.1875~7.500	0.9964
270	Pretilachlor	51218-49-6	$Y = 1.40 \times 10^6 X - 2.48 \times 10^5$	0.1250~5.000	0.9980
271	Prometon	1610-18-0	$Y = 1.47 \times 10^6 X - 3.19 \times 10^5$	0.1875–7.500	0.9988
272	Propamocarb	24579-73-5	$Y = 5.73 \times 10^5 X - 2.71 \times 10^6$	0.1875–7.500	0.9912
273	Propaphos	7292-16-2	$Y = 2.319X - 4.412 \times 10^{-1}$	0.125–5	0.9940
274	Propargite	2312-35-8	$Y = 1.43 \times 10^5 X - 1.30 \times 10^5$	0.1250–5.000	0.9995
275	Propisochlor	86763-47-5	$Y = 1.053X - 4.032 \times 10^{-1}$	0.0625–2.500	0.9970
276	Pyridaben	96489-71-3	$Y = 5.18 \times 10^6 X - 4.92 \times 10^5$	0.0625–2.500	0.9958
277	Pyrimethanil	53112-28-0	$Y = 5.84 \times 10^6 X - 4.82 \times 10^5$	0.0625–2.500	0.9980
278	Sethoxydim	74051-80-2	$Y = 2.11 \times 10^5 X + 1.26 \times 10^5$	0.5625–22.50	0.9868
279	Simetryn	1014-70-6	$Y = 2.26 \times 10^6 X - 2.94 \times 10^5$	0.1250–5.000	0.9990
280	Tetraconazole	112281-77-3	$Y = 2.72 \times 10^6 X - 5.43 \times 10^5$	0.1875~7.500	0.9992
281	Tolclofos-methyl	57018-04-9	$Y = 3.46 \times 10^6 X - 1.95 \times 10^5$	0.0625–2.500	0.9991
282	Transfluthrin	118712-89-3	$Y = 3.03 \times 10^6 X - 1.80 \times 10^5$	0.0625–2.500	0.9991
283	Triadimenol-1	55219-65-3	$Y = 1.90 \times 10^6 X - 5.96 \times 10^5$	0.1875~7.500	0.9972

TABLE 5.12 Linear Equations, Linear Ranges, and Correlation Coefficients of 567 Pesticides and Chemical Pollutants Determined by GC-MS—cont'd

No.	Compound	CAS	Linear Equation	Linear Range	Correlation Coeffient
284	Triadimenol-2	55219-65-3	$Y=1.90\times10^6X-5.96\times10^5$	0.1875~7.500	0.9972
285	Triallate	2303-17-5	$Y=8.94\times10^5X-9.82\times10^4$	0.1250–5.000	0.9993
286	Vernolate	1929-77-7	$Y=2.12\times10^6X-6.55\times10^4$	0.0625–2.500	0.9997
287	2,3,4,5-Tetrachloroaniline	634-83-3	$Y=1.61\times10^6X-1.25\times10^5$	0.1250–5.000	0.9989
288	2,3,4,5-Tetrachloroanisole	938-86-3	$Y=1.11\times10^6X-2.36\times10^4$	0.0625–2.500	0.9997
289	2,3,5,6-Tetrachloroaniline	3481-20-7	$Y=1.72\times10^6X-5.14\times10^4$	0.0625–2.500	0.9994
290	2,6-Dichlorobenzamide	2008-58-4	$Y=1.57\times10^6X-1.74\times10^5$	0.1250–5.000	0.9983
291	4,4-Dibromobenzophenone	3988-03-2	$Y=5.97\times10^5X-7.56\times10^4$	0.1250–2.500	0.9945
292	4,4-Dichlorobenzophenone	90-98-2	$Y=1.13\times10^6X-9.71\times10^4$	0.0625–2.500	0.9959
293	Athidathion	19691-80-6	$Y=1.88\times10^5X-5.69\times10^3$	0.1250–5.000	0.9995
294	Atratone	1610-17-9	$Y=2.38\times10^6X-1.24\times10^5$	0.0625–2.500	0.9988
295	Aziprotryne	4658-28-0	$Y=4.43\times10^5X-1.80\times10^5$	0.5000–20.00	0.9979
296	BDMC-1	672-99-1	$Y=2.00\times10^5X+2.44\times10^4$	0.2500–5.000	0.9978
297	BDMC-2	672-99-1	$Y=1.46\times10^6X-3.49\times10^5$	0.1250–5.000	0.9912
298	Bromfenvinfos	33399-00-7	$Y=1.20\times10^6X-1.00\times10^5$	0.0625–2.500	0.9937

Continued

299	Bromocylen	1715-40-8	$Y=5.19\times10^5X-2.05\times10^4$	0.0625–2.500	0.9990
300	Bromuconazole-1	116255-48-2	$Y=4.45\times10^5X+7.05\times10^3$	0.1250–5.000	0.9916
301	Bromuconazole-2	116255-48-2	$Y=7.98\times10^5X-5.29\times10^4$	0.1250–2.500	0.9994
302	Chlorfenethol	80-06-8	$Y=2.04\times10^6X-1.30\times10^5$	0.0625–2.500	0.9975
303	Chlorfenprop-methyl	14437-17-3	$Y=1.34\times10^6X-4.92\times10^4$	0.0625–2.500	0.9989
304	Clodinafop-propargyl	105512-06-9	$Y=9.81\times10^5X-3.07\times10^5$	0.1250–5.000	0.9820
305	Cloquintocet-mexyl	99607-70-2	$Y=4.90\times10^6X-4.73\times10^5$	0.0625–2.500	0.9915
306	Crimidine	535-89-7	$Y=1.32\times10^6X-5.33\times10^4$	0.0625–2.500	0.9987
307	Cycluron	2163-69-1	$Y=6.73\times10^5X-1.27\times10^5$	0.1875–7.500	0.9982
308	Cyproconazole	113096-99-4	$Y=1.31\times10^6X-6.22\times10^4$	0.0625–2.500	0.9983
309	Cyprodinil	121552-61-2	$Y=5.21\times10^6X-3.06\times10^5$	0.0625–2.500	0.9977
310	Dacthal	1861-32-1	$Y=4.04\times10^6X-1.08\times10^5$	0.0625–2.500	0.9995
311	DEF	78-48-8	$Y=7.30\times10^5X-1.05\times10^5$	0.1250–5.000	0.9971
312	DE-PCB 101	37680-73-2	$Y=1.74\times10^6X-3.67\times10^4$	0.0625–2.500	0.9994
313	DE-PCB 118	31508-00-6	$Y=2.10\times10^6X-6.14\times10^4$	0.0625–2.500	0.9985
314	DE-PCB 138	35065-28-2	$Y=1.31\times10^6X-3.38\times10^4$	0.0625–2.500	0.9991
315	DE-PCB 153	35065-27-1	$Y=1.60\times10^6X-4.19\times10^4$	0.0625–2.500	0.9992
316	DE-PCB 180	35065-29-3	$Y=1.16\times10^6X-3.07\times10^4$	0.0625–2.500	0.9989
317	DE-PCB 28	2012-37-5	$Y=5.60\times10^6X-5.83\times10^4$	0.0625–1.250	0.9997

TABLE 5.12 Linear Equations, Linear Ranges, and Correlation Coefficients of 567 Pesticides and Chemical Pollutants Determined by GC-MS—cont'd

No.	Compound	CAS	Linear Equation	Linear Range	Correlation Coeffient
318	DE-PCB 31	16606-02-3	$Y=5.45\times10^6X-5.64\times10^4$	0.0625–1.250	0.9997
319	DE-PCB 52	35693-99-3	$Y=1.92\times10^6X-3.05\times10^4$	0.0625–2.500	0.9997
320	Desbrom-leptophos		$Y=9.66\times10^5X-3.34\times10^4$	0.0625–2.500	0.9990
321	Desethyl-sebuthylazine	37019-18-4	$Y=3.58\times10^6X-3.93\times10^5$	0.1250–5.000	0.9985
322	Desisopropyl-atrazine	1007-28-9	$Y=8.37\times10^5X-4.16\times10^5$	0.5000–20.00	0.9975
323	Dicapthon	2463-84-5	$Y=1.75\times10^6X-1.05\times10^6$	0.3125–12.50	0.9946
324	Diclobutrazole	75736-33-3	$Y=1.90\times10^6X-5.90\times10^5$	0.2500–10.00	0.9957
325	Dimefox	115-26-4	$Y=1.34\times10^6X-1.07\times10^5$	0.1875–7.500	0.9992
326	Dimethenamid	87674-68-8	$Y=3.66\times10^6X-1.48\times10^5$	0.0625–2.500	0.9990
327	Dinobuton	973-21-7	$Y=5.32\times10^5X-1.34\times10^6$	1.250–25.00	0.9788
328	Dioxabenzofos	3811-49-2	$Y=4.35\times10^5X-2.36\times10^5$	0.6250–25.00	0.9980
329	Disulfoton sulfone	2497-06-5	$Y=1.39\times10^6X-1.94\times10^5$	0.1250–5.000	0.9966
330	Disulfoton-sulfoxide	2497-07-6	$Y=2.77\times10^5X-2.21\times10^4$	0.1250–5.000	0.9993
331	Ditalimfos	5131-24-8	$Y=2.75\times10^5X-1.67\times10^5$	0.0625–2.500	0.9976
332	Etrimfos oxon	59399-24-5	$Y=4.07\times10^5X-1.85\times10^5$	0.0625–2.500	0.9990

No.	Name	CAS	Equation	Range	R
333	Fenamiphos sulfone	31972-44-8	$Y=2.06\times10^6 X-7.52\times10^5$	0.2500–10.00	0.9954
334	Fenamiphos sulfoxide	31972-42-7	$Y=3.38\times10^5 X-2.60\times10^3$	0.2500–10.00	0.9561
335	Fenbuconazole	114369-43-6	$Y=2.96\times10^6 X+9.45\times10^5$	0.1250–5.000	0.9896
336	Fenchlorphos oxon	3983-45-7	$Y=2.18\times10^6 X-1.77\times10^5$	0.1250–5.000	0.9991
337	Fenpiclonil	74738-17-3	$Y=1.72\times10^6 X-2.58\times10^5$	0.2500–10.00	0.9981
338	Fenthion sulfone	3761-42-0	$Y=4.68\times10^5 X+1.45\times10^5$	0.2500–10.00	0.9839
339	Fenthion sulfoxide	3761-41-9	$Y=2.45\times10^5 X-1.08\times10^5$	0.2500–5.000	0.9863
340	Fluotrimazole	31251-03-3	$Y=1.55\times10^6 X-5.73\times10^4$	0.0625–2.500	0.9981
341	Fluquinconazole	136426-54-5	$Y=3.40\times10^6 X-1.22\times10^4$	0.0625–2.500	0.9991
342	Flurochloridone	61213-25-0	$Y=1.26\times10^6 X-1.66\times10^5$	0.1250–5.000	0.9979
343	Fluroxypr-1-methylheptyl ester	81406-37-3	$Y=3.14\times10^5 X-2.20\times10^4$	0.1250–2.500	0.9974
344	Flutriafol	76674-21-0	$Y=1.40\times10^6 X-1.75\times10^5$	0.1250–5.000	0.9985
345	Fuberidazole	3878-19-1			
346	Hexythiazox	78587-05-0	$Y=3.00\times10^5 X-1.65\times10^5$	0.500–20.00	0.9981
347	Isobenzan	297-78-9	$Y=5.29\times10^5 X-1.51\times10^4$	0.0625–2.500	0.9995
348	Isocarbamid	30979-48-7	$Y=2.55\times10^6 X-9.72\times10^5$	0.3125–12.50	0.9965
349	Isocarbophos	245-61-5			
350	Isodrin	465-73-6	$Y=7.73\times10^5 X-9.06\times10^3$	0.0625–2.500	0.9982
351	Isofenphos oxon	106848-93-5			

Continued

TABLE 5.12 Linear Equations, Linear Ranges, and Correlation Coefficients of 567 Pesticides and Chemical Pollutants Determined by GC-MS—cont'd

No.	Compound	CAS	Linear Equation	Linear Range	Correlation Coeffient
352	Isomethiozin	57052-04-7	$Y = 1.60 \times 10^6 X - 2.21 \times 10^5$	0.1250–5.000	0.9970
353	Lenacil	2164-08-1	$Y = 1.31 \times 10^6 X - 7.83 \times 10^5$	0.6250–25.00	0.9967
354	MCPA-butoxyethyl ester	94-81-5	$Y = 6.47 \times 10^5 X - 4.71 \times 10^4$	0.0625–2.500	0.9966
355	Mephosfolan	950-10-7	$Y = 9.51 \times 10^5 X - 2.57 \times 10^5$	0.1250–5.000	0.9890
356	Metamitron	41394-05-2	$Y = 4.59 \times 10^4 X - 4.78 \times 10^4$	2.500–25.00	0.9994
357	Methabenzthiazuron	18691-97-9	$Y = 1.30 \times 10^6 X - 1.65 \times 10^6$	0.6250–25.00	0.9930
358	Methfuroxam	2873-17-8	$Y = 1.52 \times 10^6 X - 1.33 \times 10^5$	0.0625–2.500	0.9954
359	Monalide	7287-36-7	$Y = 7.65 \times 10^5 X - 5.97 \times 10^4$	0.1250–5.000	0.9988
360	Musk ambrette	83-46-9			
361	Musk ketone	541-91-3			
362	Musk moskene	116-66-5	$Y = 9.422X - 2.217 \times 10^{-1}$	0.0625–2.500	0.9970
363	Musk tibeten	145-39-1	$Y = 2.843X - 1.358 \times 10^{-1}$	0.0625–2.500	0.9970
364	Musk xylene	81-15-2			
365	Nitralin	4726-14-1	$Y = 1.35 \times 10^6 X - 1.87 \times 10^6$	0.6250–25.00	0.9907
366	Nitrothal-isopropyl	10552-74-6	$Y = 1.80 \times 10^6 X - 4.91 \times 10^5$	0.1250–5.000	0.9911

367	Octachlorostyrene	29082-74-4	$Y = 6.84 \times 10^5 X - 1.55 \times 10^4$	0.0625–2.500	0.9997
368	Pentachloroaniline	527-20-8	$Y = 1.63 \times 10^6 X - 5.38 \times 10^4$	0.0625–2.500	0.9990
369	Pentachloroanisole	1825-21-4	$Y = 8.25 \times 10^5 X - 1.64 \times 10^4$	0.0625–2.500	0.9996
370	Pentachlorobenzene	608-93-5	$Y = 2.14 \times 10^6 X - 3.01 \times 10^4$	0.0625–2.500	0.9998
371	Perthane	72-56-0	$Y = 7.03 \times 10^6 X - 3.66 \times 10^5$	0.0625–2.500	0.9974
372	Phorate sulfone	2588-04-7	$Y = 8.24 \times 10^5 X - 5.08 \times 10^4$	0.0625–2.500	0.9946
373	Phthalic acid, benzyl butyl ester	85-68-7	$Y = 8.67 \times 10^5 X - 4.54 \times 10^4$	0.0625–2.500	0.9973
374	Prosulfocarb	52888-80-9	$Y = 1.04 \times 10^6 X - 5.13 \times 10^4$	0.0625–2.500	0.9988
375	Pyrimitate	5221-49-8	$Y = 1.657 X - 6.476 \times 10^{-2}$	0.0625–2.500	0.9990
376	Rabenzazole	69899-24-7	$Y = 5.20 \times 10^6 X - 6.05 \times 10^5$	0.0625–2.500	0.9931
377	Resmethrin-1	10453-86-8	$Y = 2.98 \times 10^5 X - 1.11 \times 10^4$	0.1250–5.000	0.9823
378	Resmethrin-2	10453-86-8	$Y = 7.98 \times 10^5 X - 1.33 \times 10^5$	0.1250–5.000	0.9953
379	Sebutylazine	7286-69-3	$Y = 3.74 \times 10^6 X - 1.70 \times 10^5$	0.0625–2.500	0.9989
380	Simeton	673-04-1	$Y = 1.93 \times 10^6 X - 2.26 \times 10^5$	0.1250–5.000	0.9984
381	Tebufenpyrad	119168-77-3	$Y = 1.69 \times 10^6 X - 8.49 \times 10^4$	0.0625–2.500	0.9965
382	Tebutam	35256-85-0	$Y = 1.22 \times 10^6 X - 8.63 \times 10^4$	0.1250–5.000	0.9990
383	Tefluthrin	79538-32-2	$Y = 5.47 \times 10^6 X - 2.89 \times 10^5$	0.0625–2.500	0.9980
384	Thionazin	297-97-2	$Y = 6.15 \times 10^5 X - 3.47 \times 10^4$	0.0625–2.500	0.9969
385	trans-Nonachlor	39765-80-5	$Y = 1.13 \times 10^6 X - 3.81 \times 10^4$	0.0625–2.500	0.9990

Continued

TABLE 5.12 Linear Equations, Linear Ranges, and Correlation Coefficients of 567 Pesticides and Chemical Pollutants Determined by GC-MS—cont'd

No.	Compound	CAS	Linear Equation	Linear Range	Correlation Coeffient
386	Triamiphos	1031-47-6	$Y = 6.059X - 5.913 \times 10^{-1}$	0.125–2.5	0.9940
387	Trichloronat	327-98-0	$Y = 1.49 \times 10^6 X - 6.67 \times 10^4$	0.0625–2.500	0.9989
388	Trietazine	1912-26-1	$Y = 2.17 \times 10^6 X - 9.66 \times 10^4$	0.0625–2.500	0.9987
389	Tri-iso-butyl phosphate	126-71-6			
390	Tri-n-butyl hosphate	126-73-8	$Y = 1.40 \times 10^6 X - 2.15 \times 10^5$	0.1250–5.000	0.9972
391	Triphenyl phosphate	115-86-6	$Y = 2.59 \times 10^6 X - 6.86 \times 10^4$	0.0625–2.500	0.9990
392	Acenaphthene	83-32-9	$Y = 5.7X + 7.57 \times 10^{-2}$	0.0625–2.500	0.9980
393	Acetochlor	34256-82-1	$Y = 1.29X - 4.33 \times 10^{-2}$	0.1250–5.000	1.0000
394	Acibenzolar-S-methyl	135158-54-2			
395	Allethrin	584-79-2	$Y = 1.88X - 3.8 \times 10^{-1}$	0.2500–10.00	0.9980
396	Benfuresate	68505-69-1	$Y = 4.25X - 1.95 \times 10^{-2}$	0.1250–5.000	1.0000
397	Benoxacor	98730-04-2	$Y = 2.74X - 2.12 \times 10^{-1}$	0.1250–5.000	0.9990
398	Bromacil	314-40-9	$Y = 1.91X - 3.65 \times 10^{-1}$	0.5000–20.00	0.9990
399	Bromobutide	74712-19-9			
400	Butafenacil	134605-64-4	$Y = 4.09X - 3.83 \times 10^{-1}$	0.0625–2.500	0.9860

Continued

401	Butamifos	8013-75-0	$Y=1.316X-1.864\times10^{-1}$	0.0625–2.500	0.9710
402	Cafenstrole	125306-83-4	$Y=2.875X-3.215$	0.5–10	0.9810
403	Captan	133-06-2	$Y=1.61\times10^{-1}X-2.67\times10^{-1}$	1.0000–40.00	0.9940
404	Chlorethoxyfos	54593-83-8	$Y=1.67X-4.32\times10^{-2}$	0.1250–5.000	1.0000
405	Chlorthal-dimethyl	1861-32-1	$Y=2.61X+6.8\times10^{-2}$	0.1250–5.000	0.9980
406	Chrysene		$Y=7.15X-3.13\times10^{-1}$	0.0625–2.500	0.9990
407	Clethodim	99129-21-2	$Y=8.75\times10^{-1}X-8.65\times10^{-2}$	0.2500–10.00	0.9970
408	Clomeprop	84496-56-0	$Y=4.83\times10^{-1}X-3.405\times10^{-1}$	0.0625–2.500	0.9970
409	Cyflufenamid	180409-60-3			
410	Demeton-S-methyl	301-12-2	$Y=5.71\times10^{-1}X-5.10\times10^{-2}$	0.2500–10.00	0.9990
411	Diallifos	10311-84-9			
412	Dibutyl succinate	141-03-7	$Y=6.82X-2.14\times10^{-1}$	0.1250–5.000	1.0000
413	Dichloran	99-30-9	$Y=8.33\times10^{-1}X-6.01\times10^{-2}$	0.1250–5.000	0.9990
414	Dimethylvinphos	2274-67-1	$Y=3.18X-3.31\times10^{-1}$	0.1250–5.000	0.9980
415	Dithiopyr	97886-45-8	$Y=2.28X-3.3\times10^{-2}$	0.0625–2.500	0.9990
416	Epoxiconazole-1	106325-08-0	$Y=6.79\times10^{-2}X+3.91\times10^{-2}$	0.5000–20.00	0.9790
417	Epoxiconazole-2	106325-08-0	$Y=2.75X-5.32\times10^{-1}$	0.5000–20.00	0.9990
418	Esprocarb	85785-20-2	$Y=2.005X-1.673\times10^{-1}$	0.125–5	0.9970
419	Etoxazole	153233-91-1			
420	Famphur	52-85-7	$Y=1.78X+9.96\times10^{-2}$	0.2500–10.00	0.9910

TABLE 5.12 Linear Equations, Linear Ranges, and Correlation Coefficients of 567 Pesticides and Chemical Pollutants Determined by GC-MS—cont'd

No.	Compound	CAS	Linear Equation	Linear Range	Correlation Coeffient
421	Fenamidone	161326-34-7	$Y = 2.255X - 7.446 \times 10^{-2}$	0.0625–2.500	0.9980
422	Fenfuram	24691-80-3	$Y = 6.91X - 6.72 \times 10^{-1}$	0.1250–5.000	0.9990
423	Fenoxanil	115852-48-7	$Y = 9.69 \times 10^{-1}X + 3.46 \times 10^{-2}$	0.1250–2.500	0.9960
424	Fenpropidin	67306-00-7	$Y = 1.21 \times 10^{1}X - 1.26$	0.1250–5.000	0.9970
425	Fenpyroximate	134098-61-6	$Y = 1.97 \times 10^{-1}X - 3.44 \times 10^{-2}$	0.5000–20.00	0.9960
426	Fipronil	120068-37-3	$Y = 8.2 \times 10^{-1}X - 2.91 \times 10^{-1}$	0.5000–20.00	0.9960
427	Flufenacet	142459-58-3	$Y = 1.95X - 6.19 \times 10^{-1}$	0.5000–20.00	0.9990
428	Fluridone	59756-60-4	$Y = 5.33X - 1.01$	0.125–5.000	0.9920
429	Flurtamone	96525-23-4	$Y = 4.89 \times 10^{-1}X - 2.29 \times 10^{-1}$	0.2500–5.000	0.9330
430	Furalaxyl	57646-30-7	$Y = 2.64X - 1.12 \times 10^{-1}$	0.1250–2.500	1.0000
431	Imazalil	35554-44-0	$Y = 9.69 \times 10^{-1}X - 1.86 \times 10^{-1}$	0.2500–10.00	0.9980
432	Imazamethabenz-methyl	81405-85-8	$Y = 5.47 \times 10^{-1}X + 2.14 \times 10^{-4}$	0.1875–7.500	0.9960
433	Imibenconazole-des-benzyl	199338-48-2			
434	Imiprothrin-1	72693-72-5	$Y = 5.69 \times 10^{-1}X - 8.96 \times 10^{-2}$	0.1250–5.000	0.9850
435	Imiprothrin-2	72693-72-5	$Y = 2.72X - 5.56 \times 10^{-1}$	0.1250–5.000	0.9900

Basic Research on Mass Spectroscopy Characteristic Parameters

No.	Name	CAS	Equation	Range	R^2
436	Iprodione	36734-19-7	$Y=1.5X-4.14\times10^{-1}$	0.2500–10.00	0.9980
437	Isoprocarb-1	2631-40-7	$Y=5.44\times10^{-1}X+1.86\times10^{-1}$	0.1250–5.000	0.9670
438	Isoprocarb-2	2631-40-7	$Y=6.46\times10^{-1}X-7.58\times10^{-1}$	0.1250–5.000	0.9670
439	Isoprothiolane	34123-59-6	$Y=7.2\times10^{-1}X-2.98\times10^{-2}$	0.1250–5.000	0.9990
440	Isoxathion	58769-20-3			
441	Lactofen	77501-63-4	$Y=3.3\times10^{-1}X-1.88\times10^{-1}$	0.5000–20.00	0.9930
442	Malaoxon	1364-78-2	$Y=1.65X-3.28$	1.0000–40.00	0.9880
443	Mefenoxam	70630-17-0	$Y=1.24X-5.51\times10^{-2}$	0.1250–5.000	1.0000
444	Mefenpyr-diethyl	135590-91-9	$Y=8.69\times10^{-1}X-8.09\times10^{-1}$	0.1875–7.500	0.9970
445	Mepanipyrim	110235-47-7	$Y=7.2X-5.17\times10^{-1}$	0.0625–2.500	0.9940
446	Methamidophos	10265-92-6	$Y=1.71\times10^{-1}X-6.58\times10^{-1}$	0.2500–10.00	0.9930
447	Methiocarb Sulfone	2178-25-1	$Y=1.69\times10^{-2}X+8.96\times10^{-4}$	0.5000–5.000	0.9990
448	Methothrin-1	34388-29-9	$Y=8.75\times10^{-1}X-8.41\times10^{-2}$	0.1250–5.000	0.9990
449	Methothrin-2	34388-29-9	$Y=2.61X-2.31\times10^{-1}$	0.1250–5.000	0.9990
450	Methyl Trithion	34388-29-9			
451	Metominostrobin	786-19-6	$Y=4.108X-7.501\times10^{-1}$	0.25–10	0.9980
452	Naled	133408-50-1	$Y=4.46\times10^{-1}X-6.33\times10^{-1}$	1.0000–40.00	0.9960
453	Naproanilide	300-76-5	$Y=1.07X-6.223\times10^{-2}$	0.0625–0.500	0.9810
454	Ofurace	52570-16-8	$Y=9.18\times10^{-1}X-9.29\times10^{-2}$	0.1875–7.500	0.9990
455	Pencycuron	58810-48-3	$Y=3.35X-2.21\times10^{-1}$	0.1250–5.000	0.9980

Continued

TABLE 5.12 Linear Equations, Linear Ranges, and Correlation Coefficients of 567 Pesticides and Chemical Pollutants Determined by GC-MS—cont'd

No.	Compound	CAS	Linear Equation	Linear Range	Correlation Coeffient
456	Phenanthrene	66063-05-6	$Y=8.74X-1.65\times10^{-2}$	0.0625–2.500	0.9990
457	Phosphamidon-1	65-01-8	$Y=1.83\times10^{-1}X-2\times10^{-1}$	0.5000–20.00	0.9770
458	Phosphamidon-2	13171-21-6	$Y=2.44X-4.03\times10^{-1}$	0.5000–20.00	0.9980
459	Picolinafen	13171-21-6	$Y=3.35X-2.06\times10^{-1}$	0.0625–2.500	0.9980
460	Picoxystrobin	137641-05-5	$Y=1.11X-3.65\times10^{-2}$	0.1250–5.000	0.9990
461	Piperophos	117428-22-5	$Y=3.04X-5.31\times10^{-1}$	0.1875–7.500	0.9990
462	Pirimicarb	24151-93-7	$Y=5.18X-1.7\times10^{-1}$	0.1250–5.000	1.0000
463	Prohydrojasmon	23103-98-2			
464	Propoxur-1	158474-72-7	$Y=1.45X+5.56\times10^{-1}$	0.1250–5.000	0.9570
465	Propoxur-2	114-26-1	$Y=6.32X-1.34$	0.1250–5.000	0.9890
466	Propyzamide	114-26-1	$Y=2.94X-2.44\times10^{-1}$	0.1250–5.000	0.9990
467	Pyraclofos	23950-58-5	$Y=7.773\times10^{-1}X-1.808\times10^{-1}$	0.5–20	0.9620
468	Pyraclostrobin	77458-01-6	$Y=4.97\times10^{-1}X-1.15$	1.5000–60.00	0.9910
469	Pyraflufen Ethyl	175013-18-0	$Y=1.18X-4.03\times10^{-2}$	0.1250–5.000	0.9990
470	Pyributicarb	129630-17-7	$Y=6.7X-6.22\times10^{-1}$	0.1250–5.000	0.9990

No.	Name	CAS	Equation	Range	R
471	Pyrifalid	88678-67-5	$Y=1.73X-9.39\times10^{-1}$	0.0625–2.500	0.9980
472	Pyrimidifen	135186-78-6	$Y=9.252X-2.558\times10^{-1}$	0.125–5	0.9720
473	Pyriminobac-Methyl	105779-78-0			
474	Pyriproxyfen	136191-64-5	$Y=2.75X-1.92\times10^{-1}$	0.0625–2.500	0.9960
475	Pyroquilon	95737-68-1	$Y=3.8X-9.6\times10^{-2}$	0.0625–2.500	1.0000
476	Quinoclamine	57369-32-1	$Y=6.31\times10^{-1}X-1.57\times10^{-1}$	0.2500–10.00	0.9980
477	Quinoxyphen	2797-51-5	$Y=3.6X-1.47\times10^{-1}$	0.0625–2.500	0.9990
478	Silafluofen	124495-18-7	$Y=6.53\times10^{-1}X-3.35\times10^{-2}$	0.0625–2.500	0.9960
479	Simeconazole	105024-66-6	$Y=3.05X-2.44\times10^{-1}$	0.1250–5.000	1.0000
480	Spirodiclofen	149508-90-7	$Y=1.95\times10^{-1}X-1.04\times10^{-1}$	0.5000–20.00	0.9950
481	TCMTB	148477-71-8			
482	Tebupirimfos	96182-53-5	$Y=8.44\times10^{-1}X-1.53\times10^{-2}$	0.1250–5.000	0.9980
483	Tebuthiuron	34014-18-1	$Y=2.9X-7.67\times10^{-1}$	0.2500–10.00	0.9960
484	Terbacil	5902-51-2			
485	Terbucarb-1	1918-11-2	$Y=6.76X-3.79\times10^{-2}$	0.1250–5.000	1.0000
486	Terbucarb-2	1918-11-2	$Y=6.76X-3.79\times10^{-2}$	0.1250–5.000	1.0000
487	Thenylchlor	96491-05-3	$Y=4.53X-2.81\times10^{-1}$	0.1250–5.000	1.0000
488	Thiamethoxam	153719-23-4	$Y=3.98\times10^{-1}X-9.13\times10^{-2}$	0.2500–5.000	0.9960
489	Thiazopyr	117718-60-2	$Y=7.9\times10^{-1}X-3.26\times10^{-2}$	0.1250–5.000	0.9990
490	Thifluzamide	130000-40-7	$Y=6.241\times10^{-1}X-3.246\times10^{-1}$	0.5–10	0.9950

Continued

TABLE 5.12 Linear Equations, Linear Ranges, and Correlation Coefficients of 567 Pesticides and Chemical Pollutants Determined by GC-MS—cont'd

No.	Compound	CAS	Linear Equation	Linear Range	Correlation Coeffient
491	Tralkoxydim	87820-88-0	$Y = 1.92X + 2.21 \times 10^{-1}$	0.5000–20.00	0.9930
492	Tridiphane	58138-08-2			
493	Trifloxystrobin	141517-21-7	$Y = 2.69X - 3.93 \times 10^{-1}$	0.2500–10.00	0.9990
494	XMC	2655-14-3	$Y = 4.751 \times 10^{-1}X + 2.739 \times 10^{-1}$	0.125–500	0.9510
495	Zoxamide	156052-68-5	$Y = 9.78 \times 10^{-1}X + 2.6 \times 10^{-2}$	0.1250–5.00	0.9990
496	2,4,5-T	93-76-5	$Y = 7.560 \times 10^{-2}X - 9.541 \times 10^{-2}$	1.25–50	0.9960
497	2,4-D	94-75-7	$Y = 7.052 \times 10^{-2}X - 6.731 \times 10^{-2}$	1.25–50	0.9930
498	3,4,5-Trimethacarb	2686-99-9		0.0625–2.500	
499	3-Phenylphenol	580-51-8	$Y = 1.865X - 4.421 \times 10^{-1}$	0.375–15	0.9990
500	4-Chlorophenoxy acetic acid	122-88-3	$Y = 2.088 \times 10^{-1}X - 5.259 \times 10^{-2}$	0.5–20	0.9940
501	Acephate	30560-19-1	$Y = 2.243 \times 10^{-1}X - 4.451 \times 10^{-1}$	1.25–50	0.9870
502	Acetamiprid	160430-64-8	$Y = 3.28 \times 10^{-1}X - 1.54 \times 10^{-1}$	0.7500–30.00	0.9930
503	Aclonifen	74070-46-5	$Y = 1.465 \times 10^{-2}X - 2.724 \times 10^{-2}$	1.25–50	0.9940
504	Azaconazole	60207-31-0			
505	Azoxystrobin	131860-33-8	$Y = 7.875 \times 10^{-1}X - 2.689 \times 10^{-1}$	0.625–25.00	0.9990

506	Bifenazate	149877-41-8	$Y = 1.849 \times 10^{-1}X - 1.115 \times 10^{-2}$	0.5000–20.00	0.9990
507	Bioresmethrin	28434-01-7	$Y = 2.138X - 1.297 \times 10^{-1}$	0.125–5.000	0.9990
508	Boscalid	188425-85-6			
509	Butralin	33629-47-9	$Y = 2.77X - 4.35 \times 10^{-1}$	0.2500–10.00	0.9940
510	Cadusafos	95465-99-9	$Y = 2.26X - 2.17 \times 10^{-1}$	0.2500–10.00	0.9990
511	Carbaryl	63-25-2	$Y = 6.292 \times 10^{-1}X + 7.786 \times 10^{-2}$	0.1875–7.500	0.9990
512	Carfentrazone-Ethyl	128621-72-7	$Y = 9.73 \times 10^{-1}X - 8.03 \times 10^{-2}$	0.1250–5.000	0.9990
513	Chlorfenapyr	122453-73-0	$Y = 2.29 \times 10^{-1}X - 1.84 \times 10^{-2}$	0.5000–20.00	0.9990
514	Cinmethylin	87818-31-3		0.0625–2.500	
515	Cyhalofop-butyl	122008-85-9			
516	Cythioate	115-93-5	$Y = 6.702 \times 10^{-3}X + 1.151 \times 10^{-3}$	1.25–10	0.9520
517	Demetom-s	126-75-0	$Y = 1.541X - 5.515 \times 10^{-1}$	0.25–10	0.9700
518	Desmedipham	13684-56-5	$Y = 2.012 \times 10^{-1}X - 1.107 \times 10^{-1}$	1.25–50	0.9950
519	Dicrotophos	141-66-2	$Y = 4.348 \times 10^{-1}X - 1.023 \times 10^{-1}$	0.5–20	0.9990
520	Diethyltoluamide	134-62-3	$Y = 9.631X - 4.478 \times 10^{-1}$	0.05–2	0.9970
521	Difenoxuron	14214-32-5	$Y = 3.536 \times 10^{-1}X - 8.184 \times 10^{-2}$	0.5–20	0.9980
522	Dimethametryn	14214-32-5	$Y = 9.04X - 3.08 \times 10^{-1}$	0.0625–2.500	1.0000
523	Dimethomorph	211867-47-9	$Y = 1.445X - 1.391 \times 10^{-1}$	0.125–5.000	0.9980
524	Dimethyl phthalate	131-11-3	$Y = 4.702 \times 10^{-1}X + 3.638 \times 10^{-1}$	0.25–5	0.9330
525	Diofenolan-1	63837-33-2	$Y = 2.13X - 2.6 \times 10^{-1}$	0.1250–5.000	0.9970

TABLE 5.12 Linear Equations, Linear Ranges, and Correlation Coefficients of 567 Pesticides and Chemical Pollutants Determined by GC-MS—cont'd

No.	Compound	CAS	Linear Equation	Linear Range	Correlation Coeffient
526	Diofenolan-2	63837-33-2	$Y=1.1X-1.04\times10^{-1}$	0.1250–5.000	0.9990
527	Dioxacarb	6988-21-2			
528	DMSA	1596-84-5	$Y=2.595\times10^{-2}X+1.340\times10^{-2}$	0.5–20	0.9910
529	Dodemorph	1593-77-7	$Y=1.102X-2.497\times10^{-2}$	0.1875–7.500	0.9990
530	Endothal	145-73-3	$Y=4.442\times10^{-2}X-4.006\times10^{-2}$	1.25–50	0.9940
531	Endrin aldehyde	7421-93-4	$Y=9.255\times10^{-2}X+8.208\times10^{-2}$	1.25–25	0.9910
532	Endrin ketone	53494-70-5	$Y=4.12\times10^{-1}X+9.06\times10^{-3}$	1–40.00	0.9980
533	Ethiofencarb	29973-13-5	$Y=3.326\times10^{-2}X+3.35\times10^{-2}$	0.625–25	0.9870
534	Fenchlorphos	3983-45-7	$Y=1.227X-9.79\times10^{-3}$	0.25–10	0.9990
535	Fenhexamid	126833-17-8	$Y=6.104\times10^{-1}X-1.226$	1.25–50	0.9940
536	Fluazinam	79622-59-6	$Y=4.636\times10^{-2}X+1.589\times10^{-2}$	0.5–20	0.9950
537	Furmecyclox	60568-05-0	$Y=2.796X-7.336\times10^{-1}$	0.1875–7.500	0.9890
538	Gamma-cyhaloterin-1	91465-08-6		0.05–2	
539	Gamma-cyhalothrin-2	91465-08-6	$Y=1.551X-2.252\times10^{-1}$	0.05–2	0.9780
540	Halfenprox	111872-58-3		0.1250–2.5	

No.	Name	CAS	Equation	Range	r
541	Halosulfuran-Methyl	100784-20-1			
542	Iprovalicarb-1	140923-17-7	$Y=3.862\times10^{-1}X-3.692\times10^{-2}$	0.25–10	0.9990
543	Iprovalicarb-2	140923-17-7	$Y=9.64\times10^{-1}X-1.342\times10^{-1}$	0.25–10	0.9990
544	Isoxadifen-Ethyl	141112-29-0	$Y=7.901\times10^{-1}X-1.039\times10^{-1}$	0.25–5	0.9970
545	Merphos	150-50-5		0.0625–2.500	
546	Metoconazole	125116-23-6	$Y=5.383\times10^{-1}X-4.572\times10^{-2}$	0.25–10	0.9980
547	Monocrotophos	6923-22-4	$Y=3.326\times10^{-2}X+3.35\times10^{-2}$	0.2500–2.500	0.9941
548	Norflurazon-desmethyl	23567-24-1	$Y=4.017\times10^{-1}X-4.393\times10^{-1}$	1.25–10	0.9880
549	Phenkapton	2275-14-1	$Y=1.409X-3.459\times10^{-1}$	0.375–3	0.9880
550	Phthalide	27355-22-2			
551	Phthalimide	5333-22-2	$Y=1.27X-1.34\times10^{-1}$	0.1250–5.000	0.9980
552	Prothoate	2275-18-5	$Y=2.235X-9.054\times10^{-1}$	0.5–4	0.9930
553	Pyrifenox-1	88283-41-4	$Y=4.07\times10^{-1}X-8.48\times10^{-2}$	0.5000–20.00	0.9990
554	Pyrifenox-2	88283-41-4	$Y=9.74\times10^{-1}X-1.01\times10^{-1}$	0.5000–20.00	0.9980
555	S 421(octachlorodipropyl ether)-1	127-90-2	$Y=1.611\times10^{-1}X+3.35\times10^{-1}$	1.25–50	0.9980
556	S 421(octachlorodipropyl ether)-2	127-90-2	$Y=1.432\times10^{-1}X-1.390\times10^{-1}$	1.25–50	0.9990
557	Sobutylazine	7286-69-3	$Y=2.04X+3.198\times10^{-1}$	0.125–5.000	0.9780
558	Spiromesifen	283594-90-1	$Y=2.709\times10^{-1}X+8.256\times10^{-2}$	0.625–5	0.9490

Continued

TABLE 5.12 Linear Equations, Linear Ranges, and Correlation Coefficients of 567 Pesticides and Chemical Pollutants Determined by GC-MS—cont'd

No.	Compound	CAS	Linear Equation	Linear Range	Correlation Coeffient
559	Spiroxamine-1	118134-30-8	$Y = 5.27X - 5.65 \times 10^{-1}$	0.1250–5.000	0.9950
560	Spiroxamine-2	118134-30-8	$Y = 1.25 \times 10^{1}X - 1.23$	0.1250–5.000	0.9980
561	Thiabendazole	148-79-8	$Y = 3.168 \times 10^{-2}X - 2.590 \times 10^{-2}$	1.25–10	0.9990
562	Tolfenpyrad	129558-76-5		0.0625–2.500	
563	Tralomethrin-1	66841-25-6			
564	Tralomethrin-2	66841-25-6			
565	Tribenuron-methyl	106040-48-6	$Y = 2.784 \times 10^{-1}X + 3.122 \times 10^{-2}$	0.0625–2.5	0.9760
566	Tricyclazole	41814-78-2	$Y = 4.369 \times 10^{-1}X + 3.83 \times 10^{-3}$	0.375–15	0.9930
567	Triflumizole	68694-11-1			

5.2.2   Linear Equations, Linear Ranges, and Correlation Coefficients of 466 Pesticides and Chemical Pollutants Determined by GC-MS-MS

See Table 5.13.

TABLE 5.13 Linear Equations, Linear Ranges, and Correlation Coefficients of 466 Pesticides and Chemical Pollutants Determined by GC-MS-MS

No.	Compound	CAS	Linear Equation	Linear Range (pg)	Correlation Coefficient
Group A					
1	Allidochlor	93-71-0	$y = 3.9877 \times 10^{-004}x - 0.0193$	90–3600	0.9996
2	Dichlormid	37764-25-3	$y = 6.6912 \times 10^{-004}x - 0.0155$	80–3200	0.9980
3	Etridiazole	2593-15-9	$y = 0.0049x - 0.0393$	20–800	0.9993
4	Chlormephos	24934-91-6	$y = 0.0031x - 0.0011$	10–400	0.9984
5	Propham	122-42-9	$y = 0.0052x - 0.0233$	10–400	0.9993
6	Thiometon	640-15-3	$y = 0.0027x - 0.0034$	30–120	0.9977
7	Cycloate	1134-23-2	$y = 0.0121x - 3.9279 \times 10^{-004}$	2.5–100	0.9994
8	Diphenylamin	122-39-4	$y = 0.0330x + 0.0030$	2–40	0.9997
9	Ethalfluralin	55283-68-6	$y = 8.1166 \times 10^{-004}x - 0.0369$	80–3200	0.9992

Continued

TABLE 5.13 Linear Equations, Linear Ranges, and Correlation Coefficients of 466 Pesticides and Chemical Pollutants Determined by GC-MS-MS—cont'd

No.	Compound	CAS	Linear Equation	Linear Range (pg)	Correlation Coefficient
10	Quintozene	82-68-8	$y=0.0011x-0.1749$	170–6800	0.9969
11	Atrazine-desethyl	6190-65-4	$y=0.0023x-0.1521$	80–3200	0.9959
12	Clomazone	81777-89-1	$y=0.0093x-0.0085$	5–200	0.9997
13	Diazinon	333-41-5	$y=0.0051x-0.0121$	10–400	0.9997
14	Fonofos	944-22-9	$y=0.0127x-0.0124$	5–200	0.9999
15	Etrimfos	38260-54-7	$y=0.0027x-0.0302$	25–1000	0.9992
16	Simazine	122-34-9	$y=0.0041x-0.3533$	13–5200	0.9988
17	Propetamphos	31218-83-4	$y=0.0031x-0.0293$	15–600	0.9981
18	Secbumeton	26259-45-0	$y=0.0033x-0.0115$	10–400	0.9996
19	Dichlofenthion	97-17-6	$y=0.0083x-0.0068$	5–200	0.9998
20	Mexacarbate	315-18-4	$y=6.4014\times10^{-004}x-0.0077$	125–2500	0.9996
21	Pronamide	23950-58-5	$y=0.0263x-0.0259$	2.5–100	0.9994
22	Dinitramine	29091-05-2	$y=2.6324\times10^{-004}x-0.1171$	160–16000	0.9978
23	Dimethoate	60-51-5	$y=0.0011x-0.0329$	50–500	0.9909
24	Ronnel	299-84-3	$y=0.0041x-0.0417$	15–600	0.9986

25	Prometryne	7287-19-6	$y=0.0039x-0.0135$	10–400	0.9995
26	Vinclozolin	50471-44-8	$y=0.0018x-0.0123$	30–1200	0.9998
27	beta-HCH	319-85-7	$y=0.0052x-0.0067$	10–400	1.0000
28	Metalaxyl	57837-19-1	$y=0.0019x-0.0109$	20–800	0.9994
29	Chlorpyrifos(ethyl)	2921-88-2	$y=0.0016x-0.0387$	20–2000	0.9995
30	Methyl-parathion	298-00-0	$y=0.0019x-0.0389$	50–500	0.9958
31	delta-HCH	319-86-8	$y=0.0026x-0.0067$	20–800	1.0000
32	Anthraquinone	84-65-1	$y=0.0142x-0.0530$	5–200	0.9983
33	Fenthion	55-38-9	$y=0.0067x-0.3743$	10–4000	0.9992
34	Malathion	121-75-5	$y=0.0041x-0.0877$	25–500	0.9907
35	Fenitrothion	122-14-5	$y=0.0024x-0.2982$	32–3200	0.9923
36	Paraoxon-ethyl	311-45-5	$y=2.6737\times10^{-004}x-0.3923$	750–30000	0.9820
37	Triadimefon	43121-43-3	$y=0.0019x-0.0209$	25–1000	0.9995
38	Pendimethalin	40318-45-4	$y=0.0040x-0.1109$	20–800	0.9920
39	Linuron	330-55-2	$y=1.1836\times10^{-004}x-0.0320$	500–5000	0.9949
40	Chlorbenside	103-17-3	$y=0.0018x-0.0152$	20–800	0.9996
41	Bromophos-ethyl	4824-78-6	$y=0.0026x-0.0347$	25–1000	0.9990
42	Quinalphos	13593-03-8	$y=0.0031x-0.0551$	20–800	0.9963
43	trans-Chlodane	5103-74-2	$y=0.0011x+0.0022$	50–2000	0.9997

Continued

TABLE 5.13 Linear Equations, Linear Ranges, and Correlation Coefficients of 466 Pesticides and Chemical Pollutants Determined by GC-MS-MS—cont'd

No.	Compound	CAS	Linear Equation	Linear Range (pg)	Correlation Coefficient
44	Metazachlor	67129-08-2	$y=0.0041x-0.0570$	20–800	0.9982
45	Prothiophos	34643-46-4	$y=0.0029x-0.0284$	20–800	0.9992
46	Folpet	133-07-3			
47	Chlorfurenol	2464-37-1	$y=0.0011x-0.2963$	100–5000	0.9861
48	Procymidone	32809-16-8	$y=0.0087x-0.0024$	5–200	0.9997
49	Methidathion	950-37-8	$y=0.0039x-0.1176$	35–700	0.9905
50	Cyanazine	21725-46-2	$y=1.6157\times10^{-004}x-0.0076$	260–5200	0.9993
51	Napropamide	15299-99-7	$y=0.0041x-0.0212$	10–400	0.9983
52	Oxadiazone	19666-30-9	$y=0.0094x-0.0084$	5–200	0.9998
53	Fenamiphos	22224-92-6	$y=6.3550\times10^{-004}x-0.0157$	80–640	0.9933
54	Tetrasul	2227-13-6	$y=0.0036x-0.0017$	10–400	0.9995
55	Bupirimate	57839-19-1	$y=0.0029x-0.0090$	10–400	0.9992
56	Carboxin	5234-68-4	$y=0.0051x-0.0181$	10–200	0.9984
57	Flutolanil	66332-96-5	$y=0.0393x-0.0133$	1.5–60	0.9986
58	p,p'-DDD	72-54-8	$y=0.0395x-0.0582$	3–120	0.9993

Continued

59	Ethion	562-12-2	$y = 1.6236 \times 10^{-004}x - 0.0420$	450–18000	0.9993
60	Etaconazole-1	1000290-09-5	$y = 0.0043x - 0.0162$	10–400	0.9988
61	Etaconazole-2	1000290-09-5	$y = 0.0040x - 0.0067$	10–400	0.9999
62	Sulprofos	35400-43-2	$y = 0.0029x - 0.0267$	15–600	0.9980
63	Myclobutanil	88671-89-0	$y = 0.0095x - 0.0038$	3.5–140	0.9997
64	Fensulfothin	115-90-2	$y = 0.0014x - 0.0252$	80–3200	0.9997
65	Propiconazole	120721-83-9	$y = 0.0033x - 0.0256$	15–600	0.9993
66	Bifenthrin	82657-04-3	$y = 0.0457x + 0.0027$	1–20	0.9993
67	Mirex	2385-85-5	$y = 0.0107x + 0.0074$	5–200	0.9994
68	Benodanil	15310-01-7	$y = 0.0023x - 0.0466$	50–1000	0.9989
69	Nuarimol	63284-71-9	$y = 0.0021x - 0.0195$	25–1000	0.9994
70	Methoxychlor	72-43-5	$y = 0.0042x - 0.0951$	20–800	0.9935
71	Oxadixyl	77732-09-3	$y = 0.0052x - 0.0516$	15–600	0.9980
72	Tetramethrin	7696-12-0	$y = 0.0058x - 0.0528$	15–300	0.9922
73	Tebuconazole	107534-96-3	$y = 0.0043x - 0.0220$	10–400	0.9988
74	Norflurazon	27314-13-2	$y = 0.0039x - 0.0279$	15–600	0.9991
75	Pyridaphenthion	119-12-0	$y = 0.0019x - 0.1754$	20–2000	0.9885
76	Phosmet	732-11-6	$y = 5.3399 \times 10^{-004}x - 0.0843$	150–3000	0.9900
77	Tetradifon	116-29-0	$y = 0.0013x - 0.0263$	50–2000	0.9989

TABLE 5.13 Linear Equations, Linear Ranges, and Correlation Coefficients of 466 Pesticides and Chemical Pollutants Determined by GC-MS-MS—cont'd

No.	Compound	CAS	Linear Equation	Linear Range (pg)	Correlation Coefficient
78	Pyrazophos	13457-18-6	$y=0.0043x-0.0620$	20–400	0.9932
79	Cypermethrin	52315-07-8	$y=0.0011x-0.1974$	250–5000	0.9948
80	Fenvalerate	51630-58-1	$y=1.3423\times10^{-004}x-0.0146$	212.5–4250	0.9974
Group B					
81	EPTC	759-94-4	$y=0.0018x-0.0102$	16–1600	0.9997
82	Butylate	2008-41-5	$y=0.0213x-0.0051$	1–100	0.9998
83	Dichlobenil	1194-65-6	$y=9.1606x+0.0429$	0.008–4	0.9999
84	Pebulate	1114-71-2	$y=0.0034x-0.0175$	20–800	0.9998
85	Nitrapyrin	1929-82-4	$y=0.0017x-0.0592$	50–2000	0.9991
86	Chloroneb	2675-77-6	$y=0.0069x-0.0037$	10–400	0.9999
87	Tecnazene	117-18-0	$y=6.1591\times10^{-004}x-0.0305$	100–2000	0.9973
88	Heptanophos	23560-59-0	$y=0.0322x-0.0335$	4–80	0.9986
89	Ethoprophos	13194-48-4	$y=0.0083x-0.0342$	10–200	0.9972
90	Hexachlorobenzene	118-74-1	$y=0.0047x-0.0022$	20–400	0.9997
91	cis-Diallate	2303-16-4	$y=0.0088x-0.0181$	10–200	0.9997

92	trans-Diallate	2303-16-4	$y=0.0088x-0.0142$	5–200	0.9996
93	Propachlor	1918-16-7	$y=0.0020x-0.0529$	60–1200	0.9983
94	Trifluralin	1582-09-8	$y=0.0046x-0.0985$	12–1200	0.9960
95	Chlorpropham	14816-20-7	$y=0.0017x-0.0690$	80–800	0.9902
96	Sulfotep	3689-24-5	$y=0.0046x-0.0337$	15–600	0.9994
97	Sulfallate	95-06-7	$y=0.0052x-0.0247$	10–400	0.9978
98	alpha-HCH	319-84-6	$y=0.0069x-0.0623$	20–400	0.9986
99	Terbufos	13071-79-9	$y=0.0094x-0.2792$	14–2800	0.9989
100	Profluralin	26399-36-0	$y=0.0023x-0.1230$	50–2000	0.9953
101	Dioxathion	78-34-2	$y=9.7800 \times 10^{-004}x-0.0366$	50–2000	0.9978
102	Propazine	2312-35-8	$y=0.0083x-0.2628$	10–2000	0.9969
103	Dicloran	99-30-9	$y=0.0020x-0.0665$	50–1000	0.9917
104	Terbuthylazine	5915-41-3	$y=0.0043x-0.1292$	20–2000	0.9975
105	Monolinuron	1746-81-2	$y=8.9084 \times 10^{-004}x-0.0922$	90–3600	0.9934
106	Chlorbufam	1967-16-4	$y=0.0011x+0.0029$	50–1000	0.9996
107	Flufenoxuron	101463-69-8	$y=6.5121 \times 10^{-004}x+0.0209$	50–1000	0.9992
108	Cyanophos	2636-26-2	$y=0.0038x-0.0511$	20–400	0.9941
109	Chlorpyrifos-methyl	5598-13-0	$y=0.0020x-0.0674$	50–1000	0.9936
110	Desmetryn	1014-69-3	$y=0.0070x-0.0181$	10–200	0.9979

Continued

TABLE 5.13 Linear Equations, Linear Ranges, and Correlation Coefficients of 466 Pesticides and Chemical Pollutants Determined by GC-MS-MS—cont'd

No.	Compound	CAS	Linear Equation	Linear Range (pg)	Correlation Coefficient
111	Dimethachloro	51218-45-2	$y=0.0092x-0.0590$	4–400	0.9958
112	Alachlor	15972-60-8	$y=0.0021x-0.0376$	10–1000	0.9967
113	Pirimiphos-methyl	29232-93-7	$y=0.0029x-0.0422$	25–1000	0.9980
114	Thiobencarb	28249-77-6	$y=0.0051x-0.0285$	10–400	0.9981
115	Terbutyrn	886-50-0	$y=0.0031x-0.0226$	15–600	0.9988
116	Dicofol	115-32-2	$y=0.0072x+0.0113$	4–400	0.9998
117	Metolachlor	51218-45-2	$y=0.0523x-0.0542$	1–100	0.9974
118	oxy-Chlordane	27304-13-8	$y=2.3915\times10^{-004}x-0.0032$	90–1800	0.9988
119	Pirimiphos-ethyl	23505-41-1	$y=0.0019x-0.0192$	25–1000	0.9975
120	Methoprene	40596-69-8	$y=0.0032x-0.1842$	40–4000	0.9979
121	Bromofos	2104-96-3	$y=8.1034\times10^{-004}x-0.1386$	150–6000	0.9946
122	Dichlofluanid	1085-98-9			
123	Ethofumesate	26225-79-6	$y=0.0100x-0.0173$	5–200	0.9985
124	Isopropalin	33820-53-0	$y=0.0049x-0.0340$	6–300	0.9915
125	Propanil	709-98-8	$y=0.0028x-0.0139$	20–400	0.9970

126	Endosulfan-1	959-98-8	$y=3.5212\times10^{-004}x-0.0103$	75–3000	0.9987
127	Crufomate	299-86-5	$y=0.0029x-0.0481$	20–400	0.9909
128	Isofenphos	25311-71-1	$y=0.0011x-0.0266$	75–600	0.9962
129	Chlorfenvinphos	470-90-6	$y=3.0999\times10^{-004}x-0.0059$	200–4000	0.9993
130	Chlorthiamid	1918-13-4	$y=8.2350\times10^{-004}x-0.0091$	40–4000	0.9998
131	cis-Chlordane	5103-71-9	$y=6.2791\times10^{-004}x+0.0117$	50–1000	0.9989
132	Tolylfluanid	731-27-1	$y=0.0044x+0.0033$	10–400	1.0000
133	p,p'-DDE	72-55-9	$y=0.0014x-0.0017$	20–800	0.9997
134	Butachlor	23184-66-9	$y=8.8088\times10^{-004}x-0.0352$	50–2000	0.9904
135	Chlozolinate	84332-86-5	$y=1.8455\times10^{-004}x-0.0564$	500–5000	0.9903
136	Iodofenphos	18181-70-9	$y=3.3615\times10^{-004}x-0.0054$	150–1200	0.9965
137	Tetrachlorvinphos	22248-79-9			
138	Chlorbromuron	13360-45-7			
139	Profenofos	41198-08-7			
140	Fluorochloridone	61213-25-0	$y=6.7979\times10^{-004}x-0.0044$	50–400	0.9980
141	Buprofenzin	69327-76-0	$y=0.0430x-0.0345$	2.5–100	0.9991
142	o,p'-DDD	53-19-0	$y=0.0244x-0.0212$	2–200	0.9996
143	Endrin	72-20-8	$y=5.6163\times10^{-004}x-0.0106$	100–4000	0.9997
144	Chlorfenson	80-33-1	$y=0.0018x-0.0358$	25–1000	0.9972

Continued

TABLE 5.13 Linear Equations, Linear Ranges, and Correlation Coefficients of 466 Pesticides and Chemical Pollutants Determined by GC-MS-MS—cont'd

No.	Compound	CAS	Linear Equation	Linear Range (pg)	Correlation Coefficient
145	Paclobutrazol	76738-62-0	$y=0.0012x+0.0104$	5–40	0.9902
146	Methoprotyne	841-06-5	$y=0.0042x-0.0479$	15–600	0.9959
147	Chlorpropylate	5836-10-2	$y=0.0034x-0.0093$	10–80	1.0000
148	Nitrofen	1836-75-5	$y=6.1661\times10^{-004}x-0.1512$	300–3000	0.9754
149	Oxyflurofen	42874-03-3	$y=3.3961\times10^{-004}x-0.0268$	125–1000	0.9864
150	Chlorthiophos	60238-56-4	$y=6.8488\times10^{-004}x-0.0277$	100–2000	0.9967
151	Endosulfan-2	115-29-7	$y=2.5386\times10^{-004}x-0.0166$	150–3000	0.9990
152	Flamprop-isopropyl	52756-22-6	$y=0.0141x-0.0172$	5–200	0.9991
153	Flamprop-methyl	52756-25-9	$y=0.0235x-0.0200$	2.5–100	0.9989
154	o,p'-DDT	789-02-6	$y=0.0322x-0.0234$	3–60	0.9992
155	p,p'-DDT	50-29-3	$y=0.0161x-0.0982$	10–100	0.9883
156	Carbofenothion	786-19-6	$y=3.1046\times10^{-004}x-0.0951$	250–10000	0.9926
157	Benalaxyl	71626-11-4	$y=0.0091x-0.0150$	5–200	0.9988
158	Edifenphos	17109-49-8	$y=5.3960\times10^{-004}x+0.0088$	50–800	0.9914
159	Cyanofenphos	13067-93-1	$y=0.0085x-0.0365$	5–200	0.9953

160	Endosulfen sulfate	1031-07-8	$y=1.5997\times10^{-004}x-0.0370$	400–4000	0.9950
161	Bromopropylate	18181-80-1	$y=0.0013x-0.0150$	20–400	0.9946
162	Benzoylprop-ethyl	22212-55-1	$y=0.0041x-0.0077$	10–400	0.9997
163	Fenpropathrin	39515-41-8	$y=6.9552\times10^{-004}x-0.0246$	100–800	0.9987
164	Leptophos	21609-90-5	$y=3.9468\times10^{-004}x-0.0517$	200–2000	0.9919
165	EPN	2104-64-5	$y=0.0018x-0.1379$	100–1000	0.9792
166	Hexazinone	51235-04-2	$y=0.0241x-0.0126$	2–80	0.9984
167	Bifenox	42576-02-3			
168	Phosalone	2310-17-0	$y=5.1168\times10^{-004}x-0.0137$	100–400	0.9960
169	Fenarimol	60168-88-9	$y=0.0018x-0.0316$	20–2000	0.9985
170	Azinphos-ethyl	2642-71-9	$y=8.5438\times10^{-004}x-0.0112$	100–400	0.9958
171	Azinphos-methyl	86-50-0	$y=4.2719\times10^{-004}x-0.0112$	200–800	0.9958
172	Prochloraz	67747-09-5	$y=4.0743\times10^{-004}x-0.0488$	200–2000	0.9918
173	Fluvalinate	102851-06-9	$y=7.1805\times10^{-004}x-0.0392$	125–1250	0.9985
174	Cyfluthrin	68359-37-5	$y=4.0432\times10^{-004}x-0.3146$	600–12000	0.9724
Group C					
175	Propamocarb	24579-73-5	$y=0.0012x-0.0969$	200–2000	0.9992
176	Vernolate	1929-77-7	$y=0.0015x-0.0045$	50–1000	0.9992
177	3,5-Dichloroaniline	626-43-7	$y=0.0035x-0.0048$	15–600	0.9998

Continued

TABLE 5.13 Linear Equations, Linear Ranges, and Correlation Coefficients of 466 Pesticides and Chemical Pollutants Determined by GC-MS-MS—cont'd

No.	Compound	CAS	Linear Equation	Linear Range (pg)	Correlation Coefficient
178	Molinate	2212-67-1	$y=0.0391x-0.0086$	1–100	0.9997
179	Methacrifos	62610-77-9	$y=0.0080x-0.0124$	5–200	0.9996
180	2-Phenylphenol	90-43-7	$y=0.0055x-0.0638$	15–300	0.9944
181	cis-1,2,3,6-Tetrahydrophthalimide	85-40-5	$y=4.7023\times10^{-004}x-0.2906$	800–4000	0.9745
182	Benfluralin	1861-40-1	$y=0.0053x-0.0882$	20–800	0.9979
183	Hexaflumuron	86479-06-3	$y=0.0022x+0.0286$	20–160	0.9952
184	Prometon	1610-18-0	$y=0.0047x-0.0203$	10–400	0.9994
185	Triallate	2303-17-5	$y=0.0032x-0.0162$	15–600	0.9998
186	pyrimethanil	53112-28-0	$y=0.0033x-0.0259$	15–600	0.9991
187	gamma-HCH	58-89-9	$y=0.0055x-0.0609$	15–600	0.9992
188	Disulfoton	298-04-4	$y=0.0151x-0.0332$	5–200	0.9991
189	Atrizine	1912-24-9	$y=0.0018x-0.1043$	50–2000	0.9964
190	Heptachlor	76-44-8	$y=0.0042x-0.0943$	25–1000	0.9982
191	Iprobenfos	26087-47-8	$y=0.0123x-0.0778$	5–200	0.9930
192	Isazofos	42509-80-8	$y=0.0011x-0.0466$	50–2000	0.9986

193	Plifenate	51366-25-7	$y=0.0022x-0.0493$	35–1400	0.9989
194	Transfluthrin	118712-89-3	$y=0.0118x-0.0163$	5–200	0.9998
195	Fenpropimorph	67306-03-0	$y=3.0233\times10^{-004}x-0.0194$	105–10500	0.9994
196	Fluchloralin	33245-39-5	$y=0.0027x-0.0719$	25–500	0.9866
197	Tolclofos-methyl	57018-04-9	$y=0.0024x-0.0457$	25–1000	0.9989
198	Ametryn	834-12-8	$y=0.0036x-0.0227$	10–400	0.9988
199	Methobromuron	40596-69-8	$y=8.3539\times10^{-005}x-0.0337$	1000–8000	0.9900
200	Metribuzin	21087-64-9	$y=0.0045x-0.1371$	20–800	0.9904
201	Dimethipin	55290-64-7	$y=1.5637\times10^{-005}x-0.0044$	1500–7500	0.9929
202	Epsilon-HCH	6108-10-7		无药	
203	Dipropetryn	4147-51-7	$y=0.0042x-0.0248$	10–400	0.9986
204	Diethofencarb	87130-20-9	$y=0.0026x-0.2342$	100–1000	0.9883
205	Dimepiperate	61432-55-1	$y=0.0883x-0.1227$	1.5–60	0.9966
206	Bitertanol	55179-31-2	$y=0.0121x-0.0567$	10–200	0.9939
207	Bioallethrin-1	584-79-2	$y=0.0168x-0.1765$	10–200	0.9937
208	o,p'-DDE	3424-82-6	$y=0.0028x-9.6395\times10^{-004}$	15–600	0.9999
209	Fenson	80-38-6	$y=0.0044x-0.0670$	20–400	0.9964
210	Diphenamid	957-51-7	$y=0.0235x-0.0316$	3–120	0.9988
211	Chlorthion	500-28-7	$y=2.2530\times10^{-004}x-1.0752$	3000–60000	0.9726

Continued

TABLE 5.13 Linear Equations, Linear Ranges, and Correlation Coefficients of 466 Pesticides and Chemical Pollutants Determined by GC-MS-MS—cont'd

No.	Compound	CAS	Linear Equation	Linear Range (pg)	Correlation Coefficient
212	Penconazole	66246-88-6	$y=0.0096x-0.0427$	10–200	0.9990
213	Mecarbam	2595-54-2	$y=6.0413\times10^{-005}x-0.0997$	3000–12000	0.9906
214	Triadimenol	55219-65-3	$y=0.0017x-0.0234$	20–400	0.9975
215	Tetraconazole	112281-77-3	$y=0.0021x-0.0456$	30–1200	0.9988
216	Flumetrialin	62924-70-3	$y=0.0071x-0.1511$	15–600	0.9902
217	Pretilachlor	51218-49-6	$y=0.0094x-0.1283$	10–400	0.9931
218	Kresoxim-methyl	143390-89-0	$y=0.0047x-0.0578$	20–400	0.9968
219	Fluazifop-butyl	79241-46-6	$y=0.0038x-0.2135$	25–2500	0.9957
220	Chlorfluazuron	71422-67-8	$y=7.7326\times10^{-004}x+0.0259$	30–1500	0.9986
221	Chlorobenzilate	510-15-6	$y=0.0042x-0.1420$	50–200	0.9832
222	Flusilazole	85509-19-9	$y=0.0047x-0.0282$	10–200	0.9947
223	Fluorodifen	15457-05-3	$y=2.1729\times10^{-004}x-0.0723$	400–3200	0.9707
224	Diniconazole	83657-24-3	$y=4.6426\times10^{-004}x-0.0277$	160–800	0.9980
225	Piperonyl butoxide	51-03-6	$y=0.0116x-0.0715$	5–200	0.9926
226	Mepronil	55814-41-0	$y=0.0467x-0.0807$	2–40	0.9924

227	Diflufenican	83164-33-4	$y=0.0035x-0.3057$	50–2000	0.9899
228	Fenazaquin	120928-09-8	$y=0.0325x-0.0660$	2.5–100	0.9983
229	Phenothrin	26002-80-2	$y=0.0249x-0.1418$	5–200	0.9944
230	Fludioxonil	131341-86-1	$y=0.0026x-0.1846$	50–2000	0.9945
231	Fenoxycarb	79127-80-3	$y=3.7510\times10^{-004}x-0.4767$	2000–10000	0.9910
232	Amitraz	33089-61-1	$y=0.0047x-0.0903$	12–1200	0.9973
233	Anilofos	64249-01-0	$y=1.7379\times10^{-004}x-0.0605$	600–2400	0.9858
234	Acrinathrin	101007-06-1	$y=7.1117\times10^{-005}x-0.0254$	800–4000	0.9901
235	Permethrin	52645-53-1	$y=0.0048x-0.0555$	15–600	0.9978
236	Mefenacet	73250-68-7	$y=8.2184\times10^{-004}x-0.0392$	75–600	0.9793
237	Lambda-cyhalothrin	91465-08-6	$y=0.0011x-0.1057$	100–800	0.9661
238	Pyridaben	96489-71-3	$y=0.0190x-0.1291$	5–200	0.9906
239	Flucythrinate	70124-77-5	$y=0.0041x-0.1492$	12–1200	0.9907
240	Fluoroglycofen-ethyl	77502-90-7			
241	Bioallethrin-2	584-79-2	$y=0.0158x-0.1429$	6–120	0.9962
242	Etofenprox	80844-07-1	$y=0.0373x-0.0538$	3–60	0.9986
243	alpha-Cypermethrin	67375-30-8	$y=0.0017x-0.7695$	250–2000	0.9913
244	Cycloxydim	101205-02-1	$y=0.0017x-0.0580$	80–400	0.9971
245	Esfenvalerate	66230-04-4	$y=1.8262\times10^{-004}x-0.2802$	2000–20000	0.9933

Continued

TABLE 5.13 Linear Equations, Linear Ranges, and Correlation Coefficients of 466 Pesticides and Chemical Pollutants Determined by GC-MS-MS—cont'd

No.	Compound	CAS	Linear Equation	Linear Range (pg)	Correlation Coefficient
246	Difenconazole	1977-6-54	$y=0.0077x-0.0486$	10–200	0.9924
247	Flumioxazin	103361-09-7	$y=5.0954\times10^{-004}x-1.0884$	1000–40000	0.9825
248	Flumiclorac-pentyl	87546-18-7	$y=2.5514\times10^{-004}x-0.0717$	500–2500	0.9957
249	Tri-iso-butyl phosphate	126-71-6		无药	
250	Dimefox	115-26-4	$y=8.9997\times10^{-005}x-0.6282$	7250–145000	0.9929
251	Disulfoton-sulfoxide	2497-07-6	$y=0.0021x-0.2586$	90–3600	0.9952
252	Pentachlorobenzen	608-93-5	$y=0.0059x-0.0329$	20–400	0.9989
253	Crimidine	535-89-7	$y=0.0053x-0.0290$	10–400	0.9985
254	BDMC-1	672-99-1	$y=0.0023x-0.3168$	95–3800	0.9946
255	Chlorfenprop-methyl	14437-17-3	$y=0.0077x-0.0899$	15–300	0.9961
256	Thionazin	297-97-2	$y=0.0088x-0.1289$	15–600	0.9959
257	2,3,5,6-Tetrachloroaniline	3481-20-7	$y=0.0045x-0.0211$	10–400	0.9993
258	Tri-n-butyl-phosphate	126-73-8	$y=0.0052x-0.0359$	10–200	0.9946
259	Pentachloroanisole	1825-21-4	$y=0.0035x-0.0292$	20–800	0.9993
260	Tebutam	35256-85-0	$y=0.0126x-0.0320$	5–200	0.9980

Continued

261	Methabenzthiazuron	18691-97-9	$y=0.0022x-0.0637$	30–300	0.9933
262	Simeton	673-04-1	$y=0.0096x-0.0647$	10–200	0.9942
263	Atratone	1610-17-9	$y=0.0051x-0.0694$	20–400	0.9923
264	Desisopropyl-atrazine	1007-28-9	$y=5.2241\times10^{-004}x-0.3629$	800–16000	0.9949
265	Tefluthrin	79538-32-2	$y=0.0236x-0.0540$	2–200	0.9967
266	Bromocylen	1715-40-8	$y=6.2377\times10^{-004}x-0.0360$	60–1200	0.9985
267	Trietazine	1912-26-1	$y=0.0070x-0.0597$	10–400	0.9965
268	DE-PCB 28	2012-37-5	$y=0.1229x-0.0204$	0.5–20	0.9996
269	DE-PCB 31	16606-02-3	$y=0.1229x-0.0204$	0.5–20	0.9996
270	Cycluron	2163-69-1	$y=0.0012x-0.0785$	50–2000	0.9921
271	2,6-Dichlorodenzamide	2008-58-4	$y=0.0090x-0.3706$	25–1000	0.9928
272	Desethyl-sebuthylazine	37019-18-4	$y=0.0020x-0.2704$	140–1400	0.9836
273	2,3,4,5-Tetrachloroaniline	634-83-3	$y=0.0030x-0.0264$	15–600	0.9971
274	Musk ambrette	83-46-9	$y=0.0015x-0.1842$	56–5600	0.9954
275	Pentachloroaniline	527-20-8	$y=0.0030x-0.0347$	20–800	0.9982
276	Aziprotryne	4658-28-0	$y=0.0022x-0.0536$	50–1000	0.9984
277	DE-PCB 52	35693-99-3	$y=0.0052x-0.0198$	6–600	0.9988
278	Sebutylazine	7286-69-3	$y=0.0036x-0.1494$	30–1200	0.9932
279	Isocarbamid	30979-48-7	$y=8.3193\times10^{-004}x-0.1867$	350–1400	0.9825

TABLE 5.13 Linear Equations, Linear Ranges, and Correlation Coefficients of 466 Pesticides and Chemical Pollutants Determined by GC-MS-MS—cont'd

No.	Compound	CAS	Linear Equation	Linear Range (pg)	Correlation Coefficient
280	Prosulfocarb	52888-80-9	$y = 0.0043x - 0.0765$	20–800	0.9952
281	Dimethenamid	87674-68-8	$y = 0.0184x - 0.0865$	5–200	0.9961
282	Monalide	7287-36-7	$y = 0.0026x - 0.0400$	20–800	0.9974
283	Octachlorostyrene	29082-74-4	$y = 0.0011x - 0.0230$	32–3200	0.9989
284	Paraoxon-methyl	950-35-6	$y = 1.9321 \times 10^{-005}x - 0.0183$	2400–12000	0.9924
285	Isobenzan	297-78-9	$y = 0.0010x - 0.0648$	40–4000	0.9970
286	Isodrin	465-73-6	$y = 0.0023x - 0.0167$	25–1000	0.9989
287	Trichloronat	327-98-0	$y = 0.0054x - 0.0314$	10–200	0.9948
288	Isomethiozin	57052-04-7	$y = 0.0035x - 0.1125$	25–1000	0.9921
289	Dacthal	1861-32-1	$y = 0.0059x - 0.0403$	6–600	0.9975
290	4,4-Dichlorobenzophenone	90-98-2	$y = 0.0074x - 0.1096$	20–400	0.9943
291	Rabenzazole	69899-24-7	$y = 0.0015x - 0.1951$	150–1200	0.9705
292	Nitrothal-isopropyl	10552-74-6	$y = 0.0032x - 0.3279$	75–1500	0.9806
293	Cyprodinil	121552-61-2	$y = 0.0086x - 0.1149$	20–400	0.9951
294	Fuberidazole	3878-19-1	$y = 0.0024x - 0.3148$	125–1000	1.0000

Continued

295	DE-PCB 101	37680-73-2	$y=0.0109x-0.0282$	4–400	0.9989
296	Methfuroxam	2873-17-8	$y=0.0074x-0.1002$	15–300	0.9957
297	Dicapthon	2463-84-5	$y=0.0019x-0.3504$	170–1700	0.9731
298	MCPA-butoxyethyl ester	94-81-5	$y=0.0017x-0.3471$	200–4000	0.9946
299	Phorate sulfone	2588-04-7			
300	Isocarbophos	245-61-5	$y=7.7768\times10^{-004}x-0.9864$	800–16000	0.9657
301	*trans*-Nonachlor	551877-74-8	$y=5.2636\times10^{-004}x-0.0317$	68–6800	0.9983
302	DEF	78-48-8	$y=0.0040x-0.1952$	40–800	0.9871
303	Flurochloridone	61213-25-0	$y=0.0015x-0.7977$	520–5200	0.9750
304	Bromfenvinfos	33399-00-7	$y=8.3731\times10^{-004}x-0.3522$	600–2400	0.9757
305	Perthane	72-56-0	$y=0.0196x-0.0973$	5–200	0.9950
306	Ditalimfos	5131-24-8	$y=0.0027x-0.0605$	40–400	0.9941
307	DE-PCB 118	31508-00-6	$y=0.0082x-0.0444$	10–400	0.9972
308	4,4-Dibromobenzophenone	3988-03-2	$y=0.0018x-0.0864$	50–1000	0.9917
309	Flutriafol	76674-21-0	$y=0.0066x-0.1150$	20–400	0.9921
310	Mephosfolan	950-10-7	$y=9.9109\times10^{-004}x-0.2800$	400–1600	0.9757
311	DE-PCB 153	35065-27-1	$y=0.0035x-0.0334$	10–1000	0.9981
312	Diclobutrazole	75736-33-3	$y=0.0036x-0.0956$	30–600	0.9903
313	Disulfoton sulfone	2497-06-5	$y=2.9827\times10^{-004}x-0.8842$	1850–37000	0.9624

TABLE 5.13 Linear Equations, Linear Ranges, and Correlation Coefficients of 466 Pesticides and Chemical Pollutants Determined by GC-MS-MS—cont'd

No.	Compound	CAS	Linear Equation	Linear Range (pg)	Correlation Coefficient
314	DE-PCB 138	35065-28-2	$y=0.0020x-0.0281$	30–1200	0.9977
315	Resmethrin-1	10453-86-8	$y=0.0018x-0.0794$	50–1000	0.9883
316	Resmethrin-2	10453-86-8	$y=0.0068x-0.2919$	20–1000	0.9828
317	Cyproconazole	113096-99-4	$y=0.0042x-0.0284$	20–200	0.9961
318	Phthalic acid, benzyl butyl ester	85-68-7	$y=0.0070x-0.1543$	20–400	0.9879
319	Clodinafop-propargyl	105512-06-9	$y=6.6802 \times 10^{-004}x-0.5025$	950–3800	0.9615
320	Fluotrimazole	31251-03-3	$y=0.0011x-0.1586$	60–6000	0.9942
321	Fluroxypr-1-methylheptyl ester	81406-37-3	$y=3.8503 \times 10^{-004}x-0.2719$	500–20000	0.9936
322	Triphenyl phosphate	115-86-6	$y=0.0030x-0.3007$	65–2600	0.9922
323	Metamitron	41394-05-2	$y=2.2177 \times 10^{-004}x-0.2271$	1400–5600	0.9776
324	DE-PCB 180	35065-29-3	$y=0.0024x-0.0272$	30–1200	0.9982
325	Tebufenpyrad	119168-77-3	$y=0.0053x-0.1447$	20–800	0.9924
326	Cloquintocet-mexyl	99607-70-2	$y=0.0112x-0.5970$	50–500	0.9726
327	Lenacil	2164-08-1	$y=0.0020x-0.0708$	55–220	0.9815

328	Bromuconazole-1	116255-48-2	$y=0.0031x-0.1701$	100–2000	0.9925
329	Bromuconazole-2	116255-48-2	$y=0.0052x-0.3923$	50–1000	0.9733
330	Nitralin	4726-14-1	$y=1.6586\times10^{-004}x-0.2160$	1600–8000	0.9629
331	Fenamiphos sulfone	31972-44-8	$y=6.1932\times10^{-004}x-0.3594$	680–3400	0.9687
332	Fenpiclonil	74738-17-3	$y=0.0017x-0.3451$	150–3000	0.9820
333	Fluquinconazole	136426-54-5	$y=0.0032x-0.1442$	40–1600	0.9933
334	Fenbuconazole	114369-43-6	$y=0.0097x-0.0632$	10–200	0.9983
Group D					
335	XMC	2655-14-3	$y=0.0083x+0.4660$	1–200	0.9956
336	Dibutyl succinate	141-03-7	$y=0.0607x+0.2400$	0.2–50	0.9940
337	Propoxur-1	114-26-1	$y=0.3627x-0.1092$	0.1–100	0.9999
338	Isoprocarb-1	2631-40-7	$y=0.0349x+0.1339$	2–200	0.9962
339	Terbucarb-1	1918-11-2	$y=0.0104x+0.1057$	10–80	0.9968
340	Chlorethoxyfos	54593-83-8	$y=7.3305\times10^{-004}x-0.0566$	120–4800	0.9991
341	Isoprocarb-2	2631-40-5	$y=0.0313x-0.1512$	5–100	0.9906
342	Tebuthiuron	34014-18-1	$y=0.0043x-0.0551$	20–400	0.9953
343	Pencycuron	66063-05-6	$y=0.0017x+0.0019$	40–800	0.9996
344	Demeton-s-methyl	919-86-8	$y=0.0025x-0.3205$	100–2000	0.9839
345	Propoxur-2	114-26-1	$y=0.0732x-0.1690$	2.5–50	0.9935

Continued

TABLE 5.13 Linear Equations, Linear Ranges, and Correlation Coefficients of 466 Pesticides and Chemical Pollutants Determined by GC-MS-MS—cont'd

No.	Compound	CAS	Linear Equation	Linear Range (pg)	Correlation Coefficient
346	Phenanthrene	65-01-8	$y=0.0024x-0.0066$	15–600	0.9995
347	Fenpyroximate	134098-61-6	$y=2.7424\times10^{-004}x+0.0150$	50–1000	0.9949
348	Tebupirimfos	96182-53-5	$y=0.0017x-0.0360$	25–1000	0.9972
349	Fenpropidin	67306-00-7	$y=0.0119x-0.0072$	5–50	0.9993
350	Fluometuron	2164-17-2	$y=1.3563\times10^{-004}x-0.1021$	1400–7000	0.9959
351	Dichloran	99-30-9	$y=0.0037x-0.1325$	45–900	0.9923
352	Pyroquilon	57369-32-1	$y=0.0099x-0.0582$	10–200	0.9973
353	Phosphamidon-1	13171-21-6	$y=7.4129\times10^{-005}x-0.5100$	6300–63000	0.9665
354	Benoxacor	98730-04-2	$y=0.0033x-0.1914$	40–1600	0.9922
355	Acetochlor	34256-82-1	$y=0.0022x-0.0388$	20–800	0.9963
356	Tridiphane	58138-08-2	$y=0.0047x-0.1484$	25–1000	0.9930
357	Propyzamide	23950-58-5	$y=0.0402x-0.1342$	2.5–100	0.9921
358	Terbucarb-2	1918-11-2	$y=0.0203x-0.0776$	4–80	0.9903
359	Fenfuram	24691-80-3	$y=0.0152x-0.1345$	10–200	0.9947
360	Acibenzolar-S-methyl	135158-54-2	$y=0.0016x-0.2730$	170–3400	0.9911

361	Benfuresate	68505-69-1	$y=0.0048x-0.0327$	15–600	0.9990
362	Mefenoxam	70630-17-0	$y=0.0020x-0.0376$	25–1000	0.9968
363	Malaoxon	1364-78-2			
364	Phosphamidon-2	13171-21-6	$y=6.5642\times10^{-004}x-1.4055$	1260–25200	0.9696
365	Simeconazole	149508-90-7	$y=0.0179x-0.1082$	5–200	0.9938
366	Chlorthal-dimethyl	1861-32-1	$y=0.0054x-0.0166$	10–400	0.9998
367	Terbacil	5902-51-2	$y=0.0018x-1.5765$	600–12000	0.9818
368	Thiazopyr	117718-60-2	$y=4.3354\times10^{-004}x-0.0246$	70–7000	0.9993
369	Dimethylvinphos	2274-67-1	$y=0.0050x-1.2963$	180–3600	0.9796
370	Zoxamide	156052-68-5	$y=0.0021x-0.0129$	30–1200	0.9994
371	Allethrin	584-79-2	$y=0.0219x-0.2610$	10–200	0.9949
372	Quinoclamine	2797-51-5	$y=0.0095x-0.7971$	800–8000	0.9698
373	Fenoxanil	115852-48-7	$y=0.0018x-0.0839$	50–1000	0.9924
374	Furalaxyl	57646-30-7	$y=0.0155x-0.0631$	5–200	0.9960
375	Thiamethoxam	153719-23-4	$y=1.0868\times10^{-004}x+0.0193$	1200–4800	0.9984
376	Thiabendazole	148-79-8	$y=0.0014x-0.0015$	60–300	0.9541
377	Bromacil	314-40-9	$y=0.0086x-0.8321$	650–13000	0.9816
378	Picoxystrobin	117428-22-5	$y=0.0029x-0.0652$	25–1000	0.9965
379	Methiocarb sulfone	2178-25-1			

Continued

TABLE 5.13 Linear Equations, Linear Ranges, and Correlation Coefficients of 466 Pesticides and Chemical Pollutants Determined by GC-MS-MS—cont'd

No.	Compound	CAS	Linear Equation	Linear Range (pg)	Correlation Coefficient
380	Butamifos	8013-75-0	$y = 0.0014x - 0.0827$	70–560	0.9705
381	TCMTB	21564-17-0	$y = 6.2982 \times 10^{-005}x - 0.0689$	1540–7700	0.9897
382	Imazalil	35554-44-0	$y = 0.0010x + 0.0031$	30–300	0.9942
383	Isoprothiolane	50512-35-1	$y = 0.0029x - 0.2128$	25–5000	0.9990
384	Cyflufenamid	180409-60-3	$y = 6.8325 \times 10^{-004}x - 0.3139$	500–10000	0.9966
385	Isoxathion	18854-01-8	$y = 0.0013x - 0.6744$	300–6000	0.9573
386	Quinoxyphen	124495-18-7	$y = 3.4451 \times 10^{-004}x - 0.0478$	200–8000	0.9985
387	Imibenconazole-des-benzyl	199338-48-2	$y = 0.0117x - 0.4611$	28–2800	0.9984
388	Trifloxystrobin	141517-21-7	$y = 0.0019x - 0.1342$	70–1400	0.9919
389	Fipronil	120068-37-3	$y = 5.4217 \times 10^{-004}x - 0.5378$	725–14500	0.9834
390	Pyraflufen ethyl	129630-17-7	$y = 3.6030 \times 10^{-005}x - 0.0051$	800–4000	0.9994
391	Thenylchlor	96491-05-3	$y = 0.0029x - 0.2018$	45–1800	0.9915
392	Clethobim	99129-21-2	$y = 4.2047 \times 10^{-004}x - 0.0589$	280–1400	0.9994
393	Mefenpyr-diethyl	135590-91-9	$y = 0.0047x - 0.1176$	20–800	0.9947
394	Chrysene	218-01-9	$y = 3.2713 \times 10^{-005}x - 0.0369$	1560–62400	0.9981
395	Epoxiconazole-1	106325-08-0	$y = 0.0144x - 0.0243$	2.5–50	0.9952

Continued

No.	Name	CAS	Equation	Range	R²
396	Epoxiconazole-2	106325-08-0	$y=0.0144x-0.0243$	2.5–50	0.9952
397	Pyriproxyfen	95737-68-1	$y=0.0096x-0.0726$	10–100	0.9905
398	Piperophos	24151-93-7	$y=4.8895\times10^{-004}x-0.5328$	1000–20000	0.9947
399	Fenamidone	161326-34-7	$y=0.0042x-0.1324$	25–1000	0.9956
400	Pyraclostrobin	175013-18-0	$y=3.2997\times10^{-004}x-0.7073$	2400–24000	0.9922
401	Tralkoxydim	87820-88-0	$y=3.6281\times10^{-004}x-0.0867$	225–9000	0.9979
402	Pyraclofos	77458-01-6	$y=4.1001\times10^{-004}x-0.5472$	1640–8200	0.9693
403	Dialifos	10311-84-9			
404	Spirodiclofen	148477-71-8	$y=0.0016x-0.0361$	260–5200	0.9939
405	Flurtamone	96525-23-4	$y=0.0020x-0.7868$	200–8000	0.9870
406	Silafluofen	105024-66-6	$y=0.0020x-0.0859$	40–1600	0.9964
407	Pyrimidifen	105779-78-0	$y=0.0087x-0.0711$	10–200	0.9886
408	Butafenacil	134605-64-4	$y=0.0160x-0.2634$	10–400	0.9868
409	Acetamiprid	160430-64-8	$y=2.5179\times10^{-004}x-0.2786$	1250–25000	0.9987
410	Fluridone	59756-60-4	$y=0.0011x-0.1573$	200–4000	0.9991
411	Tribenron-methyl	106040-48-6	$y=0.0013x-0.0784$	120–1200	0.9969
412	Dioxacarb	6988-21-2	$y=0.0032x-0.2711$	80–800	0.9948
413	Ethiofencarb	29973-13-5	$y=0.0094x-0.1289$	24–240	0.9970
414	Dimethyl phthalate	131-11-3	$y=0.0304x-0.0316$	2.5–100	0.9994
415	4-Chlorophenoxy acetic acid	40020-01-7			

TABLE 5.13 Linear Equations, Linear Ranges, and Correlation Coefficients of 466 Pesticides and Chemical Pollutants Determined by GC-MS-MS—cont'd

No.	Compound	CAS	Linear Equation	Linear Range (pg)	Correlation Coefficient
416	Phthalimide	5333-22-2	$y = 0.0033x - 2.0879$	340–6800	0.9511
417	Carbaryl	63-25-2	$y = 0.0032x - 0.2192$	150–600	0.9933
418	2,4-D	94-75-7	$y = 4.4168 \times 10^{-005}x - 0.0589$	2300–23000	0.9936
419	Cadusafos	95465-99-9	$y = 0.0208x - 0.1505$	10–200	0.9931
420	Demetom-s	126-75-0	$y = 0.0013x - 0.2322$	150–3000	0.9913
421	Dicrotophos	141-66-2	$y = 0.0019x - 0.3076$	200–2000	0.9832
422	3,4,5-Trimethacarb	2686-99-9	$y = 1.2176 \times 10^{-004}x - 0.1495$	2080–10400	0.9934
423	2,4,5-T	93-76-5	$y = 2.2384 \times 10^{-006}x - 0.0089$	7300–36500	0.9948
424	DMSA	1956-84-5	$y = 6.4955 \times 10^{-005}x - 0.0506$	1280–5120	0.9918
425	Pirimicarb	23103-98-2	$y = 0.0088x - 0.0513$	10–200	0.9959
426	Dodemorph-1	1593-77-7	$y = 0.0106x - 0.0878$	6–600	0.9962
427	Dodemorph-2	1593-77-7	$y = 0.0027x - 0.0221$	15–600	0.9983
428	Desmedipham	13684-56-5			
429	Fenchlorphos	3983-45-7	$y = 0.0028x - 0.1045$	40–800	0.9941
430	S421(octachlorodipropyl ester)-1	127-90-2	$y = 0.0031x - 0.0124$	10–80	0.9995

No.	Name	CAS	Equation	Range	R^2
431	S421(octachlorodipropyl ester)-2	127-90-2	$y=0.0055x-0.0480$	10–400	0.9984
432	Esprocarb	85785-20-2	$y=0.0261x-0.0823$	2–200	0.9960
433	Telodrin	297-78-9	$y=8.2734\times10^{-004}x-0.0215$	70–2800	0.9995
434	Difenoxuron	14214-32-5	$y=0.0014x-0.8336$	650–65000	0.9826
435	Endothal	145-73-3			
436	Butralin	33629-47-9	$y=0.0019x-0.0767$	60–1200	0.9913
437	Pyrifenox-1	88283-41-4	$y=8.5994\times10^{-004}x-0.0938$	95–3800	0.9971
438	Dimethametryn	14214-32-5	$y=0.0065x-0.0972$	20–400	0.9942
439	Flufenacet	142459-58-3	$y=0.0010x-0.2483$	320–1600	0.9846
440	Aclonifen	74070-46-5			
441	Pyrifenox-2	88283-41-4	$y=0.0023x-0.2741$	95–3800	0.9958
442	Flubenzimine	37893-02-0			
443	Azaconazole	60207-31-0	$y=0.0245x-0.0600$	5–50	0.9952
444	Iprovalicarb-1	140923-17-7	$y=0.0016x-0.3255$	280–1400	0.9885
445	Irpovalicarb-2	140923-17-7	$y=0.0018x-0.3743$	280–1400	0.9878
446	Diofenolan-1	63837-33-2	$y=0.0033x-0.1703$	50–1000	0.9901
447	Diofenolan-2	63837-33-2	$y=0.0018x-0.0909$	50–1000	0.9932
448	Chlorfenapyr	122453-73-0	$y=1.6994\times10^{-004}x-0.0334$	350–14000	0.9995
449	Bioresmethrin	28434-01-7	$y=0.0030x-0.0354$	30–120	0.9950
450	Carfentrazone-ethyl	128621-72-7	$y=0.0020x-0.4449$	200–4000	0.9944

Continued

TABLE 5.13 Linear Equations, Linear Ranges, and Correlation Coefficients of 466 Pesticides and Chemical Pollutants Determined by GC-MS-MS—cont'd

No.	Compound	CAS	Linear Equation	Linear Range (pg)	Correlation Coefficient
451	Diclofop-methyl	51338-27-3	$y = 0.0017x - 0.1474$	100–2000	0.9966
452	Endrin aldehyde	7421-93-4	$y = 1.8957 \times 10^{-004}x - 0.0214$	700–28000	0.9992
453	Halosulfuran-methyl	100784-20-1	$y = 0.0011x - 0.0811$	160–3200	0.9998
454	Carbosulfan	55285-14-8	$y = 0.0042x - 15.9704$	940–188000	0.9947
455	Fenhexamid	126833-17-8			
456	Spiromesifen	283594-90-1	$y = 0.0012x - 0.0529$	100–1000	0.9993
457	Famphur	52-85-7	$y = 0.0031x - 0.4659$	160–1600	0.9920
458	Fluazinam	79622-59-6			
459	Endrin ketone	53494-70-5	$y = 4.1812 \times 10^{-004}x - 0.0761$	250–10000	0.9998
460	Metconazole	125116-23-6	$y = 0.0028x - 0.1168$	40–800	0.9938
461	Cyhalofop-butyl	122008-85-9	$y = 0.0027x - 0.2463$	140–700	0.9923
462	Lactofen	77501-63-4			
463	Pyriftalid	135186-78-6			
464	Boscalid	188425-85-6	$y = 0.0043x - 0.3616$	80–1600	0.9936
465	Tolfenpyrad	129558-76-5	$y = 0.0016x - 0.4161$	280–2800	0.9876
466	Dimethomorph	211867-47-9	$y = 0.0036x - 0.0659$	20–400	0.9905

5.2.3 Linear Equations, Linear Ranges, and Correlation Coefficients of 284 Environmental Pollutants Determined by GC-MS-MS

See Table 5.14.

TABLE 5.14 Linear Equations, Linear Ranges, and Correlation Coefficients of 284 Environmental Pollutants Determined by GC-MS-MS

No.	Compound	Linear Equation	Linear Range	Correlation Coefficient
Heptachlor (ISTD)				
Group A				
1	2-Chlorobiphenyl	$y=2.4459x+0.0321$	1.5–120	0.9916
2	2,2′-Dichlorobiphenyl	$y=0.6702x-0.3194$	1.5–120	0.9968
3	2,4′-Dichlorobiphenyl	$y=1.3927x-0.7712$	1.5–120	0.9980
4	2,2′,6-Trichlorobiphenyl	$y=1.2178x-0.5426$	1.5–120	0.9979
5	3,4-Dichlorobiphenyl	$y=2.1949x-0.6531$	1.5–120	0.9980
6	2,3′,6-Trichlorobiphenyl	$y=1.0444x-0.0164$	1.5–120	0.9960
7	2,2′,3-Trichlorobiphenyl	$y=1.2102x-0.1932$	1.5–120	0.9952
8	2,3′,4-Trichlorobiphenyl	$y=2.2189x-1.1486$	1.5–120	0.9985
9	2,3,4-Trichlorobiphenyl	$y=1.7531x-0.3355$	1.5–120	0.9996
10	2,3,3′-Trichlorobiphenyl	$y=1.0537x-0.3765$	1.5–120	0.9985

Continued

TABLE 5.14 Linear Equations, Linear Ranges, and Correlation Coefficients of 284 Environmental Pollutants Determined by GC-MS-MS—cont'd

No.	Compound	Linear Equation	Linear Range	Correlation Coefficient
11	3,3′,5-Trichlorobiphenyl	$y = 0.8907x + 0.0022$	1.5–120	0.9932
12	2,2′,3,5-Tetrachlorobiphenyl	$y = 0.3876x - 0.4500$	4.5–120	0.9999
13	2,3,5,6-Tetrachlorobiphenyl	$y = 0.7969x - 0.2484$	1.5–120	0.9982
14	2,2′,4,6,6′-Pentachlorobiphenyl	$y = 0.5316x - 0.2535$	1.5–120	0.9958
15	2,3′,5,5′-Tetrachlorobiphenyl	$y = 0.8374x - 0.3825$	1.5–120	0.9996
16	2,2′,4,5′,6-Pentachlorobiphenyl	$y = 0.5428x + 0.1439$	1.5–120	0.9951
17	2,2′,3,4-Tetrachlorobiphenyl	$y = 0.6036x - 0.1385$	1.5–120	0.9994
18	2,3′,4,5-Tetrachlorobiphenyl	$y = 0.8405x - 0.3066$	1.5–120	0.9990
19	2,2′,3,3′-Tetrachlorobiphenyl	$y = 0.4695x - 0.1937$	1.5–120	0.9979
20	2,4,4′,5-Tetrachlorobiphenyl	$y = 0.9181x - 0.0497$	1.5–120	0.9989
21	2,2′,4,5,6′-Pentachlorobiphenyl	$y = 0.5123x + 0.0418$	1.5–120	0.9982
22	2,2′,3,5′,6-Pentachlorobiphenyl	$y = 0.6398x - 0.2194$	1.5–120	0.9963
23	2,2′,3,5,5′-Pentachlorobiphenyl	$y = 0.2195x + 0.0848$	2.5–200	0.9953
24	2,2′,4,4′,5-Pentachlorobiphenyl	$y = 0.3347x - 0.0855$	1.5–120	0.9994
25	2,2′,3,3′,6-Pentachlorobiphenyl	$y = 0.4903x + 0.1542$	1.5–120	0.9904
26	2,3,3′,4,6-Pentachlorobiphenyl	$y = 0.4238x - 0.0125$	1.5–120	0.9946

No.	Name	Equation	Range	Correlation
27	2,2',3,3',5-Pentachlorobiphenyl	$y=0.2220x-0.0859$	2.5–200	0.9989
28	2,2',3,4,5-Pentachlorobiphenyl	$y=0.5546x-0.2389$	1.5–120	0.9998
29	2',3,4,5,6-Pentachlorobiphenyl	$y=0.5407x-0.0427$	1.5–120	0.9997
30	2,2',3,4,5'-Pentachlorobiphenyl	$y=0.3698x-0.2884$	1.5–120	0.9998
31	2,3,3',4',6-Pentachlorobiphenyl	$y=0.7662x-0.9562$	1.5–120	0.9930
32	2,2',3,3',5,6-Hexachlorobiphenyl	$y=0.4681x-0.3688$	1.5–120	0.9974
33	2',3,4,5,5'-Pentachlorobiphenyl	$y=0.7963x-0.2467$	1.5–120	0.9993
34	2',3,4,4',5-Pentachlorobiphenyl	$y=0.3608x-0.3634$	1.5–120	0.9988
35	2,3',4,4',5-Pentachlorobiphenyl	$y=0.6858x-0.2333$	1.5–120	0.9978
36	2,2',3,3',5,6-Hexachlorobiphenyl	$y=0.4407x-0.2854$	1.5–120	0.9990
37	2,3,4,4',5-Pentachlorobiphenyl	$y=0.3100x-0.2190$	2.5–200	0.9993
38	2,3',4,4',5',6-Hexachlorobiphenyl	$y=0.1898x-0.0113$	1.5–120	0.9978
39	3,3',4,5,5'-Pentachlorobiphenyl	$y=0.7606x-0.0661$	1.5–120	0.9987
40	2,2',3,4,4',5-Hexachlorobiphenyl	$y=0.4780x-0.2686$	1.5–120	0.9987
41	2,3,3',4',5,6-Hexachlorobiphenyl	$y=0.7916x-0.1005$	1.5–120	0.9991
42	2,2',3,3',5,5',6-Heptachlorobiphenyl	$y=0.3163x-0.0915$	1.5–120	0.9996
43	2,2',3,4',5,5',6-Heptachlorobiphenyl	$y=0.3358x-0.1223$	1.5–120	0.9973
44	2,3,3',4',5,5'-Hexachlorobiphenyl	$y=0.6615x-0.2135$	1.5–120	0.9990
45	2,2',3,3',5,5',6,6'-Octachlorobiphenyl	$y=0.3761x-0.1960$	1.5–120	0.9992

Continued

TABLE 5.14 Linear Equations, Linear Ranges, and Correlation Coefficients of 284 Environmental Pollutants Determined by GC-MS-MS—cont'd

No.	Compound	Linear Equation	Linear Range	Correlation Coefficient
46	2,2′,3,4,4′,5,6,6′-Octachlorobiphenyl	$y = 0.3587x - 0.0924$	1.5–120	0.9990
47	2,2′,3,3′,4,4′,6,6′-Octachlorobiphenyl	$y = 0.7242x - 0.1286$	1.5–120	0.9988
48	2,3,3′,4,5,5′,6-Heptachlorobiphenyl	$y = 0.3291x - 0.0827$	1.5–120	0.9991
49	2,3,3′,4′,5,5′,6-Heptachlorobiphenyl	$y = 0.2592x - 0.0653$	1.5–120	0.9967
50	2,3,3′,4,4′,5,6-Heptachlorobiphenyl	$y = 0.6880x - 0.0956$	1.5–120	0.9989
51	3,3′,4,4′,5,5′-Hexachlorobiphenyl	$y = 0.5396x + 0.0405$	1.5–120	0.9964
52	2,2′,3,3′,4,4′,5,6-Octachlorobiphenyl	$y = 0.5381x - 0.1630$	1.5–120	0.9984
53	2,2′,3,3′,4,4′,5,5′,6-Nonachlorobiphenyl	$y = 0.2142x - 0.0274$	1.5–120	0.9980
54	2,2′,3,3′,4,4′,5,5′,6,6′-Decachlorobiphenyl	$y = 0.3092x - 0.1418$	1.5–120	0.9999
Group B				
55	3-Chlorobiphenyl	$y = 1.8077x - 0.1286$	1.5–120	0.9986
56	2,4-Dichlorobiphenyl	$y = 1.4773x + 0.4931$	1.5–120	0.9974
57	2,3-Dichlorobiphenyl	$y = 1.7668x - 0.0261$	1.5–120	0.9981
58	3,3′-Dichlorobiphenyl	$y = 1.3893x + 0.1723$	1.5–120	0.9990
59	3,4′-Dichlorobiphenyl	$y = 0.5225x + 3.6859$	2.5–200	0.9879
60	2,4′,6-Trichlorobiphenyl	$y = 1.4701x + 0.2391$	1.5–120	0.9987

Continued

61	2,4,5-Trichlorobiphenyl	$y=0.4704x-0.0694$	1.5–120	0.9962
62	2,2′,4,6-Tetrachlorobiphenyl	$y=0.4179x+0.0127$	1.5–120	0.9995
63	2,2′,5,6′-Tetrachlorobiphenyl	$y=0.3821x+0.0380$	1.5–120	0.9972
64	2,3,4′-Trichlorobiphenyl	$y=1.8027x+0.3811$	1.5–120	0.9973
65	2,3′,5′,6-Tetrachlorobiphenyl	$y=1.0349x+0.2342$	1.5–120	0.9975
66	3,4′,5-Trichlorobiphenyl	$y=0.5034x+0.0375$	1.5–120	0.9977
67	2,3,4,6-Tetrachlorobiphenyl	$y=0.8622x+0.0859$	1.5–120	0.9989
68	3,4,5-Trichlorobiphenyl	$y=0.4391x+0.2602$	2.5–200	0.9963
69	3,3′,4-Trichlorobiphenyl	$y=0.8346x-0.3293$	1.5–120	0.9992
70	2,3,4′,6-Tetrachlorobiphenyl	$y=0.6745x+0.4506$	1.5–120	0.9833
71	3,4,4′-Trichlorobiphenyl	$y=0.8905x+0.0762$	1.5–120	0.9985
72	3,3′,5,5′-Tetrachlorobiphenyl	$y=0.8679x+0.0574$	1.5–120	0.9991
73	2,3,3′,5′-Tetrachlorobiphenyl	$y=0.8447x-0.0013$	1.5–120	0.9988
74	2,3′,4,5′,6-Pentachlorobiphenyl	$y=0.3361x-0.0584$	2.5–200	0.9993
75	2,2′,3,5,6-Pentachlorobiphenyl	$y=0.6106x+0.0649$	1.5–120	0.9994
76	2,3′,4,4′-Tetrachlorobiphenyl	$y=0.8585x+0.1412$	1.5–120	0.9971
77	2,2′,3,4′,5-Pentachlorobiphenyl	$y=0.6510x+0.4744$	1.5–120	0.9971
78	2,3,3′,5′,6-Pentachlorobiphenyl	$y=0.7591x+0.3328$	1.5–120	0.9864
79	2,2′,3,4,6′-Pentachlorobiphenyl	$y=0.5683x+0.1734$	1.5–120	0.9984

TABLE 5.14 Linear Equations, Linear Ranges, and Correlation Coefficients of 284 Environmental Pollutants Determined by GC-MS-MS—cont'd

No.	Compound	Linear Equation	Linear Range	Correlation Coefficient
80	2,2′,3,5,6,6′-Hexachlorobiphenyl	$y=0.4630x+0.1136$	1.5–120	0.9960
81	2,2′,3,4,6,6′-Hexachlorobiphenyl	$y=0.2597x+0.0437$	1.5–120	0.9993
82	2,3,4,4′,6-Pentachlorobiphenyl	$y=0.7787x+0.1503$	1.5–120	0.9965
83	2,2′,4,4′,5,6′-Hexachlorobiphenyl	$y=0.4710x+0.1900$	1.5–120	0.9957
84	2,2′,3,4,4′-Pentachlorobiphenyl	$y=0.5159x-0.2218$	1.5–120	0.9985
85	2,2′,3,5,5′,6-Hexachlorobiphenyl	$y=0.3866x-0.1639$	1.5–120	0.9973
86	2,2′,3,4,4′,6-Hexachlorobiphenyl	$y=0.4978x-0.1412$	1.5–120	0.9944
87	2,2′,3,4,4′,6′-Hexachlorobiphenyl	$y=0.4286x-0.0733$	1.5–120	0.9966
88	2,3,3′,4′,5-Pentachlorobiphenyl	$y=0.4346x+0.2180$	1.5–120	0.9944
89	2,2′,3,4,5,6′-Hexachlorobiphenyl	$y=0.5288x-0.1023$	1.5–120	0.9967
90	2,2′,3,4,5,6-Hexachlorobiphenyl	$y=0.2089x-0.0063$	1.5–120	0.9975
91	2,2′,3,4′,5,5′-Hexachlorobiphenyl	$y=0.5088x+0.3019$	1.5–120	0.9957
92	2′,3,3′,4,5-Pentachlorobiphenyl	$y=0.4733x-0.1304$	1.5–120	0.9996
93	2,2′,3,4,5,5′-Hexachlorobiphenyl	$y=0.2364x+0.0591$	1.5–120	0.9991
94	2,2′,3,3′,4,5′-Hexachlorobiphenyl	$y=0.1488x+0.0030$	1.5–120	0.9944
95	2,2′,3,4,4′,5′-Hexachlorobiphenyl	$y=0.6538x+0.4058$	1.5–120	0.9964

96	2,2',3,3',4,5',6-Heptachlorobiphenyl	$y=0.3863x-0.0539$	1.5–120	0.9957
97	2,2',3,4,4',5',6-Heptachlorobiphenyl	$y=0.2910x+0.0142$	1.5–120	0.9968
98	2,3,4,4',5,6-Hexachlorobiphenyl	$y=0.2723x+0.2410$	1.5–120	0.9956
99	2,2',3,3',4,4'-Hexachlorobiphenyl	$y=0.3715x+0.1531$	1.5–120	0.9957
100	2,2',3,3',4,5',6,6'-Octachlorobiphenyl	$y=0.8084x-0.1319$	1.5–120	0.9997
101	2,2',3,3',4,5,6-Heptachlorobiphenyl	$y=0.4234x-0.0281$	1.5–120	0.9987
102	2,3,3',4,4',5'-Hexachlorobiphenyl	$y=0.7321x+0.6309$	1.5–120	0.9980
103	2,3,3',4,4',5',6-Heptachlorobiphenyl	$y=0.5642x+0.1188$	1.5–120	0.9969
104	2,2',3,4,4',5,5',6-Octachlorobiphenyl	$y=0.2451x+0.0263$	1.5–120	0.9965
105	2,2',3,3',4,5,5',6,6'-Nonachlorobiphenyl	$y=0.5763x+0.1776$	1.5–120	0.9967
106	2,2',3,3',4,4',5,5'-Octachlorobiphenyl	$y=0.1471x+0.1204$	2.5–200	0.9989
Group C				
107	4-Chlorobiphenyl	$y=2.5852x-0.7325$	1.5–120	0.9992
108	2,5-Dichlorobiphenyl	$y=1.7036x-0.9649$	1.5–120	0.9953
109	3,5-Dichlorobiphenyl	$y=3.0308x-2.0712$	1.5–120	0.9973
110	2,2',5-Trichlorobiphenyl	$y=1.0880x-1.0872$	1.5–120	0.9938
111	2,3,6-Trichlorobiphenyl	$y=0.3710x-0.4487$	1.5–120	0.9916
112	2,3,5-Trichlorobiphenyl	$y=1.0118x-0.4367$	1.5–120	0.9962
113	2,2',6,6'-Tetrachlorobiphenyl	$y=0.7109x-0.1322$	1.5–120	0.9971

Continued

TABLE 5.14 Linear Equations, Linear Ranges, and Correlation Coefficients of 284 Environmental Pollutants Determined by GC-MS-MS—cont'd

No.	Compound	Linear Equation	Linear Range	Correlation Coefficient
114	2,4′,5-Trichlorobiphenyl	$y = 0.4591x + 0.0668$	1.5–120	0.9990
115	2′,3,4-Trichlorobiphenyl	$y = 1.2614x − 0.5768$	1.5–120	0.9917
116	2,3′,4,6-Tetrachlorobiphenyl	$y = 0.5131x − 0.3672$	1.5–120	0.9945
117	2,4,4′,6-Tetrachlorobiphenyl	$y = 0.7469x − 0.2298$	1.5–120	0.9978
118	2,2′,3,6′-Tetrachlorobiphenyl	$y = 1.1978x − 0.5608$	1.5–120	0.9969
119	2,2′,4,4′-Tetrachlorobiphenyl	$y = 1.9345x − 0.7181$	1.5–120	0.9983
120	2,2′,3,5′-Tetrachlorobiphenyl	$y = 0.3093x − 0.0989$	1.5–120	0.9941
121	2,2′,3,4′-Tetrachlorobiphenyl	$y = 0.4026x − 0.1862$	1.5–120	0.9923
122	2,3′,4′,6-Tetrachlorobiphenyl	$y = 0.1522x − 0.0493$	1.5–120	0.9988
123	2,2′,3,6,6′-Pentachlorobiphenyl	$y = 0.8585x − 0.2699$	1.5–120	0.9995
124	2,2′,3,4,6-Pentachlorobiphenyl	$y = 0.3834x − 0.2149$	1.5–120	0.9902
125	2,2′,3,5,6′-Pentachlorobiphenyl	$y = 0.5321x − 0.2199$	1.5–120	0.9983
126	2,2′,3′,4,6-Pentachlorobiphenyl	$y = 0.5715x − 0.3194$	1.5–120	0.9990
127	2′,3,4,5-Tetrachlorobiphenyl	$y = 0.6350x − 0.3119$	1.5–120	0.9989
128	2,2′,3,4′,6-Pentachlorobiphenyl	$y = 0.3782x − 0.1951$	1.5–120	0.9955
129	2,2′,4,5,5′-Pentachlorobiphenyl	$y = 0.4283x − 0.1392$	1.5–120	0.9959

Continued

130	2,3,3',4'-Tetrachlorobiphenyl	$y=1.4684x-0.4890$	1.5–120	0.9973
131	2,3',4,4',6-Pentachlorobiphenyl	$y=0.5558x-0.6791$	2.5–200	0.9992
132	3,3',4,5'-Tetrachlorobiphenyl	$y=0.6847x-0.4840$	1.5–120	0.9916
133	2,3,4,5,6-Pentachlorobiphenyl	$y=0.4426x-0.3617$	1.5–120	0.9936
134	2,3,4',5,6-Pentachlorobiphenyl	$y=0.3923x-0.2829$	1.5–120	0.9970
135	3,3',4,5-Tetrachlorobiphenyl	$y=0.6307x-0.4254$	1.5–120	0.9968
136	2,2',3,3',6,6'-Hexachlorobiphenyl	$y=0.4211x-0.4115$	1.5–120	0.9956
137	2,2',3,4,5',6-Hexachlorobiphenyl	$y=0.6040x-1.2141$	1.5–120	0.9695
138	3,3',4,4'-Tetrachlorobiphenyl	$y=1.1559x-1.7888$	1.5–120	0.9868
139	2,2',3,4',5',6-Hexachlorobiphenyl	$y=0.4670x-0.4010$	1.5–120	0.9908
140	2,2',3,4,5,6,6'-Heptachlorobiphenyl	$y=0.3181x-0.3664$	1.5–120	0.9884
141	2,2',3,3',5,5'-Hexachlorobiphenyl	$y=0.3672x-0.2899$	1.5–120	0.9952
142	2,3,3',5,5',6-Hexachlorobiphenyl	$y=0.1395x-0.0626$	1.5–120	0.9963
143	2,3,3',4,5',6-Hexachlorobiphenyl	$y=0.5669x-0.7815$	1.5–120	0.9906
144	2,2',3,3',4,6'-Hexachlorobiphenyl	$y=0.6102x-0.5038$	1.5–120	0.9904
145	2,3,3',4,4'-Pentachlorobiphenyl	$y=0.5021x-0.3422$	1.5–120	0.9939
146	2,2',3,4,5,6,6'-Heptachlorobiphenyl	$y=0.3276x-0.0352$	1.5–120	0.9996
147	2,3,3',4,4',6-Hexachlorobiphenyl	$y=0.2539x-0.0976$	1.5–120	0.9987
148	2,2',3,4,4',5,6'-Heptachlorobiphenyl	$y=0.3004x-0.1752$	1.5–120	0.9982

TABLE 5.14 Linear Equations, Linear Ranges, and Correlation Coefficients of 284 Environmental Pollutants Determined by GC-MS-MS—cont'd

No.	Compound	Linear Equation	Linear Range	Correlation Coefficient
149	2,3,3′,4,5,5′-Hexachlorobiphenyl	$y=0.6772x-0.4259$	1.5–120	0.9975
150	2,3′,4,4′,5,5′-Hexachlorobiphenyl	$y=0.2552x-0.4003$	2.5–200	0.9961
151	2,2′,3,3′,4,5,6′-Heptachlorobiphenyl	$y=0.3061x-0.3686$	1.5–120	0.9933
152	2,2′,3,3′,4′,5,6-Heptachlorobiphenyl	$y=0.6042x-0.3641$	1.5–120	0.9936
153	2,3,3′,4,4′,5-Hexachlorobiphenyl	$y=0.5292x-0.084$	1.5–120	0.9979
154	2,2′,3,4,4′,5,5′-Heptachlorobiphenyl	$y=0.4919x-0.1138$	1.5–120	0.9983
155	2,2′,3,3′,4,5,5′,6-Octachlorobiphenyl	$y=0.4285x-0.2733$	1.5–120	0.9987
156	2,2′,3,3′,4,4′,5,6′-Octachlorobiphenyl	$y=0.2514x-0.1195$	1.5–120	0.9965
157	2,2′,3,3′,4,4′,5,6,6′-Nonachlorobiphenyl	$y=1.2273x-0.6215$	1.5–120	0.9965
158	2,3,3′,4,4′,5,5′,6-Octachlorobiphenyl	$y=0.7079x-0.2265$	1.5–120	0.9990
Group D				
159	2,6-Dichlorobiphenyl	$y=0.5098x+0.6470$	1.5–120	0.9948
160	2,3′-Dichlorobiphenyl	$y=0.5777x+0.7616$	2.5–200	0.9949
161	2,4,6-Trichlorobiphenyl	$y=0.9428x-0.6238$	1.5–120	0.9988
162	2,2′,4-Trichlorobiphenyl	$y=0.6599x-0.0527$	1.5–120	0.9994
163	4,4′-Dichlorobiphenyl	$y=2.6365x-0.7447$	1.5–120	0.9992

Continued

164	2′,3,5-Trichlorobiphenyl	$y=1.1703x-0.3340$	1.5–120	0.9994
165	2,3′,5-Trichlorobiphenyl	$y=1.1810x-0.4684$	1.5–120	0.9996
166	2,4,4′-Trichlorobiphenyl	$y=0.4340x-0.0469$	1.5–120	0.9989
167	2,2′,4,6′-Tetrachlorobiphenyl	$y=0.3543x-0.1073$	1.5–120	0.9991
168	2,2′,3,6-Tetrachlorobiphenyl	$y=0.4630x-0.1403$	1.5–120	0.9995
169	2,2′,5,5′-Tetrachlorobiphenyl	$y=0.5072x-0.2072$	1.5–120	0.9984
170	2,2′,4,5′-Tetrachlorobiphenyl	$y=0.5634x-0.2119$	1.5–120	0.9996
171	2,2′,4,5-Tetrachlorobiphenyl	$y=0.5424x-0.1651$	1.5–120	0.9999
172	2,3,3′,6-Tetrachlorobiphenyl	$y=0.5976x-0.0754$	1.5–120	0.9992
173	2,3′,4,5′-Tetrachlorobiphenyl	$y=0.4237x-0.1662$	1.5–120	0.9984
174	2,2′,4,4′,6-Pentachlorobiphenyl	$y=0.4722x+0.0052$	1.5–120	0.9977
175	2,3,3′,5-Tetrachlorobiphenyl	$y=0.6239x+0.1033$	1.5–120	0.9966
176	2,3,4′,5-Tetrachlorobiphenyl	$y=0.8219x+0.2707$	1.5–120	0.9974
177	2,3,4,5-Tetrachlorobiphenyl	$y=0.5322x-0.2370$	1.5–120	0.9990
178	2,2′,4,4′,6,6′-Hexachlorobiphenyl	$y=0.8670x-0.2959$	1.5–120	0.9980
179	2,3′,4′,5-Tetrachlorobiphenyl	$y=0.6048x+0.0786$	1.5–120	0.9992
180	2,3,3′,4-Tetrachlorobiphenyl	$y=0.8607x-0.3357$	1.5–120	0.9990
181	2,3,4,4′-Tetrachlorobiphenyl	$y=0.5273x-0.0476$	1.5–120	0.9980
182	2,2′,3,4′,6,6′-Hexachlorobiphenyl	$y=0.1093x+0.0513$	1.5–120	0.9966

TABLE 5.14 Linear Equations, Linear Ranges, and Correlation Coefficients of 284 Environmental Pollutants Determined by GC-MS-MS—cont'd

No.	Compound	Linear Equation	Linear Range	Correlation Coefficient
183	2,3,3',5,6-Pentachlorobiphenyl	$y = 0.8518x - 0.2858$	1.5–120	0.9993
184	2,2',3,4',5,6'-Hexachlorobiphenyl	$y = 0.3181x + 0.0355$	1.5–120	0.9968
185	2,3,3',5,5'-Pentachlorobiphenyl	$y = 0.8594x - 0.4715$	1.5–120	0.9991
186	2,2',3',4,5-Pentachlorobiphenyl	$y = 0.3223x + 0.3961$	1.5–120	0.9791
187	2,3',4,5,5'-Pentachlorobiphenyl	$y = 0.3360x - 0.1284$	1.5–120	0.9991
188	3,4,4',5-Tetrachlorobiphenyl	$y = 0.7896x - 0.2314$	1.5–120	0.9825
189	2,2',3,4',5,6-Hexachlorobiphenyl	$y = 0.2069x - 0.1182$	1.5–120	0.9991
190	2,2',3,3',4-Pentachlorobiphenyl	$y = 0.2137x + 0.0537$	1.5–120	0.9880
191	2,3,3',4,5'-Pentachlorobiphenyl	$y = 0.4724x - 0.2849$	1.5–120	0.9998
192	2,3,3',4,5-Pentachlorobiphenyl	$y = 0.2961x - 0.0953$	2.5–200	0.9984
193	2,2',3,4,4',6,6'-Heptachlorobiphenyl	$y = 0.5236x - 0.1083$	1.5–120	0.9983
194	2,2',3,3',4,6-Hexachlorobiphenyl	$y = 0.5262x - 0.3005$	1.5–120	0.9994
195	2,2',4,4',5,5'-Hexachlorobiphenyl	$y = 0.2334x - 0.1276$	2.5–200	0.9945
196	2,2',3,3',5,6,6'-Heptachlorobiphenyl	$y = 0.3483x - 0.0831$	1.5–120	0.9911
197	2,2',3,3',4,6,6'-Heptachlorobiphenyl	$y = 0.2166x + 0.0822$	1.5–120	0.9965
198	2,3,3',4,5,6-Hexachlorobiphenyl	$y = 0.7255x - 0.2075$	1.5–120	0.9973

Continued

199	2,3,3',4',5',6-Hexachlorobiphenyl	$y=0.6470x-0.1528$	1.5–120	0.9994
200	2,2',3,3',4,5-Hexachlorobiphenyl	$y=0.4375x-0.1344$	1.5–120	0.9991
201	3,3',4,4',5-Pentachlorobiphenyl	$y=0.4254x-0.0264$	1.5–120	0.9979
202	2,2',3,4,5,5',6-Heptachlorobiphenyl	$y=0.2588x+0.0625$	1.5–120	0.9998
203	2,2',3,4,4',5,6-Heptachlorobiphenyl	$y=0.6657x-0.2639$	1.5–120	0.9990
204	2,2',3,3',4,4',6-Heptachlorobiphenyl	$y=0.2589x+0.0282$	1.5–120	0.9996
205	2,2',3,3',4,5,5'-Heptachlorobiphenyl	$y=0.5493x-0.2220$	1.5–120	0.9993
206	2,2',3,3',4,5,6,6'-Octachlorobiphenyl	$y=0.2126x+0.0138$	1.5–120	0.9985
207	2,2',3,3',4,5,5',6'-Octachlorobiphenyl	$y=0.2837x-0.0349$	1.5–120	0.9992
208	2,2',3,3',4,4',5-Heptachlorobiphenyl	$y=0.2592x+0.0453$	1.5–120	0.9983
209	2,3,3',4,4',5,5'-Heptachlorobiphenyl	$y=0.8290x-0.0279$	1.5–120	0.9997
Group E				
210	Naphthalene	$y=0.1955x-0.5254$	1.0–95.6	0.9956
211	Isoprotuton	$y=0.1309x+39.3624$	126.7–10135.2	0.9920
212	Dichlorvos	$y=0.0816x-0.9668$	5.9–588.5	0.9876
213	Carbofuran	$y=0.1310x+0.1248$	4.3–432.0	0.9986
214	Methamidophos	$y=0.0240x-7.7487$	85.4–8541.5	0.9684
215	Acenaphthylene	$y=0.1953x+1.2657$	2.2–224.7	0.9875
216	Acenaphthene	$y=0.0593x+0.0296$	4.6–455.4	0.9957

TABLE 5.14 Linear Equations, Linear Ranges, and Correlation Coefficients of 284 Environmental Pollutants Determined by GC-MS-MS—cont'd

No.	Compound	Linear Equation	Linear Range	Correlation Coefficient
217	Fluorene	$y=0.7383x-3.5999$	6.6–658.8	0.9970
218	Hexachlorobenzene	$y=0.2330x+0.0854$	2.9–288.8	0.9982
219	Ethoprophos	$y=0.1949x-0.0944$	1.3–129.9	0.9967
220	Chlordimeform	$y=0.3243x+0.0058$	5.8–580.0	0.9969
221	Trifluralin	$y=0.2438x-1.6645$	4.5–451.2	0.9942
222	α-HCH	$y=0.1961x-1.0487$	4.4–437.0	0.9944
223	Omethoate	$y=0.0198x+0.7319$	18.2–1094.2	0.9969
224	Anthracene	$y=20.9615x-1.1090$	0.1–14.7	1.0000
225	Clomazone	$y=0.2356x-0.2193$	2.9–288.6	0.9969
226	Diazinon	$y=0.3229x-0.2373$	2.7–270.3	0.9983
227	Phenathrene	$y=0.1750x-0.6505$	5.3–525.0	0.9998
228	γ-HCH	$y=0.1740x-1.0058$	4.3–431.8	0.9932
229	Atrazine	$y=0.2420x-0.0580$	4.0–396.4	0.9975
230	Simazine	$y=0.0621x-0.2475$	11.8–1177.2	0.9975
231	Heptachlor	$y=0.1095x-0.8773$	6.5–649.8	0.9960
232	Pirimicarb	$y=0.4814x-1.0088$	3.1–306.0	0.9962

233	Dimethoate	$y=0.0286x-0.6293$	6.5–654.9	0.9767
234	Aldrin	$y=0.0467x+0.0240$	4.6–455.4	0.9983
235	Alachlor	$y=0.1279x-0.3246$	2.9–289.1	0.9949
236	Prometryne	$y=0.4000x-0.1287$	2.7–271.9	0.9985
237	Chlorothalonil	$y=0.0160x-180.1627$	13977.6–23296.0	0.9957
238	Phthalic acid bis-butyl ester	$y=7.3006x-52.6826$	6.6–664.4	0.9986
239	β-HCH	$y=0.1844x-0.9595$	4.5–448.0	0.9934
240	Chlorpyrifos	$y=0.1324x-0.5356$	4.5–451.5	0.9960
241	Parathion-methyl	$y=0.1515x-1.3505$	4.3–430.9	0.9901
242	Dicofol	$y=0.2491x-0.1258$	3.4–336.4	0.9986
243	Metolachlor	$y=1.3801x-1.2331$	1.1–108.0	0.9962
244	δ-HCH	$y=0.0680x-0.7771$	4.3–434.0	0.9862
245	Triadimefon	$y=0.1295x-0.3226$	4.8–481.8	0.9972
246	Fluoranthene	$y=0.1309x-0.4220$	3.0–302.6	0.9989
247	2,4′-DDE	$y=0.5787x+0.2554$	6.7–666.0	0.9985
248	cis-Chlordane	$y=0.0595x-0.1824$	6.3–625.1	0.9983
249	Phenthoate	$y=2.1028x-0.8472$	0.4–43.4	0.9921
250	trans-Chlordane	$y=0.0639x-0.1086$	6.2–618.8	0.9975
251	Pyrene	$y=0.9615x-92.2317$	5.2–524.6	0.9917

Continued

TABLE 5.14 Linear Equations, Linear Ranges, and Correlation Coefficients of 284 Environmental Pollutants Determined by GC-MS-MS—cont'd

No.	Compound	Linear Equation	Linear Range	Correlation Coefficient
252	4,4'-dde	$y=0.5374x+1.0065$	6.6–657.2	0.9990
253	Butachlor	$y=0.1201x+0.2435$	0.7–74.4	0.9900
254	Dieldrin	$y=0.0244x+0.0047$	13.2–1315.6	0.9984
255	2,4'-DDD	$y=1.1094x+1.8793$	2.7–268.8	0.9994
256	Buprofezin	$y=0.2211x+3.1681$	14.5–1452.8	0.9959
257	Endrin	$y=0.0156x+0.1034$	30–2998.6	0.9991
258	2,4'-DDT	$y=0.5975x-1.2217$	2.1–208.3	0.9955
259	Nithophen	$y=0.0629x-1.3208$	8.1–809.2	0.9875
260	Oxyfluorfen	$y=0.0775x-1.6855$	14.6–1458	0.9954
261	4,4'-DDD	$y=1.0664x-2.5332$	6.6–664.2	0.9978
262	4,4'-DDT	$y=0.4644x-5.2499$	3.0–295.0	0.9709
263	Phthalic acid benzyl butyl ester	$y=0.8329x-2.9928$	6.6–656.0	0.9984
264	Propargite	$y=0.0047x-1.1361$	127.5–12750.0	0.9958
265	Tricyclazole	$y=1.3515x-2.0591$	0.7–73.8	0.9879
266	Triazophos	$y=0.0074x-0.0245$	11.4–1142.6	0.9987
267	Mirex	$y=0.4583x-0.8473$	2.8–275.0	0.9976

268	Benzo(a)anthrancene	$y=1.0534x-0.3460$	2.0–200.0	0.9999
269	Phthalic acid bis-2-ethylhexyl ester	$y=1.7761x+16.1890$	1.3–133.2	1.0000
270	Amitraz	$y=0.0066x+0.0371$	1.6–155.9	0.9704
271	Lamba-cyhalothrin	$y=0.1814x-0.9969$	6.4–638.4	0.9956
272	Pyridaben	$y=0.8733x-5.9209$	6.7–667.0	0.9954
273	Benzo(b)fluoranthene	$y=8.4098x+0.3645$	0.1–5.4	0.9952
274	Benzo(k)fluoranthene	$y=0.3051x+0.0760$	0.2–21.8	0.9962
275	Cyfluthrin	$y=0.0272x-0.3225$	8.7–873.2	0.9801
276	Cypermethrin	$y=0.1491x-4.4244$	18.5–1854.6	0.9939
277	Benzo(a)pyrene	$y=0.8605x+1.5570$	1.8–175.5	1.0000
278	Acetamiprid	$y=0.0487x-2.0732$	27.3–2728	0.9893
279	Fenvalerate-1	$y=0.0832x-2.4583$	29.0–2904.0	0.9966
280	Fenvalerate-2	$y=0.0535x-2.4768$	29.0–2904.0	0.9949
281	Deltamethrin	$y=0.0314x-2.5842$	24.2–2418.2	0.9767
282	Indeno(1,2,3-cd)pyrene	$y=1.2713x+1.5120$	2.2–220.4	0.9992
283	Dibenzo(a,h)anthracene	$y=1.2873x-0.1164$	2.2–220.4	0.9815
284	Benzo(g,h,i)peryene	$y=0.0709x-0.1860$	2.2–220.0	0.9990

5.2.4 Linear Equations, Linear Ranges, and Correlation Coefficients of Endosulfans Determined by GC-MS(NCI)

See Table 5.15.

TABLE 5.15 Linear Equations, Linear Ranges, and Correlation Coefficients of Endosulfans Determined by GC-MS(NCI)

No.	Compound	Linear Equation	Linear Range	Correlation Coefficient
1	Endosulfan I	$y=2022.5831x +486.7318$	0.5–30.0	0.9973
2	Endosulfan II	$y=822.2465x +14.4991$	0.5–30.0	0.9995

5.2.5 Linear Equations, Linear Ranges, and Correlation Coefficients of 9 Environmental Pollutants Determined by LC-MS-MS

See Table 5.16.

TABLE 5.16 Linear Equations, Linear Ranges, and Correlation Coefficients of 9 Environmental Pollutants Determined by LC-MS-MS

No.	Compound	Linear Equation	Linear Range	Correlation Coefficient
1.	Trichlorphon	$y=1703.5808x-42.5213$	3.9–38.8	0.9991
2.	Metsulfuron-methyl	$y=45166.5928x+27852.7660$	2.3–23.0	0.9918
3.	Chlorolurons	$y=16574.7308x-12467.6596$	2.6–25.9	0.9989
4.	2,4-D	$y=940.7392x-9211.9149$	12.5–124.8	0.9939
5.	Bensulfuron-methyl	$y=45466.5694x-14917.6596$	0.6–6.0	0.9891
6.	Propanil	$y=4618.9752x+15965.5319$	4.7–47.0	0.9883
7.	Fipronil	$y=118.3954x+206.6277$	26.4–264.0	0.9962
8.	Phoxim	$y=180.0994x+275.5957$	6.6–66.2	0.9842
9.	Hexythiazox	$y=1377.9394x-19979.0476$	13–129.6	0.9874

5.2.6 Linear Equations, Linear Ranges, and Correlation Coefficients of 569 Pesticides and Chemical Pollutants Determined by LC-MS-MS

See Table 5.17.

TABLE 5.17 Linear Equations, Linear Ranges, and Correlation Coefficients of 569 Pesticides and Chemical Pollutants Determined by LC-MS-MS

No.	Compound	CAS	Linear Equation	Linear Range	Correlation Coefficient
1	Propham	122-42-9	$y = 4535.5894x + 65449.4444$	1.1000–110.0000	0.9974
2	Isoprocarb	2631-40-7	$y = 1112.6096x - 14289.0147$	0.0230–2.3000	0.9988
3	3,4,5-Trimethacarb	2686-99-9	$y = 1112.6096x - 14289.0147$	0.0034–0.3440	0.9988
4	Cycluron	2163-69-1	$y = 608.3015x + 1384.0461$	0.0021–0.2060	0.9994
5	Carbaryl	63-25-2	$y = 1526.8781x - 59353.6374$	0.1032–10.3200	0.9937
6	Propachlor	1918-16-7	$y = 626.1151x - 1686.3860$	0.0027–0.2740	0.9998
7	Rabenzazole	69899-24-7	$y = 1643.2936x - 509.7960$	0.0133–1.3320	0.9999
8	Simetryn	1014-70-6	$y = 376.7290x + 732.3834$	0.0014–0.1358	0.9997
9	Monolinuron	1746-81-2	$y = 889.4370x + 5056.1256$	0.0356–3.5600	0.9994
10	Mevinphos	7786-34-7	$y = 712.8533x + 1007.6465$	0.0157–1.5660	0.9995
11	Aziprotryne	4658-28-0	$y = 411.0509x - 4648.7270$	0.0138–1.3820	0.9965

12	Secbumeton	26259-45-0	$y=208.4258x-780.4229$	0.0007–0.0724	0.9998
13	Cyprodinil	121552-61-2	$y=545.2102x-3736.8089$	0.0074–0.7394	0.9997
14	Buturon	3766-60-7	$y=1644.4632x+46887.9697$	0.0896–8.9600	0.9913
15	Carbetamide	16118-49-3	$y=1496.3266x+12576.6376$	0.0364–3.6400	0.9975
16	Pirimicarb	23103-98-2	$y=254.9760x-200.5094$	0.0015–0.1514	0.9998
17	Clomazone	81777-89-1	$y=487.9864x+7994.2981$	0.0042–0.4220	0.9961
18	Cyanazine	21725-46-2	$y=159.3072x-630.6242$	0.0016–0.1638	0.9998
19	Prometryne	7287-19-6	$y=1435.3235x+2788.6189$	0.0016–0.1622	0.9992
20	Paraoxon methyl	950-35-6	$y=326.3815x-950.1759$	0.0076–0.7620	1.0000
21	4,4-Dichlorobenzophenone	90-98-2	$y=12.8846x+157.9041$	0.1360–13.6000	0.9940
22	Thiacloprid	111988-49-9	$y=228.2523x-976.6060$	0.0037–0.3700	0.9999
23	Imidacloprid	138261-41-3	$y=6640.2083x+289647.9396$	0.2200–22.0000	0.9957
24	Ethidimuron	30043-49-3	$y=1770.0948x+35827.9158$	0.0150–1.5000	0.9952
25	Isomethiozin	57052-04-7	$y=2900.6247x-30878.1585$	0.0107–1.0660	0.9988
26	cis- and trans-diallate	2303-16-4	$y=4752.8585x+144852.5654$	0.8920–89.2000	0.9952
27	Acetochlor	34256-82-1	$y=2.5160x+145.1775$	0.4740–47.4000	0.9966
28	Nitenpyram	150824-47-8	$y=1696.7831x+25328.4039$	0.1712–17.1200	0.9949
29	Methoprotryne	841-06-5	$y=564.6352x-1939.4407$	0.0024–0.2420	0.9999
30	Dimethenamid	87674-68-8	$y=6634.8590x+83016.8270$	0.0430–4.3008	0.9980

Continued

TABLE 5.17 Linear Equations, Linear Ranges, and Correlation Coefficients of 569 Pesticides and Chemical Pollutants Determined by LC-MS-MS—cont'd

No.	Compound	CAS	Linear Equation	Linear Range	Correlation Coefficient
31	Terbucarb	1918-11-2	$y=673.2173x+5364.1022$	0.0210–2.1000	0.9996
32	Penconazole	66246-88-6	$y=8047.2156x+53218.1036$	0.0200–2.0000	0.9996
33	Myclobutanil	88761-89-0	$y=1021.1599x+7654.9321$	0.0100–0.9960	0.9991
34	Imazethapyr	81385-77-5	$y=563.4160x+5224.5083$	0.0113–1.1260	0.9966
35	Paclobutrazol	76738-62-0	$y=1927.8071x-1979.4839$	0.0057–0.5740	0.9999
36	Fenthion sulfoxide	3761-41-9	$y=346.2671x-1334.5701$	0.0031–0.3136	0.9997
37	Triadimenol	55219-65-3	$y=7283.9170x+25541.2584$	0.1055–10.5536	0.9997
38	Butralin	33629-47-9	$y=2378.0007x+3348.4645$	0.0190–1.9000	1.0000
39	Spiroxamine	118134-30-8	$y=247.8823x-301.6067$	0.0005–0.0516	0.9997
40	Tolclofos methyl	57018-04-9	$y=1675.4622x+32850.6560$	0.6656–66.5600	0.9947
41	Desmedipham	13684-56-5	$y=-0.0024x+27.5721$	0.0403–4.0296	0.0276
42	Methidathion	950-37-8	$y=4449.6288x+70726.1570$	0.1066–10.6600	0.9947
43	Allethrin	584-79-2	$y=5878.3285x+146371.9578$	0.6040–60.4000	0.9941
44	Diazinon	333-41-5	$y=11650.2404x+3021.5872$	0.0071–0.7128	1.0000
45	Edifenphos	17109-49-8	$y=1316.5640x-20093.5375$	0.0075–0.7520	0.9951

46	Pretilachlor	51218-49-6	$y=18.4905x+175.4801$	0.0033–0.3340	0.9985
47	Flusilazole	85509-19-9	$y=1059.6021x-4263.8748$	0.0058–0.5814	0.9999
48	Iprovalicarb	140923-17-7	$y=5596.2306x-42.2949$	0.0232–2.3200	1.0000
49	Benodanil	15310-01-7	$y=554.4545x+716.5127$	0.0348–3.4800	0.9998
50	Flutolanil	66332-96-5	$y=1878.2114x+19116.7333$	0.0115–1.1460	0.9989
51	Famphur	52-85-7	$y=1898.4011x+11823.0282$	0.0360–3.6000	0.9996
52	Benalyxyl	71626-11-4	$y=1871.3401x+9394.4156$	0.0124–1.2426	0.9997
53	Diclobutrazole	75736-33-3	$y=13433.7294x+62867.5467$	0.0047–0.4680	0.9996
54	Etaconazole	1000290-09-5	$y=13433.7294x+62867.5467$	0.0178–1.7820	0.9996
55	Fenarimol	60168-88-9	$y=56.7374x-183.4200$	0.0061–0.6078	0.9997
56	Phthalic acid, dicyclobexyl ester	84-61-7	$y=0.0982x+65.3519$	0.0200–2.0000	0.7397
57	Tetramethirin	7696-12-0	$y=963.7357x-33225.4951$	0.0182–1.8200	0.9947
58	Dichlofluanid	1085-98-9	$y=0.2469x+110.8665$	0.0260–2.5999	0.7612
59	Cloquintocet mexyl	99607-70-2	$y=21105.5052x+311561.9375$	0.0188–1.8840	0.9986
60	Bitertanol	55179-31-2	$y=28096.2420x+810030.1746$	0.3340–33.4000	0.9948
61	Chlorprifos methyl	5598-13-0	$y=2301.0872x+44960.9758$	0.1600–16.0000	0.9949
62	Tepraloxydim	149979-41-9	$y=10468.1258x+182040.1498$	0.1220–12.2000	0.9978
63	Thiophanate methyl	23564-05-8	$y=9650.9738x-183448.4931$	0.2000–20.0000	0.9985
64	Azinphos ethyl	2642-71-9	$y=495.9303x+10476.6992$	1.0893–108.9280	0.9925

Continued

TABLE 5.17 Linear Equations, Linear Ranges, and Correlation Coefficients of 569 Pesticides and Chemical Pollutants Determined by LC-MS-MS—cont'd

No.	Compound	CAS	Linear Equation	Linear Range	Correlation Coefficient
65	Clodinafop propargyl	105512-06-9	$y = 18.9675x - 156.8689$	0.0244–2.4400	0.9949
66	Triflumuron	64628-44-0	$y = 1523.8721x + 2286.2687$	0.0392–3.9200	0.9999
67	Isoxaflutole	141112-29-0	$y = 1626.5635x - 38636.6348$	0.0390–3.9000	0.9975
68	Anilofos	64249-01-0	$y = 0.0827x + 90.7744$	0.0071–0.7140	0.7724
69	Thiophanat ethyl	23564-06-9	$y = 23559.9711x - 537906.3864$	0.2016–20.1600	0.9944
70	Quizalofop-ethyl	76578-14-8	$y = 478.9599x - 857.9577$	0.0068–0.6820	1.0000
71	Haloxyfop-methyl	69806-40-2	$y = 3788.3897x + 14861.3609$	0.0264–2.6400	0.9998
72	Fluazifop butyl	79241-46-6	$y = 3253.5705x - 2892.7464$	0.0026–0.2632	1.0000
73	Bromophos-ethyl	4824-78-6	$y = 408.6550x + 37887.3314$	5.6769–567.6912	0.9941
74	Bensulide	741-58-2	$y = 3183.4165x + 110541.2007$	0.3420–34.2000	0.9903
75	Triasulfuron	82097-50-5	$y = 611.6555x - 3171.1763$	0.0161–1.6089	0.9999
76	Bromfenvinfos	33399-00-7	$y = 1098.3258x + 18155.8502$	0.0302–3.0200	0.9963
77	Azoxystrobin	131860-33-8	$y = 1569.5759x - 3665.8540$	0.0045–0.4510	0.9999
78	Pyrazophos	13457-18-6	$y = 299.0403x + 2469.4351$	0.0162–1.6240	0.9991
79	Bensultap	17606-31-4		0.1428–14.2819	

No.	Name	CAS	Equation	Range	Coefficient
80	Flufenoxuron	101463-69-8	$y=808.3584x+7481.8150$	0.0317–3.1680	0.9978
81	Indoxacarb	12124-97-9	$y=1904.2440x-1532.5821$	0.0754–7.5400	0.9996
82	Emamectin benzoate	15569-91-8	$y=0.5258x+68.6057$	0.0032–0.3200	0.7457
83	Ethylene thiourea	96-45-7	$y=162.2627x+2386.2227$	0.5220–52.2000	0.9904
84	Daminozide	1596-84-5	$y=309.0319x+3823.9238$	0.0260–2.6000	0.9906
85	Dazomet	533-74-4	$y=1288.0566x+78103.3182$	1.2700–127.0000	0.9834
86	Nicotine	54-11-5	$y=454.5627x+10836.6213$	0.0220–2.2000	0.9980
87	Fenuron	101-42-8	$y=660.9790x+5338.2321$	0.0103–1.0300	0.9989
88	Cyromazine	66215-27-8	$y=563.2353x+10321.9581$	0.0724–7.2400	0.9836
89	Crimidine	535-89-7	$y=821.8290x-1452.1315$	0.0156–1.5580	0.9994
90	Acephate	30560-19-1	$y=7441.6707x+51856.2500$	0.1334–13.3400	0.9916
91	Molinate	2212-67-1	$y=715.5106x+66.1993$	0.0210–2.1000	0.9994
92	Carbendazim	10605-21-7	$y=425.4532x+1180.1779$	0.0047–0.4680	0.9950
93	6-Chloro-4-hydroxy-3-phenyl-pyridazin	40020-01-7	$y=60.1266x+701.4763$	0.0165–1.6540	0.9484
94	Propoxur	114-26-1	$y=20531.8166x+733510.9205$	0.2440–24.4000	0.9925
95	Isouron	55861-78-4	$y=347.9558x+3097.9143$	0.0041–0.4080	0.9983
96	Chlorotoluron	15545-48-9	$y=768.6574x+1232.8851$	0.0062–0.6240	0.9988
97	Thiofanox	39196-18-4	$y=2810.4270x+64595.5362$	1.5700–157.0000	0.9958

Continued

TABLE 5.17 Linear Equations, Linear Ranges, and Correlation Coefficients of 569 Pesticides and Chemical Pollutants Determined by LC–MS–MS—cont'd

No.	Compound	CAS	Linear Equation	Linear Range	Correlation Coefficient
98	Chlorbufam	1967-16-4	$y=219.5618x+5296.1947$	1.8300–183.0000	0.9941
99	Bendiocarb	22781-23-3	$y=950.7442x-12639.9461$	0.0318–3.1800	0.9801
100	Propazine	139-40-2	$y=41.9632x-247.5223$	0.0032–0.3200	0.9998
101	Terbuthylazine	5915-41-3	$y=741.5811x-7308.1745$	0.0047–0.4680	0.9999
102	Diuron	330-54-1	$y=605.2551x+8011.6022$	0.0156–1.5600	0.9919
103	Chlormephos	24934-91-6	$y=37.3290x-76.7196$	195.4000–19540.0000	0.9989
104	Carboxin	5234-68-4	$y=247.0262x+4764.3921$	0.0056–0.5560	0.9919
105	Difenzoquat-methyl sulfate	43222-48-6	$y=11.0654x+584.0573$	0.0081–0.8120	0.9747
106	Clothianidin	205510-53-8	$y=11603.0067x+184321.5874$	0.6300–63.0000	0.9932
107	Pronamide	23950-58-5	$y=5302.6716x+62927.9119$	0.1538–15.3800	0.9978
108	Dimethachloro	51218-45-2	$y=4667.1674x+70752.1903$	0.0190–1.9020	0.9950
109	Methobromuron	40596-69-8	$y=1104.1120x+29347.8000$	0.1684–16.8400	0.9938
110	Phorate	298-02-2	$y=2110.1059x+12606.8279$	3.1400–314.0000	0.9987
111	Aclonifen	74070-46-5	$y=1052.0716x+3127.7995$	0.2420–24.2000	0.9988
112	Mephosfolan	950-10-7	$y=2876.1983x+27957.4559$	0.0232–2.3200	0.9981

Continued

No.	Name	CAS	Equation	Range	R
113	Imibenzonazole-des-benzyl	199338-48-2	$y=233.1917x-1698.1113$	0.0622–6.2200	0.9993
114	Neburon	555-37-3	$y=1947.7316x+28299.1297$	0.0710–7.1000	0.9966
115	Mefenoxam	70630-17-0	$y=1975.7678x+25897.7499$	0.0154–1.5380	0.9955
116	Prothoate	2275-18-5	$y=-0.0250x+107.5127$	0.0246–2.4600	0.0248
117	Ethofume sate	26225-79-6	$y=5244.4362x+104131.4108$	3.7200–372.0000	0.9969
118	Iprobenfos	26087-47-8	$y=16837.2026x+385414.1554$	0.0828–8.2800	0.9947
119	TEPP	107-49-3	$y=516.8062x-2634.6870$	0.1040–10.4000	0.9999
120	Cyproconazole	113096-99-4	$y=643.4521x-4113.2852$	0.0073–0.7320	0.9995
121	Thiamethoxam	153719-23-4	$y=1739.2909x+28588.6451$	0.3300–33.0000	0.9959
122	Crufomate	299-86-5	$y=235.7725x-1143.2909$	0.0052–0.5180	0.9995
123	Etrimfos	38260-54-7	$y=23.3423x-177.7821$	0.1876–18.7600	0.9995
124	Coumatetralyl	5836-29-3	$y=0.0698x+81.8408$	0.0135–1.3520	0.2086
125	Cythioate	115-93-5	$y=1824.7729x+25930.7626$	0.8000–80.0000	0.9970
126	Phosphamidon	13171-21-6	$y=3107.6717x+18670.3634$	0.0388–3.8800	0.9986
127	Phenmedipham	13864-63-4	$y=0.0407x+17.6174$	0.0448–4.4800	0.6849
128	Bifenazate	149877-41-8	$y=5112.8775x+91081.7424$	0.2280–22.8000	0.9947
129	Fenhexamid	126833-17-8	$y=0.9840x+107.7399$	0.0095–0.9460	0.9694
130	Flutriafol	76674-21-0	$y=10798.2234x+101446.8458$	0.0858–8.5800	0.9983
131	Furalaxyl	57646-30-7	$y=10095.4787x+218561.3048$	0.0077–0.7700	0.9936

TABLE 5.17 Linear Equations, Linear Ranges, and Correlation Coefficients of 569 Pesticides and Chemical Pollutants Determined by LC-MS-MS—cont'd

No.	Compound	CAS	Linear Equation	Linear Range	Correlation Coefficient
132	Bioallethrin	584-79-2	$y=4863.0855x+397587.6721$	1.9800–198.0000	0.9652
133	Cyanofenphos	13067-93-1	$y=5251.2562x+112801.0922$	0.2080–20.8000	0.9941
134	Pirimiphos methyl	29232-93-7	$y=918.5729x-4161.3649$	0.0020–0.2020	0.9999
135	Buprofezin	69327-76-0	$y=1929.5826x-926.8668$	0.0088–0.8780	0.9994
136	Disulfoton sulfone	2497-06-5	$y=2108.3611x+24849.6512$	0.0246–2.4600	0.9978
137	Fenazaquin	120928-09-8	$y=42535.1395x+653348.2038$	0.0032–0.3240	0.9909
138	Triazophos	24017-47-8	$y=969.8376x-4019.1877$	0.0068–0.6800	0.9998
139	DEF	78-48-8	$y=1770.8003x-2096.7313$	0.0161–1.6140	0.9994
140	Pyriftalid	135186-78-6	$y=526.8462x-237.2240$	0.0062–0.6240	0.9997
141	Metconazole	125116-23-6	$y=927.5589x-7053.2677$	0.0132–1.3180	0.9996
142	Pyriproxyfen	95737-68-1	$y=1662.7259x+61015.3528$	0.0043–0.4300	0.9904
143	Cycloxydim	101205-02-1	$y=941.0129x-9065.1397$	0.0254–2.5400	0.9999
144	Isoxaben	82558-50-7	$y=491.9346x+8078.7678$	0.0019–0.1860	0.9969
145	Flurtamone	96525-23-4	$y=593.3652x-1122.0940$	0.0044–0.4440	0.9996
146	Trifluralin	1582-09-8	$y=27.5179x+729.2464$	12.4000–1240.0000	0.9958

	Name	CAS	Equation	Range	R
147	Flamprop methyl	52756-25-9	$y=23678.0875x+800578.1693$	0.2020–20.2000	0.9919
148	Bioresmethrin	28434-01-7	$y=1403.9813x+28065.8045$	0.0742–7.4200	0.9950
149	Propiconazole	60207-90-1	$y=2664.8397x-19906.5546$	0.0176–1.7580	1.0000
150	Chlorpyrifos	2921-88-2	$y=5894.1313x+206296.7456$	0.5380–53.8000	0.9792
151	Fluchloralin	33245-39-5	$y=311.8411x+8518.8396$	4.8800–488.0000	0.9922
152	Chlorsulfuron	64902-72-3	$y=39.1361x-497.4055$	0.0274–2.7400	0.9966
153	Clethodim	99129-21-2	$y=830.2219x-21325.9773$	0.0208–2.0800	0.9970
154	Flamprop isopropyl	52756-22-6	$y=247.2410x-1177.0420$	0.0043–0.4340	0.9999
155	Tetrachlorvinphos	22248-79-9	$y=539.9792x+197.7294$	0.0222–2.2200	0.9998
156	Propargite	2312-35-8	$y=5072.8013x+96338.0050$	0.6860–68.6000	0.9909
157	Bromuconazole	116255-48-2	$y=22.4850x-21.6017$	0.0314–3.1400	0.9936
158	Picolinafen	137641-05-5	$y=961.6737x-36816.8027$	0.0073–0.7260	0.9901
159	Fluthiacet methyl	117337-19-6	$y=183.3926x-2207.7119$	0.0530–5.3000	0.9991
160	Trifloxystrobin	141517-21-7	$y=5285.1909x+30372.2004$	0.0200–2.0000	0.9991
161	Chlorimuron ethyl	90982-32-4	$y=41.7499x-348.4110$	0.3040–30.4000	0.9995
162	Hexaflumuron	86479-06-3	$y=1448.3638x+5274.8438$	0.2520–25.2000	0.9975
163	Novaluron	116714-46-6	$y=1263.0591x+31714.6404$	0.0804–8.0400	0.9904
164	Hydramethylnon	67485-29-4	$y=1627.9241x+4747.8670$	0.0172–1.7160	0.9989
165	Flurazuron	86811-58-7	$y=4291.9442x+27513.8511$	0.2680–26.8000	0.9977

Continued

TABLE 5.17 Linear Equations, Linear Ranges, and Correlation Coefficients of 569 Pesticides and Chemical Pollutants Determined by LC-MS-MS—cont'd

No.	Compound	CAS	Linear Equation	Linear Range	Correlation Coefficient
166	Maleic hydrazide	123-33-1	y=198.2946x+8947.9613	0.8000–80.0000	0.9869
167	Methamidophos	10265-92-6	y=495.2414x+20947.8577	0.0493–4.9300	0.9853
168	EPTC	759-94-4	y=201.6454x+1242.4048	0.3734–37.3380	0.9990
169	Diethyltoluamide	134-62-3	y=1273.0515x+12442.7309	0.0055–0.5500	0.9998
170	Monuron	150-68-5	y=7916.4313x+496481.8838	0.3474–34.7360	0.9902
171	Pyrimethanil	53112-28-0	y=297.8113x−442.7539	0.0068–0.6800	1.0000
172	Fenfuram	24691-80-3	y=495.4576x+6248.3192	0.0078–0.7800	0.9982
173	Quinoclamine	2797-51-5	y=190.9102x+1060.3676	0.0792–7.9200	0.9997
174	Fenobucarb	3766-81-2	y=4559.3338x+57553.4853	0.0590–5.9000	0.9984
175	Ethirimol	23947-60-6	y=197.6192x+2562.2084	0.0056–0.5600	0.9962
176	Propanil	709-98-8	y=1862.7776x+12581.9564	0.2159–21.5900	0.9994
177	Carbofuran	1563-66-2	y=9973.4801x+241072.6078	0.1306–13.0600	0.9949
178	Acetamiprid	160430-64-8	y=654.6118x+13642.1111	0.0144–1.4400	0.9969
179	Mepanipyrim	110235-47-7	y=357.9960x−1061.8607	0.0032–0.3200	0.9999
180	Prometon	1610-18-0	y=1303.1226x+3859.3918	0.0013–0.1310	0.9996

181	Methiocarb	2032-65-7	$y=-0.0044x+192.8365$	0.4120–41.2000	0.0003
182	Metoxuron	19937-59-8	$y=299.5215x+4631.8144$	0.0064–0.6372	0.9964
183	Dimethoate	60-51-5	$y=4064.9134x+205809.1447$	0.0760–7.6000	0.9868
184	Methfuroxam	2873-17-8	$y=1028.0700x+4428.8175$	0.0027–0.2704	0.9997
185	Fluometuron	2164-17-2	$y=589.0483x+16489.4690$	0.0092–0.9200	0.9922
186	Dicrotophos	141-66-2	$y=718.2417x+2790.5141$	0.0114–1.1440	0.9985
187	Monalide	7287-36-7	$y=511.8486x+1710.3270$	0.0120–1.2000	0.9992
188	Diphenamid	957-51-7	$y=1481.4691x-2901.4019$	0.0014–0.1414	0.9999
189	Ethoprophos	13194-48-4	$y=1637.4591x+13357.7575$	0.0276–2.7648	0.9992
190	Fonofos	944-22-9	$y=5273.8108x+88034.9413$	0.0746–7.4580	0.9976
191	Etridiazol	2593-15-9	$y=225.3888x+2635.2119$	1.0042–100.4210	0.9958
192	Furmecyclox	60568-05-0	$y=1048.2963x+2500.1511$	0.0083–0.8320	0.9998
193	Hexazinone	51235-04-2	$y=584.4002x+1578.2577$	0.0012–0.1190	0.9992
194	Dimethametryn	14214-32-5	$y=515.7551x-3828.4887$	0.0011–0.1100	0.9998
195	Trichlorphon	52-68-6	$y=-0.0480x+223.3850$	0.0112–1.1224	0.0416
196	Demeton(O+S)	126-75-0	$y=595.8260x+2922.4572$	0.0677–6.7704	0.9991
197	Benoxacor	98730-04-2	$y=715.7650x+1637.8850$	0.0690–6.9000	0.9997
198	Bromacil	314-40-9	$y=1746.0330x+15437.5364$	0.2360–23.6000	0.9989
199	Phorate sulfoxide	2588-03-6	$y=9.7368x-69.8236$	3.6828–368.2800	0.9991

Continued

TABLE 5.17 Linear Equations, Linear Ranges, and Correlation Coefficients of 569 Pesticides and Chemical Pollutants Determined by LC-MS-MS—cont'd

No.	Compound	CAS	Linear Equation	Linear Range	Correlation Coefficient
200	Brompyrazon	3042-84-0	$y=282.9273x+7986.4013$	0.0360–3.6000	0.9936
201	Oxycarboxin	5259-88-1	$y=501.8658x-25816.0131$	0.0090–0.8960	0.9905
202	Mepronil	55814-41-0	$y=363.2365x+4086.5943$	0.0038–0.3780	0.9981
203	Disulfoton	298-04-4	$y=10446.1164x+226111.6989$	4.6970–469.6960	0.9934
204	Fenthion	55-38-9	$y=3870.7884x+89147.7230$	0.5200–52.0000	0.9952
205	Metalaxyl	57837-19-1	$y=1189.1167x+9947.9627$	0.0050–0.5000	0.9986
206	Ofurace	58810-48-3	$y=508.6080x+10677.4384$	0.0100–1.0000	0.9933
207	Dodemorph	1593-77-7	$y=836.6801x+584.1797$	0.0040–0.4000	0.9989
208	Fosthiazate	98886-44-3	$y=1.1626x+137.4950$	0.00568–0.5680	0.7209
209	Imazamethabenz-methyl	81405-85-8	$y=1314.6426x+5059.2474$	0.0016–0.1638	0.9986
210	Disulfoton-sulfoxide	2497-07-6	$y=1.1239x+3280.9409$	0.0284–2.8440	0.4886
211	Isoprothiolane	50512-35-1	$y=2638.4855x+39192.8681$	0.0185–1.8480	0.9967
212	Imazalil	35554-44-0	$y=951.7959x+7371.4414$	0.0200–2.0000	0.9994
213	Phoxim	14816-18-3	$y=9911.7799x+604149.2113$	0.8280–82.8000	0.9935
214	Quinalphos	13593-03-8	$y=865.7880x+9658.7584$	0.0200–1.9980	0.9976

No.	Name	CAS	Equation	Range	R^2
215	Ditalimfos	5131-24-8	y=18167.3659x−1128108.9193	0.6721−67.2100	0.9981
216	Fenoxycarb	79127-80-3	y=30.1865x+3656.8194	0.1827−18.2700	0.8951
217	Pyrimitate	5221-49-8	y=728.3243x−2829.8820	0.0017−0.1740	0.9999
218	Fensulfothin	115-90-2	y=1075.4355x+9987.5546	0.0200−2.0013	0.9980
219	Fluorochloridone	61213-25-0	y=814.8037x+18697.5800	0.1378−13.7800	0.9965
220	Butachlor	23184-66-9	y=5342.1104x+122190.1493	0.2007−20.0660	0.9937
221	Imazaquin	81335-37-7	y=941.3703x+2564.5663	0.0289−2.8880	0.9998
222	Kresoxim-methyl	143390-89-0	y=3067.5518x+180350.1522	1.0058−100.5800	0.9850
223	Triticonazole	131983-72-7	y=589.2076x−3536.9922	0.0302−3.0200	0.9996
224	Fenamiphos sulfoxide	31972-42-7	y=832.6923x+12375.6069	0.0074−0.7392	0.9970
225	Thenylchlor	96491-05-3	y=42333.7439x+1799969.6460	0.2414−24.1400	0.9918
226	Fenoxanil	115852-48-7	y=23939.2280x+813656.6825	0.3940−39.4000	0.9942
227	Fluridone	59756-60-4	y=331.9877x+3675.5578	0.0018−0.1800	0.9995
228	Epoxiconazole	106325-08-0	y=240.0921x+3999.2990	0.0406−4.0560	0.9925
229	Chlorphoxim	14816-20-7	y=7137.1858x+238332.6184	0.7757−77.5740	0.9916
230	Fenamiphos sulfone	31972-44-8	y=803.4927x+22984.8341	0.0045−0.4452	0.9921
231	Fenbuconazole	114369-43-6	y=1081.0112x+2554.7208	0.0165−1.6490	0.9998
232	Isofenphos	25311-71-1	y=49.1653x+11611.1212	2.1867−218.6720	0.9659
233	Phenothrin	26002-80-2	y=20245.2913x+121159.4061	3.3920−339.2000	0.9957

Continued

TABLE 5.17 Linear Equations, Linear Ranges, and Correlation Coefficients of 569 Pesticides and Chemical Pollutants Determined by LC-MS-MS—cont'd

No.	Compound	CAS	Linear Equation	Linear Range	Correlation Coefficient
234	Fentin-chloride	6369-58-7	y=981.9434x+10764.6836	0.1725–17.2500	0.9985
235	Piperophos	24151-93-7	y=14716.3683x+451792.1536	0.0924–9.2400	0.9952
236	Piperonyl butoxide	51-03-6	y=9283.1656x+209027.3863	0.0113–1.1316	0.9952
237	Oxyflurofen	42874-03-3	y=689.9970x+21833.5155	0.5855–58.5480	0.9883
238	Coumaphos	56-72-4	y=4970.7879x+43543.6013	0.0210–2.1000	0.9991
239	Flufenacet	142459-58-3	y=290.9839x+1018.3351	0.0530–5.3000	0.9996
240	Phosalone	2310-17-0	y=1640.4725x+38388.1640	0.4804–48.0408	0.9946
241	Methoxyfenozide	161050-58-4	y=12590.3849x+245653.8657	0.0370–3.7000	0.9979
242	Prochloraz	67747-09-5	y=2978.0386x+1934.6194	0.0207–2.0698	0.9998
243	Aspon	3244-90-4	y=682.2165x+19697.2170	0.0173–1.7300	0.9908
244	Ethion	562-12-2	y=1927.1075x+87171.8652	0.0296–2.9562	0.9839
245	Diafenthiuron	80060-09-9	y=330.5676x−4802.6808	0.0028–0.2800	0.9983
246	Thifensulfuron-methyl	79277-27-3	y=498.3365x−1835.7614	0.2140–21.4000	0.9999
247	Ethoxysulfuron	126801-58-9	y=102.6484x+331.3256	0.0458–4.5820	0.9998
248	Dithiopyr	97886-45-8	y=599.2316x+16816.3808	0.1040–10.4000	0.9917

249	Spirodiclofen	148477-71-8	$y=2576.7006x+48597.6639$	0.0991–9.9060	0.9952
250	Fenpyroximate	134098-61-6	$y=15606.6519x+129482.2293$	0.0136–1.3600	0.9991
251	Flumiclorac-pentyl	87546-18-7	$y=3704.8939x-73082.8121$	0.1061–10.6080	0.9976
252	Temephos	3383-96-8	$y=124.3151x-853.2678$	0.0122–1.2150	0.9996
253	Butafenacil	134605-64-4	$y=5435.4055x+109585.4037$	0.0950–9.5000	0.9976
254	Spinosad	131929-63-0	$y=0.0268x+37.7938$	0.0057–0.5684	1.0000
255	Mepiquat chloride	24307-26-4	$y=61.1769x+1284.4492$	0.0090–0.9000	0.9956
256	Allidochlor	93-71-0	$y=2852.2896x+73242.4805$	0.4104–41.0400	0.9951
257	Propamocarb	24579-73-5	$y=10.7676x+1187.4636$	0.0009–0.0876	0.9289
258	Tricyclazole	41814-78-2	$y=0.1725x+130.1401$	0.1248–12.4800	0.5466
259	Thiabendazole	148-79-8	$y=194.3938x+572.0249$	0.0049–0.4880	0.9971
260	Metamitron	41394-05-2	$y=837.4865x+5143.0632$	0.0636–6.3600	0.9993
261	Isoproturon	34123-59-6	$y=116.6237x+326.4895$	0.0014–0.1356	0.9999
262	Atratone	1610-17-9	$y=543.7094x+622.4082$	0.0018–0.1832	0.9999
263	Oesmetryn	1014-69-3	$y=248.0989x+1451.9357$	0.0017–0.1704	0.9994
264	Metribuzin	21087-64-9	$y=2.7769x-19.4039$	0.0054–0.5400	0.9652
265	DMST	66840-71-9	$y=9474.6497x+323916.1902$	0.4000–40.0000	0.9920
266	Cycloate	1134-23-2	$y=1848.8387x+27511.3811$	0.0444–4.4400	0.9971
267	Atrazine	1912-24-9	$y=173.1963x+1836.1567$	0.0036–0.3604	0.9987

Continued

TABLE 5.17 Linear Equations, Linear Ranges, and Correlation Coefficients of 569 Pesticides and Chemical Pollutants Determined by LC-MS-MS—cont'd

No.	Compound	CAS	Linear Equation	Linear Range	Correlation Coefficient
268	Butylate	2008-41-5	y=32234.8508x+1353075.8181	3.0200–302.0000	0.9927
269	Pymetrozin	123312-89-0	y=3462.3035x+1602385.7207	0.3428–34.2800	0.5514
270	Chloridazon	1968-60-8	y=219.8168x+1316.0376	0.0233–2.3280	0.9989
271	Sulfallate	95-06-7	y=4542.9259x+93406.5102	2.0720–207.2000	0.9962
272	Ethiofencarb	29973-13-5	y=4.8106x+171.2435	0.0492–4.9200	0.9981
273	Terbumeton	33693-04-8	y=926.5814x+1209.2664	0.0010–0.0960	0.9997
274	Cyprazine	22936-86-3	y=2575.4123x+6962.3458	0.00428-0.4280	0.9997
275	Ametryn	834-12-8	y=2575.4123x+6962.3458	0.0096–0.9600	0.9997
276	Tebuthiuron	34014-18-1	y=362.7670x−959.1598	0.0022–0.2168	0.9999
277	Trietazine	1912-26-1	y=1406.3615x+4295.4245	0.0060–0.6040	0.9998
278	Sebutylazine	7286-69-3	y=891.9600x+1390.9694	0.0031–0.3140	0.9998
279	Dibutyl succinate	141-03-7	y=44854.7853x+2538025.4845	2.2240–222.4000	0.9933
280	Tebutam	35256-85-0	y=2024.1474x+13998.5326	0.0014–0.1360	0.9995
281	Thiofanox-sulfoxide	39184-27-5	y=99.8573x+533.4822	0.0829–8.2940	0.9970
282	Cartap hydrochloride	15263-52-2	y=1786.6346x−240134.7510	20.8000–2080.0000	0.7874

283	Methacrifos	62610-77-9	$y=103041.2938x+4175464.4145$	24.2370–2423.6960	0.9936
284	Terbutryn	886-50-0	$y=169.3932x-525.0790$	0.0002-0.0020	0.9999
285	Triazoxide	72459-58-6	$y=1255.9644x+13370.1869$	0.0800–8.0000	0.9985
286	Thionazin	297-97-2	$y=4444.9530x+41025.6897$	0.2268–22.6800	0.9990
287	Linuron(a)	330-55-2	$y=1767.3063x+2794.6827$	0.1163–11.6340	0.9999
288	Heptanophos	23560-59-0	$y=2589.8453x+36451.4087$	0.0584–5.8400	0.9971
289	Prosulfocarb	52888-80-9	$y=527.3550x+670.3112$	0.0037–0.3668	0.9997
290	Dipropetryn	4147-51-7	$y=793.8462x-220.0440$	0.0027–0.2700	1.0000
291	Thiobencarb	28249-77-6	$y=6110.6145x+90820.8254$	0.0330–3.3000	0.9971
292	Tri-iso-butyl phosphate	126-71-6	$y=6619.1341x+122612.7095$	0.3576-35.7600	0.9983
293	Tri-n-butyl phosphate	126-73-8	$y=6619.1341x+122612.7095$	0.0037–0.3740	0.9983
294	Diethofencarb	87130-20-9	$y=821.6445x+4418.7161$	0.0200–2.0000	0.9995
295	Alachlor	15972-60-8	$y=3014.3541x+47669.6833$	0.0740–7.4000	0.9979
296	Cadusafos	95465-99-9	$y=5.1788x+373.2663$	0.0115–1.1520	0.9971
297	Metazachlor	67129-08-2	$y=2339.9685x+18152.3547$	0.0098–0.9800	0.9994
298	Propetamphos	31218-83-4	$y=42.3058x+224.2160$	0.5400–54.0000	0.9997
299	Terbufos	13071-79-9	$y=17.0585x+1089.5581$	22.4000–2240.0043	0.8747
300	Simeconazole	149508-90-7	$y=3902.4474x+13977.0751$	0.0294–2.9400	0.9998

Continued

TABLE 5.17 Linear Equations, Linear Ranges, and Correlation Coefficients of 569 Pesticides and Chemical Pollutants Determined by LC-MS-MS—cont'd

No.	Compound	CAS	Linear Equation	Linear Range	Correlation Coefficient
301	Triadimefon	43121-43-3	$y=2850.1366x+8752.3272$	0.0788–7.8800	0.9998
302	Phorate sulfone	2588-04-7	$y=5037.9609x+74316.6949$	0.4200–42.0000	0.9955
303	Tridemorph	24602-86-6	$y=448.1759x-5222.6051$	0.0260–2.6040	0.9982
304	Mefenacet	73250-68-7	$y=1375.1645x+5984.8963$	0.0221–2.2080	0.9998
305	Azaconazole	60207-31-0	$y=0.1764x+103.5656$	0.0081–0.8064	0.9635
306	Fenamiphos	22224-92-6	$y=0.0113x+21.7285$	0.0021–0.2068	0.9785
307	Fenpropimorph(c)	67306-03-0	$y=806.4761x-1630.0388$	0.0018–0.1840	1.0000
308	Tebuconazole	107534-96-3	$y=2944.7220x+10025.3093$	0.0223–2.2320	0.9998
309	Isopropalin	33820-53-0	$y=1570.8561x+91301.8669$	0.3000–30.0000	0.9906
310	Nuarimol	63284-71-9	$y=0.0438x+69.7575$	0.0100–0.9960	0.9998
311	Bupirimate	57839-19-1	$y=167.1571x+1276.3810$	0.0070–0.7000	0.9993
312	Azinphos-methyl	86-50-0	$y=17995.0170x+951026.8043$	11.0433–1104.3340	0.9969
313	Tebupirimfos	96182-53-5	$y=1625.8675x+31691.6334$	0.0013–0.1292	0.9946
314	Phenthoate	2597-03-7	$y=9469.7346x+448551.7108$	0.9235–92.3520	0.9907
315	Sulfotep	3689-24-5	$y=2213.2598x+49462.9003$	0.0260–2.6000	0.9945

Continued

316	Sulprofos	35400-43-2	$y=664.1067x+18530.7053$	0.0584–5.8400	0.9976
317	EPN	2104-64-5	$y=2571.7229x+57254.8564$	0.3300–33.0000	0.9947
318	Azamethiphos	35575-96-3	$y=350.5892x-1529.6606$	0.0081–0.8080	0.9964
319	Diniconazole	83657-24-3	$y=2606.3991x+11653.5772$	0.0134–1.3440	0.9997
320	Flumetsulam	98967-40-9	$y=173.6748x+565.0047$	0.0030–0.2968	1.0000
321	Sethoxydim	74051-80-2	$y=0.0455x+65.5173$	8.9600–896.0000	0.6459
322	Pencycuron	66063-05-6	$y=518.3200x+1111.6428$	0.0027–0.2732	0.9995
323	Mecarbam	2595-54-2	$y=16297.3708x+404245.7279$	0.1960–19.6000	0.9934
324	Tralkoxydim	87820-88-0	$y=78.7504x+589.4615$	0.0032–0.3208	0.9980
325	Malathion	121-75-5	$y=4822.1235x+128759.6046$	0.0564–5.6442	0.9935
326	PYRIBUTICARB	88678-67-5	$y=634.3592x-1694.2849$	0.0034–0.3388	0.9996
327	Pyridaphenthion	119-12-0	$y=1497.8111x+967.7513$	0.0087–0.8720	0.9998
328	Pirimiphos-ethyl	23505-41-1	$y=1403.1428x+12257.1289$	0.00476–0.476	0.9982
329	Thiodicarb	59669-26-0	$y=217.7714x+1082.3544$	0.3937–39.3680	
330	Pyraclofos	77458-01-6	$y=10823.8581x+390570.1748$	0.0100–1.0040	0.9996
331	Picoxystrobin	117428-22-5	$y=1420.3457x+4189.6734$	0.0844–8.4400	0.9936
332	Tetraconazole	112281-77-3	$y=31730.8228x+1397070.9903$	0.0172–1.7200	0.9996
333	Mefenpyr-diethyl	135590-91-9	$y=2441.5194x+16050.0118$	0.1256–12.5600	0.9918
334	Profenefos	41198-08-7		0.0202–2.0160	0.9990

TABLE 5.17 Linear Equations, Linear Ranges, and Correlation Coefficients of 569 Pesticides and Chemical Pollutants Determined by LC-MS-MS—cont'd

No.	Compound	CAS	Linear Equation	Linear Range	Correlation Coefficient
335	Pyraclostrobin	175013-18-0	y=294.2071x+1012.6109	0.0051–0.5051	0.9993
336	Dimethomorph	211867-47-9	y=−0.1179x+63.0441	0.0035–0.3524	0.8778
337	Kadethrin	58769-20-3	y=468.4548x+5319.2106	0.0333–3.3280	0.9974
338	Thiazopyr	117718-60-2	y=657.5481x+3978.0302	0.0196–1.9600	0.9981
339	Benfuracarb-methyl	82560-54-1	y=7849.0751x+119127.7299	0.1638–16.3760	0.9985
340	Cinosulfuron	94593-91-6	y=837.8124x−1767.6328	0.0112–1.1240	1.0000
341	Pyrazosulfuron-ethyl	93699-74-6	y=863.7984x−1231.2501	0.0684–6.8400	0.9999
342	Metosulam	139528-85-1	y=118.7489x+466.4379	0.0440–4.4000	0.9999
343	Chlorfluazuron	71422-67-8	y=1350.1950x+52504.4409	0.0868–8.6800	0.9958
344	4-Aminopyridine	504-24-5	y=1.1687x+194.9642	0.0087–0.8680	1.0000
345	Chlormequat	999-81-5	y=109.4692x+1469.0886	0.0012–0.1210	0.9949
346	Methomyl	16752-77-5	y=1104.7884x+30018.8117	0.0956–9.5600	0.9906
347	Pyroquilon	57369-32-1	y=845.8417x+22604.6008	0.0348–3.4800	0.9922
348	Fuberidazole	3878-19-1	y=2610.2411x+20460.3733	0.0189–1.8900	0.9907
349	Isocarbamid	30979-48-7	y=1023.8102x+15078.1656	0.0170–1.6980	0.9960

Continued

350	Butocarboxim	34681-10-2	$y=82.9344x+1370.6113$	0.0157–1.5700	0.9962
351	Chlordimeform	6164-98-3	$y=42.2670x+548.3823$	0.0133–1.3320	0.9992
352	Cymoxanil	57966-95-7	$y=521.0033x-15836.1579$	0.5560–55.6000	0.9982
353	Vernolate	1929-77-7	$y=5.9720x+133.8317$	0.0026–0.2580	0.9905
354	Chlorthiamid	1918-13-4	$y=-0.1856x+261.8419$	0.0882–8.8200	0.6681
355	Aminocarb	2032-59-9	$y=6816.0957x+353887.1140$	0.1642–16.4200	0.9901
356	Dimethirimol	5221-53-4	$y=163.3937x+1124.7516$	0.0012–0.1246	0.9942
357	Omethoate	1113-02-6	$y=796.3507x+30866.6902$	0.0965–9.6500	0.9947
358	Ethoxyquin	91-53-2	$y=38.1282x-496.7101$	0.0352–3.5200	0.9999
359	Dichlorvos	62-73-7	$y=10668.6148x+28771.7328$	0.0055–0.5480	0.9975
360	Aldicarb sulfone	1646-88-4	$y=236.2511x+6970.3190$	0.2140–21.4000	0.9906
361	Dioxacarb	6988-21-2	$y=10027.2028x+402269.4998$	0.0336–3.3600	0.9939
362	Benzyladenine	1214-39-7	$y=25472.3781x+760346.7160$	0.7080–70.8000	0.9937
363	Demeton-s-methyl	919-86-8	$y=354.2901x+13956.8399$	0.0530–5.3000	0.9903
364	Ethiofencarb-sulfoxide	53380-22-6	$y=46444.9107x+335958.4554$	2.2400–224.0000	0.9965
365	Cyanohos	2636-26-2	$y=-0.1934x+765.7837$	0.1012–10.1200	0.2539
366	Thiometon	640-15-3	$y=26.5917x+589.8780$	5.7800–578.0000	0.9975
367	Folpet	133-07-3	$y=16.1809x+814.4634$	1.3860–138.6000	0.9971
368	Demeton-s-methyl sulfone	17040-19-6	$y=3149.4204x+186365.6020$	0.1976–19.7600	0.9897

TABLE 5.17 Linear Equations, Linear Ranges, and Correlation Coefficients of 569 Pesticides and Chemical Pollutants Determined by LC-MS-MS—cont'd

No.	Compound	CAS	Linear Equation	Linear Range	Correlation Coefficient
369	Dimepiperate	61432-55-1	$y=-0.0233x+103.0486$	37.8000–3780.0000	0.0770
370	Fenpropidin	67306-00-7	$y=355.8065x-2541.6466$	0.0018–0.1830	0.9996
371	Amidithion	919-76-7	$y=4.0419x+88.2751$	6.5800–658.0000	0.9963
372	Paraoxon-ethyl	311-45-5	$y=429.7212x+2144.1489$	0.0047–0.4740	0.9986
373	Aldimorph	1704-28-5	$y=46.0093x+14185.8345$	0.0316–3.1600	0.9902
374	Vinclozolin	50471-44-8	$y=17.6782x-169.1461$	0.0254–2.5400	0.9985
375	Uniconazole	83657-17-4	$y=1817.6256x-9941.6046$	0.0240–2.4000	0.9994
376	Pyrifenox	88283-41-4	$y=324.1458x+2009.4549$	0.0027–0.2660	0.9953
377	Chlorthion	500-28-7	$y=0.0861x+174.0543$	1.3360–133.6000	0.0710
378	Dicapthon	2463-84-5	$y=-0.2023x+494.4527$	0.0024–0.2380	0.3815
379	Clofentezine	74115-24-5	$y=63.7269x+2467.3521$	0.0076–0.7640	0.9938
380	Norflurazon	27314-13-2	$y=192.1462x+795.9883$	0.0026–0.2580	0.9997
381	Triallate	2303-17-5	$y=1278.8007x+15922.2273$	0.4620–46.2000	0.9960
382	Ziram	137-30-4	$y=82.0717x+50446.3136$	61.2000–6120.0000	0.9983
383	Quinoxyphen	124495-18-7	$y=98306.9981x+2120128.8256$	1.5340–153.4000	0.9871

384	Fenthion sulfone	3761-42-0	$y=4341.6469x+165245.4231$	0.1746–17.4600	0.9934
385	Flurochloridone	61213-25-0	$y=320.8582x+380.8278$	0.0129–1.2900	0.9916
386	Phthalic acid, benzyl butyl ester	85-68-7	$y=18.3720x+2803.8831$	6.3200–632.0000	0.9949
387	Isazofos	42509-80-8	$y=554.2332x+1798.2357$	0.0018–0.1784	0.9988
388	Dichlofenthion	97-17-6	$y=991.6565x+10620.2658$	0.3020–30.2000	0.9967
389	Vamidothion sulfone	70898-34-9	$y=-14.3629x+9771.7890$	4.7600–476.0000	0.9780
390	Terbufos sulfone	56070-16-7	$y=27447.2830x+1244932.1309$	0.8860–88.6000	0.9957
391	Dinitramine	29091-05-2	$y=42.2483x-274.4581$	0.0179–1.7920	0.9991
392	Cyazofamid	120116-88-3	$y=4.7986x-488.2846$	0.0450–4.5000	0.9578
393	Trichloronat	327-98-0	$y=196.4102x-1978.8775$	0.6680–66.8000	0.9995
394	Resmethrin-2	10453-86-8	$y=703.6100x+39091.8906$	0.0030–0.3000	0.9984
395	Boscalid	188425-85-6	$y=2213.6037x+1839.6788$	0.0476–4.7600	0.9990
396	Nitralin	4726-14-1	$y=615.2840x+2287.1067$	0.3440–34.4000	0.9975
397	Fenpropathrin	39515-41-8	$y=6141.8629x+337059.4024$	2.4500–245.0000	0.9926
398	Hexythiazox	78587-05-0	$y=7708.0484x+62615.2640$	0.2360–23.6000	0.9950
399	Florasulam	145701-23-1	$y=0.1932x+40.2449$	0.1740–17.4000	0.8538
400	Benzoximate	29104-30-1	$y=1411.7030x+17393.7979$	0.1966–19.6600	0.9861
401	Benzoylprop-ethyl	22212-55-1	$y=93645.7583x+777253.3565$	3.0800–308.0000	0.9927
402	Pyrimidifen	105779-78-0	$y=13.5375x+773.1731$	0.1400–14.0000	0.8789

Continued

TABLE 5.17 Linear Equations, Linear Ranges, and Correlation Coefficients of 569 Pesticides and Chemical Pollutants Determined by LC-MS-MS—cont'd

No.	Compound	CAS	Linear Equation	Linear Range	Correlation Coefficient
403	Furathiocarb	65907-30-4	$y=3757.9696x-90830.8254$	0.0192–1.9180	0.9948
404	*trans*-Permethin	551877-74-8	$y=144940.3566x$ $+13586835.5807$	0.0480–4.8000	0.9773
405	Etofenprox	80844-07-1	$y=52422.6895x+993176.1053$	22.8000–2280.0000	0.9833
406	Pyrazoxyfen	71561-11-0	$y=159.9437x-450.6158$	0.0033–0.3260	0.9997
407	Flubenzimine	37893-02-0	$y=9.1937x+233.7501$	0.0778–7.7800	0.9955
408	Zeta cypermethrin	52315-07-8	$y=0.0113x+308.8147$	0.0068–0.6780	0.0084
409	Haloxyfop-2-ethoxyethyl	87237-48-7	$y=2276.8566x+23031.1273$	0.0250–2.5000	0.9971
410	Esfenvalerate	66230-04-4	$y=-0.0278x+65.1524$	41.0000–4100.0000	0.2962
411	Fluoroglycofen-ethyl	77502-90-7	$y=66.6293x+365.1939$	0.0500–5.0000	0.9986
412	Tau-fluvalinate	102851-06-9	$y=1079.4595x+29675.2916$	2.3000–230.0000	0.9922
413	Acrylamide	79-06-1	$y=78.5401x+2720.1502$	0.1780–17.8000	0.9938
414	Tert-butylamine	75-64-9	$y=92.9714x+4142.8598$	0.3895–38.9500	0.9966
415	Hymexazol	10004-44-1	$y=1043.3346x-17597.5831$	2.2414–224.1360	0.9993
416	Chlormequat chloride	999-81-5	$y=86.8613x+4466.3742$	0.0070–0.7040	0.9937
417	Phthalimide	5333-22-2	$y=274.3572x+1545.7599$	0.4300–43.0000	0.9988

418	Dimefox	115-26-4	y=8261.6586x+309298.1906	0.6820–68.2000	0.9963
419	Metolcarb	1129-41-5	y=21.1101x+978.5273	0.2540–25.4000	0.9926
420	Diphenylamin	122-39-4	y=367.9085x+4817.7280	0.0041–0.4140	0.9991
421	1-Naphthy acetamide	86-86-2	y=513.1340x+4096.6725	0.0081–0.8100	0.9989
422	Atrazine-desethyl	6190-65-4	y=255.4311x+2598.5443	0.0062–0.6200	0.9979
423	2,6-Dichlorobenzamide	2008-58-4	y=577.5666x+25445.5813	0.0450–4.5000	0.9914
424	Aldicarb	116-06-3	y=6988.0771x+84055.6763	2.6100–261.0000	0.9792
425	Dimethyl phthalate	131-11-3		0.1320–13.2000	
426	Chlordimeform hydrochloride	19750-95-9	y=2.9006x+195.5954	0.0264–2.6400	0.9923
427	Simeton	673-04-1	y=3002.0893x+69986.0397	0.0110–1.1040	0.9954
428	Dinotefuran	165252-70-0	y=28.0192x+1010.6177	0.1018–10.1800	0.9940
429	Pebulate	1114-71-2	y=189.2352x+696.6051	0.0340–3.4000	0.9991
430	Acibenzolar-S-methyl	135158-54-2	y=33.6084x+55.5798	0.0308–3.0800	0.9998
431	Dioxabenzofos	3811-49-2	y=-0.0456x+111.0720	0.1384–13.8400	0.0984
432	Oxamyl	23135-22-0	y=26448.9708x+424443.7157	5.4806–548.0608	0.9832
433	Thidiazuron PESTANAL	41118-83-6	y=48.4079x+227.6618	0.0029–0.2940	0.9991
434	Methabenzthiazuron	18691-97-9	y=48.9268x+276.3932	0.0007–0.0734	0.9999
435	Butoxycarboxim	34681-23-7	y=235.2117x+9946.1018	0.2660–26.6000	0.9980
436	Mexacarbate	315-18-4	y=0.1301x+86.1724	0.0094–0.9400	0.9997

Continued

TABLE 5.17 Linear Equations, Linear Ranges, and Correlation Coefficients of 569 Pesticides and Chemical Pollutants Determined by LC-MS-MS—cont'd

No.	Compound	CAS	Linear Equation	Linear Range	Correlation Coefficient
437	Demeton-S-methyl sulfoxide	301-12-2	$y=609.6832x+5113.5306$	0.0392–3.9200	0.9992
438	Thiofanox sulfone	39184-59-3	$y=72.6762x-2810.5989$	0.2408–24.0800	0.9935
439	Phosfolan	950-10-7	$y=193.4480x-1771.9920$	0.0049–0.4860	0.9994
440	Triclopyr	55335-06-3	$y=0.0544x+41.8416$	0.0020–0.2000	0.6225
441	Demeton-S	8065-48-3	$y=0.0625x+1012.4209$	0.8000–80.0000	0.0015
442	Imazapyr	81334-34-1	$y=0.4753x+368.8999$	0.1028–10.2800	0.9988
443	Fenthion oxon	6552-12-1	$y=1105.6217x+10901.9584$	0.0119–1.1880	0.9992
444	Napropamide	15299-99-7	$y=2258.4391x+52089.9185$	0.0127–1.2740	0.9935
445	Fenitrothion	122-14-5	$y=400.0687x+1614.7608$	0.2680–26.8000	0.9998
446	Phthalic acid, dibutyl ester	84-74-2	$y=147722.9118x+15556210.4944$	0.3960–39.6000	0.9950
447	Metolachlor	51218-45-2	$y=3123.9033x+29312.3918$	0.0039–0.3900	0.9987
448	Procymidone	32809-16-8	$y=1960.3625x+70236.7827$	0.8660–86.6000	0.9863
449	Vamidothion	2265-23-2	$y=10319.0861x+387250.5245$	0.0456–4.5600	0.9935
450	Chloroxuron	1982-47-4	$y=0.3445x+9.9829$	0.0044–0.4440	0.9927
451	Triamiphos	1031-47-6	$y=1.6144x-121.5512$	0.001–0.0100	0.8937

452	Prallethrin	23031-36-9			
453	Cumyluron	99485-76-4	$y=1724.7489x+2011.9108$	0.0132–1.3180	0.9999
454	Imazamox	114311-32-9	$y=-0.1650x+400.9340$	0.0180–1.8000	0.3015
455	Warfarin	00081-81-2	$y=110.9884x+3333.2446$	0.0268–2.6800	0.9935
456	Phosmet	732-11-6	$y=24748.5772x+657561.5828$	0.1772–17.7200	0.9933
457	Ronnel	8003-34-7	$y=35.2124x+381.8608$	0.1313–13.1300	0.9962
458	Pyrethrin	8003-34-7	$y=1168.5641x+1160.1158$	0.3580–35.8000	0.9995
459	Phthalic acid, biscyclohexyl ester	84-61-7	$y=1594.5382x-17600.5078$	0.0068–0.6780	0.9996
460	carpropamid	104030-54-8	$y=1376.3912x+13487.6023$	0.0520–5.2000	0.9979
461	Tebufenpyrad	119168-77-3	$y=11383.0704x+303944.2068$	0.0025–0.2546	0.9930
462	Tebufenozide	112410-23-8	$y=99802.4374x+3668232.4095$	0.2780–27.8000	0.9900
463	Iminoctadine triacetate	39202-40-9		0.0061–0.6080	
464	Chlorthiophos	60238-56-4	$y=1297.8765x+54114.4596$	0.3180–31.8000	0.9956
465	Naled	300-76-5	$y=0.1685x+24.8709$	1.4820–148.2000	0.9725
466	Dialifos	10311-84-9	$y=1687.9268x+115808.1608$	1.5700–157.0000	0.9900
467	Cinidon-ethyl	142891-20-1	$y=1036.1571x+44597.7280$	0.1458–14.5800	0.9940
468	Rotenone	83-79-4	$y=562.9032x-2063.2506$	0.0232–2.3200	1.0000
469	Imibenconazole	86598-92-7	$y=675.6232x-17887.5593$	0.1026–10.2600	0.9931
470	Propaquizafop	111479-05-1	$y=513.7900x-572.6864$	0.0124–1.2360	1.0000

Continued

TABLE 5.17 Linear Equations, Linear Ranges, and Correlation Coefficients of 569 Pesticides and Chemical Pollutants Determined by LC-MS-MS—cont'd

No.	Compound	CAS	Linear Equation	Linear Range	Correlation Coefficient
471	Lactofen	77501-63-4	$y=13102.3167x+166244.0259$	0.6200–62.0000	0.9905
472	2,3,4,5-Tetrachloroaniline	634-83-3	$y=-0.0309x+181.0508$	0.5360–53.6000	0.0328
461	Benzofenap	82692-44-2	$y=30.5235x-72.0631$	0.0008–0.0800	0.9998
474	Dinoseb acetate	2813-95-8	$y=-22.6178x+34782.1204$	0.4128–41.2800	0.9988
475	Imazapic	104098-49-9	$y=2992.9311x+35232.0039$	0.0168–1.6800	0.9983
476	pyriminobac-methyl(Z)	136191-64-5	$y=0.0130x+52.9569$	0.0008–0.0800	0.4420
477	Propisochlor	86763-47-5	$y=64.5185x+2075.4492$	0.0080–0.8000	0.9955
478	Silafluofen,	105024-66-6	$y=0.0336x+71.5216$	6.0800–608.0000	0.1262
479	Triphenyl phosphate	115-86-6			
480	Etobenzanid	79540-50-4	$y=84.0634x+320.0916$	0.0080–0.8000	0.9992
481	Fentrazamide	158237-07-1	$y=4169.3848x+181909.9937$	0.1240–12.4000	0.9943
482	Pentachloroaniline	527-20-8	$y=0.6614x+63.3572$	0.0374–3.7440	0.9035
483	Carbosulfan	55285-14-8			
484	Cyphenothrin	39515-40-7	$y=129.1536x+3499.6625$	0.1680–16.8000	0.9981
485	Dieldrin	60-57-1	$y=0.1336x+92.4975$	1.6160–161.6000	0.3399

486	Dimefuron	34205-21-5	$y=5.7831x+65.1268$	0.0400–4.0000	0.9984
487	Etoxazole	153233-91-1	$y=0.2379x+85.9116$	0.0087–0.8720	0.2673
488	Malaoxon	1364-78-2	$y=0.0544x+98.4962$	0.0469–4.6880	0.2845
489	Chlorbenside sulfone	7082-99-7			
490	Dodine	2439-10-3	$y=-1.0355x+2826.9902$	0.0800–8.0000	0.9762
491	Propylene thiourea	2122-19-2	$y=1443.7894x+55229.6801$	0.3008–30.0800	0.9953
492	Dalapon	17040-19-6	$y=16.5435x+4215.5432$	2.3074–230.7400	0.9926
493	Ethephon	16672-87-0	$y=7.7376x+237.2904$	0.9384–93.8400	0.9930
494	Flupropanate	756-09-2	$y=0.6295x+27.2309$	0.2298–22.9824	0.9767
495	2,6-Difluorobenzoic acid	385-00-2	$y=0.0175x+29.0027$	17.0408–1704.0800	0.3388
496	Trichloroacetic acid sodium salt	650-51-1	$y=0.0035x+20.9276$	2.8158–281.5800	0.0148
497	Tert-butyl-4-hydroxyanisole	25013-16-5	$y=84.3912x-1338.2358$	0.0023–0.2300	
498	2-Phenylphenol	90-43-7	$y=84.3912x-1338.2358$	1.6988–169.8800	0.9965
499	3-Phenylphenol	580-51-8	$y=87.4363x-658.0829$	0.0400–4.0032	0.9950
500	Clopyralld	1702-17-6	$y=1.6393x+137.8885$	2.8000–280.0000	0.9976
501	DNOC	534-52-1	$y=55.9063x+115.5717$	0.0260–2.6000	0.9997
502	Cloprop	101-10-0	$y=152.6101x+1428.1018$	0.1140–11.4000	0.9970
503	Dicloran	99-30-9	$y=14.0866x-125.6277$	0.4856–48.5560	0.9990
504	Aminopyralid	150114-71-9			

Continued

TABLE 5.17 Linear Equations, Linear Ranges, and Correlation Coefficients of 569 Pesticides and Chemical Pollutants Determined by LC-MS-MS—cont'd

No.	Compound	CAS	Linear Equation	Linear Range	Correlation Coefficient
505	Chlorpropham	101-21-3	$y=38.5725x-709.0287$	0.1577–15.7680	0.9956
506	Mecoprop	94596-45-9	$y=92.3581x-703.9107$	0.0490–4.8960	0.9996
507	Terbacil	5902-51-2	$y=10.5604x-100.5870$	0.0088–0.8778	0.9990
508	2,4-D	94-75-7	$y=243.6842x+7685.5105$	0.1186–11.8600	0.9913
509	Dicamba	1918-00-9	$y=3.6383x+516.9783$	12.6592–1265.920	0.9903
510	MCPB	94-81-5	$y=60.4809x+1024.1458$	0.1418–14.1800	0.9987
511	Fenaminosulf	140-56-7		2.2540–225.4000	
512	Dichlorprop	15165-67-0		0.0147–1.4706	
513	Picloram	1918-02-1	$y=56.7467x+1061.2832$	5.3411–534.1060	0.9990
514	Bentazone	25057-89-0	$y=19.2706x-258.9725$	0.0103–1.0336	0.9966
515	Dinoseb	88-85-7	$y=508.1906x+18314.5557$	0.0040–0.3960	0.9935
516	Dinoterb	1420-07-1	$y=70.4148x+618.6087$	0.0024–0.2400	0.9987
517	Forchlorfenuron	68157-60-8	$y=204.8374x-665.9657$	0.1140–11.4000	1.0000
518	2,4-DB		$y=0.0209x+67.2938$	21.3978–2139.776	0.0733
519	Fludioxonil	131341-86-1	$y=4.4075x-43.4941$	0.6216–62.1600	0.9972

No.	Name	CAS	Equation	Range	r
520	Trinexapac-ethyl	95266-40-3	$y=0.1451x+4.3252$	0.7069–70.6860	0.8088
521	2,4,5-T	93-76-5	$y=1460.7370x+4977.4311$	0.1748–17.4800	0.9986
522	Fluroxypyr	69377-81-7	$y=1460.7370x+4977.4311$	1.9206–192.0600	0.9986
523	Chlorfenethol	80-06-8	$y=-11.8956x+3375.5347$	1.6430–164.3000	0.9236
524	Fenoprop	93-72-1		0.0654–6.5372	
525	Cyclanilide	113136-77-9		0.0344–3.4400	
526	Bromoxynil	1689-84-5	$y=10.1402x-163.2630$	0.0180–1.7992	0.9983
527	Pentachlorophenol	87-86-5		0.0039–0.3910	
528	Isocarbophos	245-61-5		0.0004–0.0360	
529	Naptalam	132-66-1	$y=12.7926x+77.4575$	0.0195–1.9456	0.9946
530	Chlorobenzuron	57160-47-1	$y=7.1922x-1.8859$	0.2040–20.4000	0.9996
531	Chloramphenicolum	56-75-7	$y=14.3455x+255.1415$	0.0388–3.8800	0.9955
532	Alloxydim-sodium	66003-55-2	$y=27.8662x-497.1785$	0.0020–0.1994	0.9994
533	Pyrithlobac sodium	123343-16-8		—	
534	Dinobuton	973-21-7		0.0043–0.4284	
535	Fluorodifen	15457-05-3		2.7331–273.3120	
536	Dimehypo	7772-98-7	$y=2.2972x+15.7850$	4.0020–400.2000	0.9910
537	Fempxaprop-ethyl	66441-23-4	$y=30.1361x-333.5816$	0.0490–4.8960	0.9964
538	Diflufenzopyr-sodium	109293-98-3		0.3080–30.8000	

Continued

TABLE 5.17 Linear Equations, Linear Ranges, and Correlation Coefficients of 569 Pesticides and Chemical Pollutants Determined by LC-MS-MS—cont'd

No.	Compound	CAS	Linear Equation	Linear Range	Correlation Coefficient
539	Sulfanitran	122-16-7	$y=6.2932x+243.0328$	0.0304–3.0400	0.9908
540	Mesotrion	104206-82-8	$y=316.5324x+8345.1880$	23.0056–2300.560	0.9942
541	Oryzalin	19044-88-3	$y=24.7555x+268.9195$	0.0491–4.9140	0.9963
542	Gibberellic acid	77-06-5	$y=32.6442x+318.8949$	0.6634–66.3400	0.9920
543	Acifluorfen	50594-66-6	$y=304.2937x+12432.6715$	1.1800–118.0000	0.9936
544	Heptachlor	76-44-8	$y=0.0066x+22.7959$	0.0002–0.0216	0.1134
545	Plifenate	51366-25-7		0.0002–0.0226	
546	Ioxynil	689-83-4	$y=9.6066x-107.7810$	0.0062–0.6154	0.9996
547	Famoxadone	131807-57-3	$y=304.2686x-461.0252$	0.4529–45.2880	0.9973
548	Metsulfuron-methyl	74223-64-6		5.6700–567.0000	
549	Sulfentrazone	122836-35-5			
550	Diflufenican	83164-33-4	$y=312.4554x-5410.5163$	0.2827–28.2720	0.9977
551	Ethiprole	181587-01-9	$y=899.3631x-2846.9036$	0.3985–39.8520	1.0000
552	Propoxycarbzone-sodium	181274-15-7			
553	Flazasulfuron	104040-78-0		2.9810–298.1000	

No.	Name	CAS	Equation	Range	R^2
554	Flusulfamide	106917-52-6	$y=41.4627x+81.5874$	0.0041–0.4140	0.9997
555	Endosulfan-sulfate	1031-07-8		0.0418–4.1797	0.9959
556	Cyclosulfamuron	136849-15-5	$y=238.2037x+11649.3806$	3.4368–343.6800	
557	Triforine	26644-46-2			
558	Halosulfuron-methyl	100784-20-1		0.0980–9.7970	0.9997
559	Fomesafen	72178-02-0	$y=25.9945x+306.6959$	0.0202–2.0200	
560	Tecloftalam	34014-18-1		0.0009–0.0880	0.9916
561	Fluazinam	79622-59-6	$y=3859.6694x+238735.5505$	0.7060–70.6000	0.9983
562	Fluazuron	86811-58-7	$y=60.8451x+469.0285$	0.0002–0.0200	0.9995
563	Iodosulfuron-methyl sodium	144550-36-7	$y=293.2463x+3471.0234$	0.2120–21.2000	
564	Lufenuron	103055-07-8		0.0002–0.0200	
565	Thifluzamide	130000-40-7		!	0.9928
566	Kelevan	4234-79-1	$y=46.7960x+1932.9938$	964.0000–96400.0	0.9995
567	Acrinathrin	101007-06-1	$y=21.9525x-253.3594$	0.0808–8.0800	0.9995
568	Iodosulfuron-methyl	1225-60-61	$y=293.2463x+3471.0234$	0.6660–66.6000	
569	Octachlorostyrene	29082-74-4		0.0336–3.3600	

5.3 GPC ANALYTICAL PARAMETERS OF PESTICIDES AND CHEMICAL POLLUTANTS

5.3.1 GPC Analytical Parameters of 744 Pesticides and Chemical Pollutants

See Table 5.18.

TABLE 5.18 GPC Analytical Parameters of 744 Pesticides and Chemical Pollutants

No.	Compound	Collection Time (min)
1	2,3,4,5-Tetrachloroaniline	24–30
2	2,3,4,5-Tetrachloroanisole	23–35
3	2,3,5,6-Tetrachloroaniline	23–33
4	2,4-D	22–33
5	2,4-DB	22–33
6	2,6-Dichlorobenzamide	22–34
7	2,6-Difluorobenzoic acid	23–30
8	2-Phenylphenol	22–33
9	3,5-Dichloroaniline	22–30
10	3,5-Xylyl methylcarbamate	23–27
11	3,4-Trimethacarb	22–33
12	3-Hydroxycarbofuran	22–33
13	3-phenylphenol	24–31
14	4,4-Dibromobenzophenone	29–38
15	4,4-Dichlorobenzophenone	27–36
16	4-Bromo-3-dimethylphenyl N-methylcarbamate	22–33
17	4-Chlorophenoxyacetic acid	24–35
18	6-Chloro-4-hydroxy-3-phenyl-pyridazin	25–32
19	Acephate	22–33
20	Acequinocyl	20–30
21	Acetamiprid	26–34
22	Acetochlor	23–29

TABLE 5.18 GPC Analytical Parameters of 744 Pesticides and Chemical Pollutants—cont'd

No.	Compound	Collection Time (min)
23	Acifluorfen	17–30
24	Aclonifen	26–35
25	Acrinathrin	15–21
26	Alachlor	24–31
27	Aldicarb	23–32
28	Aldicarb sulfone	23–33
29	Aldicarb sulfoxide	24–32
30	Aldoxycarb	24–30
31	Aldrin	22–33
32	Allethrin mixture	20–29
33	Allidochlor	23–31
34	Alloxydim-sodium	21–29
35	Alpha-cypermethrin	21–27
36	Ametryn	22–29
37	Amidithion	22–36
38	Amidosulfuron	22–33
39	Aminocarb	24–30
40	Amitraz	23–36
41	Amitrole	23–32
42	Amobam	22–35
43	Anilazine	22–30
44	Anilofos	24–31
45	Anthraquinone	20–42
46	Aramite	18–27
47	Aroclor 1221	26–35
48	Aroclor 1232	26–36
49	Aroclor 1242	26–35
50	Aroclor 1254	25–35

Continued

TABLE 5.18 GPC Analytical Parameters of 744 Pesticides and Chemical Pollutants—cont'd

No.	Compound	Collection Time (min)
51	Aroclor 1260	25–35
52	Aroclor 1262	25–35
53	Aroclor 1268	26–32
54	Aspon	23–34
55	Asulam	21–29
56	Athidathion	19–33
57	Atratone	22–33
58	Atrazine	20–27
59	Atrazine-desethyl	20–26
60	Azaconazole	23–34
61	Azamethiphos	33–39
62	Azinphos-ethyl	25–35
63	Azinphos-methyl	24–38
64	Aziprotryne	23–30
65	Azoxystrobin	24–35
66	Benalaxyl	23–33
67	Benazolin	23–40
68	Bendiocarb	24–33
69	Benfluralin	18–28
70	Benfuracarb	20–30
71	Benfuresate	22–31
72	Benodanil	24–33
73	Benoxacor	26–33
74	Bensulfuron-methyl	25–32
75	Bensulide	20–33
76	Bensultap	24–33
77	Bentazone	23–31
78	Benzofenap	23–33

TABLE 5.18 GPC Analytical Parameters of 744 Pesticides and Chemical Pollutants—cont'd

No.	Compound	Collection Time (min)
79	Benzoximate	25–33
80	Benzoylprop-ethyl	23–30
81	Bifenazate	22–30
82	Bifenox	22–34
83	Bifenthrin	21–34
84	Bioallethrin	19–33
85	Bioresmethrin	21–33
86	Biphenyl	23–36
87	Bitertanol	19–27
88	Bromacil	22–29
89	Bromfenvinfos	23–31
90	Bromobutide	22–36
91	Bromocylen	22–32
92	Bromophos (-methyl)	23–33
93	Bromophos-ethyl	24–31
94	Bromopropylate	22–33
95	Bromoxynil	25–32
96	Brompyrazon	22–32
97	Bromuconazole	22–33
98	Bupirimate	21–28
99	Buprofezin	22–31
100	Butachlor	22–37
101	Butafenacil	19–25
102	Butamifos	22–38
103	Butocarboxim	26–33
104	Butocarboxim-sulfoxide	25–32
105	Butoxycarboxim	24–33
106	Butoxycarboxim-sulfoxid	25–32

Continued

TABLE 5.18 GPC Analytical Parameters of 744 Pesticides and Chemical Pollutants—cont'd

No.	Compound	Collection Time (min)
107	Butralin	21–28
108	Buturon	21–27
109	Butylate	22–30
110	Cadusafos	22–34
111	Cafenstrol	23–32
112	Captafol	22–33
113	Captan	23–35
114	Carbaryl	26–33
115	Carbendazim	16–33
116	Carbetamide	20–34
117	Carbofuran	24–31
118	Carbophenothion	23–35
119	Carbosulfan	19–28
120	Carboxin	28–37
121	Carfentrazone-ethyl	20–29
122	Cartap	22–37
123	Chlorbenside	27–34
124	Chlorbenside sulfone	25–33
125	Chlorbromuron	23–30
126	Chlorbufam	21–28
127	Chlordecone	22–32
128	Chlordimeform	26–34
129	Chlorfenapyr	20–26
130	Chlorfenethol	22–34
131	Chlorfenprop-methyl	23–34
132	Chlorfenson	23–33
133	Chlorfenvinphos	22–32
134	Chlorfluazuron	17–23

TABLE 5.18 GPC Analytical Parameters of 744 Pesticides and Chemical Pollutants—cont'd

No.	Compound	Collection Time (min)
135	Chlorflurenol-methyl	25–34
136	Chlorfurenol	25–34
137	Chloridazon	23–33
138	Chlorimuron-ethyl	20–31
139	Chlormephos	24–32
140	Chlormequat	20–28
141	Chlorobenzilate	22–28
142	Chloroneb	23–35
143	Chloropicrin	23–33
144	Chloropropylate	21–33
145	Chlorothalonil	27–35
146	Chlorotoluron	23–31
147	Chlorphoxim	23–29
148	Chlorpropham	21–33
149	Chlorpyrifos (-ethyl)	22–31
150	Chlorpyrifos-methyl	23–33
151	Chlorsulfuron	23–32
152	Chlorthal-dimethyl	23–33
153	Chlorthiamid	22–32
154	Chlorthion	25–33
155	Chlorthiophos	24–33
156	Chlozolinate	20–28
157	Chromafenozide	22–37
158	Chrysene	37–47
159	Cinosulfuron	17–33
160	cis-Permethrin	22–33
161	cis-1,2,3,6-Tetrahydrophthalimide	18–33
162	cis-Chlordane	22–32

Continued

TABLE 5.18 GPC Analytical Parameters of 744 Pesticides and Chemical Pollutants—cont'd

No.	Compound	Collection Time (min)
163	Clethodim	25–32
164	Clodinafop-propargyl	21–28
165	Clomazone	24–33
166	Clomeprop	22–35
167	Cloprop	22–33
168	Clopyralld	23–34
169	Cloquintocet-mexyl	23–30
170	Cloransulam-methyl	23–31
171	Clothianidin	22–29
172	Coumaphos	23–33
173	Coumatetralyl	26–34
174	Crimidine	28–35
175	Crotoxyphos	22–31
176	Crufomate	22–33
177	Cyanazine	19–33
178	Cyanofenphos	22–32
179	Cyanophos	20–28
180	Cyclanilide	23–33
181	Cycloate	23–34
182	Cycloprothrin	23–36
183	Cycloxydim	20–35
184	Cycluron	23–33
185	Cyfluthrin	19–26
186	Cymoxanil	22–30
187	Cypermethrin	20–33
188	Cyprazine	21–27
189	Cyproconazole	24–30
190	Cyprodinil	26–33

TABLE 5.18 GPC Analytical Parameters of 744 Pesticides and Chemical Pollutants—cont'd

No.	Compound	Collection Time (min)
191	Cyromazine	23–34
192	Cythioate	22–30
193	Dacthal(DCPA)	27–33
194	Dalapon	22–33
195	Daminozide	21–35
196	Dazomet	27–41
197	o,p'-DDD	23–33
198	p,p'-DDD	22–32
199	o,p'-DDE	25–32
200	p,p'-DDE	24–33
201	o,p'-DDT	26–32
202	p,p'-DDT	23–33
203	DEF	22–29
204	Deltamethrin	19–33
205	Demephion(Tinox)	20–33
206	Demeton (O+S)	23–32
207	Demeton-S	23–30
208	Demeton-S-methyl	24–33
209	Demeton-S-methyl sulfoxide	22–30
210	Demeton-S-methyl suphone	23–33
211	deneton-s-methyl	23–34
212	DE-PCB 28 2,4,4'-Trichlorobiphenyl	23–34
213	DE-PCB 31 2,4',5-Trichlorobiphenyl	23–35
214	DE-PCB 52 2,2',5,5'-Tetrachlorobiphenyl	23–34
215	DE-PCB 101 2,2',4,5,5'-Pentachlorobiphenyl	23–34
216	DE-PCB 118 2,3',4,4',5-Pentachlorobiphenyl	27–35
217	DE-PCB 138 2,2',3,4,4',5'-Hexachlorobiphenyl	23–34

Continued

TABLE 5.18 GPC Analytical Parameters of 744 Pesticides and Chemical Pollutants—cont'd

No.	Compound	Collection Time (min)
218	DE-PCB 153 2,2′,4,4′,5,5′-Hexachlorobiphenyl	23–34
219	DE-PCB 180 2,2′,3,4,4′,5,5′-Heptachlorobiphenyl	22–35
220	Desethyl-sebuthylazine	20–25
221	Desisopropyl-atrazine	21–27
222	Desmedipham	21–27
223	Desmethylformamido-pirimicarb	23–37
224	Desmethyl-norflurazon	19–33
225	Desmethyl-pirimicarb	22–29
226	Desmetryn	23–32
227	Diallate	22–34
228	Diafenthiuron	19–26
229	Dialifos	24–31
230	Diazinon	21–32
231	Dibutyl succinate	23–30
232	Dicamba	23–29
233	Dicapthon	26–33
234	Dichlobenil	23–34
235	Dichlofluanid	22–31
236	Dichlone	29–38
237	Dichlormid	23–32
238	Dichlorprop	20–33
239	Dichlorvos	22–33
240	Diclobutrazole	22–36
241	Diclofenthion	23–32
242	Diclofluanid	24–32
243	Diclofop-methyl	22–34
244	Dicloran	24–32
245	Dicofol	23–33

TABLE 5.18 GPC Analytical Parameters of 744 Pesticides and Chemical Pollutants—cont'd

No.	Compound	Collection Time (min)
246	Dicrotophos	26–35
247	Dieldrin	23–34
248	Dienochlor	23–33
249	Diethofencarb	20–30
250	Diethyltoluamide	26–34
251	Difenconazole	22–33
252	Difenoxuron	25–33
253	Difenzoquat-methyl sulfate	22–37
254	Diflufenican	19–28
255	Diflufenzopyr-sodium	21–35
256	Dimefox	23–33
257	Dimefuron	19–27
258	Dimepiperate	22–33
259	Dimethachlor	23–32
260	Dimethametryn	19–32
261	Dimethenamid	24–33
262	Dimethipin	21–33
263	Dimethirimol	23–29
264	Dimethoate	23–34
265	Dimethomorph	29–37
266	Dimethylaminosulfanilide	22–33
267	Dimethylaminosulfotoluidide	22–33
268	Diniconazole	19–30
269	Dinitramine	20–33
270	Dinobuton	21–28
271	Dinoseb	23–30
272	Dinoseb acetate	22–29
273	Dinotefuran	22–33

Continued

TABLE 5.18 GPC Analytical Parameters of 744 Pesticides and Chemical Pollutants—cont'd

No.	Compound	Collection Time (min)
274	Dinoterb	23–30
275	Diofenolan	24–32
276	Dioxabenzofos	28–35
277	Dioxacarb	22–32
278	Dioxathion	22–33
279	Diphenamid	17–22; 28–35
280	Diphenylamine	27–33
281	Dipropetryn	20–27
282	Diquat dibromide hydrate	23–34
283	Disufoton	26–31
284	Disulfoton sulfone	22–34
285	Disulfoton sulfoxide	22–34
286	Ditalimfos	23–34
287	Dithiopyr	19–26
288	Diuron	22–31
289	DMSA	23–30
290	DMST	22–29
291	DNOC	27–34
292	Dodemorph	19–29
293	Edifenphos	27–37
294	Endosulfan sulfate	22–32
295	Endosulfan, (alpha beta)	22–34
296	Endothal	26–32
297	Endrin	22–34
298	Endrin ketone	18–33
299	EPN	24–33
300	Epoxiconazole	23–30

TABLE 5.18 GPC Analytical Parameters of 744 Pesticides and Chemical Pollutants—cont'd

No.	Compound	Collection Time (min)
301	EPTC	19–28
302	Erbon	22–31
303	Esfenvalerate	20–33
304	Etaconazole	23–34
305	Ethalfluralin	19–33
306	Ethephon	21–33
307	Ethidimuron	24–31
308	Ethiofencarb	24–32
309	Ethiofencarbsulfon	23–34
310	Ethiofencarbsulfoxid	23–34
311	Ethion	23–31
312	Ethirimol	22–30
313	Ethofumesate	22–32
314	Ethoprophos	24–32
315	Ethoxysulfuron	22–33
316	Ethylene thiourea	14–37
317	Etofenprox	23–29
318	Etridiazole	27–33
319	Etrimfos	22–33
320	Etrimfos oxon	24–30
321	Famoxadone	22–29
322	Famphur	26–33
323	Fenamidone	22–30
324	Fenamiphos	22–29
325	Fenamiphos sulfone	21–34
326	Fenamiphos sulfoxide	22–34
327	Fenarimol	23–33
328	Fenazaflor	20–34

Continued

TABLE 5.18 GPC Analytical Parameters of 744 Pesticides and Chemical Pollutants—cont'd

No.	Compound	Collection Time (min)
329	Fenazaquin	22–33
330	Fenbuconazole	22–34
331	Fenbutatin oxide	22–31
332	Fenchlorphos oxon	22–34
333	Fenfuram	24–30
334	Fenhexamid	21–33
335	Fenitorthion	26–33
336	Fenitrothion	24–33
337	Fenobucarb	22–28
338	fenothiocarb	23–33
339	Fenoxanil	22–31
340	Fenoxaprop-ethyl	23–33
341	Fenoxycarb	22–30
342	Fenpiclonil	21–33
343	Fenpropathrin	20–28
344	Fenpropidin	20–25
345	Fenpropimorph	18–33
346	fenpyroximate	21–29
347	Fenson	21–33
348	Fensulfothion	23–33
349	Fenthion	27–37
350	Fenthion PO-sulfone	22–34
351	Fenthion PO-sulfoxide	23–37
352	Fenthion PS-sulfone	23–33
353	Fentin-chloride	21–34
354	Fenuron	26–33
355	Fenvalerate	16–27
356	Fipronil	16–21

TABLE 5.18 GPC Analytical Parameters of 744 Pesticides and Chemical Pollutants—cont'd

No.	Compound	Collection Time (min)
357	Flamprop-isopropyl	20–33
358	Flamprop-methyl	23–31
359	Flazasulfuron	21–33
360	florasulam	23–33
361	Fluazifop-butyl	18–26
362	Fluazinam	17–23
363	Fluazuron	17–23
364	Flubenzimine	18–24
365	Fluchloralin	21–27
366	Flucythrinate	20–30
367	Fludioxonil	20–26
368	Flufenacet	20–26
369	Flufenoxuron	16–24
370	Flumethrin	18–26
371	Flumetralin	20–26
372	Flumetsulam	23–31
373	Flumiclorac-pentyl	22–31
374	Flumioxazin	23–35
375	Fluometuron	21–26
376	Fluorochloridone	20–27
377	Fluorodifen	26–34
378	Fluoroglycofen-ethyl	20–26
379	Fluoroimide	19–30
380	Fluotrimazole	22–33
381	Fluquinconazole	24–32
382	Fluridone	23–30
383	Flurochloridone	20–27
384	Fluroxypr 1-methylheptyl	18–24

Continued

TABLE 5.18 GPC Analytical Parameters of 744 Pesticides and Chemical Pollutants—cont'd

No.	Compound	Collection Time (min)
385	Fluroxypyr	21–36
386	Flurtamone	21–28
387	Flusilazole	22–28
388	Flusulfamide	21–27
389	Fluthiacet-Methyl	28–36
390	Flutolanil	19–26
391	Flutriafol	22–33
392	Folpet	28–36
393	Fomesafen	18–28
394	Fonofos	26–34
395	Fonofos, O-analogue	26–33
396	Forchlorfenuron	23–35
397	Formothion	22–37
398	Fosthiazate	22–36
399	Fuberidazole	25–32
400	Furalaxyl	26–33
401	Furathiocarb	22–32
402	Furmecyclox	25–32
403	gamma-Cyhalothrin	23–38
404	Gibberellic acid	18–28
405	Halosulfuran-methyl	22–40
406	Haloxyfop-ehyoxyethyl	19–25
407	Haloxyfop-methyl	19–33
408	Alpha-HCH	23–34
409	Beta-HCH	23–34
410	Delta-HCH	23–33
411	Epsilon-HCH	22–30
412	Heptachlor	23–34

TABLE 5.18 GPC Analytical Parameters of 744 Pesticides and Chemical Pollutants—cont'd

No.	Compound	Collection Time (min)
413	Heptachlor epoxide, *cis*-	23–33
414	Heptachlor epoxide, *trans*-	20–34
415	Heptanophos	22–34
416	Hexachlorobenzene	25–37
417	Hexaconazole	22–33
418	Hexaflumuron	16–33
419	Hexazinone	28–36
420	Hexythiazox	25–33
421	Hydramethylnon	16–30
422	Imazalil	24–33
423	Imazamethabenz-methyl	22–28
424	Imazamox	19–31
425	Imazapyr	23–32
426	Imazaquin	24–31
427	Imazethapyr	23–29
428	Imibenconazole	27–33
429	Imidacloprid	25–35
430	Imiprothrin	22–31
431	Indoxacarb	20–26
432	Iodofenphos	23–34
433	Iodosulfuron-methyl	23–32
434	Ioxynil	26–37
435	Iprobenfos	22–29
436	Iprodione	18–23
437	Iprovalicarb	19–24
438	Isazofos	22–33
439	Isobenzan	23–33
440	Isocarbamid	23–34

Continued

TABLE 5.18 GPC Analytical Parameters of 744 Pesticides and Chemical Pollutants—cont'd

No.	Compound	Collection Time (min)
441	Isocarbophos	26–33
442	Isodrin	23–34
443	Isofenphos	20–33
444	Isomethiozin	26–36
445	Isoprocarb	22–29
446	Isopropalin	19–31
447	Isoprothiolane	27–34
448	Isoproturon	22–31
449	Isouron	20–33
450	Isoxaben	21–33
451	Isoxaflutole	21–33
452	Isoxathion	23–33
453	Kadethrin	23–33
454	Kelevan	23–34
455	Kresoxim-methyl	22–32
456	Lactofen	18–27
457	Lambda-cyhalothrin	18–33
458	Lenacil	23–34
459	Leptophos	27–34
460	Leptophos desbrom	27–34
461	Leptophos oxon	17–27
462	Lindane	22–31
463	Linuron	22–30
464	Lufenuron	16–19
465	Malaoxon	22–30
466	Malathion	22–30
467	Maleic hydrazide	23–31
468	MCPA butoxyethyl ester	22–34

TABLE 5.18 GPC Analytical Parameters of 744 Pesticides and Chemical Pollutants—cont'd

No.	Compound	Collection Time (min)
469	MCPB	19–31
470	Mecarbam	22–33
471	Mefenacet	23–38
472	Mefenoxam	23–33
473	Mefenpyr-diethyl	23–29
474	Mepanipyrim	26–34
475	Mephosfolan	27–36
476	Mepiquat chloride	23–33
477	Mepronil	21–30
478	Merphos	22–29
479	Mesotrion	26–40
480	Metalaxyl	25–33
481	Metamitron	27–34
482	Metamitron	23–33
483	Metazachlor	23–33
484	Metconazole	22–33
485	Methabenzthiazuron	28–39
486	Methacrifos	24–33
487	Methamidophos	25–34
488	Methfuroxam	23–33
489	Methidathion	28–36
490	Methiocarb	23–32
491	Methiocarb sulfoxide	22–32
492	Methomyl	23–36
493	Methoprene	18–28
494	Methoprotryne	21–34
495	Methothrin	20–32
496	Methoxychlor	26–34

Continued

TABLE 5.18 GPC Analytical Parameters of 744 Pesticides and Chemical Pollutants—cont'd

No.	Compound	Collection Time (min)
497	Methoxyfenozide	21–30
498	Metobromuron	24–33
499	Metolachlor	23–33
500	Metolcarb	22–32
501	Metominostrobin-(E)	23–32
502	Metominostrobin-(Z)	20–32
503	Metosulam	24–33
504	Metoxuron	25–33
505	Metribuzin	23–34
506	Metsulfuron-methyl	23–32
507	Mevinphos	23–34
508	Mexacarbate	22–34
509	Mirex	22–32
510	Molinate	21–36
511	Monalide	20–33
512	Monocrotophos	22–33
513	Monolinuron	23–31
514	Monuron	20–27
515	Musk ambrette	23–32
516	Musk Ketone	22–30
517	Musk moskene	22–32
518	Musk tibetene	23–33
519	Musk xylene	23–33
520	Myclobutanil	22–34
521	Naled (Dibrom)	23–34
522	Napropamide	22–32
523	Neburon	21–27
524	Nicosulfuron	24–34

TABLE 5.18 GPC Analytical Parameters of 744 Pesticides and Chemical Pollutants—cont'd

No.	Compound	Collection Time (min)
525	Nicotine	28–34
526	Nitenpyram	26–35
527	Nitralin	20–34
528	Nitrapyrin	24–33
529	Nitrofen	27–33
530	Nitrothal-isopropyl	21–33
531	Nonachlor, *trans*	22–34
532	Norflurazon	21–33
533	Novaluron	15–20
534	Nuarimol	23–30
535	Octachlorostyrene	23–34
536	Ofurace	24–34
537	Omethoate	22–33
538	Oryzalin	17–34
539	Oxadiazon	21–28
540	Oxadixyl	23–35
541	Oxamyl	23–29; 35–37
542	Oxycarboxin	26–34
543	Oxy-chlordane	22–30
544	Oxyfluorfen	21–33
545	Paclobutrazol	21–33
546	Paraoxon	23–31
547	Paraoxon-methyl	23–34
548	Paraquat Dichloride	23–40
549	Parathion-ethyl	22–33
550	Parathion-methyl	25–32
551	Pebulate	22–33

Continued

TABLE 5.18 GPC Analytical Parameters of 744 Pesticides and Chemical Pollutants—cont'd

No.	Compound	Collection Time (min)
552	Penconazole	21–33
553	pencycuron	24–32
554	Pendimethalin	23–33
555	Pentachloroaniline	25–33
556	Pentachloroanisole	28–35
557	Pentachlorobenzene	23–35
558	Pentachlorophenol	23–34
559	Permethrin	23–32
560	Perthane	21–34
561	Phenkapton	23–33
562	phenmedipham	21–28
563	Phenothrin	20–30
564	Phenthoate	24–33
565	Phorate	22–33
566	Phorate sulfone	22–34
567	Phorate sulfoxide	22–35
568	Phosalone	23–34
569	Phosmet	23–28; 29–37
570	Phosmet, O-analogue	23–36
571	Phosmet-oxon	23–33
572	Phosphamidon	24–31
573	phoxim	23–30
574	Phthalic acid, benzyl butyl ester	23–33
575	Phthalic acid, di-(2-ethylhexyl) ester	18–24
576	Phthalic acid, dibutyl ester	22–33
577	Phthalic acid, dicyclohexyl ester	22–34
578	Picloram	23–30
579	Picolinafen	20–29

TABLE 5.18 GPC Analytical Parameters of 744 Pesticides and Chemical Pollutants—cont'd

No.	Compound	Collection Time (min)
580	Picoxystrobin	22–28
581	Piperonyl butoxide	22–28
582	Piperophos	23–29
583	Pirimicarb	23–36
584	Pirimiphos-ethyl	21–33
585	Pirimiphos-methyl	22–33
586	Pirmicarb-desmethyl	23–24
587	Plifenate	22–33
588	Prallethrin	21–30
589	Pretilachlor	22–29
590	primisulfuron-methyl	18–24
591	Probenazole	22–30
592	Prochloraz	23–33
593	Procymidone	23–29
594	Profenofos	23–33
595	Profluralin	19–33
596	Prohydrojasmon	23–32
597	Promecarb	21–33
598	Prometon	22–28
599	Prometryne	21–34
600	Pronamide	20–33
601	Propachlor	23–32
602	Propamocarb	22–34
603	Propanil	22–28
604	Propaphos	22–33
605	Propargite	20–34
606	Propazine	19–33
607	Propetamphos	20–27

Continued

TABLE 5.18 GPC Analytical Parameters of 744 Pesticides and Chemical Pollutants—cont'd

No.	Compound	Collection Time (min)
608	Propham	24–34
609	Propiconazole	22–32
610	Propisochlor	22–30
611	Propoxur	23–33
612	Propylene thiourea	26–34
613	Propyzamide	21–30
614	Prosulfocarb	25–33
615	Prosulfuron	19–33
616	Prothiofos	23–33
617	Prothoate	28–33
618	Pymetrozin	26–34
619	Pyraclofos	29–40
620	Pyraclostrobin	25–33
621	Pyraflufen Ethyl	23–30
622	Pyrazolynate	22–27
623	Pyrazophos	23–33
624	Pyrazosulfuron-ethyl	21–32
625	Pyrethrin	20–27
626	Pyributicarb	23–31
627	Pyridaben	18–32
628	Pyridaphenthion	23–33
629	Pyridate	23–30
630	Pyrifenox	26–33
631	Pyriftalid	27–35
632	Pyrimethanil	24–33
633	Pyrimidifen	26–32
634	Pyrimitate	22–33
635	Pyriproxyfen	24–32

TABLE 5.18 GPC Analytical Parameters of 744 Pesticides and Chemical Pollutants—cont'd

No.	Compound	Collection Time (min)
636	Pyrithlobac sodium	23–32
637	Pyroquilon	35–44
638	Quinalphos	25–33
639	Quinmerac	27–36
640	Quinoclamine	27–35
641	Quinoxyphen	28–35
642	Quintozene	30–36
643	Quizalofop-ethyl	24–31
644	Rabenzazole	25–32
645	Resmethrin	22–33
646	Rimsulfuron	25–36
647	Ronnel	23–33
648	S421(Octachlorodipropyl ether)	26–34
649	Sebutylazine	20–26
650	Secbumeton	22–33
651	Sethoxydim	15–32
652	Simazine	22–34
653	Simeconazole	20–31
654	Simeton	24–33
655	Simetryn	24–33
656	Spinosad	17–24
657	Spirodiclofen	21–29
658	Spiroxamine	21–34
659	Sulfallate	25–37
660	Sulfanitran	18–26
661	Sulfentrazone	21–33
662	Sulfotep	22–33
663	Sulprofos	24–35

Continued

TABLE 5.18 GPC Analytical Parameters of 744 Pesticides and Chemical Pollutants—cont'd

No.	Compound	Collection Time (min)
664	Tau-fluvalinate	14–25
665	TCA-sodium	22–33
666	TCMTB	26–37
667	Tebuconazole	21–33
668	Tebufenozide	19–29
669	Tebufenpyrad	21–28
670	Tebupirimfos	20–27
671	Tebutam	23–33
672	Tebuthiuron	24–32
673	Tecnazene	23–33
674	Teflubenzuron	20–26
675	Tefluthrin	18–23; 24–33
676	Temephos	24–33
677	TEPP	17–32
678	Tepraloxydim	24–32
679	Terbacil	22–33
680	Terbucarb	20–31
681	Terbufos	23–29
682	Terbufos sulfone O-analogue	23–33
683	Terbumeton	22–33
684	Terbutryne	22–33
685	Terbutylazine	21–32
686	Tetrachlorvinphos	25–32
687	Tetraconazole	19–30
688	Tetradifon	23–33
689	Tetrahydrophthalimide	23–33
690	Tetramethrin	23–34
691	Tetrasul	27–34

TABLE 5.18 GPC Analytical Parameters of 744 Pesticides and Chemical Pollutants—cont'd

No.	Compound	Collection Time (min)
692	Thenylchlor	26–34
693	Thiabendazole	27–35
694	Thiacloprid	28–36
695	Thiamethoxam	27–37
696	Thiazopyr	20–27
697	Thifensulfuron-methyl	23–35
698	Thiobencarb	26–33
699	Thiodicarb	28–38
700	Thiofanox	24–30
701	Thiofanox sulfone	23–34
702	Thiofanox-sulfoxid	23–34
703	Thiometon	18–34
704	Thionazin	26–33
705	Thiophanat-methyl	22–30
706	Thiram	38–50
707	Tolclofos-methyl	21–31
708	Tolfenpyrad	22–31
709	Tolylfluanide	19–33
710	Tralkoxydim	23–30
711	*trans*-Chlordane	25–34
712	Transfluthrin	19–28
713	*trans*-permethrin	22–29
714	Triadimefon	22–32
715	Triadimenol	20–34
716	Triallate	23–30
717	Triamiphos	22–33
718	Triasulfuron	22–33
719	Triazophos	23–33

Continued

TABLE 5.18 GPC Analytical Parameters of 744 Pesticides and Chemical Pollutants—cont'd

No.	Compound	Collection Time (min)
720	Triazoxide	32–40
721	Tribenuron-methyl	26–33
722	Trichlorfon	23–30
723	Trichloronat	22–31
724	Tricyclazole	33–43
725	Tridemorph	22–30
726	Trietazine	22–29
727	Trifloxystrobin	21–27
728	Triflumizole	20–26
729	Triflumuron	18–25
730	Trifluralin	18–34
731	Triflusulfuron-methyl	18–33
732	Triforine	18–26
733	Tri-isobutyl phosphate	22–31
734	Trimethylsulfonium iodide	22–31
735	Tri-n-butyl phosphate	23–34
736	Triphenyl phosphate	23–32
737	Triticonazole	22–33
738	Uniconazole	22–28
739	Vamidothion	22–35
740	Vamidothion sulfone	23–33
741	Vamidothion sulfoxide	25–33
742	Vernolate	22–32
743	Vinclozolin	22–30
744	Zoxamide	20–27

5.3.2   GPC Analytical Parameters of 107 Pesticides and Chemical Pollutants

See Table 5.19.

TABLE 5.19 GPC Analytical Parameters of 107 Pesticides and Chemical Pollutants

No.	Compound	Start Collection Time (min)	Stop Collection Time (min)
1	2,4′−DDD	27	39
2	2,4′−DDE	29	38
3	2,4-D	24	36
4	2,4′-DDT	24	34
5	4,4′−DDD	26	38
6	4,4′−DDE	28	39
7	4,4′-DDT	24	34
8	Acenaphthene	30	43
9	Acenaphthylene	29	41
10	Acetamiprid	26	36
11	Alachlor	24	34
12	Aldrin	27	36
13	Amitraz	24	33
14	Anthracene	30	45
15	Atrazine	21	35
16	Bensulfuron methyl	23	34
17	Benzo[a]anthracene	38	48
18	Benzo[a]pyrene	43	56
19	Benzo[b]fluoranthene	35	51
20	Benzo[g,h,i]peryene	−	−
21	Benzo[k]fluoranthene	38	50
22	Buprofezin	23	33
23	Butachlor	22	34
24	Carbofuran	25	34
25	Chlordimeform	27	35
26	Chlorolurons	25	35

Continued

TABLE 5.19 GPC Analytical Parameters of 107 Pesticides and Chemical Pollutants—cont'd

No.	Compound	Start Collection Time (min)	Stop Collection Time (min)
27	Chlorothalonil	28	37
28	Chlorpyrifos	24	35
29	cis-Chlordane	–	–
30	Clomazone	27	35
31	Cyfluthrin	20	29
32	Cypermethrin	21	35
33	Deltamethrin	22	35
34	Diazinon	23	34
35	Dibenzo[a,h] anthracene	40	53
36	Dichlorvos	29	34
37	Dicofol	23	29
38	Dieldrin	25	36
39	Dimethoate	23	34
40	Endosulfuran I	22	35
41	Endrin	28	35
42	Ethoprophos	23	34
43	Fenvalerate-1	21	28
44	Fipronil	16	34
45	Fluoranthene	35	46
46	Fluorene	29	43
47	Heptachlor	23	34
48	Hexachlorobenzene	29	33
49	Hexythiazox	25	34
50	Indeno[1,2,3-cd]pyrene	26	40
51	Isoproturon	22	33
52	Lamba cyhalothrin	18	27
53	Methamidophos	23	33
54	Metolachlor	23	34
55	Metsulfuron-methyl	24	34

TABLE 5.19 GPC Analytical Parameters of 107 Pesticides and Chemical Pollutants—cont'd

No.	Compound	Start Collection Time (min)	Stop Collection Time (min)
56	Mirex	23	34
57	Naphthalene	30	38
58	Nithophen	27	35
59	Omethoate	24	42
60	Oxyfluorfen	21	29
61	Parathion-methyl	25	35
62	Phenathrene	29	45
63	Phenthoate	23	36
64	Phoxim	23	34
65	Phthalic acid benzylbutyl ester	24	34
66	Phthalic acid bis-2-ethylhexyl ester	18	34
67	Phthalic acid bis-butyl ester	22	35
68	Pirimicarb	23	37
69	Prometryne	22	29
70	Propanil	22	29
71	Propargite	22	33
72	Pyrene	29	50
73	Pyridaben	21	34
74	Simazine	23	33
75	*trans*-Chlordane	20	35
76	Triadimefon	22	35
77	Triazophos	24	34
78	Trichlorfon	23	37
79	Tricyclazole	35	45
80	Trifluralin	19	27
81	α-HCH	24	36
82	β-HCH	23	35
83	γ-HCH	23	35

Continued

TABLE 5.19 GPC Analytical Parameters of 107 Pesticides and Chemical Pollutants—cont'd

No.	Compound	Start Collection Time (min)	Stop Collection Time (min)
84	δ-HCH	23	35
85	Aroclor 28	27	35
86	Aroclor 52	23	34
87	Aroclor 101	25	35
88	Aroclor 118	27	36
89	Aroclor 138	26	35
90	Aroclor 153	24	35
91	Aroclor 180	24	35
92	Aroclor 1016	28	36
93	Aroclor 1221	29	36
94	Aroclor 1232	26	41
95	Aroclor 1242	28	36
96	Aroclor 1248	28	36
97	Aroclor 1254	29	34
98	Aroclor 1260	26	35
99	PCB-1,2,3	20	38
100	PCB-4,5,8,10,12	20	38
101	PCB-16,19,20,23,24	20	37
102	PCB-48,51,52,59,60	20	36
103	PCB-108,109,111,112,113	20	36
104	PCB-133,134,137,138,139	25	36
105	PCB-170,172,179,190,192	20	36
106	PCB-194,195,197,198,200	20	35
107	PCB-206,207,208	20	35

Appendix A

Index of 1039 Pesticides and Chemical Pollutants Determined by GC-MS, GC-MS-MS and LC-MS-MS

No.	Compound	GC-MS									GC-MS/MS	LC-MS-MS									
		Fruit and Vegetable 500ᵃ	Cereals 475	Honey, Juice and Juice Wine 497	Tea 519	Edible Fungi 503	Chin. Herb. Med. 488	Animal Muscle 478	Puffer, Sea Eel and Prawn 485	Milk Powder 511	Animal Muscle 295	Fruit and Vegetable 450	Cereals 486	Honey, Juice and Juice Wine 512	Tea 448	Edible Fungi 440	Chin. Herb. Medi. 413	Animal Muscle 461	Honey 486	Puffer, Sea Eel and Prawn 450	Milk Powder 493
1.	Acenaphthene	✓	✓	✓			✓	✓	✓	✓	✓										
2.	Acenaphthylene										✓										
3.	Acephate	✓	✓	✓		✓	✓	✓	✓	✓	✓		✓	✓		✓		✓	✓	✓	✓
4.	Acetamiprid	✓	✓	✓	✓	✓	✓	✓	✓	✓	✓	✓	✓	✓	✓	✓	✓	✓	✓	✓	✓
5.	Acetochlor	✓	✓		✓	✓		✓	✓	✓		✓	✓	✓	✓	✓	✓	✓	✓	✓	✓
6.	Acibenzolar-s-methyl				✓	✓				✓		✓	✓	✓	✓	✓	✓	✓	✓	✓	✓
7.	Acifluorfen				✓		✓	✓	✓				✓	✓				✓	✓		✓
8.	Aclonifen	✓	✓	✓		✓				✓		✓	✓	✓	✓	✓	✓			✓	✓
9.	Acrinathrin					✓	✓			✓				✓							✓
10.	Acrylamide			✓								✓			✓	✓	✓			✓	✓

Continued

No.	Compound																				
11.	Alachlor	✓	✓	✓	✓	✓	✓	✓	✓	✓							✓	✓	✓	✓	✓
12.	Aldicarb		✓	✓	✓	✓	✓	✓	✓	✓							✓	✓	✓	✓	✓
13.	Aldicarb sulfone			✓	✓	✓	✓	✓	✓	✓							✓	✓	✓	✓	✓
14.	Aldimorph				✓	✓	✓	✓	✓	✓		✓					✓	✓	✓	✓	✓
15.	Aldrin	✓	✓	✓	✓	✓	✓	✓	✓	✓		✓	✓	✓	✓	✓	✓	✓	✓	✓	✓
16.	Allethrin	✓	✓	✓	✓	✓	✓	✓	✓	✓		✓	✓	✓	✓	✓	✓	✓	✓	✓	✓
17.	Allidochlor	✓	✓	✓	✓	✓	✓	✓	✓	✓			✓	✓	✓	✓	✓	✓	✓	✓	✓
18.	Alloxydim-sodium	✓	✓	✓			✓		✓				✓	✓	✓	✓	✓	✓	✓		✓
19.	Alpha-cypermethrin	✓	✓	✓	✓	✓	✓	✓	✓	✓	✓	✓	✓	✓	✓	✓	✓	✓	✓	✓	✓
20.	Alpha-HCH	✓	✓	✓	✓	✓	✓	✓	✓	✓		✓	✓	✓	✓	✓	✓	✓	✓	✓	✓
21.	Ametryn	✓										✓		✓	✓	✓	✓	✓	✓	✓	✓
22.	Amidithion																✓	✓	✓	✓	✓
23.	Aminocarb																✓	✓	✓	✓	✓
24.	Aminopyralid																			✓	
25.	Amitraz						✓	✓	✓	✓		✓		✓	✓	✓	✓				✓
26.	Anilofos	✓	✓	✓	✓	✓	✓	✓	✓	✓		✓	✓	✓	✓	✓	✓	✓	✓	✓	✓
27.	Anthracene	✓	✓	✓	✓	✓					✓										
28.	Anthraquinone	✓	✓	✓	✓	✓	✓	✓	✓					✓	✓	✓				✓	✓
29.	Aramite	✓	✓	✓										✓	✓					✓	

30.	Arboxin	✓	✓	✓	✓	✓	✓		✓	✓			✓	✓	✓	✓	✓	✓	✓			✓		✓				
31.	Aspon	✓	✓	✓	✓	✓	✓	✓	✓			✓	✓	✓	✓	✓		✓		✓	✓	✓	✓	✓	✓				
32.	Athidathion	✓	✓	✓	✓		✓		✓	✓			✓		✓	✓	✓	✓		✓			✓						
33.	Atratone	✓	✓	✓	✓	✓	✓	✓	✓	✓	✓	✓	✓	✓	✓	✓	✓	✓	✓	✓	✓	✓	✓	✓	✓				
34.	Atrazine	✓	✓	✓	✓	✓	✓		✓	✓	✓	✓	✓	✓	✓	✓	✓	✓	✓	✓	✓	✓	✓	✓					
35.	Atrazine-desethyl	✓	✓	✓	✓	✓	✓	✓				✓	✓	✓	✓	✓	✓	✓	✓	✓	✓	✓	✓						
36.	Atrizine	✓	✓	✓	✓	✓	✓		✓	✓	✓			✓	✓	✓	✓	✓	✓	✓	✓	✓	✓		✓				
37.	Azaconazole										✓			✓	✓	✓	✓	✓	✓	✓	✓	✓							
38.	Azamethiphos	✓	✓	✓	✓	✓	✓	✓	✓	✓	✓	✓	✓	✓	✓	✓	✓	✓	✓	✓	✓	✓	✓						
39.	Azinphos ethyl	✓	✓	✓	✓	✓	✓	✓	✓	✓	✓	✓	✓	✓	✓	✓	✓	✓	✓	✓	✓	✓	✓						
40.	Azinphos-methyl	✓	✓	✓	✓	✓	✓	✓	✓	✓	✓	✓	✓	✓	✓	✓	✓	✓	✓	✓	✓	✓	✓						
41.	Aziprotryne	✓	✓	✓	✓	✓	✓	✓	✓	✓	✓		✓	✓	✓	✓	✓	✓	✓	✓	✓	✓							
42.	Azoxystrobin	✓	✓	✓	✓	✓	✓	✓	✓	✓	✓			✓	✓	✓	✓	✓	✓	✓	✓	✓							
43.	BDMC-1	✓	✓	✓	✓	✓	✓	✓	✓					✓	✓	✓	✓	✓	✓	✓	✓	✓							
44.	BDMC-2	✓	✓	✓	✓	✓	✓	✓	✓					✓	✓	✓	✓	✓	✓	✓	✓	✓							
45.	Bediocarb													✓	✓	✓	✓	✓	✓		✓	✓							
46.	Benalaxyl	✓	✓	✓	✓	✓	✓	✓	✓					✓	✓	✓	✓	✓	✓	✓	✓	✓	✓	✓	✓				
47.	Bendiocarb	✓	✓	✓	✓	✓	✓	✓	✓					✓	✓	✓	✓	✓	✓	✓	✓	✓	✓	✓	✓				
48.	Benfluralin	✓	✓	✓	✓	✓	✓	✓	✓	✓				✓	✓	✓	✓	✓	✓				✓	✓	✓				
49.	Benfuracarb-methyl	✓	✓	✓	✓	✓	✓	✓	✓					✓	✓	✓	✓	✓	✓				✓	✓	✓				

50.	Benfuresate	
51.	Benodanil	
52.	Benoxacor	
53.	Bensulfuron-methyl	
54.	Bensulide	
55.	Bentazone	
56.	Benzo(a)anthrancene	
57.	Benzo(a)pyrene	
58.	Benzo(b)fluoranthene	
59.	Benzo(g,h,i)peryene	
60.	Benzo(k)fluoranthene	
61.	Benzofenap	
62.	Benzoximate	
63.	Benzoylprop-ethyl	
64.	Benzyladenine	
65.	Beta-HCH	
66.	Bifenazate	
67.	Bifenox	
68.	Bifenthrin	
69.	Bioallethrin-1	

Continued

No.	Compound
70.	Bioallethrin-2
71.	Bioresmethrin
72.	Biphenyl
73.	Bispyribacsodium
74.	Bitertanol
75.	Boscalid
76.	Bromacil
77.	Bromfenvinfos
78.	Bromobutide
79.	Bromocylen
80.	Bromofos
81.	Bromophos-ethyl
82.	Bromopropylate
83.	Bromoxynil
84.	Brompyrazon
85.	Bromuconazole-1
86.	Bromuconazole-2
87.	Bupirimate
88.	Buprofezin
89.	Butachlor

#	Name	1	2	3	4	5	6	7	8	9	10	11	12	13	14	15	16	17	18	19	20
90.	Butafenacil	√	√	√	√	√	√	√	√	√		√	√	√	√	√	√	√	√		√
91.	Butamifos	√	√	√	√	√	√	√	√	√											
92.	Butocarboxim												√	√	√	√	√	√	√	√	√
93.	Butoxycarboxim											√	√	√	√	√	√	√	√	√	√
94.	Butralin	√	√	√	√	√	√	√	√	√		√	√	√	√	√	√	√	√		
95.	Buturon											√	√	√	√	√	√	√	√	√	√
96.	Butylate	√	√	√		√	√	√		√		√	√	√	√	√	√	√	√	√	√
97.	Cadusafos	√	√	√	√	√	√	√	√	√		√		√	√	√	√	√	√	√	√
98.	Cafenstrole	√	√	√	√			√	√										√		
99.	Captafol			√			√			√											
100.	Captan	√					√														
101.	Carbaryl				√	√	√		√	√		√	√	√	√	√	√	√	√	√	√
102.	Carbendazim											√	√	√	√	√		√	√	√	√
103.	Carbetamide											√	√	√	√	√	√	√	√	√	√
104.	Carbofenothion	√	√	√	√	√	√	√	√	√											
105.	Carbofuran										√	√	√	√	√	√	√	√	√	√	√
106.	Carbosulfan			√	√		√	√	√						√						
107.	Carboxin	√	√	√	√	√	√	√	√	√		√		√	√	√	√	√			√
108.	Carfentrazone-ethyl	√	√	√	√	√	√	√	√	√									√		
109.	Carpropamid											√	√	√	√	√	√	√	√	√	√

Continued

No.	Compound
110.	Cartap hydrochloride
111.	Chloramben
112.	Chloramphenicolum
113.	Chlorbenside
114.	Chlorbenside sulfone
115.	Chlorbromuron
116.	Chlorbufam
117.	Chlordimeform
118.	Chlordimeform hydrochloride
119.	Chlorethoxyfos
120.	Chlorfenapyr
121.	Chlorfenethol
122.	Chlorfenprop-methyl
123.	Chlorfenson
124.	Chlorfenvinphos
125.	Chlorfluazuron
126.	Chlorflurenol
127.	Chloridazon
128.	Chlorimuron ethyl
129.	Chlormephos

		1	2	3	4	5	6	7	8	9	10	11	12	13	14	15	16	17	18
130.	Chlormequat	✓				✓		✓	✓										
131.	Chlormequat chloride					✓		✓		✓									
132.	Chlorobenzilate	✓	✓	✓	✓	✓		✓	✓			✓	✓	✓	✓	✓	✓		✓
133.	Chlorobenzuron	✓	✓	✓	✓	✓		✓	✓			✓	✓	✓	✓	✓			✓
134.	Chlorolurons																		
135.	Chloroneb											✓	✓	✓	✓	✓	✓		✓
136.	Chloropropylate	✓										✓	✓	✓	✓		✓		✓
137.	Chlorothalonil									✓				✓					✓
138.	Chlorotoluron	✓	✓	✓	✓	✓	✓	✓	✓			✓	✓	✓	✓	✓	✓		✓
139.	Chloroxuron	✓	✓	✓	✓	✓	✓	✓	✓		✓	✓	✓	✓	✓	✓	✓		✓
140.	Chlorphoxim	✓	✓	✓	✓	✓	✓	✓	✓		✓	✓	✓	✓	✓	✓	✓		✓
141.	Chlorpropham	✓	✓	✓	✓	✓	✓	✓	✓		✓	✓	✓	✓	✓	✓	✓		✓
142.	Chlorpyrifos	✓	✓	✓	✓	✓	✓	✓	✓		✓	✓	✓	✓	✓	✓	✓		✓
143.	Chlorpyrifos methyl	✓	✓	✓	✓	✓	✓		✓			✓	✓	✓	✓	✓	✓		✓
144.	Chlorsulfuron	✓	✓	✓	✓		✓		✓		✓	✓	✓	✓	✓	✓	✓		✓
145.	Chlorthal-dimethyl											✓	✓	✓	✓	✓	✓		✓
146.	Chlorthiamid	✓	✓	✓	✓	✓	✓	✓	✓	✓		✓	✓	✓	✓	✓	✓		✓
147.	Chlorthion	✓	✓	✓	✓									✓		✓	✓		✓
148.	Chlorthiophos	✓	✓	✓	✓	✓	✓	✓	✓			✓	✓	✓	✓	✓	✓		✓
149.	Chlortoluron											✓	✓	✓	✓	✓	✓		✓

Continued

		1	2	3	4	5	6	7	8	9	10	11	12	13	14	15	16	17	18	19	20	21	22	23	24
150.	Chlozolinate	✓	✓	✓	✓	✓	✓	✓															✓		
151.	Chromafenozide	✓	✓	✓	✓	✓	✓	✓																	
152.	Chrysene			✓	✓																				
153.	Cinidon-ethyl					✓		✓						✓	✓	✓	✓	✓	✓	✓		✓	✓	✓	✓
154.	Cinmethylin				✓																				
155.	Cinosulfuron																								
156.	cis and trans-Chlorfenvinphos		✓								✓			✓	✓	✓	✓	✓	✓						
157.	cis and trans-Diallate																								
158.	cis and trans-Diallate		✓				✓																		
159.	cis-Chlordane	✓	✓	✓	✓	✓	✓	✓						✓	✓	✓	✓	✓	✓	✓	✓	✓	✓	✓	
160.	cis-1,2,3,6-Tetrahydrophthalimide																					✓			
161.	cis-Chlordane																								
162.	cis-Diallate	✓	✓	✓	✓	✓	✓	✓	✓	✓		✓		✓	✓	✓	✓	✓	✓	✓					
163.	cis-Permethrin	✓	✓	✓	✓	✓	✓	✓	✓	✓		✓		✓	✓	✓	✓	✓	✓	✓					
164.	Clethodim	✓	✓	✓	✓	✓	✓	✓		✓		✓		✓	✓	✓	✓	✓	✓	✓					
165.	Clodinafop propargyl	✓	✓	✓	✓	✓	✓	✓						✓	✓	✓	✓	✓	✓	✓	✓	✓	✓	✓	✓
166.	Clofentezine	✓	✓	✓	✓	✓	✓	✓						✓	✓	✓	✓	✓	✓	✓	✓	✓	✓	✓	✓
167.	Clomazone	✓	✓	✓	✓	✓	✓	✓						✓	✓	✓	✓	✓	✓	✓	✓	✓	✓	✓	✓
168.	Clomeprop		✓	✓	✓	✓	✓	✓						✓	✓	✓	✓	✓	✓	✓	✓	✓	✓	✓	✓

169.	Cloprop							✓	✓	✓			✓								✓	✓
170.	Clopyralid						✓	✓					✓								✓	✓
171.	Cloquintocet mexyl	✓			✓	✓	✓	✓	✓	✓	✓	✓	✓	✓	✓	✓		✓	✓	✓	✓	✓
172.	Clothianidin			✓	✓	✓	✓	✓	✓	✓	✓	✓	✓	✓	✓	✓		✓	✓	✓	✓	✓
173.	Coumaphos	✓			✓	✓	✓	✓	✓	✓	✓	✓	✓	✓		✓		✓	✓	✓	✓	✓
174.	Coumatetralyl			✓	✓	✓	✓					✓											
175.	Crimidine	✓		✓	✓		✓	✓	✓	✓	✓	✓	✓	✓	✓	✓		✓	✓	✓	✓	✓
176.	Crotoxyphos	✓		✓																			
177.	Crufomate	✓	✓	✓	✓	✓	✓	✓	✓	✓	✓	✓	✓	✓	✓	✓		✓	✓	✓	✓	✓
178.	Cumyluron		✓	✓	✓	✓	✓	✓	✓	✓	✓	✓	✓	✓	✓	✓		✓	✓	✓	✓	✓
179.	Cyanazine	✓	✓	✓	✓	✓					✓	✓	✓	✓	✓	✓		✓	✓	✓	✓	✓
180.	Cyanofenphos	✓	✓	✓							✓	✓	✓	✓	✓	✓		✓	✓	✓	✓	✓
181.	Cyanophos	✓	✓								✓	✓	✓	✓	✓	✓				✓	✓	✓
182.	Cyazofamid						✓	✓	✓	✓					✓	✓		✓		✓		✓	✓
183.	Cyclanilide																						
184.	Cycloate	✓		✓	✓	✓	✓				✓	✓	✓	✓	✓	✓		✓	✓	✓	✓	✓
185.	Cyclosulfamuron	✓		✓	✓	✓	✓												✓		✓	✓
186.	Cycloxydim	✓	✓	✓	✓	✓	✓											✓		✓	✓	✓
187.	Cycluron	✓		✓	✓																✓	✓
188.	Cyflufenamid	✓		✓	✓	✓	✓											✓	✓	✓	✓	✓

Continued

No.	Compound																				
189.	Cyfluthrin	✓	✓									✓	✓	✓	✓	✓		✓	✓	✓	✓
190.	Cyhalofop-butyl	✓	✓								✓	✓	✓	✓	✓	✓				✓	✓
191.	Cymoxanil	✓	✓	✓	✓	✓	✓	✓			✓	✓	✓	✓	✓	✓		✓	✓	✓	✓
192.	Cypermethrin	✓	✓	✓	✓	✓	✓	✓		✓	✓	✓	✓	✓	✓	✓		✓	✓	✓	✓
193.	Cyprazine	✓	✓	✓	✓	✓	✓	✓			✓	✓	✓	✓	✓	✓		✓	✓	✓	✓
194.	Cyproconazole	✓	✓	✓	✓	✓	✓	✓			✓	✓	✓	✓	✓	✓		✓	✓	✓	✓
195.	Cyprodinil	✓	✓	✓	✓	✓	✓				✓	✓	✓	✓	✓			✓	✓	✓	✓
196.	Cyromazine																				
197.	Cythioate																				
198.	Dacthal	✓	✓	✓	✓	✓	✓	✓		✓	✓	✓	✓	✓	✓	✓		✓	✓	✓	✓
199.	Daimuron																				
200.	Dalapon																				
201.	Daminozide																				
202.	Dazomet																				
203.	Def	✓	✓	✓	✓	✓	✓	✓		✓	✓	✓	✓	✓	✓	✓		✓	✓	✓	✓
204.	delta-HCH	✓	✓	✓	✓	✓	✓	✓		✓	✓	✓	✓	✓	✓	✓		✓	✓	✓	✓
205.	Deltamethrin	✓	✓	✓	✓	✓	✓	✓			✓	✓	✓	✓	✓					✓	✓
206.	Demetom-s	✓	✓			✓					✓	✓	✓								
207.	Demeton(o+s)																				
208.	Demeton-s											✓									
209.	Demeton-s-methyl	✓									✓	✓	✓	✓	✓			✓	✓	✓	✓

No.	Compound																				
210.	Demeton-s-methyl sulfone											✓	✓	✓	✓	✓	✓	✓	✓	✓	
211.	Demeton-s-methyl sulfoxide											✓	✓	✓	✓	✓	✓	✓	✓	✓	
212.	de-PCB 101	✓	✓	✓	✓	✓	✓	✓													
213.	de-PCB 118	✓	✓	✓	✓	✓	✓	✓													
214.	de-PCB 138	✓	✓	✓	✓	✓	✓	✓													
215.	de-PCB 153	✓	✓	✓	✓	✓	✓	✓													
216.	de-PCB 180	✓	✓	✓	✓	✓	✓	✓													
217.	de-PCB 28	✓	✓	✓	✓	✓	✓	✓													
218.	de-PCB 31	✓	✓	✓	✓	✓	✓	✓													
219.	de-PCB 52	✓	✓	✓	✓	✓	✓	✓													
220.	Desbrom-leptophos	✓	✓	✓	✓	✓															
221.	Desethyl-sebuthylazine	✓	✓	✓		✓	✓														
222.	Desisopropyl-atrazine	✓	✓			✓	✓														
223.	Desmedipham	✓	✓	✓	✓	✓	✓			✓	✓	✓	✓	✓			✓	✓	✓	✓	
224.	Desmetryn	✓	✓	✓	✓	✓	✓										✓	✓	✓	✓	
225.	Diafenthiuron	✓	✓	✓						✓	✓	✓	✓	✓	✓	✓	✓	✓	✓	✓	
226.	Dialifos	✓	✓	✓	✓					✓	✓	✓	✓	✓	✓	✓	✓	✓	✓	✓	
227.	Diallate	✓	✓	✓	✓	✓				✓	✓	✓	✓	✓	✓	✓	✓	✓	✓	✓	
228.	Diazinon	✓	✓	✓	✓	✓	✓		✓	✓	✓	✓	✓	✓	✓	✓	✓	✓	✓	✓	

Continued

No.	Compound
229.	Dibenzo[a,h]anthracene
230.	Dibutyl succinate
231.	Dicamba
232.	Dicapthon
233.	Dichlobenil
234.	Dichlofenthion
235.	Dichlofluanid
236.	Dichloran
237.	Dichlormid
238.	Dichlorprop
239.	Dichlorvos
240.	Diclobutrazole
241.	Diclofop-methyl
242.	Dicloran
243.	Dicofol
244.	Dicrotophos
245.	Dieldrin
246.	Diethofencarb
247.	Diethyltoluamide
248.	Difenconazole-1
249.	Difenconazole-2

No.	Compound																		
250.	Difenoxuron	✓										✓				✓	✓		
251.	Difenzoquat-methyl sulfate		✓	✓	✓	✓		✓	✓	✓		✓	✓		✓	✓	✓	✓	✓
252.	Diflufenican	✓	✓	✓	✓			✓	✓	✓	✓	✓	✓	✓	✓	✓	✓	✓	✓
253.	Diflufenzopyr-sodium			✓	✓		✓				✓	✓	✓	✓		✓	✓		✓	
254.	Dimefox	✓	✓	✓	✓	✓	✓	✓	✓	✓	✓	✓	✓	✓	✓	✓	✓	✓	✓
255.	Dimefuron	✓	✓	✓	✓	✓	✓			✓		✓	✓		✓			✓	
256.	Dimehypo			✓	✓	✓				✓			✓		✓	✓	✓		✓	
257.	Dimepiperate	✓	✓	✓	✓	✓		✓	✓	✓	✓	✓	✓	✓	✓	✓	✓	✓	✓
258.	Dimethachlor	✓	✓	✓	✓	✓		✓	✓	✓	✓	✓	✓	✓	✓	✓	✓	✓	✓
259.	Dimethametryn	✓	✓	✓	✓	✓		✓	✓	✓	✓	✓	✓	✓	✓	✓	✓	✓	✓
260.	Dimethenamid	✓	✓		✓	✓		✓		✓	✓	✓	✓	✓	✓	✓	✓	✓	✓
261.	Dimethipin	✓	✓	✓	✓	✓		✓	✓	✓	✓		✓	✓					
262.	Dimethirimol			✓	✓	✓		✓	✓	✓	✓	✓	✓	✓		✓	✓	✓	✓
263.	Dimethoate		✓	✓	✓	✓	✓	✓	✓	✓	✓	✓	✓	✓	✓	✓	✓	✓	✓
264.	Dimethomorph	✓	✓	✓	✓	✓		✓	✓	✓	✓	✓	✓	✓		✓	✓	✓	✓
265.	Dimethyl phthalate		✓	✓	✓		✓			✓		✓	✓	✓		✓	✓	✓	✓
266.	Dimethylvinphos	✓	✓	✓	✓	✓		✓	✓	✓									
267.	Diniconazole	✓	✓	✓	✓	✓		✓	✓	✓	✓	✓	✓	✓	✓	✓	✓	✓	✓
268.	Dinitramine	✓	✓	✓	✓	✓		✓	✓	✓	✓	✓	✓	✓	✓	✓	✓	✓	✓
269.	Dinobuton	✓	✓		✓	✓		✓			✓	✓	✓	✓		✓	✓	✓	✓

Continued

#	Compound																	
270.	Dinoseb	✓	✓	✓	✓		✓		✓	✓								
271.	Dinoseb acetate	✓	✓	✓	✓	✓	✓	✓	✓	✓								
272.	Dinotefuram	✓	✓	✓	✓	✓	✓	✓	✓	✓								
273.	Dinotefuran	✓	✓	✓	✓	✓	✓	✓	✓	✓		✓	✓	✓				
274.	Dinoterb	✓	✓	✓	✓	✓	✓	✓	✓	✓		✓	✓				✓	✓
275.	Diofenolan-1	✓	✓	✓	✓	✓	✓	✓	✓	✓		✓	✓	✓	✓	✓	✓	✓
276.	Diofenolan-2	✓	✓	✓	✓	✓	✓	✓	✓	✓		✓	✓	✓	✓	✓	✓	✓
277.	Dioxabenzofos	✓	✓	✓	✓	✓	✓	✓	✓	✓			✓	✓	✓	✓	✓	✓
278.	Dioxacarb	✓	✓	✓	✓	✓	✓	✓	✓	✓		✓	✓	✓	✓	✓	✓	✓
279.	Dioxathion	✓	✓	✓	✓	✓	✓	✓	✓	✓		✓	✓	✓	✓	✓	✓	✓
280.	Diphenamid	✓	✓	✓	✓	✓	✓	✓	✓	✓		✓	✓	✓	✓	✓	✓	✓
281.	Diphenylamine	✓	✓	✓	✓	✓	✓	✓	✓	✓		✓	✓	✓	✓	✓	✓	✓
282.	Dipropetryn	✓	✓	✓	✓	✓	✓	✓	✓	✓		✓	✓	✓	✓	✓	✓	✓
283.	Disulfoton	✓	✓	✓	✓	✓	✓	✓	✓	✓		✓	✓	✓	✓	✓	✓	✓
284.	Disulfoton sulfone	✓	✓	✓	✓	✓		✓	✓	✓		✓	✓	✓	✓	✓		
285.	Disulfoton-sulfoxide	✓	✓	✓	✓	✓		✓	✓	✓		✓						
286.	Ditalimfos	✓	✓	✓	✓	✓	✓	✓	✓	✓		✓		✓				
287.	Dithianon			✓											✓			
288.	Dithiopyr	✓	✓	✓	✓	✓	✓	✓	✓	✓		✓	✓		✓			
289.	Diuron	✓	✓	✓	✓	✓	✓	✓	✓	✓								
290.	DMSA	✓		✓	✓	✓						✓	✓	✓				

Continued

No.	Compound
291.	DMST
292.	DNOC
293.	Dodemorph
294.	Dodine
295.	Edifenphos
296.	Emamectin benzoate
297.	Endosulfan-2
298.	Endosulfan-1
299.	Endosulfan-sulfate
300.	Endothal
301.	Endrin
302.	Endrin aldehyde
303.	Endrin ketone
304.	EPN
305.	Epoxiconazole-2
306.	Epoxiconazole-1
307.	Epsilon-HCH
308.	EPTC
309.	Erbon
310.	Esfenvalerate

#	Compound																		
311.	Esprocarb										✓	✓	✓	✓	✓	✓		✓	✓
312.	Etaconazole-1	✓	✓	✓	✓	✓	✓	✓	✓	✓	✓	✓	✓	✓				✓	✓
313.	Etaconazole-2	✓	✓	✓	✓	✓	✓	✓	✓	✓	✓	✓	✓	✓				✓	✓
314.	Ethalfluralin	✓	✓	✓	✓						✓	✓	✓	✓				✓	✓
315.	Ethidimuron	✓	✓	✓	✓	✓	✓	✓	✓	✓	✓	✓	✓	✓				✓	✓
316.	Ethiofencarb	✓		✓		✓	✓	✓	✓	✓	✓	✓	✓	✓				✓	✓
317.	Ethiofencarb-sulfoxide										✓	✓	✓	✓					
318.	Ethion	✓	✓	✓	✓	✓	✓	✓	✓	✓	✓	✓	✓	✓				✓	✓
319.	Ethiprole	✓	✓	✓	✓	✓	✓	✓	✓	✓	✓	✓	✓	✓				✓	✓
320.	Ethirimol										✓	✓	✓	✓					✓
321.	Ethofume sate	✓	✓	✓	✓	✓	✓	✓	✓	✓	✓	✓	✓	✓				✓	✓
322.	Ethoprophos	✓	✓	✓	✓	✓	✓	✓	✓	✓	✓	✓	✓	✓				✓	✓
323.	Ethoxyquin										✓	✓	✓	✓					
324.	Ethoxysulfuron												✓						
325.	Ethylene thiourea						✓												
326.	Etobenzanid	✓	✓	✓	✓	✓	✓	✓	✓	✓	✓	✓	✓	✓	✓	✓		✓	✓
327.	Etofenprox	✓	✓	✓	✓	✓	✓	✓	✓	✓	✓	✓	✓	✓				✓	✓
328.	Etoxazole	✓	✓	✓	✓	✓	✓	✓	✓	✓	✓	✓	✓	✓				✓	✓
329.	Etridiazol	✓	✓	✓	✓	✓	✓	✓	✓	✓	✓	✓	✓	✓				✓	✓
330.	Etrimfos	✓	✓	✓	✓	✓	✓	✓	✓	✓	✓	✓	✓	✓				✓	✓
331.	Etrimfos oxon	✓	✓	✓					✓										

No.	Compound																
332.	Famoxadone	✓									✓	✓		✓	✓	✓	✓
333.	Famphur	✓	✓	✓	✓	✓				✓	✓	✓		✓	✓	✓	✓
334.	Fempxaprop-ethyl						✓				✓	✓	✓					
335.	Fenamidone	✓	✓	✓	✓	✓	✓	✓	✓		✓				✓			
336.	Fenaminosulf	✓	✓	✓	✓		✓	✓	✓						✓	✓	✓	✓
337.	Fenamiphos	✓	✓	✓	✓	✓	✓	✓	✓	✓	✓	✓	✓		✓	✓	✓	✓
338.	Fenamiphos sulfone	✓	✓	✓	✓	✓	✓	✓	✓	✓	✓	✓	✓		✓	✓	✓	✓
339.	Fenamiphos sulfoxide	✓	✓	✓	✓	✓	✓	✓	✓	✓	✓	✓	✓		✓	✓	✓	✓
340.	Fenarimol	✓	✓	✓	✓	✓	✓		✓		✓	✓	✓					
341.	Fenazaquin	✓	✓	✓	✓	✓	✓		✓		✓	✓	✓					
342.	Fenbuconazole	✓	✓	✓	✓	✓	✓		✓		✓	✓	✓					
343.	Fenchlorphos	✓		✓	✓	✓	✓	✓	✓									
344.	Fenchlorphos oxon	✓	✓	✓	✓	✓			✓									
345.	Fenfuram	✓	✓	✓	✓	✓	✓		✓		✓	✓	✓		✓	✓	✓	✓
346.	Fenhexamid	✓	✓		✓	✓	✓	✓	✓		✓	✓	✓		✓	✓	✓	✓
347.	Fenitrothion	✓	✓	✓	✓	✓	✓	✓	✓		✓	✓	✓		✓	✓	✓	✓
348.	Fenobucarb	✓	✓	✓	✓	✓	✓	✓	✓		✓	✓	✓		✓	✓	✓	✓
349.	Fenoprop					✓	✓											
350.	Fenothiocarb	✓	✓	✓	✓	✓	✓	✓	✓		✓							
351.	Fenoxanil	✓	✓	✓	✓	✓	✓	✓	✓		✓	✓	✓		✓	✓	✓	✓

Continued

No.	Compound																	
352.	Fenoxaprop	✓	✓		✓	✓				✓								✓
353.	Fenoxycarb	✓	✓	✓	✓	✓	✓	✓	✓	✓	✓	✓	✓				✓	✓
354.	Fenpiclonil	✓	✓	✓	✓	✓	✓	✓	✓	✓	✓	✓	✓				✓	✓
355.	Fenpropathrin	✓	✓	✓	✓	✓	✓	✓	✓	✓	✓	✓	✓				✓	✓
356.	Fenpropidin	✓	✓	✓	✓	✓	✓	✓	✓	✓	✓	✓	✓				✓	✓
357.	Fenpropimorph	✓	✓	✓	✓	✓	✓	✓	✓	✓	✓	✓	✓				✓	✓
358.	Fenpyroximate	✓	✓	✓	✓	✓	✓	✓	✓	✓	✓	✓	✓				✓	✓
359.	Fenson	✓	✓	✓	✓	✓	✓	✓	✓	✓	✓		✓				✓	
360.	Fensulfothion	✓	✓	✓	✓	✓	✓	✓	✓	✓	✓	✓	✓				✓	✓
361.	Fenthion	✓	✓	✓	✓	✓	✓	✓	✓	✓	✓	✓	✓				✓	✓
362.	Fenthion oxon		✓			✓												
363.	Fenthion sulfone	✓	✓	✓	✓	✓	✓	✓	✓	✓	✓	✓	✓				✓	✓
364.	Fenthion sulfoxide	✓	✓	✓	✓	✓	✓	✓	✓	✓	✓	✓	✓				✓	✓
365.	Fentin-chloride											✓	✓					
366.	Fentrazamide	✓	✓	✓	✓	✓	✓	✓	✓	✓	✓	✓	✓	✓			✓	✓
367.	Fenuron	✓	✓	✓	✓	✓	✓	✓	✓	✓	✓	✓	✓				✓	✓
368.	Fenvalerate-1	✓	✓	✓	✓	✓	✓		✓	✓	✓	✓	✓	✓	✓		✓	✓
369.	Fenvalerate-2	✓	✓	✓	✓	✓	✓		✓	✓	✓	✓	✓	✓	✓	✓	✓	✓
370.	Fipronil	✓	✓	✓	✓		✓		✓	✓	✓	✓	✓	✓	✓		✓	✓
371.	Flamprop acid																	
372.	Flamprop-isopropyl	✓	✓	✓	✓	✓	✓	✓	✓	✓	✓	✓	✓				✓	✓

373.	Flamprop-methyl	✓	✓	✓	✓								✓	✓			✓	✓	✓	✓
374.	Florasulam	✓	✓	✓									✓	✓			✓	✓		✓
375.	Fluazifop			✓																	
376.	Fluazifop butyl	✓	✓	✓	✓								✓	✓	✓	✓	✓	✓	✓	✓
377.	Fluazinam			✓									✓			✓				
378.	Fluazuron																				
379.	Flubenzimine	✓	✓	✓									✓	✓	✓	✓	✓	✓	✓	✓
380.	Fluchloralin	✓	✓	✓	✓	✓							✓	✓	✓		✓	✓		✓
381.	Flucythrinate-1	✓	✓	✓	✓	✓							✓	✓	✓	✓	✓	✓	✓	✓
382.	Flucythrinate-2	✓	✓	✓	✓								✓	✓		✓				
383.	Fludioxonil	✓	✓		✓	✓							✓	✓	✓	✓	✓	✓	✓	✓
384.	Flufenacet	✓	✓	✓	✓	✓							✓	✓	✓	✓	✓	✓	✓	✓
385.	Flufenoxuron	✓	✓	✓	✓	✓							✓	✓	✓	✓	✓	✓	✓	✓
386.	Flumetralin	✓	✓	✓	✓	✓							✓	✓			✓			
387.	Flumetsulam	✓	✓	✓	✓								✓	✓	✓	✓	✓	✓	✓	✓
388.	Flumiclorac-pentyl	✓	✓	✓	✓	✓							✓	✓	✓	✓	✓	✓	✓	✓
389.	Flumioxazin	✓	✓																		✓
390.	Fluometuron												✓			✓	✓	✓	✓	✓	✓
391.	Fluoranthene								✓												
392.	Fluorene								✓												

Continued

No.	Compound
393.	Fluorochloridone
394.	Fluorodifen
395.	Fluoroglycofen-ethyl
396.	Fluotrimazole
397.	Flupropanate
398.	Fluquinconazole
399.	Flurazuron
400.	Fluridone
401.	Flurochloridone
402.	Fluroxypyr-1-methylheptyl ester
403.	Fluroxypyr
404.	Flurtamone
405.	Flusilazole
406.	Flusulfamide
407.	Fluthiacet methyl
408.	Flutolanil
409.	Flutriafol
410.	Fluvalinate
411.	Folpet
412.	Fomesafen

413.	Fonofos	✓	✓	✓	✓	✓	✓	✓			✓	✓	✓	✓	✓		✓	✓	✓	✓	✓	✓	✓	✓
414.	Forchlorfenuron	✓	✓	✓	✓	✓	✓	✓			✓										✓	✓	✓	✓
415.	Formothion	✓	✓								✓	✓	✓	✓										✓
416.	Fosthiazate	✓	✓	✓	✓	✓	✓	✓			✓	✓	✓	✓	✓	✓	✓	✓			✓	✓	✓	✓
417.	Fuberidazole	✓	✓	✓	✓	✓					✓	✓	✓	✓	✓	✓	✓	✓	✓	✓	✓	✓	✓	✓
418.	Furalaxyl	✓	✓	✓	✓	✓	✓	✓			✓	✓	✓	✓	✓	✓	✓	✓	✓	✓	✓	✓	✓	✓
419.	Furathiocarb	✓	✓	✓	✓	✓	✓	✓			✓	✓	✓	✓	✓	✓	✓	✓		✓	✓	✓	✓	✓
420.	Furmecyclox	✓	✓	✓	✓	✓	✓	✓			✓	✓	✓	✓	✓	✓				✓	✓	✓	✓	✓
421.	Gamma-cyhaloterin-1						✓	✓			✓	✓	✓	✓										
422.	Gamma-cyhalothrin-2						✓	✓			✓	✓	✓	✓										
423.	Gamma-HCH	✓	✓	✓	✓	✓	✓	✓	✓		✓	✓	✓	✓		✓								
424.	Gibberellic acid	✓									✓						✓							✓
425.	Halfenprox	✓	✓	✓	✓	✓	✓	✓			✓	✓	✓	✓										
426.	Halosulfuran-methyl						✓	✓			✓									✓				
427.	Haloxyfop			✓																				
428.	Haloxyfop-2-ethoxyethyl						✓	✓			✓	✓	✓	✓	✓	✓	✓	✓	✓	✓	✓	✓	✓	✓
429.	Haloxyfop-methyl	✓	✓	✓	✓	✓	✓	✓			✓	✓	✓	✓	✓	✓	✓	✓	✓	✓	✓	✓	✓	✓
430.	Heptachlor	✓	✓	✓	✓	✓	✓	✓	✓		✓	✓	✓	✓										
431.	Heptachlor (ISTD)								✓															
432.	Heptanophos	✓	✓	✓	✓	✓	✓	✓			✓	✓	✓	✓	✓	✓	✓	✓	✓	✓	✓	✓	✓	✓

Continued

433.	Hexachlorobenzene
434.	Hexaconazole
435.	Hexaflumuron
436.	Hexazinone
437.	Hexythiazox
438.	Hydramethylnon
439.	Hymexazol
440.	Imazalil
441.	Imazamethabenz-methyl
442.	Imazamox
443.	Imazapic
444.	Imazapyr
445.	Imazaquin
446.	Imazethapyr
447.	Imibenconazole
448.	Imibenconazole-des-benzyl
449.	Imidacloprid
450.	Iminoctadine triacetate
451.	Imiprothrin-1
452.	Imiprothrin-2

No.	Compound
453.	Indeno(1,2,3-cd)pyrene
454.	Indoxacarb
455.	Iodofenphos
456.	Iodosulfuron-methyl
457.	Iodosulfuron-methyl sodium
458.	Ioxynil
459.	Iprobenfos
460.	Iprodione
461.	Iprovalicarb-1
462.	Iprovalicarb-2
463.	Isazofos
464.	Isobenzan
465.	Isocarbamid
466.	Isocarbophos
467.	Isodrin
468.	Isofenphos
469.	Isofenphos oxon
470.	Isomethiozin
471.	Isoprocarb-2
472.	Isoprocarb-1

Continued

No.	Compound
473.	Isopropalin
474.	Isoprothiolane
475.	Isoproturon
476.	Isouron
477.	Isoxaben
478.	Isoxadifen-ethyl
479.	Isoxaflutole
480.	Isoxathion
481.	Kadethrin
482.	Kelevan
483.	Kresoxim-methyl
484.	Lactofen
485.	Lambda-cyhalothrin
486.	Lenacil
487.	Leptophos
488.	Linuron
489.	lufenuron
490.	Malaoxon
491.	Malathion
492.	Maleic hydrazide
493.	MCPA

#	Name																					
494.	mcpa-Butoxyethyl ester	✓	✓						✓	✓	✓					✓				✓		
495.	MCPB	✓	✓	✓	✓	✓	✓	✓	✓	✓	✓					✓				✓	✓	
496.	Mecarbam	✓	✓	✓	✓	✓	✓	✓		✓	✓			✓		✓	✓	✓	✓	✓	✓	✓
497.	Mecoprop	✓	✓	✓	✓	✓	✓		✓	✓	✓							✓	✓	✓		
498.	Mefenacet	✓	✓	✓	✓	✓	✓	✓	✓	✓	✓			✓		✓	✓	✓	✓	✓	✓	✓
499.	Mefenoxam	✓	✓	✓	✓	✓	✓	✓	✓	✓	✓			✓		✓	✓	✓	✓	✓	✓	✓
500.	Mefenpyr-diethyl	✓	✓	✓	✓	✓		✓	✓	✓	✓			✓		✓	✓	✓	✓	✓	✓	✓
501.	Mepanipyrim	✓	✓	✓	✓	✓	✓		✓	✓	✓			✓		✓	✓	✓	✓	✓	✓	✓
502.	Mephosfolan	✓	✓	✓				✓		✓	✓			✓		✓		✓	✓		✓	✓
503.	Mepiquat chloride	✓								✓	✓					✓	✓	✓	✓			
504.	Mepronil	✓	✓	✓	✓	✓	✓	✓	✓	✓	✓			✓		✓	✓	✓	✓	✓	✓	✓
505.	Merphos															✓				✓		
506.	Mesotrion													✓		✓	✓	✓	✓			
507.	Metalaxyl	✓	✓	✓	✓	✓	✓	✓	✓	✓	✓			✓		✓	✓	✓	✓	✓	✓	✓
508.	Metamitron	✓	✓	✓	✓	✓	✓	✓	✓	✓	✓					✓	✓	✓	✓	✓	✓	✓
509.	Metazachlor	✓	✓	✓	✓	✓	✓	✓	✓	✓	✓			✓		✓	✓	✓	✓	✓	✓	✓
510.	Metconazole	✓	✓	✓	✓	✓	✓	✓	✓	✓	✓					✓	✓	✓	✓	✓	✓	✓
511.	Methabenzthiazuron	✓	✓	✓	✓	✓	✓	✓	✓	✓	✓			✓		✓	✓	✓	✓	✓	✓	✓
512.	Methacrifos	✓	✓	✓	✓	✓	✓	✓						✓		✓	✓	✓	✓	✓	✓	✓
513.	Methamidophos	✓						✓		✓				✓		✓	✓	✓	✓	✓	✓	✓

Continued

No.	Compound
514.	Methfuroxam
515.	Methidathion
516.	Methiocarb
517.	Methiocarb sulfone
518.	Methobromuron
519.	Methomyl
520.	Methoprene
521.	Methoprotryne
522.	Methothrin-1
523.	Methothrin-2
524.	Methoxychlor
525.	Methoxyfenozide
526.	Methyl-parathion
527.	Metobromuron
528.	Metoconazole
529.	Metolachlor
530.	Metolcarb
531.	Metominostrobin-1
532.	Metominostrobin-2
533.	Metosulam
534.	Metoxuron

No.	Compound
535.	Metribuzin
536.	Metsulfuron-methyl
537.	Mevinphos
538.	Mexacarbate
539.	Mirex
540.	Molinate
541.	Monalide
542.	Monocrotophos
543.	Monolinuron
544.	Monuron
545.	Musk ambrette
546.	Musk ketone
547.	Musk moskene
548.	Musk tibeten
549.	Musk xylene
550.	Myclobutanil
551.	NAA
552.	Naled
553.	Naphthalene
554.	Naproanilide

Continued

No.	Compound
555.	Napropamide
556.	Naptalam
557.	Neburon
558.	Nicotine
559.	Nitenpyram
560.	Nitralin
561.	Nitrapyrin
562.	Nitrofen
563.	Nitrothal-isopropyl
564.	Norflurazon
565.	Norflurazon-desmethyl
566.	Novaluron
567.	Nuarimol
568.	o,p'-DDD
569.	o,p'-DDT
570.	o,p'-DDE
571.	Octachlorostyrene
572.	Oesmetryn
573.	Ofurace
574.	OH-ioxynil
575.	Omethoate

No.	Name
576.	Oryzalin
577.	Oxabetrinil
578.	Oxadiazone
579.	Oxadixyl
580.	Oxamyl
581.	Oxamyl-oxime
582.	Oxycarboxin
583.	Oxy-chlordane
584.	oxyfluorfen
585.	p,p'-DDE
586.	p,p'-DDT
587.	p,p'-DDD
588.	Paclobutrazol
589.	Paraoxon ethyl
590.	Paraoxon methyl
591.	Parathion
592.	Pebulate
593.	Penconazole
594.	Pencycuron
595.	Pendimethalin

Continued

No.	Compound
596.	Pentachloroaniline
597.	Pentachloroanisole
598.	Pentachlorobenzene
599.	Pentachlorphenol
600.	Permethrin
601.	Perthane
602.	Phenanthrene
603.	Phenkapton
604.	Phenmedipham
605.	Phenothrin
606.	Phenthoate
607.	Phorate
608.	Phorate sulfone
609.	Phorate sulfoxide
610.	Phosalone
611.	Phosfolan
612.	Phosmet
613.	Phosphamidon-1
614.	Phosphamidon-2
615.	Phoxim

No.	Compound
616.	Phthalic acid benzyl butyl ester
617.	Phthalic acid bis-2-ethylhexyl ester
618.	Phthalic acid bis-butyl ester
619.	Phthalic acid, biscyclohexyl ester
620.	Phthalic acid, dibutyl ester
621.	Phthalic acid, benzyl butyl ester
622.	Phthalic acid, dicyclohexyl ester
623.	Phthalide
624.	Phthalimide
625.	Picloram
626.	Picolinafen
627.	Picoxystrobin
628.	Piperonyl butoxide
629.	Piperophos
630.	Pirimicarb
631.	Pirimiphos ethyl
632.	Pirimiphos methyl

Continued

633.	Plifenate
634.	Prallethrin
635.	Pretilachlor
636.	Prochloraz
637.	Procymidone
638.	Profenofos
639.	Profluralin
640.	Prohydrojasmon
641.	Promecarb
642.	Prometon
643.	Prometryne
644.	Pronamide
645.	Propachlor
646.	Propamocarb
647.	Propanil
648.	Propaphos
649.	Propaquizafop
650.	Propargite
651.	Propazine
652.	Propetamphos
653.	Propham

No.	Compound																		
654.	Propiconazole-1	✓	✓	✓	✓	✓	✓	✓	✓	✓							✓	✓	✓
655.	Propiconazole-2	✓	✓	✓	✓	✓											✓	✓	
656.	Propisochlor	✓	✓	✓	✓	✓										✓	✓	✓	✓
657.	Propoxur-1	✓	✓	✓	✓	✓	✓									✓	✓	✓	✓
658.	Propoxur-2	✓	✓	✓	✓	✓													
659.	Propylene thiourea														✓	✓	✓	✓	✓
660.	Propyzamide	✓	✓	✓	✓	✓	✓							✓					
661.	Prosulfocarb	✓	✓	✓	✓	✓	✓							✓	✓	✓	✓	✓	✓
662.	Prothiophos	✓	✓	✓	✓	✓								✓	✓	✓	✓	✓	✓
663.	Prothoate													✓	✓	✓	✓	✓	✓
664.	Pymetrozin	✓	✓	✓	✓									✓	✓	✓	✓	✓	✓
665.	Pyraclofos	✓	✓	✓	✓	✓								✓	✓	✓	✓	✓	✓
666.	Pyraclostrobin	✓	✓	✓	✓	✓	✓							✓	✓	✓	✓	✓	✓
667.	Pyraflufen ethyl	✓	✓	✓	✓									✓	✓	✓	✓	✓	✓
668.	Pyrazophos	✓	✓	✓										✓		✓	✓	✓	✓
669.	Pyrazosulfuron ethyl											✓							
670.	Pyrazoxyfen													✓	✓	✓	✓	✓	✓
671.	Pyrene													✓	✓	✓	✓	✓	✓
672.	Pyrethrin																✓	✓	✓
673.	Pyrethrins															✓	✓		

Continued

	1	2	3	4	5	6	7	8	9	10	11	12	13	14	15	16	17	18	19	20	21	22	23	24	25
674. Pyributicarb	✓	✓	✓	✓	✓	✓	✓	✓	✓		✓	✓	✓	✓	✓					✓	✓	✓	✓	✓	✓
675. Pyridaben	✓	✓	✓						✓		✓	✓	✓	✓						✓	✓	✓		✓	✓
676. Pyridaphenthion	✓	✓	✓					✓	✓		✓	✓	✓	✓	✓		✓			✓	✓	✓	✓	✓	✓
677. Pyridate									✓			✓	✓	✓							✓				✓
678. Pyrifenox-2	✓		✓	✓	✓	✓	✓	✓	✓			✓	✓	✓	✓					✓	✓	✓	✓	✓	✓
679. Pyrifenox-1	✓	✓	✓	✓	✓	✓	✓	✓	✓		✓	✓	✓	✓	✓					✓	✓	✓	✓	✓	✓
680. Pyriftalid	✓	✓	✓	✓	✓	✓	✓	✓	✓		✓	✓	✓	✓	✓					✓	✓	✓	✓	✓	✓
681. Pyrimethanil	✓	✓	✓	✓	✓	✓	✓	✓	✓		✓	✓	✓	✓	✓					✓	✓	✓	✓	✓	✓
682. Pyrimidifen	✓	✓	✓	✓					✓		✓	✓	✓	✓	✓					✓	✓	✓	✓	✓	✓
683. Pyriminobac methyl	✓	✓	✓	✓					✓		✓	✓	✓	✓	✓					✓					
684. Pyrimitate	✓	✓	✓	✓	✓	✓	✓	✓	✓		✓	✓	✓	✓	✓		✓			✓	✓	✓	✓	✓	✓
685. Pyriproxyfen	✓	✓	✓	✓	✓	✓	✓	✓	✓		✓	✓	✓	✓	✓					✓	✓	✓	✓	✓	✓
686. Pyrithlobac sodium	✓		✓	✓	✓	✓			✓			✓	✓	✓	✓										
687. Pyroquilon	✓	✓	✓	✓	✓	✓	✓	✓	✓		✓	✓	✓	✓	✓					✓	✓	✓	✓	✓	✓
688. Quinalphos	✓	✓	✓	✓	✓	✓	✓	✓	✓		✓	✓	✓	✓	✓					✓	✓	✓		✓	✓
689. Quinclorac	✓	✓	✓	✓	✓	✓			✓		✓		✓	✓	✓										
690. Quinoclamine	✓	✓	✓	✓	✓	✓	✓	✓	✓		✓	✓	✓	✓	✓					✓	✓	✓	✓	✓	✓
691. Quinoxyphen	✓	✓	✓	✓	✓	✓	✓	✓	✓		✓	✓	✓	✓	✓					✓	✓	✓	✓	✓	✓
692. Quintozene	✓	✓	✓	✓					✓		✓		✓	✓	✓										
693. Quizalofop		✓	✓	✓					✓		✓		✓	✓											
694. Quizalofop ethyl									✓			✓	✓	✓						✓	✓	✓	✓	✓	✓

#	Compound	1	2	3	4	5	6	7	8	9	10	11	12	13	14	15	16	17	18	19	20	21
695.	Rabenzazole	√	√	√	√	√	√	√	√	√		√	√	√	√	√	√	√	√	√	√	
696.	Resmethrin-1	√		√	√	√	√	√	√	√											√	
697.	Resmethrin-2	√		√	√	√	√	√	√	√		√	√	√	√	√	√	√	√		√	
698.	Ronnel	√	√	√	√	√	√	√	√	√		√	√	√	√	√	√	√	√	√	√	
699.	Rotenone											√	√	√	√	√	√	√	√	√	√	
700.	s421(octachlorodipropyl ether)-1				√	√	√		√	√												
701.	s421(octachlorodipropyl ether)-2				√	√	√		√	√												
702.	Sebutylazine	√	√	√	√	√	√	√	√	√		√	√	√	√	√	√	√	√	√	√	
703.	Secbumeton	√	√	√	√	√	√	√	√	√		√	√	√	√	√	√	√	√	√	√	
704.	Sethoxydim	√	√	√		√	√	√					√	√	√	√			√			
705.	Silafluofen	√	√	√	√	√	√	√	√	√			√	√	√	√	√	√			√	
706.	Simazine	√	√	√				√			√											
707.	Simeconazole	√	√	√	√	√	√	√	√	√		√	√	√	√	√	√	√	√		√	
708.	Simeton	√	√	√	√	√	√	√	√	√		√	√	√	√	√	√	√	√	√	√	
709.	Simetryn	√		√		√	√	√	√	√		√	√	√	√	√	√	√	√	√	√	
710.	Sobutylazine				√	√	√		√	√												
711.	Spinosad											√		√	√	√		√	√			√
712.	Spirodiclofen	√	√	√	√	√	√	√	√	√		√	√	√	√	√		√	√	√	√	
713.	Spiromesifen				√	√	√			√												

Continued

No.	Name	1	2	3	4	5	6	7	8	9	10	11	12	13	14	15	16	17	18	19
714.	Spiroxamine-1	√		√	√	√	√		√	√		√	√	√	√	√	√	√	√	√	
715.	Spiroxamine-2	√		√	√	√	√		√												
716.	Sulfallate	√	√	√	√	√	√	√		√		√	√	√	√	√	√	√	√	√	√
717.	Sulfanitran											√	√	√			√			√	√
718.	Sulfentrazone											√				√	√			√	
719.	Sulfotep	√	√	√	√	√	√	√	√	√		√	√	√	√	√	√	√	√	√	√
720.	Sulprofos	√	√	√	√	√	√	√	√	√		√	√	√	√	√	√	√	√	√	√
721.	Tau-fluvalinate					√				√		√	√	√	√					√	
722.	TCMTB	√	√	√		√	√	√	√	√											
723.	Tebuconazole	√	√	√	√	√	√	√	√	√		√	√	√	√	√	√	√	√	√	√
724.	Tebufenozide											√	√	√		√	√	√	√		√
725.	Tebufenpyrad	√	√	√	√	√	√	√	√	√		√	√	√	√	√	√	√	√	√	
726.	Tebupirimfos	√	√	√	√	√	√	√	√	√		√	√	√	√	√	√	√	√	√	√
727.	Tebutam	√	√	√	√	√	√	√	√	√		√	√	√	√	√	√	√	√	√	√
728.	Tebuthiuron	√	√	√	√	√	√	√	√	√		√	√	√	√	√	√	√	√	√	√
729.	Tecnazene	√	√	√	√	√	√	√	√	√											
730.	Tefluthrin	√	√	√	√	√	√		√	√											
731.	Temephos											√	√	√	√	√	√	√	√	√	√
732.	TEPP											√	√	√	√			√	√	√	√
733.	Tepraloxydim											√	√				√			√	√
734.	Terbacil		√	√	√	√		√	√			√	√	√	√	√	√	√	√	√	√
735.	Terbucarb-1	√	√	√	√	√	√	√		√					√						

No.	Compound	1	2	3	4	5	6	7	8	9	10	11	12	13	14	15	16	17	18	19	20
736.	Terbucarb-2				✓	✓	✓			✓											
737.	Terbufos	✓	✓	✓	✓	✓	✓	✓	✓	✓		✓	✓	✓	✓			✓	✓	✓	✓
738.	Terbufos sulfone	✓	✓	✓								✓	✓	✓	✓	✓	✓	✓	✓	✓	✓
739.	Terbumeton	✓	✓	✓		✓		✓	✓	✓		✓	✓	✓	✓	✓	✓	✓	✓	✓	✓
740.	Terbuthylazine	✓	✓	✓	✓	✓	✓	✓	✓	✓		✓	✓	✓	✓	✓	✓	✓	✓	✓	✓
741.	Terbutryn	✓	✓	✓	✓	✓	✓	✓	✓			✓	✓	✓		✓	✓	✓	✓		✓
742.	Terrbucarb											✓	✓	✓	✓	✓		✓	✓		✓
743.	Tert-butylamine											✓	✓	✓	✓	✓	✓	✓	✓	✓	✓
744.	Tetrachlorvinphos	✓	✓	✓	✓	✓	✓	✓	✓			✓	✓	✓	✓	✓	✓	✓	✓	✓	✓
745.	Tetraconazole	✓	✓	✓	✓	✓	✓	✓	✓			✓	✓	✓	✓	✓	✓	✓	✓	✓	✓
746.	Tetradifon	✓	✓	✓	✓	✓	✓	✓	✓	✓											
747.	Tetrahydrophthalimide	✓	✓	✓				✓													
748.	Tetramethirn	✓	✓	✓	✓	✓	✓	✓		✓		✓	✓	✓	✓	✓	✓	✓	✓	✓	✓
749.	Tetrasul	✓	✓	✓	✓	✓	✓	✓	✓												
750.	Thenylchlor	✓	✓	✓	✓	✓	✓	✓	✓			✓	✓	✓	✓	✓	✓	✓	✓	✓	✓
751.	Thiabendazole				✓	✓			✓	✓		✓	✓	✓	✓	✓		✓	✓	✓	✓
752.	Thiacloprid											✓	✓	✓	✓	✓	✓	✓	✓	✓	✓
753.	Thiamethoxam	✓	✓			✓		✓	✓	✓		✓	✓	✓	✓	✓	✓	✓	✓		✓
754.	Thiazopyr	✓	✓	✓	✓	✓	✓	✓	✓	✓		✓	✓	✓	✓	✓	✓	✓	✓	✓	✓
755.	Thidiazuron											✓	✓			✓		✓			
756.	Thifensulfuron methyl											✓	✓			✓			✓	✓	

Continued

		1	2	3	4	5	6	7	8	9	10	11	12	13	14	15	16	17	18	19	20
757.	Thifluzamide	√	√																√		
758.	Thiobencarb	√	√	√	√	√	√	√	√	√		√	√	√	√	√	√	√	√	√	√
759.	Thiodicarb												√	√	√	√			√	√	√
760.	Thiofanox											√	√	√	√	√	√	√	√	√	√
761.	Thiofanox sulfone											√	√	√	√	√	√	√	√		
762.	Thiofanox sulfoxide											√	√	√	√	√	√	√	√	√	√
763.	Thiometon	√	√	√	√	√	√	√	√	√		√	√	√	√	√	√	√	√	√	√
764.	Thionazin	√	√	√	√	√	√	√	√	√		√	√	√	√	√	√	√	√	√	√
765.	Thiophanat ethyl											√	√		√			√		√	
766.	Thiophanate methyl											√	√					√		√	
767.	Tolclofos methyl	√	√	√	√	√	√	√	√			√	√	√	√	√	√	√	√	√	√
768.	Tolfenpyrad											√						√			
769.	Tolylfluanide	√	√	√	√	√	√	√	√	√											
770.	Tralkoxydim	√	√	√		√	√	√	√	√		√	√	√	√	√	√	√	√	√	√
771.	Tralomethrin-1				√				√												
772.	Tralomethrin-2				√				√												
773.	Trans chlordane	√	√	√	√	√	√	√	√	√	√										
774.	Trans diallate	√	√		√	√	√	√	√	√											
775.	Trans nonachlor	√	√	√	√	√	√	√	√	√											
776.	Trans permethin	√	√	√	√	√	√	√	√	√		√	√	√	√	√	√	√	√	√	√
777.	Transfluthrin	√	√	√	√	√	√	√	√	√											

No.	Compound
778.	Triadimefon
779.	Triadimenol-1
780.	Triadimenol-2
781.	Triallate
782.	Triamiphos
783.	Triasulfuron
784.	Triazophos
785.	Triazoxide
786.	Tribenuron-methyl
787.	Trichloroacetic acid sodium salt
788.	Trichloronat
789.	Trichlorphon
790.	Triclopyr
791.	Tricyclazole
792.	Tridemorph
793.	Tridiphane
794.	Trietazine
795.	Trifloxystrobin
796.	Triflumizole
797.	Triflumuron
798.	Trifluralin

Continued

No.	Compound	1	2	3	4	5	6	7	8	9	10	11	12	13	14	15	16	17	18
799.	Triforine			✓				✓	✓	✓									
800.	tri-iso-Butyl phosphate	✓					✓	✓	✓	✓						✓	✓	✓	✓
801.	tri-N-Butyl phosphate	✓	✓	✓	✓	✓	✓	✓	✓	✓		✓	✓	✓	✓	✓	✓	✓	✓
802.	Trinexapac-ethyl						✓					✓	✓	✓	✓	✓	✓	✓	✓
803.	Triphenyl phosphate											✓	✓	✓	✓	✓	✓	✓	✓
804.	Triticonazole			✓	✓	✓	✓	✓	✓	✓		✓	✓	✓	✓	✓	✓	✓	✓
805.	Uniconazole		✓	✓	✓	✓	✓	✓	✓	✓		✓	✓	✓	✓	✓	✓	✓	✓
806.	Vamidothion	✓	✓	✓	✓	✓	✓	✓	✓	✓									
807.	Vamidothion Sulfone	✓	✓	✓	✓	✓	✓	✓	✓	✓									
808.	Vernolate	✓	✓		✓	✓	✓	✓	✓	✓		✓	✓	✓	✓	✓	✓	✓	✓
809.	Vinclozolin	✓	✓	✓	✓	✓	✓	✓		✓		✓	✓	✓	✓	✓	✓	✓	✓
810.	Warfarin																		
811.	XMC											✓	✓	✓	✓	✓	✓	✓	✓
812.	Zeta cypermethrin	✓	✓	✓	✓	✓	✓	✓	✓	✓									
813.	Ziram	✓	✓																
814.	Zoxamide			✓	✓	✓	✓	✓		✓		✓	✓	✓	✓	✓	✓	✓	✓
815.	1-Naphthy acetamide	✓	✓	✓	✓	✓	✓	✓	✓	✓									
816.	2,2',3,3',4,4',5,5',6,6'-Decachlorobiphenyl										✓								
817.	2,2',3,3',4,4',5,5',6-Nonachlorobiphenyl										✓								

Continued

818.	2,2',3,3',4,4',5, 5'-Octachlorobiphenyl	✓
819.	2,2',3,3',4,4',5,6, 6'-Nonachlorobiphenyl	✓
820.	2,2',3,3',4,4',5, 6'-Octachlorobiphenyl	✓
821.	2,2',3,3',4,4',5, 6-Octachlorobiphenyl	✓
822.	2,2',3,3',4,4', 5-Heptachlorobiphenyl	✓
823.	2,2',3,3',4,4',6, 6'-Octachlorobiphenyl	✓
824.	2,2',3,3',4,4', 6-Heptachlorobiphenyl	✓
825.	2,2',3,3',4, 4'-Hexachlorobiphenyl	✓
826.	2,2',3,3',4,5,5',6, 6'-Nonachlorobiphenyl	✓
827.	2,2',3,3',4,5,5', 6'-Octachlorobiphenyl	✓
828.	2,2',3,3',4,5,5', 6-Octachlorobiphenyl	✓
829.	2,2',3,3',4,5, 5'-Heptachlorobiphenyl	✓
830.	2,2',3,3',4,5,6, 6'-Octachlorobiphenyl	✓

831.	2,2',3,3',4,5',6,6'-Octachlorobiphenyl	∨
832.	2,2',3,3',4',5,6-Heptachlorobiphenyl	∨
833	2,2',3,3',4,5',6-Heptachlorobiphenyl	∨
834.	2,2',3,3',4,5,6'-Heptachlorobiphenyl	∨
835.	2,2',3,3',4,5,6-Heptachlorobiphenyl	∨
836.	2,2',3,3',4,5'-Hexachlorobiphenyl	∨
837.	2,2',3,3',4,5-Hexachlorobiphenyl	∨
838.	2,2',3,3',4,6,6'-Heptachlorobiphenyl	∨
839.	2,2',3,3',4,6'-Hexachlorobiphenyl	∨
840.	2,2',3,3',4,6-Hexachlorobiphenyl	∨
841.	2,2',3,3',4-Pentachlorobiphenyl	∨
842.	2,2',3,3',5,5',6,6'-Octachlorobiphenyl	∨
843.	2,2',3,3',5,5',6-Heptachlorobiphenyl	∨

Continued

844.	2,2',3,3',5, 5'-Hexachlorobiphenyl	✓
845.	2,2',3,3',5,6, 6'-Heptachlorobiphenyl	✓
846.	2,2',3,3',5, 6'-Hexachlorobiphenyl	✓
847.	2,2',3,3',5, 6-Hexachlorobiphenyl	✓
848.	2,2',3,3', 5-Pentachlorobiphenyl	✓
849.	2,2',3,3',6, 6'-Hexachlorobiphenyl	✓
850.	2,2',3,3', 6-Pentachlorobiphenyl	✓
851.	2,2',3, 3'-Tetrachlorobiphenyl	✓
852.	2,2',3,4,4',5,5', 6-Octachlorobiphenyl	✓
853.	2,2',3,4,4',5, 5'-Heptachlorobiphenyl	✓
854.	2,2',3,4,4',5,6, 6'-Octachlorobiphenyl	✓
855.	2,2',3,4,4',5', 6-Heptachlorobiphenyl	✓
856.	2,2',3,4,4',5, 6'-Heptachlorobiphenyl	✓

No.	Compound		✓
857.	2,2',3,4,4',5,6-Heptachlorobiphenyl		✓
858.	2,2',3,4,4',5'-Hexachlorobiphenyl		✓
859.	2,2',3,4,4',5-Hexachlorobiphenyl		✓
860.	2,2',3,4,4',6,6'-Heptachlorobiphenyl		✓
861.	2,2',3,4,4',6'-Hexachlorobiphenyl		✓
862.	2,2',3,4,4',6-Hexachlorobiphenyl		✓
863.	2,2',3,4,4'-Pentachlorobiphenyl		✓
864.	2,2',3,4,5,5'-Heptachlorobiphenyl		✓
865.	2,2',3,4',5,5'-Heptachlorobiphenyl		✓
866.	2,2',3,4,5,5'-Hexachlorobiphenyl		✓
867.	2,2',3,4',5,5'-Hexachlorobiphenyl		✓
868.	2,2',3,4,5,6,6'-Heptachlorobiphenyl		✓
869.	2,2',3,4',5,6,6'-Heptachlorobiphenyl		✓

Continued

No.	Name	
870.	2,2',3,4,5,6-Hexachlorobiphenyl	✓
871.	2,2',3,4,5',6-Hexachlorobiphenyl	✓
872.	2,2',3,4',5',6-Hexachlorobiphenyl	✓
873.	2,2',3,4,5,6'-Hexachlorobiphenyl	✓
874.	2,2',3,4',5,6'-Hexachlorobiphenyl	✓
875.	2,2',3,4',5,6-Hexachlorobiphenyl	✓
876.	2,2',3',4,5-Pentachlorobiphenyl	✓
877.	2,2',3,4',5-Pentachlorobiphenyl	✓
878.	2,2',3,4,5'-Pentachlorobiphenyl	✓
879.	2,2',3,4,5-Pentachlorobiphenyl	✓
880.	2,2',3,4,6'-Hexachlorobiphenyl	✓
881.	2,2',3,4',6,6'-Hexachlorobiphenyl	✓
882.	2,2',3,4,6-Pentachlorobiphenyl	✓

No.	Name	
883.	2,2',3',4,6-Pentachlorobiphenyl	∿
884.	2,2',3,4',6-Pentachlorobiphenyl	∿
885.	2,2',3,4,6'-Pentachlorobiphenyl	∿
886.	2,2',3,4'-Tetrachlorobiphenyl	∿
887.	2,2',3,4-Tetrachlorobiphenyl	∿
888.	2,2',3,5,5'-Hexachlorobiphenyl	∿
889.	2,2',3,5,5'-Pentachlorobiphenyl	∿
890.	2,2',3,5,6-Hexachlorobiphenyl	∿
891.	2,2',3,5',6-Pentachlorobiphenyl	∿
892.	2,2',3,5,6'-Pentachlorobiphenyl	∿
893.	2,2',3,5,6-Pentachlorobiphenyl	∿
894.	2,2',3,5'-Tetrachlorobiphenyl	∿
895.	2,2',3,5-Tetrachlorobiphenyl	∿

Continued

896.	2,2′,3,6, 6′-Pentachlorobiphenyl	
897.	2,2′,3, 6′-Tetrachlorobiphenyl	✓
898.	2,2′,3, 6-Tetrachlorobiphenyl	✓
899.	2,2′,3-Trichlorobiphenyl	✓
900.	2,2′,4,4′,5, 5′-Hexachlorobiphenyl	✓
901.	2,2′,4,4′,5, 6′-Hexachlorobiphenyl	✓
902.	2,2′,4,4′, 5-Pentachlorobiphenyl	✓
903.	2,2′,4,4′,6, 6′-Hexachlorobiphenyl	✓
904.	2,2′,4,4′, 6-Pentachlorobiphenyl	✓
905.	2,2′,4, 4′-Tetrachlorobiphenyl	✓
906.	2,2′,4,5, 5′-Pentachlorobiphenyl	✓
907.	2,2′,4,5′, 6-Pentachlorobiphenyl	✓
908.	2,2′,4,5, 6′-Pentachlorobiphenyl	✓

No.	Compound	
909.	2,2',4,5'-Tetrachlorobiphenyl	✓
910.	2,2',4,5-Tetrachlorobiphenyl	✓
911.	2,2',4,6,6'-Pentachlorobiphenyl	✓
912.	2,2',4,6'-Tetrachlorobiphenyl	✓
913.	2,2',4,6-Tetrachlorobiphenyl	✓
914.	2,2',4-Trichlorobiphenyl	✓
915.	2,2',5,5'-Tetrachlorobiphenyl	✓
916.	2,2',5,6'-Tetrachlorobiphenyl	✓
917.	2,2',5-Trichlorobiphenyl	✓
918.	2,2',6,6'-Tetrachlorobiphenyl	✓
919.	2,2',6-Trichlorobiphenyl	✓
920.	2,2'-Dichlorobiphenyl	✓
921.	2,3,3',4,4',5,5'-Octachlorobiphenyl	✓
922.	2,3,3',4,4',5,5'-Heptachlorobiphenyl	✓

Continued

		✓
923.	2,3,3',4,4',5,6-Heptachlorobiphenyl	✓
924.	2,3,3',4,4',5',6-Heptachlorobiphenyl	✓
925.	2,3,3',4,4',5'-Hexachlorobiphenyl	✓
926.	2,3,3',4,4',5-Hexachlorobiphenyl	✓
927.	2,3,3',4,4',6-Hexachlorobiphenyl	✓
928.	2,3,3',4,4'-Pentachlorobiphenyl	✓
929.	2,3,3',4,5,5',6-Heptachlorobiphenyl	✓
930.	2,3,3',4',5,5',6-Heptachlorobiphenyl	✓
931.	2,3,3',4,5,5'-Hexachlorobiphenyl	✓
932.	2,3,3',4',5,5'-Hexachlorobiphenyl	✓
933.	2,3,3',4,5,6-Hexachlorobiphenyl	✓
934.	2,3,3',4',5,6-Hexachlorobiphenyl	✓
935.	2,3,3',4,5',6-Hexachlorobiphenyl	✓

936.	2,3,3',4',5,6-Hexachlorobiphenyl	✓	
937.	2,3,3',4,5-Pentachlorobiphenyl	✓	
938.	2',3,3',4,5-Pentachlorobiphenyl	✓	
939.	2,3,3',4',5-Pentachlorobiphenyl	✓	
940.	2,3,3',4,5'-Pentachlorobiphenyl	✓	
941.	2,3,3',4,6-Pentachlorobiphenyl	✓	
942.	2,3,3',4',6-Pentachlorobiphenyl	✓	
943.	2,3,3',4'-Tetrachlorobiphenyl	✓	
944.	2,3,3',4-Tetrachlorobiphenyl	✓	
945.	2,3,3',5,5'-Hexachlorobiphenyl	✓	
946.	2,3,3',5,5'-Pentachlorobiphenyl	✓	
947.	2,3,3',5,6-Pentachlorobiphenyl	✓	
948.	2,3,3',5',6-Pentachlorobiphenyl	✓	

Continued

		✓
949.	2,3,3',5'-Tetrachlorobiphenyl	✓
950.	2,3,3',5-Tetrachlorobiphenyl	✓
951.	2,3,3',6-Tetrachlorobiphenyl	✓
952.	2,3,3'-Trichlorobiphenyl	✓
953.	2,3',4,4',5,5'-Hexachlorobiphenyl	✓
954.	2,3,4,4',5,6-Hexachlorobiphenyl	✓
955.	2,3',4,4',5',6-Hexachlorobiphenyl	✓
956.	2,3,4,4',5-Pentachlorobiphenyl	✓
957.	2',3,4,4',5-Pentachlorobiphenyl	✓
958.	2,3',4,4',5-Pentachlorobiphenyl	✓
959.	2,3,4,4',6-Pentachlorobiphenyl	✓
960.	2,3',4,4',6-Pentachlorobiphenyl	✓
961.	2,3,4,4'-Tetrachlorobiphenyl	✓

No.	Compound										
962.	2,3',4,4'-Tetrachlorobiphenyl						✓				
963.	2',3,4,5,5'-Pentachlorobiphenyl						✓				
964.	2,3',4,5,5'-Pentachlorobiphenyl						✓				
965.	2',3,4,5,6'-Pentachlorobiphenyl						✓				
966.	2,3,4,5,6-Pentachlorobiphenyl						✓				
967.	2,3',4,5',6-Pentachlorobiphenyl						✓				
968.	2,3,4',5,6-Pentachlorobiphenyl	✓	✓	✓	✓	✓	✓				
969.	2,3,4,5-Tetrachloroaniline	✓	✓	✓	✓	✓		✓			✓
970.	2,3,4,5-Tetrachloroanisole	✓	✓	✓	✓	✓		✓			
971.	2,3,4,5-Tetrachlorobiphenyl						✓				
972.	2',3,4,5-Tetrachlorobiphenyl						✓				
973.	2,3,4',5-Tetrachlorobiphenyl						✓				
974.	2,3',4',5-Tetrachlorobiphenyl						✓				

	Compound						
975.	2,3',4,5'-Tetrachlorobiphenyl					✓	
976.	2,3',4,5-Tetrachlorobiphenyl					✓	
977.	2,3,4,6-Tetrachlorobiphenyl					✓	
978.	2,3',4,6-Tetrachlorobiphenyl					✓	
979.	2,3,4',6-Tetrachlorobiphenyl					✓	
980.	2,3',4',6-Tetrachlorobiphenyl					✓	
981.	2',3,4-Trichlorobiphenyl					✓	
982.	2,3',4-Trichlorobiphenyl					✓	
983.	2,3,4'-Trichlorobiphenyl					✓	
984.	2,3,4-Trichlorobiphenyl					✓	
985.	2,3',5,5'-Tetrachlorobiphenyl					✓	
986.	2,3,5,6-Tetrachloroaniline	✓	✓	✓	✓	✓	✓
987.	2,3,5,6-Tetrachlorobiphenyl					✓	
988.	2,3',5',6-Tetrachlorobiphenyl					✓	
989.	2,3,5-Trichlorobiphenyl					✓	
990.	2',3,5-Trichlorobiphenyl					✓	

Continued

	1	2	3	4	5	6	7	8	9	10	11	12	13	14	15	16	17	18	19	20	21	22	23	24	25	26
991.	2,3′,5-Trichlorobiphenyl												✓									✓			✓	
992.	2,3,6-Trichlorobiphenyl												✓										✓		✓	✓
993.	2,3′,6-Trichlorobiphenyl												✓												✓	
994.	2,3′-Dichlorobiphenyl							✓					✓							✓					✓	
995.	2,3-Dichlorobiphenyl												✓												✓	
996.	2,4,4′,5-Tetrachlorobiphenyl												✓												✓	
997.	2,4,4′,6-Tetrachlorobiphenyl											✓													✓	
998.	2,4,4′-Trichlorobiphenyl												✓												✓	
999.	2,4,5-T										✓	✓	✓			✓	✓									
1000.	2,4,5-Trichlorobiphenyl												✓													
1001.	2,4′,5-Trichlorobiphenyl												✓					✓			✓					
1002.	2,4,6-Trichlorobiphenyl												✓					✓	✓							
1003.	2,4′,6-Trichlorobiphenyl												✓													
1004.	2,4-D							✓		✓	✓	✓	✓			✓	✓			✓						
1005.	2,4-DB															✓										
1006.	2,4′-Dichlorobiphenyl												✓													
1007.	2,4-Dichlorobiphenyl												✓													
1008.	2,5-Dichlorobiphenyl												✓			✓					✓					
1009.	2,6-Dichlorobenzamide			✓	✓	✓	✓	✓	✓				✓	✓	✓	✓				✓		✓	✓	✓	✓	
1010.	2,6-Dichlorobiphenyl												✓													

Continued

1011.	2,6-Difluorobenzoic acid								✓			
1012.	2-Chlorobiphenyl	✓	✓				✓		✓	✓	✓	✓
1013.	2-Phenylphenol	✓	✓	✓	✓	✓	✓		✓	✓	✓	✓
1014.	3,3',4,4',5,5'-Hexachlorobiphenyl					✓						
1015.	3,3',4,4',5-Pentachlorobiphenyl					✓						
1016.	3,3',4,4'-Tetrachlorobiphenyl					✓	✓					
1017.	3,3',4,5,5'-Pentachlorobiphenyl					✓						
1018.	3,3',4,5'-Tetrachlorobiphenyl					✓						
1019.	3,3',4,5-Tetrachlorobiphenyl					✓						
1020.	3,3',4-Trichlorobiphenyl					✓						
1021.	3,3',5,5'-Tetrachlorobiphenyl					✓						
1022.	3,3',5-Trichlorobiphenyl					✓						
1023.	3,3'-Dichlorobiphenyl					✓						
1024.	3,4,4',5-Tetrachlorobiphenyl					✓						
1025.	3,4,4'-Trichlorobiphenyl					✓						
1026.	3,4,5-Trichlorobiphenyl					✓						
1027.	3,4',5-Trichlorobiphenyl					✓						

No.	Compound	1	2	3	4	5	6	7	8	9	10	11	12	13	14	15	16	17	18	19
1028.	3,4,5-Trimethacarb				✓							✓	✓	✓	✓	✓	✓	✓	✓	✓
1029.	3,4′-Dichlorobiphenyl										✓									
1030.	3,4-Dichlorobiphenyl										✓									
1031.	3,5-Dichloroaniline	✓	✓	✓	✓	✓	✓	✓	✓	✓										
1032.	3,5-Dichlorobiphenyl										✓									
1033.	3.4.5-Trimethacarb								✓	✓										
1034.	3-Chlorobiphenyl										✓									
1035.	3-Phenylphenol				✓	✓	✓		✓	✓		✓	✓	✓	✓	✓	✓	✓	✓	✓
1036.	4,4-Dibromobenzophenone	✓	✓	✓	✓	✓	✓	✓	✓	✓										
1037.	4,4-Dichlorobenzophenone	✓	✓	✓	✓	✓	✓	✓	✓	✓		✓	✓	✓	✓	✓			✓	✓
1038.	4,4′-Dichlorobiphenyl										✓									
1039.	4-Aminopyridine											✓	✓	✓	✓	✓	✓	✓	✓	
1040.	4-Chlorobiphenyl										✓									
1041.	4-Chlorophenoxy acetic acid				✓	✓	✓			✓										
1042.	4-CPA				✓															
1043.	6-Chloro-4-hydroxy-3-phenyl-pyridazin											✓	✓	✓	✓	✓	✓		✓	✓

*a*Pesticide quantities

Appendix B

Solvent Selected and Concentration of Mixed Standard Solution of Pesticides and Chemical Pollutants

TABLE B.1 Solvent Selected and Concentration of Mixed Standard Solution of 587 Pesticides and Chemical Pollutants Analyzed by GC-MS

No.	Compound	Solvent	Concentration of Mixed Standard Solution (mg/L)
1.	Trans-Diallate	Toluene	5
2.	(tau-)Fluvalinate	Toluene	30
3.	2,3,4,5-Tetrachloroaniline	Toluene	5
4.	2,3,4,5-Tetrachloroanisole	Toluene-acetone(8:2)	2.5
5.	2,3,5,6-Tetrachloroaniline	Toluene	2.5
6.	2,4,5-T	Toluene	50
7.	2,4'-DDD	Toluene	2.5
8.	2,4-D	Toluene	50
9.	2,4'-DDE	Toluene	2.5
10.	2,4'-DDT	Toluene	5
11.	2,6-Dichlorobenzamide	Toluene	5

Continued

TABLE B.1 Solvent Selected and Concentration of Mixed Standard Solution of 587 Pesticides and Chemical Pollutants Analyzed by GC-MS—cont'd

No.	Compound	Solvent	Concentration of Mixed Standard Solution (mg/L)
12.	2.4′-DDE	Toluene	2.5
13.	2-Phenylphenol	Toluene	2.5
14.	3,4,5-Trimethacarb	Toluene	20
15.	3,5-Dichloroaniline	Toluene	2.5
16.	3.4.5-Trimethacarb	Toluene	20
17.	3-Phenylphenol	Toluene	15
18.	4,4′-DDE	Toluene	2.5
19.	4,4′-DDT	Toluene	5
20.	4,4′-DDD	Toluene	2.5
21.	4,4-Dibromobenzophenone	Toluene	2.5
22.	4-Chlorophenoxy acetic acid	Toluene	1.3
23.	Acenaphthene	Cyclohexane	2.5
24.	Acephate	Toluene	50
25.	Acetamiprid	Toluene	10
26.	Acetochlor	Toluene	5
27.	Acibenzolar-s-methyl	Toluene	5
28.	Aclonifen	Toluene	50
29.	Acrinathrin	Toluene	5
30.	Alachlor	Toluene	7.5
31.	Aldrin	Toluene	5
32.	Allethrin	Toluene	10
33.	Allidochlor	Toluene	5
34.	Alpha-cypermethrin	Toluene	5
35.	Alpha-HCH	Toluene	2.5
36.	Ametryn	Toluene	7.5
37.	Amitraz	Toluene	7.5

TABLE B.1 Solvent Selected and Concentration of Mixed Standard Solution of 587 Pesticides and Chemical Pollutants Analyzed by GC-MS—cont'd

No.	Compound	Solvent	Concentration of Mixed Standard Solution (mg/L)
38.	Anilofos	Toluene	5
39.	Anthraquinone	Dichlormethane	2.5
40.	Aramite	Dichlormethane	2.5
41.	Aspon	Toluene	5
42.	Athidathion	Toluene	5
43.	Atratone	Toluene	2.5
44.	Atrazine-desethyl	Toluene-acetone(8:2)	2.5
45.	Atrizine	Toluene-acetone(9:1)	2.5
46.	Azaconazole	Toluene	10
47.	Azinphos-ethyl	Toluene	5
48.	Azinphos-methyl	Toluene	15
49.	Aziprotryne	Toluene	20
50.	Azoxystrobin	Toluene	25
51.	BDMC-1	Toluene-acetone(8:2)	5
52.	BDMC-2	Toluene	5
53.	Benalaxyl	Toluene	2.5
54.	Benfluralin	Toluene	2.5
55.	Benfuresate	Toluene	5
56.	Benodanil	Toluene	7.5
57.	Benoxacor	Cyclohexane	5
58.	Bensulide	Toluene	20
59.	Benzoylprop-ethyl	Toluene	7.5
60.	Beta-HCH	Toluene	2.5
61.	Bifenazate	Toluene	20
62.	Bifenox	Toluene	5
63.	Bifenthrin	n-Hexane	2.5

Continued

TABLE B.1 Solvent Selected and Concentration of Mixed Standard Solution of 587 Pesticides and Chemical Pollutants Analyzed by GC-MS—cont'd

No.	Compound	Solvent	Concentration of Mixed Standard Solution (mg/L)
64.	Bioallethrin-1	Toluene	10
65.	Bioallethrin-2	Toluene	10
66.	Bioresmethrin	Toluene	5
67.	Biphenyl	Toluene	2.5
68.	Bitertanol	Toluene	7.5
69.	Boscalid	Toluene	10
70.	Bromacil	Toluene	5
71.	Bromfenvinfos	Toluene-acetone(8:2)	2.5
72.	Bromobutide	Cyclohexane	2.5
73.	Bromocylen	Toluene	2.5
74.	Bromofos	Toluene	5
75.	Bromophos-ethyl	Toluene	2.5
76.	Bromopropylate	Toluene	5
77.	Bromuconazole-1	Toluene	5
78.	Bromuconazole-2	Toluene	5
79.	Bupirimate	Toluene	2.5
80.	Buprofezin	Toluene	5
81.	Butachlor	Toluene	5
82.	Butafenacil	Acetonitrile	2.5
83.	Butamifos	Cyclohexane	2.5
84.	Butralin	Toluene	10
85.	Butylate	Toluene	7.5
86.	Cadusafos	Toluene	10
87.	Cafenstrole	Acetonitrile	10
88.	Captafol	Toluene-acetone(8:2)	45
89.	Captan	Toluene	40

TABLE B.1 Solvent Selected and Concentration of Mixed Standard Solution of 587 Pesticides and Chemical Pollutants Analyzed by GC-MS—cont'd

No.	Compound	Solvent	Concentration of Mixed Standard Solution (mg/L)
90.	Carbaryl	Toluene	7.5
91.	Carbofenothion	Toluene	5
92.	Carbosulfan	Toluene	7.5
93.	Carboxin	Toluene	7.5
94.	Carfentrazone-ethyl	Toluene	5
95.	Chlorbenside	Toluene	5
96.	Chlorbenside sulfone	Toluene	5
97.	Chlorbromuron	Toluene	60
98.	Chlorbufam	Toluene	5
99.	Chlordimeform	n-Hexane	2.5
100.	Chlorethoxyfos	Toluene	5
101.	Chlorfenapyr	Toluene	20
102.	Chlorfenethol	Toluene	2.5
103.	Chlorfenprop-methyl	Toluene	2.5
104.	Chlorfenson	Toluene	5
105.	Chlorfenvinphos	Toluene	7.5
106.	Chlorfluazuron	Toluene	7.5
107.	Chlorflurenol	Toluene-acetone(9:1)	7.5
108.	Chlormephos	Toluene	5
109.	Chlorobenzilate	Toluene	2.5
110.	Chloroneb	Toluene	2.5
111.	Chloropropylate	Toluene	2.5
112.	Chlorothalonil	Toluene	5
113.	Chlorpropham	Toluene	5
114.	Chlorpyrifos (-ethyl)	Toluene	2.5
115.	Chlorpyrifos-methyl	Toluene	2.5
116.	Chlorthal-dimethyl	Toluene	5

Continued

TABLE B.1 Solvent Selected and Concentration of Mixed Standard Solution of 587 Pesticides and Chemical Pollutants Analyzed by GC-MS—cont'd

No.	Compound	Solvent	Concentration of Mixed Standard Solution (mg/L)
117.	Chlorthion	Toluene	5
118.	Chlorthiophos	Toluene	7.5
119.	Chlozolinate	Toluene	5
120.	Chromafenozide	Toluene	20
121.	Chrysene	Acetonitrile	2.5
122.	Cinmethylin	Toluene	5
123.	cis and trans-Chlorfenvinphos	Toluene	7.5
124.	cis and trans-Diallate	Toluene	5
125.	cis-Chlordane	Toluene	5
126.	cis-Diallate	Toluene	5
127.	cis-1,2,3,6-Tetrahydrophthalimide	Methanol	7.5
128.	cis-Permethrin	Toluene	2.5
129.	Clethodim	Cyclohexane	10
130.	Clodinafop-propargyl	Toluene	5
131.	Clomazone	Toluene	2.5
132.	Clomeprop	Acetonitrile	2.5
133.	Cloquintocet-mexyl	Toluene	2.5
134.	Coumaphos	Toluene	15
135.	Crimidine	Toluene	2.5
136.	Crotoxyphos	Toluene	15
137.	Crufomate	Toluene	15
138.	Cyanazine	Toluene-acetone(8:2)	7.5
139.	Cyanofenphos	Toluene	2.5
140.	Cyanophos	Toluene	5
141.	Cycloate	Toluene	2.5
142.	Cycloxydim	Toluene	30

TABLE B.1 Solvent Selected and Concentration of Mixed Standard Solution of 587 Pesticides and Chemical Pollutants Analyzed by GC-MS—cont'd

No.	Compound	Solvent	Concentration of Mixed Standard Solution (mg/L)
143.	Cycluron	Toluene	7.5
144.	Cyflufenamid	Toluene	40
145.	Cyfluthrin	Toluene	30
146.	Cyhalofop-butyl	Toluene	5
147.	Cypermethrin	Toluene	7.5
148.	Cyprazine	Toluene-acetone(9:1)	2.5
149.	Cyproconazole	Toluene	2.5
150.	Cyprodinil	Toluene	2.5
151.	Cythioate	Toluene	50
152.	Dacthal	Toluene	2.5
153.	DEF	Toluene	5
154.	delta-HCH	Toluene	5
155.	Deltamethrin	Toluene	15
156.	Demetom-s	Toluene	10
157.	Demeton-s-methyl	Toluene	10
158.	de-PCB 101	Toluene	2.5
159.	de-PCB 118	Toluene	2.5
160.	de-PCB 138	Toluene	2.5
161.	de-PCB 153	Toluene	2.5
162.	de-PCB 180	Toluene	2.5
163.	de-PCB 28	Toluene	2.5
164.	de-PCB 31	Toluene	2.5
165.	De-PCB 52	Toluene	2.5
166.	Desbrom- leptophos	Toluene	2.5
167.	Desethyl-sebuthylazine	Toluene	5
168.	Desethyl-Sebuthylazine	Toluene-acetone(8:2)	5

Continued

TABLE B.1 Solvent Selected and Concentration of Mixed Standard Solution of 587 Pesticides and Chemical Pollutants Analyzed by GC-MS—cont'd

No.	Compound	Solvent	Concentration of Mixed Standard Solution (mg/L)
169.	Desisopropyl-atrazine	Toluene-acetone(8:2)	20
170.	Desmedipham	Toluene	50
171.	Desmetryn	Toluene	2.5
172.	Dialifos	Toluene	80
173.	Diazinon	Toluene	2.5
174.	Dibutyl succinate	Toluene	5
175.	Dicapthon	Toluene	12.5
176.	Dichlobenil	Toluene	0.5
177.	Dichlofenthion	Toluene	2.5
178.	Dichlofluanid	Toluene	15
179.	Dichloran	Toluene	5
180.	Dichlormid	Toluene	2.5
181.	Dichlorvos	Toluene	15
182.	Diclobutrazole	Toluene	10
183.	Diclofop-methyl	Toluene	2.5
184.	Dicloran	Toluene	5
185.	Dicofol	Toluene	5
186.	Dicrotophos	Toluene	20
187.	Dieldrin	Toluene	5
188.	Diethofencarb	Toluene	15
189.	Diethyltoluamide	Toluene	2
190.	Difenoconazole	Toluene	15
191.	Difenonazole	Toluene	15
192.	Difenoxuron	Toluene	20
193.	Diflufenican	Toluene	2.5
194.	Dimefox	Toluene	7.5
195.	Dimefuron	Toluene	10

TABLE B.1 Solvent Selected and Concentration of Mixed Standard Solution of 587 Pesticides and Chemical Pollutants Analyzed by GC-MS—cont'd

No.	Compound	Solvent	Concentration of Mixed Standard Solution (mg/L)
196.	Dimepiperate	Toluene	5
197.	Dimethachlor	Toluene	7.5
198.	Dimethametryn	Toluene	2.5
199.	Dimethenamid	Methanol	2.5
200.	Dimethipin	Toluene	7.5
201.	Dimethoate	Toluene	10
202.	Dimethomorph	Toluene	5
203.	Dimethyl phthalate	Toluene	10
204.	Dimethylvinphos	Toluene	5
205.	Diniconazole	Toluene	7.5
206.	Dinitramine	Toluene	10
207.	Dinobuton	Toluene	25
208.	Dinoterb	Toluene	10
209.	Diofenolan -1	Toluene	5
210.	Dioxabenzofos	Toluene	25
211.	Dioxacarb	Toluene	20
212.	Dioxathion	Toluene	10
213.	Diphenamid	Toluene	2.5
214.	Diphenylamine	Toluene	2.5
215.	Dipropetryn	Toluene	2.5
216.	Disulfoton	Toluene	2.5
217.	Disulfoton sulfone	Toluene	5
218.	Disulfoton-sulfoxide	Toluene	5
219.	Ditalimfos	Toluene	2.5
220.	Dithiopyr	Toluene	2.5
221.	DMSA	Toluene	20
222.	Dodemorph	Toluene	7.5
223.	Edifenphos	Toluene	5

Continued

TABLE B.1 Solvent Selected and Concentration of Mixed Standard Solution of 587 Pesticides and Chemical Pollutants Analyzed by GC-MS—cont'd

No.	Compound	Solvent	Concentration of Mixed Standard Solution (mg/L)
224.	Endosulfan -1	Toluene	15
225.	Endosulfan -2	Toluene	15
226.	Endosulfan-sulfate	Toluene	7.5
227.	Endothal	Toluene	50
228.	Endrin	Toluene	30
229.	Endrin aldehyde	Toluene	50
230.	Endrin Ketone	Toluene	10
231.	EPN	Toluene	10
232.	Epoxiconazole -1	Toluene	20
233.	Epoxiconazole -2	Toluene	20
234.	Epsilon-HCH	Methanol	5
235.	EPTC	Toluene	7.5
236.	Erbon	Toluene	5
237.	Esfenvalerate	Toluene	10
238.	Esprocarb	Toluene	5
239.	Etaconazole-1	Toluene	7.5
240.	Etaconazole-2	Toluene	7.5
241.	Ethalfluralin	Toluene	10
242.	Ethiofencarb	Toluene	25
243.	Ethion	Toluene	5
244.	Ethofumesate	Toluene	5
245.	Ethoprophos	Toluene	7.5
246.	Etofenprox	Toluene	2.5
247.	Etoxazole	Toluene	15
248.	Etridiazol	Toluene	7.5
249.	Etrimfos	Toluene	2.5
250.	Etrimfos oxon	Toluene	2.5
251.	Famphur	Toluene	10

TABLE B.1 Solvent Selected and Concentration of Mixed Standard Solution of 587 Pesticides and Chemical Pollutants Analyzed by GC-MS—cont'd

No.	Compound	Solvent	Concentration of Mixed Standard Solution (mg/L)
252.	Fenamidone	Toluene	2.5
253.	Fenamiphos	Toluene	7.5
254.	Fenamiphos sulfone	Toluene-acetone(8:2)	10
255.	Fenamiphos sulfoxide	Toluene	10
256.	Fenarimol	Toluene	5
257.	Fenazaquin	Toluene	2.5
258.	Fenbuconazole	Toluene-acetone(8:2)	5
259.	Fenchlorphos	Toluene	10
260.	Fenchlorphos Oxon	Toluene	5
261.	Fenfuram	Toluene	5
262.	Fenhexamid	Toluene	50
263.	Fenitrothion	Toluene	5
264.	Fenobucarb	Toluene	5
265.	Fenothiocarb	Acetone	5
266.	Fenoxanil	Toluene	5
267.	Fenoxycarb	Toluene	15
268.	Fenpiclonil	Toluene	10
269.	Fenpropathrin	Toluene	5
270.	Fenpropidin	Toluene	5
271.	Fenpropimorph	Toluene	2.5
272.	Fenpyroximate	Toluene	20
273.	Fenson	Toluene	2.5
274.	Fensulfothion	Toluene	5
275.	Fenthion	Toluene	2.5
276.	Fenthion sulfone	Toluene	10
277.	Fenthion sulfoxide	Toluene	10

Continued

TABLE B.1 Solvent Selected and Concentration of Mixed Standard Solution of 587 Pesticides and Chemical Pollutants Analyzed by GC-MS—cont'd

No.	Compound	Solvent	Concentration of Mixed Standard Solution (mg/L)
278.	Fenvalerate-1	Toluene	10
279.	Fenvalerate-2	Toluene	10
280.	Fipronil	Toluene	20
281.	Flamprop-isopropyl	Toluene	2.5
282.	Flamprop-methyl	Toluene-acetone(8:2)	2.5
283.	Fluazifop-butyl	Cyclohexane	2.5
284.	Fluazinam	Toluene	20
285.	Fluchloralin	Cyclohexane	10
286.	Flucythrinate-1	Toluene-acetone(8:2)	5
287.	Flucythrinate-2	Toluene	5
288.	Fludioxonil	Toluene	2.5
289.	Flufenacet	Toluene	20
290.	Flufenoxuron	Toluene	7.5
291.	Flumetralin	Cyclohexane	5
292.	Flumiclorac-pentyl	Toluene	5
293.	Flumioxazin	Toluene	5
294.	Fluorochloridone	Toluene	5
295.	Fluorodifen	Toluene	2.5
296.	Fluoroglycofen-ethyl	Toluene	30
297.	Fluotrimazole	Methanol	2.5
298.	Fluquinconazole	Toluene	2.5
299.	Fluridone	Toluene	5
300.	Flurochloridone	Toluene	5
301.	Fluroxypr-1-methylheptyl ester	Toluene	2.5
302.	Flurtamone	Toluene	5
303.	Flusilazole	Toluene	7.5

TABLE B.1 Solvent Selected and Concentration of Mixed Standard Solution of 587 Pesticides and Chemical Pollutants Analyzed by GC-MS—cont'd

No.	Compound	Solvent	Concentration of Mixed Standard Solution (mg/L)
304.	Flutolanil	Toluene	2.5
305.	Flutriafol	Toluene	5
306.	Fluvalinate	Toluene	30
307.	Folpet	Toluene	30
308.	Fonofos	Toluene	2.5
309.	Formothion	Methanol	5
310.	Fuberidazole	Toluene	12.5
311.	Furalaxyl	Toluene	5
312.	Furmecyclox	Cyclohexane	7.5
313.	gamma-Cyhaloterin-1	Toluene	2
314.	gamma-Cyhalothrin-2	Toluene	2
315.	gamma-HCH	Toluene	5
316.	Halfenprox	Toluene	5
317.	Halosulfuran-methyl	Toluene	50
318.	HCH, epsilon-	Methanol	5
319.	Heptachlor	Toluene	7.5
320.	Heptanophos	Toluene	7.5
321.	Hexachlorobenzene	Toluene	2.5
322.	Hexaconazole	Toluene	15
323.	Hexaflumuron	Toluene	15
324.	Hexazinone	Toluene	7.5
325.	Hexythiazox	Toluene	20
326.	Imazalil	Toluene	10
327.	Imazamethabenz-methyl	Toluene	7.5
328.	Imibenconazole-des-benzyl	Toluene-acetone(8:2)	10
329.	Imiprothrin-1	Toluene	5
330.	Imiprothrin-2	Toluene	5

Continued

TABLE B.1 Solvent Selected and Concentration of Mixed Standard Solution of 587 Pesticides and Chemical Pollutants Analyzed by GC-MS—cont'd

No.	Compound	Solvent	Concentration of Mixed Standard Solution (mg/L)
331.	Iodofenphos	Toluene	5
332.	Iprobenfos	Toluene	7.5
333.	Iprodione	Toluene	10
334.	Iprovalicarb-1	Toluene	10
335.	Iprovalicarb-2	Toluene	10
336.	Isazofos	Toluene	5
337.	Isobenzan	Toluene	2.5
338.	Isocarbamid	Toluene	12.5
339.	Isocarbophos	Toluene	5
340.	Isodrin	Toluene	2.5
341.	Isofenphos	Toluene	5
342.	Isofenphos oxon	Toluene	5
343.	Isomethiozin	Toluene	5
344.	Isoprocarb -1	Toluene	5
345.	Isoprocarb -2	Toluene	5
346.	Isopropalin	Toluene	5
347.	Isoprothiolane	Toluene	5
348.	Isoxadifen-ethyl	Toluene	5
349.	Isoxathion	Toluene	20
350.	Kresoxim-methyl	Toluene	2.5
351.	Lactofen	Toluene	20
352.	Lambda-cyhalothrin	Toluene	2.5
353.	Lenacil	Toluene	25
354.	Leptophos	Toluene	5
355.	Linuron	Toluene-acetone(9:1)	10
356.	Malaoxon	Toluene	40
357.	Malathion	Toluene	10

TABLE B.1 Solvent Selected and Concentration of Mixed Standard Solution of 587 Pesticides and Chemical Pollutants Analyzed by GC-MS—cont'd

No.	Compound	Solvent	Concentration of Mixed Standard Solution (mg/L)
358.	MCPA-Butoxyethyl ester	Toluene	2.5
359.	Mecarbam	Toluene	10
360.	Mefenacet	Toluene	7.5
361.	Mefenoxam	Toluene	5
362.	Mefenpyr-diethyl	Toluene	7.5
363.	Mepanipyrim	Toluene	2.5
364.	Mephosfolan	Toluene	5
365.	Mepronil	Toluene	2.5
366.	Metalaxyl	Toluene	7.5
367.	Metamitron	Toluene	25
368.	Metazachlor	Toluene	7.5
369.	Methabenzthiazuron	Toluene	25
370.	Methacrifos	Toluene	2.5
371.	Methamidophos	Toluene	10
372.	Methfuroxam	Toluene	2.5
373.	Methidathion	Toluene	5
374.	Methiocarb sulfone	Toluene	80
375.	Methoprene	Toluene	10
376.	Methoprotryne	Toluene	7.5
377.	Methothrin-1	Toluene	5
378.	Methothrin-2	Acetonitrile	5
379.	Methoxychlor	Toluene	20
380.	Methyl-parathion	Toluene	10
381.	Metobromuron	Toluene	15
382.	Metoconazole	Toluene	10
383.	Metolachlor	Toluene	2.5
384.	Metominostrobin-1	Acetonitrile	10
385.	Metominostrobin-2	Acetonitrile	10

Continued

TABLE B.1 Solvent Selected and Concentration of Mixed Standard Solution of 587 Pesticides and Chemical Pollutants Analyzed by GC-MS—cont'd

No.	Compound	Solvent	Concentration of Mixed Standard Solution (mg/L)
386.	Metribuzin	Toluene	7.5
387.	Mevinphos	Toluene	5
388.	Mexacarbate	Toluene	7.5
389.	Mirex	Toluene	2.5
390.	Molinate	Toluene	2.5
391.	Monalide	Toluene	5
392.	Monocrotophos	Toluene	20
393.	Monolinuron	Toluene	10
394.	Musk ambrette	Toluene	2.5
395.	Musk ketone	Toluene	2.5
396.	Musk moskene	Toluene	2.5
397.	Musk Tibeten	Toluene	2.5
398.	Musk xylene	Toluene	2.5
399.	Myclobutanil	Toluene	2.5
400.	Naled	Toluene	40
401.	Naproanilide	Acetonitrile	2.5
402.	Napropamide	Toluene	7.5
403.	Nitralin	Toluene	25
404.	Nitrapyrin	Toluene	7.5
405.	Nitrofen	Toluene	15
406.	Nitrothal-isopropyl	Toluene	5
407.	Norflurazon	Toluene-acetone(9:1)	2.5
408.	Norflurazon-desmethyl	Toluene	10
409.	Nuarimol	Toluene-acetone(9:1)	5
410.	o,p′-DDD	Toluene	2.5
411.	o,p′-DDT	Toluene	5

TABLE B.1 Solvent Selected and Concentration of Mixed Standard Solution of 587 Pesticides and Chemical Pollutants Analyzed by GC-MS—cont'd

No.	Compound	Solvent	Concentration of Mixed Standard Solution (mg/L)
412.	o,p'-DDE	Toluene	2.5
413.	Octachlorostyrene	Toluene	2.5
414.	Ofurace	Toluene	7.5
415.	Oxadiazone	Toluene	2.5
416.	Oxadixyl	Toluene	2.5
417.	Oxycarboxin	Toluene-acetone(9:1)	15
418.	oxy-Chlordane	Toluene	2.5
419.	Oxyfluorfen	Toluene	10
420.	p,p'-DDE	Toluene	2.5
421.	p,p'-DDT	Toluene	5
422.	p,p'-DDD	Toluene	2.5
423.	Paclobutrazol	Toluene	7.5
424.	Paraoxon-ethyl	Toluene	10
425.	Paraoxon-Methyl	Toluene	5
426.	Parathion	Toluene	10
427.	Pebulate	Toluene	7.5
428.	Penconazole	Toluene	7.5
429.	Pencycuron	Toluene	10
430.	Pendimethalin	Toluene	10
431.	Pentachloroaniline	Toluene	2.5
432.	Pentachloroanisole	Toluene	2.5
433.	Pentachlorobenzene	Toluene	2.5
434.	Permethrin	Toluene	5
435.	Perthane	Toluene	2.5
436.	Phenanthrene	Toluene	2.5
437.	Phenkapton	Toluene	15
438.	Phenothrin	Toluene	2.5

Continued

TABLE B.1 Solvent Selected and Concentration of Mixed Standard Solution of 587 Pesticides and Chemical Pollutants Analyzed by GC-MS—cont'd

No.	Compound	Solvent	Concentration of Mixed Standard Solution (mg/L)
439.	Phenthoate	Toluene	5
440.	Phorate	Toluene	2.5
441.	Phorate sulfone	Toluene	2.5
442.	Phosalone	Toluene	5
443.	Phosmet	Toluene	5
444.	Phosphamidon -1	Toluene	20
445.	Phosphamidon -2	Toluene	20
446.	Phthalic acid,benzyl butyl ester	Toluene	2.5
447.	Phthalide	Acetone	10
448.	Phthalimide	Toluene	5
449.	Picolinafen	Toluene	2.5
450.	Picoxystrobin	Toluene	5
451.	Piperonyl butoxide	Toluene	2.5
452.	Piperophos	Toluene	7.5
453.	Pirimicarb	Toluene	5
454.	Pirimiphos-ethyl	Toluene	5
455.	Pirimiphos-methyl	Toluene	2.5
456.	Plifenate	Toluene	5
457.	Prallethrin	Toluene	7.5
458.	Pretilachlor	Toluene	5
459.	Prochloraz	Toluene	15
460.	Procymidone	Toluene	2.5
461.	Profenofos	Toluene	15
462.	Profluralin	Toluene	10
463.	Prohydrojasmon	Cyclohexane	10
464.	Prometon	Toluene	7.5
465.	Prometryne	Toluene	2.5

TABLE B.1 Solvent Selected and Concentration of Mixed Standard Solution of 587 Pesticides and Chemical Pollutants Analyzed by GC-MS—cont'd

No.	Compound	Solvent	Concentration of Mixed Standard Solution (mg/L)
466.	Pronamide	Toluene-acetone(9:1)	2.5
467.	Propachlor	Toluene	7.5
468.	Propamocarb	Toluene	7.5
469.	Propanil	Toluene	5
470.	Propaphos	Toluene	5
471.	Propargite	Toluene	5
472.	Propazine	Toluene	2.5
473.	Propetamphos	Toluene	2.5
474.	Propham	Toluene	2.5
475.	Propiconazole-1	Toluene	7.5
476.	Propiconazole-2	Toluene	7.5
477.	Propisochlor	Toluene	2.5
478.	Propoxur -1	Toluene	5
479.	Propoxur -2	Toluene	5
480.	Propyzamide	Toluene	5
481.	Prosulfocarb	Toluene	2.5
482.	Prothiophos	Toluene	2.5
483.	Pyraclofos	Toluene	20
484.	Pyraclostrobin	Toluene	60
485.	Pyraflufen ethyl	Toluene	5
486.	Pyrazophos	Toluene	5
487.	Pyributicarb	Acetonitrile	5
488.	Pyridaben	Toluene	2.5
489.	Pyridaphenthion	Toluene	2.5
490.	Pyrifenox -1	Toluene	20
491.	Pyrifenox -2	Toluene	20
492.	Pyriftalid	Cyclohexane	2.5

Continued

TABLE B.1 Solvent Selected and Concentration of Mixed Standard Solution of 587 Pesticides and Chemical Pollutants Analyzed by GC-MS—cont'd

No.	Compound	Solvent	Concentration of Mixed Standard Solution (mg/L)
493.	Pyrimethanil	Toluene	2.5
494.	Pyrimidifen	Toluene	5
495.	Pyriminobac-methyl	Toluene	10
496.	Pyrimitate	Toluene	2.5
497.	Pyriproxyfen	Toluene	5
498.	Pyroquilon	Toluene	2.5
499.	Quinalphos	Toluene	2.5
500.	Quinoclamine	Toluene	10
501.	Quinoxyphen	Toluene	2.5
502.	Quintozene	Toluene	5
503.	Rabenzazole	Toluene	2.5
504.	Resmethrin-1	Toluene	5
505.	Resmethrin-2	Toluene	5
506.	Ronnel	Toluene	5
507.	s421(Octachlorodipropyl ether)-1	Toluene	50
508.	s421(Octachlorodipropyl ether)-2	Toluene	50
509.	Sebutylazine	Toluene	2.5
510.	Secbumeton	Toluene	2.5
511.	Sethoxydim	Toluene	22.5
512.	Silafluofen	Toluene	2.5
513.	Simazine	Methanol	2.5
514.	Simeconazole	Toluene	5
515.	Simeton	Toluene	5
516.	Simetryn	Toluene	5
517.	Sobutylazine	Toluene	5
518.	Spirodiclofen	Toluene	20

TABLE B.1 Solvent Selected and Concentration of Mixed Standard Solution of 587 Pesticides and Chemical Pollutants Analyzed by GC-MS—cont'd

No.	Compound	Solvent	Concentration of Mixed Standard Solution (mg/L)
519.	Spiromesifen	Toluene	25
520.	Spiroxamine-1	Toluene	5
521.	Spiroxamine-2	Toluene	5
522.	Sulfallate	Toluene	5
523.	Sulfotep	Toluene	2.5
524.	Sulprofos	Toluene	5
525.	tau-Fluvalinate	Toluene	30
526.	TCMTB	Toluene	40
527.	Tebuconazole	Toluene	7.5
528.	Tebufenpyrad	Toluene	2.5
529.	Tebupirimfos	Toluene	5
530.	Tebutam	Toluene	5
531.	Tebuthiuron	Toluene	10
532.	Tecnazene	Toluene	5
533.	Tefluthrin	Toluene	2.5
534.	Terbacil	Toluene	5
535.	Terbucarb-1	Toluene	5
536.	Terbucarb-2	Toluene	5
537.	Terbufos	Toluene	5
538.	Terbufos Sulfone	Toluene	2.5
539.	Terbumeton	Toluene	7.5
540.	Terbuthylazine	Toluene	2.5
541.	Terbutryn	Toluene	5
542.	Tetrachlorvinphos	Toluene	7.5
543.	Tetraconazole	Toluene	7.5
544.	Tetradifon	Toluene	2.5
545.	Tetrahydrophthalimide	Toluene	7.5
546.	Tetramethirn	Toluene	5

Continued

TABLE B.1 Solvent Selected and Concentration of Mixed Standard Solution of 587 Pesticides and Chemical Pollutants Analyzed by GC-MS—cont'd

No.	Compound	Solvent	Concentration of Mixed Standard Solution (mg/L)
547.	Tetrasul	Toluene	2.5
548.	Thenylchlor	Toluene	5
549.	Thiabendazole	Toluene	50
550.	Thiamethoxam	Toluene	10
551.	Thiazopyr	Toluene	5
552.	Thifluzamide	Acetonitrile	20
553.	Thiobencarb	Toluene	5
554.	Thiometon	Toluene	2.5
555.	Thionazin	Toluene	2.5
556.	Tolclofos-methyl	Toluene	2.5
557.	Tolylfluanide	Toluene	60
558.	Tralkoxydim	Toluene	20
559.	Tralomethrin-1	Toluene	2.5
560.	Tralomethrin-2	Toluene	2.5
561.	trans-Chlordane	Toluene	2.5
562.	trans-Diallate	Toluene	5
563.	Transfluthrin	Toluene	2.5
564.	trans-Nonachlor	Toluene	2.5
565.	trans-Permethrin	Toluene	2.5
566.	Triadimefon	Toluene	5
567.	Triadimenol-1	Toluene	7.5
568.	Triadimenol-2	Toluene	7.5
569.	Triallate	Toluene	5
570.	Triamiphos	Toluene	5
571.	Tribenuron-methyl	Toluene	2.5
572.	Trichloronat	Toluene	2.5
573.	Tricyclazole	Toluene	15
574.	Tridiphane	Isooctane	10

TABLE B.1 Solvent Selected and Concentration of Mixed Standard Solution of 587 Pesticides and Chemical Pollutants Analyzed by GC-MS—cont'd

No.	Compound	Solvent	Concentration of Mixed Standard Solution (mg/L)
575.	Trietazine	Toluene	2.5
576.	Trifloxystrobin	Toluene-acetone(8:2)	10
577.	Triflumizole	Toluene	10
578.	Trifluralin	Toluene	5
579.	tri-iso-Butyl phosphate	Toluene	2.5
580.	tri-N-Butyl phosphate	Toluene	5
581.	Triphenyl phosphate	Toluene	2.5
582.	Uniconazole	Cyclohexane	5
583.	Vernolate	Toluene	2.5
584.	Vinclozolin	Toluene	2.5
585.	XMC	Toluene	5
586.	Zoxamide	Toluene-acetone(8:2)	5

TABLE B.2 Solvent Selected and Concentration of Mixed Standard Solution of 584 Pesticides and Chemical Pollutants Analyzed by LC-MS/MS

No.	Compound	Solvent	Concentration of Mixed Standard Solution (mg/L)
1.	Propham	Toluene	11.00
2.	isoprocarb	Methanol	0.23
3.	3,4,5-Trimethacarb	Methanol	0.03
4.	Cycluron	Methanol	0.02
5.	Carbaryl	Methanol	1.03
6.	Propachlor	Methanol	0.03

Continued

TABLE B.2 Solvent Selected and Concentration of Mixed Standard Solution of 584 Pesticides and Chemical Pollutants Analyzed by LC-MS/MS—cont'd

No.	Compound	Solvent	Concentration of Mixed Standard Solution (mg/L)
7.	Rabenzazole	Methanol	0.13
8.	Simetryn	Methanol	0.01
9.	Monolinuron	Methanol	0.36
10.	Mevinphos	Toluene	0.16
11.	Aziprotryne	Methanol	0.14
12.	Secbumeton	Methanol	0.01
13.	Cyprodinil	Methanol	0.07
14.	Buturon	Methanol	0.90
15.	Carbetamide	Methanol	0.36
16.	Pirimicarb	Methanol	0.02
17.	Clomazone	Methanol	0.04
18.	Clomazone dimethazone	Methanol	20.00
19.	Cyanazine	Methanol	0.02
20.	Prometryne	Methanol	0.02
21.	Paraoxon methyl	Methanol	0.08
22.	4,4-Dichlorobenzophenone	Methanol	1.36
23.	Thiacloprid	Methanol	0.04
24.	Ethidimuron	Methanol	64.00
25.	Imidacloprid	Methanol	2.20
26.	Diallate	Methanol	8.92
27.	Isomethiozin	Methanol	0.11
28.	Acetochlor	Methanol	4.74
29.	cis and trans Diallate	Methanol	8.92
30.	Methoprotryne	Methanol	0.02
31.	Nitenpyram	Methanol	1.71
32.	Dimethenamid	Methanol	0.43
33.	Terrbucarb	Methanol	0.21

TABLE B.2 Solvent Selected and Concentration of Mixed Standard Solution of 584 Pesticides and Chemical Pollutants Analyzed by LC-MS/MS—cont'd

No.	Compound	Solvent	Concentration of Mixed Standard Solution (mg/L)
34.	Myclobutanil	Methanol	0.10
35.	Penconazole	Methanol	0.20
36.	Imazethapyr	Methanol	0.11
37.	Fenthion sulfoxide	Methanol	0.03
38.	Paclobutrazol	Methanol	0.06
39.	Butralin	Methanol	0.19
40.	Triadimenol	Methanol	1.06
41.	Spiroxamine	Methanol	0.01
42.	Tolclofos methyl	Methanol	6.66
43.	Desmedipham	Methanol	0.40
44.	Methidathion	Methanol	94.00
45.	Allethrin	Methanol	6.04
46.	Benodanil	Methanol	0.35
47.	Diazinon	Toluene	0.07
48.	Edifenphos	Methanol	12.00
49.	Flutolanil	Methanol	0.11
50.	Pretilachlor	Methanol	0.03
51.	Flusilazole	Methanol	0.06
52.	Iprovalicarb	Methanol	0.23
53.	Famphur	Methanol	0.36
54.	Benalyxyl	Methanol	0.12
55.	Dichlofluanid	Toluene	0.26
56.	Diclobutrazole	Methanol	0.05
57.	Etaconazole	Methanol	0.18
58.	Fenarimol	Methanol	0.06
59.	Phthalic acid, dicyclobexyl ester	Methanol	0.20

Continued

TABLE B.2 Solvent Selected and Concentration of Mixed Standard Solution of 584 Pesticides and Chemical Pollutants Analyzed by LC-MS/MS—cont'd

No.	Compound	Solvent	Concentration of Mixed Standard Solution (mg/L)
60.	Tetramethirn	Methanol	0.18
61.	Triflumuron	Methanol	0.39
62.	Bitertanol	Methanol	3.34
63.	Cloquintocet mexyl	Methanol	0.19
64.	Chlorprifos methyl	Methanol	1.60
65.	Tepraloxydim	Methanol	1.22
66.	Thiophanat ethyl	Methanol	2.02
67.	Azinphos ethyl	Methanol	10.89
68.	Clodinafop propargyl	Methanol	0.24
69.	Haloxyfop-methyl	Methanol	0.26
70.	Fluazifop butyl	Methanol	0.03
71.	Thiophanate methyl	Methanol	2.00
72.	Isoxaflutole	Methanol	0.39
73.	Anilofos	Methanol	0.07
74.	Quizalofop-ethyl	Methanol	0.07
75.	Azoxystrobin	Methanol	0.05
76.	Bensulide	Methanol	3.42
77.	Bromophos-ethyl	Methanol	56.77
78.	Bromfenvinfos	Methanol	0.30
79.	Triasulfuron	Methanol	0.16
80.	Nicotine	Methanol	0.22
81.	Pyrazophos	Methanol	0.16
82.	Indoxacarb	Methanol	0.75
83.	Daminozide	Methanol	0.26
84.	Acephate	Methanol	1.33
85.	Dazomet	Methanol	12.70
86.	Ethylene thiourea	Methanol	5.22

TABLE B.2 Solvent Selected and Concentration of Mixed Standard Solution of 584 Pesticides and Chemical Pollutants Analyzed by LC-MS/MS—cont'd

No.	Compound	Solvent	Concentration of Mixed Standard Solution (mg/L)
87.	Fenuron	Methanol	0.10
88.	Flufenoxuron	Methanol	0.32
89.	Emamectin benzoate	Methanol	0.03
90.	Cyromazine	Methanol	0.72
91.	Crimidine	Methanol	0.16
92.	Carbendazim	Methanol	0.05
93.	6-Chloro-4-hydroxy-3-phenyl-pyridazin	Methanol	0.17
94.	Chlorotoluron	Methanol	0.06
95.	Isouron	Methanol	0.04
96.	Molinate	Methanol	0.21
97.	Propoxur	Methanol	2.44
98.	Chlorbufam	Methanol	18.30
99.	Terbuthylazine	Methanol	0.05
100.	Bendiocarb	Methanol	0.32
101.	Propazine	Methanol	0.03
102.	Thiofanox	Methanol	15.70
103.	Carboxin	Methanol	0.06
104.	Difenzoquat-methyl sulfate	Methanol	0.08
105.	Clothianidin	Methanol	6.30
106.	Diuron	Methanol	0.16
107.	Chlormephos	Methanol	44.80
108.	Methobromuron	Toluene	1.68
109.	Pronamide	Methanol	1.54
110.	Dimethachloro	Methanol	0.19
111.	Arboxin	Methanol	0.06
112.	Mephosfolan	Methanol	0.23

Continued

TABLE B.2 Solvent Selected and Concentration of Mixed Standard Solution of 584 Pesticides and Chemical Pollutants Analyzed by LC-MS/MS—cont'd

No.	Compound	Solvent	Concentration of Mixed Standard Solution (mg/L)
113.	Aclonifen	Methanol	2.42
114.	Imibenzonazole-des-benzyl	Methanol	0.62
115.	Phorate	Methanol	31.40
116.	Ethofume sate	Methanol	37.20
117.	Neburon	Methanol	0.71
118.	Mefenoxam	Methanol	0.15
119.	Prothoate	Methanol	0.25
120.	Thiamethoxam	Methanol	3.30
121.	Crufomate	Methanol	0.05
122.	Iprobenfos	Methanol	0.83
123.	Coumatetralyl	Methanol	0.14
124.	Cyproconazole	Methanol	0.07
125.	TEPP	Methanol	1.04
126.	Cythioate	Methanol	8.00
127.	Phenmedipham	Methanol	0.45
128.	Bifenazate	Methanol	2.28
129.	Phosphamidon	Methanol	0.39
130.	Fenhexamid	Methanol	0.09
131.	Flutriafol	Methanol	0.86
132.	Furalaxyl	Methanol	0.08
133.	Bioallethrin	Methanol	19.80
134.	Etrimfos	Methanol	1.88
135.	Pirimiphos methyl	Methanol	0.02
136.	Cyanofenphos	Methanol	2.08
137.	Disulfoton sulfone	Methanol	0.25
138.	Fenazaquin	Methanol	0.03
139.	DEF	Methanol	0.16

TABLE B.2 Solvent Selected and Concentration of Mixed Standard Solution of 584 Pesticides and Chemical Pollutants Analyzed by LC-MS/MS—cont'd

No.	Compound	Solvent	Concentration of Mixed Standard Solution (mg/L)
140.	Metconazole	Methanol	0.13
141.	Triazophos	Toluene	0.07
142.	Pyriproxyfen	Methanol	0.04
143.	Pyriftalid	Methanol	0.06
144.	Cycloxydim	Methanol	0.25
145.	Flurtamone	Methanol	0.04
146.	Buprofezin	Methanol	0.09
147.	Trifluralin	Toluene	33.48
148.	Bioresmethrin	Methanol	0.74
149.	Chlorpyrifos	Methanol	5.38
150.	Propiconazole	Methanol	0.18
151.	Chlorsulfuron	Methanol	0.27
152.	Clethodim	Methanol	0.21
153.	Flamprop isopropyl	Methanol	0.04
154.	Tetrachlorvinphos	Methanol	0.22
155.	Isoxaben	Methanol	0.02
156.	Fluchloralin	Methanol	3268.00
157.	Propargite	Toluene	6.86
158.	Bromuconazole	Methanol	0.31
159.	Flamprop methyl	Methanol	2.02
160.	Fluthiacet methyl	Methanol	0.53
161.	Trifloxystrobin	Methanol	0.20
162.	Hexaflumuron	Methanol	2.52
163.	Methamidophos	Methanol	0.49
164.	Chlorimuron ethyl	Methanol	3.04
165.	EPTC	Methanol	3.73
166.	Novaluron	Methanol	0.80

Continued

TABLE B.2 Solvent Selected and Concentration of Mixed Standard Solution of 584 Pesticides and Chemical Pollutants Analyzed by LC-MS/MS—cont'd

No.	Compound	Solvent	Concentration of Mixed Standard Solution (mg/L)
167.	Flurazuron	Methanol	2.68
168.	Maleic hydrazide	Methanol	8.00
169.	Diethyltoluamide	Methanol	0.06
170.	Monuron	Methanol	3.47
171.	Picolinafen	Methanol	0.07
172.	Carbofuran	Methanol	1.31
173.	Hydramethylnon	Methanol	0.17
174.	Pyrimethanil	Methanol	0.07
175.	Fenfuram	Methanol	0.08
176.	Quinoclamine	Methanol	0.79
177.	Fenobucarb	Methanol	0.59
178.	Propanil	Methanol	2.16
179.	Acetamiprid	Methanol	0.14
180.	Mepanipyrim	Methanol	0.03
181.	Dimethoate	Methanol	0.76
182.	Methiocarb	Methanol	4.12
183.	Prometon	Methanol	0.01
184.	Metoxuron	Methanol	0.06
185.	Fluometuron	Methanol	0.09
186.	Monalide	Methanol	0.12
187.	Methfuroxam	Methanol	20.00
188.	Dicrotophos	Methanol	0.11
189.	Ethirimol	Methanol	0.06
190.	Hexazinone	Methanol	0.01
191.	Dimethametryn	Methanol	0.01
192.	Furmecyclox	Methanol	0.08
193.	Diphenamid	Methanol	0.01

TABLE B.2 Solvent Selected and Concentration of Mixed Standard Solution of 584 Pesticides and Chemical Pollutants Analyzed by LC-MS/MS—cont'd

No.	Compound	Solvent	Concentration of Mixed Standard Solution (mg/L)
194.	Ethoprophos	Methanol	0.28
195.	Etridiazol	Methanol	10.04
196.	Fonofos	Methanol	46.00
197.	Bromacil	Methanol	2.36
198.	Trichlorphon	Methanol	0.11
199.	Benoxacor	Methanol	0.69
200.	Brompyrazon	Methanol	0.36
201.	Mepronil	Methanol	0.04
202.	Demeton(o+s)	Methanol	0.68
203.	Disulfoton	Methanol	46.97
204.	Metalaxyl	Methanol	0.05
205.	Phorate sulfoxide	Methanol	34.65
206.	Dodemorph	Methanol	0.04
207.	Fenthion	Methanol	5.20
208.	Imazamethabenz-methyl	Methanol	0.02
209.	Oxycarboxin	Methanol	0.09
210.	Fosthiazate	Methanol	0.01
211.	Disulfoton-sulfoxide	Methanol	0.28
212.	Ofurace	Methanol	0.10
213.	Phoxim	Methanol	8.28
214.	Imazalil	Methanol	0.20
215.	Isoprothiolane	Methanol	0.18
216.	Fenoxycarb	Methanol	1.83
217.	Pyrimitate	Methanol	0.02
218.	Quinalphos	Methanol	0.20
219.	Kresoxim-methyl	Methanol	10.06
220.	Butachlor	Methanol	2.01

Continued

TABLE B.2 Solvent Selected and Concentration of Mixed Standard Solution of 584 Pesticides and Chemical Pollutants Analyzed by LC-MS/MS—cont'd

No.	Compound	Solvent	Concentration of Mixed Standard Solution (mg/L)
221.	Fensulfothin	Methanol	0.20
222.	Ditalimfos	Methanol	6.72
223.	Triticonazole	Isooctane	0.30
224.	Fenamiphos sulfoxide	Methanol	0.07
225.	Fluorochloridone	Methanol	282.00
226.	Fenamiphos sulfone	Methanol	0.04
227.	Fluridone	Methanol	0.02
228.	Imazaquin	Methanol	0.29
229.	Thenylchlor	Methanol	2.41
230.	Chlorphoxim	Methanol	7.76
231.	Fenoxanil	Methanol	3.94
232.	Epoxiconazole	Methanol	0.41
233.	Fenbuconazole	Methanol	0.16
234.	Isofenphos	Methanol	21.87
235.	Phenothrin	Methanol	33.92
236.	Piperonyl butoxide	Methanol	0.11
237.	Methoxyfenozide	Methanol	0.37
238.	Piperophos	Methanol	0.92
239.	Ethion	Methanol	0.30
240.	Diafenthiuron	Methanol	0.03
241.	Oxyflurofen	Methanol	5.85
242.	Phosalone	Methanol	4.80
243.	Coumaphos	Methanol	0.21
244.	Fentin-chloride	Methanol	1.73
245.	Ethoxysulfuron	Methanol	0.46
246.	Prochloraz	Methanol	0.21
247.	Aspon	Methanol	0.17

TABLE B.2 Solvent Selected and Concentration of Mixed Standard Solution of 584 Pesticides and Chemical Pollutants Analyzed by LC-MS/MS—cont'd

No.	Compound	Solvent	Concentration of Mixed Standard Solution (mg/L)
248.	Spinosad	Methanol	0.06
249.	Temephos	Methanol	0.12
250.	Flumiclorac-pentyl	Methanol	1.06
251.	Thifensulfuron-methyl	Methanol	2.14
252.	Diafenthiuron pestanal	Methanol	134.00
253.	Allidochlor	Methanol	4.10
254.	Spirodiclofen	Methanol	0.99
255.	Fenpyroximate	Methanol	0.14
256.	Flufenacet	Methanol	0.53
257.	Metamitron	Methanol	0.64
258.	Oesmetryn	Methanol	0.02
259.	Tricyclazole	Methanol	0.01
260.	Butafenacil	Methanol	0.95
261.	Thiabendazole	Methanol	0.05
262.	Isoproturon	Methanol	0.01
263.	Atrazine	Methanol	0.04
264.	Propamocarb	Methanol	0.01
265.	Dithiopyr	Methanol	1.04
266.	DMST	Methanol	4.00
267.	Pymetrozin	Methanol	3.43
268.	Mepiquat chloride	Methanol	0.09
269.	Sulfallate	Toluene	20.72
270.	Atratone	Methanol	0.02
271.	Desmetryn	Methanol	0.02
272.	Ethiofencarb	Methanol	0.49
273.	Metribuzin	Toluene	0.05
274.	Cycloate	Methanol	0.44

Continued

TABLE B.2 Solvent Selected and Concentration of Mixed Standard Solution of 584 Pesticides and Chemical Pollutants Analyzed by LC-MS/MS—cont'd

No.	Compound	Solvent	Concentration of Mixed Standard Solution (mg/L)
275.	Tebuthiuron	Methanol	0.02
276.	Butylate	Methanol	30.20
277.	Cyprazine	Methanol	0.00
278.	Sebutylazine	Methanol	0.03
279.	Tebutam	Methanol	0.01
280.	thiofanox-sulfoxide	Methanol	0.83
281.	Methacrifos	Methanol	242.37
282.	Terbumeton	Methanol	0.01
283.	Ametryn	Methanol	0.10
284.	Triazoxide	Methanol	0.80
285.	Cartap hydrochloride	Methanol	208.00
286.	Heptanophos	Methanol	0.58
287.	Terbutryn	Methanol	2.27
288.	Thionazin	Methanol	2.27
289.	Prosulfocarb	Methanol	0.04
290.	Trietazine	Methanol	0.06
291.	Dibutyl succinate	Methanol	22.24
292.	Chloridazon	Methanol	0.23
293.	Dipropetryn	Methanol	0.03
294.	Thiobencarb	Methanol	0.33
295.	Diethofencarb	Methanol	0.20
296.	Propyzamide	Methanol	0.70
297.	Linuron	Methanol	202.00
298.	Cadusafos	Methanol	0.12
299.	tri-n-Butyl phosphate	Methanol	0.04
300.	tri-iso-Butyl phosphate	Methanol	0.40
301.	Triadimefon	Methanol	0.79

TABLE B.2 Solvent Selected and Concentration of Mixed Standard Solution of 584 Pesticides and Chemical Pollutants Analyzed by LC-MS/MS—cont'd

No.	Compound	Solvent	Concentration of Mixed Standard Solution (mg/L)
302.	Metazachlor	Methanol	0.10
303.	Fenamiphos	Methanol	0.02
304.	Alachlor	Methanol	0.74
305.	Fenpropimorph	Methanol	0.02
306.	Tebuconazole	Methanol	0.22
307.	Tridemorph	Methanol	0.26
308.	Propetamphos	Methanol	5.40
309.	Pyraclostrobin	Methanol	0.05
310.	Simeconazole	Methanol	0.29
311.	Terbufos	Methanol	1216.00
312.	Tebupirimfos	Methanol	0.01
313.	Azaconazole	Methanol	0.08
314.	Phorate sulfone	Methanol	4.20
315.	Sulfotep	Methanol	0.26
316.	Sulprofos	Toluene	0.58
317.	Isopropalin	Methanol	3.00
318.	Phenthoate	Methanol	9.24
319.	Nuarimol	Methanol	0.10
320.	Sethoxydim	Methanol	9.96
321.	Azinphos-methyl	Methanol	110.43
322.	Mecarbam	Methanol	1.96
323.	Pyributicarb	Methanol	0.03
324.	Tralkoxydim	Methanol	0.03
325.	Azamethiphos	Methanol	0.08
326.	pyridaphenthion	Methanol	0.09
327.	Mefenacet	Methanol	0.22
328.	Pirimiphos-ethyl	Methanol	0.02

Continued

TABLE B.2 Solvent Selected and Concentration of Mixed Standard Solution of 584 Pesticides and Chemical Pollutants Analyzed by LC-MS/MS—cont'd

No.	Compound	Solvent	Concentration of Mixed Standard Solution (mg/L)
329.	Pyraclofos	Methanol	0.10
330.	Diniconazole	Methanol	0.13
331.	Tetraconazole	Methanol	0.17
332.	EPN	Methanol	266.00
333.	Flumetsulam	Methanol	0.03
334.	Pencycuron	Methanol	0.03
335.	Malathion	Methanol	0.56
336.	Picoxystrobin	Methanol	0.84
337.	Bupirimate	Methanol	0.07
338.	Mefenpyr-diethyl	Methanol	1.26
339.	Thiazopyr	Methanol	0.20
340.	4-Aminopyridine	Methanol	0.09
341.	Dimethomorph	Methanol	0.04
342.	Thiodicarb	Methanol	3.94
343.	Chlorfluazuron	Methanol	0.87
344.	Methomyl	Methanol	0.96
345.	Profenefos	Methanol	0.20
346.	Pyroquilon	Methanol	0.35
347.	Benfuracarb-methyl	Methanol	1.64
348.	Chlormequat	Methanol	0.01
349.	Cinosulfuron	Methanol	0.11
350.	Pyrazosulfuron-ethyl	Methanol	0.68
351.	Omethoate	Methanol	0.97
352.	Aminocarb	Methanol	1.64
353.	Aldicarb sulfone	Methanol	2.14
354.	Fuberidazole	Methanol	0.19
355.	Promecarb	Methanol	0.86

TABLE B.2 Solvent Selected and Concentration of Mixed Standard Solution of 584 Pesticides and Chemical Pollutants Analyzed by LC-MS/MS—cont'd

No.	Compound	Solvent	Concentration of Mixed Standard Solution (mg/L)
356.	Dioxacarb	Methanol	0.34
357.	Thiometon	Methanol	57.80
358.	Ethiofencarb-sulfoxide	Methanol	22.40
359.	Oxamyl-oxime	Methanol	10.00
360.	Trichloronate	Methanol	6.68
361.	Amidithion	Methanol	65.80
362.	Butocarboxim	Methanol	0.16
363.	Chlorthiamid	Methanol	0.88
364.	Dichlorvos	Methanol	0.05
365.	Chlordimeform	Methanol	0.13
366.	Imazapic	Methanol	0.59
367.	Cymoxanil	Methanol	5.56
368.	Dimethirimol	Methanol	0.01
369.	Kadethrin	Methanol	0.33
370.	Aldimorph	Methanol	0.32
371.	Vernolate	Methanol	0.03
372.	Ethoxyquin	Methanol	0.35
373.	Oxabetrinil	Methanol	4.00
374.	Vinclozolin	Methanol	0.25
375.	Paraoxon-ethyl	Methanol	0.05
376.	Uniconazole	Methanol	0.24
377.	Cyanohos	Methanol	1.01
378.	Chlortoluron	Methanol	0.03
379.	Demeton-s-methyl sulfone	Methanol	1.98
380.	Etridiazole	Methanol	0.10
381.	Metosulam	Methanol	0.44
382.	Pyrifenox	Methanol	0.03

Continued

TABLE B.2 Solvent Selected and Concentration of Mixed Standard Solution
of 584 Pesticides and Chemical Pollutants Analyzed by LC-MS/MS—cont'd

No.	Compound	Solvent	Concentration of Mixed Standard Solution (mg/L)
383.	Clofentezine	Methanol	0.08
384.	benzyladenine	Methanol	7.08
385.	Folpet	Methanol	13.86
386.	Demeton-s-methyl	Methanol	0.53
387.	Dimepiperate	Methanol	378.00
388.	Triallate	Methanol	4.62
389.	Phthalic acid, benzyl butyl ester	Methanol	63.20
390.	Quinoxyphen	Methanol	15.34
391.	Chlorthion	Methanol	13.36
392.	Vamidothion sulfone	Methanol	47.60
393.	Methoprene	Methanol	0.52
394.	Terbufos sulfone	Methanol	8.86
395.	Cyazofamid	Acetonitrile	0.45
396.	Nitralin	Methanol	3.44
397.	Trichloronat	Methanol	6.68
398.	Boscalid	Methanol	0.48
399.	Benzoximate	Methanol	1.97
400.	Benzoylprop-ethyl	Methanol	30.80
401.	Dicapthon	Methanol	0.02
402.	Fenthion sulfone	Methanol	1.75
403.	Pyrimidifen	Methanol	1.40
404.	Norflurazon	Methanol	0.03
405.	Hexythiazox	Methanol	2.36
406.	Isazofos	Methanol	0.02
407.	Ziram	Methanol	7.84
408.	Fenpropidin	Methanol	0.02

TABLE B.2 Solvent Selected and Concentration of Mixed Standard Solution of 584 Pesticides and Chemical Pollutants Analyzed by LC-MS/MS—cont'd

No.	Compound	Solvent	Concentration of Mixed Standard Solution (mg/L)
409.	Florasulam	Acetonitrile	1.74
410.	Resmethrin	Methanol	0.03
411.	Dinitramine	Toluene	0.18
412.	Etofenprox	Methanol	228.00
413.	Flurochloridone	Methanol	0.13
414.	Pyridaben	Methanol	1.22
415.	Dichlofenthion	Methanol	3.02
416.	Resmethrin-2	Methanol	0.03
417.	Pyridate	Methanol	7.98
418.	Acrylamide	Methanol	1.78
419.	Fenpropathrin	Methanol	24.50
420.	Fluoroglycofen-ethyl	Methanol	0.50
421.	Zeta cypermethrin	Methanol	0.07
422.	Dimefox	Methanol	6.82
423.	Esfenvalerate	Methanol	41.60
424.	Metolcarb	Methanol	2.54
425.	1-Naphthy acetamide	Methanol	0.08
426.	Furathiocarb	Methanol	0.19
427.	Aldicarb	Methanol	26.10
428.	Atrazine-desethyl	Methanol	0.06
429.	Phthalimide	Methanol	4.30
430.	2,6-Dichlorobenzamide	Methanol	0.45
431.	Flubenzimine	Methanol	0.78
432.	Dinotefuran	Methanol	1.02
433.	trans-Permethin	Methanol	0.48
434.	Dimethyl phthalate	Methanol	1.32
435.	Pebulate	Methanol	0.34

Continued

TABLE B.2 Solvent Selected and Concentration of Mixed Standard Solution of 584 Pesticides and Chemical Pollutants Analyzed by LC-MS/MS—cont'd

No.	Compound	Solvent	Concentration of Mixed Standard Solution (mg/L)
436.	Pyrazoxyfen	Methanol	0.03
437.	Thiofanox sulfone	Methanol	2.41
438.	Haloxyfop-2-ethoxyethyl	Methanol	0.25
439.	Methabenzthiazuron	Methanol	0.01
440.	Butoxycarboxim	Methanol	2.66
441.	Demeton-s	Methanol	8.00
442.	tau-Fluvalinate	Methanol	23.00
443.	Demeton-s-methyl sulfoxide	Methanol	0.39
444.	Napropamide	Methanol	0.13
445.	tert-Butylamine	Methanol	3.90
446.	Phosfolan	Cyclohexane	0.05
447.	Silafluofen	Methanol	60.80
448.	Vamidothion	Methanol	0.46
449.	Chlordimeform hydrochloride	Methanol	0.26
450.	Diphenylamine	Methanol	0.04
451.	Triclopyr	Methanol	0.02
452.	Imazapyr	Methanol	1.03
453.	Metolachlor	Methanol	0.04
454.	Procymidone	Methanol	8.66
455.	Tebufenpyrad	Methanol	0.03
456.	Triamiphos	Methanol	0.00
457.	Pyrethrins	Methanol	3.58
458.	Simeton	Methanol	0.11
459.	Tebufenozide	Methanol	2.78
460.	Thidiazuron pestanal	Methanol	0.03
461.	Cumyluron	Methanol	0.13

TABLE B.2 Solvent Selected and Concentration of Mixed Standard Solution of 584 Pesticides and Chemical Pollutants Analyzed by LC-MS/MS—cont'd

No.	Compound	Solvent	Concentration of Mixed Standard Solution (mg/L)
462.	Dialifos	Methanol	15.70
463.	Imazamox	Methanol	0.18
464.	Acibenzolar-s-methyl	Methanol	0.31
465.	Hymexazol	Methanol	22.41
466.	Chlormequat chloride	Methanol	0.07
467.	Dioxabenzofos	Methanol	1.38
468.	Dithianon	Methanol	0.85
469.	Chlorthiophos	Methanol	3.18
470.	Oxamyl	Methanol	54.81
471.	Lactofen	Methanol	6.20
472.	Propaquizafop	Methanol	0.12
473.	Pyrethrin	Methanol	3.58
474.	Fenitrothion	Methanol	2.68
475.	Propisochlor	Methanol	0.08
476.	Daimuron	Methanol	26.00
477.	Cinidon-ethyl	Methanol	1.46
478.	Zoxamide	Methanol	0.45
479.	Etobenzanid	Methanol	0.08
480.	Phthalic acid, dibutyl ester	Methanol	3.96
481.	Imibenconazole	Methanol	1.03
482.	Naled	Methanol	14.82
483.	Prallethrin	Methanol	0.13
484.	Tolfenpyrad	Methanol	0.01
485.	Dicofol	Methanol	0.18
486.	Chloroxuron	Methanol	18.00
487.	Runnel	Methanol	1.31
488.	Cloprop	Methanol	1.14

Continued

TABLE B.2 Solvent Selected and Concentration of Mixed Standard Solution of 584 Pesticides and Chemical Pollutants Analyzed by LC-MS/MS—cont'd

No.	Compound	Solvent	Concentration of Mixed Standard Solution (mg/L)
489.	Chlorpropham	Methanol	1.58
490.	Malaoxon	Methanol	0.47
491.	Phosmet	Methanol	1.77
492.	Phthalic acid, biscyclohexyl ester	Methanol	0.07
493.	Carpropamid	Methanol	0.52
494.	Terbacil	Methanol	0.09
495.	Thidiazuron	Methanol	0.03
496.	Bentazone	Methanol	0.10
497.	Dinoterb	Methanol	0.02
498.	DNOC	Methanol	0.26
499.	Mexacarbate	Methanol	0.09
500.	Dimefuron	Methanol	0.40
501.	Rotenone	Methanol	0.23
502.	Trinexapac-ethyl	Methanol	7.07
503.	Athidathion	Methanol	40.00
504.	Chlorbenside sulfone	Methanol	0.08
505.	Chlorobenzuron	Methanol	2.04
506.	Chloramphenicolum	Methanol	0.39
507.	Fenthion oxon	Methanol	0.12
508.	Benzofenap	Methanol	0.01
509.	Dalapon	Methanol	23.07
510.	Oryzalin	Methanol	0.49
511.	2-Phenylphenol	Methanol	16.99
512.	Dinoseb acetate	Methanol	4.13
513.	3-Phenylphenol	Methanol	0.40
514.	Diflufenican	Methanol	2.83

TABLE B.2 Solvent Selected and Concentration of Mixed Standard Solution of 584 Pesticides and Chemical Pollutants Analyzed by LC-MS/MS—cont'd

No.	Compound	Solvent	Concentration of Mixed Standard Solution (mg/L)
515.	2,3,4,5-Tetrachloroaniline	Methanol	5.36
516.	MCPB	Methanol	1.42
517.	2,4-DB	Methanol	213.98
518.	Iminoctadine triacetate	Methanol	110.00
519.	Dimehypo	Methanol	40.02
520.	Dieldrin	Methanol	16.16
521.	Forchlorfenuron	Methanol	1.14
522.	2,4-D	Methanol	1.19
523.	Fluotrimazole	Methanol	8066.00
524.	Pentachloroaniline	Methanol	0.37
525.	Propylene thiourea	Methanol	3.01
526.	Carfentrazone—ethyl	Methanol	48.00
527.	Cyphenothrin	Methanol	1.68
528.	Warfarin	Methanol	0.27
529.	Flupropanate	Methanol	2.30
530.	Etoxazole	Methanol	0.09
531.	cis-1,2,3,6-Tetrahydrophthalimide	Methanol	620.00
532.	Dodine	Methanol	0.80
533.	Fempxaprop-ethyl	Methanol	0.49
534.	2,6-Difluorobenzoic acid	Methanol	170.41
535.	Dicloran	Methanol	4.86
536.	Mesotrion	Methanol	230.06
537.	Aminopyralid	Methanol	36.60
538.	Trichloroacetic acid sodium salt	Methanol	28.16
539.	Mecoprop	Methanol	0.49
540.	Clopyralld	Methanol	28.00

Continued

TABLE B.2 Solvent Selected and Concentration of Mixed Standard Solution of 584 Pesticides and Chemical Pollutants Analyzed by LC-MS/MS—cont'd

No.	Compound	Solvent	Concentration of Mixed Standard Solution (mg/L)
541.	Sulfentrazone	Methanol	8.96
542.	Dicamba	Methanol	126.59
543.	Ethiprole	Methanol	3.99
544.	Fenaminosulf	Methanol	1600.00
545.	Picloram	Methanol	53.41
546.	Iodosulfuron-methyl sodium	Methanol	2.12
547.	Fentrazamide	Methanol	1.24
548.	Dinoseb	Methanol	0.04
549.	2,4,5-T	Methanol	1.75
550.	Chlorfenethol	Methanol	16.43
551.	Fluorodifen	Methanol	2080.00
552.	Diflufenzopyr-sodium	Methanol	54.00
553.	Fluroxypyr	Methanol	19.21
554.	Cyclanilide	Methanol	0.34
555.	Bromoxynil	Methanol	0.18
556.	Naptalam	Methanol	0.19
557.	Alloxydim-sodium	Methanol	0.02
558.	Sulfanitran	Methanol	0.30
559.	Endosulfan-sulfate	Methanol	16.00
560.	Pyrithlobac sodium	Methanol	138.20
561.	Halosulfuran-methyl	Methanol	8.00
562.	Ioxynil	Methanol	0.06
563.	Famoxadone	Methanol	4.53
564.	Gibberellic acid	Methanol	6.63
565.	Lufenuron	Methanol	40.00
566.	Acifluorfen	Methanol	11.80

TABLE B.2 Solvent Selected and Concentration of Mixed Standard Solution of 584 Pesticides and Chemical Pollutants Analyzed by LC-MS/MS—cont'd

No.	Compound	Solvent	Concentration of Mixed Standard Solution (mg/L)
567.	Thifluzamide	Methanol	56.00
568.	Bediocarb	Methanol	30.00
569.	Bromocylen	Methanol	70.00
570.	Flusulfamide	Methanol	0.04
571.	Dimethipin	Methanol	170.00
572.	Benfuresate	Methanol	550.00
573.	Dichlorprop	Methanol	0.15
574.	Fomesafen	Methanol	0.20
575.	Nitrofen	Methanol	60.00
576.	Bifenox	Methanol	110.00
577.	Cyclosulfamuron	Methanol	34.37
578.	Iodosulfuron-methyl	Methanol	6.66
579.	Fludioxonil	Methanol	6.22
580.	Fluazinam	Methanol	7.06
581.	Acrinathrin	Methanol	0.81
582.	Heptachlor	Methanol	0.00
583.	Triforine	Methanol	42.00
584.	Kelevan	Methanol	962.27

Appendix C

Physicochemical Properties of 900 Pesticides and Chemical Pollutants

No.	Compound	CAS	Toxicity	Formula	Characteristic Atom[1]	Effect[2]	MW	WS (mg/L)	Solubility of Organic Solvent (g/L)
(1) *Organochlorine pesticides*									
1.	1,2-Dibromo-3-chloropropane[a]	96-12-8	Medium	$C_3H_5Br_2Cl$	–Br	N/F	236.32	1000	Diffluent in fatty hydrocarbon, arene, isopropanol and trichloro ethylene, etc.
2.	2,3,4,5-Tetrachloroanisole[a]	938-86-3		$C_7H_4Cl_4O$	–Cl	I/A	245.92	Insoluble	Soluble in multiple organic solvents
3.	Acifluorifen[a]	50594-66-6	Low	$C_4H_7ClF_3NO_5$	–F	H	241.45	120	Acetone, toluene 500, benzene 10, chloroform, hexane, $CH_2Cl_2 <10$
4.	Aldrin[a]	309-00-2	Medium	$C_{12}H_8Cl_6$	–Cl	I	364.93	0.027	Benzene 830, acetone 660, ethanol 150
5.	Aroclor 1221[a]	11104-28-2	–Cl	PCB mixture	Cl		200.7	40	Soluble in various organic solvents
6.	Aroclor 1232[a]	11141-16-5	–Cl		Cl		232.2	407	Soluble in various organic solvents
7.	Aroclor 1242[a]	53469-21-9	–Cl		Cl		266.5	0.23	Soluble in various organic solvents
8.	Aroclor 1248[a]	12672-29-6	–Cl		Cl		299.5	0.054	Soluble in various organic solvents
9.	Aroclor 1254[a]	11097-69-1	–Cl		Cl		328.4	0.031	Soluble in various organic solvents
10.	Aroclor 1260[a]	11096-82-5	–Cl		Cl		375.7	0.0027	Soluble in various organic solvents
11.	Aroclor 1262[a]	37324-23-5	–Cl		Cl				Soluble in various organic solvents
12.	Aroclor 1268[a]	11100-14-4	–Cl		Cl				Soluble in various organic solvents

No.	Name	CAS number		Molecular formula			MW	Water solubility	Solubility
13.	Bromocyclen[a]	1715-40-8	Low	$C_8H_3Cl_6Br$	-Cl	A/I	391.9	Insoluble	Soluble in benzene
14.	Camphechlor[a]	8001-35-2	Medium	$C_{10}H_{10}Cl_8$	-Cl	I	414	3	Hexane 7200, acetone 6000, benzene 5000, methanol 150
15.	Carbon tetrachloride[a]	56-23-5	Low	CCl_4	-Cl	I	153.8	800	Miscible with ethanol, benzene, chloroform, and petroleum ether etc.
16.	Carpropamid[a]	104030-54-8	Low	$C_{15}H_{18}Cl_3NO$	Cl/N	F	334.67	1.7	Acetone 153, methanol 106, toluene 38, hexane 0.9
17.	Chlorbenside[a]	103-17-3	Low	$C_{13}H_7Cl_2S$	-Cl	I/A	269.19	Insoluble	Toluene 10780, benzene 1110, acetone 920, methanol 400
18.	Chlorbenside sulfone[a]	7082-99-7	Low	$C_{13}H_{10}Cl_2O_2S$	-S	I/A	301.19	Insoluble	Soluble in various organic solvents
19.	Chlorbenside sulfoxide[a]	7047-28-1	Low	$C_{13}H_7Cl_2SO$	-Cl	I/A	285.19	Insoluble	Soluble in various organic solvents
20.	Chlordecone[a]	143-50-0	Medium	$C_{10}Cl_{10}O$	-Cl	I/F	490.64	4000	Diffluent in acetone, slightly soluble in benzene
21.	Chlorfenethol[a]	80-06-8	Low	$C_{14}H_{12}Cl_2O$	-Cl	A	267.14	Insoluble	Soluble in various organic solvents
22.	Chlorfenprop-methyl[a]	14437-17-3	Low	$C_{10}H_{10}Cl_2O_2$	-Cl	H	233.10	40	Soluble in acetone, aromatic hydrocarbon, and diethyl ether
23.	Chlorthal-dimethyl[a]	1861-32-1	Low	$C_{10}H_6Cl_4O_4$	-Cl	H	331.99	0.5	Benzene 250, toluene 170, xylene 140, acetone 100, dioxane 120

Continued

No.	Compound	CAS	Toxicity	Formula	Characteristic Atom[1]	Effect[2]	MW	WS (mg/L)	Solubility of Organic Solvent (g/L)
24.	Chloropicrin[a]	76-06-2	Medium	Cl_3CNO_2	-Cl	I	164.35	1620	Soluble in benzene, methanol, ethanol and petroleum ether
25.	Chlorobenzilate[a]	510-15-6	Low	$C_{16}H_{14}Cl_2O_3$	-Cl	A	325.21	Insoluble	Soluble in various organic solvents
26.	cis-Chlordane[a]	5103-71-9	Low	$C_{10}H_6Cl_8$	-Cl	I	409.83	Insoluble	Soluble in various organic solvents
27.	trans-Chlordane[a]	5103-74-2	Low	$C_{10}H_6Cl_8$	-Cl	I	409.83	Insoluble	Soluble in various organic solvents
28.	Chlorfenson[a]	80-33-1	Low	$C_8H_5Cl_3O_2$	-Cl	A	239.49	Indissolvable	Acetone 1300, xylene 780
29.	Chlorfenethol	80-06-8	Low	$C_{14}H_{12}Cl_2O$	-Cl	A	267.04	Indissolvable	Soluble in various organic solvents
30.	Chlorfenprop-methyl[a]	14437-17-3	Low	$C_{10}H_{10}Cl_2O_2$	-Cl	H	233	40	Soluble in acetone, diethyl ether, aromatic hydrocarbon and fatty hydrocarbon
31.	Chlorflurenol-methyl[a]	2536-31-4	Low	$C_{15}H_{11}ClO_3$	-Cl	PGR	274.72	22	Acetone 260, methanol 150, benzene 70
32.	Chloroneb[a]	2675-77-6	Low	$C_8H_8Cl_2O_2$	-Cl	F	207	8	CH_2Cl_2 133, acetone 115, xylene 89
33.	Chloropropylate[a]	5836-10-2	Low	$C_{17}H_{16}Cl_2O_3$	-Cl	A	339.07	Insoluble	Soluble in various organic solvents
34.	Chlorothalonil[a]	1897-45-6	Low	$C_8Cl_4N_2$	-Cl	F	265.92	0.6	Xylene 80, acetone 20
35.	Chlordane[a]	57-74-9	Medium	$C_{10}H_6Cl_8$	Cl	I	409.83	Insoluble	Soluble in various organic solvents

#	Name	CAS	Volatility	Formula			MW	Solubility (water)	Solubility
36.	Chlozolinate[a]	84332-86-5	Low	$C_{13}H_{10}Cl_2NO_5$	–Cl	F	332.1	32	Acetone, chloroform and CH_2Cl_2 30, methanol 1, hexane 3
37.	Clophen A30[a]	1336-36-3	Low	$C_{12}H_xCl_{10-x}$	–Cl			Insoluble	Soluble in various organic solvents
38.	Clophen A40[a]	1336-36-3	Low	$C_{12}H_xCl_{10-x}$	–Cl			Insoluble	Soluble in various organic solvents
39.	Clophen A50[a]	1336-36-3	Low	$C_{12}H_xCl_{10-x}$	–Cl			Insoluble	Soluble in various organic solvents
40.	Clophen A60[a]	1336-36-3	Low	$C_{12}H_xCl_{10-x}$	–Cl			Insoluble	Soluble in various organic solvents
41.	Cloprop[a]	101-10-0	Low	$C_9H_9ClO_3$	–Cl	PGR	200.6	350	Soluble in various organic solvents
42.	Cyflufenamid[a]	180409-60-3	Low	$C_{20}H_{17}F_5N_2O_2$	–N	F	412.35	0.52	logKow=4.7, CH_2Cl_2 902, acetone 920, xylene 658, acetonitrile 943, methanol 653, ethanol 500, ethyl acetate 808, n-hexane 18.6
43.	Dalapon[a]	17040-19-6	Low	$C_3H_4Cl_2O_2$	–Cl	H	142.96	7.5×10^5	Methanol 827, ethanol 185, indissolvable in other solvents
44.	o,p′-DDD[a]	53-19-0	Medium	$C_{14}H_{10}Cl_4$	–Cl	I	320.04		Diffluent in aromatic hydrocarbon and chlorinated hydrocarbon
45.	p,p′-DDD[a]	72-54-8	Medium	$C_{14}H_{10}Cl_4$	–Cl	I	320.04	0.1	Diffluent in aromatic hydrocarbon and chlorinated hydrocarbon

Continued

No.	Compound	CAS	Toxicity	Formula	Characteristic Atom[1]	Effect[2]	MW	WS (mg/L)	Solubility of Organic Solvent (g/L)
46.	o,p'-DDE[a]	3424-82-6	Medium	$C_{14}H_8Cl_4$	–Cl	I	318.04	Insoluble	Diffluent in aromatic hydrocarbon and chlorinated hydrocarbon
47.	p,p'-DDE[a]	72-55-9	Medium	$C_{14}H_8Cl_4$	–Cl	I	318.04	Insoluble	Diffluent in aromatic hydrocarbon and chlorinated hydrocarbon
48.	o,p'-DDT[a]	789-02-6	Medium	$C_{14}H_9Cl_5$	–Cl	I	354.5	Insoluble	Diffluent in aromatic hydrocarbon and chlorinated hydrocarbon
49.	p,p-DDT[a]	50-29-3	Medium	$C_{14}H_9Cl_5$	–Cl	I	354.5	Insoluble	Diffluent in aromatic hydrocarbon and chlorinated hydrocarbon
50.	4,4-Dibromobenzophenone[a]	3988-03-2		$C_7H_6Br_2O$	–Br	Chemical materials	266.27	Insoluble	Soluble in various organic solvents
51.	Dicamba[a]	1918-00-9	Low	$C_8H_6Cl_2N$	–Cl	H	221.04	4500	Ethanol 922, xylene 78
52.	Dichlobenil[a]	1194-65-6	Low	$C_7H_3Cl_2N$	–Cl	H	172.02	18	Acetone, benzene, ethanol and toluene 50
53.	Dichlone[a]	117-80-6	Low	$C_{10}H_4Cl_2O_2$	–Cl	F	227.06	0.1	Benzene and chloroform 30, acetone 20, acetic acid 9, diethyl ether 6, ethanol 2
54.	3,5-Dichloroaniline[a]	626-43-7	Low	$C_6H_5Cl_2N$	–Cl	F	161.96	Insoluble	Soluble in various organic solvents
55.	Diclofop methyl[a]	51338-27-3	Low	$C_{16}H_{14}Cl_2O_4$	–Cl	H	341.19	0.3	Soluble in acetone and xylene
56.	Dicloran[a]	99-30-9	Low	$C_6H_4Cl_2N_2O_2$	Cl/N	F	207.06	Insoluble	Acetone 34, chloroform 12, ethyl acetate 9

No.	Name	CAS number		Molecular formula	-Cl	Chemical materials	Molecular weight	Water solubility	Solubility
57.	4,4-Dichlorobenzophenone[a]	90-98-2	Low	$C_7H_6Cl_2NO$	-Cl		176.97	Insoluble	Soluble in various organic solvents
58.	1,2-Dichloro ethane[a]	107-06-2	Low	$C_2H_4Cl_2$	-Cl	I	98.96	8.69	Soluble in various organic solvents
59.	Dichlorprop[a]	15165-67-0	Medium	$C_9H_8Cl_2O_3$	-Cl	H	235.06	710	Acetone 595, isopropanol 510, benzene 85, toluene 69, xylene 51
60.	Dicofol[a]	115-32-2	Low	$C_{14}H_9Cl_5O$	-Cl	A	370.51	Insoluble	Soluble in aliphatic and aromatic solvents
61.	Dieldrin[a]	60-57-1	Medium	$C_{12}H_8Cl_6O$	-Cl	I	380.93	0.186	Benzene 0.75, xylene 0.520
62.	Dienochlor[a]	2227-17-0	Low	$C_{10}Cl_{10}$	-Cl	A	474.64	Insoluble	Slightly soluble in aromatic hydrocarbon, slightly dissolved in hot ethanol, acetone and aliphatic hydrocarbon
63.	α-Endosulfan[a]	959-98-8	High	$C_9H_6Cl_6O_3S$	-Cl	I	406.92	Insoluble	Soluble in various organic solvents
64.	β-Endosulfan[a]	33213-65-9	High	$C_9H_6Cl_6O_3S$	-Cl	I	406.92	Insoluble	Soluble in various organic solvents
65.	Endosulfan sulfate[a]	1031-07-8	High	$C_9H_6Cl_6O_4S$	-Cl	I	422.9	0.22	logKow=0.05
66.	Endrin[a]	72-20-8	High	$C_{12}H_8Cl_6O$	-Cl	I	381.20	Insoluble	Soluble in benzene, xylene and acetone, slightly dissolved in alcohol and petroleum hydrocarbon
67.	Endrin aldehyde[a]	7421-93-4	Low	$C_{12}H_8Cl_6O$	-Cl	I	381.20	50	Soluble in benzene, xylene and acetone, slightly soluble in alcohol and petroleum hydrocarbon. lgKow=1.4×10^3

Continued

No.	Compound	CAS	Toxicity	Formula	Characteristic Atom[1]	Effect[2]	MW	WS (mg/L)	Solubility of Organic Solvent (g/L)
68.	Endrin ketone[a]	53494-70-5	Low	$C_{12}H_8Cl_6O$	–Cl	I	381.20	Insoluble	Soluble in benzene, xylene and acetone, slightly dissolved in alcohol and petroleum hydrocarbon
69.	Erbon[a]	136-25-4	Low	$C_{11}H_9Cl_3O_3$	–Cl	H	295.46	Insoluble	Soluble in acetone, ethanol, benzene, xylene
70.	Fenoprop[a]	93-72-1	Low	$C_9H_7O_3Cl_3$	–Cl	H	269.51	140	Acetone 180, Methanol 134, diethyl ether 98, benzene 16.8, CCl_4 0.95, heptane 0.86
71.	Fluazuron[a]	86811-58-7	Low	$C_{20}H_{10}Cl_2F_5N_3O_3$	–N	IGR/A	506.2	0.02	Methanol 2.4, isopropanol 0.9
72.	Fluoroglycofen-ethyl[a]	77502-90-7	Low	$C_{18}H_{13}ClF_3NO_7$	–F	H	447.8	1	Most solvents >100
73.	Fluoroimide[a]	41205-21-4	Low	$C_{10}H_4FCl_2NO_2$	–Cl	F	260.04	5.9	logKow=2.3, acetone 1.92, methanol 0.84, indissolvable in hexane etc.
74.	Fluorodifen[a]	15457-05-3	Low	$C_{13}H_7F_3N_2O_5$	–F	H	328.13		
75.	Halfenprox[a]	111872-58-3	Medium	$C_{24}H_{23}BrF_2O_3$	–F	A/I	477.3	5×10^{-5}	Soluble in various organic solvents
76.	α-HCH[a]	319-84-6	Low	$C_6H_{12}Cl_6$	–Cl	I	296.78	1.13	Petroleum ether 13, benzene 62, chloroform 63
77.	β-HCH[a]	319-85-7	Low	$C_6H_{12}Cl_6$	–Cl	I	296.78	0.02	Petroleum ether 2, benzene 13, chloroform 3
78.	γ-HCH[a]	58-89-9	Medium	$C_6H_{12}Cl_6$	–Cl	I	296.78	5.75	Petroleum ether 35, benzene 289, chloroform 240

No.	Name	CAS	Persistence	Molecular formula	Substituent	Type	MW	Water solubility	Organic solubility
79.	δ-HCH[a]	319-86-8	Low	$C_6H_{12}Cl_6$	-Cl	I	296.78	20.3	Benzene 411, chloroform 137
80.	ε-HCH[a]	58-89-9	Low	$C_6H_{12}Cl_6$	-Cl	I	296.78	Insoluble	Soluble in various organic solvents
81.	Heptachlor[a]	76-44-8	Medium	$C_{10}H_5Cl_7$	-Cl	I	373.74	Insoluble	Benzene 1260, xylene 1020, acetone 750, ethanol 45
82.	Heptachlor epoxide,Cis-[a]	1024-57-3	High	$C_{10}H_{10}Cl_7O$	-Cl	I	384.74	Insoluble	Soluble in various organic solvents
83.	Heptachlor epoxide, trans-[a]	1024-57-3	High	$C_{10}H_{10}Cl_7O$	-Cl	I	384.74	Insoluble	Soluble in various organic solvents
84.	Hexachlorobenzene[a]	118-74-1	Low	C_6Cl_6	-Cl	I	284.81	Insoluble	Soluble in hot benzene, slightly soluble in diethyl ether
85.	Isobenzan[a]	297-78-9	High	C_9Cl_8O	-Cl	I	1164.32	Insoluble	Soluble in acetone and benzene etc.
86.	Isodrin[a]	465-73-6	High	$C_{12}H_6Cl_6$	-Cl	I	362.82	Insoluble	Soluble in benzene and xylene etc.
87.	Kelevan[a]	4234-79-1	Medium	$C_{17}H_{12}Cl_{10}O_4$	-Cl	I/A	634.83	5.5	Difluent in organic solvents
88.	Lindan[a]	58-89-9	Medium	$C_6H_6Cl_6$	-Cl	I	290.85	10	Acetone 435, benzene 289, toluene 276, xylene 247, chloroform 240, methanol 74, ethanol 64
89.	Lufenuron[a]	103055-07-8	Low	$C_{17}H_8Cl_2F_8N_2O_3$	-F	IGR/A	511.15	0.06	Methanol 41,acetone 460, toluene 72,n-octyl alcohol 8.9,n-hexane 0.13
90.	MCPA butoxyethyl ester[a]	19480-43-4		$C_{15}H_{24}ClO_4$	-Cl	H	300.8	Insoluble	Soluble in various organic solvents

Continued

No.	Compound	CAS	Toxicity	Formula	Characteristic Atom[1]	Effect[2]	MW	WS (mg/L)	Solubility of Organic Solvent (g/L)
91.	MCPA[a]	94-74-6	Low	$C_9H_9ClO_3$	–Cl	H	200.6	82.5	Diethyl ether 770, ethanol 153, toluene 6.2, xylene 4.9
92.	MCPB[a]	94-81-5	Low	$C_{11}H_{13}ClO_3$	–Cl	H	228.68	44	Acetone 200, ethanol 150, slightly soluble in benzene and CCl_4
93.	Mecoprop-P[a]	94596-45-9	Low	$C_{10}H_{11}ClO_3$	–Cl	H	214.6	860	Acetone, diethyl ether and ethanol > 1000, CH_2Cl_2 968, hexane 9, toluene 330
94.	Methoxychlor[a]	72-43-5	Low	$C_{16}H_{15}Cl_3O_2$	–Cl	—	345.65	Insoluble	Difluent in aromatic hydrocarbon, slightly soluble in hydrocarbon and ethanol
95.	Mirex[a]	2385-85-5	Medium	$C_{12}H_{12}Cl_{12}$	–Cl	I	581.56	Insoluble	Xylene 143, benzene 122
96.	Nitrofen[a]	1836-75-5	Low	$C_{12}H_7Cl_2NO_3$	–Cl	H	284.10	1.2	Acetone, ethanol and xylene 250
97.	Nonachlor,trans[a]	39765-80-5	Low	$C_{10}H_5Cl_9$	–Cl	—	319.05	Insoluble	Soluble in various organic solvents
98.	Nonachlor,cis[a]	5103-73-1	Low	$C_{10}H_5Cl_9$	–Cl	—	319.05	Insoluble	Soluble in various organic solvents
99.	Octachlorodipropyl ether[a]	127-90-2	Low	$C_6H_6Cl_8O$	–Cl	Synergist	377.73	Insoluble	Miscible with various solvents
100.	Octachlorostyrene[a]	29082-74-4	Low	C_8Cl_8	–Cl		I/A	379.68	Soluble in various organic solvents
101.	Oxychlordane[a]	27304-13-8	Low	$C_{10}H_4Cl_8O$	–Cl	—	424.14	Insoluble	Soluble in various organic solvents

No.	Name	CAS							Soluble in various solvents
102.	Oxyfluorfen[a]	42874-03-3	Low	$C_{15}H_{11}ClF_3NO_4$	–F	H	361.7	0.1	Soluble in various solvents
103.	DE-PCB 28 2,4,4'-Trichlorobiphenyl[a]	2012-37-5	Low	$C_{12}H_7Cl_3$	–Cl		257.47	Insoluble	Soluble in various organic solvents
104.	DE-PCB 31 2,4',5-trichlorobiphenyl[a]	16606-02-3	Low	$C_{12}H_7Cl_3$	–Cl		357.47	Insoluble	Soluble in various organic solvents
105.	DE-PCB 52 2,2',5,5'-Tetrachlorobiphenyl[a]	35693-99-3	Low	$C_{12}H_6Cl_4$	–Cl		291.92	Insoluble	Soluble in various organic solvents
106.	DE-PCB 101 2,2',4,5,5'-Pentachlorobiphenyl[a]	37680-73-2	Low	$C_{12}H_5Cl_5$	–Cl		326.37	Insoluble	Soluble in various organic solvents
107.	DE-PCB 118 2,3',4,4',5-Pentachlorobiphenyl[a]	31508-00-6	Low	$C_{12}H_5Cl_5$	–Cl		326.37	Insoluble	Soluble in various organic solvents
108.	DE-PCB 138 2,2',3,4,4',5'-Hexachlorobiphenyl[a]	35065-28-2	Low	$C_{12}H_4Cl_6$	–Cl		360.82	Insoluble	Soluble in various organic solvents
109.	DE-PCB 153 2,2',4,4',5,5'-Hexachlorobiphenyl[a]	35065-27-1	Low	$C_{12}H_4Cl_6$	–Cl		360.82	Insoluble	Soluble in various organic solvents
110.	DE-PCB 180[a] 2,2',3,4,4',5,5'-Heptachlorobiphenyl[a]	35065-29-3	Low	$C_{12}H_3Cl_7$	–Cl		395.27	Insoluble	Soluble in various organic solvents
111.	Pentachloroanline[a]	527-20-8	Low	C_6Cl_5N	–Cl	F	265.31	Insoluble	Soluble in various organic solvents
112.	Pentachloroanisole[a]	1825-21-4	Low	$C_7H_3Cl_5O$	–Cl	F	268.31	Insoluble	Soluble in various organic solvents

Continued

No.	Compound	CAS	Toxicity	Formula	Characteristic Atom[1]	Effect[2]	MW	WS (mg/L)	Solubility of Organic Solvent (g/L)
113.	Pentachlorobenzene[a]	608-93-5	Low	C_6HCl_5	–Cl	F	250.31	Insoluble	Soluble in various organic solvents
114.	Pentachlorophenol[a]	87-86-5	High	C_6HCl_5O	–Cl	F	266.31	20	Soluble in various organic solvents, acetone 215
115.	Perthane[a]	72-56-0	Low	$C_{18}H_{20}Cl_2$	–Cl	I	307.25	Insoluble	Soluble in aromatic hydrocarbon and CH_2Cl_2
116.	Plifenate[a]	51366-25-7	Low	$C_{10}H_7Cl_5O_2$	–Cl	I	336.43	50	Toluene and cyclohexanone 600, isopropanol 10
117.	Quinclorac[a]	84087-01-4	Low	$C_{10}H_5Cl_2NO_2$	–Cl	H	242.1	0.065	logKow=–1.15; acetone and ethanol 2, diethyl ether and ethyl acetate 1, indissolvable in toluene, acetonitrile, hexane and CH_2Cl_2
118.	Quinoxyphen[a]	124495-18-7	Low	$C_{15}H_8FCl_2NO$	–Cl	F	308.1	0.116	logKow=4.66, CH_2Cl_2 589, toluene 272, acetone 116, n-octyl alcohol 37.9, hexane 9.6
119.	Quintozene[a]	82-68-8	Low	$C_6H_5NO_2$	–Cl	F	295.34	0.1	logKow=5.1, toluene 1140, methanol 20, heptane 30
120.	Spirodiclofen[a]	148477-71-8	Low	$C_{21}H_{24}Cl_2O_4$	–Cl	A	411.32	50	Dichloromethane 250, xylene 250, isopropanol 47, n-hexane 20

No.	Name	CAS number	Toxicity	Molecular formula			Molecular weight	Water solubility	Organic solvent solubility
121.	2,4,5-T[a]	93-76-5	Medium	$C_8H_5Cl_3O_3$	–Cl	H	255.49	238	Acetone and ethanol 0.59 (50°C)
122.	Tecloftalam[a]	76280-91-6	Low	$C_{14}H_5Cl_6NO_3$	–Cl	F	447.9	14	Acetone 25.6, ethanol 19.2, methanol 5.4, benzene 0.95, toluene 0.16
123.	Tecnazene	117-18-0	Low	$C_{14}H_6Cl_4NO_2$	–Cl	F	260.96	Insoluble	Diffluent in benzene and chloroform, ethanol 40
124.	2,3,4,5-Tetrachloroaniline[a]	634-83-3	Low	$C_6H_3Cl_4$	–Cl	F	230.86	Slight soluble	Soluble in various organic solvents
125.	2,3,4,5-Tetrachloroanisole[a]	938-86-3	Low	$C_7H_4Cl_4O$	–Cl	F	245.87	Slight soluble	Soluble in various organic solvents
126.	2,3,5,6-Tetrachloroaniline[a]	3481-20-7	Low	$C_6H_3Cl_4N$	–Cl	F	230.86	Slight soluble	Soluble in various organic solvents
127.	Tetradifon[a]	116-29-0	Low	$C_{12}H_6Cl_4O_2S$	–Cl	I	356.06	200	Chloroform 255, benzene 148, toluene 135, xylene 115, acetone 82, methanol 10
128.	Tetrasul[a]	2227-13-6	Low	$C_{12}H_6Cl_4S$	–Cl	A	324.06	Insoluble	Diffluent in chlorinated hydrocarbon, aromatic hydrocarbon and chloroform
129.	Triclopyr[a]	55335-06-3	Low	$C_7H_4Cl_3NO_3$	–Cl	H	256.5	440	Acetone 989, n-octyl alcohol 307, acetonitrile 126, xylene 27.9, benzene and chloroform 27.3, hexane 0.41
130.	Tridiphane[a]	58138-08-2	Low	$C_{10}H_5Cl_5O$	–Cl	H	320.4	1.8	Acetone 9.1, CH_2Cl_2 718, methanol 980, xylene 4.6

Continued

(2) Organophosphorus pesticide

No.	Compound	CAS	Toxicity	Formula	Characteristic Atom[1]	Effect[2]	MW	WS (mg/L)	Solubility of Organic Solvent (g/L)
131.	Acephate^b	30560-19-1	Low	$C_4H_{10}NO_3PS$	P/S	I	183.16	7.9×10^5	CH_2Cl_2, ethanol and acetone 151, benzene 16, hexane 0.1
132.	Amidithion^b	919-76-7	High	$C_7H_{16}PS_2NO_4$	P/S	I/A	273.12	2×10^4	Diffluent in organic solvents
133.	Anilofos^b	64249-01-0	Low	$C_{13}H_{19}ClNO_3PS_2$	P/S	H	367.8	13.6	Acetone, chloroform and toluene 1000, benzene, ethanol, ethyl acetate and CH_2Cl_2 20, hexane 12
134.	Athidathion^b	19691-80-6		$C_6H_{15}N_2O_4PS_3$	P/S	I	306.21	Insoluble	Soluble in various organic solvents
135.	Azamethiphos^b	35575-96-3	Low	$C_9H_{10}ClN_2O_5PS$	P/S	I	324.67	1100	Benzene 13, methanol 10, CH_2Cl_2 6.1
136.	Azinphos-ethyl^b	2642-71-9	High	$C_{12}H_{16}N_3O_3PS_2$	P/S	I/A	345.36	Insoluble	Soluble in most organic solvents except for petroleum ether
137.	Azinphos-methyl^b	86-50-0	High	$C_{10}H_{12}N_3O_3PS_2$	P/S	I/A	317.33	33	Soluble in various organic solvents
138.	Bensulide^b	741-58-2	Low	$C_{14}H_{24}NO_4PS_3$	P/S	H	397.54	25	Diffluent in acetone and ethanol, xylene 250
139.	Bromfenvinfos^b	33399-00-7	Low	$C_{12}H_{18}BrCl_2O_4P$	−Cl	I/A	404.02		Soluble in various organic solvents
140.	Bromophos^b	2104-96-3	Low	$C_8H_8BrCl_2O_3PS$	−Cl	I	366	40	Diffluent in toluene and dithyl ether
141.	Bromophos-ethyl^b	4824-78-6	Medium	$C_{10}H_{12}BrCl_2O_3PS$	−Cl	I	394	2	Miscible with common organic solvents

#	Name	CAS No.		Molecular formula		H			
142.	Butamifos[b]	8013-75-0	Low	$C_{13}H_{21}N_2O_3PS$	P/S	H	316.34	5.1	Diffluent in methanol and acetone, etc.
143.	Cadusafos[b]	95465-99-9	High	$C_{10}H_{23}O_2PS_2$	P/S	I/N	270.4	248	Miscible with acetone, acetonitrile, methanol and CH_2Cl_2, etc.
144.	Carbophenothion[b]	786-19-6	High	$C_{11}H_{16}ClO_2PS_3$	P/S	I	342.96	40	Soluble in various organic solvents
145.	Chlorethoxyfos[b]	54593-83-8	High	$C_6H_{11}Cl_4O_3PS$	P/S	I	336.0	1	Soluble in acetonitrile, chloroform, ethanol, n-hexane and xylene
146.	Chlorthion[b]	500-28-7	Low	$C_8H_9ClNO_5PS$	P/S	I	297.56		Soluble in various solvents
147.	Chlorfenvinphos[b]	470-90-6	High	$C_{12}H_{14}Cl_3O_4P$	-Cl	I	359.5	145	Miscible with acetone, ethanol and xylene, etc.
148.	Chlormephos[b]	24934-91-6	High	$C_5H_{12}ClO_2PS_2$	P/S	I	234.71	60	Miscible with most organic solvents
149.	Chlorphoxim[b]	14816-20-7	Low	$C_{12}H_{14}ClN_2O_3PS$	P/S	I	332.75	1.7	Toluene 400
150.	Chlorpyrifos[b]	2921-88-2	Medium	$C_9H_{11}Cl_3NO_3PS$	P/S	I	350.62	2	Soluble in various organic solvents
151.	Chlorpyrifos, O-Nanlogue[b]	2921-88-2	Medium	$C_8H_{11}Cl_3NO_3PS$	P/S	I	350.47	12	Soluble in various solvents
152.	Chlorpyrifos-methyl[b]	5598-13-0	Medium	$C_7H_7Cl_3NO_3PS$	P/S	I	322.5	5	Diffluent in acetone, benzene, diethyl ether and chloroform
153.	Chlorpyrifos-methyl-oxn[b]	5598-13-0	Medium	$C_7H_7Cl_3NO_3PS$	P/S	I	322.5	4	Soluble in various solvents
154.	Chlorthiophos[b]	60238-56-4	Medium	$C_{11}H_{15}Cl_2O_3PS$	P/S	I	361.25	Insoluble	Soluble in benzene, acetone and ethanol

Continued

No.	Compound	CAS	Toxicity	Formula	Characteristic Atom[1]	Effect[2]	MW	WS (mg/L)	Solubility of Organic Solvent (g/L)
155.	Coumaphos[b]	56-72-4	High	$C_{14}H_{16}ClO_5PS$	P/S	I	362.62	1.5	Diffluent in ketone and aromatic hydrocarbon
156.	Crotoxyphox[b]	7700-17-6	Medium	$C_{14}H_{19}O_6P$	–P	I	314.28	1×10^5	Soluble in acetone, chloroform, ethanol and chlorinated hydrocarbon
157.	Crufomate[b]	299-86-5	Medium	$C_{12}H_{19}ClNO_3 P$	–N	I	291.69	5000	Soluble in acetone, acetonitrile and methanol
158.	Cyanofenphos[b]	13067-93-1	High	$C_{14}H_{14}NO_2PS$	P/S	I	291.17	0.6	Diffluent in acetone and benzene
159.	Cythioate[b]	115-93-5	Medium	$C_8H_{12}NO_5PS_2$	P/S	I	297.17	Insoluble	Soluble in acetone, benzene, diethyl ether and ethanol
160.	DEF[b]	78-48-8	Medium	$C_{12}H_{27}OPS_2$	P/S	H	314.51	2.3	Soluble in acetone, ethanol, benzene, xylene, hexane, kerosene, diesel, naphtha and methylnaphthalene
161.	Demephion[b]	2587-90-8		$C_5H_{13}O_3PS_2$	P/S	I/A	216.14	Insoluble	Soluble in various organic solvents
162.	Demeton (O+S)[b]	126-75-0	High	$C_8H_{19}O_3PS_2$	P/S	I	258.34	2×10^5	Miscible with various solvents
163.	Demeton-O[b]	126-75-0	High	$C_8H_{19}O_3PS_2$	P/S	I	258.34	60	Soluble in various organic solvents
164.	Demeton-S[b]	8065-48-3	High	$C_8H_{19}O_3PS_2$	P/S	I	258.34	2×10^5	Soluble in most organic solvents
165.	Demeton-S-methyl[b]	919-86-8	Medium	$C_6H_{15}O_3PS_2$	P/S	I	230.29	330	Soluble in most organic solvents

No.	Name	CAS		Formula			MW	Solubility	Description
166.	Demeton-S-methyl sulphone[b]	17040-19-6	High	$C_6H_{15}O_5PS_2$	P/S	I	262.26	3300	Diffluent in alcohol, indissolvable in aromatic hydrocarbon
167.	Demeton-S-methyl sulfoxide[b]	301-12-2		$C_6H_{15}O_4PS_2$	P/S	I	246.29	Insoluble	Soluble in various organic solvents
168.	Demeton-S-sulfoxide[b]	2496-92-6		$C_8H_{19}O_4PS_2$	P/S	I	274.34	Insoluble	Soluble in various organic solvents
169.	Dialifos[b]	10311-84-9	High	$C_{14}H_{17}ClNO_4PS_2$	P/S	A/I	393.85	1000	Diffluent in acetone, chloroform and xylene
170.	Diazinon[b]	333-41-5	Medium	$C_{12}H_{21}N_2O_3PS$	P/S	I	304.35	4	Soluble in most organic solvents
171.	Dicapthon[b]	2463-84-5	Medium	$C_8H_9ClO_5PS$	P/S	I/A	283.56	35	Soluble in acetone, slightly soluble in cyclohexanone, ethyl acetate, toluene and xylene
172.	Dichlofenthrion[b]	97-17-6	Low	$C_{10}H_{13}Cl_2O_3PS$	P/S	I	315.17	0.245	Soluble in most organic solvents
173.	Dichlorvos[b]	62-73-7	Medium	$C_4H_7Cl_2O_4P$	-P	I	220.91	1000	Soluble in most organic solvents
174.	Dicrotophos[b]	141-66-2	High	$C_8H_{16}NO_5P$	-P	I/A	237.21	Miscible	Miscible with acetone and ethanol etc.
175.	Dimethoate	60-51-5	Medium	$C_5H_{12}NO_3PS_2$	P/S	I	229.28	2.5×10^4	Diffluent in methanol and benzene
176.	Dimefox	115-26-4	High	$C_4H_{12}FN_2OP$	-N	I/A	154.13	Diffluent	Soluble in various solvents
177.	Dimethylvinphos	2274-67-1	Medium	$C_{10}H_{10}Cl_3O_4P$	-Cl	I	331.51	130	Acetone and trichloroethane 500, xylene 300

Continued

No.	Compound	CAS	Toxicity	Formula	Characteristic Atom[1]	Effect[2]	MW	WS (mg/L)	Solubility of Organic Solvent (g/L)
178.	Dioxabenzofos[b]	3811-49-2	Medium	$C_9H_9O_3P$	–P	I	184.05	58	Soluble in acetone, benzene, ethanol and diethyl ether, slightly soluble in toluene and xylene
179.	Dioxathion[b]	78-34-2	High	$C_{12}H_{26}O_6P_2S_4$	P/S	A/I	456.6	Insoluble	Soluble in most organic solvents
180.	Disulfoton[b]	298-04-4	High	$C_8H_{19}O_2PS_3$	P/S	I/A	270.40	25	Soluble in most organic solvents
181.	Disulfoton sulfone[b]	2497-06-5	High	$C_8H_{19}O_4PS_3$	P/S	I/A	306.4	Insoluble	Soluble in various organic solvents
182.	Disulfoton sulfoxide[b]	2497-07-6		$C_8H_{19}O_3PS_3$	P/S	I/A	290.4	Insoluble	Soluble in various organic solvents
183.	Ditalimfos[b]	5131-24-8	Low	$C_{12}H_{14}NO_4PS$	P/S	F	299.29	133	Soluble in hexane, ethanol and cyclohexane, diffluent in benzene, xylene and ethyl acetate
184.	Dithianon[b]	3347-22-6	Medium	$C_{14}H_4N_2O_2S_2$	N/S	F	296.33	Insoluble	Soluble in chloroform and chlorobenzene
185.	Edifenphos[b]	17109-49-8	Medium	$C_{14}H_{15}O_2PS_2$	P/S	F	310.23	5	Soluble in methanol, acetone and chloroform, etc.
186.	EPN[b]	2104-64-5	High	$C_{14}H_{14}NO_4PS$	P/S	I	323.17	Insoluble	Soluble in most organic solvents
187.	Ethephon[b]	16672-87-0	Low	$C_2H_6ClO_3P$	–Cl	PGR	144.50	Diffluent	Diffluent in ethanol and propanediol, slightly soluble in aromatic solvents
188.	Ethion[b]	562-12-2	Medium	$C_9H_{22}O_4P_2S_4$	P/S	I/A	384.48	Slight soluble	Soluble in most organic solvents

No.	Name	CAS	Persistence	Formula	Type	I/N	MW	Solubility (value)	Solubility
189.	Ethoprophos[b]	13194-48-4	Medium	$C_8H_{19}O_2PS_2$	P/S	I/N	242.3	750	Soluble in various organic solvents
190.	Etrimfos[b]	38260-54-7	Low	$C_{10}H_{17}N_2O_4PS$	P/S	I	292.29	40	Soluble in acetone, diethyl ether, ethanol and xylene
191.	Etrimfos oxon[b]	59399-24-5		$C_{10}H_{17}N_2O_4PS$	P/S	I	292.29	Insoluble	Soluble in most organic solvents
192.	Famphur[b]	52-85-7	Medium	$C_{10}H_{16}NO_5PS_2$	P/S	I	293.32	100	Diffluent in chloroform, CCl_4, slightly soluble in polar solvents, insoluble in hexane and heptane, etc.
193.	Fenamiphos[b]	22224-92-6	High	$C_{13}H_{22}NO_3PS$	P/S	N	303.30	700	Soluble in various organic solvents
194.	Fenamiphos sulfone[b]	31972-44-8		$C_{13}H_{22}NO_5PS$	P/S	N	335.3	Insoluble	Soluble in most organic solvents
195.	Fenamiphos sulfoxide[b]	31972-42-7		$C_{13}H_{22}NO_4PS$	P/S	N	319.3	Insoluble	Soluble in most organic solvents
196.	Fenchlorphos[b]	3983-45-7	Low	$C_8H_8Cl_3O_3PS$	Cl/P	I	321.56	44	Acetone 908, toluene 592, chloroform 347, methanol 25
197.	Fenchlorphos oxon[b]	3983-45-7		$C_8H_8Cl_3O_4P$	Cl/P	I / A	305.48	Insoluble	Soluble in various organic solvents
198.	Fenitrothion[b]	122-14-5	Low	$C_9H_{12}NO_5PS$	P/S	I	277.24	30	Diffluent in acetone and benzene
199.	Fensulfothion[b]	115-90-2	High	$C_{11}H_{17}O_4PS_2$	P/S	I/N	308.35	1540	Soluble in various organic solvents
200.	Fenthion[b]	55-38-9	Medium	$C_{10}H_{15}O_3PS_2$	P/S	I	278.33	55	Soluble in methanol, ethanol, chlorinated hydrocarbon and aromatic hydrocarbon

Continued

No.	Compound	CAS	Toxicity	Formula	Characteristic Atom[1]	Effect[2]	MW	WS (mg/L)	Solubility of Organic Solvent (g/L)
201.	Fenthion-oxon[b]	6552-12-1		$C_{10}H_{15}O_4PS$	P/S	I	262.13		Soluble in various organic solvents
202.	Fenthion PO-sulfone[b]		Medium	$C_{10}H_{15}O_5PS_2$	P/S	I	310.33	Insoluble	Soluble in various organic solvents
203.	Fenthion PO-sulfoxide[b]		Medium	$C_{10}H_{15}O_4PS_2$	P/S	I	294.33	Insoluble	Soluble in various organic solvents
204.	Fenthion oxon sulfone[b]	14086-35-2	Medium	$C_{10}H_{15}O_6PS$	P/S	I	294.13	Insoluble	Soluble in various organic solvents
205.	Fenthion oxon sulfoxide[b]	6552-13-2	Medium	$C_{10}H_{15}O_5PS$	P/S	I	278.13	Insoluble	Soluble in various organic solvents
206.	Fenthion sulfone[b]	3761-42-0	Medium	$C_{10}H_{15}O_5PS_2$	P/S	I	310.34	Insoluble	Soluble in most organic solvents
207.	Fenthion sulfoxide[b]	3761-41-9	Medium	$C_{10}H_{15}O_4PS_2$	P/S	I	294.3	Insoluble	Soluble in various organic solvents
208.	Fonofos[b]	944-22-9	High	$C_{10}H_{15}OPS_2$	P/S	I	246.34	13	Miscible with xylene and acetone, etc.
209.	Formothion[b]	2540-82-1	Medium	$C_6H_{12}NO_4PS_2$	P/S	I/A	257.27	2600	Miscible with ethanol, chloroform, diethyl ether and benzene, etc.
210.	Fosetyl-aluminium[b]	39148-24-8	Low	$C_6H_{18}AlO_9P_3$	–P	F	354.1	1.2×10^5	Indissolvable in organic solvents, acetonitrile < 0.08
211.	Fosthiazate[b]	98886-44-3	Medium	$C_9H_{18}NO_3PS_2$	P/S	I/N	283.3	9850	Soluble in various organic solvents
212.	Fosthietan[b]	21548-32-3	High	$C_6H_{12}NO_3PS_2$	P/S	N/I	241.3	5×10^4	Soluble in acetone, chloroform, methanol and toluene, etc.

No.	Name	CAS		Formula			MW	Water sol.	Organic solvent solubility
213.	Glgphosate[b]	1071-83-6	Low	$C_3H_8NO_5P$	-P	H	169.08	1.28×10^4	Insoluble in common organic solvents
214.	Glufosinate ammonium[b]	77182-82-2	Low	$C_5H_{15}N_2O_4P$	-P	H	198.2	Diffluent	Indissolvable
215.	Heptenophos[b]	23560-59-0	Medium	$C_9H_{11}ClO_4P$	-P	I	250.62	2500	Acetone, methanol and xylene 1000, hexane 130
216.	Heterophos[b]	40626-35-5	Medium	$C_9H_{12}ClO_4P$	-P	I	250.62	2500	Acetone, methanol and xylene 1000, hexane 130
217.	Iodofenphos[b]	18181-70-9	Low	$C_8H_8Cl_2IO_3PS$	P/S	I	413.0	2	CH_2Cl_2 860, benzene 610, acetone 480, hexane 33
218.	Iprobenfos[b]	26087-47-8	Low	$C_{13}H_{21}O_3PS$	P/S	F	288.32	1000	Soluble in various organic solvents
219.	Isazofos[b]	42509-80-8	Medium	$C_9H_{17}ClN_3O_3PS$	-N	I	313.17	250	Soluble in benzene, chloroform, hexane and methanol
220.	Isocarbophos[b]	245-61-5	High	$C_{11}H_{16}NO_4PS$	P/S	I	288.0	Insoluble	Soluble in petroleum ether, acetone, benzene and ethyl acetate
221.	Isofenphos[b]	25311-71-1	High	$C_{15}H_{24}NO_4PS$	P/S	I	345.40	20	Benzene 600
222.	Isofenphos oxon[b]	106848-93-5	Medium	$C_{15}H_{24}NO_5P$	-P	I/A	329.34	Insoluble	Soluble in various organic solvents
223.	Isoxathion[b]	18854-01-8	Medium	$C_{13}H_{16}NO_4PS$	P/S	I	313.16	1.9	Soluble in various organic solvents
224.	Leptophos[b]	21609-90-5	High	$C_{13}H_{10}Cl_2BrO_2PS$	P/S	I	412.16	2.4	Acetone 170, cyclohexane 142, benzene 3
225.	Leptophos desbrom[b]		High	$C_{13}H_{11}Cl_2O_2PS$	P/S	I/A	333.06	Insoluble	Soluble in various organic solvents

Continued

No.	Compound	CAS	Toxicity	Formula	Characteristic Atom[1]	Effect[2]	MW	WS (mg/L)	Solubility of Organic Solvent (g/L)
226.	Leptophos oxon[b]	25006-32-0	High	$C_{13}H_{10}Cl_2BrO_3P$	–P	I	396	2.4	Soluble in acetone, benzene, cyclohexane, heptane and propanol etc.
227.	Malaoxon[b]	1364-78-2	Medium	$C_{10}H_{19}O_7PS$	P/S	I	314.3	Insoluble	Soluble in various organic solvents
228.	Malathion[b]	121-75-5	Low	$C_{10}H_{19}O_6PS_2$	P/S	I	330.36	145	Diffluent in ether, acetone and aromatic hydrocarbon, slightly soluble in petroleum ether
229.	Mecarbam[b]	2595-54-2	High	$C_{10}H_{20}NO_5PS_2$	P/S	A/I	329.19	<1000	Fatty hydrocarbon <50, miscible with organic solvents such as alcohol, ester, ketone, aromatic hydrocarbon and chlorinated hydrocarbon, etc.
230.	Mephosfolan[b]	950-10-7	High	$C_8H_{16}NO_3PS_2$	P/S	I	269.3	5.7×10^4	Soluble in acetone, ethanol and benzene, etc.
231.	Merphos[b]	150-50-5	Medium	$C_{12}H_{27}PS_3$	P/S	Defoliant	298.27	Slight soluble	Soluble in various organic solvents
232.	Methacrifos[b]	62610-77-9	Low	$C_7H_{13}O_5PS$	P/S	I/A	240.22	400	Diffluent in benzene, CH_2Cl_2 and methanol
233.	Methamidophos[b]	10265-92-6	High	$C_2H_8NO_2PS$	P/S	I/A	141.13	2×10^6	Benzene and xylene 100, chloroform, CH_2Cl_2 and diethyl ether 20, hexane 10
234.	Methidathion[b]	950-37-8	High	$C_6H_{11}N_2O_4PS_3$	P/S	I	302.33	240	Diffluent in acetone, benzene and methanol

No.	Name	CAS	Persistence	Formula			MW	Water solubility	Solubility
235.	Mevinphos[b]	7786-34-7	High	$C_7H_{13}O_6P$	-P	I	224.04	Miscible	Miscible with acetone, methanol, ethanol, benzene, toluene and xylene
236.	Monocrotophos[b]	6923-22-4	High	$C_7H_{14}NO_5P$	-P	I/A	223.2	Miscible	Soluble in acetone and ethanol, slightly soluble in toluene
237.	Naled[b]	300-76-5	Medium	$C_4H_7Br_2Cl_2O_4P$	-Cl	I	380.84	Insoluble	Diffluent in aromatic hydrocarbon, chlorinated hydrocarbon, hexane 5
238.	Omethoate[b]	1113-02-6	High	$C_5H_{12}NO_4PS$	P/S	I/A	213.19	Diffluent	Diffluent in ethanol and acetone, insoluble in petroleum ether
239.	Oxydemeton-methyl[b]	301-12-2	Medium	$C_6H_{15}O_4PS_2$	P/S	I/A	246.15	Miscible	Soluble in various solvents except for petroleum ether
240.	Paraoxon[b]	311-45-5	High	$C_{10}H_{14}NO_6P$	-P	I	203.24	2.5×10^4	Soluble in various solvents
241.	Paraoxon-methyl[b]	950-35-6		$C_8H_{10}NO_6P$	-P	I	215.05	Insoluble	Soluble in various organic solvents
242.	Parathion[b]	56-38-2	High	$C_{10}H_{14}NO_5PS$	P/S	I/A	291.27	24	Diffluent in ethanol, acetone, benzene and chloroform
243.	Parathion-methyl[b]	298-00-0	High	$C_8H_{10}NO_5PS$	P/S	I/A	263.21	60	Diffluent in ethanol, acetone, benzene and chloroform
244.	Phenkapton[b]	2275-14-1	Medium	$C_{11}H_{15}Cl_2PS_3$	P/S	A	377.33	Insoluble	Miscible with various organic solvents
245.	Phenthoate[b]	2597-03-7	Medium	$C_{12}H_{17}O_5PS_3$	P/S	I	320.37	11	Diffluent in methanol, ethanol, acetone and benzene, hexane 120

Continued

No.	Compound	CAS	Toxicity	Formula	Characteristic Atom[1]	Effect[2]	MW	WS (mg/L)	Solubility of Organic Solvent (g/L)
246.	Phorate[b]	298-02-2	High	$C_7H_{17}O_2PS_3$	P/S	I/A/N	260.22	50	Miscible with ethanol, ketone, ether, ester and chlorinated hydrocarbon, etc.
247.	Phorate, O-analogue[b]	2600-69-3	High	$C_7H_{17}O_3PS_2$	P/S	I/A	244.16	Insoluble	Soluble in various organic solvents
248.	Phorate Sulfone[b]	2588-04-7	High	$C_7H_{17}O_4PS_3$	P/S	I/A	292.22	Insoluble	Soluble in various organic solvents
249.	Phorate sulfoxide[b]	2588-03-6	High	$C_7H_{17}O_3PS_3$	P/S	I/A	276.22	Insoluble	Soluble in various organic solvents
250.	Phosalone[b]	2310-17-0	Medium	$C_{12}H_{15}ClNO_4PS_2$	P/S	I/A	367.82	100	Diffluent in acetone, acetonitrile, benzene, chloroform, CH_2Cl_2, toluene and xylene, methanol and ethanol 200
251.	Phosfolan[b]	950-10-7	High	$C_7H_{14}NO_3PS_2$	P/S	I/A	255.06	Insoluble	Soluble in various organic solvents
252.	Phosmet[b]	732-11-6	Medium	$C_{11}H_{12}NO_4PS_2$	P/S	I/A	317.33	25	Soluble in methanol, ethanol, CH_2Cl_2, benzene, toluene and xylene, acetone 100
253.	Phosmet, O-analogue[b]	3735-33-9	Medium	$C_{11}H_{12}NO_5PS$	P/S	I/A	301.27	Insoluble	Soluble in various organic solvents
254.	Phosmet-oxon[b]	732-11-6	Medium	$C_{11}H_{13}NO_4PS$	P/S	I/A	286.25	Insoluble	Soluble in various organic solvents
255.	Phosphamidon[b]	13171-21-6	High	$C_{10}H_{19}ClNO_5P$	–P	I	299.69	Miscible	Soluble in various solvents

No.	Name	CAS		Formula	P/S		MW	Sol.	Solubility
256.	Phoxim[b]	14816-18-3	Low	$C_{12}H_{15}N_2O_3PS$	P/S	I	298.30	7	Soluble in alcohol, ketone and aromatic hydrocarbon, slightly soluble in petroleum ether
257.	Piperophos[b]	24151-93-7	Medium	$C_{14}H_{33}NO_3PS_2$	P/S	H	358.50	25	Soluble in benzene, n-hexane, acetone and dichloromethane
258.	Pirimiphos-ethyl[b]	23505-41-1	Medium	$C_{13}H_{24}N_3O_3PS$	–N	I	333.16	1	Soluble in various organic solvents
259.	Pirimiphos-methyl[b]	29232-93-7	Low	$C_{11}H_{20}N_3O_3PS$	–N	I/A	305.34	5	Soluble in various organic solvents
260.	Profenofos[b]	41198-08-7	Medium	$C_{11}H_{15}BrClO_3PS$	P/S	I	373.6	20	Soluble in various organic solvents
261.	Propaphos[b]	7292-16-2	Medium	$C_{12}H_{21}O_4PS$	P/S	I	292.09	125	Soluble in various solvents
262.	Prothoate[b]	2275-18-5	High	$C_9H_{20}NO_3PS_2$	P/S	A/I	285.08	2500	Miscible with various solvents, hexane 30, petroleum ether 20
263.	Prothiofos[b]	34643-46-4	Low	$C_{11}H_{15}Cl_2O_2PS_2$	P/S	I	345.25	1.7	Isopropanol, toluene and CH_2Cl_2 1200
264.	Pyraclofos[b]	77458-01-6	Medium	$C_{14}H_{18}ClN_2O_3PS$	P/S	I	360.8	33	Mutually soluble with various solvents, slightly soluble in n-hexane
265.	Pyrazophos[b]	13457-18-6	Medium	$C_{14}H_{20}N_3O_3PS$	P/S	F	373.37	3300	Diffluent in benzene, toluene, xylene, ethanol, ethyl acetate and CH_2Cl_2
266.	Pyrazophos, O-analogue[b]		Medium	$C_{14}H_{20}N_3O_6P$	P/S	F	357.11	Insoluble	Soluble in various organic solvents

Continued

No.	Compound	CAS	Toxicity	Formula	Characteristic Atom[1]	Effect[2]	MW	WS (mg/L)	Solubility of Organic Solvent (g/L)
267.	Pyridaphenthion[b]	119-12-0	Low	$C_{14}H_{17}N_2O_4PS$	P/S	I	340.17	Indissolvable	Difiluent in acetone, methanol and diethyl ether, etc.
268.	Pyrimitate[b]	5221-49-8	Low	$C_{11}H_{20}N_3O_3PS$	P/S	I/A	305.14	Insoluble	Soluble in various organic solvents
269.	Quinalphos[b]	13593-03-8	Medium	$C_{12}H_{15}N_2O_3PS$	P/S	I	298.3	22	Soluble in toluene, xylene, ethyl acetate, diethyl ether, acetone, acetonitrile, methanol and ethanol
270.	Schradan[b]	152-16-9	High	$C_8H_{24}N_4O_3P_2$	–N	I/A	286.24	Miscible	Miscible with benzene and acetone, etc., indissolvable in alkane
271.	Sulfotep[b]	3689-24-5	High	$C_8H_{10}O_5P_2S_2$	P/S	I/A	322.33	25	Miscible with various solvents, slightly soluble in petroleum ether
272.	Sulprofos[b]	35400-43-2	Medium	$C_{12}H_{19}O_2PS_3$	P/S	I	322.45	50	Cyclohexanone 120, isopropanol 60
273.	Tebupirimfos[b]	96182-53-5	High	$C_{13}H_{23}N_2O_3PS$	P/S	I	318.4	5.5	Soluble in various solvents
274.	Temephos[b]	3383-96-8	Low	$C_{16}H_{20}O_6P_2S_3$	P/S	I	466.46	Insoluble	Soluble in various solvents
275.	TEPP[b]	107-49-3	High	$C_8H_{20}O_7P_2$	–P	A/I	290.02	Miscible	Soluble in various organic solvents
276.	Terbufos[b]	13071-79-9	High	$C_9H_{21}O_2PS_3$	P/S	I	288.43	4.5	Aromatic hydrocarbon, chlorinated hydrocarbon, ethanol and acetone 30
277.	Terbufos sulfone[b]	56070-16-7	High	$C_9H_{21}O_4PS_3$	P/S	I	320.43	Insoluble	Soluble in various organic solvents

278.	Terbufos sulfone O-analogue[b]	56070-15-6	High	$C_9H_{21}O_4PS_3$	P/S	I	320.42	Insoluble	Soluble in various organic solvents
279.	Tetrachlorvinphos[b]	22248-79-9	Low	$C_{10}H_9Cl_4O_4P$	–Cl	N/A	365.97	11	Chloroform and CH_2Cl_2 400, acetone 200, xylene 150
280.	Thiometon[b]	640-15-3	Medium	$C_6H_{15}O_2PS_3$	P/S	I	246.35	200	Soluble in various organic solvents
281.	Thionazin[b]	297-97-2	High	$C_8H_{13}N_2O_3PS$	P/S	I/N	284.24	1140	Soluble in various solvents
282.	Tolclofos-methyl[b]	57018-04-9	Low	$C_9H_{11}Cl_2O_3PS$	P/S	F	301.1	0.4	Acetone 502, cyclohexanone 537, cyclohexane 498
283.	Riazophos[b]	24017-47-8	Medium	$C_{12}H_{16}N_3O_3PS$	–N	I	313.31	39	Toluene, ethyl acetate, acetone and ethanol 300, hexane 7
284.	Triamiphos[b]	1031-47-6	High	$C_{12}H_{19}N_6OP$	–N	F	294.09	250	Soluble in various solvents
285.	Trichlorfon[b]	52-68-6	Medium	$C_4H_8Cl_3O_4P$	–P	I	257.44	1.54×10^5	Chloroform 750, ethyl acetate 170, toluene 152, hexane 0.8
286.	Trichloronate[b]	327-98-0	High	$C_{10}H_{12}Cl_3O_2PS$	–Cl	I	333.42	50	Soluble in various solvents
287.	Tri-n-butyl phosphate[b]	126-73-8	Low	$C_{12}H_{27}PO_4$	–P	N/A	266.09	Insoluble	Soluble in various solvents
288.	Tri-isobutyl phosphate[b]	126-71-6	Low	$C_{12}H_{27}PO_4$	–P	N/A	266.09	Insoluble	Soluble in various solvents
289.	Triphenyl phosphate[b]	115-86-6	Low	$C_{18}H_{15}PO_4$	–P	N/A	326.15	Insoluble	Soluble in various solvents
290.	Vamidothion[b]	2265-23-2	Medium	$C_8H_{18}NO_4PS_2$	P/S	I	287.35	4×10^6	Benzene, toluene, ethyl acetate, CH_2Cl_2 and chloroform 1000, xylene 125

Continued

No.	Compound	CAS	Toxicity	Formula	Characteristic Atom[1]	Effect[2]	MW	WS (mg/L)	Solubility of Organic Solvent (g/L)
291.	Vamidothion sulfone[b]	70898-34-9	Medium	$C_8H_{18}NO_6PS_2$	P/S	I	319.35	Insoluble	Soluble in various organic solvents
292.	Vamidothion sulfoxide[b]	20300-00-9	Medium	$C_8H_{18}NO_5PS_2$		I	303.35	Insoluble	Soluble in various organic solvents
(3) Carbamates pesticide									
293.	Alanycarb[d]	83130-01-2	Medium	$C_{17}H_{25}N_3O_4S_2$	–N	I	399.5	20	Soluble in methanol, acetone, benzene, xylene and CH_2Cl_2; logKow = 3.43
294.	Aldicarb[d]	116-06-3	High	$C_7H_{14}N_2O_2S$	–S		190.17	6000	Acetone 430, benzene 240, chloroform 440, toluene 120
295.	Aldicarb sulfone (aldoxycarb)[d]	1646-88-4	High	$C_7H_{14}N_2O_3S$	–S	I	206.17	9000	Acetone 50, acetonitrile 75, CH_2Cl_2 41
296.	Aldicarb sulfoxide[d]	1646-87-3	High	$C_7H_{14}N_2O_4S$	–S	I	222.17	8000	Acetone 430, benzene 240, chloroform 440, toluene 120
297.	Aminocarb[d]	2032-59-9	High	$C_{11}H_{13}N_2O_2$	–N	I	205.11	Slightly soluble	Diffluent in methanol, slightly soluble in aromatic solvents
298.	Asulam[d]	3337-71-1	Low	$C_8H_{10}N_2O_4S$	–N	H	230	5000	Acetone 300, methanol 200, hydrocarbon and chlorinated hydrocarbon 20
299.	Barban[d]	101-27-9	Low	$C_{11}H_9Cl_2NO_2$	–Cl	H	258.19	11	Benzene 370, toluene 295, xylene 206, hexane 2
300.	BDMC[d]	672-99-1		$C_{10}H_{12}BrNO_2$	–N	I/A	258.11	Insoluble	Soluble in various organic solvents

No.	Name	CAS No.	Toxicity	Formula			MW	Water solubility	Organic solubility
301.	Bendiocarb[d]	22781-23-3	Medium	$C_{11}H_{23}NO_4$	-N	I	223.11	4×10^4	Chloroform and acetone 200, ethanol and benzene 40
302.	Benfuracarb[d]	82560-54-1	Medium	$C_{20}H_{30}N_2O_5S$	-N	I	410.5	8.1	logKow=4.3, benzene, CH_2Cl_2, ethanol, acetone, n-hexane, xylene and ethyl acetate >1000
303.	Butocarboxim[d]	34681-10-2	Medium	$C_7H_{14}N_2O_2S$	-N	I	190.27	3.5×10^4	Diffluent in aromatic hydrocarbon and ketone
304.	Butcocarboxim-sulfoxide[d]	34681-24-8	Medium	$C_7H_{14}N_2O_4S$	-N	I	206.27	Insoluble	Diffluent in aromatic hydrocarbon and ketone
305.	Butoxycarboxim-sulfone[d]	34681-23-7		$C_7H_{14}N_2O_4S$	-N	I	238.13		Diffluent in aromatic hydrocarbon and ketone
306.	Butoxycarboxim[d]	34681-23-7	Low	$C_7H_{14}N_2O_4S$	-N	I/A	222.25	2×10^5	Chloroform 186, acetone 172, isopropanol 101, toluene 29
307.	Carbaryl[d]	63-25-2	Low	$C_{12}H_{11}NO_2$	-N	I	201.23	120	Acetone 200, cyclohexane 200
308.	Carbetamide[d]	16118-49-3	Low	$C_{12}H_{16}N_2O_3$	-N	H	236.27	3.5×10^5	Soluble in acetone, methanol and CH_2Cl_2
309.	Carbofuran[d]	1563-66-2	High	$C_{12}H_{15}NO_3$	-N	I/A/N	221.25	700	Acetone 150, acetonitrile 140, benzene 40
310.	Carbosulfan[d]	55285-14-8	Medium	$C_{20}H_{32}N_2O_3S$	N/S	I	380.55	0.3	Soluble in various organic solvents
311.	Cartap[d]	15263-52-2	Medium	$C_7H_{15}N_3O_2S_2$	N/S	I/F	237.19	2×10^5	Slightly soluble in methanol
312.	Cartap hydrochloride[d]	15263-52-2	Medium	$C_7H_{15}N_3O_2S_2 \cdot HCl$	N/S	I	237.3	2×10^5	Insoluble in acetone, diethyl ether, chloroform, hexane and benzene, etc.

Continued

No.	Compound	CAS	Toxicity	Formula	Characteristic Atom[1]	Effect[2]	MW	WS (mg/L)	Solubility of Organic Solvent (g/L)
313.	Chlorbufam[d]	1967-16-4	Low	$C_{11}H_{10}ClNO_2$	-N	H	223.56	540	Methanol 286, acetone 280, ethanol 95
314.	Chlorpropham[d]	101-21-3	Low	$C_{10}H_{12}ClNO_2$	-N	H/PGR	213.67	89	Diffluent in aromatic hydrocarbon and ethanol, miscible with acetone
315.	Cycloate[d]	1134-23-2	Low	$C_{11}H_{21}NOS$	N/S	H	215.37	85	Miscible with acetone, benzene, ethanol and xylene
316.	Desmedipham[d]	13684-56-5	Low	$C_{16}H_{16}N_2O_4$	-N	H	300.32	7	Acetone 400, methanol 180, chloroform 80
317.	DesmethylformamidO-pirimicarb[d]	27218-04-8	Medium	$C_9H_{13}N_3O$	-N	I	195.24	Insoluble	Soluble in various organic solvents
318.	Diallate[d]	2303-16-4	Medium	$C_{10}H_{17}Cl_2NOS$	-Cl	H	270.21	14	Miscible with ethanol, acetone and benzene
319.	Diethofencarb[d]	87130-20-9	Low	$C_{14}H_{21}NO_4$	-N	F	267.3	2.66×10^4	Hexane 1.3, methanol 101
320.	Dimepiperate[d]	61432-55-1	Low	$C_{15}H_{21}NOS$	-N	H	263.4	20	Acetone 620, chloroform 580, xylene 10, hexane 200
321.	Dioxacarb[d]	6988-21-2	Medium	$C_{11}H_{13}NO_4$	-N	I	223.23	6000	Acetone 280, ethanol 60, CH_2Cl_2 345
322.	EPTC[d]	759-94-4	Low	$C_9H_{19}NOS$	-N	H	189.32	365	Miscible with benzene, toluene, xylene, methanol and acetone
323.	Erbon[d]	136-25-4	Low	$C_{11}H_9Cl_3O_3$	-Cl	H	295.46	Insoluble	Soluble in acetone, ethanol, benzene and xylene

No.	Name	CAS		Formula			MW	Water solubility	Solubility in organic solvents
324.	Esprocarb[d]	85785-20-2	Low	C$_{15}$H$_{23}$NOS	–N	H	265.4	4.9	Acetone, ethanol, acetonitrile, chlorobenzene and xylene 1000
325.	Ethiofencarb[d]	29973-13-5	Medium	C$_{11}$H$_{15}$NO$_2$S	N/S	I	225.31	1820	Acetone, CH$_2$Cl$_2$ and toluene 600
326.	Ethiofencarb sulfon[d]	53380-23-7		C$_{11}$H$_{15}$NO$_4$S	–N	I	257.31	1820	Soluble in various organic solvents
327.	Ethiofencarb sulfoxid[d]	53380-22-6		C$_{11}$H$_{15}$NO$_3$S	–N	I	241.31	Insoluble	Soluble in various organic solvents
328.	Fenobucarb[d]	3766-81-2	Medium	C$_{12}$H$_{17}$NO$_2$	–N	I	207.28	Slight soluble	Soluble in acetone, methanol, benzene, toluene and xylene, etc.
329.	Fenoxycarb[d]	79127-80-3	Low	C$_{17}$H$_{19}$NO$_4$	–N	I	301.4	5.7	Acetone, chloroform, diethyl ether, ethyl acetate, toluene and methanol 250, hexane 5
330.	Fenothiocarb[d]	62850-32-2	Low	C$_{13}$H$_{19}$NO$_2$S	N/S	A	253.4	30	Ethanol 3120, acetone 2530, xylene 2460, methanol 1430
331.	Furathiocarb[d]	65907-30-4	Medium	C$_{18}$H$_{26}$N$_2$OPS	–N	I	382.5	10	Soluble in acetone, hexane and methanol, etc.
332.	3-Hydroxycarb-ofuran[d]	16655-82-6		C$_{12}$H$_{15}$NO$_4$	–N	I	237.25	Insoluble	Soluble in various organic solvents
333.	Iprovalicarb[d] (SS)	140923-17-7	Low	C$_{18}$H$_{28}$N$_2$O$_3$	–N	F	320.43	6.8	CH$_2$Cl$_2$ 35, toluene 2.4, acetone 19, hexane 0.04
334.	Iprovalicarb[d] (SR)	140923-17-7	Low	C$_{18}$H$_{28}$N$_2$O$_3$	–N	F	320.43	11	CH$_2$Cl$_2$ 97, toluene 2.9, acetone 22, hexane 0.06

Continued

No.	Compound	CAS	Toxicity	Formula	Characteristic Atom[1]	Effect[2]	MW	WS (mg/L)	Solubility of Organic Solvent (g/L)
335.	Isoprocarb[d]	2631-40-7	Medium	$C_{11}H_{15}NO_2$	–N	I	193.11	265	Acetone 400, methanol 125, xylene 50
336.	Methiocarb[d]	2032-65-7	Medium	$C_{11}H_{15}NO_2S$	–N	I/A	234.17	27	Isopropanol 350, cyclohexanone 268, toluene 52.5
337.	Methiocarb sulfoxide[d]	2635-10-1	Medium	$C_{11}H_{15}NO_3S$	–N	I	250.17	Insoluble	Soluble in various organic solvents
338.	Methiocarb sulfone[d]	2178-25-1	Medium	$C_{11}H_{16}N_4OS$	–N	I/A	252.33	Insoluble	Soluble in various organic solvents
339.	Methomyl[d]	16752-77-5	High	$C_5H_{10}N_2O_2S$	–N	I	162.21	5.8×10^4	Methanol 1000, acetone 730, ethanol 420, toluene 30
340.	Metolcarb[d]	1129-41-5	Low	$C_9H_{11}NO_2$	–N	I	165.2	2600	Soluble in various solvents
341.	Mexacarbate[d]	315-18-4	High	$C_{12}H_{18}N_2O_2$	–N	I/A	218.12	100	Soluble in various solvents
342.	Molinate[d]	2212-67-1	Low	$C_9H_{17}NOS$	–N	H	187.32	800	Miscible with acetone, methanol, benzene and xylene
343.	Oxamyl[d]	23135-22-0	High	$C_7H_{13}N_3O_3S$	–N	I/A/N	219.36	2.8×10^5	Methanol 1440, acetone 670, ethanol 330, isopropanol 110, toluene 10
344.	Oxamyl-oxime[d]	30558-43-1	High	$C_7H_{12}N_2O_2S$	–N	I/A/N	188.36	2.8×10^5	Soluble in various solvents
345.	Pebulate[d]	1114-71-2	Low	$C_{10}H_{21}NOS$	–N	H	203.36	92	Miscible with acetone, benzene, toluene, xylene and methanol

No.	Name	CAS	Toxicity	Formula			MW	Water solubility	Solubility
346.	Phenmedipham[d]	13864-63-4	Low	$C_{16}H_{16}N_2O_4$	$-N$	H	300.32	10	Acetone 200, methanol 150, chloroform 20, benzene 5, hexane 0.5
347.	Pirimicarb[d]	23103-98-2	Medium	$C_{11}H_{18}N_4O_2$	$-N$	I	238.3	2700	Acetone 400, ethanol 250, methanol 230
348.	Pirimicarb-desmethyl[d]	30614-22-3	Medium	$C_{10}H_{15}N_4O_2$	$-N$	I	223.3	Aqueous	Soluble in solvents like acetone, ethanol, xylene and chloroform, etc.
349.	Promecarb[d]	2631-37-0	Medium	$C_{12}H_{17}NO_2$	$-N$	I	207.26	92	Acetone 600, methanol 400, xylene 200
350.	Propamocarb[d]	24579-73-5	Low	$C_9H_{20}N_2O_2$	$-N$	F	188.28	5×10^5	Methanol 500, CH_2Cl_2 450, ethyl acetate 23, benzene 0.1
351.	Propham[d]	122-42-9	Low	$C_{10}H_{13}NO_2$	$-N$	H	179.22	250	Soluble in various solvents
352.	Propoxur[d]	114-26-1	Medium	$C_{11}H_{15}NO_3$	$-N$	I	209.25	2000	Soluble in various solvents
353.	Prosulfocarb[d]	52888-80-9	Low	$C_{14}H_{21}NOS$	$-N$	H	251.4	13.2	Soluble in acetone, ethanol and xylene
354.	Pyributicarb[d]	88678-67-5	Low	$C_{18}H_{22}N_2O_2S$	$-N$	H	330.4	0.32	Acetone 780, xylene 580, ethyl acetate 560, chloroform 390, ethanol 33, methanol 28
355.	Sulfallate[d]	95-06-7	Low	$C_8H_{14}ClNS_2$	$-S$	H	223.79	100	Soluble in ether, acetone, benzene, chloroform and ethanol, etc.
356.	Thiodicarb[d]	59669-26-0	Medium	$C_{10}H_{18}N_4O_4S_3$	$-N$	I	354.46	300	Soluble in acetone, CH_2Cl_2, acetonitrile and diethyl ether

Continued

No.	Compound	CAS	Toxicity	Formula	Characteristic Atom[1]	Effect[2]	MW	WS (mg/L)	Solubility of Organic Solvent (g/L)
357.	Thiofanox^d	39196-18-4	High	$C_9H_{18}N_2O_2S$	–N	I/A	218.15	5200	Diffluent in chlorinated hydrocarbon, aromatic hydrocarbon, ketone and nonpolar solvents, slightly soluble in fatty hydrocarbon
358.	Thiofanox-sulfoxide^d	39184-27-5	High	$C_9H_{18}N_2O_3S$	–N	I/A	234.32	5200	Diffluent in chlorinated hydrocarbon, aromatic hydrocarbon and ketone, slightly soluble in fatty hydrocarbon
359.	Thiofanox-sulfon^d	53380-23-7	High	$C_9H_{18}N_2O_4S$	–N	I/A	250.32	Insoluble	Diffluent in chlorinated hydrocarbon, aromatic hydrocarbon and ketone, slightly soluble in fatty hydrocarbon
360.	3,4,5-Trimethacarb^d	2686-99-9	Medium	$C_{11}H_{15}NO_2$	–N	I	193.24	58	Soluble in various solvents
361.	Vernolate^d	1929-77-7	Low	$C_{10}H_{21}NOS$	–N	H	203.36	90	Miscible with various solvents
(4) Pyrethroids pesticide									
362.	Acrinathrin^c	101007-06-1	Low	$C_{26}H_{21}F_6NO_5$	–F	A	541.4	0.02	Acetone, chloroform, CH_2CL_2 and ethyl acetate 500, ethanol 40, hexane 10
363.	Allethrin^c	584-79-2	Medium	$C_{19}H_{26}O_3$	/	I	302.4	Insoluble	Diffluent in organic solvents
364.	Alpha-cypermethrin^c	67375-30-8	Medium	$C_{22}H_{19}Cl_2NO_3$	–Cl	I	416.3	0.1	Acetone, chloroform, ethanol and xylene 450

No.	Name	CAS Number	Volatility	Molecular Formula	Cl/F		Mol. Wt.	Water Solubility	Solvent Solubility
365.	Bifenthrin[c]	82657-04-3	Medium	$C_{23}H_{22}ClF_3O_2$	—	I	422.71	0.1	Soluble in common organic solvents
366.	Bioallethrin[c]	584-79-2	Medium	$C_9H_{25}O_3$	/	I	302.42	Insoluble	Miscible with acetone, benzene, toluene, ethanol and hexane
367.	Bioresmethrin[c]	28434-01-7	Low	$C_{22}H_{26}O_3$	/	I	338.45	Insoluble	Soluble in common organic solvents
368.	cis Permethrin[c]	52645-53-1	Low	$C_{21}H_{20}Cl_2O_3$	–Cl	I	391.11	0.07	Acetone, methanol, ethanol, diethyl ether, xylene and $CH_2Cl_2 > 500$
369.	Cyfluthrin[c]	68359-37-5	Low	$C_{23}H_{18}Cl_2FNO_3$	–Cl	I	434.12	Slightly soluble	Diffluent in acetone, toluene and CH_2Cl_2
370.	gamma-Cyhalothrin[c]	91465-08-6	Medium	$C_{23}H_{19}F_3ClNO_3$	–F	I/A	449.9	Insoluble	Acetone, methanol, ethyl acetate toluene >500
371.	lambda-Cyhalothrin[c]	91465-08-6	Medium	$C_{23}H_{19}ClF_3NO_3$	–F	I	449.68	Insoluble	Soluble in most solvents
372.	Cypermethrin[c]	52315-07-8	Medium	$C_{22}H_{19}Cl_2NO_3$	–Cl	I	416.3	0.2	Acetone, ethanol, xylene and chloroform 450
373.	Cyphenothrin[c]	39515-40-7	Medium	$C_{24}H_{25}NO_3$	/	I	375.5	0.01	Hexane, methanol and xylene 500
374.	Cycloprothrin[c]	6993-38-6	Low	$C_{26}H_{21}Cl_2NO_4$	–Cl	I	482.4	0.091	Soluble in most organic solvents, but indissolvable in fatty hydrocarbon
375.	Deltamethrin[c]	52918-63-5	Medium	$C_{12}H_{19}Br_2NO_3$	–Br	I	505.1	0.002	Acetone 500, benzene 450, ethyl acetate 300, xylene 250, ethanol 15
376.	S-Esfenvalerate[c]	66230-04-4	Medium	$C_{25}H_{22}ClNO_3$	–Cl	I	419.9	0.3	Acetone, ethyl acetate, chloroform, acetonitrile and xylene 600; methanol 100, hexane 50

Continued

No.	Compound	CAS	Toxicity	Formula	Characteristic Atom[1]	Effect[2]	MW	WS (mg/L)	Solubility of Organic Solvent (g/L)
377.	Etofenprox^c	80844-07-1	Low	$C_{25}H_{28}O_2$	/	-	360.5	1	Chloroform 9000, acetone 7800, ethyl acetate 6000, ethanol 150, methanol 66
378.	Fenpropathrin^c	39515-41-8	Medium	$C_{22}H_{23}NO_3$	-N	-	349.22	Insoluble	Soluble in cyclohexnae and xylene, etc.
379.	Fenvalarate^c	51630-58-1	Low	$C_{25}H_{22}ClNO_3$	-Cl	I/A	419.91	1	Soluble in most organic solvents >450
380.	Flucythrinate^c	70124-77-5	Medium	$C_{26}H_{23}F_2NO_4$	-F	-	451.26	0.5	Xylene 1810, acetone 820, hexane 90
381.	Flumethrin^c	69770-45-2	Low	$C_{28}H_{22}FCl_2NO_2$	-Cl	-	493.37	Insoluble	Soluble in toluene, acetone and cyclohexane etc.
382.	Imiprothrin^c	72693-72-5	Low	$C_{17}H_{22}N_2O_4$	-N	-	318.4	93.5	LogKow=2.9, soluble in various organic solvents
383.	Kadethrin^c	58769-20-3	Medium	$C_{23}H_{24}O_4S$	-S	-	396.51	Insoluble	Soluble in acetone, toluene and CH_2Cl_2
384.	Methothrin^c	34388-29-9	Low	$C_{19}H_{26}O_3$		-	302.4	Insoluble	Difluent in alcohol, acetone, benzene and toluene
385.	Permethrin^c	52645-53-1	Medium	$C_{21}H_{20}Cl_2O_3$	Cl	-	391.11	0.07	Acetone, ethanol, methanol, diethyl ether, xylene and CH_2Cl_2 > 500
386.	cis-Permethrin^c	74774-45-7	Medium	$C_{21}H_{20}Cl_2O_3$	Cl	-	391.11	0.07	
387.	trans-Permethrin^c	551877-74-8	Medium	$C_{21}H_{20}Cl_2O_3$	Cl	-	391.11	0.07	
388.	Phenothrin^c	26002-80-2	Low	$C_{23}H_{26}O_3$	/	-	350.5	2	Soluble in various solvents
389.	Prallethrin^c	23031-36-9	Low	$C_{19}H_{24}O_3$	/	-	300.4	8.03	Soluble in various organic solvents

No.	Name	CAS number		Molecular formula			Molecular weight	Water solubility	Solubility
390.	Pyrethrins[c]	8003-34-7	Low	$C_{21}H_{28}O_3$-$C_{22}H_{30}O_5$	/	I	328.43-360.43	Insoluble	Diffluent in ethanol and chlorinated hydrocarbon, etc.
391.	Resmethrin[c]	10453-86-8	Low	$C_{22}H_{26}O_3$	/	.I	338.45	500	Acetone and CH_2Cl_2 500, ethanol 80, isopropanol 70
392.	tau-Fluvalinate[c]	102851-06-9	Medium	$C_{26}H_{22}ClF_3N_2O_3$	–F	I	502.9	0.002	Acetone and chloroform 1000, methanol 760, miscible with aromatic hydrocarbon, diethyl ether and CH_2Cl_2
393.	Tefluthrin[c]	79538-32-2	High	$C_{17}H_{14}ClF_7O_2$	–F	I	418.7	0.02	Acetone, ethyl acetate, hexane, toluene and CH_2Cl_2 500, methanol 263
394.	Tetramethrin[c]	7696-12-0	Low	$C_{19}H_{25}NO_4$	–N	I	331.42	Insoluble	Benzene xylene 500, toluene and acetone 400, ethyl acetate 350, hexane 20
395.	Tralomethrin[c]	66841-25-6	Medium	$C_{22}H_{19}Br_4NO_3$	–Br	I	665.0	70	Acetone, toluene, xylene and CH_2Cl_2 1000, ethanol 180
396.	Transfluthrin[c]	118712-89-3	Low	$C_{15}H_{12}Cl_2F_4O_2$	–F	I	371.15	0.057	Soluble in various solvents >200
397.	zeta Cypermethrin[c]	52315-07-8	Medium	$C22H19Cl2NO3$	–Cl	I	416.3	0.2	Acetone, ethanol, xylene and chloroform 450
(5) Organonitrogen pesticide									
398.	6-Chloro-4hydroxy-3-phenyl-pyridazin[e]	40020-01-7		$C_{10}H_7ClN_2O$	–N	I/A	206.63	Insoluble	Soluble in various organic solvents
399.	Acetamiprid[e]	160430-64-8	Low	$C_{10}H_{11}ClN_4$	–N	I	222.68	4200	Soluble in acetone, methanol, ethanol, CH_2Cl_2 and chloroform, etc.

Continued

No.	Compound	CAS	Toxicity	Formula	Characteristic Atom[1]	Effect[2]	MW	WS (mg/L)	Solubility of Organic Solvent (g/L)
400.	Acetochlor^e	34256-82-1	Low	$C_{14}H_{20}ClNO_2$	-N	H	269.7	223	Soluble in acetone, benzene, ethanol and ethyl acetate, etc.
401.	Acibenzolar^e	135158-54-2	Low	$C_8H_6N_2OS_2$	-N	F	210.28	7.7	logKow=3.1, methanol 4.2, ethyl acetate 25, n-hexane 1.3, toluene 36, n-octyl alcohol 5.4, acetone 28, CH_2Cl_2 160
402.	Acrylamide^e	79-06-1	Medium	C_3H_5NO	-N	H	71.08	Insoluble	Soluble in various organic solvents
403.	Alachlor^e	15972-60-8	Low	$C_{14}H_{20}ClNO_2$	-Cl	H	269.7	148	Soluble in acetone, ethanol, ethyl acetate and diethyl ether
404.	Aldimorph^e	1704-28-5	Low	$C_{18}H_{37}NO$	-N	F	283.48	20	Soluble in various organic solvents
405.	Allidochlor^e	93-71-0	Medium	$C_8H_{12}ClNO$	Cl/N	H	173.63	1.97×10^4	Chloroform and ethanol 500, hexane and xylene 200
406.	Ametryn^e	834-12-8	Low	$C_9H_{17}N_5S$	-N	H	227.15	185	Soluble in organic solvents
407.	Amidosulfuron^e	120923-37-7	Low	$C_9H_{15}N_5O_7S_2$	-N	H	369.37	9	Hexane 0.1, acetone 8.1, toluene 0.256, CH_2Cl_2 6.9
408.	Amitraz^e	33089-61-1	Low	$C_{19}H_{23}N_3$	-N	I/A	293.19	1	Acetone and xylene 300
409.	Amitrole^e	61-82-5	Low	CH_4N_4	-N	H	72.01	2.8×10^5	Ethanol 26, insoluble in diethyl ether, acetone and nonpolar solvents
410.	Anilazine^e	101-05-3	Low	$C_4H_5N_4Cl_3$	Cl/N	F	215.39	Insoluble	Acetone 10, toluene 5, xylene 4

No.	Name		Molecular formula			Molecular weight	Water solubility	Soluble in organic solvents
411.	Atratone[e]	Low	$C_9H_{17}N_5O$	-N	H	211.27	1800	Soluble in organic solvents
412.	Atrazine[e]	Low	$C_8H_{14}ClN_5$	N	H	215.69	33	Chloroform 52, ethyl acetate 28, methanol 18, diethyl ether 12
413.	Atrazine,des ethyl[e]	Low	$C_6H_{10}ClN_5$	-N	H	187.68	Insoluble	Soluble in various organic solvents
414.	Azaconazole[e]	Medium	$C_{12}H_{11}Cl_3N_3O_2$	-N	F	300.1	300	Acetone 160, methanol 150, toluene 79, n-hexane 0.8
415.	Azimsulfuron[e]	Low	$C_{13}H_{16}N_{10}O_5S$	-N	H	424.4	1050	logKow=0.646, acetonitrile 0.014, acetone 0.026, methanol 0.002, toluene 0.002, n-hexan 0.0002, ethyl acetate 0.0013, CH_2Cl_2 0.066
416.	Aziprotryne[e]	Low	$C_7H_{11}N_7S$	-N	H	225.3	55	Soluble in various organic solvents
417.	Azocyclotin[e]	Medium	$C_{12}H_{35}N_3Sn$	-N	A	436.21	1	Toluene, CH_2Cl_2 and isopropanol 10
418.	Azoxystrobin[e]	Low	$C_{22}H_{17}N_3O_5$	-N	F	403.4	60	Difiluent in ethyl acetate and CH_2Cl_2;Soluble in methanol and toluene; Slightly soluble in hexane and n-octyl alcohol
419.	Benalaxyl[e]	Low	$C_{20}H_{23}NO_3$	-N	F	325.4	37	Acetone, chloroform and CH_2Cl_2 500, xylene 300, hexane 50
420.	Benfluralin[e]	Low	$C_{13}H_{16}F_3N_3O_4$	F/N	H	335.13	70	Acetone 650, xylene 420, ethanol 24

Continued

No.	Compound	CAS	Toxicity	Formula	Characteristic Atom[1]	Effect[2]	MW	WS (mg/L)	Solubility of Organic Solvent (g/L)
421.	Benodanil^e	15310-01-7	Low	$C_{13}H_{10}INO$	-N	F	323.1	20	Acetone 401, ethyl acetate 120, chloroform 77
422.	Benomyl^e	17804-35-2	Low	$C_{14}H_{18}N_4O_3$	-N	F	290.14	3.8	Chloroform 90
423.	Benoxacor^e	98730-04-2	Low	$C_{11}H_{11}Cl_3NO_2$	-N	H	260.1	20	logKow=2.70; CH_2Cl_2 400, cyclohexanone 300, acetone 230, toluene 90, methanol 3
424.	Bensulfuron-methyl^e	83055-99-6	Low	$C_{16}H_{18}N_4O_7$	-N	H	410.4	120	CH_2Cl_2 11.72, acetonitrile 5.38, ethyl acetate 1.66, acetone 1.38
425.	Benzofenap^e	82692-44-2	Low	$C_{22}H_{20}Cl_3N_2O_3$	-N	H	431.3	0.13	log Kow=4.69; chloroform 920, acetone 73, xylene 90, hexane 0.46
426.	Bifenazate^e	149877-41-8	Low	$C_{17}H_{20}N_2O_3$	-N	I	300.35	3.76	Soluble in various organic solvents
427.	Benzyladenine^e	1214-39-7	Low	$C_{12}H_{11}N_5$	-N	PGR	225.2	Indissolvable	Indissolvable in common solvents, soluble in hot ethanol
428.	Bitertanol^e	55179-31-2	Low	$C_{20}H_{23}N_3O_2$	-N	F	337.4	8	Diffluent in isopropanol, toluene and CH_2Cl_2
429.	Boscalid^e	188425-85-6	Low	$C_{18}H_{12}Cl_2N_2O_2$	-N	F	343.21	4.6	Acetone 160-200, methanol 40-50, n-heptane 正庚烷 <10
430.	Bromacil^e	314-40-9	Low	$C_9H_{13}BrN_2O_2$	-N	H	261.1	815	Acetone 167, ethanol 134, acetonitrile 71, xylene 32
431.	Bromobutide^e	74712-19-9	Low	$C_{15}H_{22}BrNO$	-N	H	312.3	3.54	Toluene 35, xylene 4.7, hexane 0.5

No.	Name	CAS	Level	Formula		Type	MW	Sol.	Solubility
432.	Bromuconazole[e]	116255-48-2	Medium	$C_{13}H_{12}Cl_2BrN_3O$	–N	F	377.13	50	Slightly soluble in organic solvents
433.	Brompyrazon[e]	3042-84-0	Low	$C_{10}H_8BrN_3O$	–Br	H	266.2	Insoluble	Soluble in various organic solvents
434.	Bupirimate[e]	57839-19-1	Low	$C_{13}H_{24}N_4O_3S$	–N	F	316.43	22	logKow=3.9, soluble in most organic solvents
435.	Buprofezin[e]	69327-76-0	Low	$C_{16}H_{23}N_3O_3$	–N	I	305.4	0.9	Benzene 370, toluene 320, acetone 240
436.	Butachlor[e]	23184-66-9	Low	$C_{17}H_{26}Cl_2NO_2$	Cl/N	H	311.5	23	Diffluent in aromatic hydrocarbon and ketone, soluble in acetone, diethyl ether, benzene, ethanol and ethyl acetate
437.	Butafenacil[e]	134605-64-4	Low	$C_{20}H_{18}F_3ClN_2O_6$	–F	H	474.82	10	Kow3.2
438.	Butralin[e]	33629-47-9	Low	$C_{14}H_{22}N_3O_4$	–N	PGR	296.14	1	Methanol 125, acetone 4.48, benzene 2.7, xylene 3.88
439.	Buturon[e]	3766-60-7	Low	$C_{12}H_{13}ClN_2O$	–N	H	236.7	30	Acetone 279, methanol 128, benzene 9.8
440.	Butylate[e]	2008-41-5	Low	$C_{11}H_{23}NOS$	N/S	H	217.38	45	Miscible with acetone, xylene and ethanol
441.	Cafenstrole[e]	125306-83-4	Low	$C_{16}H_{22}N_4O_3S$	–N	H	350.45	2.5	logKow=3.21
442.	Captafol[e]	2425-06-1	Low	$C_{10}H_9Cl_4NO_2S$	–Cl	F	348.96	1.4	Toluene 400
443.	Captan[e]	133-06-2	Low	$C_9H_8Cl_3NO_2S$	–Cl	F	300.61	3.3	Chloroform 70, acetone 21, xylene 20
444.	Carbendazim[e]	10605-21-7	Low	$C_9H_9N_3O_2$	–N	F	191.19	5.8	Acetone and ethanol 0.3, CH_2Cl_2 0.068, benzene 0.036

Continued

No.	Compound	CAS	Toxicity	Formula	Characteristic Atom[1]	Effect[2]	MW	WS (mg/L)	Solubility of Organic Solvent (g/L)
445.	Carfentrazone-ethyl[e]	128621-72-7	Low	$C_{15}H_{14}F_3Cl_2N_3O_3$	–F	H	412.2	0.012	Soluble in methanol and acetone, etc.
446.	Chloramphenicol[e]	2787-09-9	Low	$C_{11}H_{12}Cl_2N_2O_7$	Cl/N	F	355.01	2500	Diffluent in ethanol, acetone and diethyl ether
447.	Chlorbenzuron[e]	57160-47-1	Low	$C_{14}H_{10}Cl_2N_2O_2$	Cl/N	I	309.1	0.17	Acetone and chloroform 5, CH_2Cl_2 1, benzene and toluene 0.5, methanol and ethanol 0.3
448.	Chlordimeform[e]	6164-98-3	Medium	$C_{10}H_{13}ClN$	–N	I	182.55	250	Acetone, benzene, chloroform, ethyl acetate, methanol and hexane >200
449.	Chlorfluazuron[e]	71422-67-8	Low	$C_{20}H_8F_5Cl_3N_3O_3$	Cl/N	I	539.58	0.017	Methanol 2.6, acetone 55, xylene 4.2, CH_2Cl_2 15.9
450.	Chlorfenapyr[e]	122453-73-0	Medium	$C_{15}H_{11}BrClF_3N_2O$	F	I/A	407.6	Insoluble	Soluble in acetone, diethyl ether, acetonitrile, dimethylsulfoxide and alcohol, etc.
451.	Chlorotoluron[e]	15545-48-9	Low	$C_{10}H_{13}ClN_2O$	–N	H	212.55	70	Acetone 50, C_2Cl_2 43, benzene 24
452.	Chlorimuron-ethyl[e]	90982-32-4	Low	$C_{15}H_{15}ClN_4O_6S$	–N	H	414.8	1200	logKow=0.11
453.	Chlorbromuron[e]	13360-45-7	Low	$C_9H_{10}BrClN_2O_2$	–Cl	H	293.5	35	Soluble in acetone and dimethylformamide, slightly soluble in acetone
454.	Chlordimeform hydrochloride[e]	19750-95-9	Medium	$C_{10}H_{13}Cl_2N_2$	Cl/N	A/I	196.68	5×10^5	Methanol 300, chloroform 10

No.	Name	CAS		Formula			MW		Solubility
455.	Chloridazon[e]	1968-60-8	Low	$C_{10}H_8ClN_3O$	–N	H	221.56	400	Soluble in various organic solvents
456.	Chloroxuron[e]	1982-47-4	Low	$C_{15}H_{15}ClN_2O_2$	–N	H	290.75	3.7	Soluble in acetone and chloroform, slightly soluble in benzene and ethanol
457.	Chlorsulfuron[e]	64902-72-3	Low	$C_{12}H_{12}ClN_5O_4S$	–N	H	357.8	2.7×10^4	Slightly soluble in acetone, methanol and CH_2Cl_2
458.	Chlortoluron[e]	15545-48-9	Low	$C_{10}H_{13}ClN_2O$	–N	H	213.55	70	Acetone 50, CH_2Cl_2 43, benzene 24
459.	Chlorthiamid[e]	1918-13-4	Low	$C_7H_5Cl_2NS$	–N	H	206.1	950	Aromatic hydrocarbon and chlorinated hydrocarbon 50-100
460.	Cinosulfuron[e]	94593-91-6	Low	$C_{15}H_9N_5O_7S$	–N	H	413.42	3700	CH_2Cl_2 9.5, acetone 36, ethanol 19, toluene 5.4
461.	Clethodim[e]	99129-21-2	Low	$C_{17}H_{26}ClNO_3S$	–N	H	359.9	Insoluble	Soluble in various organic solvents
462.	Clethodim sulfone[e]		Low	$C_{17}H_{26}ClNO_5S$	–Cl	H	391.9	Insoluble	Soluble in various organic solvents
463.	Clethodim sulfoxide[e]		Low	$C_{17}H_{26}ClNO_4S$	–Cl	H	375.9	Insoluble	Soluble in various organic solvents
464.	Clethodim-imin sulfone[e]		Low	$C_{17}H_{26}ClNO_5S$	–Cl	H	376.9	Insoluble	Soluble in various organic solvents
465.	Clodinafop-propargyl[e]	105512-06-9	Low	$C_{17}H_{13}ClFNO_4$	–F	H	349.62	4	Acetone 500, methanol 180, hexane 7.5, n-octyl alcohol 21
466.	Clofentezine[e]	74115-24-5	Low	$C_{14}H_8Cl_2N_4$	–N	A	303.1	Indissolvable	Chloroform 50, acetone 9.3, benzene 2.5, ethanol 0.5

Continued

No.	Compound	CAS	Toxicity	Formula	Characteristic Atom[1]	Effect[2]	MW	WS (mg/L)	Solubility of Organic Solvent (g/L)
467.	Clomazone[e]	81777-89-1	Low	$C_{12}H_{14}ClNO_2$	-N	H	239.58	1100	Soluble in various solvents
468.	Clomeprop[e]	84496-56-0	Low	$C_{16}H_{15}Cl_2NO_2$	-N	H	324.2	0.032	Acetone 33, cyclohexane 9, dimethylformamide 20, xylene 17
469.	Clopyralid[e]	1702-17-6	Low	$C_6H_3Cl_2N_2O_3$	-Cl	H	222.09	1000	Acetone 153, acetonitrile 121, hexane 6, methanol 104
470.	Cloransulam-methyl[e]	147150-35-4	Low	$C_{15}H_{13}ClFN_5O_5S$	-N	H	429.8	184	logKow=0.268,
471.	Crimidine[e]	535-89-7	High	$C_7H_{10}ClN_3$	-N	R	171.64	Insoluble	Soluble in acetone, benzene, chloroform and diethyl ether
472.	Cumyluron[e]	99485-76-4	Low	$C_{17}H_{18}ClN_2$	N	H	302.8	1	Methanol 21.5, ethanol 19.4, acetone 14.5, acetonitrile 7.5, benzene 0.8, hexane 0.4
473.	Cyazofamid[e]	120116-88-3	Low	$C_{13}H_{12}ClN_4O_2S$	-N	F	324.78	0.121	logKow=3.2
474.	Cyanazine[e]	21725-46-2	Medium	$C_9H_{13}ClN_6$	-N	H	240.7	171	Chloroform 210, ethanol 45, benzene and hexane 15
475.	Cyclanilide[e]	113136-77-9	Medium	$C_{11}H_9Cl_2NO_3$	-Cl	H	274.1	Insoluble	logKow=3.25
476.	Cyclosulfamuron[e]	136849-15-5	Low	$C_{17}H_{19}N_5O_6S$	-N	H	421.4	6.52	logKow=1.41,
477.	Cycluron[e]	2163-69-1	Low	$C_{11}H_{22}N_2O$	-H	H	198.31	1000	Methanol 500, acetone 67, benzene 55
478.	Cymoxanil[e]	57966-95-7	Low	$C_7H_{10}N_4O_3$	-N	F	198.18	100	Acetone 10.5, CH_2Cl_2 10.3, methanol 4.1, benzene 0.2, hexane 0.1

479.	Cyprazine[e]	22936-86-3	Low	C$_9$H$_{13}$ClN$_5$	–N	H	226.54	6.9	Soluble in acetone, slightly soluble in chloroform and methanol
480.	Cyproconazole[e]	113096-99-4	Low	C$_{15}$H$_{18}$ClN$_3$O	–N	F	291.8	1400	Acetone 2300, dimethyl sulfoxide 180, xylene 200
481.	Cyprodinil[e]	121552-61-2	Low	C$_{14}$H$_{15}$N$_3$	–N	F	225.3	13	Ethanol 160, acetone 610, toluene 460, n-octyl alcohol 160, n-hexane 30
482.	Cyromazine[e]	66215-27-8	Low	C$_6$H$_{10}$N$_6$	–N	I	166.2	1.3×10^4	Methanol 22, acetone 1.7, CH$_2$Cl$_2$ 0.025, toluene 0.015, hexane 0.0002
483.	Dazomet[e]	533-74-4	Medium	C$_5$H$_8$N$_2$S$_2$	N/S	F/N	160.17	3000	Chloroform 391, acetone 173, benzene 51, ethanol 15
484.	Desamino-metamitron[e]	41394-05-2	Low	C$_{10}$H$_8$N$_3$	–N	H	186.2	Insoluble	Soluble in various organic solvents
485.	Desethyl-sebuthylazine[e]	37019-18-4	Low	C$_7$H$_{12}$ClN$_5$	–N	H	201.49	Insoluble	Soluble in various organic solvents
486.	Desisopropyl-atrazine[e]	1007-28-9	Low	C$_5$H$_8$ClN$_5$	–N	H	173.5	Insoluble	Soluble in various organic solvents
487.	Desmethyl-norflurazon[e]	23576-24-1	Low	C$_{11}$H$_7$ClF$_3$N$_3$O	–F	H	289.56	Insoluble	Soluble in various organic solvents
488.	Desmetryne[e]	1014-69-3	Low	C$_8$H$_{15}$N$_5$S	–N	H	213.31	600	Soluble in most organic solvents
489.	Diafenthiuron[e]	80060-09-9	Low	C$_{23}$H$_{32}$N$_2$OS	–N	I/A	384.6	0.05	CH$_2$Cl$_2$ 600, cyclohexanone 380, toluene 320, acetone 280, xylene 210, hexane 8

Continued

No.	Compound	CAS	Toxicity	Formula	Characteristic Atom[1]	Effect[2]	MW	WS (mg/L)	Solubility of Organic Solvent (g/L)
490.	2,6-Dichlorobenza mide[e]	2008-58-4		$C_7H_5Cl_2NO$	–N	F	189.97	Insoluble	Soluble in various organic solvents
491.	Dichloran[e]	99-30-9	Low	$C_6H_4Cl_2N_2O_2$	–N	F	207.06	Insoluble	Dioxane 40, acetone 34, ethyl acetate 19, chloroform 12, benzene 5.3, ethanol 2
492.	Dichlormid[e]	37764-25-3	Low	$C_8H_{11}Cl_2NO$	–N	H	208.09	500	Kerosene 0.15l, miscible with various organic solvents such as acetone, ethanol and xylene, etc.
493.	Diclobutrazol[e]	75736-33-3	Low	$C_{13}H_{17}Cl_2N_3O$	–N	F	302.03	Insoluble	Soluble in various organic solvents
494.	Diclosulam[e]	145701-21-9	Low	$C_{13}H_{10}Cl_2FN_5O_3S$	–N	H	406.2	0.006	logKow=0.85, acetone 7.97, acetonitrile 4.59, CH_2Cl_2 2.17, ethyl acetate 1.45, methanol 0.81, toluene 0.588
495.	Diethyltoluamide[e]	134-62-3	Low	$C_{12}H_{17}NO$	–N	I	191.12	Insoluble	Miscible with ethanol, benzene and diethyl ether
496.	Difenoconazole[e]	1977-6-54	Low	$C_{19}H_{17}Cl_2N_3O_3$	–N	F	406.3	3.3	Diffluent in solvents
497.	Difenoxuron[e]	14214-32-5	Low	$C_{16}H_{18}N_2O_3$	–N	H	286.33	20	CH_2Cl_2 156, acetone 63, isopropanol 10
498.	Difenzoquat-methyl sulfate[e]	43222-48-6	Medium	$C_{18}H_{20}N_2O_4S$	–N	H	360.44	7.6×10^5	Xylene 0.1
499.	Diflubenzuron[e]	35367-38-5	Low	$C_{14}H_9F_2ClN_2O_2$	F/N	IGR	310.7	0.2	Acetone 650, acetonitrile 200, methanol 100

No.	Name	CAS		Formula	F/N		MW	Solubility	Solubility notes
500.	Diflufenican[e]	83164-33-4	Low	$C_{19}H_{11}F_5N_2O_2$	F/N	H	394.3	0.05	Acetone and dimethylformamide 100, xylene 20, cyclohexanone 50
501.	Diflufenzopyr[e]	109293-97-2	Low	$C_{15}H_{12}F_2N_4O_3$	–N	H	394.3	5850	logKow=0.037,
502.	Dimethachlor[e]	51218-45-2	Low	$C_{13}H_{18}Cl_5NO_2$	–Cl	H	255.74	50	Soluble in polar solvents
503.	Dimethomorph[e]	211867-47-9	Low	$C_{21}H_{22}ClNO_4$	–N	F	387.9	50	Acetone 15, CH_2Cl_2 315
504.	Dimethenamid[e]	87674-68-8	Low	$C_{12}H_{18}ClNO_2S$	–N	H	275.79	1200	n-Heptane 282, isooctanol 220, diethyl ether and ethanol > 500
505.	Dimethylaminosulfanilide[e]	1596-84-5	Low	$C_6H_{12}N_2O_3$	–N	PGR	160.17	1×10^5	Methanol 50, acetone 25. Insoluble in hydrocarbon
506.	Dimefuron[e]	34205-21-5	Low	$C_{13}H_{19}ClN_4O_3$	–N	H	314.58	Insoluble	Soluble in various solvents
507.	Dimethametryn[e]	14214-32-5	Low	$C_{12}H_{20}N_5S$	–N	H	266.38	50	Soluble in polar solvents
508.	Dimethirimol	5221-53-4	Low	$C_{11}H_{19}N_3O$	–N	F	209.3	1200	Chloroform 1200, xylene 360, ethanol 65, acetone 45
509.	Dimuron[e]	42609-52-9	Low	$C_{17}H_{20}N_2O$	–N	H	268.36	1.3	Methanol and ethanol 10, diethyl ether 100, dimethylformamide 182, dimethyl sulfoxide 200
510.	Diniconazole[e]	83657-24-3	Medium	$C_{14}H_6Cl_2N_3O$	–N	F	303.04		Soluble in various solvents
511.	Dinitramine[e]	29091-05-2	Low	$C_{11}H_{13}F_3N_3O_4$	–N	H	314.58	1.1	Acetone 640, benzene 490, chloroform 600, hexane 140
512.	Dinocap[e]	39300-45-3	Medium	$C_{18}H_{24}N_2O_6$	–N	A/F	364.41	Insoluble	Soluble in various organic solvents

Continued

No.	Compound	CAS	Toxicity	Formula	Characteristic Atom[1]	Effect[2]	MW	WS (mg/L)	Solubility of Organic Solvent (g/L)
513.	Dinoseb[e]	88-85-7	High	$C_{10}H_{12}N_2O_5$	–N	A/H	240.42	52	Ethanol 480, hexane 270, miscible with diethyl ether, toluene and xylene
514.	Dinotefuran[e]	165252-70-0	High	$C_7H_{14}N_4O_3$	–N	I	202.19	Insoluble	Soluble in various organic solvents
515.	Dinoterb[e]	1420-07-1	High	$C_{10}H_{12}N_2O_5$	–N	H	240.42	Insoluble	Ethyl acetate, cyclohexanone and dimethyl sulfoxide 200, glycol and fatty hydrocarbon 100, soluble in alcohol
516.	Diphenamid[e]	957-51-7	Low	$C_{16}H_{17}NO$	–N	H	239.3	260	Acetone 198, xylene 50
517.	Diphenylamine[e]	122-39-4	Medium	$C_{12}H_{11}N$	–N	I	169.12	Insoluble	Soluble in acetone, ethanol, benzene and diethyl ether
518.	Dipropetryn[e]	4147-51-7	Low	$C_{11}H_{21}N_5S$	–N	H	255.17	16	Acetone and benzene 540, ethanol 180, xylene 220
519.	Diquat[e]	6385-62-2	Medium	$C_{12}H_{12}Br_2N_2$	–N	H	344.06	7×10^5	Slightly soluble in ethanol, insoluble in nonpolar solvents
520.	Diuron[e]	330-54-1	Low	$C_9H_{10}Cl_2N_2O$	N/Cl	H	233.10	42	Acetone 53, benzene 1.4
521.	DNOC[e]	534-52-1	High	$C_7H_6N_2O_5$	–N	H	198.13	130	Soluble in various organic solvents, chloroform 372, ethanol 43
522.	Dodemorph[e]	1593-77-7	Low	$C_{18}H_{35}NO$	–N	F	281.5	100	Chloroform >1000, ethyl acetate 185, acetone 57, ethanol 50

No.	Name	CAS number		Molecular formula			MW	Soluble in hot water	Soluble in other solvents
523.	Dodine[e]	2439-10-3	Low	$C_{15}H_{33}N_3O_2$	-N	F	287.44		Soluble in ethanol, slightly soluble in other solvents
524.	Epoxiconazole[e]	106325-08-0	Low	$C_{17}H_{13}ClFN_3O$	-N	F	329.76	6.63	logKow=3.1
525.	Etaconazole[e]	1000290-09-5	Low	$C_{14}H_{15}Cl_2N_3O_2$	-N	F	328	Indissolvable	Diffluent in organic solvents
526.	Ethalfluralin[e]	55283-68-6	Low	$C_{13}H_{14}F_3N_3O_4$	-F	H	333.13	0.2	Acetone, acetonitrile, chloroform and xylene 500, methanol 100
527.	Ethametsulfuron-methyl[e]	97780-06-8	Low	$C_{15}H_{18}N_6O_6S$	-N	H	410.4	50	CH_2Cl_2 3.9, acetone 1.6, methanol 0.35, ethyl acetate 0.68, acetonitrile 0.8
528.	Ethidimuron[e]	30043-49-3	Low	$C_7H_{12}N_4O_3S_2$	-N	H	264.33	2960	CH_2Cl_2 0.02
529.	Ethirimol[e]	23947-60-6	Low	$C_{11}H_{19}N_3O$	-N	F	209.3	200	Chloroform 150, ethanol 24, acetone 5
530.	Ethoxysulfuron[e]	126801-58-9	Low	$C_{15}H_{18}N_4O_7S$	-N	H	398.4	1353	logKow=0.004
531.	Etobenzanid[e]	79540-50-4	Low	$C_{16}H_{15}Cl_2NO_3$	-N	H	340	0.92	Acetone >100, methanol 22.4, n-hexane 2.42
532.	Etridiazole[e]	2593-15-9	Low	$C_5H_5Cl_3N_2OS$	-Cl	F	247.53	50	Soluble in acetone, diethyl ether, ethanol, benzene and xylene
533.	Famoxadone[e]	131807-57-3	Low	$C_{22}H_{18}N_2O_4$	-N	F	374.4	52	logKow=4.65
534.	Fenamidone[e]	161326-34-7	Low	$C_{17}H_{17}N_3O_2S$	-N	F	311.4	7.8	logKow=2.8
535.	fenaminosulf[e]	140-56-7	Medium	$C_8H_{10}N_3O_3SNa$	-N	F	251.14	3000	Soluble in methanol, insoluble in diethyl ether and benzene

Continued

No.	Compound	CAS	Toxicity	Formula	Characteristic Atom[1]	Effect[2]	MW	WS (mg/L)	Solubility of Organic Solvent (g/L)
536.	Fenarimol[e]	60168-88-9	Low	$C_{17}H_{12}Cl_2N_2O$	Cl/N	F	331.20	13.7	Diffluent in acetone, acetonitrile, benzene, chloroform and methanol, slightly soluble in hexane
537.	Fenazaflor[e]	14255-88-0	Medium	$C_{15}H_7Cl_2F_3N_2O_2$	–N	A	375.14	1	Slightly soluble in organic solvents except for acetone, benzene and dioxane
538.	Fenazaquin[e]	120928-09-8	Medium	$C_{20}H_{22}N_2O$	–N	A	306.4	0.22	Chloroform 500, acetone 400, ethanol and toluene 50, acetonitrile and toluene 33
539.	Fenbuconazole[e]	114369-43-6	Low	$C_{19}H_{17}ClN_4$	–N	F	336.8	0.2	Soluble in acetone, ethanol and diethyl ether, insoluble in fatty hydrocarbon
540.	Fenfuram[e]	24691-80-3	Low	$C_{12}H_{11}NO_2$	–N	F	201.22	100	Cyclohexanone 340, acetone 300, methanol 145
541.	Fenhexamid[e]	126833-17-8	Low	$C_{14}H_{17}Cl_2NO_2$	–Cl	F	302.2	20	logKow=3.51, soluble in various organic solvents
542.	Fenoxanil[e]	115852-48-7	Low	$C_{15}H_{18}Cl_2N_2O_2$	–N	F	329.23	30.7	Soluble in various organic solvents
543.	Fenoxaprop-ethyl[e]	66441-23-4	Low	$C_{18}H_{16}ClNO_5$	–N	H	361.8	0.9	Diffluent in acetone, toluene and ethyl acetate, soluble in ethanol, cyclohexane and octanol
544.	Fenoxaprop-P-ethyl[e]	66441-23-4	Low	$C_{18}H_{16}ClNO_5$	Cl/N	H	361.8	0.7	Acetone 200, toluene 200, ethyl acetate > 200, ethanol 24

No.	Name	CAS	Mobility	Formula			MW	Solubility	Solvents
545.	Fenpiclonil[e]	74738-17-3	Low	$C_{11}H_6Cl_2N_2$	–N	F	237.1	4.8	logKow=3.86, ethanol 73, acetone 360, toluene 7.2, n-hexane 0.26, n-octyl alcohol 41
546.	Fenpropidin[e]	67306-00-7	Low	$C_{19}H_{31}N$	–N	F	273.5	350	Acetone, chloroform, ethanol, ethyl acetate, heptane and xylene >250
547.	Fenpropimorph[e]	67306-03-0	Low	$C_{20}H_{33}NO$	–N	F	303.5	4.3	Acetone, chloroform, cyclohexane, diethyl ether, ethanol, ethyl acetate and toluene >1000
548.	Fenpyroximate[e]	134098-61-6	Medium	$C_{24}H_{27}N_3O_4$	–N	A	421.5	0.015	Diffluent in chloroform, benzene, toluene, acetone and methanol, etc.
549.	Fentrazamide[e]	158237-07-1	Low	$C_{16}H_{20}ClN_5O_2$	–N	H	349.8	2.3	LogKow=4.01,isopropanol 32,xylene 250
550.	Fenuron[e]	101-42-8	Low	$C_9H_{12}N_2O$	–N	H	164.21	3850	Diffluent in alcohol, ketone and halohydrocarbon, indissolvable in alkane
551.	Fipronil[e]	120068-37-3	Medium	$C_{12}H_{14}F_6Cl_2N_4OS$	–F	I	437.2	1.9	Acetone 546, CH_2Cl_2 22.3, methanol 137.5, hexane and toluene 30
552.	Flamprop-isopropyl[e]	52756-22-6	Low	$C_{19}H_{19}ClFNO_3$	–Cl	H	363.82	18	Acetone and xylene 500
553.	Flamprop-methyl[e]	52756-25-9	Low	$C_{17}H_{15}ClFNO_3$	–Cl	H	335.77	35	Acetone 500, xylene 258, ethanol 135, hexane 7
554.	Flazasulfuron[e]	104040-78-0	Low	$C_{13}H_{12}F_3N_5O_5S$	–N	H	407.3	2100	Methanol 4.2, acetone 22.7, acetonitrile 8.7, toluene 0.56

Continued

No.	Compound	CAS	Toxicity	Formula	Characteristic Atom[1]	Effect[2]	MW	WS (mg/L)	Solubility of Organic Solvent (g/L)
555.	Florasulam[e]	145701-23-1	Low	$C_{12}H_8F_3N_5O_3S$	-N	H	359.3	6360	logKow=-1.22
556.	Fluazifop-P-butyl[e]	79241-46-6	Low	$C_9H_{20}F_3NO_4$	-F	H	383.4	1	Miscible with methanol, acetone, hexane, xylene and CH_2Cl_2
557.	Fluazinam[e]	79622-59-6	Low	$C_{13}H_4Cl_2F_6N_4O_4$	-F	F	465.1	1.7	Ethyl acetate 680, acetone 470, toluene 410, CH_2Cl_2 330, ethanol 150, hexane 12
558.	Fluazifop-butyl[e]	79241-46-6	Low	$C_{19}H_{20}F_3NO_4$	-N	H	383.36	2	Miscible with toluene, acetone, cyclohexanone, hexane, xylene and CH_2Cl_2
559.	Flubenzimine[e]	37893-02-0	Low	$C_{17}H_{10}F_6N_4S$	-F	A	424.18	1.6	CH_2Cl_2 0.2, toluene 0.01
560.	Flucarbazone-sodium[e]	181274-17-9	Low	$C_{12}H_{10}F_3N_4NaO_6S$	-N	H	418.27	44000	logKow=-1.85
561.	Fluchloralin[e]	33245-39-5	Low	$C_{12}H_{13}ClF_3N_3O_4$	F/N	H	355.7	1	Acetone, benzene, chloroform, diethyl ether and ethyl acetate > 1000, ethanol 177
562.	Fludioxonil[e]	131341-86-1	Low	$C_{12}H_6F_2N_2O_2$	-F	F	248.2	1.8	Acetone 0.19, methanol 0.044, hexane 0.0078(ppm)
563.	Flufenacet[e]	142459-58-3	Low	$C_{14}H_{13}F_4N_3O_2S$	-N	H	363.3	56	Acetone, dimethylformamide, CH_2Cl_2 and toluene > 200, hexane 8.7
564.	Flufenoxuron[e]	101463-69-8	Low	$C_{21}H_{11}ClF_6N_2O_3$	-F	I/A	488.66	0.04	Acetone 82, CH_2Cl_2 24, xylene 6

No.	Name	CAS		Formula		PGR	MW	Solubility (water)	Solubility
565.	Flumetralin[e]	62924-70-3	Low	C₁₆H₁₂ClF₄N₃O₄	–F	PGR	421.61	0.1	CH₂Cl₂ 800, benzene 550, methanol 250, hexane 13
566.	Flumetsulam[e]	98967-40-9	Low	C₁₂H₉F₂N₅O₂S	–N	H	325.3	5600	Acetone < 0.016, methanol < 0.04, insoluble in toluene and hexane
567.	Flumiclorac-pentyl[e]	87546-18-7	Low	C₂₁H₂₃ClFNO₅	–F	H	423.9	0.189	Acetone 590, methanol 47.8, n-octyl alcohol 16, hexane 3.28
568.	Flumioxazin[e]	103361-09-7	Low	C₁₉H₁₅FN₂O₄	–N	H	354.34	1790	Soluble in common organic solvents
569.	Fluridone[e]	59756-60-4	Low	C₁₉H₁₄F₃NO	–F	H	329.3	12	Methanol, chloroform and diethyl ether > 10, ethyl acetate > 5, n-hexane < 0.5
570.	Fluroxypyr[e]	69377-81-7	Low	C₇H₅Cl₂FN₂O₃	–Cl	H	254.97	91	Acetone 41.6
571.	Fluometuron[e]	2164-17-2	Low	C₁₀H₁₀F₃N₂O	–F	H	231.2	105	Soluble in acetone, dimethylformamide, isopropanol, ethanol and propanediol, etc.
572.	Fluotrimazole[e]	31251-03-3	Low	C₂₁H₁₅F₃N₃	–N	F	366.21	0.0015	CH₂Cl₂ 400, cyclopropanone 200, toluene 100, acetone 50
573.	Fluquinconazole[e]	136426-54-5	Medium	C₁₆H₈Cl₂FN₅O	–Cl	F	376.17	1	Ethanol 3, acetone 50, toluene 10, dimethylsulfoxide 200
574.	Fluridone[e]		Low	C₁₈H₁₄F₃NO	–F	H	317.31	12	Methanol and chloroform > 10, ethyl acetate > 0.5, diethyl ether > 1, acetone < 0.5

Continued

No.	Compound	CAS	Toxicity	Formula	Characteristic Atom[1]	Effect[2]	MW	WS (mg/L)	Solubility of Organic Solvent (g/L)
575.	Flurochloridone^e	61213-25-0	Low	$C_{12}H_{10}Cl_2F_3NO$	–F	H	312.1	28	Acetone, xylene and ethanol 100-150
576.	Flurtamone^e	96525-23-4	Medium	$C_{18}H_{14}F_3NO_2$	–F	H	317.3	35	Soluble in acetone, methanol and CH_2Cl_2, slightly soluble in isopropanol
577.	Flutolanil^e	66332-96-5	Low	$C_{18}H_{16}F_3NO_2$	–F	F	323.17	9.6	Acetone 642, methanol 480, chloroform 341, toluene 56, xylene 29
578.	Fluroxypy 1-methylheptyl^e	81406-37-3	Low	$C_8H_7Cl_2FN_2O_3$	–Cl	H	369.2	0.9	CH_2Cl_2 896, acetone 867, ethyl acetate 792, toluene 735, xylene 642, methanol 469, hexane 45
579.	Flupyrsulfuron-methyl-sodium^e	144740-54-5	Low	$C_{15}H_{13}F_3N_5NaO_7S$	–N	H	487.3	63	logKow=0.10, acetonitrile 4.3, acetone 3, ethyl acetate 0.49, CH_2Cl_2 0.6, n-hexane 0.001
580.	Fluquinconazole^e	607-68-1	Low	$C_{16}H_8Cl_2FN_5O$	–Cl	F	376.17	Insoluble	Soluble in various organic solvents
581.	Flusilazole^e	85509-19-9	Low	$C_{16}H_{15}F_2N_3Si$	–F	F	315.25	45	Acetone and petroleum ether, etc. >2000
582.	Flusulfamide^e	106917-52-6	Medium	$C_{13}H_7F_3Cl_2N_2O_4S$	–F	F	415.2	2.9	Tetrahydrofunan 592, acetone 314, methanol 24, xylene 14
583.	Fluthiacet-methyl^e	117337-19-6	Low	$C_{15}H_{15}ClN_3O_3S_2$	–N	H	403.9	0.78	Methanol 4.4, acetone 100, toluene 84, acetonitrile 68.7, ethyl acetate 73.5

Continued

No.	Name	CAS		Formula			MW	Water solubility	Organic solubility
584.	Flutolanil[e]	66332-96-5	Low	$C_{17}H_{16}F_3NO_2$	-F	F	323.17	9.6	Acetone 642, methanol 480, chloroform 341, toluene 56, xylene 29
585.	Flutriafol[e]	76674-21-0	Low	$C_{16}H_{14}F_2N_3O$	-N	F	301.3	130	Acetone 190, CH_2Cl_2 150, methanol 69, xylene 12, hexane 0.3
586.	Folpet[e]	133-07-3	Low	$C_9H_4Cl_3NO_2S$	-Cl	F	296.56	1	Slightly soluble in organic solvents
587.	Fomesafen[e]	72178-02-0	Low	$C_{15}H_{10}F_3ClN_2O_6S$	-F	H	438.7	60%	Soluble in various organic solvents
588.	Formetanate hydrochloride[e]	23422-53-9	Low	$C_{11}H_{16}ClN_3O_2$	-N	A/I	257.72	1000	Methanol 200, acetone and chloroform 100
589.	Forchlorfenuron[e]	68157-60-8	Low	$C_{12}H_{10}ClN_3O$	-N	PGR	247.7	65	Diffluent in acetone, ethanol and dimethylsulfoxide
590.	Fuberidazole[e]	3878-19-1	Low	$C_{11}H_8N_2O$	-N	F	184.20	78	Isopropanol 50, CH_2Cl_2, xylene and petroleum ether 10
591.	Furalaxyl[e]	57646-30-7	Low	$C_{17}H_{19}NO_4$	-N	F	301.34	230	CH_2Cl_2 600, acetone 520, methanol 500, hexane 4
592.	Furmecyclox[e]	60568-05-0	Low	$C_{14}H_{21}NO_3$	-N	I/A	251.32	Insoluble	Soluble in various organic solvents
593.	Guazatine acetate[e]	39202-40-9	Medium	$C_{24}H_{53}N_7O_6$	-N	F	535.74	3×10^6	Methanol 300, ethanol 200
594.	Halosulfuron-methyl[e]	100784-20-1	Low	$C_{13}H_{15}ClN_6O_7S$	-N	H	434.8	1650	logKow=−0.0186, methanol 1.62
595.	Haloxyfop-etotyl[e]	87237-48-7	Medium	$C_{19}H_{19}F_3ClNO_5$	-F	H	433.8	0.58	Acetone, CH_2Cl_2, toluene and xylene 1000, methanol 233, hexane 44

No.	Compound	CAS	Toxicity	Formula	Characteristic Atom[1]	Effect[2]	MW	WS (mg/L)	Solubility of Organic Solvent (g/L)
596.	Haloxyfop-methyl[e]	69806-40-2	Medium	$C_{16}H_{13}F_3ClNO_4$	–F	H	375.7	43.3	Acetone and methanol 1000, ethyl acetate 518, CH_2Cl_2 459, toluene 118, xylene 74, hexane 0.17
597.	Hexaconazole[e]	79983-71-4	Low	$C_{14}H_{17}Cl_2N_3O$	–N	F	314.2	0.018	Acetone 164, toluene 59, hexane 0.8
598.	Hexaflumuron[e]	86479-06-3	Low	$C_{16}H_8Cl_2F_6N_2O_3$	–F	I	461.1	0.7	Methanol 11.3, xylene 5.2
599.	Hexazinone[e]	51235-04-2	Low	$C_{12}H_{20}N_4O_2$	–N	H	252.32	3.3×10^4	Chloroform 3880, methanol 2650, benzene 940, CH_2Cl_2 836, hexane 3
600.	Hexythiazox[e]	78587-05-0	Low	$C_{17}H_{21}ClN_2O_2S$	–N	A	352.9	0.5	Chloroform 1380, xylene 362, acetone 160, acetonitrile 28.6, methanol 20.6, hexane 3.9
601.	Hydramethylnon[e]	67485-29-4	Low	$C_{25}H_{24}F_6N_4$	–F	I	494.5	0.005	log Kow=2.31, acetone 360, chlorobenzol 390, methanol 230, xylene 94, ethanol 72
602.	Hymexazol[e]	10004-44-1	Low	$C_4H_5NO_2$	–N	F	99.15	8.5×10^4	Soluble in methanol, acetone and ethanol, etc.
603.	Imazalil[e]	35554-44-0	Medium	$C_{14}H_{14}Cl_2N_2O$	–N	F	297.18	1400	Methanol, ethanol, benzene and xylene 500
604.	Imazamethabenz-methyl[e]	81405-85-8	Low	$C_{16}H_{20}N_2O_3$	–N	H	288.3	1.3	CH_2Cl_2 300, methanol 242, dimethylsulfoxide 238, acetone 1.82, hexane 0.4
605.	Imazamox[e]	114311-32-9	Low	$C_{15}H_{19}N_3O_4$	–N	H	305.3	Diffluent	CH_2Cl_2 0.143, methanol 0.0668

No.	Name	CAS		Formula			MW		Solubility
606.	Imazapic[e]	104098-49-9	Low	$C_{14}H_{17}N_3O_4$	–N	H	275.3	2150	Acetone 18.9
607.	Imazapyr[e]	81334-34-1	Low	$C_{13}H_{15}N_3O_3$	–N	H	261.3	1.13	Methanol 230, ethanol and CH_2Cl_2 72, acetone 6
608.	Imazaquin[e]	81335-37-7	Low	$C_{17}H_{17}N_3O_3$	–N	H	311.34	60	CH_2Cl_2 14, toluene 0.4
609.	Imazethapyr[e]	81385-77-5	Low	$C_{15}H_{19}N_3O_3$	–N	H	289.3	1400	CH_2Cl_2 185, methanol 105, acetone 48.2, toluene 5
610.	Imibenconazole[e]	86598-92-7	Low	$C_{17}H_{13}Cl_3N_4S$	N/Cl	F	411.7	1790	Acetone 1063, benzene 580, xylene 250, methanol 120
611.	Imibenconazole-des-benzyl[e]	199338-48-2	Low	$C_{10}H_{17}Cl_3N_4S$	–N	F	321.58	Insoluble	Soluble in various organic solvents
612.	Imidacloprid[e]	138261-41-3	Medium	$C_9H_{10}ClN_5O_2$	–N	I	255.66	510	Hexane 0.1, isopropanol 1-2
613.	Iminoctadine triacetate[e]	39202-40-9	Medium	$C_{24}H_{53}N_7O_6$	–N	F	535.7	7.64×10^5	Ethanol 117
614.	Iodosulfuron-methyl[e]	1225-60-61	Low	$C_{14}H_{14}IN_5O_6S$	–N	H	506.24	Insoluble	Soluble in various organic solvents
615.	Iodosulfuron-methyl sodium[e]	144550-36-7	Low	$C_{14}H_{12}IN_5NaO_6S$	–N	H	528.24	2.5×10^4	logKow=–0.70
616.	Indoxacarb[e]	12124-97-9		$C_{22}H_{17}ClF_3N_3O_7$	–N	I	527.8	0.2	Acetone 250, acetonitrile 139, methanol 103
617.	Iprodione[e]	36734-19-7	Low	$C_{13}H_{13}Cl_2N_3O_3$	–Cl	F	330.17	13	CH_2Cl_2 500, acetone 300, benzene 200, xylene and acetonitrile 150, ethanol 20
618.	Isocarbamid[e]	30979-48-7	Low	$C_8H_{15}N_3O_2$	–N	H	185.23	1300	CH_2Cl_2 281, cyclohexanone 130
619.	Isomethiozin[e]	57052-04-7	Low	$C_{12}H_{20}N_4OS$	–N	H	268.38	10	CH_2Cl_2 152, cyclohexanone 103

Continued

No.	Compound	CAS	Toxicity	Formula	Characteristic Atom[1]	Effect[2]	MW	WS (mg/L)	Solubility of Organic Solvent (g/L)
620.	Isopropalin[e]	33820-53-0	Low	$C_{15}H_{23}N_3O_4$	$-N$	H	309.41	0.1	Acetone, acetonitrile, benzene, chloroform, diethyl ether, methanol and propane >1000
621.	Isoproturon[e]	34123-59-6	Low	$C_{12}H_{18}N_2O$	$-N$	H	206.29	70	CH_2Cl_2 63, methanol 56, benzene 5, hexane 0.1
622.	Isouron[e]	55861-78-4	Low	$C_{10}H_{17}N_3O_2$	$-N$	H	211.26	300	Ethanol 357, acetone 270, xylene 240
623.	Isoxaben[e]	82558-50-7	Low	$C_{18}H_{24}N_2O_4$	$-N$	H	332.39	1.42	Methanol, ethyl acetate and CH_2Cl_2 50-100, acetonitrile 30-50, toluene 5, ethane 0.7
624.	Isoxaflutole[e]	141112-29-0	Low	$C_{15}H_{12}F_3NO_4S$	$-F$	H	359.32	6.2	Soluble in various organic solvents
625.	Kresoxim-methyl[e]	143390-89-0	Low	$C_{18}H_{19}NO_4$	$-N$	F	313.35	2	Diffluent in various solvents
626.	Lactofen[e]	77501-63-4	Low	$C_{19}H_{15}F_3ClNO_7$	$-F$	H	581.9	1	Propanol 192, soluble in xylene
627.	Lenacil[e]	2164-08-1	Low	$C_{13}H_{18}N_2O_2$	$-N$	H	234.29	6	Diffluent in pyridine, other organic solvents < 10
628.	Linuron[e]	330-55-2	Low	$C_9H_{10}Cl_2N_2O_2$	$-N$	H	249.10	75	Diffluent in acetone, chloroform and diethyl ether
629.	Lmibenconazole[e]	86598-92-7	Low	$C_{17}H_{13}Cl_3N_4S$	$-N$	F	411.7	1.7	Acetone 1030, methanol 120, xylene 50
630.	Maleic hydrazide[e]	123-33-1	Low	$C_4H_4N_2O_2$	$-N$	PGR	112.10	6000	Acetone, ethanol and xylene 10

No.	Name	CAS number		Formula		Type	MW	Value	Solubility
631.	Mefenacet[e]	73250-68-7	Low	$C_{16}H_{14}N_2O_2S$	–N	H	298.4	4	CH_2Cl_2 200, acetone 100, acetonitrile 60, toluene 50
632.	Mefenoxam[e]	70630-17-0	Low	$C_{15}H_{21}NO_4$	–N	F	279.33	2.6×10^4	n-Hexane 59, mutually soluble with acetone, ethyl acetate, methanol, dichloromethane, toluene and n-octyl alcohol etc.
633.	Mefenpyr-diethyl[e]	135590-91-9	Low	$C_{16}H_{18}Cl_2N_2O_4$	–N	H	373.2	20	logKow=3.83, acetone > 500, ethyl acetate >400, toluene >400, methanol >400
634.	Mepanipyrim[e]	110235-47-7	Low	$C_{14}H_{13}N_3$	–N	F	223.3	5.58	Soluble in various organic solvents, logKow=3.41
635.	Mepronil[e]	55814-41-0	Low	$C_{17}H_{19}NO_2$	–N	F	279.0	12.7	Acetone and methanol 500, benzene 28.2, hexane 1.1
636.	Mepiquat chloride[e]	24307-26-4	Low	$C_7H_{16}NCl$	–N	PGR	149.7	100	Ethanol 162, acetone 1
637.	Mesosulfuron-methyl[e]	208465-21-8	Low	$C_{17}H_{21}N_5O_9S_2$	–N	H	503.5	21.4	Acetone 13.66, hexane 0.0002
638.	Metalaxyl[e]	57837-19-1	Low	$C_{15}H_{21}NO_4$	–N	F	279.34	7100	CH_2Cl_2 750, methanol 650, benzene 550, hexane 9.1
639.	Metamitron[e]	41394-05-2	Low	$C_{10}H_{10}N_4O$	–N	H	202.22	1800	CH_2Cl_2 and cyclohexanol 10-50
640.	Metazachlor[e]	67129-08-2	Low	$C_{14}H_{16}ClN_3$	–N	H	277.76	17	Acetone and chloroform 1000, ethyl acetate 590, ethanol 200
641.	Metconazole[e]	125116-23-6	Low	$C_{17}H_{17}ClN_3O$	–N	F	319.8	15	logKow=3.85, methanol 235, acetone 239

Continued

No.	Compound	CAS	Toxicity	Formula	Characteristic Atom[1]	Effect[2]	MW	WS (mg/L)	Solubility of Organic Solvent (g/L)
642.	Methabenthiazuron[e]	18691-97-9	Low	$C_{10}H_{11}N_3OS$	–N	H	221.29	59	Acetone 116, dimethylformamide 100, methanol 65.9
643.	Methfuroxam[e]	2873-17-8	Low	$C_{14}H_{15}NO_2$	–N	F	229.27	10	Dimethyl formamide 412, acetone 125, methanol 64, benzene 36
644.	Methoxyfenozide[e]	161050-58-4	Low	$C_{22}H_{28}N_2O_3$	–N	IGR	368.47	<1	Dimethylsulfoxide 110, cyclohexanone 99, acetone 90
645.	Methoprotryne[e]	841-06-5	Low	$C_{11}H_{21}N_5OS$	–N	H	271.39	320	Benzene and cyclohexane 230, acetone 200, xylene 140
646.	Metobromuron[e]	3060-89-7	Low	$C_9H_{11}BrN_2O_2$	–N	H	259.11	330	Diffluent in acetone, chloroform and ethanol
647.	Metolachlor[e]	51218-45-2	Low	$C_{15}H_{22}ClNO_2$	–N	H	283.80	530	Miscible with toluene and xylene
648.	Metominostrobin-(E)[e]	133408-50-1	Low	$C_{18}H_{19}NO_3$	–N	F	313.35	128	logKow=2.32, CH_2Cl_2 1380, chloroform 1280
649.	Metominostrobin-(Z)[e]	133408-50-1	Low	$C_{18}H_{19}N_2O_3$	–N	F	313.35	128	logKow=2.32, CH_2Cl_2 1380, chloroform 1280
650.	Metoxuron[e]	19937-59-8	Low	$C_{10}H_{13}ClN_2O_2$	–N	H	228.68	678	Soluble in acetone, cyclopropanone and hot ethanol, appropriately soluble in benzene and ethanol, insoluble in petroleum ether

No.	Name	CAS		Formula		Type	MW	Water sol.	Solubility
651.	Metribuzin[e]	21087-64-9	Low	$C_8H_{14}N_4OS$	–N	H	214.29	1200	Methanol 450, CH_2Cl_2 333, toluene 130
652.	Metosulam[e]	139528-85-1	Low	$C_{14}H_{13}Cl_2N_5O_4S$	–N	H	418.3	200	logKow=0.9778, acetone, acetonitrile and CH_2Cl_2 1 >0.5, n-octyl alcohol, hexane and toluene <0.5
653.	Metsulfuron-methyl[e]	74223-64-6	Low	$C_{14}H_{15}N_5O_6S$	–N	H	381.4	9500	CH_2Cl_2 121, acetone 36, methanol 7.3, ethanol 2.3, xylene 0.58, hexane 0.79
654.	Monalide[e]	7287-36-7	Low	$C_{13}H_{13}ClFNO$	–N	H	253.58	22.8	Cyclohexanone 500, xylene 100, petroleum ether 10
655.	Monolinuron[e]	1746-81-2	Low	$C_9H_{11}ClN_2O_2$	–N	H	214.65	735	Soluble in acetone, ethanol, benzene and xylene
656.	Monuron[e]	150-68-5	Low	$C_9H_{11}ClN_2O$	–N	H	198.66	230	Acetone 52, benzene 2
657.	Myclobutanil[e]	88671-89-0	Low	$C_{15}H_{17}ClN_4$	–N	F	288.78	142	Soluble in various solvents, insoluble in aliphatic hydrocarbon
658.	1-Naphthyl acetic acid[e]	86-87-3	Low	$C_{12}H_{10}O_2$	/	PGR	186.12	Soluble in hot water	Diffluent in acetone, diethyl ether and chloroform, etc.
659.	1-Naphthy acetamide[e]	86-86-2	Low	$C_{12}H_{11}NO$	–N	Rooting agent	185	Aqueous	Soluble in various organic solvents
660.	Napropamide[e]	15299-99-7	Low	$C_{17}H_{21}NO_2$	–N	H	271	73	Acetone and ethanol >1000, xylene 505, hexane 1
661.	Naproanilide[e]	52570-16-8	Low	$C_{19}H_{17}NO_2$	–N	H	291.3	0.74	Acetone 171, benzene and toluene 46, ethanol 17

Continued

No.	Compound	CAS	Toxicity	Formula	Characteristic Atom[1]	Effect[2]	MW	WS (mg/L)	Solubility of Organic Solvent (g/L)
662.	Naptalam^e	132-66-1	Low	$C_{18}H_{13}NO_3$	–N	H	291.29	200	Acetone 5, isopropanol 2.11, insoluble in benzene, xylene and hexane
663.	NC-330^e	114874-05-4				H			
664.	Neburon^e	555-37-3	Low	$C_{12}H_{16}Cl_2N_2O$	–N	H	275.18	4.8	Slightly soluble in organic solvents, insoluble in aliphatic hydrocarbon
665.	Nicosulfuron^e	111991-09-4	Low	$C_{15}H_{18}N_6O_6S$	–N	H	410.4	1.2×10^5	CH_2Cl_2 160, chloroform 64, acetonitrile 23, acetone 18, ethanol 4.5
666.	Nicotine^e	54-11-5	Medium	$C_{10}H_{14}N_2$	–N	I	162.23	Miscible	Soluble in various solvents, insoluble in aliphatic hydrocarbon
667.	Nitenpyram^e	150824-47-8	Low	$C_{11}H_{15}ClN_4O_2$	–N	I	270.71	8.4×10^5	Chloroform 700, acetone 290, xylene 4.5
668.	Nitralin^e	4726-14-1	Low	$C_{13}H_{19}N_3O_6S$	–N	H	345.38	0.6	Acetone 360, dimethylsulfoxide 330, benzene 125, methanol 11, isopropanol 1.8
669.	Nitrapyrin^e	1929-82-4	Low	$C_6H_3Cl_4N$	–Cl	F	230.86	Insoluble	Soluble in acetone, ethanol and benzene
670.	Norflurazon^e	27314-13-2	Low	$C_{12}H_9ClF_3N_3O$	–F	H	303.57	28	Ethanol 142, propanol 50, xylene 2.5
671.	Novaluron^e	116714-46-6	Low	$C_{17}H_9ClF_8N_2O_4$	–F	I	492.7	0.05	Soluble in various organic solvents

672.	Nuarimol[e]	Medium	$C_{17}H_{12}ClFN_2O$	–Cl	F	314.75	26	Soluble in acetone, benzene, acetonitrile, chloroform and methanol
673.	Ofurace[e]	Low	$C_{14}H_{16}ClNO_3$	–N	F	281.7	140	CH_2Cl_2 255, cyclohexanone 141, ethyl acetate 44
674.	Olaquindox[e]	Low	$C_{12}H_{13}N_3O_4$	–N	F	263.25	Slight soluble	Insoluble in methanol, ethanol and chloroform
675.	Oryzalin[e]	Low	$C_{12}H_8N_4O_6S$	–N	H	346.36	25	Acetone 500, acetonitrile 150, methanol 50, CH_2Cl_2 30, benzene 4, xylene 2
676.	Oxabetrinil[e]	Low	$C_{12}H_{11}N_2O_3$	–N	F	232.2	20	CH_2Cl_2 450, cyclohexane 300, acetone 250, toluene 220, xylene 150, methanol 30, hexane 5.6
677.	Oxadiazon[e]	Low	$C_{15}H_{18}Cl_2N_2O_3$	–N	H	345.23	0.7	Benzene, toluene and chloroform 1000, acetone and isopropanol 600, methanol and ethanol 100
678.	Oxadixyl[e]	Low	$C_{14}H_{28}N_2O_4$	–N	F	288.18	3400	Soluble in polar solvents
679.	Oxasulfuron[e]	Low	$C_{17}H_{18}N_4O_6S$	–N	H	406.4	1700	logKow=–0.81, acetone 9.3, CH_2Cl_2 6.9, methanol 1.5, toluene 0.32, ethyl acetate 0.23, n-hexane 0.002
680.	Oxine-Copper[e]	Low	$C_{18}H_{12}CuN_2O_2$	–N	F	351.8	Indissolvable	Soluble in chloroform, indissolvable in various organic solvents

Continued

No.	Compound	CAS	Toxicity	Formula	Characteristic Atom[1]	Effect[2]	MW	WS (mg/L)	Solubility of Organic Solvent (g/L)
681.	Oxolinic acid^e	14698-29-4	Low	$C_{13}H_{11}NO_5$	–N	F	261.2	0.003	Methanol, acetone and ethyl acetate <10
682.	Oxycarboxin^e	5259-88-1	Low	$C_{12}H_{13}NO_4S$	–N	F	267.30	1000	Acetone 360, methanol 70, benzene 34, ethanol 30
683.	Paclobutrazol^e	76738-62-0	Low	$C_{18}H_{20}ClN_3O$	–N	PGR	293.8	35	Methanol 150, acetone 110, CH_2Cl_2 100, xylene 60, hexane 10
684.	Paraquat dichloride^e	4685-14-7	Medium	$C_{12}H_{14}N_2 \cdot Cl_2$	–N	H	257.04	Diffluent	Slightly soluble in low alcohols, insoluble in varsol
685.	Penconazole^e	66246-88-6	Low	$C_{13}H_{15}Cl_2N_3$	–N	F	284.2	73	Propanol 770, ethanol 730, toluene 610, n-octyl alcohol 400, n-hexane 2
686.	Pencycuron^e	66063-05-6	Low	$C_{19}H_{21}ClN_2O$	–N	F	328.82	Insoluble	Soluble in CH_2Cl_2, slightly soluble in most organic solvents
687.	Pendimethalin^e	40318-45-4	Low	$C_{13}H_{19}N_3O_4$	–N	H	281.31	0.3	Diffluent in ethanol, acetone, benzene, toluene, chloroform and CH_2Cl_2
688.	Phthalimide^e	5333-22-2	Low	$C_8H_5NO_2$	–N	F	147.12	Slight soluble in cold water	Soluble in alcohol and acetic acid, slightly soluble in benzene and chloroform
689.	Picloram^e	1918-02-1	Low	$C_6H_3Cl_3N_2O_2$	–Cl	H	241.48	430	Acetone 19.8, ethanol 10.5, acetonitrile 1.6, benzene 0.2, CH_2Cl_2 0.6
690.	Picolinofen^e	137641-05-5	Low	$C_{19}H_{12}F_4N_2O_2$	–F	H	376.31	0.047	Acetone 557, CH_2Cl_2 764, ethyl acetate 464, methanol 30.4

No.	Name	CAS		Formula			MW		Solubility
691.	Pretilachlor[e]	51218-49-6	Low	$C_{17}H_{25}ClNO_2$	–N	H	310.62	50	Methanol and hexane 500
692.	Primisulfuron-methyl[e]	86209-51-0	Low	$C_{15}H_{12}F_4N_4O_7S$	–N	H	468.3	390	Acetone 45, toluene 5.79, n-octyl alcohol 0.13, n-hexane 0.001
693.	Probenazole[e]	27605-76-1	Low	$C_{10}H_9NO_3S$	–S	F	223.2	150	Diffluent in acetone and chloroform, soluble in benzene, ethanol and methanol, slightly soluble in hexane
694.	Prochloraz[e]	67747-09-5	Low	$C_{15}H_{16}Cl_3N_3O_2$	–N	F	376.5	34	Acetone and ethanol 600, xylene 0.6, hexane 0.0075
695.	Procymidone[e]	32809-16-8	Low	$C_{13}H_{11}Cl_2NO_2$	–Cl	F	284.1	47.5	Acetone 3500, chloroform, diethyl ether and xylene 2500
696.	Profluralin[e]	26399-36-0	Low	$C_{13}H_{16}F_3N_3O_5$	–N	H	351.13	5.1	Acetone, xylene, ethanol, hexane, ketone, aromatic hydrocarbon and aliphatic hydrocarbon 500
697.	Prometone[e]	1610-18-0	Low	$C_{10}H_{19}N_5O$	–N	H	225.92	750	Acetone and methanol 500, benzene 250
698.	Prometryne[e]	7287-19-6	Low	$C_{10}H_{19}N_5S$	–N	H	241.37	48	Soluble in various solvents
699.	Propachlor[e]	1918-16-7	Low	$C_{11}H_{14}ClNO$	–N	H	211.69	700	Benzene 500, acetone 309, ethanol 290, xylene 193
700.	Propanil[e]	709-98-8	Low	$C_9H_9Cl_2NO$	–Cl	H	218.08	225	Toluene and xylene 250, diffluent in methanol, ethanol and benzene
701.	Propaquizafop[e]	111479-05-1	Low	$C_{22}H_{22}ClN_3O_5$	–N	I/A	443.9	Insoluble	Soluble in various organic solvents

Continued

No.	Compound	CAS	Toxicity	Formula	Characteristic Atom[1]	Effect[2]	MW	WS (mg/L)	Solubility of Organic Solvent (g/L)
702.	Propazine[e]	139-40-2	Low	$C_9H_{16}ClN_5$	-N	H	230.09	8.6	Indissolvable
703.	Propetamphos[e]	31218-83-4	Medium	$C_{10}H_{20}N_4PS$	-N	I	281.3	110	Soluble in various solvents
704.	Propiconazol[e]	60207-90-1	Low	$C_{15}H_{17}Cl_2N_3O_2$	-N	F	342.2	110	Soluble in various solvents
705.	Propisochlor[e]	86763-47-5	Low	$C_{15}H_{22}ClNO_2$	-N	H	283.8	184	logKow=3.5, soluble in methanol, hexane and CH_2Cl_2 etc.
706.	Propoxycarbazone-sodium[e]	181274-15-7	Low	$C_{15}H_{17}N_4NaO_7S$	-N	H	420.4	4.2×10^4	CH_2Cl_2 1.5, n-heptane, xylene and isopropanol <0.1
707.	Propyzamide[e]	23950-58-5	Low	$C_{12}H_{11}Cl_2NO$	-N	H	256.13	15	Dimethyl sulfoxide 330, cyclohexane 200, methanol and isopropanol 150
708.	Prosulfuron[e]	94125-34-5	Low	$C_{15}H_{16}F_3N_5O_4S$	-N	H	419.38	Insoluble	Soluble in various organic solvents
709.	Pymetrozin[e]	123312-89-0	Low	$C_{10}H_{11}N_5O$	-N	PGR	217.1	270	Ethanol 2.25, hexane 0.01
710.	Pyraclostrobin[e]	175013-18-0	Low	$C_{19}H_{17}F_4N_2O_2$	-F	F	387.82	1.9	logKow=3.99
711.	Pyraflufen ethyl[e]	129630-17-7	Low	$C_{15}H_{13}F_3Cl_2N_2O_4$	-N	H	413.2	0.082	Xylene 41.7-43.5, acetone 167-182, methanol 7.39, ethyl acetate 105-111
712.	Pyrazolate[e]	58011-68-0	Low	$C_{19}H_{16}Cl_2N_2O_4S$	-N	H	439.3	0.056	Chloroform 587, benzene 205, ethylacetate 118, ethanol 14, hexane 0.6
713.	Pyrazosulfuron-ethyl[e]	93699-74-6	Low	$C_{14}H_{18}N_6O_7S$	-N	H	414.4	14.5	Chloroform 234, acetone 31.7, methanol 0.7, hexane 0.2

No.	Name		CAS	Formula		Type	MW	Water solubility	Solubility in solvents
714.	Pyrazoxyfen[e]	Low	71561-11-0	$C_{20}H_{16}Cl_2N_2O_3$	–N	H	403.3	900	Chloroform 1068, n-hexane 900, benzene 325, acetone 223, toluene 200, xylene 116, ethanol 14
715.	Pyridaben[e]	Low	96489-71-3	$C_{19}H_{25}ClN_2OS$	–N	I/A	364.9	0.012	Benzene 110, ethanol 57, hexane 10, acetone 0.46
716.	Pyridate[e]	Low	55512-33-9	$C_{19}H_{23}ClN_2O_2S$	–N	H	378.11	90	Diffluent in various solvents
717.	Pyrimethanil[e]	Low	53112-28-0	$C_{12}H_{13}N_3$	–N	F	199.26	3.5	Diffluent in methanol and acetone etc.
718.	Pyrimidifen[e]	Medium	105779-78-0	$C_{20}H_{28}ClN_3O_2$	–N	A/I	377.9	2.17	Soluble in various organic solvents
719.	Pyridinitril[e]	Low	1086-02-8	$C_{13}H_5Cl_2N_3$	–N	F	274.14	Insoluble	Slightly soluble in acetone, benzene, chloroform and CH_2Cl_2
720.	Pyrifenox[e]	Low	88283-41-4	$C_{14}H_{12}Cl_2N_2O$	–N	F	295.2	115	Ethyl acetate, chloroform and toluene 25, acetone and diethyl ether 20, hexane 1
721.	cis-Pyriminobac-methyl[e]	Low	136191-64-5	$C_{17}H_{19}N_3O_6$	–N	H	361.4	0.9	Methanol 14.6
722.	trans-Pyriminobac-methyl[e]	Low	136191-64-5	$C_{17}H_{19}N_3O_6$	–N	H	361.4	17.5	Methanol 14.6
723.	Pyriproxyfen[e]	Low	95737-68-1	$C_{20}H_{29}NO_3$	–N	IGR	331.5	Insoluble	Xylene 500, hexane 400, methanol 200
724.	Pyroquilon[e]	Medium	57369-32-1	$C_{11}H_{11}NO$	–N	F	173.2	4000	CH_2Cl_2 580, methanol 240, benzene 200, acetone 125, isopropanol 85
725.	Quinmerac[e]	Low	90717-03-6	$C_{11}H_8ClNO_2$	–Cl	H	221.6	22	Acetone and CH_2Cl_2 2, ethanol 1

Continued

No.	Compound	CAS	Toxicity	Formula	Characteristic Atom[1]	Effect[2]	MW	WS (mg/L)	Solubility of Organic Solvent (g/L)
726.	Quizalofop-ethyl[e]	76578-14-8	Low	$C_{19}H_{17}ClN_2O_4$	–N	H	372.8	0.4	Acetone 650, xylene 60, ethanol 20, hexane 5
727.	Rabenzazole[e]	69899-24-7	Low	$C_{12}H_{12}N_4$	–N	I/A	212.25	Slight soluble	Soluble in various organic solvents
728.	Rimsulfuron[e]	122931-48-0	Low	$C_{14}H_{17}N_5O_7S_2$	–N	H	431.4	10	Soluble in various organic solvents
729.	Sebutylazine[e]	7286-69-3	Low	$C_9H_{16}ClN_5$	–N	H	229.54	40	Soluble in various organic solvents
730.	Secbumeton[e]	26259-45-0	Low	$C_{10}H_{19}N_5O$	–N	H	225.30	620	Diffluent in ethanol, acetone, chloroform and aromatic hydrocarbon
731.	Simazine[e]	122-34-9	Low	$C_7H_{12}ClN_5$	–N	H	201.66	5	Chloroform 90, methanol 40, ethyl acetate 30
732.	Simeconazole[e]	149508-90-7	Low	$C_{14}H_{20}FN_3OSi$	–N	F	293.41	57.5	Soluble in most organic solvents
733.	Simetone[e]	673-04-1	Low	$C_7H_{14}N_5O$	–N	H	184.07	3200	Soluble in methanol, etc.
734.	Simetryn[e]	1014-70-6	Low	$C_8H_{15}N_5S$	–N	H	213.14	Indissolvable	Soluble in methanol, ethanol and chloroform, etc.
735.	Spinosad[e]	131929-60-7	Low	$C_{42}H_{67}NO_{16}$	–N	I	746.00	235	logKow = 4
736.	Spinosad[e]	131929-63-0	Low	$C_{41}H_{65}NO_{16}$	–N	I	731.98	0.332	logKow = 4.5
737.	Spiroxamine(A)[e]	118134-30-8	Low	$C_{18}H_{35}NO_2$	–N	F	297.48	470(A)	logKow = 2.79, diffluent in various organic solvents
738.	Spiroxamine(B)[e]	118134-30-8	Low	$C_{18}H_{35}NO_2$	–N	F	297.48	340(B)	logKow = 2.79, diffluent in various organic solvents

No.	Name		CAS		Formula			MW	Water sol.	Solubility
739.	Sulfentrazone[e]	Low	122836-35-5	$C_{11}H_{10}Cl_2F_2N_4O_3S$	-F	H	387.2	0.78	Soluble in polar solvents such as acetone, etc.
740.	Sulfosulfuron[e]	Low	141776-32-1	$C_{15}H_{16}N_4O_5S$	-N	H	470.5	1627	logKow <1
741.	Sulfometuron-methyl[e]	Low	74222-97-2	$C_{16}H_{18}N_6O_7S$	-N	H	364.4	244	Acetone 3.3, acetonitrile 1.8, ethyl acetate 0.65, diethyl ether 0.06, methanol 0.550, octanol 0.140, CH_2Cl_2 1.5, dimethyl sulfoxide 32, toluene 0.240, hexane 0.001
742.	Sulfanitran[e]	Low	122-16-7	$C_{14}H_{13}N_3O_5S$	-N	F	335.34	Insoluble	Soluble in most organic solvents
743.	Tebuconazole[e]	Low	107534-96-3	$C_{16}H_{22}ClN_3O$	-N	F	307.8	32	CH_2Cl_2 200, toluene 100, hexane 0.1
744.	Tebufenozide[e]	Low	112410-23-8	$C_{22}H_{28}N_2O_2$	-N	I	352.52	1	Slightly soluble in organic solvents
745.	Tebufenpyrad[e]	Low	119168-77-3	$C_{18}H_{24}ClN_3O$	-N	A	333.8	2.8	Soluble in acetone, methanol, acetonitrile, benzene and chloroform, etc.
746.	Tebuthiuron[e]	Low	34014-18-1	$C_9H_{16}N_4OS$	-S	H	228.32	2300	Chloroform 250, methanol 170, acetone 70, acetonitrile 60, hexane 6.1, benzene 3.7
747.	Teflubenzuron[e]	Low	83121-18-0	$C_{14}H_6Cl_2F_4N_2O_2$	-F	I	381.06	0.02	Acetone 10, ethanol 1.4
748.	Terbacil[e]	Low	5902-51-2	$C_9H_{13}ClN_2O_2$	-N	H	216.66	710	Difluent in cyclohexanone, slightly soluble in xylene, almost insoluble in aliphatic hydrocarbon

Continued

No.	Compound	CAS	Toxicity	Formula	Characteristic Atom[1]	Effect[2]	MW	WS (mg/L)	Solubility of Organic Solvent (g/L)
749.	Terbucarb[e]	1918-11-2	Low	$C_{17}H_{27}NO_2$	-N	H	277.45	Insoluble	Soluble in most organic solvents
750.	Tebutam[e]	35256-85-0	Low	$C_{15}H_{23}NO$	-N	H	233.36	1030	Soluble in acetone, hexane, methanol, ethanol, benzene, toluene and chloroform etc.
751.	Terbumeton[e]	33693-04-8	Medium	$C_{10}H_{19}N_5O$	-N	H	225.3	130	Isopropanol and xylene 200
752.	Terbuthylazine[e]	5915-41-3	Low	$C_9H_{16}ClN_5$	-N	H	229.72	8.5	Dimethylformamide 100, ethyl acetate 40, isopropanol and xylene 10
753.	Terbutryne[e]	886-50-0	Low	$C_{10}H_{19}N_5S$	-N	H	241.36	58	Soluble in most solvents
754.	tert-Butylamine[e]	75-64-9	Medium	$C_4H_{11}N$	N	I	72.14	Aqueous	Soluble in ethanol and acetone
755.	Tetraconazole[e]	112281-77-3	Low	$C_{13}H_{11}Cl_2F_4N_3O$	-N	F	372.1	150	Mutually soluble with CH_2Cl_2 40, acetone and methanol
756.	Tetrahydrophthalimide[e]	85-40-5	Low	$C_8H_9NO_2$	-N	I/A	151.16	Insoluble	Soluble in most organic solvents
757.	Thenylchlor[e]	96491-05-3	Low	$C_{16}H_{18}ClNO_2$	-N	H	323.8	11	logKow=3.53
758.	Thidiazuron[e]	41118-83-6	Low	$C_9H_8N_4OS$	-N	PGR	220.2	50	Dimethylformamide 500, cyclohexane 21, methanol 4.5, acetone 8
759.	Thifluzamide[e]	130000-40-7	Low	$C_{13}H_6Br_2F_6N_2O_2S$	-F	F	528.06	1.6	logKow=4.1

No.	Name	CAS		Formula			MW	Solubility	Solubility in solvents
760.	Thifensulfuron-methyl[e]	79277-27-3	Low	$C_{12}H_{13}N_5O_6S_2$	–N	H	387.4	2.48	CH_2Cl_2 27.5, acetone 11.9, acetonitrile 7.3, methanol 2.6, ethanol 0.9
761.	Thiacloprid[e]	111988-49-9	Low	$C_{10}H_9ClN_4S$	–N	I	252.8	185	Soluble in most organic solvents
762.	Thiabendazole[e]	148-79-8	Low	$C_{10}H_7N_3S$	–N	F	201.3	50	Methanol 9.3, ethanol 7.9, acetone 4.2, chloroform 0.8, benzene 0.23
763.	Thiazopyr[e]	117718-60-2	Low	$C_{16}H_{17}F_5N_2O_2S$	–F	H	396.4	2.5	$logKow=3.89$
764.	Thiobencarb[e]	28249-77-6	Low	$C_{12}H_{16}ClNOS$	–N	H	257.63	27.5	Diffluent in xylene, alcohol and acetone
765.	Thiophanate-methyl[e]	23564-05-8	Low	$C_{12}H_{14}N_4O_4S_2$	–N	F	342.40	3.5	Acetone 58.1, methanol 29.2, chloroform 26.2, acetonitrile 24.4
766.	Tiamulin-fumarate[e]	55297-96-6	Medium	$C_{32}H_{51}NO_8$	–N	I/A	609.82	6×10^5	Soluble in most solvents
767.	Tolfenpyrad[e]	129558-76-5	Medium	$C_{21}H_{22}ClN_3O_2$	–N	I	383.91	Insoluble	Soluble in most organic solvents
768.	Tolylfluanid[e]	731-27-1	Low	$C_{10}H_{13}FCl_2N_2O_2S_2$	–Cl	F/A	347.26	40	Benzene 570, xylene 230, methanol 46
769.	trans-diallate[e]	2303-16-4	Medium	$C_{10}H_{17}Cl_2NOS$	–Cl	H	270.21	14	Miscible with ethanol, acetone, benzene and xylene, etc.
770.	Triadimefon[e]	43121-43-3	Medium	$C_{14}H_{16}ClN_3O_2$	–N	F	293.76	260	CH_2Cl_2 1200, toluene 600, isopropyl-ketone 400
771.	Triadimenol[e]	55219-65-3	Low	$C_{14}H_{18}ClN_3O_2$	–N	F	295.8	95	CH_2Cl_2 2000, toluene 300, hexane 10
772.	tri-Allate[e]	2303-17-5	Low	$C_{10}H_{16}Cl_3NOS$	–Cl	H	304.66	4	Soluble in various solvents

Continued

No.	Compound	CAS	Toxicity	Formula	Characteristic Atom[1]	Effect[2]	MW	WS (mg/L)	Solubility of Organic Solvent (g/L)
773.	Triasulfuron[e]	82097-50-5	Low	$C_{14}H_{16}ClN_5O_5S$	–N	H	401.8	1500	CH_2Cl_2 and acetone 0.18, xylene 0.166
774.	Triazoxide[e]	72459-58-6	Medium	$C_{10}H_6ClN_5O$	–N	F	247.6	34	CH_2Cl_2 32, xylene 6.9, isopropanol 1.8, n-hexane 0.05
775.	Triflusulfuron-methyl[e]	126535-15-7	Low	$C_{17}H_{19}F_3N_6O_6S$	–N	H	492.4	11	logKow=0.96
776.	Tricyclazole[e]	41814-78-2	Medium	$C_9H_7N_3S$	–N	F	189.2	1.6	Chloroform 500, methanol and ethanol 35, acetone and acetonitrile 10.4
777.	Trietazine[e]	1912-26-1	Low	$C_9H_{16}ClN_5$	–N	H	229.7	20	Chloroform 500, benzene 200, acetone 170, ethanol 30
778.	Trifloxystrobin[e]	141517-21-7	Low	$C_{20}H_{19}F_3N_2O_4$	–F	F	408.37	0.61	logKow=4.5
779.	Triflumizole[e]	68694-11-1	Low	$C_{15}H_{15}ClF_3N_3O$	–F	F	345.1	1.25×10^4	Chloroform 2220, xylene 639, hexane 17.6
780.	Triflumuron[e]	64628-44-0	Low	$C_{15}H_{10}ClF_3N_2O_2$	–F	I	538.7	0.025	CH_2Cl_2 50, toluene 5, isopropanol 2
781.	Trifluralin[e]	1582-09-8	Low	$C_{13}H_{16}F_3N_3O_4$	–N	H	335.28	1	Xylene 580, acetone 400
782.	Triforine[e]	26644-46-2	Low	$C_{10}H_{14}Cl_6N_4O_2$	–F	F	434.97	9	logKow=2.2, dimethylformamide 330, dimethylsulfoxide 476, acetone 11, methanol 10, CH_2Cl_2 1
783.	Triticonazole[e]	131983-72-7	Low	$C_{17}H_{20}ClN_3O$	–N	F	317.8	9.3	logKow=3.29

No.	Name	CAS No.		Formula		MW	Solubility	Type	Solubility notes
784.	Tribenuron-methyl[e]	106040-48-6	Low	$C_{16}H_{17}N_5O_5S$	–N	395.4	280	H	Acetonitrile 0.054.2, acetone 0.0438, ethyl acetate 0.017, methanol 0.00339, hexane 0.000028
785.	Uniconazole[e]	83657-17-4	Low	$C_{15}H_{18}ClN_3O$	–N	291.8	8.41	PGR	Methanol 0.88, toluene 0.7, hexane 0.01
786.	Vinclozoline[e]	50471-44-8	Low	$C_{12}H_9Cl_2NO_3$	–Cl	286.11	1000	F	Acetone 435, chloroform 319, ethyl acetate 253, benzene 146, diethyl ether 63, ethanol 14
787.	XMC[e]	2655-14-3	Low	$C_{10}H_{13}NO_2$	–N	179.2	470	I	Acetone 5.74, ethanol 3.52, ethyl acetate 2.77, benzene 2.04
788.	Zoxamide[e]	156052-68-5	Low	$C_{14}H_{16}Cl_3NO_2$	–Cl	336.64	0.68	F	logKow=3.76
(6) Organophosphorus pesticides									
789.	Amobam[f]	3566-10-7	Medium	$C_4H_{14}N_4S_4$	N/S	246.28	Aqueous	F	Slightly soluble in ethanol and acetone, insoluble in benzene
790.	Aramite[f]	140-57-8	Low	$C_{13}H_{20}ClSO_4$	Cl/S	307.58	Insoluble	A	Diffluent in acetone, benzene and hexane
791.	Aspon[f]	3244-90-4	Medium	$C_{12}H_{28}P_2S_2NO$	P/S	314.06	1600	I	Miscible with various solvents, indissolvable in petroleum ether
792.	Benazolin-ethyl[f]	3813-05-6	Low	$C_9H_6ClNO_3S$	–N	243.7	600	H	Acetone 132, ethanol 111
793.	Benfuresate[f]	68505-69-1	Low	$C_{12}H_{16}O_4S$	–S	256.3	190	H	logKow=2.46, diffluent in acetone, benzene, chloroform, ethanol, cyclohexane 51, hexane 12

Continued

No.	Compound	CAS	Toxicity	Formula	Characteristic Atom[1]	Effect[2]	MW	WS (mg/L)	Solubility of Organic Solvent (g/L)
794.	Bensultap[f]	17606-31-4	Low	$C_{17}H_{21}NO_4S_4$	–S	I	431.6	Insoluble	Soluble in most organic solvents
795.	Bentazone[f]	25057-89-0	Low	$C_{10}H_{12}N_2O_3S$	–N	H	240.28	500	Acetone 1507, ethanol 861, ethyl acetate 650, diethyl ether 16, chloroform 180, benzene 33
796.	Benthiazole[f]	21564-17-0	Low	$C_9H_6N_2S_3$	–N	F	281.5	Insoluble	Soluble in most organic solvents
797.	Benzoylprop-ethyl[f]	22212-55-1	Low	$C_{18}H_{17}Cl_2NO_3S$	–Cl	H	366.25	20	Acetone 750
798.	Bisultap(dimthypo)[f]	7772-98-7	Low	$C_5H_{11}NO_6S_4Na_2$	–S	I	419.8	Diffluent	Soluble in hot ethanol, methanol, dimethylformamide, dimethyl sulfoxide; Slightly soluble in acetone; Insoluble in ethyl acetate and diethyl ether
799.	Butoxycarboxim-sulfoxid[f]	34681-24-8	Low	$C_9H_{11}NO_5S$	–N	I	238.24	Insoluble	Soluble in most organic solvents
800.	Carbon disulphide[f]	75-15-0	Low	CS_2	–S	I/A	76.13	220	Miscible with ethanol, diethyl ether and chloroform, etc.
801.	Carbonyl sulfide[f]	463-58-1		COS	–S	F	60.07	Soluble	Soluble in ethanol and toluene
802.	Carboxin[f]	5234-68-4	Low	$C_{12}H_{13}NO_2S$	N/S	F	235.31	170	Acetone 600, ethanol 210, benzene 150, ethanol 100

No.	Name	Leaching	Formula		A/F	MW	Water solubility	Solubility in organic solvents
803.	Chinomethionat[f]	Low	$C_{10}H_6N_2OS_2$	-N		234.3	1	logKow=3.78, toluene 25, dichloromethane 4, hexane 18, isopropanol 0.9, cyclohexanone 18, petroleum ether 4, dimethylformamide 10
804.	Clothianidin[f]		$C_6H_8ClN_5O_2S$	-N	I	249.63	Insoluble	Soluble in most organic solvents
805.	Cycloxydim[f]	Low	$C_{17}H_{27}NO_3S$	S	H	325.46	85	diffluent in most solvents
806.	Dichlofluanid[f]	Low	$C_9H_{11}FCl_2N_2O_2S_2$	-N	F	329.22	Insoluble	Soluble in acetone, xylene 70, methanol 15
807.	Dimethipin[f]	Low	$C_6H_{10}O_4S_2$	-S	H/PGR	210.3	4600	Acetonitrile 180, xylene 8.979, methanol 10.7
808.	Dithiopyr[f]	Low	$C_{15}H_{16}F_5NO_2S_2$	-F	H	401.4	1.38	logKow=4.75, soluble in various organic solvents
809.	Ethofumesate[f]	Low	$C_{13}H_{18}O_5S$	-S	H	286.3	110	Acetone, chloroform and benzene 400, ethanol 100, hexane 4
810.	Ethylene thiourea[f]		$C_3H_6N_2S$	-S	Metabolite	102.09	2×10^5	Soluble in methanol, etc.
811.	Fenson[f]	Low	$C_{12}H_9ClO_3S$	-Cl	A	268.63	Insoluble	Soluble in most organic solvents
812.	Isoprothiolane[f]	Low	$C_{12}H_{18}O_4S_2$	-S	F	290.4	48	Acetone 4000, ethyl acetate, chloroform, toluene and xylene 2300, methanol and ethanol 1500, hexane 40
813.	Mancozeb[f]	Low	$C_4H_6N_2S_4MnZn$	-S	F	330.67	Slight soluble	Insoluble in common solvents

Continued

No.	Compound	CAS	Toxicity	Formula	Characteristic Atom[1]	Effect[2]	MW	WS (mg/L)	Solubility of Organic Solvent (g/L)
814.	Maneb[f]	84070-12-2	Low	$C_4H_6N_2S_4Mn$	–S	F	265.29	Insoluble	Insoluble in common solvents
815.	Mesotrione[f]	104206-82-8	Low	$C_{14}H_{13}NO_7S$	–S	H	339.31	1.5×10^4	Indissolvable in organic solvents
816.	Methyl isothiocyanate[f]	556-61-6	Medium	C_2H_3NS	–N	F	73.11	7600	Diffluent in most organic solvents
817.	Metiram[f]	9006-42-2	Low	$C_{16}H_{33}N_{11}S_{16}Zn_3$	–S	F	1088.7	Insoluble	Insoluble
818.	Propargite[f]	2312-35-8	Low	$C_{19}H_{24}O_4S$	–S	A	348.19	0.5	Soluble in acetone, benzene and ethanol, etc.
819.	Propineb[f]	120721-83-9	Low	$(C_5H_8N_2S_4Zn)x$	–S	F	289.76	Insoluble	logKow=–0.26, toluene, n-hexane, dichloromethane 1<0.1
820.	Propylene thiourea[f]	2122-19-2	Low	$C_4H_8N_2S$	–S	Metabolite	116.18	Insoluble	Soluble in various organic solvents
821.	Pyriftalid[f]	135186-78-6	Low	$C_{15}H_{14}N_2O_4S$	–N	PGR	318.34	Insoluble	Soluble in various organic solvents
822.	Pyrithiobac-sodium[f]	123343-16-8	Low	$C_{13}H_{10}ClN_2NaO_4S$	–N	H	348.7	7.05×10^5	logKow=–0.84, methanol 270, acetone 0.812, dichloromethane 0.00838, n-hexane 0.01
823.	Sethoxydim[f]	74051-80-2	Low	$C_{17}H_{29}NO_3S$	–N	H	327.5	24.5	Soluble in various solvents
824.	Thiamethoxam[f]	153719-23-4	Low	$C_8H_{10}ClN_5O_3S$	N	I	291.72	4100	Acetone 48, ethyl acetate 7, methanol 13, dichloromethane 110, toluene 0.68, hexane 0.01

No.	Name	CAS		Formula			MW	Solubility	Solvent solubility
825.	Thiocyclam hydrogenoxalate^f	31895-21-3	Medium	$C_7H_{13}NO_4S_3$	–S	I	271.4	8.4%	Dimethyl sulfoxide 92, methanol 17, ethanol 1.9, acetonitrile 1.2, acetone 0.5
826.	Thiram^f	137-26-8	Low	$C_6H_{12}N_2S_4$	–N	F	240.44	30	Soluble in chloroform and acetone, slightly soluble in ethanol and ethers
827.	Trimethylsulfonium iodide^f	2181-42-2	Low	C_3H_9SI	–S	F	204.07	Aqueous	Soluble in various solvents
828.	Zineb^f	12122-67-7	Low	$C_4H_6N_2S_4Zn$	–S	F	275.74	10	Slightly soluble in pyridine, insoluble in common solvents
829.	Ziram^f	137-30-4	Low	$C_6H_{12}N_2S_4Zn$	–S	F	305.81	65	Soluble in chloroform and diffluent in acetone, insoluble in ethanol and diethyl ether
830.	Abamectin^g	71757-41-2	High	$C_{43}H_{72}O_{14}$/ $C_{47}H_{70}O_{16}$	/	I	872/858	0.0078	Diffluent in toluene, acetone, chloroform, ethanol and methanol
831.	Acenaphthene^g	83-32-9	Low	$C_{12}H_{10}$	/	A	154.21	Insoluble	Chloroform 40, benzene 20, ethanol 3, methanol 2
832.	Acequinocyl^g	57960-19-7	Low	$C_{24}H_{32}O_4$	/	A	384.5	0.007	Dichloromethane 620, toluene 450, acetone 220, dimethyl sulfoxide 190, n-hexane 44, methanol 7.8
833.	Aclonifen^g	74070-46-5	Low	$C_{12}H_9ClN_2O_3$	–Cl	H	264.7	4.9	Miscible with organic solvents such as methanol, acetone, toluene, n-hexane, n-octyl alcohol etc.

Continued

No.	Compound	CAS	Toxicity	Formula	Characteristic Atom[1]	Effect[2]	MW	WS (mg/L)	Solubility of Organic Solvent (g/L)
834.	Alloxydim-sodium[g]	66003-55-2	Low	$C_{17}H_{24}O_5Na$	–O	I	345.4	2×10^6	Dimethylformamide 1000, methanol 619, ethanol 50, acetone 14, xylene and ethyl acetate 4
835.	Aluminum phosphide[g]	20859-73-8	High	H_6OPAl	–P	I	79.95	26%	Insoluble in organic solvents
836.	Anthraquinone[g]	84-65-1	Low	$C_{13}H_8O_2$	/	Bird attractant	196.13	Insoluble	Chloroform 6.1, benzene 2.6, ethanol and diethyl ether 1
837.	Benzoximate[g]	29104-30-1	Low	$C_{18}H_{18}ClNO_5$	–N	A	363.8	Insoluble	Dimethylformamide 1460, acetone 980, xylene 710, benzene 650, ethanol 70
838.	Bifenox[g]	42576-02-3	Low	$C_{14}H_9Cl_2NO_5$	–Cl	H	342.1	0.35	Acetone 400, xylene 300, ethanol 50
839.	Binapacryl[g]	485-31-4	Medium	$C_{15}H_{18}N_2O_6$	–N	I/A/F	322.15	Insoluble	Acetone 780, xylene 700, ethanol 110
840.	Diphenyl[g]	92-52-4	Low	$C_{12}H_{10}$	–O	F	154.21	Insoluble	Diffluent in solvents like alcohol ether etc.
841.	Bromoxynil[g]	1689-84-5	Medium	$C_7H_3Br_2NO$	–Br	H	277	130	Dimethylformamide 610, tetrachlorofunan 410, acetone 170, methanol 90, benzene 10
842.	Bromopropylate[g]	18181-80-1	Low	$C_{17}H_8Br_2O_3$	–Br	A	428.1	5	Soluble in most organic solvents
843.	Bromoxynil octanoate[g]	1689-99-2	Medium	$C_{15}H_{17}Br_2NO_2$	–Br	H	403.35	Insoluble	Xylene 700, acetone, methanol 100

No.	Name		Formula		PGR/H	MW	Water solubility	Solubility
844.	Chlorfurenol[g]	Low	$C_{14}H_9ClO_3$	–Cl	PGR/H	260.68	Insoluble	Soluble in various organic solvents
845.	Chlormequat[g]	Low	$C_5H_{13}Cl_2N$	–Cl	PGR	158.07	Insoluble	Soluble in various organic solvents
846.	Cloquintocet-mexyl[g]	Low	$C_{11}H_8ClNO_3$	–Cl	S	335.8	0.59	Toluene 0.36, acetone 0.34, ethanol 0.19, n-octyl alcohol 0.011, hexane 0.0014
847.	Coumatetralyl[g]	High	$C_{19}H_{16}O_3$	/	R	292.3	Insoluble	Soluble in acetone and ethanol
848.	Cyhexatin[g]	Low	$C_{18}H_{34}OSn$	/	A	385.16	1000	Chloroform 216, methanol 37, acetone 1.3
849.	1,2-Dibromo ethane[g]	Low	$C_2H_4Br_2$	–Br	I	187.88	Insoluble	Soluble in various organic solvents
850.	Dibutyl succinate[g]	Low	$C_{13}H_{22}O_4$	–O	M	242.32	Insoluble	Miscible with various organic solvents
851.	Dimethyl phthalate[g]	Low	$C_{10}H_{10}O_4$	/	Insect attractant	194.19	4300	Miscible with ethanol and diethyl ether etc.
852.	Dinobuton[g]	Medium	$C_{14}H_{18}N_2O_7$	–N	A/F	326.31	Insoluble	Soluble in aliphatic hydrocarbon and ethanol, diffluent in aliphatic ketone and aromatic hydrocarbon
853.	Dinoseb acetate[g]	Medium	$C_{12}H_{14}N_2O_6$	–N	H	282.26	2200	Soluble in aromatic solvents
854.	Diofenolan[g]	Medium	$C_{18}H_{20}O_4$	/	I	300.35	Insoluble	Soluble in various organic solvents
855.	Emamectin benzoate [B1a][g]	Medium	$C_{49}H_{75}NO_{13} \cdot C_7H_6O_2$	/	I	1008.6	Slight soluble	Soluble in acetone and toluene, insoluble in hexane

Continued

No.	Compound	CAS	Toxicity	Formula	Characteristic Atom[1]	Effect[2]	MW	WS (mg/L)	Solubility of Organic Solvent (g/L)
856.	Emamectin benzoate [B1b]g	15569-91-8	Low	$C_{48}H_{73}NO_{13}$ · $C_7H_6O_2$	/	I	994.23	Slight soluble	Soluble in acetone and toluene, insoluble in hexane
857.	Endothalg	145-73-3	High	$C_8H_{10}O_5$		PGR/H	186.2	1×10^5	Methanol 280, dioxane 76, acetone 70, benzene 0.1
858.	Ethoxyquing	91-53-2	Low	$C_{14}H_{19}NO$	–N	F	217.31	Insoluble	Soluble in various organic solvents
859.	Fenbutatin oxideg	13356-08-6	Low	$C_{60}H_{78}OSn_2$	/	A	1052.66	0.005	CH_2Cl_2 380, benzene 140, acetone 6
860.	Fentin-chlorideg	6369-58-7	Medium	$C_{18}H_{15}ClSn$		F/I	384.72	Slight soluble	Slightly soluble in various organic solvents
861.	Fentin acetateg	900-95-8	Medium	$C_{20}H_{18}O_2Sn$		F/I	409.04	28	Slightly soluble in various organic solvents
862.	Gibberellic acidg	77-06-5	Low	$C_{19}H_{22}O_6$	/	PGR	346.4	5000	Soluble in ethyl acetate, methanol, ethanol and acetone, insoluble in benzene and chloroform
863.	Inorganic bromideg	24959-67-9	Low	Br_2	–Br		160.2		
864.	Ioxynilg	689-83-4	Medium	$C_7H_3NOI_2$	–N	H	370.92	50	Tetrachlorofuran 340, acetone 70, methanol 20, benzene 11, CCl_4 0.14
865.	Metaldehydeg	108-62-3	Low	$C_8H_{16}O_4$		Molluscicide	176.2	200	Diffluent in benzene and chloroform, slightly soluble in diethyl ether
866.	Methopreneg	40596-69-8	Low	$C_{19}H_{34}O_3$	/	IGR	310.19	1.39	Soluble in organic solvents

	Name	CAS	Volatility	Formula			MW	Water solubility	Solubility
867.	Methyl bromide[g]	74-83-9	High	CH_3Br	−Br	I	94.95	1.34×10^4	Soluble in various solvents
868.	Milbemectin A3[g]	51596-10-2	Medium	$C_{29}H_{67}O_7$	/	A/I	528.7	0.88	Benzene 143, ethyl acetate 69.5, acetone 66.1, methanol 64.8, ethanol 41.9, n-hexane 1.4
869.	Milbemectin A4[g]	51596-11-3	Medium	$C_{29}H_{67}O_7$	/	A/I	542.7	0.88	Benzene 143, ethyl acetate 69.5, acetone 66.1, methanol 64.8, ethanol 41.9, n-hexane 1.4
870.	Musk ambrette[g]	83-46-9	Low	$C_{12}H_{16}N_2O_5$	−N	I/A	268.12	Insoluble	Soluble in various organic solvents
871.	Musk ketone[g]	541-91-3	Low	$C_{16}H_{30}O$	/	I/A	238.41	Slight soluble	Soluble in ethanol
872.	Musk moskene[g]	116-66-5	Low	$C_{14}H_{18}N_2O_4$	−N	I/A	278.34	Insoluble	Soluble in various organic solvents
873.	Musk tibetene[g]	145-39-1	Low	$C_{13}H_{18}N_2O_4$	−n	I/A	266.33	Insoluble	Soluble in various organic solvents
874.	Musk xylene[g]	81-15-2	Low	$C_{12}H_{15}N_3O_6$	−N	I/A	297.27	Insoluble	Soluble in various organic solvents
875.	Nitrothal-isopropyl[g]	10552-74-6	Low	$C_{14}H_{17}NO_6$	−N	F	295.3	Insoluble	Diffluent in various solvents
876.	Ortho-phenylphenol[g]	90-43-7	Low	$C_{12}H_{10}O$	/	F	170.2	Slight soluble	Soluble in various solvents
877.	Phenanthrene[g]	65-01-8	Low	C_4H_{10}	/	I/A	178.23	Insoluble	Soluble in various organic solvents
878.	2-Phenylphenol[g]	90-43-7	Low	$C_{12}H_{10}O$	/	F	170.12	700	Soluble in various solvents
879.	3-Phenylphenol[g]	580-51-8	Low	$C_{12}H_{10}O$	/	F	170.2	Slight soluble	Soluble in various solvents

Continued

No.	Compound	CAS	Toxicity	Formula	Characteristic Atom[1]	Effect[2]	MW	WS (mg/L)	Solubility of Organic Solvent (g/L)
880.	Phthalic acid, benzylbutyl ester[g]	85-68-7	Low	$C_{19}H_{20}O_4$	/	I/A	312.19	Insoluble	Soluble in various organic solvents
881.	Phthalic acid,di-(2-ethylhexyl)ester[g]	117-81-7	Low	$C_{16}H_{20}O_4$	/	I/A	276.16	Insoluble	Soluble in various organic solvents
882.	Phthalic acid,dibutyl ester[g]	84-74-2	Low	$C_{16}H_{22}O_4$	/	I/A	278.16	11.2	logKow=4.45
883.	Phthalic acid, discyclohexyl ester[g]	84-61-7	Low	$C_{20}H_{26}O_4$	/	I/A	330.2	Insoluble	Soluble in various organic solvents
884.	Phthalide[g]	27355-22-2	Low	$C_8H_2Cl_4O_2$	—Cl	F	271.9	2.49	Tetrahydrofuran 19.3, benzene 14.1, acetone 8.3, ethanol 1.1
885.	Picoxystrobin[g]	117428-22-5	Low	$C_{18}H_{16}F_3NO_4$	–F	F	367.32	128	logKow=3.6, diffluent in various organic solvents
886.	Piperonyl butoxide[g]	51-03-6	Low	$C_{19}H_{30}O_5$	/	Synergist	338.43	Insoluble	Soluble in various solvents
887.	prohexadione-calcium[g]	124537-28-6	Low	$C_{10}H_{10}CaO_3$	/	PGR	250.3	168	Soluble in various organic solvents
888.	Prohydrojasmon[g]	158474-72-7	Low	$C_{15}H_{26}O_3$	/	PGR	254.37	60	Acetone, acetonitrile, chloroform, ethyl acetate, methanol >100
889.	propylene oxide[g]	75-56-9	Low	C_3H_6O	/		58.08	4.05×10^5	Mutual soluble with ethanol, diethyl ether, acetone, methanol
890.	Pyriproxyfen[g]	95737-68-1	Low	$C_{20}H_{29}NO_3$	–N	IGR	331.5	Insoluble	Xylene 500, hexane 400, methanol 200
891.	Pyroquilon[g]	57369-32-1	Medium	$C_{11}H_{11}NO$	–N	F	173.2	4000	CH_2Cl_2 580, methanol 240, benzene 200, acetone 125, isopropanol 85

No.	Name	CAS	Toxicity	Formula	[1]	[2]	MW	Solubility (water)	Solubility
892.	Quinoclamine[g]	2797-51-5	Low	$C_{10}H_6ClNO_2$	-Cl	H	207.6		logKow = 1.48, nitrobenzol 37, acetic acid 16, acetone 13, chlorobenzol 5,
893.	Rotenone[g]	83-79-4	Medium	$C_{23}H_{22}O_6$		I	394.43	15	Slightly soluble in CCl_4 and polar solvents
894.	Silafluofen[g]	105024-66-6	Low	$C_{25}H_{29}FO_2Si$	-F	I	408.59	0.001	Soluble in various organic solvents
895.	Sulfur[g]	774-94-7	Low	S		F	32.06	Insoluble	Slightly soluble in ethanol and diethyl ether
896.	TCA-sodium[g]	650-51-1	Medium	$C_2Cl_3O_2Na$	-Cl	H	185.4	Diffluent	Indissolvable in organic solvents
897.	Tepraloxydim[g]	149979-41-9	Low	$C_{17}H_{24}ClNO_4$	-N	H	341.8	0.14	Soluble in most organic solvents
898.	Tralkoxydim[g]	87820-88-0	Low	$C_{20}H_{27}NO_3$	-N	H	3292	6	CH_2Cl_2 500, toluene 213, ethyl acetate 110, acetone 89, methanol 25, hexane 18
899.	Tridemorph[g]	24602-86-6	Low	$C_{19}H_{39}NO$	-N	F	297.5	100	Miscible with acetone, benzene, chloroform, ethanol, cyclohexane, etc.
900.	Warfarin[g]	00081-81-2	Medium	$C_{19}H_{16}O_4$	/	R	308.4	Insoluble	Soluble in methanol, ethanol, isopropanol; dioxane 100, acetone 65, chloroform 56

Note: [1] Characteristic atom; [2] Classification for pesticide functions: I, Insecticide; F, Bactericide; H, Herbicide; A, Acaricide; N, Nematicide; R, Rodenticide, IGR, Insect growth regulator; PGR, Plant growth regulator, S Safener; [3] Classification for pesticides according to the compound: [a] Organic halogen pesticide, [b] Organophosphorus pesticide, [c] Pyrethroid pesticide, [d] Carbamate pesticide, [e] Organic nitrogen pesticide, [f] Organosulphorous pesticide, [g] Other pesticides.

FURTHER READING

[1] Feng J, Gu Q, Bo Y. Handbook of Pesticide terms. 3rd ed. Beijing, China: Chemical industry Press; 2009.

[2] Shi D. Chinese Pesticide Dictionary. Beijing, China: Chemical Industry Press; 2008.

[3] Zhu Y, Wang Z, Li B. Encyclopedia of agricultural chemical. Beijing, China: Three Gorges Publishing House/Agricultural Technology and Education Press; 2006.

[4] Zhang M. New Handbook of Pesticide Commodities. Beijing, China: Chemical industry Press; 2006.

[5] Liu C. World Pesticide Encyclopedia-for Bactericide. Beijing, China: Chemical industry Press; 2006.

[6] Liu C. World Pesticide Encyclopedia-for Herbicide. Beijing, China: Chemical industry Press; 2002.

[7] Liu Y. Encyclopedia of pesticide application. Beijing, China: Agriculture Press; 1989.

[8] Royal Society of Chemistry. The Agrochemicals Handbook; 1983.

[9] Shi D, Hu X, Cao W. English-Chinese Pesticide Dictionary. Beijing, China: Petroleum and Chemical industry Press; 1978.

[10] Martin H, Worthing CR. Pesticide manual. Issued by the British Crop Protection Council; 1968.

Appendix D

MRM Chromatograms of 493 Pesticides and Related Chemicals

MRM chromatograms of pesticides and related chemicals divided into seven groups, i.e., A, B, C, D, E, F and G groups

A Group

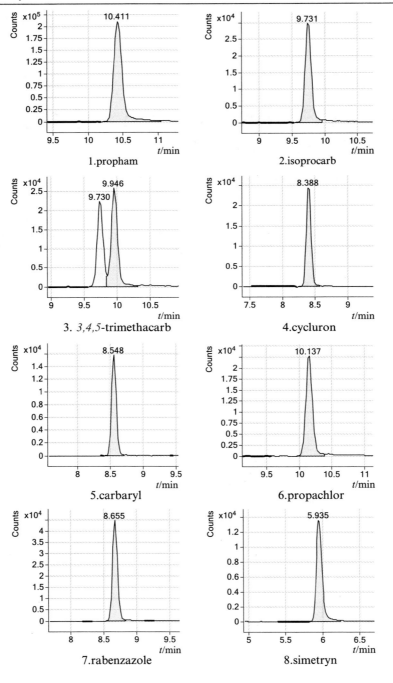

1.propham

2.isoprocarb

3. *3,4,5*-trimethacarb

4.cycluron

5.carbaryl

6.propachlor

7.rabenzazole

8.simetryn

A Group—cont'd

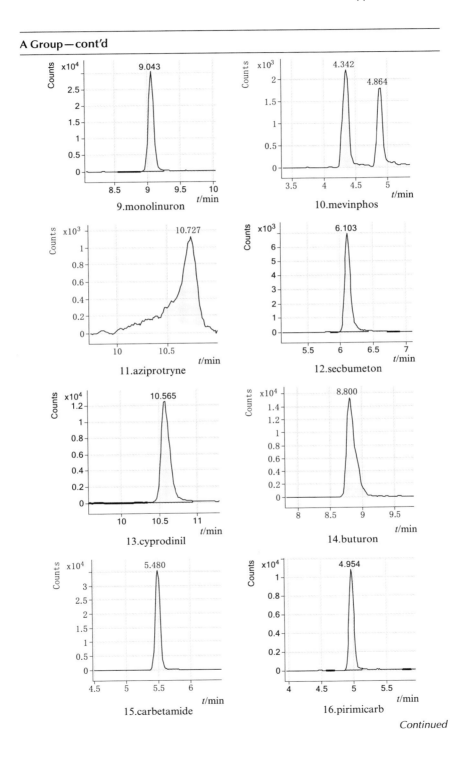

9.monolinuron

10.mevinphos

11.aziprotryne

12.secbumeton

13.cyprodinil

14.buturon

15.carbetamide

16.pirimicarb

Continued

A Group—cont'd

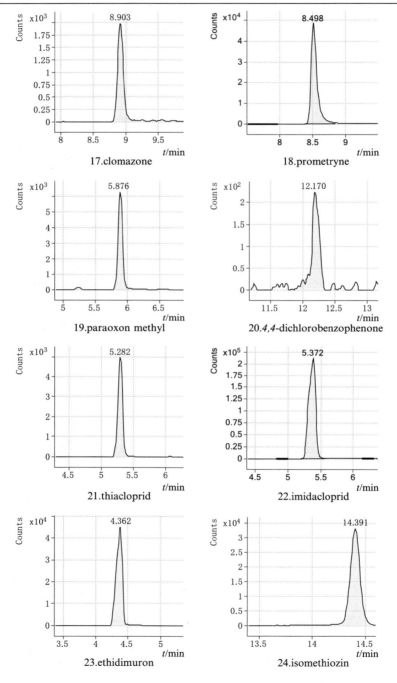

17.clomazone

18.prometryne

19.paraoxon methyl

20.*4,4*-dichlorobenzophenone

21.thiacloprid

22.imidacloprid

23.ethidimuron

24.isomethiozin

A Group—cont'd

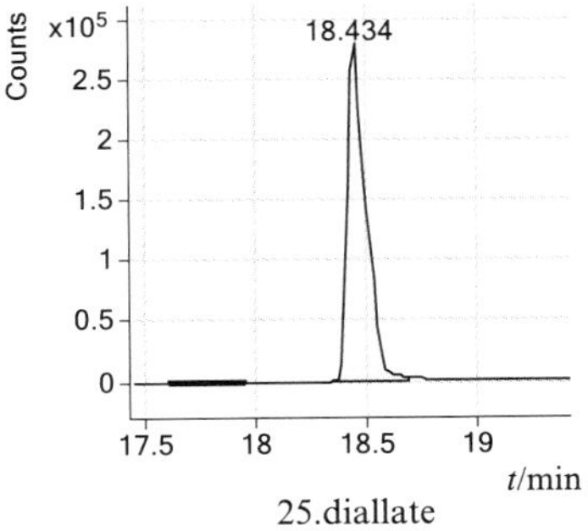

25.diallate

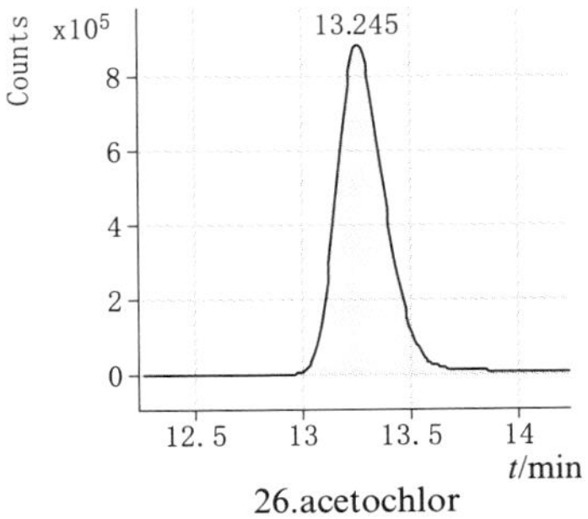

26.acetochlor

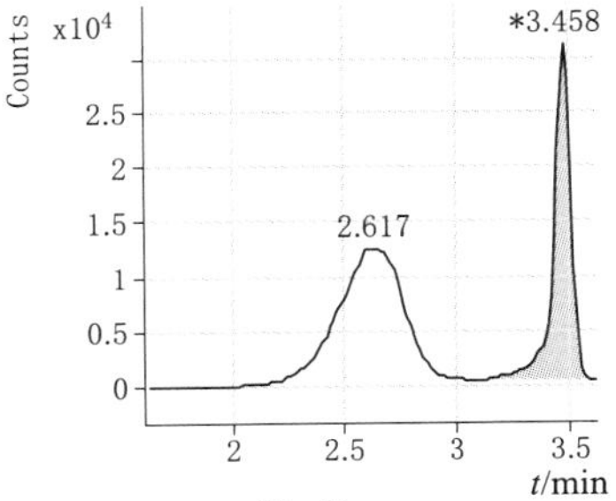

27.nitenpyram

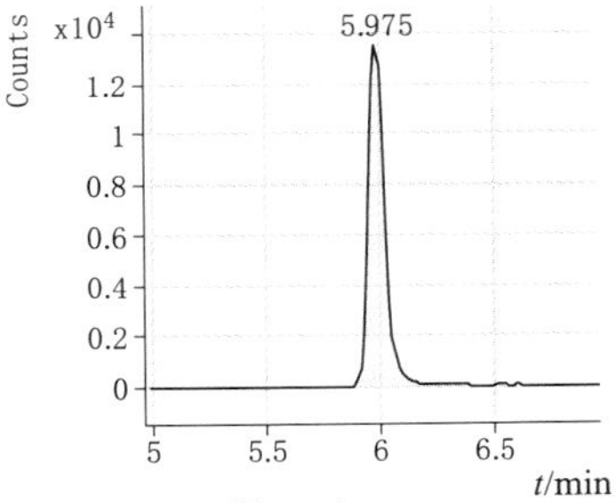

28.methoprotryne

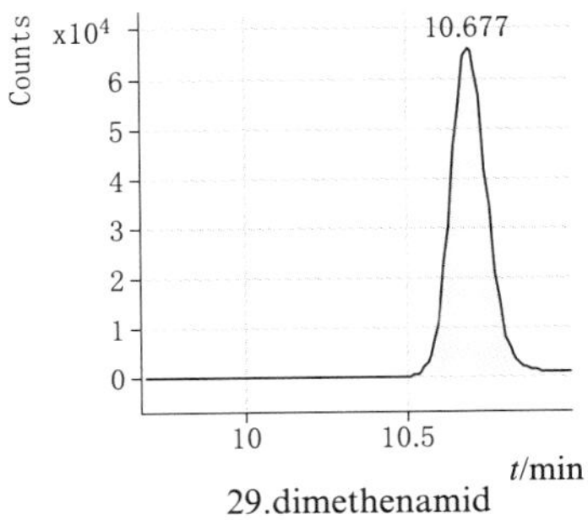

29.dimethenamid

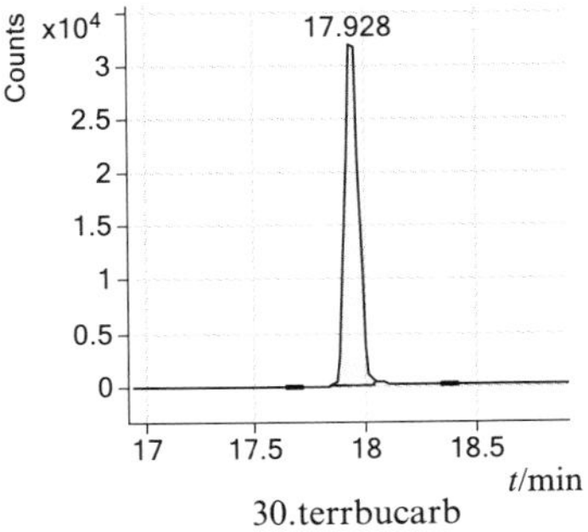

30.terrbucarb

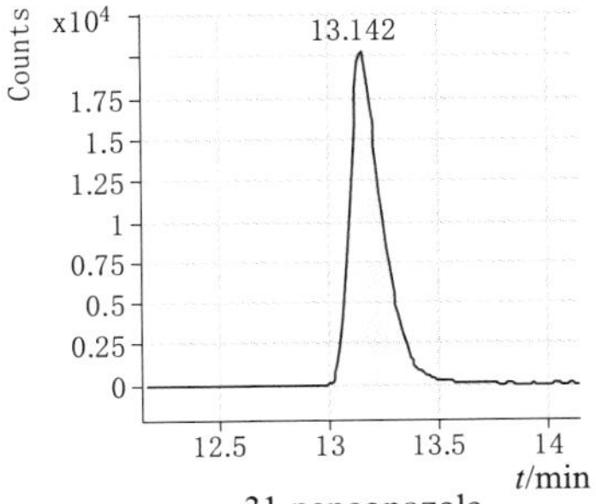

31.penconazole

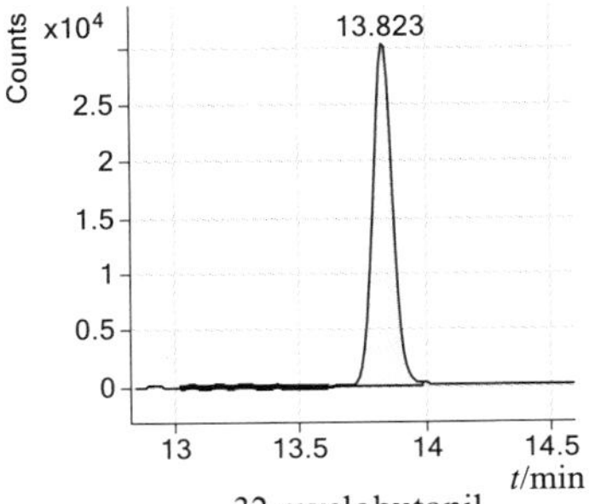

32.myclobutanil

Continued

A Group—cont'd

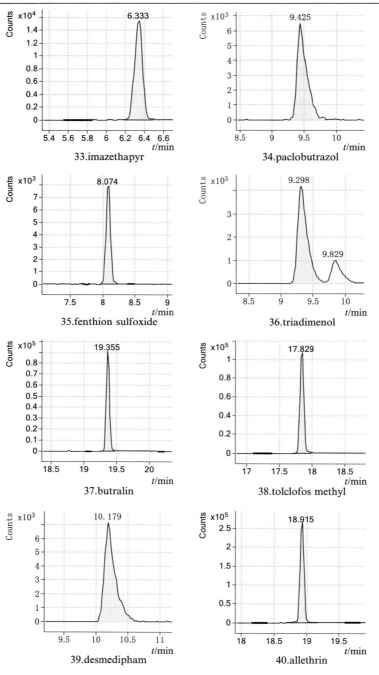

A Group—cont'd

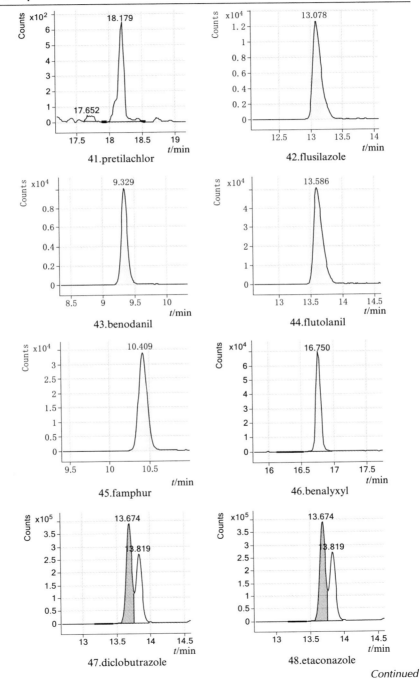

41.pretilachlor

42.flusilazole

43.benodanil

44.flutolanil

45.famphur

46.benalyxyl

47.diclobutrazole

48.etaconazole

Continued

A Group—cont'd

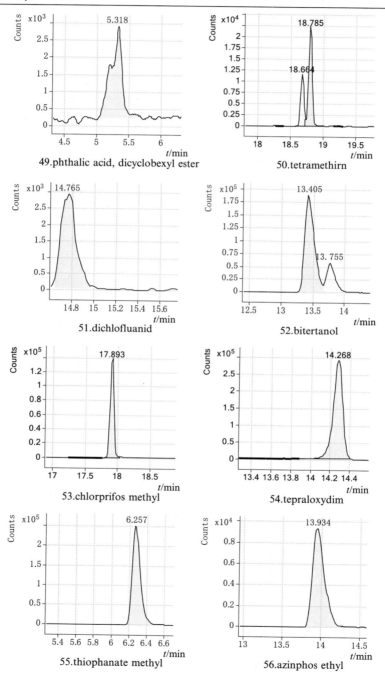

49.phthalic acid, dicyclobexyl ester

50.tetramethirn

51.dichlofluanid

52.bitertanol

53.chlorprifos methyl

54.tepraloxydim

55.thiophanate methyl

56.azinphos ethyl

A Group—cont'd

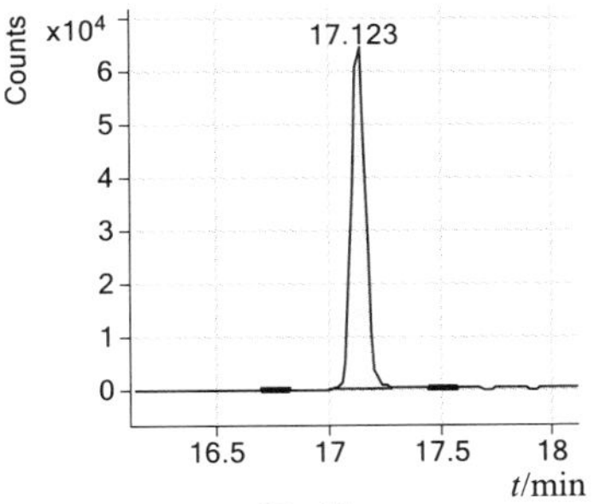

57.triflumuron

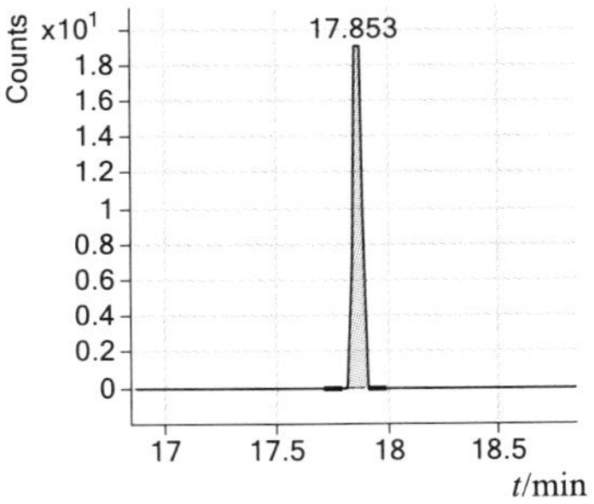

58.anilofos

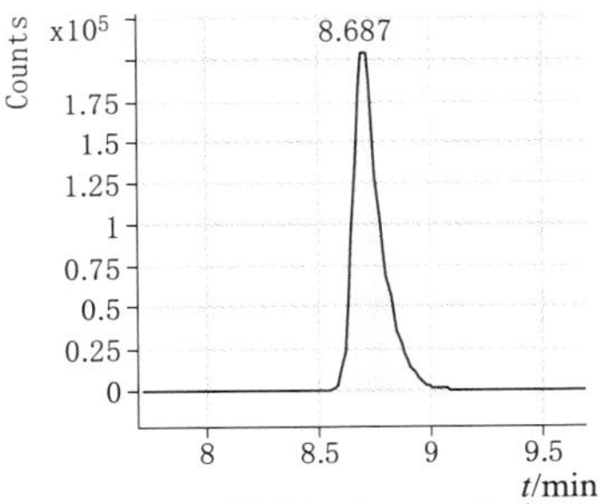

59.thiophanat ethyl

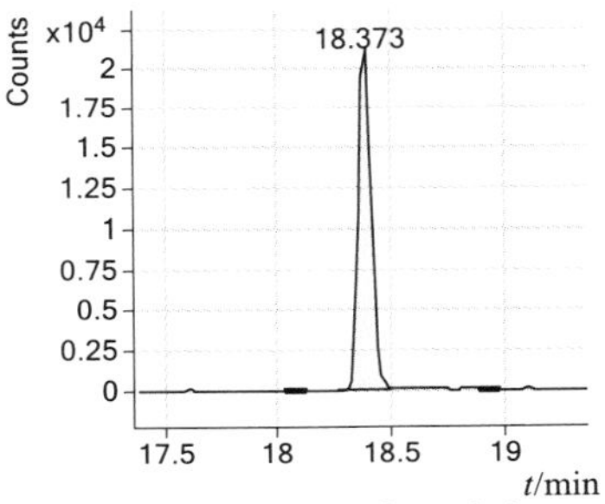

60.quizalofop-ethyl

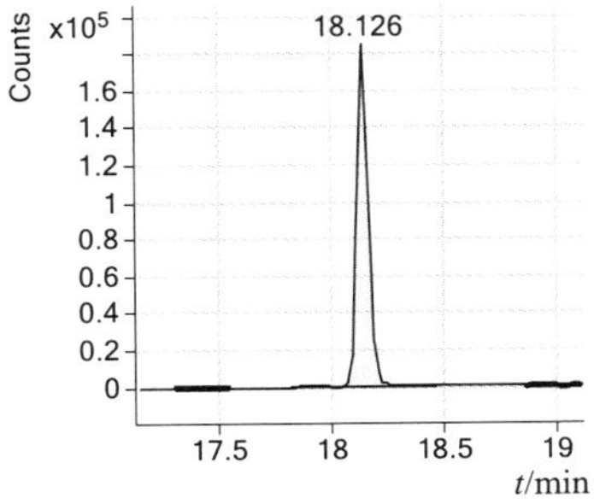

61.haloxyfop-methyl

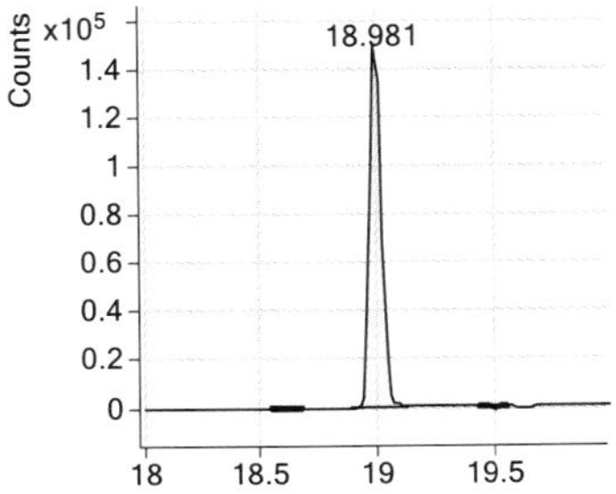

62.fluazifop butyl

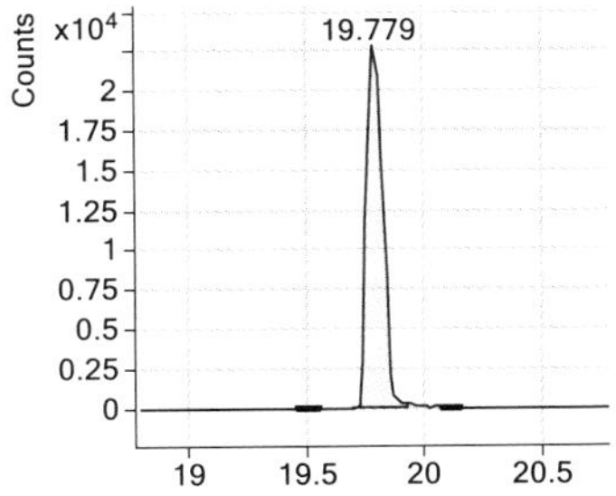

63.bromophos-ethyl

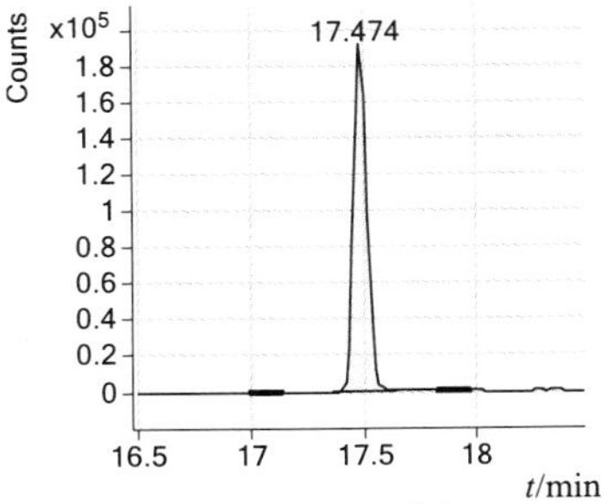

64.bensulide

Continued

A Group—cont'd

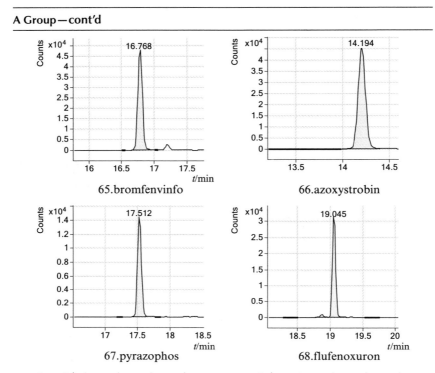

65.bromfenvinfo

66.azoxystrobin

67.pyrazophos

68.flufenoxuron

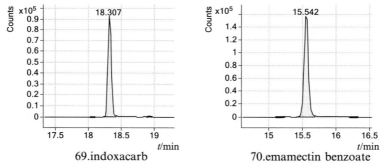

69.indoxacarb

70.emamectin benzoate

B Group

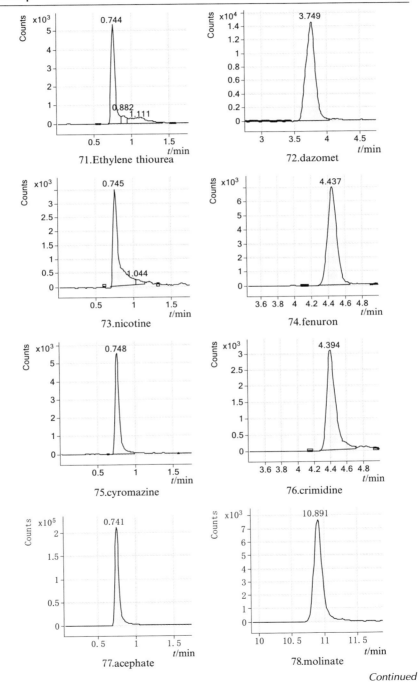

71.Ethylene thiourea

72.dazomet

73.nicotine

74.fenuron

75.cyromazine

76.crimidine

77.acephate

78.molinate

Continued

B Group—cont'd

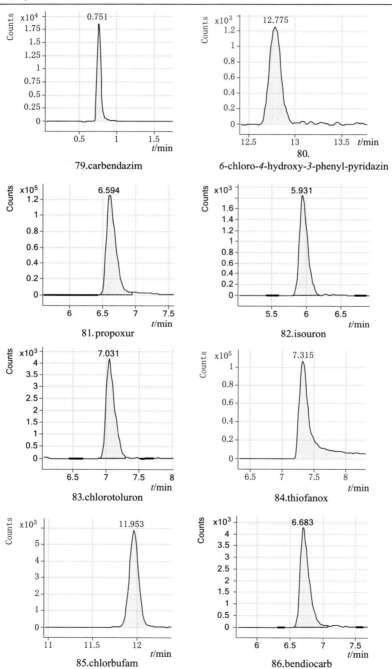

79.carbendazim

80.
6-chloro-4-hydroxy-3-phenyl-pyridazin

81. propoxur

82.isouron

83.chlorotoluron

84.thiofanox

85.chlorbufam

86.bendiocarb

B Group—cont'd

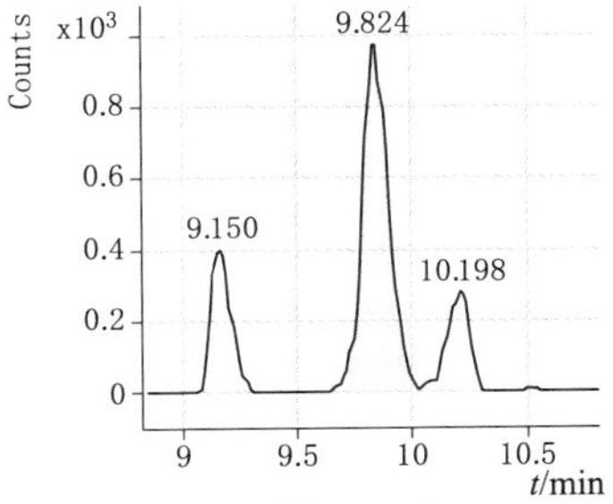

87.propazine

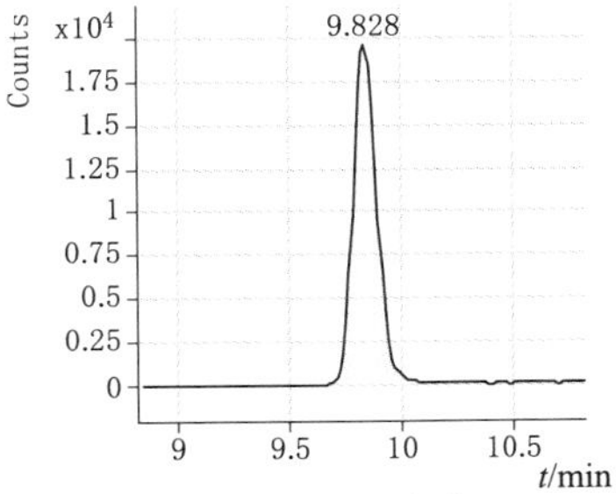

88.terbuthylazine

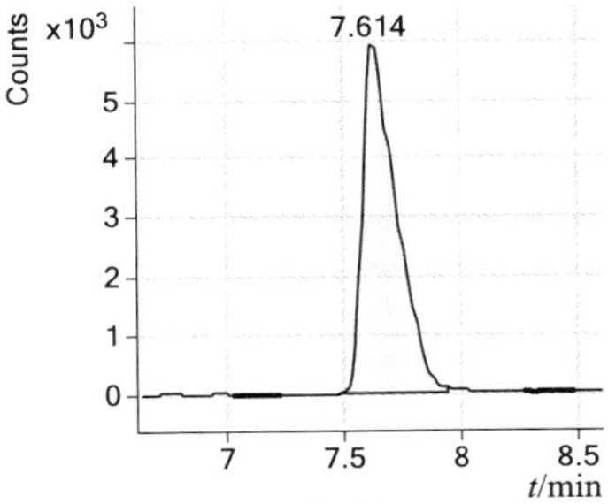

89.diuron

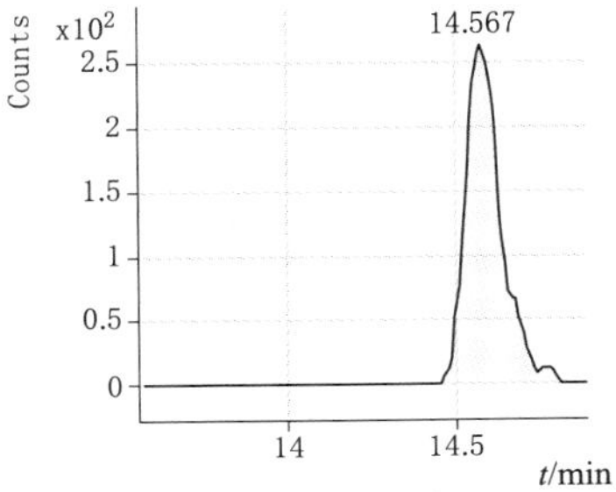

90.chlormephos

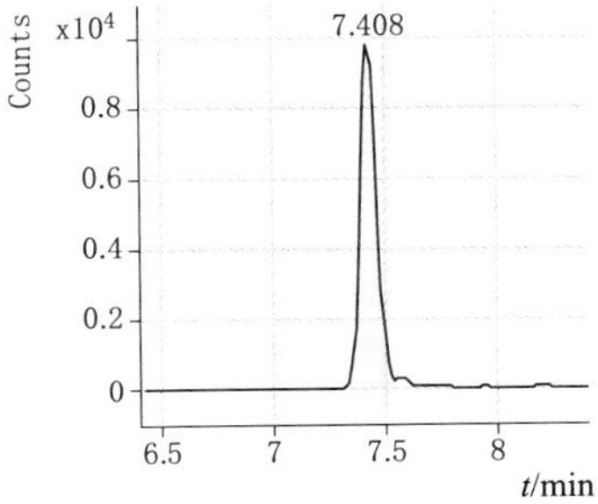

91.carboxin

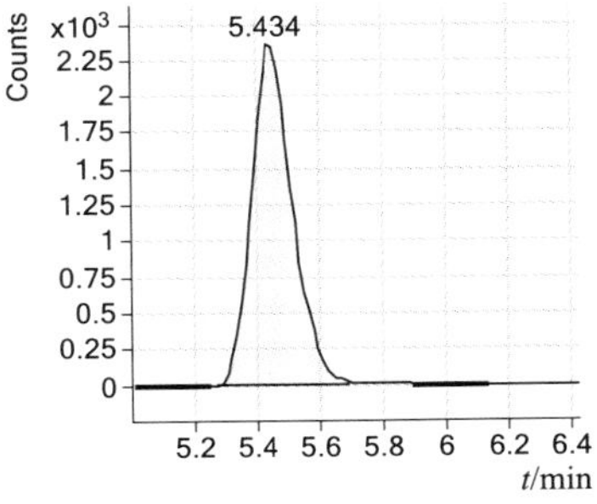

92.difenzoquat-methyl sulfate

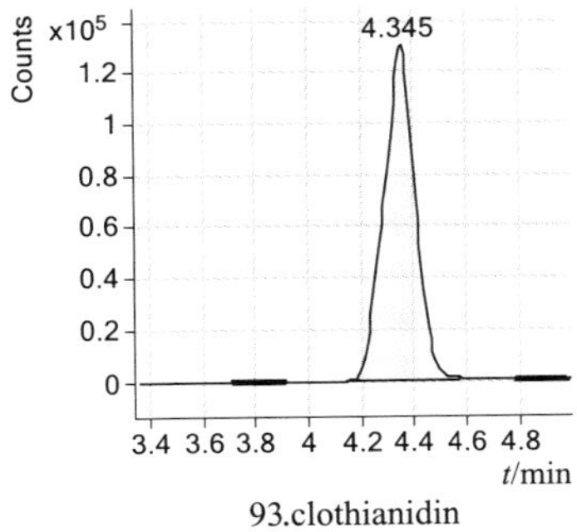

93.clothianidin

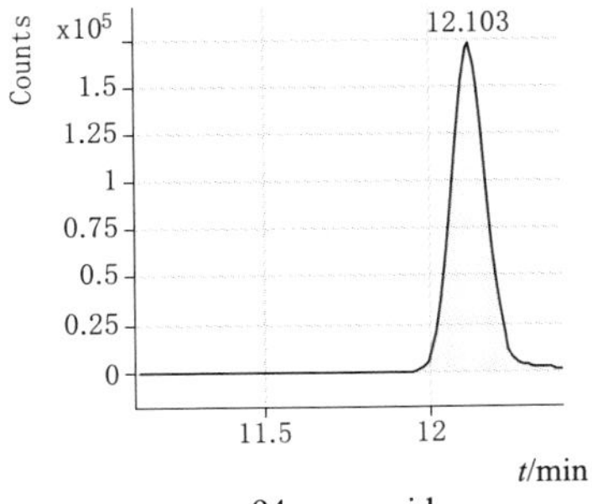

94.pronamide

Continued

B Group—cont'd

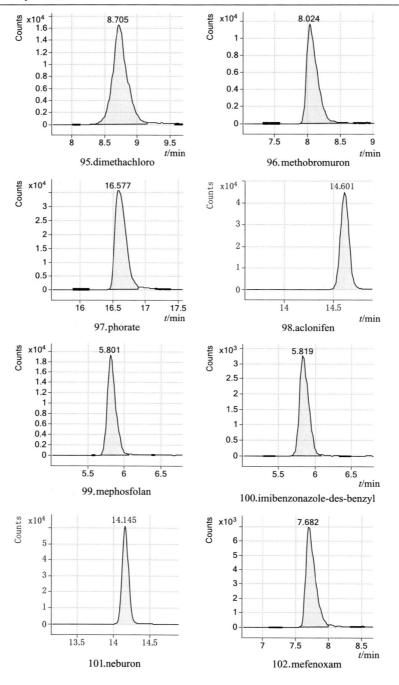

95. dimethachloro

96. methobromuron

97. phorate

98. aclonifen

99. mephosfolan

100. imibenzonazole-des-benzyl

101. neburon

102. mefenoxam

B Group—cont'd

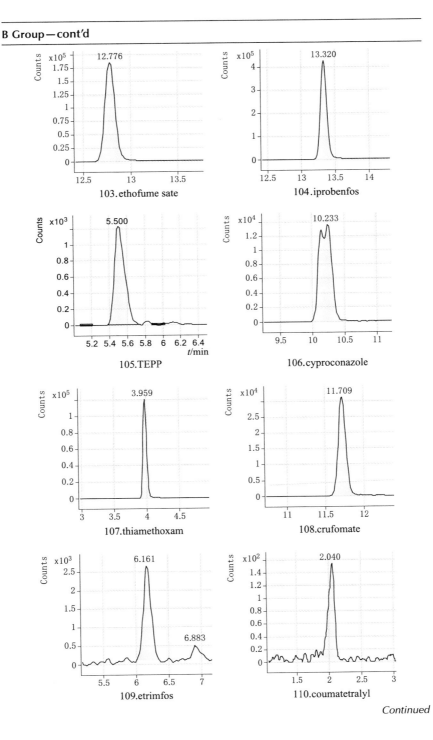

103. ethofume sate

104. iprobenfos

105. TEPP

106. cyproconazole

107. thiamethoxam

108. crufomate

109. etrimfos

110. coumatetralyl

Continued

B Group—cont'd

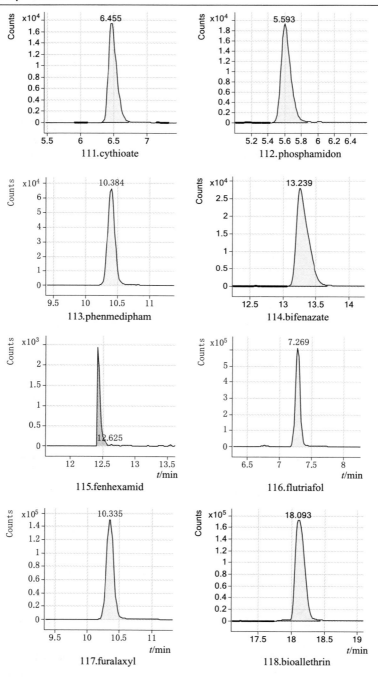

B Group—cont'd

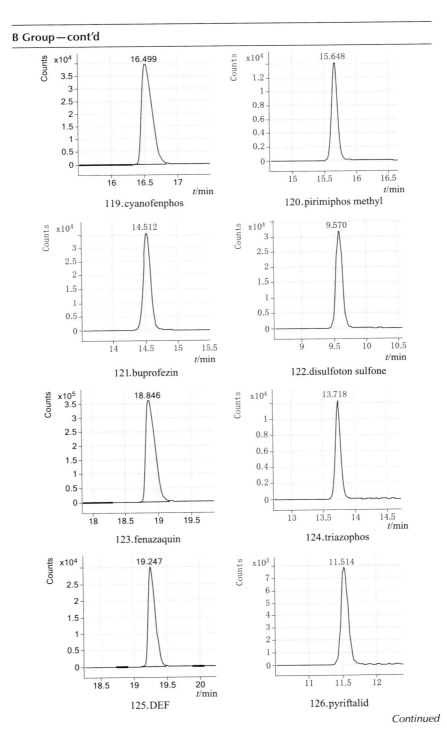

119.cyanofenphos

120.pirimiphos methyl

121.buprofezin

122.disulfoton sulfone

123.fenazaquin

124.triazophos

125.DEF

126.pyriftalid

Continued

B Group—cont'd

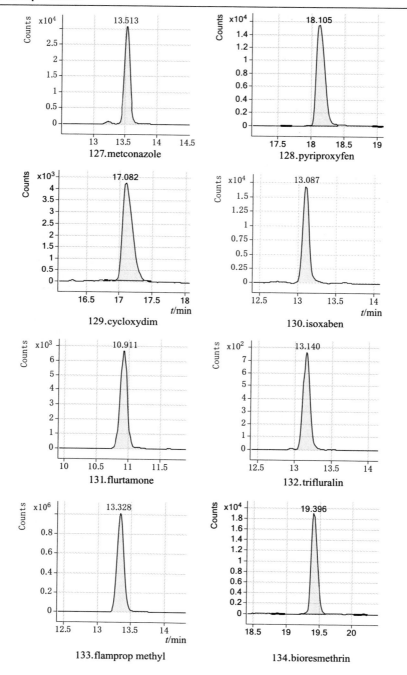

127. metconazole

128. pyriproxyfen

129. cycloxydim

130. isoxaben

131. flurtamone

132. trifluralin

133. flamprop methyl

134. bioresmethrin

B Group—cont'd

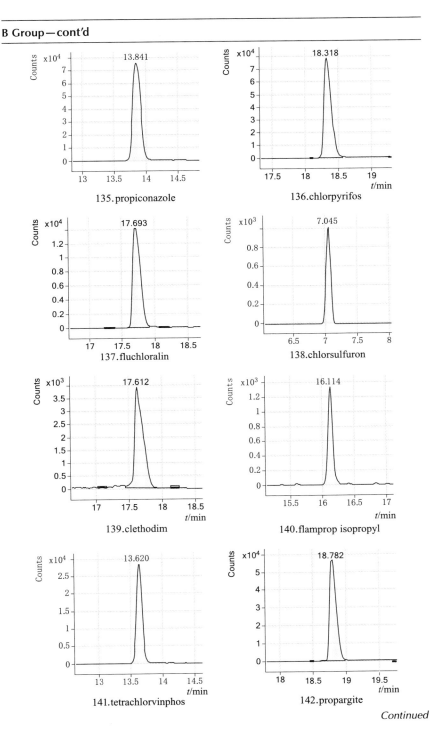

135. propiconazole

136. chlorpyrifos

137. fluchloralin

138. chlorsulfuron

139. clethodim

140. flamprop isopropyl

141. tetrachlorvinphos

142. propargite

Continued

B Group—cont'd

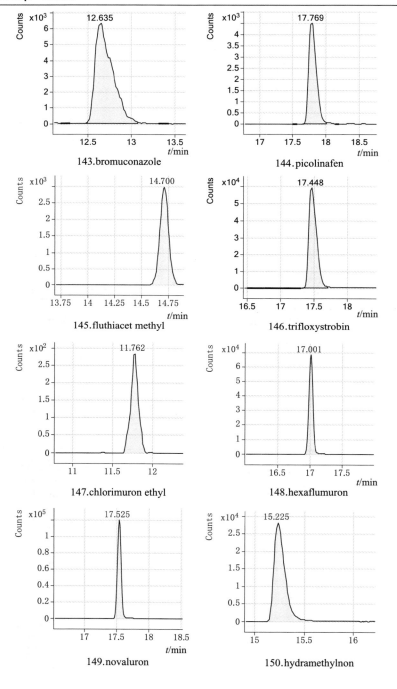

143.bromuconazole

144.picolinafen

145.fluthiacet methyl

146.trifloxystrobin

147.chlorimuron ethyl

148.hexaflumuron

149.novaluron

150.hydramethylnon

B Group—cont'd

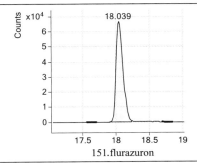

151.flurazuron

C Group

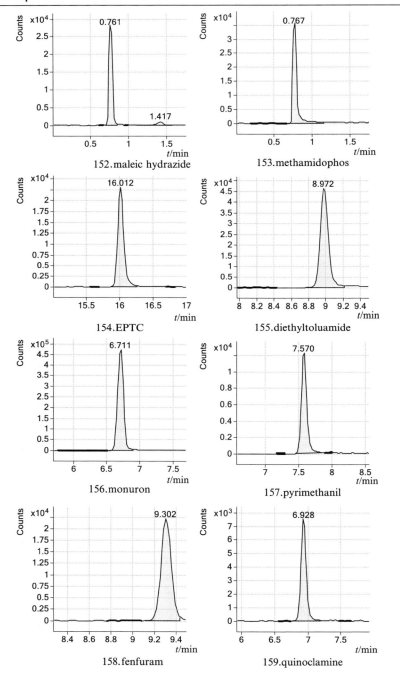

152. maleic hydrazide

153. methamidophos

154. EPTC

155. diethyltoluamide

156. monuron

157. pyrimethanil

158. fenfuram

159. quinoclamine

C Group—cont'd

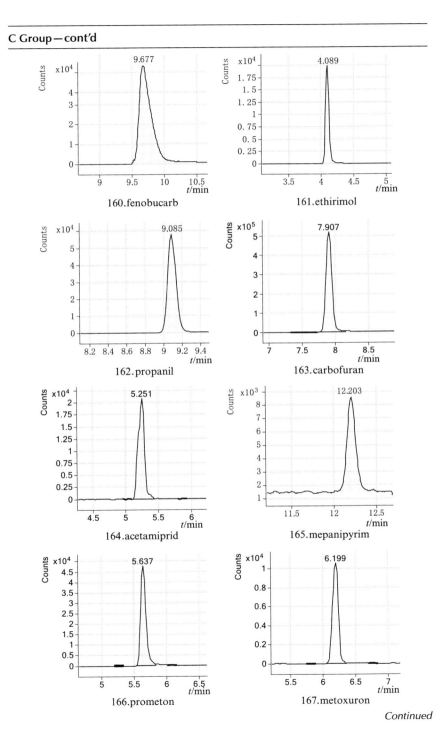

160.fenobucarb

161.ethirimol

162.propanil

163.carbofuran

164.acetamiprid

165.mepanipyrim

166.prometon

167.metoxuron

Continued

C Group—cont'd

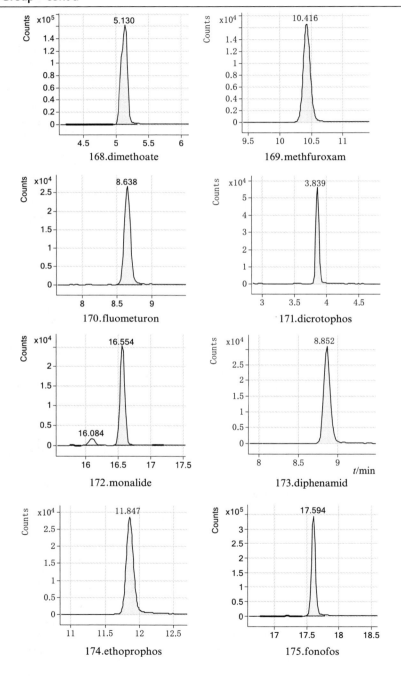

168.dimethoate

169.methfuroxam

170.fluometuron

171.dicrotophos

172.monalide

173.diphenamid

174.ethoprophos

175.fonofos

C Group—cont'd

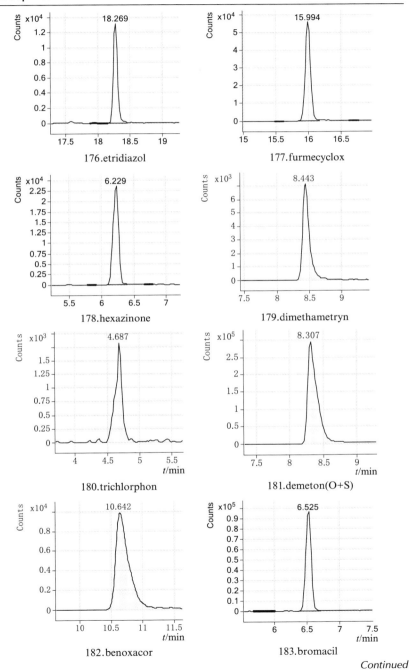

176.etridiazol

177.furmecyclox

178.hexazinone

179.dimethametryn

180.trichlorphon

181.demeton(O+S)

182.benoxacor

183.bromacil

Continued

C Group—cont'd

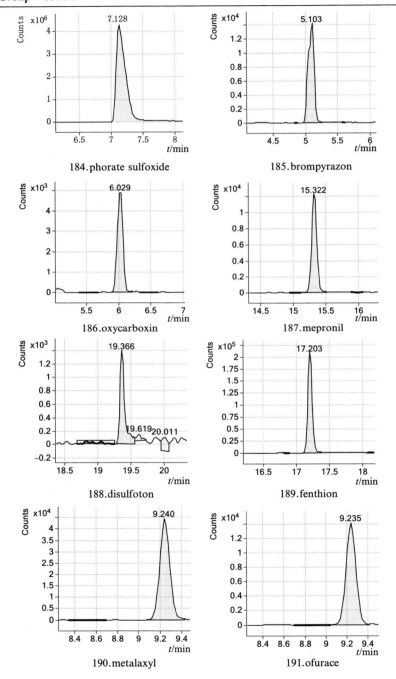

184.phorate sulfoxide

185.brompyrazon

186.oxycarboxin

187.mepronil

188.disulfoton

189.fenthion

190.metalaxyl

191.ofurace

C Group—cont'd

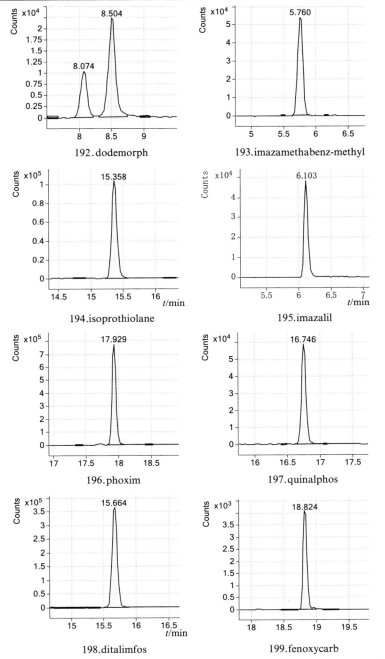

192. dodemorph

193. imazamethabenz-methyl

194. isoprothiolane

195. imazalil

196. phoxim

197. quinalphos

198. ditalimfos

199. fenoxycarb

Continued

C Group—cont'd

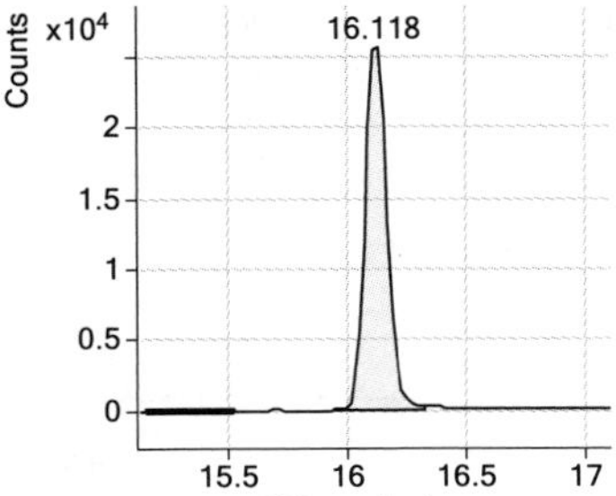

200. pyrimitate

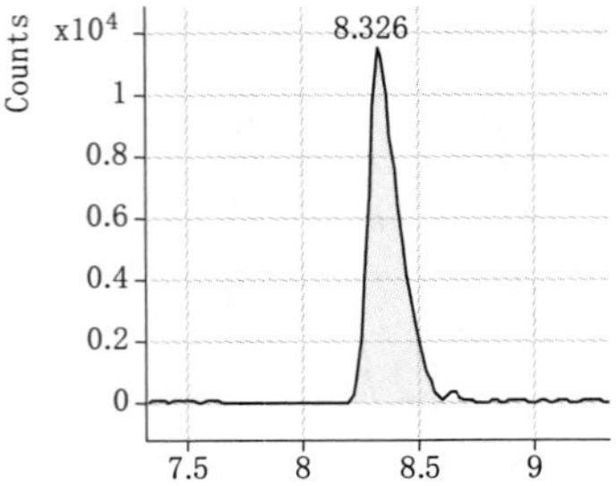

201. fensulfothin

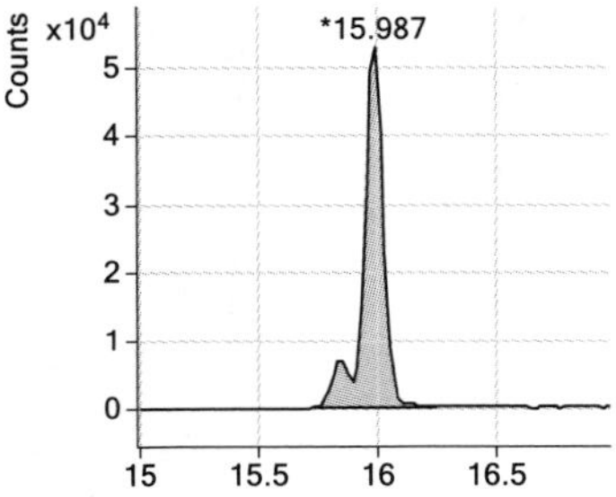

202. fluorochloridone

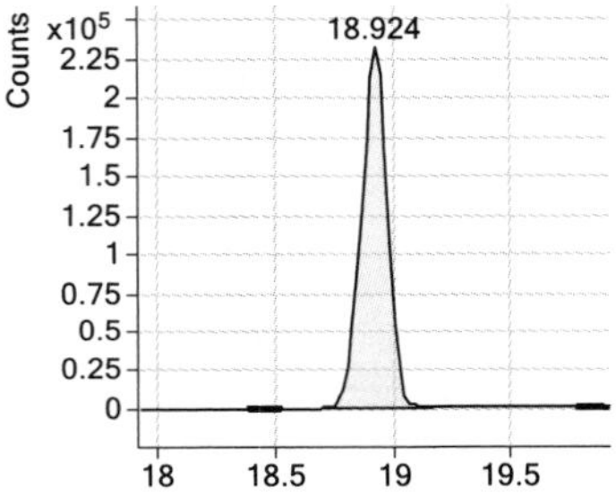

203. butachlor

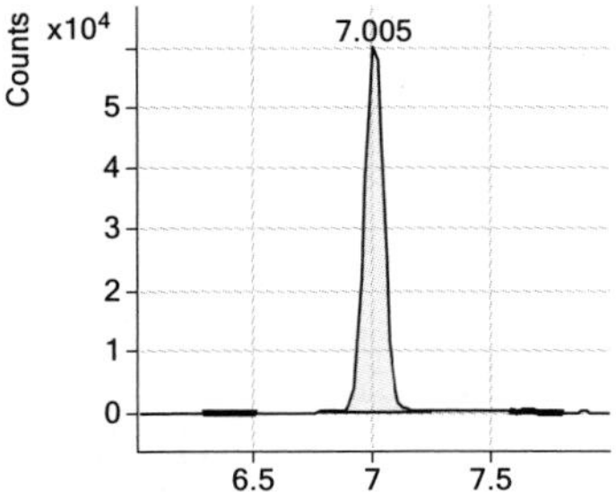

204. imazaquin

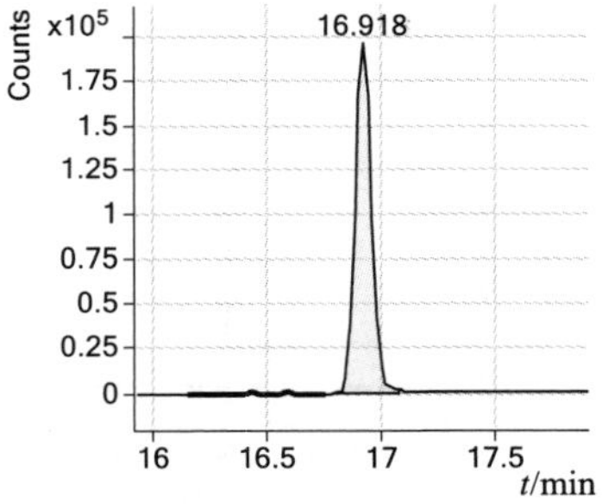

205. kresoxim-methyl

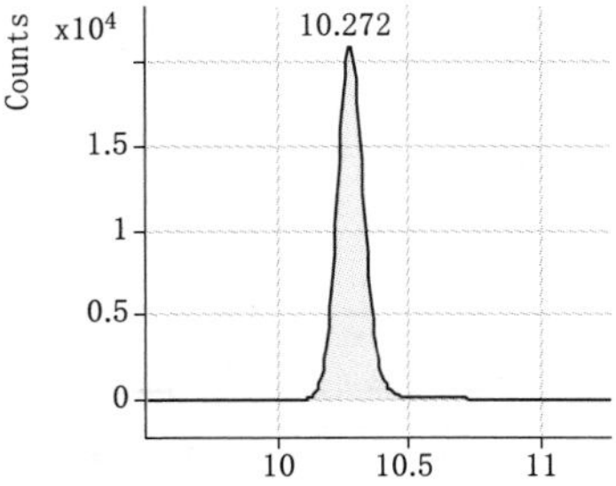

206. triticonazole

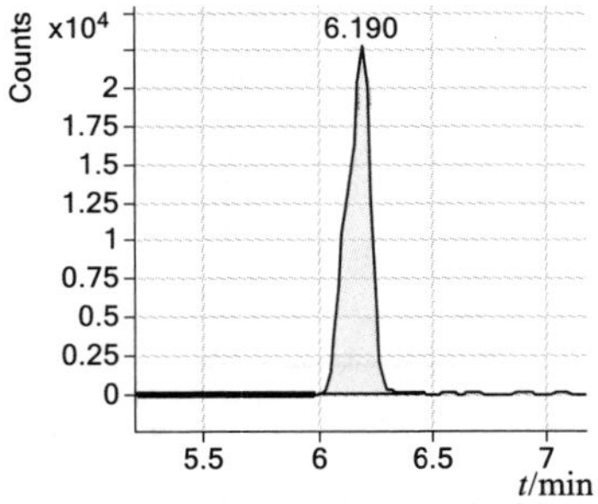

207. fenamiphos sulfoxide

C Group—cont'd

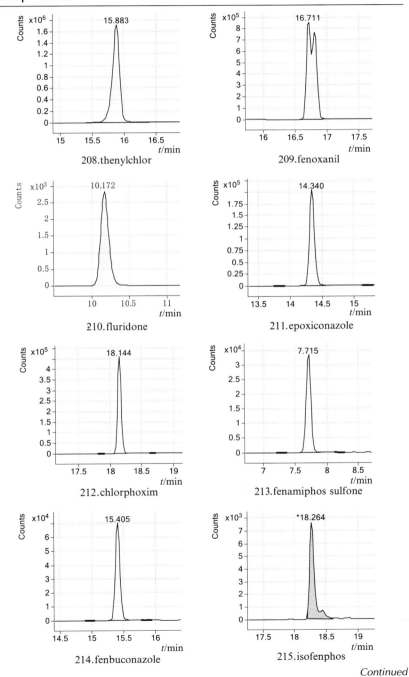

208.thenylchlor

209.fenoxanil

210.fluridone

211.epoxiconazole

212.chlorphoxim

213.fenamiphos sulfone

214.fenbuconazole

215.isofenphos

Continued

C Group—cont'd

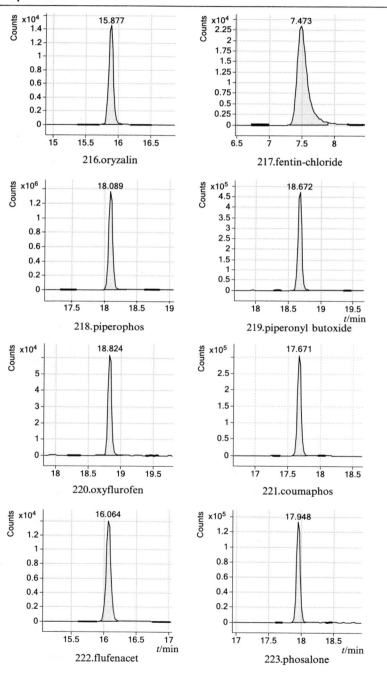

216.oryzalin

217.fentin-chloride

218.piperophos

219.piperonyl butoxide

220.oxyflurofen

221.coumaphos

222.flufenacet

223.phosalone

C Group—cont'd

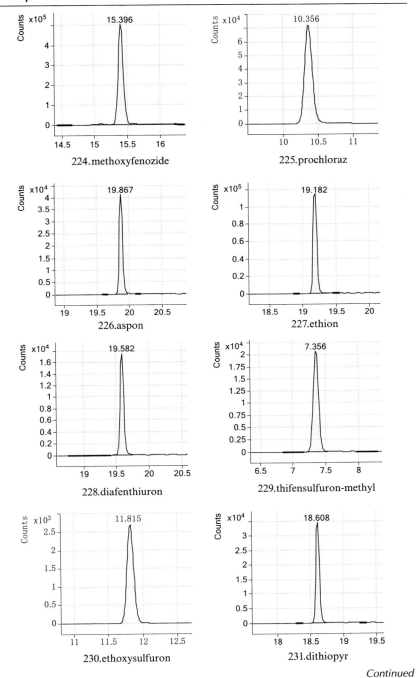

224. methoxyfenozide

225. prochloraz

226. aspon

227. ethion

228. diafenthiuron

229. thifensulfuron-methyl

230. ethoxysulfuron

231. dithiopyr

Continued

C Group—cont'd

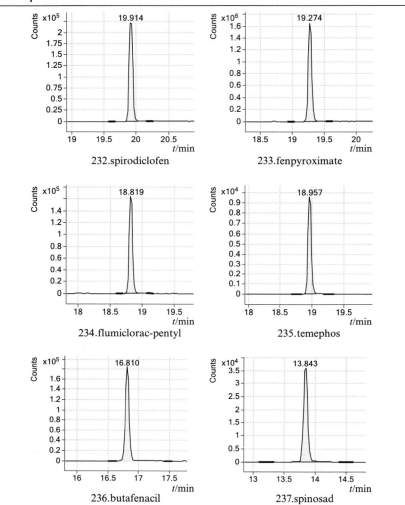

232.spirodiclofen

233.fenpyroximate

234.flumiclorac-pentyl

235.temephos

236.butafenacil

237.spinosad

D Group

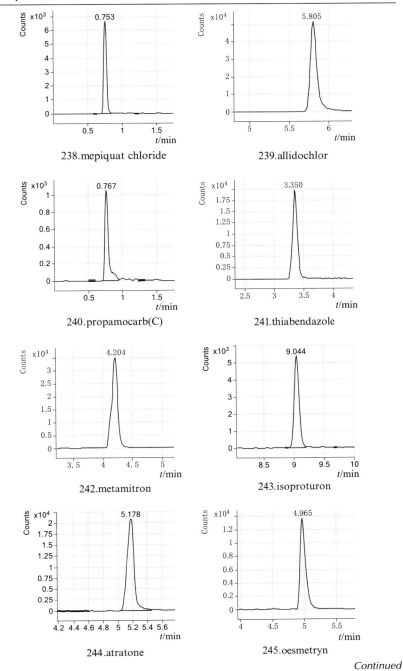

238.mepiquat chloride

239.allidochlor

240.propamocarb(C)

241.thiabendazole

242.metamitron

243.isoproturon

244.atratone

245.oesmetryn

Continued

D Group—cont'd

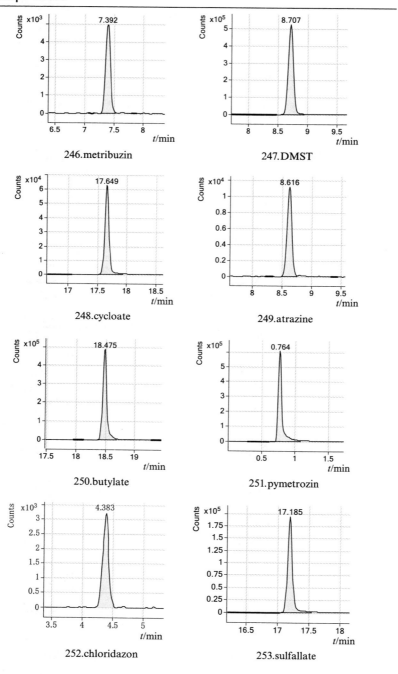

246.metribuzin

247.DMST

248.cycloate

249.atrazine

250.butylate

251.pymetrozin

252.chloridazon

253.sulfallate

D Group—cont'd

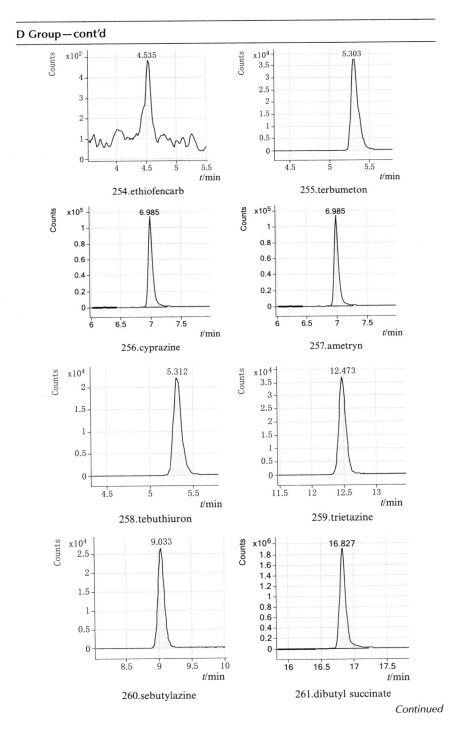

254.ethiofencarb

255.terbumeton

256.cyprazine

257.ametryn

258.tebuthiuron

259.trietazine

260.sebutylazine

261.dibutyl succinate

Continued

D Group—cont'd

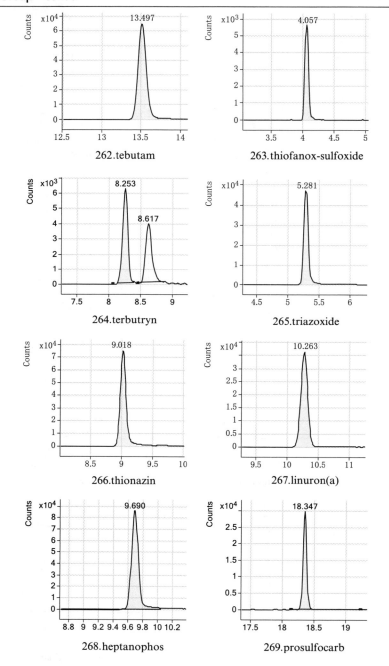

262.tebutam

263.thiofanox-sulfoxide

264.terbutryn

265.triazoxide

266.thionazin

267.linuron(a)

268.heptanophos

269.prosulfocarb

D Group—cont'd

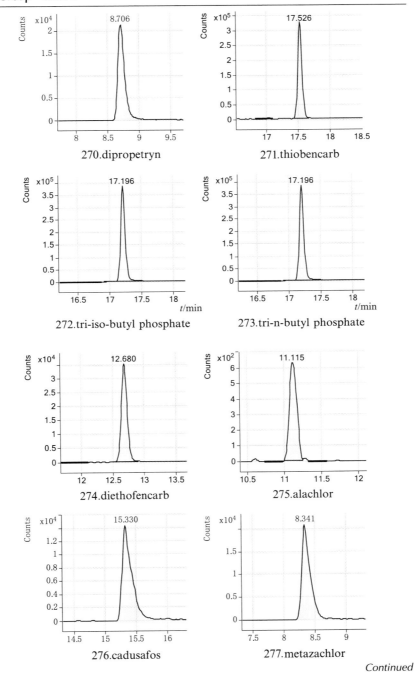

270.dipropetryn

271.thiobencarb

272.tri-iso-butyl phosphate

273.tri-n-butyl phosphate

274.diethofencarb

275.alachlor

276.cadusafos

277.metazachlor

Continued

D Group—cont'd

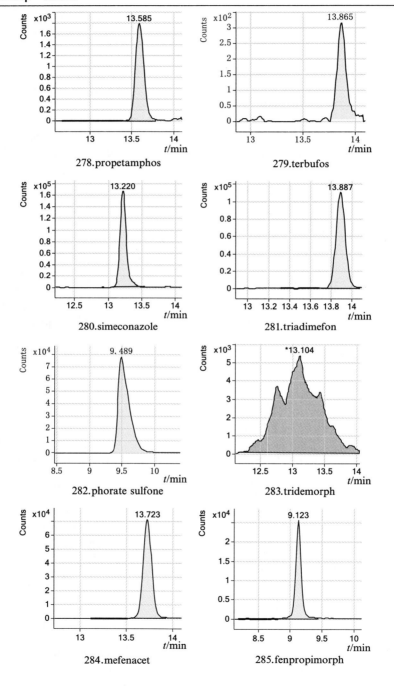

278. propetamphos

279. terbufos

280. simeconazole

281. triadimefon

282. phorate sulfone

283. tridemorph

284. mefenacet

285. fenpropimorph

D Group—cont'd

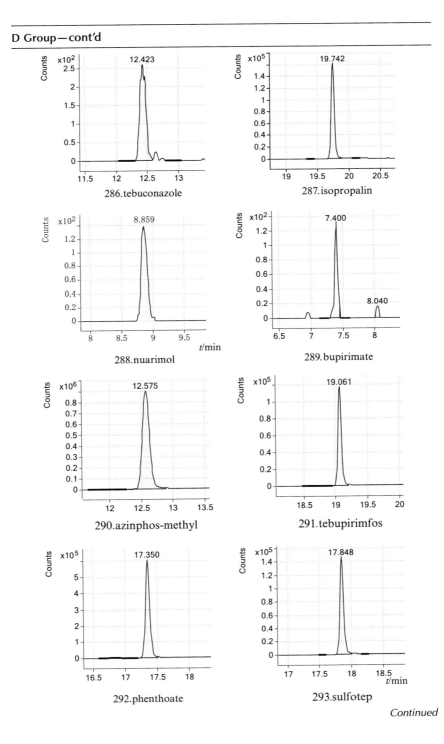

286. tebuconazole

287. isopropalin

288. nuarimol

289. bupirimate

290. azinphos-methyl

291. tebupirimfos

292. phenthoate

293. sulfotep

Continued

D Group—cont'd

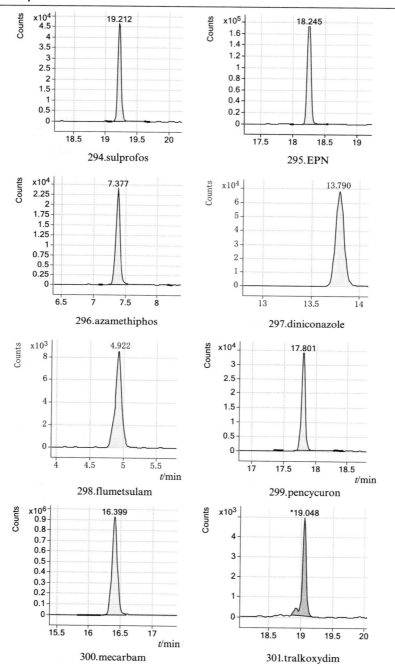

294.sulprofos

295.EPN

296.azamethiphos

297.diniconazole

298.flumetsulam

299.pencycuron

300.mecarbam

301.tralkoxydim

D Group—cont'd

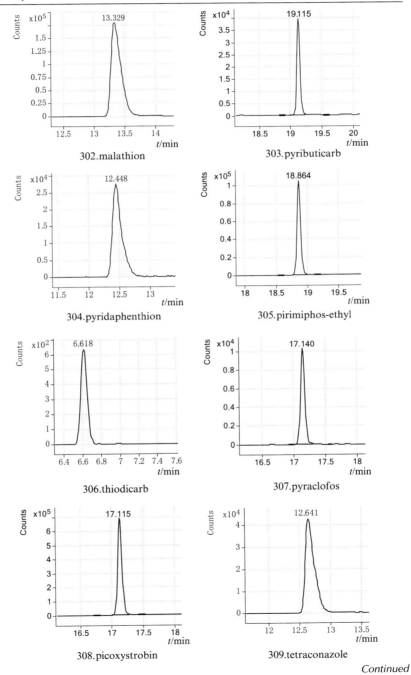

302. malathion

303. pyributicarb

304. pyridaphenthion

305. pirimiphos-ethyl

306. thiodicarb

307. pyraclofos

308. picoxystrobin

309. tetraconazole

Continued

D Group—cont'd

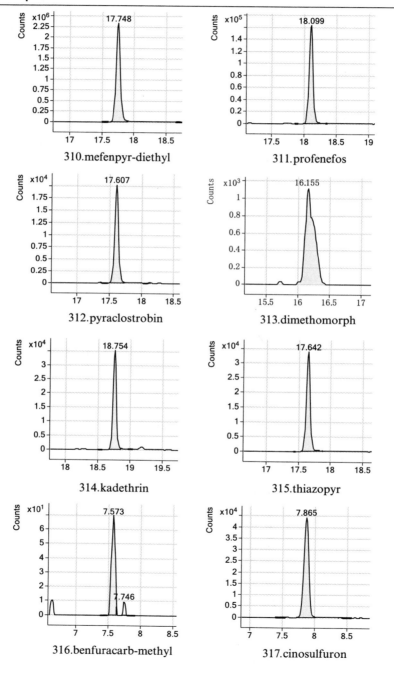

310.mefenpyr-diethyl

311.profenefos

312.pyraclostrobin

313.dimethomorph

314.kadethrin

315.thiazopyr

316.benfuracarb-methyl

317.cinosulfuron

D Group—cont'd

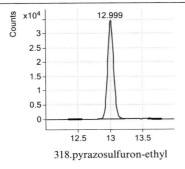

318.pyrazosulfuron-ethyl

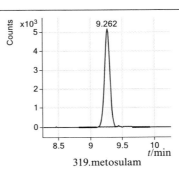

319.metosulam

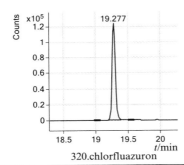

320.chlorfluazuron

E Group

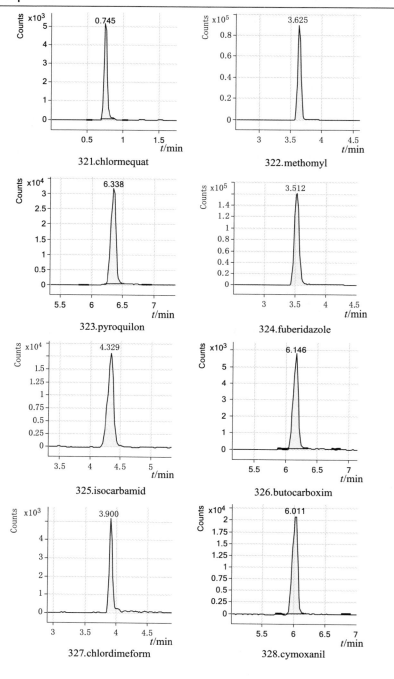

321.chlormequat

322.methomyl

323.pyroquilon

324.fuberidazole

325.isocarbamid

326.butocarboxim

327.chlordimeform

328.cymoxanil

E Group—cont'd

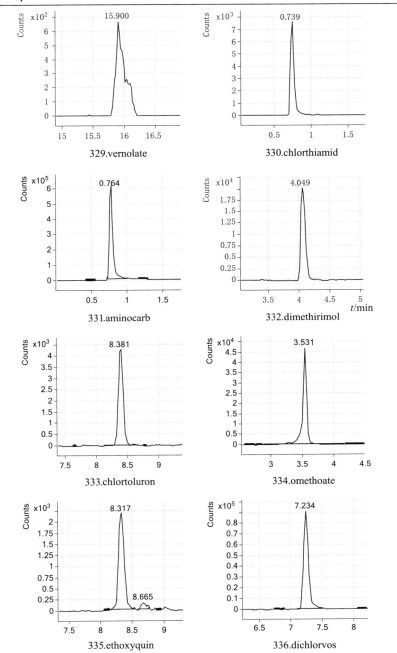

329.vernolate

330.chlorthiamid

331.aminocarb

332.dimethirimol

333.chlortoluron

334.omethoate

335.ethoxyquin

336.dichlorvos

Continued

E Group—cont'd

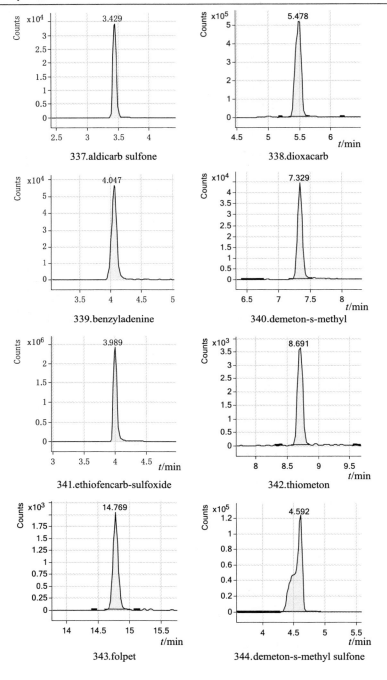

337.aldicarb sulfone

338.dioxacarb

339.benzyladenine

340.demeton-s-methyl

341.ethiofencarb-sulfoxide

342.thiometon

343.folpet

344.demeton-s-methyl sulfone

E Group—cont'd

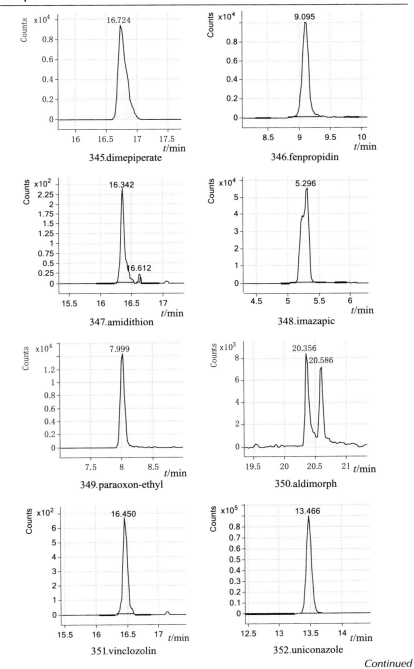

345.dimepiperate

346.fenpropidin

347.amidithion

348.imazapic

349.paraoxon-ethyl

350.aldimorph

351.vinclozolin

352.uniconazole

Continued

E Group—cont'd

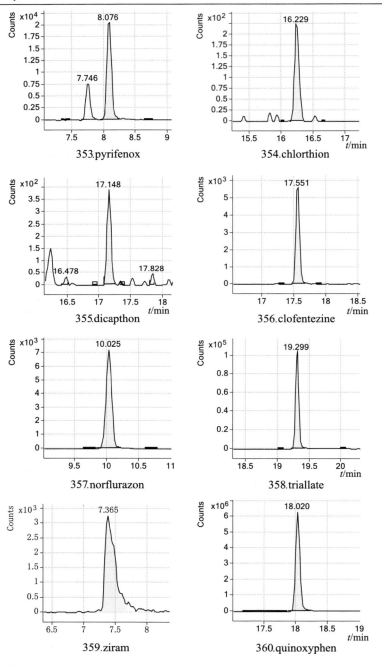

353.pyrifenox

354.chlorthion

355.dicapthon

356.clofentezine

357.norflurazon

358.triallate

359.ziram

360.quinoxyphen

E Group—cont'd

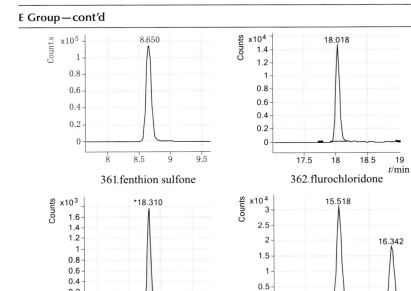

361.fenthion sulfone

362.flurochloridone

363.phthalic acid,benzyl butyl ester

364.isazofos

365.dichlofenthion

366.vamidothion sulfone

367.terbufos sulfone

368.dinitramine

Continued

E Group—cont'd

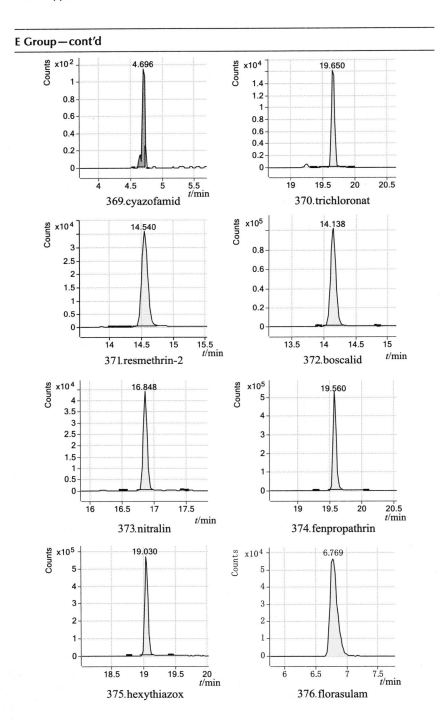

369. cyazofamid

370. trichloronat

371. resmethrin-2

372. boscalid

373. nitralin

374. fenpropathrin

375. hexythiazox

376. florasulam

E Group—cont'd

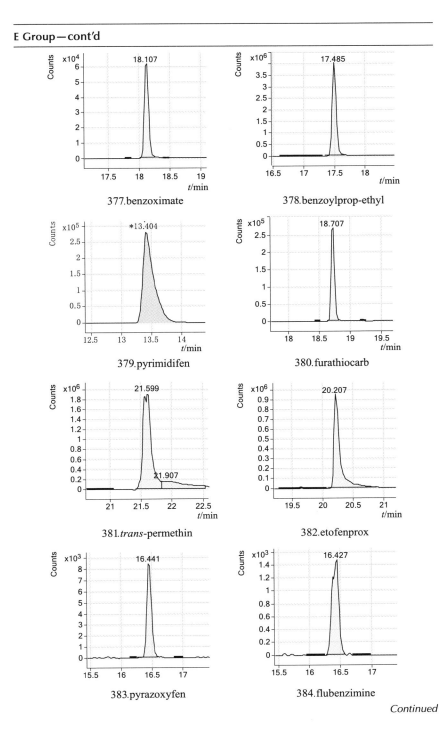

377.benzoximate

378.benzoylprop-ethyl

379.pyrimidifen

380.furathiocarb

381.*trans*-permethin

382.etofenprox

383.pyrazoxyfen

384.flubenzimine

Continued

E Group—cont'd

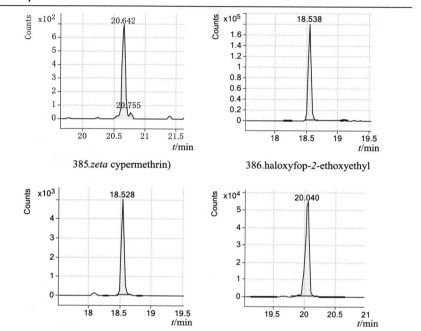

385.*zeta* cypermethrin)

386.haloxyfop-*2*-ethoxyethyl

387.fluoroglycofen-ethyl

388.*tau*-fluvalinate

F Group

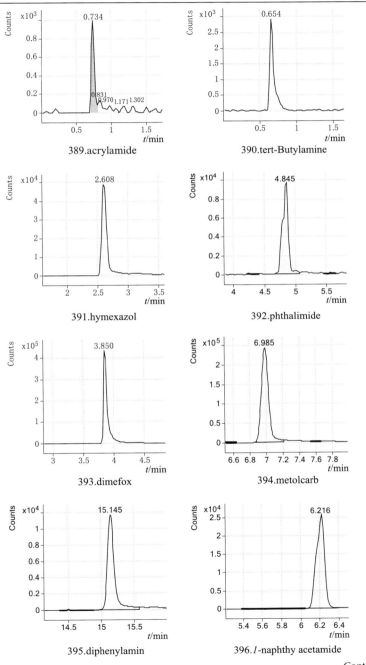

389.acrylamide

390.tert-Butylamine

391.hymexazol

392.phthalimide

393.dimefox

394.metolcarb

395.diphenylamin

396.*1*-naphthy acetamide

Continued

F Group—cont'd

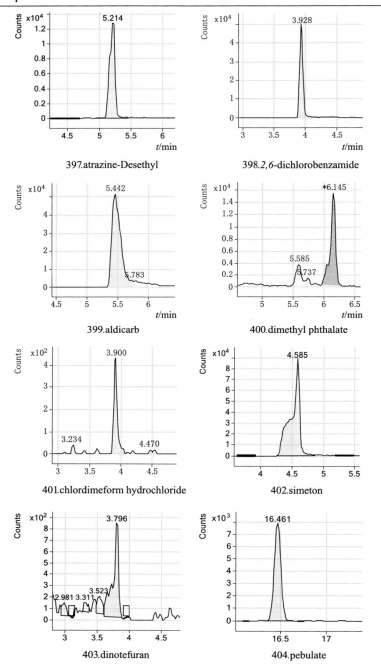

397.atrazine-Desethyl

398.2,6-dichlorobenzamide

399.aldicarb

400.dimethyl phthalate

401.chlordimeform hydrochloride

402.simeton

403.dinotefuran

404.pebulate

F Group—cont'd

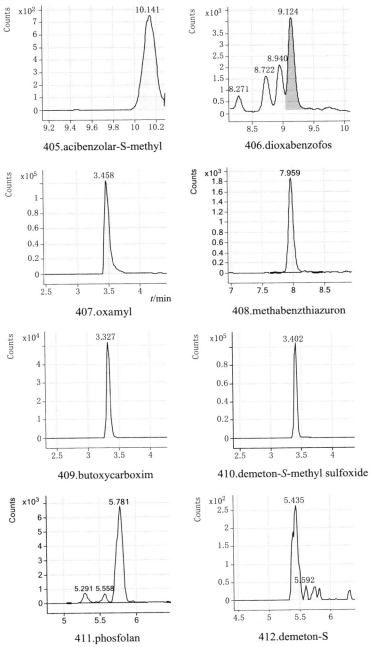

405.acibenzolar-S-methyl

406.dioxabenzofos

407.oxamyl

408.methabenzthiazuron

409.butoxycarboxim

410.demeton-S-methyl sulfoxide

411.phosfolan

412.demeton-S

Continued

F Group—cont'd

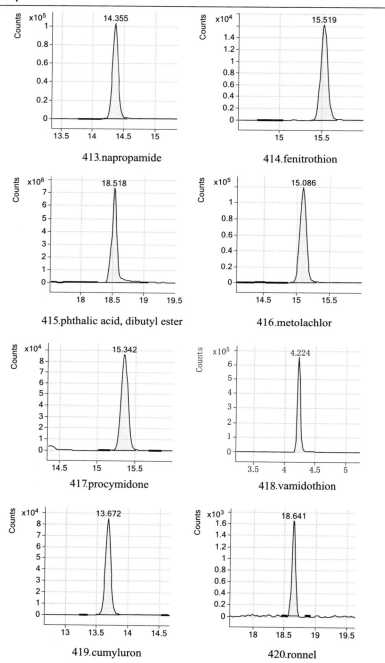

413. napropamide

414. fenitrothion

415. phthalic acid, dibutyl ester

416. metolachlor

417. procymidone

418. vamidothion

419. cumyluron

420. ronnel

F Group—cont'd

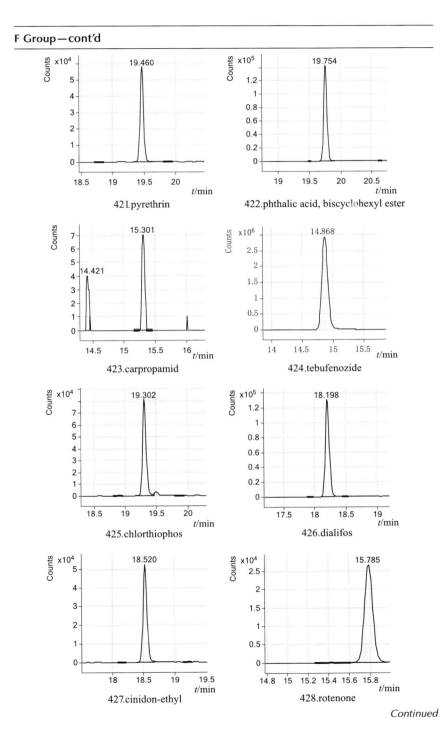

421.pyrethrin

422.phthalic acid, biscyclohexyl ester

423.carpropamid

424.tebufenozide

425.chlorthiophos

426.dialifos

427.cinidon-ethyl

428.rotenone

Continued

F Group—cont'd

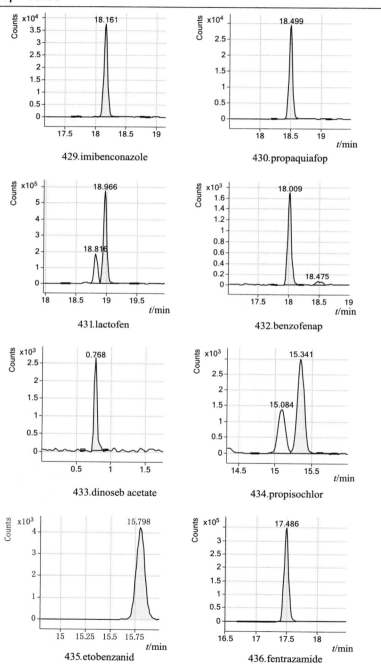

429. imibenconazole

430. propaquiafop

431. lactofen

432. benzofenap

433. dinoseb acetate

434. propisochlor

435. etobenzanid

436. fentrazamide

F Group—cont'd

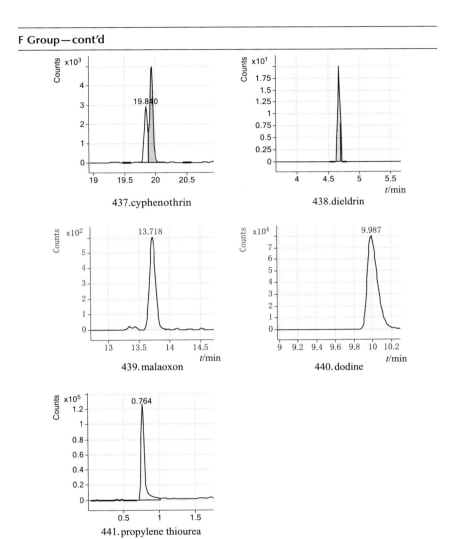

437. cyphenothrin

438. dieldrin

439. malaoxon

440. dodine

441. propylene thiourea

G Group

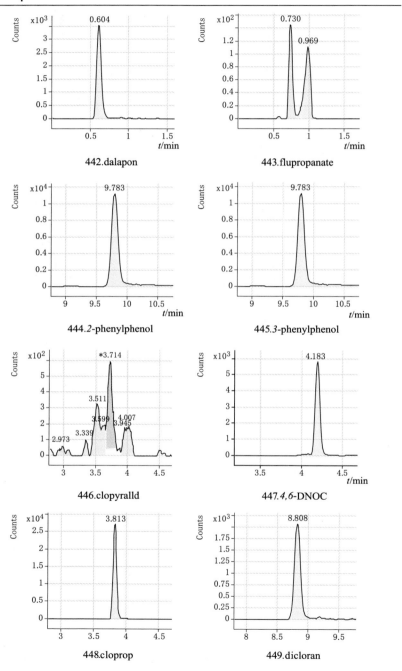

442.dalapon

443.flupropanate

444.*2*-phenylphenol

445.*3*-phenylphenol

446.clopyralld

447.*4,6*-DNOC

448.cloprop

449.dicloran

G Group—cont'd

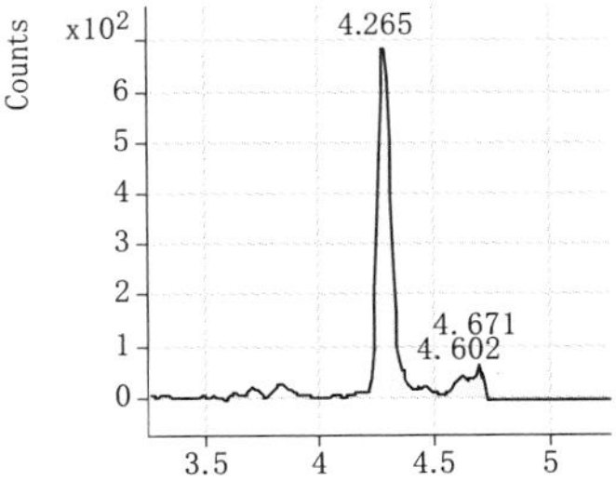

450.aminopyralid

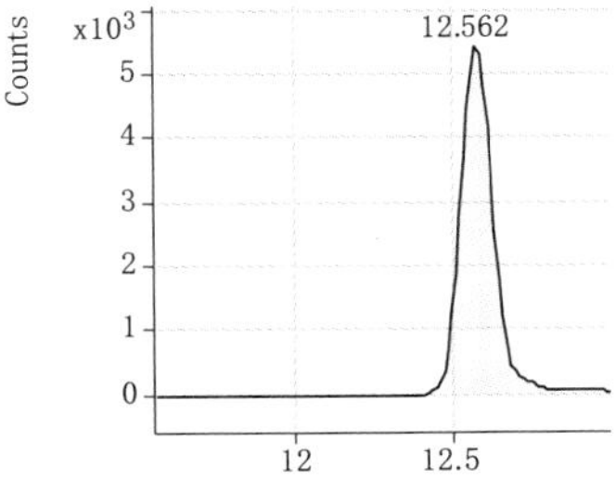

451.chlorpropham

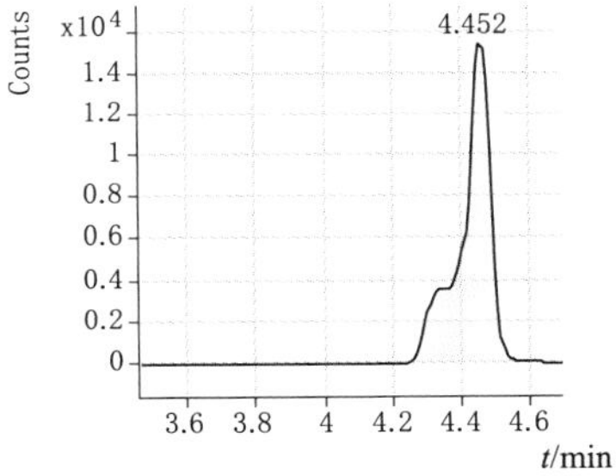

452.*2-4*-mecoprop

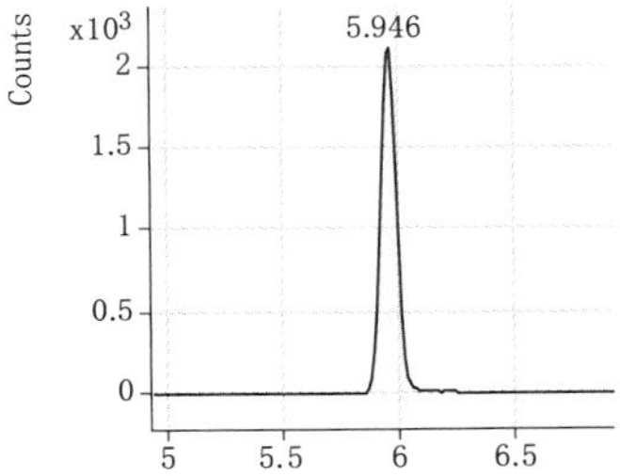

453.terbacil

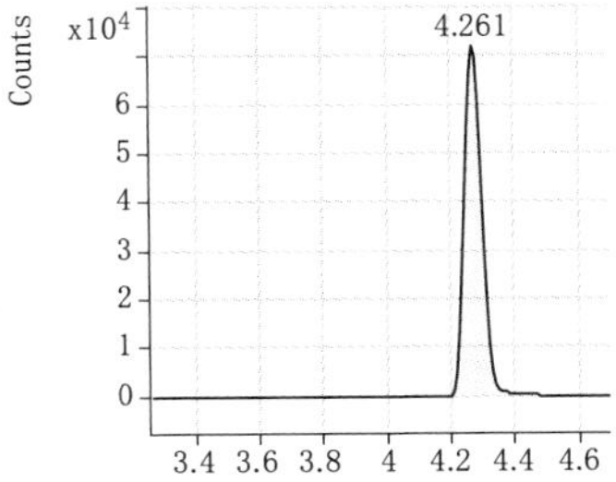

454.*2,4*-D

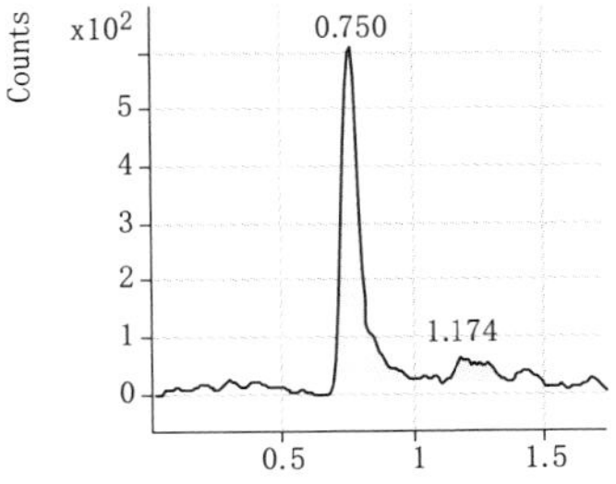

455.dicamba

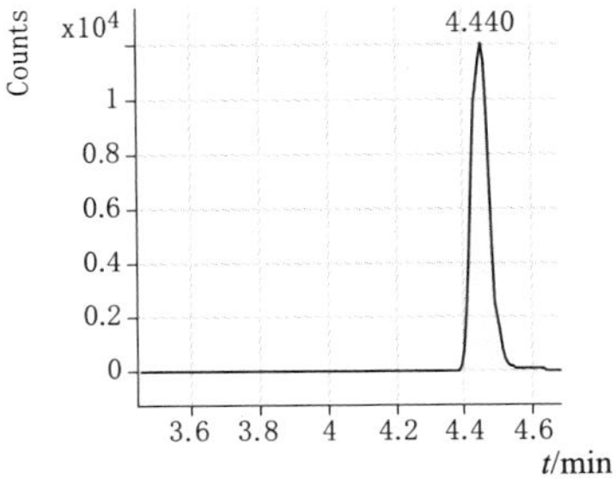

456.MCPB

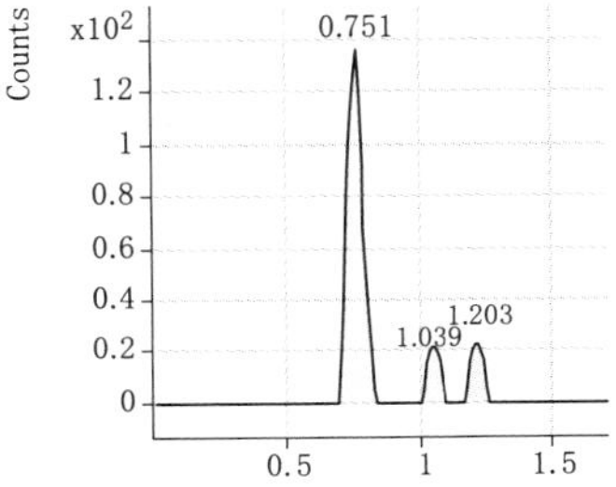

457.fenaminosulf

Continued

G Group—cont'd

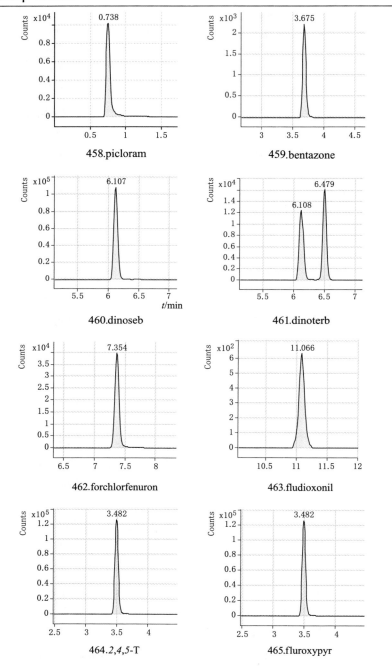

458.picloram

459.bentazone

460.dinoseb

461.dinoterb

462.forchlorfenuron

463.fludioxonil

464.2,4,5-T

465.fluroxypyr

G Group—cont'd

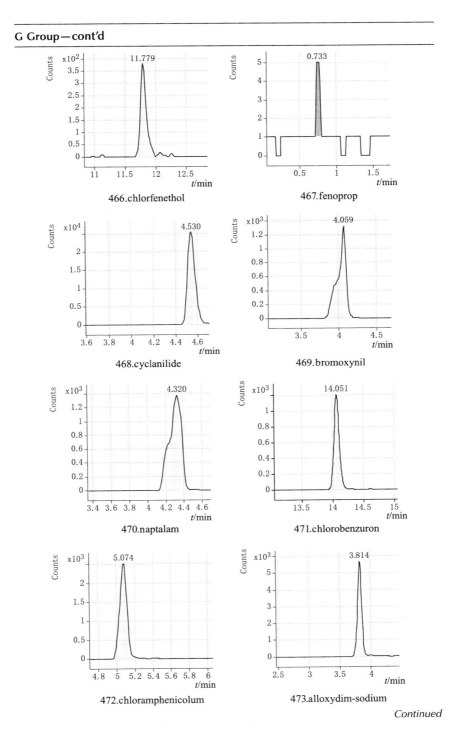

466.chlorfenethol

467.fenoprop

468.cyclanilide

469.bromoxynil

470.naptalam

471.chlorobenzuron

472.chloramphenicolum

473.alloxydim-sodium

Continued

G Group—cont'd

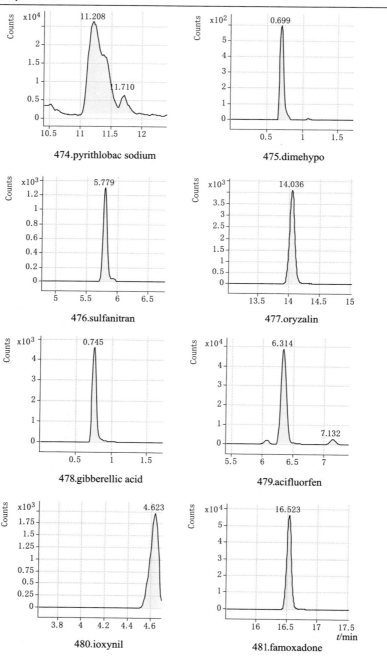

474.pyrithlobac sodium

475.dimehypo

476.sulfanitran

477.oryzalin

478.gibberellic acid

479.acifluorfen

480.ioxynil

481.famoxadone

G Group—cont'd

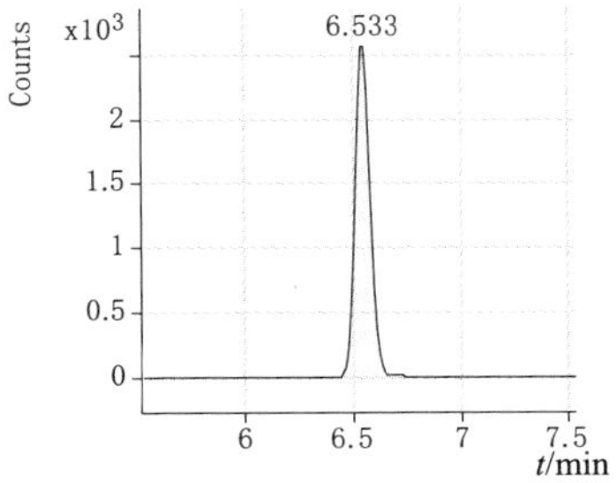

482.sulfentrazone

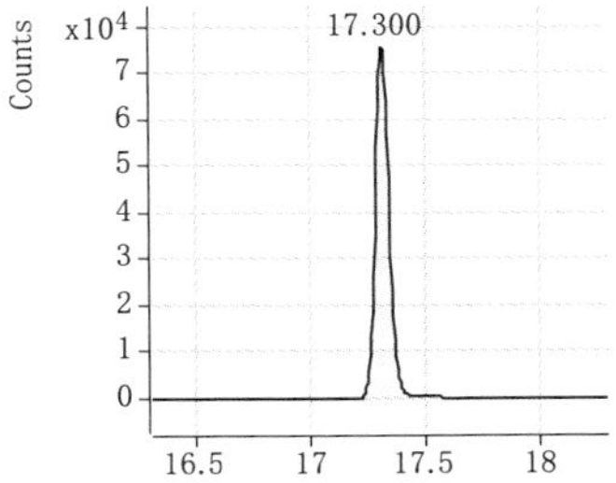

483.diflufenican

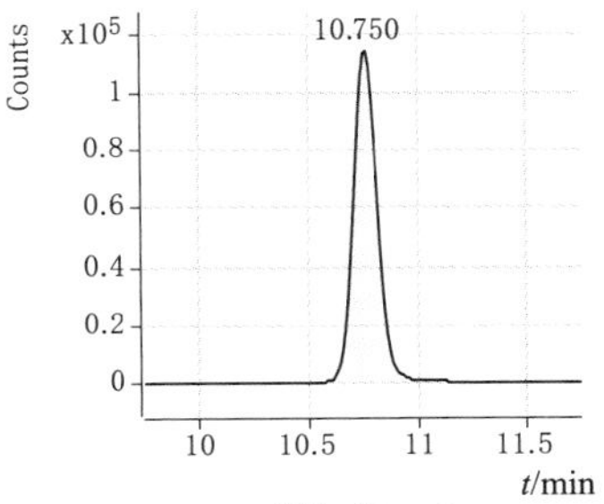

484.ethiprole

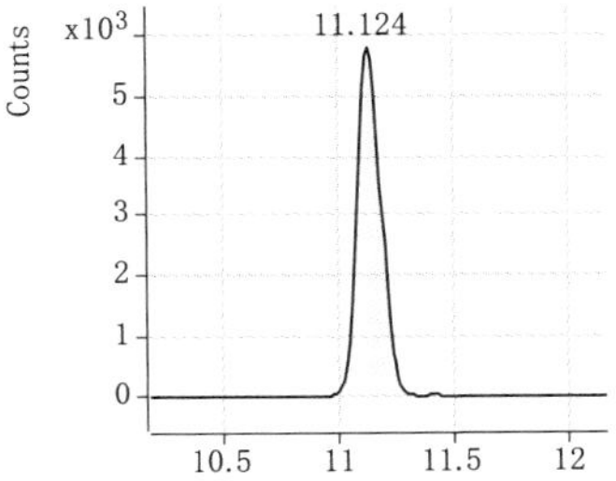

485.flusulfamide

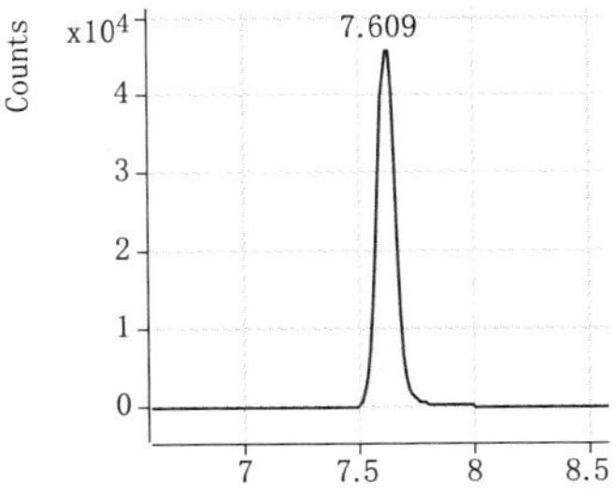

486.cyclosulfamuron

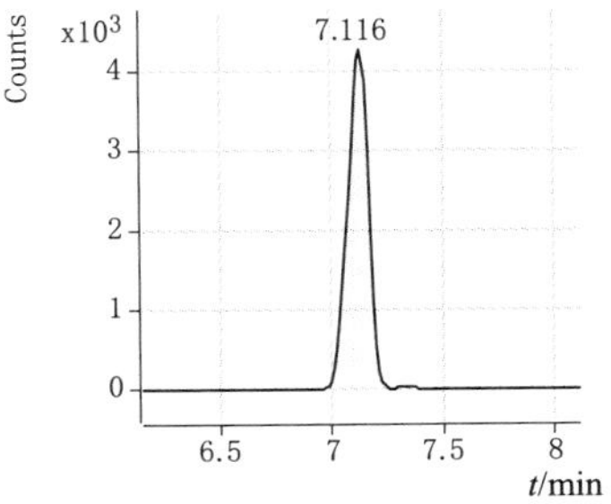

487.fomesafen

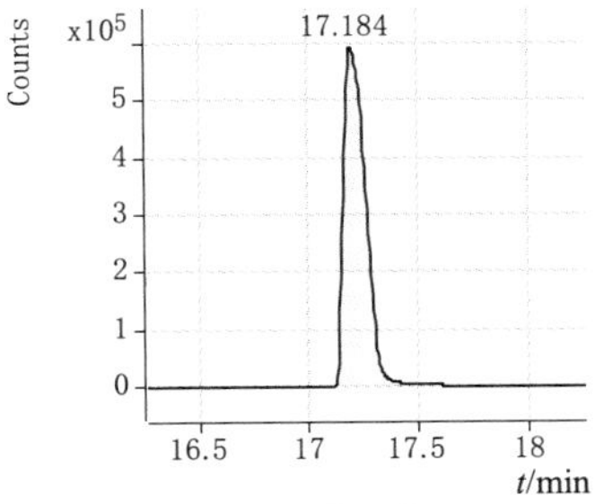

488.fluazinam

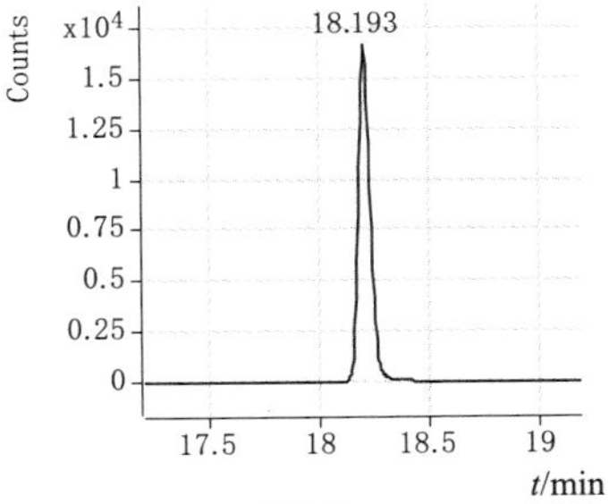

489.fluazuron

Continued

G Group—cont'd

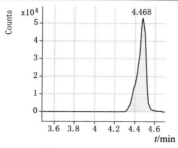

490.iodosulfuron-methyl sodium

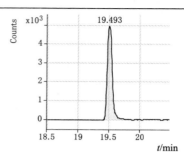

491.kelevan

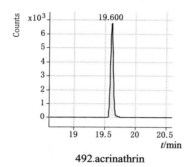

492.acrinathrin

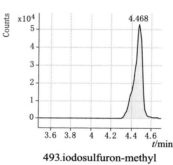

493.iodosulfuron-methyl

Index

Note: Page numbers followed by *t* indicate tables.